Numerical Linear Approximation in C

CHAPMAN & HALL/CRC
Numerical Analysis and Scientific Computing

Aims and scope:
Scientific computing and numerical analysis provide invaluable tools for the sciences and engineering. This series aims to capture new developments and summarize state-of-the-art methods over the whole spectrum of these fields. It will include a broad range of textbooks, monographs and handbooks. Volumes in theory, including discretisation techniques, numerical algorithms, multiscale techniques, parallel and distributed algorithms, as well as applications of these methods in multi-disciplinary fields, are welcome. The inclusion of concrete real-world examples is highly encouraged. This series is meant to appeal to students and researchers in mathematics, engineering and computational science.

Proposals for the series should be submitted to one of the series editors above or directly to:
CRC Press, Taylor & Francis Group
24-25 Blades Court
Deodar Road
London SW15 2NU
UK

Numerical Linear Approximation in C

Nabih N. Abdelmalek, Ph.D.
William A. Malek, M.Eng.

CRC Press is an imprint of the
Taylor & Francis Group, an **informa** business
A CHAPMAN & HALL BOOK

CRC Press
Taylor & Francis Group
6000 Broken Sound Parkway NW, Suite 300
Boca Raton, FL 33487-2742

First issued in paperback 2019

ISBN-13: 978-1-58488-978-6 (hbk)
ISBN-13: 978-0-367-38731-0 (pbk)

Library of Congress Cataloging-in-Publication Data

Abdelmalek, Nabih N.
Numerical linear approximation in C / Nabih Abdelmalek and William A. Malek.
p. cm. -- (CRC numerical analysis and scientific computing)
Includes bibliographical references and index.
ISBN 978-1-58488-978-6 (alk. paper)
1. Chebyshev approximation. 2. Numerical analysis. 3. Approximation theory. I. Malek, William A. II. Title. III. Series.

QA297.A23 2008
518--dc22 2008002447

Visit the Taylor & Francis Web site at
http://www.taylorandfrancis.com

and the CRC Press Web site at
http://www.crcpress.com

Contents

PART 1

Preliminaries and Tutorials

Chapter 1 Applications of Linear Approximation

PART 2

The L_1 Approximation

Chapter 5 Linear L_1 Approximation

Chapter 6 One-Sided L_1 Approximation

Chapter 7 L_1 Approximation with Bounded Variables

Chapter 8 L_1 Polygonal Approximation of Plane Curves

Chapter 9 Piecewise L_1 Approximation of Plane Curves

PART 3

The Chebyshev Approximation

Chapter 10 Linear Chebyshev Approximation

Chapter 11 One-Sided Chebyshev Approximation

Chapter 14 Strict Chebyshev Approximation

Chapter 15 Piecewise Chebyshev Approximation

Chapter 16 Solution of Linear Inequalities

PART 4

The Least Squares Approximation

Chapter 17 Least Squares and Pseudo-Inverses of Matrices

Chapter 18 Piecewise Linear Least Squares Approximation

Chapter 19 Solution of Ill-Posed Linear Systems

PART 5

Solution of Underdetermined Systems Of Linear Equations

Chapter 20 L_1 Solution of Underdetermined Linear Equations

Chapter 21 Bounded and L_1 Bounded Solutions of Underdetermined Linear Equations

Chapter 22 Chebyshev Solution of Underdetermined Linear Equations

Chapter 23 Bounded Least Squares Solution of Underdetermined Linear Equations

Appendices

List of Figures

Preface

Discrete linear approximation is one of the most frequently used techniques in all areas of science and engineering. Linear approximation of a continuous function is typically done by digitizing the function, then applying discrete linear approximation to the resulting data. Discrete linear approximation is equivalent to the solution of an overdetermined system of linear equations in an appropriate measure or norm, with or without some constraints.

This book describes algorithms for the solution of overdetermined and underdetermined systems of linear equations. Software implementations of the algorithms and test drivers, all written in the programming language C, are provided at the end of each chapter. Also included are piecewise linear approximations to plane curves in the three main norms, as well as solutions of overdetermined linear inequalities and of ill-posed linear systems.

It is assumed that the reader has basic knowledge of elementary functional analysis and the programming language C.

All the algorithms in this book are based on linear programming techniques, except the least squares ones. In addition, Chapter 23 uses a quadratic programming technique.

Linear programming proved to be a powerful tool in solving linear approximation problems. The solution, if it exists, is obtained in a finite number of iterations, and no conditions are imposed on the coefficient matrix, such as the Haar condition or the full rank condition. The issue of the uniqueness of the solution is determined by methods that use linear programming. Characterization of the solution may also be deduced from the final tableau of the linear programming problem.

The algorithms presented in this book are compared with existing algorithms and are found to be the best or among the best of them.

The first author developed and published almost all the algorithms presented in this book while he was with the National Research Council of Canada (NRC), in Ottawa, Canada. While at the NRC, the algorithms were applied successfully in several engineering projects. Since then, the software was converted to FORTRAN 77, and then to C for this book.

The second author is the son of the first author. He has been instrumental in converting the software from FORTRAN 77 to C. He has worked for over 25 years as a software engineer in the high-tech industry in Ottawa, and has spent 4 years teaching courses in software engineering at Carleton University in Ottawa, Canada.

This book may be considered a companion to the books, *Numerical Recipes in C, the Art of Scientific Computing*, second edition, W.H. Press et. al, Cambridge University Press, 1992 and *A Numerical Library in C for Scientists and Engineers*, H.T. Lau, CRC Press, 1994. Except for solving the least squares approximation by the normal equation and by the Singular Value Decomposition methods (Section 15.4 in the former and 3.4.2 in the latter), they lack the subject of Numerical Linear Approximation. Also, the generic routine on linear programming and the simplex method in the former reference (Section 10.8) has little or no application to linear approximation. Standard Subroutine Packages, such as LINPACK, LAPACK, IMSL and NAG, may contain few linear approximation routines, but not in C.

This book is divided into 5 main parts. Part 1 includes introductory chapters 1 and 2, and tutorial chapters 3 and 4. Chapter 1 describes some diverse applications of linear approximation. In Chapter 2, the relationship between Discrete Linear Approximation in a certain norm and the solution of overdetermined linear equations in the same norm is established. The L_1, the least squares (L_2) and the Chebyshev (minimax or L_∞) norms are considered. A comparison is made between the approximations in the three norms, using a simple example. An explanation of single- and double-precision computation, tolerance parameters and the programmatic representation of vectors and matrices is also given. The chapter concludes with a study of outliers or odd points in data, how to identify them, and some methods of dealing with them. Chapter 3 gives an overall description of the subject of linear programming and the simplex method. Also, the (ordinary or two-sided) L_1 and the (ordinary or two-sided) Chebyshev approximation problems are formulated as linear programming problems. Chapter 4 starts with the familiar subject of vector and matrix norms and some relevant theorems. Then, elementary matrices, which are used to perform elementary operations on a matrix equation, are described. The solution of square linear systems using the Gauss **LU** factorization method with complete pivoting and the Householder's **QR** factorization method with pivoting follows. A note on the Gauss-Jordan elimination method for a set of underdetermined system of linear equations is given. This chapter ends with a presentation of rounding error analysis for simple and extended arithmetic operations. Chapter 4 serves as an introduction to Chapter 17 on the pseudo-inverses of matrices and the solution of linear least squares problems.

Part 2 of this book includes chapters 5 through 9. Chapters 5, 6 and 7 present the ordinary L_1 approximation problem, the one-sided L_1 approximation, and the L_1 approximation with bounded variables, respectively. Chapters 8 and 9 present, respectively, the L_1 polygonal approximation and the linear L_1 piecewise approximation of plane curves.

Part 3 contains chapters 10 through 16. Chapters 10 through 15 present, respectively, the ordinary Chebyshev approximation problem, the one-sided Chebyshev approximation, the Chebyshev approximation with bounded variables, the Restricted Chebyshev approximation, the Strict Chebyshev approximation and the piecewise Chebyshev approximation of plane curves. Chapter 16 presents the solution of overdetermined linear inequalities and its application to pattern classification, using the one-sided Chebyshev and the one-sided L_1 approximation algorithms.

Part 4 contains chapters 17 through 19. Chapter 17 introduces the pseudo-inverses of matrices and the minimal length least squares solution. It then describes the least squares approximation by the Gauss **LU** factorization method with complete pivoting and by Householder's **QR** factorization method with pivoting. An introduction of linear spaces and the pseudo-inverses is given next. The interesting subject of multicollinearity, or the ill-conditioning of the coefficient matrix, is presented. This leads to the subject of dealing with multicollinearity via the Principal Components Analysis (PCA), the Partial Least Squares (PLS) method and the Ridge equation technique. Chapter 18 presents the piecewise approximation of plane curves in the L_2 norm. Chapter 19 presents the solution of ill-posed systems such as those arising from the discretization of the Fredholm integral equation of the first kind.

Part 5 contains Chapters 20 through 23. They present the solution of underdetermined systems of consistent linear equations subject to different constraints on the elements of the unknown solution vector (not on the residuals, since the residuals are zeros). Chapter 20 constitutes the L_1 approximation problem for the elements of the solution vector. Chapter 21 describes (a) the solution of underdetermined linear equations subject to bounds on the elements of the solution vector and (b) the L_1 approximation problem for these elements that satisfy bounds on them. Chapter 22 describes the L_∞ approximation problem for the elements of the solution vector. Finally, Chapter 23 describes the bounded L_2 approximation problem for the elements of the solution vector.

Chapters 5 through 23 include the C functions that implement the linear approximation algorithms, along with sample drivers. Each driver contains a number of test case examples. The results of one or more of the test cases are given in the text. Most of these chapters also include a numerical example solved in detail.

To use the software provided in this book, it is not necessary to understand the theory behind each one of the functions. The examples given in each driver act as a guide to their usage. The code was compiled and tested using Microsoft™ Visual C++™ 6.0, Standard Edition. To ensure maximum portability, only the most basic features of the ANSI C standard were used, and no C++ features were employed.

It is hoped that this book will be of benefit to scientists, engineers and university students.

Acknowledgments

The first author wishes to thank his friends and colleagues at the Institute for Informatics of the National Research Council of Canada, particularly, Stuart Baxter, Brice Wightman, Frank Farrell, Bert Lewis and Tom Bach for their unfailing help during his stay with the Institute. From the Institute for Information Technology, he wishes to thank his friend and colleague Tony Kasvand, who got him interested in the fields of image processing and computer vision to which numerical techniques were applied.

We also wish to sincerely thank Robert B. Stern, Executive Editor at Taylor & Francis Group, Chapman and Hall, and CRC Press for his support and interest in this book. Our thanks also go to the reviewers for their constructive comments.

Last but not least, we thank our families for their love and encouragement.

Warranties

The authors and the publisher of this book do not assume any responsibility for any damage to person or property caused directly or indirectly by the use, misuse or any other application of any of the algorithms contained in this book.

We also make no warranties, in any way or form that the programs contained in this book are free of errors or are consistent or inconsistent with any type of machine.

About the authors

Nabih N. Abdelmalek

The first author was born in Egypt in 1929. He received a B.Sc. in electrical engineering in 1951, a B.Sc. in mathematics in 1954, both from Cairo University, Egypt, and a Ph.D. in mathematical physics in 1958 from Manchester University, England.

From 1959 to 1965 he was an associate professor of mathematics in the faculty of engineering, Cairo University, Egypt. From 1965 to 1967 he was a member of Scientific Staff at Northern Electric (now Nortel Networks) in Ottawa, Canada. From 1967 until his retirement in 1990, he was with the National Research Council of Canada, in Ottawa, where he was a senior research officer. His interest was in the application of numerical analysis techniques to image processing, particularly image restoration, data compression, pattern classification and segmentation of 3-D range images.

William A. Malek

The second author was born in Egypt in 1960. He received a B.Eng. in electrical engineering in 1982 and an M.Eng. in computer engineering in 1984, both from Carleton University in Ottawa Canada.

From 1984 to 1988 he was a lecturer in the Department of Systems and Computer Engineering at Carleton University, teaching undergraduate and graduate courses in software engineering and real-time systems. Since 1982, he has worked in Ottawa as a software engineer for networking, avionics and military applications.

"Every good gift and every perfect gift is from above, and comes down from the Father of lights."
(James 1:17 NKJV)

PART 1

Preliminaries and Tutorials

Chapter 1

Applications of Linear Approximation

1.1 Introduction

In this chapter, we present a number of everyday life problems whose solutions need the application of linear approximation, which is the subject of this book. The problems presented here occur in social sciences, economics, industry and digital image processing. In the following, we assume that mathematical models have been agreed upon, that observed data have been collected constituting the columns of a matrix **A**, and that the response data constitute vector **b**. The solution of the problem, vector **x**, is the solution vector of the **matrix equation**

$$\mathbf{Ax} = \mathbf{b}$$

The n by m, n > m, matrix **A** consists of n observations in m parameters. Because of measuring errors in the elements of matrix **A** and in the response n-vector **b**, a larger number of observations than the number of parameters are collected. Thus system **Ax** = **b** is an overdetermined system of linear equations.

Equation **Ax** = **b** may also be written as

$$x_1\mathbf{a}_1 + x_2\mathbf{a}_2 + \ldots + x_m\mathbf{a}_m = \mathbf{b}$$

where $\mathbf{a}_1, \mathbf{a}_2, \ldots, \mathbf{a}_m$ are the m columns of matrix **A** and the (x_i) are the elements of the solution m-vector **x**.

In many instances, the mathematical model represented by **Ax** = **b** is an approximate or even incorrect one. In other instances, matrix **A** is badly conditioned and needs to be stabilized. A badly conditioned matrix results in a solution vector **x** that would be inaccurate or even wrong.

Since $\mathbf{Ax} = \mathbf{b}$ is an overdetermined system, in most cases it has no exact solution, but it may have an approximate solution. For calculating an approximate solution, usually, the residual, or the error vector

$$\mathbf{r} = \mathbf{Ax} - \mathbf{b}$$

is minimized in a certain norm. The p-vector norm of the residual $\mathbf{r}$ is given by

$$\| \mathbf{r} \|_p = \left[\sum_{i=1}^{n} |r(x_i)|^p \right]^{1/p}, \quad 1 \le p \le \infty$$

There are three main vector-norms, namely for $p = 1$ (the L_1 approximation), $p = 2$ (the L_2 or the least squares approximation) and for $p = \infty$ (the L_∞, the minimax or the Chebyshev approximation). This notion of norms is repeated in Chapter 2.

The approximate solution of $\mathbf{Ax} = \mathbf{b}$ may be achieved as follows:

(a) One may minimize the norm of the **residual vector r**.

(b) One may also minimize the norm of the residual vector $\mathbf{r}$ subject to the condition that all the elements of $\mathbf{r}$ are either non-negative; that is, $r_i \ge 0$, or non-positive; $r_i \le 0$, $i = 1, 2, \ldots, n$. This is known as the **linear one-sided approximation** problem.

(c) The solution of $\mathbf{Ax} = \mathbf{b}$ may also be obtained by minimizing the norm of $\mathbf{r}$ subject to constraints on the solution vector $\mathbf{x}$. The elements (x_i) may be required to be bounded

$$b_i \le x_i \le c_i, \quad i = 1, 2, \ldots, m$$

where (b_i) and (c_i) are given parameters. This problem is known as the **linear approximation with bounded variables**. If all the c_i are very large positive numbers and all the b_i are 0's, then the approximation has a **non-negative solution** vector **a**.

(d) One may also minimize the norm of the residual vector $\mathbf{r}$ in the L_∞ norm subject to the constraints that the l.h.s. $\mathbf{Ax}$ is bounded; that is

$$\mathbf{l} \le \mathbf{Ax} \le \mathbf{u}$$

where $\mathbf{l}$ and $\mathbf{u}$ are given n-vectors. This approximation is known as the **restricted Chebyshev approximation**. If vector $\mathbf{l}$ is $\mathbf{0}$ and all the elements of vector $\mathbf{u}$ are very large numbers, the approximation is known to be with **non-negative functions**.

(e) Instead of solving an overdetermined system of linear equations, we may have an overdetermined system of **linear inequalities** in the form

$$\mathbf{Ax} \geq \mathbf{b} \text{ or } \mathbf{Ax} \geq \mathbf{0}$$

where $\mathbf{0}$ is a zero n-vector. Such systems of linear inequalities have applications to digital pattern classification.

(f) Once more, in some cases, particularly in control theory, equation $\mathbf{Ax} = \mathbf{b}$ is an underdetermined system; that is, $n < m$. The residual vector $\mathbf{r} = \mathbf{0}$ and system $\mathbf{Ax} = \mathbf{b}$ has an infinite number of solutions. In this case, one minimizes the norm of the solution vector $\mathbf{x}$ (not the residual vector $\mathbf{r}$, since $\mathbf{r} = \mathbf{0}$), and may set bounds on the elements of $\mathbf{x}$.

In Section 1.2, applications to social sciences and economics are presented. In Section 1.3, we present two applications to industrial problems and in Section 1.4, applications are presented to digital image processing.

For all the problems in Section 1.2, the matrix equations are overdetermined and are solved in the least squares sense. However, they may also be solved in the L_1 or in the Chebyshev sense. These problems are taken from published literature.

1.2 Applications to social sciences and economics

Let the response vector be $\mathbf{b}$ and the columns of matrix $\mathbf{A}$ be $(\mathbf{a}_i)$, $i = 1, 2, \ldots, m$. As mentioned before, in all of these applications, the linear model $\mathbf{Ax} = \mathbf{b}$ is a convenient way to formulate the problem mathematically. In no way is this model assumed to be the right one. Whenever possible, for each problem, comments are provided on the given data and/or on the obtained solution.

Note 1.1

The minimum L_2 norm $\|\mathbf{r}\|_2$ of a given problem is of importance in assessing the quality of the least squares fit. Obviously, if this norm is zero, there is a perfect fit by the approximating curve (surface) for the given data. On the other hand, if $||\mathbf{r}||_2$ is too large, there might be large errors in the measured data that cause this norm to be too high, or the used model may not be suitable for the given data and a new model is to be sought.

1.2.1 Systolic blood pressure and age

The data of this problem is given in Kleinbaum et al. ([18], p. 52). It gives the observed systolic blood pressure (SBP) and the age for a sample of n = 30 individuals.

$\mathbf{b}$ = The observed systolic blood pressure (SBP)
$\mathbf{a}_1$ = $\mathbf{1}$
$\mathbf{a}_2$ = The ages

Equation $\mathbf{Ax} = \mathbf{b}$ is of the form

$$x_0\mathbf{a}_1 + x_1\mathbf{a}_2 = \mathbf{b}$$

where $\mathbf{1}$ is an n-column of 1's and x_0 and x_1 are the elements of $\mathbf{x}$.

The straight line fit reflects the trend that the SBP increases with age. It is observed that one of the given points seems to be an odd point to the other points. This is known as an **outlier**. Because outliers can affect the solution parameters x_0 and x_1, it is important to decide if the outlier should be removed from the data and a new least squares solution be calculated. In this problem, the outlier was removed and the new line fit was slightly below the one obtained by using all the data.

On the other hand, if this example is solved in the L_1 norm instead of the least squares, the fitted straight line would have ignored the odd point and would have interpolated (passed through) two of the given data points (Section 2.3).

1.2.2 Annual teacher salaries

The data of this problem is given in Gunst and Mason ([16], p. 221) and it gives the average annual salaries of particular teachers in the USA during the 11 year periods from 1964-65 to 1974-75. By examining the average salaries vs. time, three different curve fitting methods were used. Here, we consider only the first two;

a linear fit and a quadratic fit

$$x_0\mathbf{1} + x_1\mathbf{t} = \mathbf{b}$$

and

$$x_0\mathbf{1} + x_1\mathbf{t} + x_2\mathbf{t}^2 = \mathbf{b}$$

where $\mathbf{b}$ is the average salary, $\mathbf{a}_1 = \mathbf{1}$, an n-column of 1's and $\mathbf{a}_2 = (t_1, t_2, \ldots, t_{11})^T$, represent the time (years). For the quadratic fit, $\mathbf{b}$, $\mathbf{a}_1$ and $\mathbf{a}_2$ are the same and $\mathbf{a}_3 = (t_1^2, t_2^2, \ldots, t_{11}^2)^T$.

The quadratic equation gives the better fit since the residuals are the smallest for almost all the points and are randomly centered around zero. The linear fit is not as good as it offers larger residuals and are not distributed as randomly well around the zero value. Hence, the quadratic fit seems the best. However, it is observed that the quadratic curve is an inverted parabola. It reaches a maximum value and decreases after this value. This is unrealistic, as after this maximum, the salaries of teachers decrease in the latest years!

1.2.3 Factors affecting survival of island species

The Galapagos, 29 islands off the coast of Ecuador, provide data for studying the factors that influence the survival of different life species. The data of this problem is given in Weisberg ([26], pp. 224, 225).

$\mathbf{b}$ = Number of species
$\mathbf{a}_1$ = Endemics
$\mathbf{a}_2$ = Area of the island in square km
$\mathbf{a}_3$ = Elevation of the island in meters
$\mathbf{a}_4$ = Distance in km from nearest island
$\mathbf{a}_5$ = Distance in km from Santa Cruz
$\mathbf{a}_6$ = Area of adjacent island in square km

In this data, one complicating factor is that the elevations of 6 of the islands (elements of $\mathbf{a}_3$) were not recorded, so provisions must be made for this; delete these 6 islands from the data or substitute a plausible value for the missing data.

1.2.4 Factors affecting fuel consumption

For a whole year, the fuel consumption in millions of gallons was measured in 48 states in the United States as a function of fuel tax, average income and other factors. The data of this problem is given in Weisberg ([26], pp. 35, 36). The model $\mathbf{Ax} = \mathbf{b}$ is as follows:

$\mathbf{b}$ = Fuel consumption in millions of gallons
$\mathbf{a}_1$ = $\mathbf{1}$, an n-column of 1's
$\mathbf{a}_2$ = Fuel tax in cents per gallon
$\mathbf{a}_3$ = Percentage of population with drivers licenses
$\mathbf{a}_4$ = Average income in thousands of dollars
$\mathbf{a}_5$ = Length in miles of the highway network

In this problem all the variables have been scaled to be roughly of the same magnitude. This scaling does not affect the relationship between the measured values. For example, it does not matter if the elements of column $\mathbf{a}_3$ are expressed as fractions or as percentages.

1.2.5 Examining factors affecting the mortality rate

In 1973, a study was conducted to examine the effect of air pollution, environmental and other factors, 15 of them, on the death rate for (60) sixty metropolitan areas in the United States. The data is given in Gunst and Mason ([16], Appendix B, pp. 368-372). In this problem

$\mathbf{b}$ = Age adjusted mortality rate, or death per 100,000 population
$\mathbf{a}_1$ = Average annual precipitation in inches
$\mathbf{a}_2$ = Average January temperature in degrees F
$\mathbf{a}_3$ = Average July temperature in degrees F
$\mathbf{a}_4$ = Percentage of 1960 population 65 years and older
$\mathbf{a}_5$ = Population per household, 1960
$\mathbf{a}_6$ = Median school years completed to those over 22 years
$\mathbf{a}_7$ = Percentage of housing units that are sound
$\mathbf{a}_8$ = Population per square mile in urbanized area, 1960
$\mathbf{a}_9$ = Percentage of 1960 urbanized non-white population
$\mathbf{a}_{10}$ = Percentage employed in white collar in 1960
$\mathbf{a}_{11}$ = Percentage of families with annual income < \$3000

$\mathbf{a}_{12}$ = Relative pollution potential of Hydrocarbons
$\mathbf{a}_{13}$ = Relative pollution potential of Oxides of Nitrogen
$\mathbf{a}_{14}$ = Relative pollution of Sulfur Dioxide
$\mathbf{a}_{15}$ = Percentage relative humidity, annual average at 1 pm

One purpose of this study was to determine whether the pollution variables defined by $\mathbf{a}_{12}$, $\mathbf{a}_{13}$ and $\mathbf{a}_{14}$ were influential in the results, once the other factors (other $\mathbf{a}_i$) were included.

1.2.6 Effects of forecasting

This example displays a case of **mild multicollinearity** (ill-conditioning of matrix **A**). For the subject of multicollinearity, see Section 17.8. The problem discusses the factors of domestic production, stock formation and domestic consumption of the imports, all measured in millions of French francs, from the years 1949 through 1966. The data for this problem is given in Chatterjee and Price ([14], p. 152).

$\mathbf{b}$ = Imports
$\mathbf{a}_1$ = **1**, an n-column of 1's
$\mathbf{a}_2$ = Domestic production
$\mathbf{a}_3$ = Stock formation
$\mathbf{a}_4$ = Domestic consumption

Statistically, this model is not well specified. The solution of this problem gives a negative value to x_1, the coefficient of $\mathbf{a}_2$. The reason is that the problem has collinear data. In other words, matrix **A** is badly conditioned. It means that one or more of the columns of matrix **A** are mildly dependent on the others.

This example is solved again as Example 17.5 in Section 17.11, using the Ridge equation.

1.2.7 Factors affecting gross national products

This example displays case of **strong multicollinearity**. The gross national products for 49 countries were presented as a function of six socioeconomic factors. The data for this problem is given in Gunst and Mason ([16], Appendix A, p. 358).

$\mathbf{b}$ = Gross national products

$\mathbf{a}_1$ = Infant dearth rate
$\mathbf{a}_2$ = Physician/population ratio
$\mathbf{a}_3$ = Population density
$\mathbf{a}_4$ = Density as a function of agricultural land area
$\mathbf{a}_5$ = Literacy measure
$\mathbf{a}_6$ = An index of higher education

The results of this data indicate a strong multicollinearity (dependence) between the population density and density as a function of agricultural land area (columns $\mathbf{a}_3$ and $\mathbf{a}_4$).

Mason and Gunst ([16], pp. 253-259) solved this example and gave the results for few options, of which we report the following:

(1) They obtained the least squares solution for the full data by ignoring the strong multicollinearity (dependence) between $\mathbf{a}_3$ and $\mathbf{a}_4$, and as a result, matrix $(\mathbf{A}^T\mathbf{A})$ is nearly singular.

(2) They used a **Ridge technique** (Section 17.11) to overcome the near singularity of matrix $(\mathbf{A}^T\mathbf{A})$; reasonable results were obtained.

(3) Two points that influenced the outcome of the results were deleted. These two points were outliers (odd points) and their influences masked one another in one of the statistical tests. When these two points were deleted from the data, satisfactory results for certain estimates were obtained. See Mason and Gunst [20].

Once more, solving this example in the L_1 sense instead of the least squares sense would resolve the issue of the odd points.

1.3 Applications to industry

1.3.1 Windmill generating electricity

A research engineer was investigating the use of a windmill to generate DC (direct current) electricity. The data is the DC output vs. wind velocity and is given in Montgomery and Peck ([21], p. 92). The suggested model was

$$x_1\mathbf{1} + x_2\mathbf{V} = \mathbf{DC} \tag{1.3.1}$$

where **DC** is the current output, **1** is an n-column of 1's and **V** is the wind velocity.

Inspection of the plot of **DC** (the y-axis) and **V** (the x-axis) suggests that as the velocity of the wind increases, the DC output approaches an upper limit and the linear model of (1.3.1) is not suitable.

A quadratic model (inverted parabola) was then suggested, namely

$$x_0\mathbf{1} + x_1\mathbf{V} + x_2\mathbf{V}^2 = \mathbf{DC}$$

However, this model is also not suitable because the inverted parabola will eventually reach a maximum point and will bend downwards.

A third model was suggested which is

$$x_0\mathbf{1} + x_1\mathbf{V}' = \mathbf{DC}$$

Here, $\mathbf{V}' = (1/\mathbf{V})$. A plot of **DC** (y-axis) and $\mathbf{V}'$ (x-axis) is a straight line with negative slope and it seems more appropriate.

1.3.2 A chemical process

This example displays a case of **severe multicollinearity**. This example concerns the percentage of conversion of n-Heptane to acetylene P. It is given in Montgomery and Peck ([21], pp. 311, 312). There are three main factors; reactor temperature T, ratio of H_2 to n-Heptane H and contact time C. Data is collected for 16 observations. The given model is quadratic, as follows

$$\text{(1.3.2)} \qquad \mathbf{P} = x_0\mathbf{1} + x_1\mathbf{T} + x_2\mathbf{H} + x_3\mathbf{C} + x_4\mathbf{TH} + x_5\mathbf{TC} + x_6\mathbf{HC} + x_7\mathbf{T}^2 + x_8\mathbf{H}^2 + x_9\mathbf{C}^2$$

That is, in (1.3.2), $\mathbf{b} = \mathbf{P}$, $\mathbf{a}_1 = \mathbf{1}$, $\mathbf{a}_2 = \mathbf{T}$, $\mathbf{a}_3 = \mathbf{H}$, …, etc. **TH** means each element of **T** is multiplied by the corresponding element of **H**. Similarly, $\mathbf{T}^2$ means each element of **T** is squared, etc. In this example, matrix **A** is highly ill-conditioned, or one says, there is severe multicollinearity due to strong near linear dependency in the data. Column $\mathbf{a}_2 = \mathbf{T}$ is highly interrelated with $\mathbf{a}_4 = \mathbf{C}$, the longer the time of operation, the higher is the temperature. There are also quadratic terms and cross product terms.

Model (1.3.2) was replaced by 5 other models which are 5 subsets of (1.3.2). We cite two of them here: (a) Linear model, where the first 4 columns of **A** are kept and the other 6 columns were eliminated and (b) All the columns with **C** were eliminated and the equation is full quadratic in **T** and **H** only. The first model still maintains multicollinearity. The second model appears more satisfactory, since the causes of ill-conditioning of matrix **A** are eliminated. This problem is studied further in Example 17.6 of Section 17.11, using the Ridge equation.

1.4 Applications to digital images

1.4.1 Smoothing of random noise in digital images

Digital images are usually corrupted by random noise. A gray level image may be enhanced by a smoothing operation. Each pixel in the image is replaced by the average of the pixel values of all the pixels inside a square window or a mask centered at the pixel. Let the window be of size L by L, where L is an odd integer.

We show here that the smoothing operation is equivalent to fitting a plane in the least squares sense to all the pixels in the window. The ordinate of the fitting plane at the centre of the window is the new pixel value in the enhanced image, which equals the average of the pixel values inside the window.

This idea was given first by Graham [15] and was rediscovered later by Haralick and Watson [17]. We present here this idea and elaborate on it.

Let $N = L^2$ be the number of pixels inside the window. Let us take $N = 9$, i.e., $L = 3$. Let the coordinate of the pixel at hand be (0, 0). Let the pixels around the centre have coordinates as shown in Figure 1-1. Let these pixels be labeled 1 to 9 as shown in Figure 1-2.

$$\begin{array}{cccc} & \rightarrow j & & \\ \downarrow & (-1,-1) & (-1,0) & (-1,1) \\ i & (0,-1) & (0,0) & (0,1) \\ & (1,-1) & (1,0) & (1,1) \end{array}$$

Figure 1-1: Relative coordinates of pixels inside a 3 by 3 window

$$\begin{array}{ccc} 1 & 2 & 3 \\ 4 & 5 & 6 \\ 7 & 8 & 9 \end{array}$$

Figure 1-2: Pixel labels inside the 3 by 3 window

Let the pixel values inside the window be given by z_k, $k = 1, 2, \ldots, 9$ and let the fitting plane for the pixels be

(1.4.1) $$x_1\mathbf{1} + x_2\mathbf{i} + x_3\mathbf{j} = \mathbf{z}$$

where **1** is an 9-column of 1's, **i** and **j** are the coordinates of the pixels, and x_1, x_2 and x_3 are to be calculated. By substituting the pixel coordinates from Figures 1.1 and 1.2, into (1.4.1) we get the equation

(1.4.2) $$\mathbf{Ax} = \mathbf{z}$$

which is

$$\begin{bmatrix} 1 & -1 & -1 \\ 1 & -1 & 0 \\ 1 & -1 & 1 \\ 1 & 0 & -1 \\ 1 & 0 & 0 \\ 1 & 0 & 1 \\ 1 & 1 & -1 \\ 1 & 1 & 0 \\ 1 & 1 & 1 \end{bmatrix} \begin{bmatrix} x_1 \\ x_2 \\ x_3 \end{bmatrix} = \begin{bmatrix} z_1 \\ z_2 \\ z_3 \\ z_4 \\ z_5 \\ z_6 \\ z_7 \\ z_8 \\ z_9 \end{bmatrix}$$

The least squares solution of this equation is obtained by solving the so-called **normal equation** to equation (1.4.2), which is obtained by pre-multiplying equation (1.4.2) by $\mathbf{A}^T$, the transpose of $\mathbf{A}$ (equation (17.2.3)), namely ($\mathbf{A}^T\mathbf{Ax} = \mathbf{A}^T\mathbf{z}$), which gives

$$\begin{bmatrix} 9 & 0 & 0 \\ 0 & 6 & 0 \\ 0 & 0 & 6 \end{bmatrix} \begin{bmatrix} x_1 \\ x_2 \\ x_3 \end{bmatrix} = \begin{bmatrix} Z_1 \\ Z_2 \\ Z_3 \end{bmatrix}$$

where

$$Z_1 = (z_1 + z_2 + \ldots + z_9)$$
$$Z_2 = (-z_1 - z_2 - z_3 + z_7 + z_8 + z_9)$$
$$Z_3 = (-z_1 + z_3 - z_4 + z_6 - z_7 + z_9)$$

From equation (1.4.1), x_1 is the value at the centre of the window, (0, 0), which is

$$\sum_{k=1}^{9} z_k/9$$

or for any other size of window $N = L^2$

$$\sum_{k=1}^{N} z_k/N$$

The obtained result proves the argument concerning smoothing of random noise in digital images.

Smoothing random noise by applying non-square masks, such as pentagonal and hexagonal windows were suggested by Nagao and Matsuyama [22]. Such operations are explained in a similar manner to the above in [3].

1.4.2 Filtering of impulse noise in digital images

Another type of noise that contaminates a gray level digital image is the impulse noise known as salt and/or pepper noise. The noisy image, in this case, has white and/or black dots, investing the whole image. Filtering the impulse noise may be done by applying a simple technique using L_1 approximation. The noise salt and pepper dots are identified as follows. If the gray level intensity of the image be measured from 0 to 256. A white (salt) noise pixel has the intensity zero and a black (pepper) noise pixel has the intensity 256.

The L_1 approximation has an interesting property that the L_1 approximating curve to a given point set interpolates (passes through) a number of points of the given point set. The number of interpolated points equals at least the number of the unknown parameters of the approximating curve. As a result, the L_1 approximation of a given point set that contains an odd (wild) point, almost entirely ignores the odd point. See Figure 2-1. Each of the salt and the pepper points in the gray image is an odd or a wild point.

Let us have a set of three points in the x-y plane of which the middle point is a wild point. Let us fit the straight line $\mathbf{y} = a_1 + a_2\mathbf{x}$ in the L_1 norm to these three points. According to the property described above, the best fit in the L_1 norm would be the straight line that interpolates the first and the third points. In this case, the approximated y value on the straight line of the middle (wild) point $= (y_1 + y_3)/2$, where y_1 and y_3 are the y-coordinates of the first and third points respectively. This idea is now utilized to filter the impulse noise corrupting a digital gray level image.

Let the image in the x-y plane be of size N by M pixels. We first scan the digital noisy image horizontally, one row at a time and identify the impulse noise pixels (that have gray intensity zero or 256). We scan the image from row 1 to row N, starting from the second pixel to the last pixel but one in each row. A horizontal window of size 1 by 3 is centered at each noise pixel. Let the three pixel values inside the window be denoted in succession by z_1, z_2 and z_3. An L_1 fit of the three pixel values would interpolate pixels 1 and 3 of the window. The approximated pixel value of the middle (noise) pixel would be $\underline{z}_2 = (z_1 + z_3)/2$. This would filter the impulse noise pixel.

At the end of this process, the noisy image is only partially filtered as the impulse noise pixels may not all be removed. Any two adjacent white pixels or any two adjacent black pixels in any row would not be filtered.

We then reapply the same process to the partially filtered image in a column-wise manner. At the end of the second process, most of the impulse noise pixels are filtered. Experimental results [3] show that this technique is powerful in filtering impulse noise.

1.4.3 Applications to pattern classification

We solve here overdetermined systems of linear inequalities. This problem is discussed in detail in Chapter 16. We describe it here for completion. Given is a class **A** of s patterns and a class **B** of t patterns, where each pattern is a point in an m-dimensional Euclidean space. Let $n = (s + t)$. Usually $n >> m$.

It is required to find a surface in the m-dimensional space such that all points of class **A** be on one side of this surface and all points of class **B** be on the other side of the surface. Let the equation of this separating surface be

$$a_1\phi_1(x) + a_2\phi_2(x) + \ldots + a_{m+1}\phi_{m+1}(x) = 0$$

where $\mathbf{a} = (a_1, a_2, \ldots a_{m+1})^T$ is a vector of parameters or weights to be calculated and $\{\phi_1(x), \phi_2(x), \ldots, \phi_{m+1}(x)\}$ is a set of linearly independent functions to be specified according to the geometry of the problem. Following Tou and Gonzalez ([25], pp. 40-41, 48-49), a decision function d(x) of the form

$$d(x) = a_1\phi_1(x) + a_2\phi_2(x) + \ldots + a_{m+1}\phi_{m+1}(x)$$

is established. This function has the property that

(1.4.3a) $$d(x_i) < 0, \quad x_i \in \mathbf{A}$$

(1.4.3b) $$d(x_i) > 0, \quad x_i \in \mathbf{B}$$

By multiplying the first set of inequalities by a –ve signs, we get

(1.4.3c) $$-d(x_i) > 0, \quad x_i \in \mathbf{A}$$

The problem may now be posed as follows. Using (1.4.3c) and (1.4.3b), let **C** be an n by (m + 1) matrix whose i^{th} row $\mathbf{C}_i$ is

$$\mathbf{C}_i = (-\phi_1(x_i), -\phi_2(x_i), \ldots, -\phi_m(x_i), -\phi_{m+1}(x)), \; 1 \le i \le s$$

and

$$\mathbf{C}_i = (\phi_1(x_i), \phi_2(x_i), \ldots, \phi_m(x_i), \phi_{m+1}(x)), \quad (s + 1) \le i \le n$$

It is required to calculate the elements of the (m + 1)-vector **a** that satisfies the inequalities

$$\mathbf{Ca} > \mathbf{0}$$

We now have a system of linear inequalities and is solved by linear one-sided approximation algorithms [2]. See Chapter 16.

1.4.4 Restoring images with missing high-frequency components

We solve here a constrained underdetermined system of linear equations. Let the algebraic image restoration problem in the one-dimension case be defined by the system of linear equations

(1.4.4) $$\mathbf{Ax} = \mathbf{b}$$

where **A** is the discrete Fourier transform low-pass filter matrix and **b** is the observed image vector. Matrix $\mathbf{A}$ is an N by L matrix, $N \geq L$.

Equation (1.4.4) is an inconsistent equation and thus has no exact solution as vector $\mathbf{b}$ is often contaminated with noise. By inconsistent, we mean $\text{rank}(\mathbf{A}|\mathbf{b}) > \text{rank}(\mathbf{A})$. It is made consistent by pre-multiplying it by matrix $\mathbf{A}^T$ the transpose of $\mathbf{A}$. We get

$$(1.4.5) \qquad \mathbf{A}^T\mathbf{A}\mathbf{x} = \mathbf{A}^T\mathbf{b}$$

This equation may be written as

$$\mathbf{C}\mathbf{x} = \mathbf{c}$$

where $\mathbf{C} = (\mathbf{A}^T\mathbf{A})$ and $\mathbf{c} = (\mathbf{A}^T\mathbf{b})$.

If matrix $\mathbf{A}$ is of rank L, equation (1.4.5) is the normal equation and its solution is the least squares solution of $\mathbf{A}\mathbf{x} = \mathbf{b}$ and is given by $\mathbf{x} = (\mathbf{A}^T\mathbf{A})^{-1}\mathbf{A}^T\mathbf{b}$ (Chapter 17).

However, in general, matrix $\mathbf{A}$ is a rank deficient matrix and is of rank k, $k \leq L$. As a result, matrix $(\mathbf{A}^T\mathbf{A})$ is singular and has no inverse. In other words, system (1.4.5) or equation $\mathbf{C}\mathbf{x} = \mathbf{c}$ has $k \leq L$ linearly independent equations and $(L - k)$ redundant equations. Let these k linearly independent equations be given by (see also equation (1.4.13) in Section 1.4.7)

$$(1.4.5a) \qquad \mathbf{C}_{(k)}\mathbf{x} = \mathbf{c}_{(k)}$$

which is an underdetermined system of k equations in L unknowns.

One may obtain equation (1.4.5a) from equation $\mathbf{C}\mathbf{x} = \mathbf{c}$ by applying Gauss-Jordan elimination steps (Section 4.6) to matrix $\mathbf{C}$ and its updates with partial pivoting. Since $\mathbf{C} = (\mathbf{A}^T\mathbf{A})$ is symmetric positive semi-definite, one may pivot on the diagonal elements of matrix $\mathbf{C}$ and its updates and exchange (permute) the equations of $\mathbf{C}\mathbf{x} = \mathbf{c}$ and the columns of $\mathbf{C}$. After $k \leq L$ steps, the pivot elements become small enough and would be replaced by 0's. We thus get the k linearly independent equations (1.4.5a).

The physical problem identifies the solution vector $\mathbf{x}$ of (1.4.5a) as having non-negative and bounded elements

$$(1.4.6) \qquad 0 \leq x_i \leq x_{max}, \quad i = 1, 2, \ldots, L$$

The underdetermined system (1.4.5a) may now be solved by minimizing its solution vector $\mathbf{x}$ in the L_1 norm subject to the constraints (1.4.6). This problem is known as the **minimum fuel problem for discrete linear admissible control systems** and it is solved in [9] by the algorithm described in Chapter 21.

One may also solve equation (1.4.5a) subject to the constraints (1.4.6) in the least squares sense (L_2 norm) [1]. This problem is known as the **minimum energy problem for discrete linear admissible control systems** and it is solved by the algorithm described in Chapter 23. See Section 1.4.7.

It was found, however, that solving equation (1.4.5a) in the L_1 norm subject to the constraints (1.4.6) is much faster than obtaining the solution in the L_2 norm [10].

1.4.5 De-blurring digital images using the Ridge equation

In the linear model for image restoration of digital images, the problem is described by the Fredholm integral equation of the first kind (Chapter 19). The discretization of this equation gives a system of linear equations of the form

(1.4.7) $$[\mathbf{H}]\mathbf{f} = \mathbf{g}$$

where **g** is a stacked real m-vector representing the known or given degraded (blurred) image, **f** is a stacked real n-vector representing the unknown or un-degraded image. [**H**] is an n by m real matrix.

If the degraded image **g** is represented by an I by J matrix, $n = I \times J$. Also, if the unknown un-degraded image **f** is represented by a K by L matrix, $m = K \times L$. Without loss of generality, we assume that $n \geq m$.

It is common to solve equation (1.4.7) in the least squares sense. However, this equation is ill-posed in the sense that small error in vector **g** may cause large error in the solution vector **f**. A successful technique for overcoming the ill-posedness of equation (1.4.7) is to regularize or dampen its least squares solution. This method is also known as the **Ridge technique** (Section 17.11). Variations of the Ridge equation were given by some authors, such as Phillips [23].

The dampened least squares solution to system (1.4.7) is obtained from the normal equation

(1.4.8) $$([\mathbf{H}]^T[\mathbf{H}] + \varepsilon[\mathbf{I}])\mathbf{f} = [\mathbf{H}]^T\mathbf{g}$$

$[\mathbf{H}]^T$ is the transpose of [**H**] and [**I**] is an m-unit matrix. The parameter ε, $0 \leq \varepsilon \leq 1$, is known as the **regularization parameter**. Adding the term $\varepsilon[\mathbf{I}]$ in the above equation results in adding the positive quantity

ε to each of the diagonal elements of $[\mathbf{H}]^T[\mathbf{H}]$ which are themselves non-negative and real. Also it means adding the positive quantity ε to each of the eigenvalues of $[\mathbf{H}]^T[\mathbf{H}]$.

Hence, assuming that the matrix on the l.h.s. of (1.4.8) is non-singular, an approximate solution to (1.4.8) is given by

$$\underline{\mathbf{f}} = ([\mathbf{H}]^T[\mathbf{H}] + \varepsilon[\mathbf{I}])^{-1}[\mathbf{H}]^T\mathbf{g}$$

The parameter ε is increased or decreased, and a new solution is calculated each time. This is usually done a few times until a visually acceptable solution is obtained. The cost of these repeated solutions in terms of computation time is prohibitive if the problem is to be solved from scratch each time. However, in this problem, the regularization parameter for the best or near-best solution may be obtained by an inverse interpolation method. This is explained in detail in [8]. The inverse interpolation method is described by Ralston ([24], p. 657).

1.4.6 De-blurring images using truncated eigensystem

As in the previous example, in the linear model for image restoration of digital images, the problem is described by the Fredholm integral equation of the first kind. The discretization of this equation gives a system of linear equations of the form

$$[\mathbf{H}]\mathbf{f} = \mathbf{g} \tag{1.4.9}$$

The definitions and the dimensions of matrix $[\mathbf{H}]$ and vectors $\mathbf{g}$ and $\mathbf{f}$ are exactly as are given at the beginning of the previous example. We consider here the case where matrix $[\mathbf{H}]$ is the discretization of the so-called space-invariant point-spread function (SIPSF).

Equation (1.4.9) is ill-posed in the sense that small changes in vector $\mathbf{g}$ result in large changes in the solution vector $\mathbf{f}$. There are two main approaches for obtaining a numerically stable and physically acceptable solution to this equation, using direct (non-iterative) methods. The first approach is demonstrated by the Ridge technique described in Section 1.4.5.

The second approach is the technique of a truncated singular value decomposition (SVD – Chapter 17), or truncated eigensystem of matrix $[\mathbf{H}]$. A major drawback to using the SVD expansion is the high cost in terms of the number of arithmetic operations of computing the singular value system, especially for large size matrices.

However, in this example, an efficient method is used since matrix [**H**] has a structure similar to that of the so-called circulant matrix. Circulant matrices have special properties [13]. Hence, [**H**] may be replaced by its approximate circulant matrix. As a result, the border regions are the only regions in the unknown image that are affected. The obtained system of linear equations may be solved efficiently by using the Fast Fourier Transform (FFT) techniques. See for example, Andrews and Hunt ([11], Chapter 8).

Matrix [**H**] in (1.4.9) should be a square and symmetric one, so as to have real eigenvalues. If [**H**] is not square and symmetric, we pre-multiply (1.4.9) by $[\mathbf{H}]^T$, and get a square symmetric coefficient matrix of the equation $[\mathbf{H}]^T[\mathbf{H}]\mathbf{f} = [\mathbf{H}]^T\mathbf{g}$.

Without loss of generality, we assume that the coefficient matrix in (1.4.9) is m by m and symmetric. We now replace matrix [**H**] in (1.4.9) by its approximate circulant matrix, denoted by $[\mathbf{H}_c]$. We get

$$[\mathbf{H}_c]\mathbf{f} = \mathbf{g}$$

Once more, following the approach of Baker et al. [12], matrix $[\mathbf{H}_c]$ may be approximated by another circulant matrix of smaller rank r. This is done by retaining the largest r diagonal elements in absolute value, in the eigensystem of $[\mathbf{H}_c]$, $r \leq m$, and replacing the remaining $(m - r)$ diagonal elements by 0's. Hence, this is called the **truncated eigensystem technique**.

A method for finding out the optimum value of the rank r of matrix $[\mathbf{H}_c]$ is described in detail in [7]. Also in [7], analysis is provided for the case when the point-spread function is spatially-separable. That is, matrix $[\mathbf{H}_c]$ may be written in the form $[\mathbf{H}_c] = [\mathbf{A}]\otimes[\mathbf{B}]$, where $\otimes$ is the Kronecker product operator.

We note here that some experimentation was done with a truncated **LU** factorization for restoring blurred images [5]. However, results were not as successful as with the truncated eigensystem technique.

1.4.7 De-blurring images using quadratic programming

As in the previous two examples, in the linear model for image restoration of digital images, the problem is described by the

Fredholm integral equation of the first kind. The discretization of this equation gives a system of linear equations of the form

$$[\mathbf{H}]\mathbf{f} = \mathbf{g} \tag{1.4.10}$$

The definitions and the dimensions of matrix $[\mathbf{H}]$ and vectors $\mathbf{g}$ and $\mathbf{f}$ are exactly as those given in the previous two examples. However, in this example, for a realistic solution, equation (1.4.10) is solved under the conditions that the elements of vector $\mathbf{f}$ satisfy

$$0 \le f_j \le f_{max} \tag{1.4.11}$$

where f_{max} is a specified pixel value, usually equal to 256.

Equation (1.4.10) is ill-posed. As in the previous example, the ill-posedness of equation (1.4.10) is dealt with by one of two ways. The first is by a dampening technique such as that demonstrated by the Ridge technique described in Section 1.4.5. The second approach is by using a rank reduction technique to matrix $[\mathbf{H}]$ or equivalently to matrix $[\mathbf{H}]^T[\mathbf{H}]$. One way of doing this is to use a truncated eigensystem of matrix $[\mathbf{H}]$, as demonstrated by the previous example. In this method, we apply another rank reduction technique to matrix $[\mathbf{H}]^T[\mathbf{H}]$. That is besides taking into account the physical conditions (1.4.11).

To start, convert system (1.4.10) to a consistent system of linear equations by pre-multiplying by $[\mathbf{H}]^T$. We get

$$[\mathbf{H}]^T[\mathbf{H}]\mathbf{f} = [\mathbf{H}]^T\mathbf{g} \tag{1.4.12}$$

Let $[\mathbf{D}] = [\mathbf{H}]^T[\mathbf{H}]$ and $\mathbf{b} = [\mathbf{H}]^T\mathbf{g}$. Then (1.4.12) becomes

$$[\mathbf{D}]\mathbf{f} = \mathbf{b} \tag{1.4.13}$$

Assume the m by m matrix $[\mathbf{H}]$ in (1.4.12) is of rank r. Then system (1.4.13) has r linearly independent equations and $(m - r)$ dependent (redundant) equations.

Assume that the equations in (1.4.13) are permuted such that the first r equations are linearly independent and the last $(m - r)$ are the linearly dependent equations. Let the first r equations be given by

$$[\mathbf{C}]\mathbf{f} = \mathbf{p} \tag{1.4.14}$$

where $[\mathbf{C}]$ is an r by m matrix $r \le m$, and $\mathbf{p}$ is an r-vector. Equation (1.4.14) is a consistent underdetermined system of linear equations and thus the residual $\mathbf{r} = [\mathbf{C}]\mathbf{f} - \mathbf{p} = \mathbf{0}$ and it has an infinite number of solutions (Chapter 4).

In order to obtain a unique least squares solution, the problem is formulated as find **f** that minimizes (**f**, **f**), subject to the condition $[\mathbf{C}]\mathbf{f} - \mathbf{p} = \mathbf{0}$. This is known as the **minimal length least squares solution** of (1.4.14) (Theorem 17.3).

Let us now take condition (1.4.11) into account. Let

$$a_i = f_i/f_{max}, \quad i = 1, 2, \ldots, m$$

and

$$d_j = p_j/f_{max}, \quad j = 1, 2, \ldots, r$$

where a_i and d_j are the elements of vectors **a** and **d** respectively.

This problem is now formulated as a quadratic programming problem with bounded variables, as follows

$$\text{minimize } \mathbf{a}^T\mathbf{a}$$

subject to

$$0 \leq a_i \leq 1, \quad i = 1, 2, \ldots, m$$

and

$$[\mathbf{C}]\mathbf{a} = \mathbf{d}$$

As in Section 1.4.4, this problem is also known as the Minimum energy problem for discrete linear admissible control systems (Chapter 23). In the current example, for estimating the rank r of system $[\mathbf{C}]\mathbf{a} = \mathbf{d}$, which gives a best or near-best solution, we may use a special technique, based on the knowledge of the unbiased estimate of the variance and the mean of the noise in the blurred image. The detail of this method is given in [6].

The above cited applications are a small fraction of those in the field of linear approximation. The author also applied linear approximation to segmentation of 3-D range images [4, 19] and similar problems.

References

1. Abdelmalek, N.N., Restoration of images with missing high-frequency components using quadratic programming, *Applied Optics*, 22(1983)2182-2188.
2. Abdelmalek, N.N., Linear one-sided approximation algorithms for the solution of overdetermined systems of linear inequalities, *International Journal of Systems Science*, 15(1984)1-8.
3. Abdelmalek, N.N., Noise filtering in digital images and approximation theory, *Pattern Recognition*, 19(1986)417-424.
4. Abdelmalek, N.N., Heuristic procedure for segmentation of 3-D range images, *International Journal of Systems Science*, 21(1990)225-239.
5. Abdelmalek, N.N. and Kasvand, T., Image restoration by Gauss **LU** decomposition, *Applied Optics*, 18(1979)1684-1686.
6. Abdelmalek, N.N. and Kasvand, T., Digital image restoration using quadratic programming, *Applied Optics*, 19(1980)3407-3415.
7. Abdelmalek, N.N., Kasvand, T. and Croteau, J.P., Image restoration for space invariant pointspread functions, *Applied Optics*, 19(1980)1184-1189.
8. Abdelmalek, N.N., Kasvand, T., Olmstead, J. and Tremblay, M.M., Direct algorithm for digital image restoration, *Applied Optics*, 20(1981)4227-4233.
9. Abdelmalek, N.N. and Otsu, N., Restoration of Images with missing high-frequency components by minimizing the L_1 norm of the solution vector, *Applied Optics*, 24(1985)1415-1420.
10. Abdelmalek, N.N. and Otsu, N., Speed comparison among methods for restoring signals with missing high-frequency components using two different low-pass-filter matrix dimensions, *Optics Letters*, 10(1985)372-374.
11. Andrews, H.C. and Hunt, B.R., *Digital Image Restoration*, Prentice-Hall, Englewood Cliffs, NJ, 1977.

12. Baker, C.T.H., Fox, L., Mayers, D.F. and Wright, K., Numerical solution of Fredholm integral equations of the first kind, *Computer Journal*, 7(1964)141-148.
13. Bellman, R., *Introduction to Matrix Analysis*, McGraw-Hill, New York, 1970.
14. Chatterjee, S. and Price, B., *Regression Analysis by Example*, John Wiley & Sons, New York, 1977.
15. Graham, R.E., Snow removal – A noise-stripping process for picture signals, *IEEE Transactions on Information Theory*, IT-8(1962)129-144.
16. Gunst, R.F. and Mason, R.L., *Regression Analysis and its Application: A Data Oriented Approach*, Marcel Dekker, Inc., New York, 1980.
17. Haralick, M.R. and Watson, L., A facet model for image data, *Computer Graphics and Image Processing*, 15(1981)113-129.
18. Kleinbaum, D.G., Kupper, L.L. and Muller, K.E., *Applied Regression Analysis and Other Multivariate Methods*, Second Edition, PWS-Kent Publishing Company, Boston, 1988.
19. Kurita, T. and Abdelmalek, N.N., An edge based approach for the segmentation of 3-D range images of small industrial-like objects, *International Journal of Systems Science*, 23(1992)1449-1461.
20. Mason, R.L. and Gunst, R.F., Outlier-induced collinearities, *Technometrics*, 27(1985)401-407.
21. Montgomery, D.C. and Peck, E.A., *Introduction to Linear Regression Analysis*, John Wiley & Sons, New York, 1992.
22. Nagao, M. and Matsuyama, T., Edge preserving smoothing, *Computer Graphics and Image Processing*, 9(1979)394-407.
23. Phillips, D.L., A technique for the numerical solution of certain integral equations, *Journal of ACM*, 9(1962)84-97.
24. Ralston, A., *A First Course in Numerical Analysis*, McGraw-Hill, New York, 1965.
25. Tou, J.L. and Gonzalez, R.C., *Pattern Recognition Principles*, Addison-Wesley, Reading, MA, 1974.
26. Weisberg, S., *Applied Linear Regression*, John Wiley & Sons, Second Edition, New York, 1985.

Chapter 2

Preliminaries

2.1 Introduction

Each algorithm described in this book is implemented by a C function that typically has several child functions to perform sub-tasks in the computation. Every algorithm is accompanied by a driver function that provides examples of how to use the algorithm. All code is provided at the end of each chapter.

For naming convention, algorithm and child function names are prefixed with "LA_" (for Linear Approximation) and driver function names are prefixed with "DR_". Source file naming follows a similar convention. For example, the Linear L_1 approximation algorithm (Chapter 5) is implemented by LA_L1() and its child functions, all of which are in the source file LA_L1.c. Similarly, DR_L1() is in the source file DR_L1.c. For ease of use, the header file hierarchy is kept to a minimum. An application that exercises these algorithms must include the header file LA_Prototypes.h to access all the algorithms.

In this book, we adopt vector-matrix notation and use bold letters for vectors and matrices. In this chapter, we discuss the following issues:

(1) To start with, we show that the discrete linear approximation in a certain residual (error) norm is equivalent to the solution of an overdetermined system of linear equations in the same norm [1, 2].

(2) A comparison is made between approximation in the 3 main residual norms, the L_1, the L_2 and the L_∞, using a simple practical example. We also describe some main properties concerning the L_1 and the L_∞ approximations.

(3) In dealing with approximation problems using digital computers with finite word length, we have to specify tolerance parameters to account for round-off error. The values of the tolerance parameters are set according to the word lengths of single- and double-precision numbers.

(4) The C implementation of vectors and matrices in this book facilitates array indexing from 1 to n, instead of from 0 to (n – 1), the latter being the convention in C. This allows indexing, such as that found in “for” loops (“DO” loops in FORTRAN), to match the mathematical convention used to describe the algorithms. The algorithms dynamically allocate vectors and matrices according to the size of the data passed to them, then relinquish that memory upon termination.

(5) Outlier, spikes, wild or odd points in a given data are identified and some methods of dealing with them are described.

2.2 Discrete linear approximation and solution of overdetermined linear equations

To clarify the relationship between discrete linear approximation and the solution of overdetermined systems of linear equations, consider the following example.

Let us have the set of 8 points in the x-y plane: (1, 2), (2, 2.5), (3, 2), (4, 6.5), (5, 3.5), (6, 4.5), (7, 6), (8, 7). It is required to approximate this discrete set of points by the vertical parabola

$$y = a_1 + a_2x + a_3x^2 \tag{2.2.1}$$

where a_1, a_2 and a_3 are unknowns to be calculated. If we substitute the 8 given points in equation (2.2.1), we get

$$\begin{aligned}
a_1 + a_2 + a_3 &= 2 \\
a_1 + 2a_2 + 4a_3 &= 2.5 \\
a_1 + 3a_2 + 9a_3 &= 2 \\
a_1 + 4a_2 + 16a_3 &= 6.5 \\
a_1 + 5a_2 + 25a_3 &= 3.5 \\
a_1 + 6a_2 + 36a_3 &= 4.5 \\
a_1 + 7a_2 + 49a_3 &= 6 \\
a_1 + 8a_2 + 64a_3 &= 7
\end{aligned} \tag{2.2.2}$$

In vector-matrix form, the set of equations in (2.2.2) is written as

(2.2.3) $$\mathbf{Ca} = \mathbf{f}$$

C is the matrix on the l.h.s. of (2.2.2), **a** is the solution vector and **f** is the vector on the right hand side r.h.s., as in

$$\begin{bmatrix} 1 & 1 & 1 \\ 1 & 2 & 4 \\ 1 & 3 & 9 \\ 1 & 4 & 16 \\ 1 & 5 & 25 \\ 1 & 6 & 36 \\ 1 & 7 & 49 \\ 1 & 8 & 64 \end{bmatrix} \begin{bmatrix} a_1 \\ a_2 \\ a_3 \end{bmatrix} = \begin{bmatrix} 2 \\ 2.5 \\ 2 \\ 6.5 \\ 3.5 \\ 4.5 \\ 6 \\ 7 \end{bmatrix}$$

This set of 8 equations in the 3 the unknowns (a_1, a_2, a_3) has no exact solution and can only be solved approximately. The approximate solution is done with respect to a certain measure or norm of the error vector. The error, or the **residual vector**, is the difference between the l.h.s. and the r.h.s. in (2.2.3)

$$\mathbf{r} = \mathbf{Ca} - \mathbf{f}$$

As indicated in Chapter 1, the (approximate) solution vector **a** of problem (2.2.2) or (2.2.3) requires that the norm of vector **r** be as small as possible. The p-measure or p-norm of vector **r**, is denoted by L_p or by $\|\mathbf{r}\|_p$ and is given by

$$z = \|\mathbf{r}\|_p = \left[\sum_{i=1}^{n} |r(x_i)|^p\right]^{1/p}, \quad 1 \leq p \leq \infty$$

where n is the number of elements of vector **r**.

Let $\mathbf{r} = (r(x_i))$, where $r(x_i)$ are the elements of vector **r**. In (2.2.2) it is given by

$$r(x_i) = a_1 + a_2x_i + a_3x_i^2 - y_i, \quad i = 1, 2, \ldots, n$$

Then for p = 1, the approximate solution of (2.2.2) or (2.2.3) requires that the norm z

$$z = \sum_{i=1}^{n} |r(x_i)| = \sum_{i=1}^{n} |a_1 + a_2x_i + a_3x_i^2 - y_i|$$

be as small as possible, where for this example $n = 8$. We call the obtained (approximate) solution vector **a**, the **L_1 solution** of system (2.2.2).

For $p = 2$, the approximate solution of (2.2.2) require function z be as small as possible

$$z = \text{sqrt}\left[\sum_{i=1}^{n} [r(x_i)]^2\right] = \text{sqrt}\left[\sum_{i=1}^{n} [a_1 + a_2x_i + a_3x_i^2 - y_i]^2\right]$$

The obtained solution vector **a** is the **L_2** or the **least squares solution** of system (2.2.2).

For $p = \infty$, we require that

$$z = \max|r(x_i)| = \max|a_1 + a_2\, x_i + a_3\, x_i^2 - y_i|, \quad i = 1, 2, \ldots, n$$

be as small as possible. We call the solution vector **a** the **Chebyshev,** the **L_∞**, or the **minimax** solution to system (2.2.2) or (2.2.3).

Now consider the following two problems and assume that all the functions are real valued.

Problem (a)

Let f(x) be a given function defined on a finite subset $\mathbf{X} = \{x_1, x_2, \ldots, x_n\}$ of an interval **I** on the real line. Let the set of arbitrary linearly independent continuous functions $\{\phi_1(\mathbf{x}), \phi_2(\mathbf{x}), \ldots, \phi_m(\mathbf{x})\}$, $m < n$, be defined on **I**. We define the polynomial $\mathbf{P}(a_1, a_2, \ldots, a_m, \mathbf{x})$ as

(2.2.4) $$\mathbf{P}(a_1, a_2, \ldots, a_m, \mathbf{x}) = a_1\phi_1(\mathbf{x}) + \ldots + a_m\phi_m(\mathbf{x})$$

or simply the function $\mathbf{P}(\mathbf{a}, \mathbf{x})$, where **a** denotes the parameter vector $(a_1, \ldots, a_m)^T$ in the E_m space. By comparing (2.2.4) and (2.2.1), $\phi_1(\mathbf{x}) = \mathbf{1}$, an n-vector of 1's, $\phi_2(\mathbf{x}) = \mathbf{x}$ and $\phi_3(\mathbf{x}) = \mathbf{x}^2$.

The linear L_1 approximation problem for f(**x**) on **X** is to determine vector **a** that minimizes the function

$$z = \sum_{i=1}^{n} |r(x_i)|$$

where the residuals $r(x_i)$ are given by

(2.2.5) $$r(x_i) = \mathbf{P}(\mathbf{a}, x_i) - f(x_i), \quad i = 1, 2, \ldots, n$$

The linear L_2 or the linear least squares approximation problem for $f(\mathbf{x})$ on $\mathbf{X}$ is to determine vector $\mathbf{a}$ that minimizes the function

$$z = \text{sqrt}\left[\sum_{i=1}^{n} [r(x_i)]^2\right]$$

The linear Chebyshev, L_∞ or minimax approximation problem for $f(\mathbf{x})$ on $\mathbf{X}$ is to determine vector $\mathbf{a}$ that minimizes the function

$$z = \max|r(x_i)|, \quad i = 1, 2, \ldots, n$$

Problem (b)

Consider now the overdetermined system of linear equations

(2.2.6) $$\begin{array}{l} c_{11}a_1 + c_{12}a_2 + \ldots + c_{1m}a_m = f_1 \\ \ldots \qquad \ldots \qquad \ldots \\ c_{n1}a_1 + c_{n2}a_2 + \ldots + c_{nm}a_m = f_n \end{array}$$

where $\mathbf{C} = (c_{ij})$ is an n by m constant matrix of rank m, m < n, and $\mathbf{f} = (f_i)$ and $\mathbf{a} = (a_i)$ are n- and m-vectors in the Euclidean m- and n-spaces respectively.

The linear L_1 solution to system (2.2.6) is to determine vector $\mathbf{a}$ that minimizes the function

$$z = \sum_{i=1}^{n} |r_i|$$

where

(2.2.7) $$r_i = c_{i1}a_1 + \ldots + c_{im}a_m - f_i, \quad i = 1, 2, \ldots, n$$

In the same manner, the L_2 and the L_∞ approximations for the system of linear equations (2.2.6) are defined.

The symbols used for problem (b) are chosen to match those of problem (a). In (2.2.6), $\mathbf{f} = (f_i)$ and $\mathbf{a} = (a_i)$ correspond to $(f(x_i))$ and (a_i) of problem (a) respectively. Also, matrix $\mathbf{C} = (c_{ij})$ corresponds to $(\phi_j(x_i))$. Consequently (r_i) of (2.2.7) corresponds to $(r(x_i))$ of (2.2.5).

It is clear, as demonstrated, that problem (a) is equivalent to problem (b); that is, the discrete linear approximation problem in a certain norm is equivalent to the solution of an overdetermined system of linear equations in the same norm. Throughout this book, the above-mentioned 3 norms are used.

Note 2.1

The approximation for each of the 3 norms is called **linear** because the residuals in (2.2.5) or in (2.2.7) depend linearly on the solution vector **a**.

Note 2.2

In most algorithms in this book, we define the residuals $r(x_i)$, as in (2.2.5), or r_i, as in (2.2.7). In some algorithms, the residuals are defined as the negative of the r.h.s. of (2.2.5) or (2.2.7). This choice is arbitrary and does not affect the analysis of the algorithm. Hence, if $\mathbf{r} = \mathbf{Ca} - \mathbf{f}$, then $\mathbf{Ca} = \mathbf{f} + \mathbf{r}$, and if $\mathbf{r} = \mathbf{f} - \mathbf{Ca}$, then $\mathbf{Ca} = \mathbf{f} - \mathbf{r}$.

Note 2.3

From now on, the expressions discrete linear approximation in a certain norm and the solution of an overdetermined system of linear equations in the same norm, would be used alternatively.

2.3 Comparison between the L_1, the L_2 and the L_∞ norms by a practical example

The motivation behind this section is summarized by the following. Suppose we are given a set of experimental data points in the x-y plane that contains spikes, odd or wild points. Let this data be approximated by a curve in the L_p norm, where $p \geq 1$.

It is known that the L_1 norm is recommended over other norms for fitting a curve to a data that contains odd or wild points [11]. We are given the set of 8 points of Section 2.2, which contains the wild point (4, 6.5). This set is approximated by vertical parabolas of the form $y = a_1 + a_2x + a_3x^2$ in the L_1, the L_2 and the L_∞ norms. The parameters a_1, a_2 and a_3 are calculated for each case. The results are shown in Figure 2-1.

We observe in this figure that the L_1 approximation almost entirely ignores the wild point (4, 6.5), while the L_2 and especially the L_∞ approximation curves bend towards the wild point.

Barrodale [4] presented numerical evidence showing that the L_1 approximation is superior to the L_2 approximation for data that contains some very inaccurate points. Rosen et al. [12], in a practical application of signal identification, also illustrated the robustness of a solution based on minimizing the L_1 error norm over minimizing the L_2 error norm.

The wild point (4, 6.5) is called an **outlier**, since its residual in the L_1 approximation is very large, compared with the residuals of the other points. As explained in the next section, the fitting curve in the L_1 norm passes through at least m points of the given set, where $m = \text{rank}(\mathbf{C})$. On the other hand, the Chebyshev approximation curve bends towards the wild point since $(m + 1)$ residuals in the L_∞ approximation, have the same absolute maximum L_∞ norm. The least squares approximation curve also curves towards the wild point but not as much as the Chebyshev approximation curve.

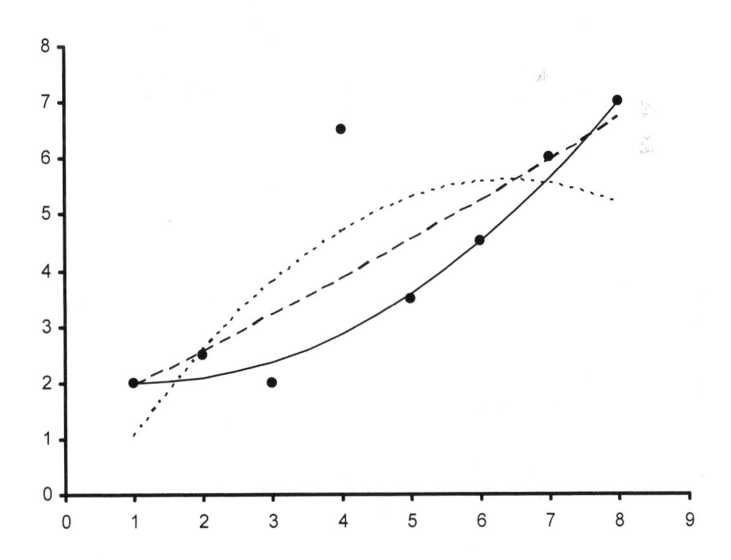

Figure 2-1: Curve fitting a set of 8 points, including a wild point, using the L_1, L_2 and L_∞ norms

In Figure 2-1, the solid curve is the L_1 approximation. The dashed curve is the L_2 (least squares) approximation and the dotted curve is the L_∞ (Chebyshev) approximation.

2.3.1 Some characteristics of the L_1 and the Chebyshev approximations

(a) The L_1 approximation

A main property is that the L_1 approximation curve (see Figure 2-1) passes through m of the given points, where m is the number of unknowns of the approximating curve. Here, $y = a_1 + a_2x + a_3x^2$, $m = 3$. It passes through points 1, 6 and 8. This is known as the **interpolatory characteristic of the L_1 approximation**. This property is explained further in Chapter 5. If the coefficient matrix in (2.2.2) is of rank k, $k < m$, the fitting curve will pass through at least k point. In fact, the interpolatory characteristic of the L_1 approximation results in making the L_1 approximation curve be far away from the wild point(s). See also Sposito et al. [14].

(b) The Chebyshev approximation

A main property of the Chebyshev approximation is that each of the $(m + 1)$ residuals for the given points equals $\pm z$, where z is the optimum Chebyshev norm and m is the number of unknowns of the approximating curve, which equals 3 in this example. We observe in Figure 2-1 that the 4 residuals for the points 3, 4, 5 and 8 are equal in magnitude and oscillating in sign; that is, $r_3 = +z$, $r_4 = -z$, $r_5 = +z$ and $r_8 = -z$. This characteristic is known as the **equioscillatory characteristic of the Chebyshev approximation**. This property will be discussed further in Chapter 10.

2.4 Error tolerances in the calculation

Since the programs in this book are for linear approximation, two tolerance parameters named "PREC" and "EPS" are specified in LA_Defs.h.

PREC stands for precision, or the round-off error that occurs in simple floating-point arithmetic operations [15]. Single-precision (s.p.) floating-point numbers typically have a precision of around 6 decimal places. This means that the figure after the 6th decimal place is either truncated if it is < 5, or rounded up if it is ≥ 5. Hence, for s.p. computation we set PREC = 1.0E–06.

EPS is the tolerance to be used during computation. A calculated number x is considered to be 0 if |x| < EPS. For s.p. computation we set EPS = 1.0E–04.

Double-precision (d.p.) floating-point numbers typically have a precision of around 16 decimal places. Thus the figure after the 16th decimal place is either truncated or rounded up. For d.p. computation we set PREC = 1.0E–16 and EPS = 1.0E–11. See Section 4.7.1 for a description of normalized floating-point representation.

The values of the parameters PRES and EPS are controlled by conditional compile variable DOUBLE_PRECISION, which is used to toggle PREC and EPS between single- and double-precision values.

The formats of s.p. and d.p numbers can vary between computer systems, so the values of PREC and EPS need to be adjusted accordingly.

Floating-point processors perform all simple arithmetic operations in at-least double-precision (often higher than d.p.), regardless of whether the result is stored in single- or double-precision.

All computation in our C implementation is performed in d.p. To simulate s.p. computation, as indicated in several sample problems in the book, we change PREC and EPS to s.p. values.

Note 2.4

Some programs in this book do not employ PREC, but all employ EPS.

2.5 Representation of vectors and matrices in C

Arrays in C are **zero-offset**, or **zero-origin**, meaning that an array of size n is indexed from [0] to [n – 1]). In FORTRAN (and the mathematical convention of linear approximation), arrays are indexed from [1] to [n]. In C, a vector $\mathbf{v} = (1, 2, 3, 4)^T$ of length 4 is accessed as v[0] = 1, v[1] = 2, v[2] = 3 and v[3] = 4, whereas in FORTAN (and the mathematical notation), it is accessed as v[1] = 1, …, v[4] = 4.

In order to counteract the zero-origin nature of C, after memory for a vector variable has been allocated (via malloc()), the pointer to the memory is decremented to allow indexing to start from 1 instead of 0. Statically-initialized vectors (used only for initial data in drivers) cannot have their pointers manipulated, so the value at the 0th index is set to 0 and ignored, and data initialization is from indices [1] to [n] such that $\mathbf{v} = (0, 1, 2, 3, 4)^T$, or v[1] = 1, … v[4] = 4. See also [8] and [10].

An n by m matrix variable is allocated using similar pointer manipulation such that it can be indexed from [1][1] to [n][m]. However, unlike statically-initialized vectors, statically-initialized matrices (also used only for initial data in drivers) are not padded with 0 elements, as this would necessitate the entire first column and the entire first row of each matrix to be set to 0's. Instead, a statically-initialized matrix, indexed from [0][0] to [n – 1][m – 1], is copied to a dynamically-allocated matrix variable indexed from [1][1] to [n][m].

LA_Utils.c contains a collection of utility functions for dynamic allocation, deallocation, copying and printing of vector and matrix data. See any driver function for examples of how these utilities are used.

2.6 Outliers and dealing with them

In Section 2.2, we have given 8 data points, with one of them being an odd point, or outlier. When the L_1 approximation was used to fit this data (Figure 2-1), the residual of the fourth point was much larger than the residuals of the other 7 points, so we denoted this point as an outlier. The L_1 curve fitting almost entirely ignores the outliers. As mentioned earlier, the reason is that in the L_1 curve fitting, the curve has to interpolate (passes through) at least k of the given points, k is the rank of the coefficient matrix.

On the other hand, in the L_∞ (Chebyshev) approximation for the same data, (k + 1) residuals have the same absolute maximum value and the curve bends towards this outlier. The L_2 (the Least squares) approximation also leans towards the outlier, not as much as the L_∞ approximation curve. As a result, the identification of these outliers is difficult as their residuals in the L_2 norm have been made small.

Hence, for the L_1 approximation, the fourth point of this data set is an outlier, while for the L_∞ approximation, this fourth point is not, and thus the notion of outliers depends on the fitting strategy. Statisticians mostly use the L_2 (the Least squares) approximation for which an outlier has a larger residual, compared with the residuals of the other points. They would like to use statistical measures to detect the outliers and discard them before the curve fitting takes place.

It is important to detect outliers, but it is equally important to realize that outliers should be discarded only if it is determined that they are not valid data points. Outliers that are valid data points usually suggest that the used model is incorrect. If the model is correct, an outlier that is a valid point might be given a smaller weight to suit the model. Deleting odd points as outliers from the data should be done if their presence can be attributed to a provable error in the data collection.

Ryan ([13], p. 408) gives an example of 10 data points that were fitted by a least squares method, in which the last point was an outlier. When using only the first 9 points, the residual estimator was reduced by 50%. The last point caused high correlation (dependence) between two of the columns of the coefficient matrix.

However, it is reasonable to discard the outlier that does not suggest the unsuitability of the model. Consider the fourth point in Figure 2-1. This point is not extreme in either its x-coordinate nor in its y-coordinate, but it has an influence on the equation of the L_2 and the L_∞ fit. In fact, when this outlier is removed, both the L_2 and the L_∞ curve fits for the remaining 7 points, almost coincide on the L_1 curve fit (see an almost identical example in [3], p. 420).

According to Ryan ([13], p. 350), there are 5 types of outliers:

(1) Regression outlier: a point that deviates from the rest of the points or at least from the majority of them.

(2) Residual outlier: a point that has a large calculated residual,

(3) x-outlier: a point that is outlying in regards to the x-coordinate only.

(4) y-outlier: a point that is outlying in regards to its y-coordinate only.

(5) x- and y-outlier: a point that is outlying in both coordinates.

Each one of these types of outliers may severely distort the coefficients of the approximating curve. These types of outliers may also be detected by statistical tests, which are outside the scope of this book. For statistical study of the problem, see for example, Belsley et al. [5], Cook and Weisberg [6] and Ryan [13].

2.6.1 Data editing and residual analysis

As suggested by Gunst and Mason [7], data screening before approximation calculations may get rid of costly errors that computer calculation may not detect. Because the approximating curve attempts to fit all the data points, the residuals might all be large, while if one or two odd points were eliminated before the calculation, that would give a better fit with smaller residuals.

One of the easiest ways for spotting irregularities in a given data is to visually scan the given data. Also, by routinely plotting the given data, one may specify the approximating curve, would it be a straight line or maybe better a quadratic, or a higher degree polynomial. By plotting the given data points, one may distinguish the outlier in the data and determine if it belongs to any one of the types described above.

One important point when examining plots of data points is to look at the overall trend of the data, not at small perturbations of few points. In some problems, plotting the data points does not clearly identify the pattern of the data. Eliminating the variability of the data may be done by a smoothing technique. This results in enabling the recognition of the trend of the data.

Gunst and Mason ([7], pp. 39-41) presented an example where they did just that. Given is the raw data of a set of points; the y-value of each data point (apart of the first and the last ones) is replaced by the median of the y-values of three points. These are the point itself and the points before and after it. This kind of smoothing technique may be repeated once or twice. Even after smoothing, one should examine the overall trend of the smoothed data rather than the localized trends. Median smoothing is not the only type of smoothing to reduce the variability of the raw data. Moving averages and exponential smoothing have also been used effectively.

Residual analysis is an important task in the approximation of given data. It means examining the differences between the residuals of the scanned or the plotted given data and the calculated residuals from the fitted curves. This kind of examination assists the detection of large residuals such as those of the outliers. A certain pattern may be observed and suggests actions to be taken, such as using a different approximation equation, eliminating some points from, or adding other points to the data.

Montgomery and Peck ([9], pp. 74-79) suggest plotting the residuals against the calculated values of y (or against x). If the plot indicates that the residuals are contained in a horizontal band, then there are no obvious model defects. Otherwise there might be symptomatic model deficiencies. Also, the effect of outliers on the fitting equation may be easily checked by dropping the outliers and re-calculating the fitting equation again. If the coefficients of the fitting equation are over sensitive to the removal of the outliers, then the chosen fitting equation may not be suitable for the given data, or that the outliers are to be removed to give a better fit for the remaining data points.

References

1. Abdelmalek, N.N., Linear L_1 approximation for a discrete point set and L_1 solutions of overdetermined linear equations, *Journal of ACM*, 18(1971)41-47.
2. Abdelmalek, N.N., On the discrete L_1 approximation and L_1 solutions of overdetermined linear equations, *Journal of Approximation Theory*, 11(1974)38-53.
3. Abdelmalek, N.N., Noise filtering in digital images and approximation theory, *Pattern Recognition*, 19(1986)417-424.
4. Barrodale, I., L_1 approximation and analysis of data, *Applied Statistics*, 17(1968)51-57.
5. Belsley, D.A., Kuh, E. and Welch, R.E., *Regression Diagnostics Identifying Influential Data and Sources of Collinearity*, John Wiley & Sons, New York, 1980.
6. Cook, R.D. and Weisberg, S., *Residuals and Influence in Regression*, Chapman-Hall, London, 1982.
7. Gunst, R.F. and Mason, R.L., *Regression Analysis and its Application: A Data Oriented Approach*, Marcel Dekker, Inc., New York, 1980.
8. Lau, H.T., *A Numerical Library in C for Scientists and Engineers*, CRC Press, Ann Arbor, 1995.
9. Montgomery, D.C. and Peck, E.A., *Introduction to Linear Regression Analysis*, John Wiley & Sons, New York, 1992.

10. Press, W.H., Flannery, B.P., Teukolsky, S.A. and Vetterling, W.T., *Numerical Recipes in C, The Art of Scientific Computing*, Second Edition, Cambridge University Press, Cambridge, 1992.
11. Rice, J.R. and White, J.S., Norms for smoothing and estimation, *SIAM Review*, 6(1964)243-256.
12. Rosen, J.B., Park, H. and Glick, J., Signal identification using a least L_1 norm algorithm, *Optimization and Engineering*, 1(2000)51-65.
13. Ryan, T.P., *Modern Regression Methods*, John Wiley & Sons, New York, 1997.
14. Sposito, V.A., Kennedy, W.J. and Gentle, J.E., Useful generalized properties of L_1 estimators, *Communications in Statistics-Theory and Methods*, A9(1980)1309-1315.
15. Wilkinson, J.H., *Rounding Errors in Algebraic Processes*, Prentice-Hall, Englewood Cliffs, NJ, 1963.

Chapter 3

Linear Programming and the Simplex Algorithm

3.1 Introduction

This chapter is a tutorial one. Its purpose is to introduce the reader to the subject of Linear Programming, which is used as the main tool for solving all the approximation problems in this book, with the exception of some least squares approximation problems.

In general, linear programs deal with optimization problems in real life, such as those found in industry, transportation, economics, numerical analysis and other fields [6, 8, 10, 11].

A linear programming problem is an optimization problem in which a linear function of a set of variables is to be maximized (or minimized) subject to a set of linear constraints. It may be formulated as follows.

Find the n variables $x_1, x_2, \ldots, x_n$ that maximize the linear function

$$z = c_1x_1 + c_2x_2 + \ldots + c_nx_n \tag{3.1.1}$$

subject to the m linear constraints (conditions)

$$\begin{array}{ccccccccc} a_{11}x_1 & + & a_{12}x_2 & + & \ldots & + & a_{1n}x_n & \eta_1 & b_1 \\ a_{21}x_1 & + & a_{22}x_2 & + & \ldots & + & a_{2n}x_n & \eta_2 & b_2 \\ \ldots & & & & \ldots & & & & \ldots \\ a_{m1}x_1 & + & a_{m2}x_2 & + & \ldots & + & a_{mn}x_n & \eta_m & b_m \end{array} \tag{3.1.2}$$

where η_i, $i = 1, 2 \ldots, m$, is a $\leq$, $\geq$, or $=$ sign. The following n constraints are also specified

$$x_1 > 0, \ x_2 > 0, \ \ldots, \ x_n > 0 \tag{3.1.3}$$

The following is a common linear programming problem that

occurs in industry. The so-called **simplex method** solves the problem via **Gauss-Jordan elimination processes**.

Example 3.1

A firm has 3 workshops, one for parts, one for wiring and one for testing. The firm produces 2 different products A and B. Each product has to undergo an operation in each workshop consecutively.

One unit of product A requires 1, 1, and 2 hours in the 3 workshops respectively. One unit of product B requires 1, 3, and 1 hours in the 3 workshops respectively. The workshops are available 20, 50 and 30 hours per week respectively.

The profit in selling one unit of product A is \$20 and in selling one unit of product B is \$30. What is the number of weekly output units x_1 and x_2 of products A and B respectively that will maximize profit?

The formulation of the problem is

$$\text{maximize } z = 20x_1 + 30x_2$$

subject to the constraints

$$\begin{aligned} x_1 + x_2 &\le 20 \\ x_1 + 3x_2 &\le 50 \\ 2x_1 + x_2 &\le 30 \end{aligned} \tag{3.1.4}$$

and

$$x_1 \ge 0 \text{ and } x_2 \ge 0 \tag{3.1.5}$$

Conditions (3.1.5) indicate that we cannot produce a negative number of goods.

This problem has two independent variables x_1 and x_2 and may be solved graphically. It is known that a straight line, say $ax_1 + bx_2 + c = 0$, divides the Euclidean plane into two halves. The quantity $(ax_1 + bx_2 + c) < 0$, in one half of the plane and $(ax_1 + bx_2 + c) > 0$ in the other half.

The intersection of the five half planes satisfying the above five inequalities (3.1.4) and (3.1.5) constitute the **feasibility region** for the solution (x_1, x_2). For this problem, it is a polygonal region that has 5 sides and 5 corners as shown in Figure 3-1.

The value $z = 20x_1 + 30x_2$ defines a straight line that moves parallel to itself as z increases or decreases. The maximum value of z, z_{max}, will be obtained when this line touches the region of feasible solution. That is when the line $z = 20x_1 + 30x_2$ passes through the furthest corner of this region. The solution of the problem is $(x_1, x_2) = (5, 15)$ and $z_{max} = 550$, depicted in Figure 3-1.

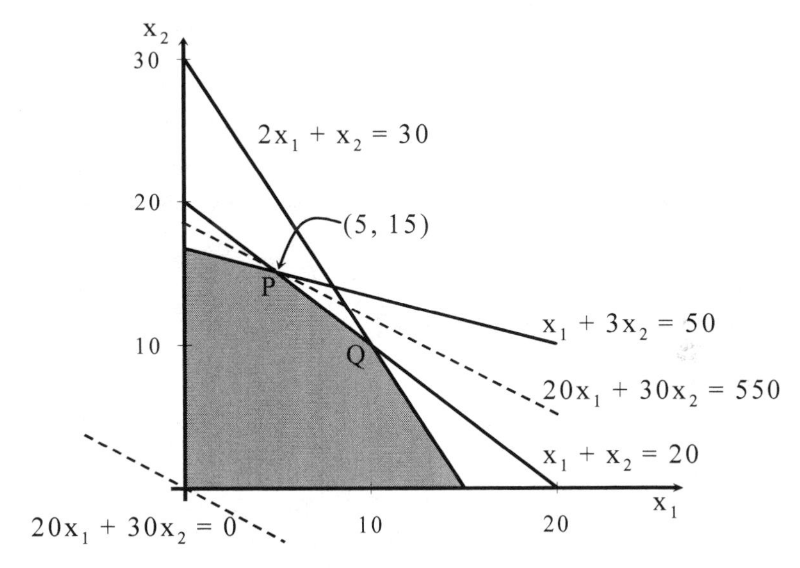

Figure 3-1: Feasibility Region

In this example, the region of feasible solution has as boundaries the lines given by the set of constraints (3.1.4) and (3.1.5). One vertex (corner) of this region is the optimal solution. This result is always true for any general linear programming problem with the number of variables ≥ 2.

We see from this example that since all the functions in linear programming problems are linear, the feasible region defined by problem (3.1.1-3) is a convex set. Thus the linear programming problem is convex. Again, the optimizer of linear programming must lie on the boundary of the feasible region.

In this problem the solution is said to be feasible and unique. However, there are some exceptional cases, which are discussed next.

3.1.1 Exceptional linear programming problems

Various possibilities may arise in linear programming problems. A problem may have a unique solution, an infinite number of solutions or no solution at all. In the last case, the problem may have an unbounded solution or it may have inconsistent constraints. In both cases, we say that the problem has no solution. Consider the following examples.

Example 3.2: (non-unique solution)

Example 3.1, illustrated by Figure 3-1, has the unique solution $(x_1, x_2) = (5, 15)$. Consider the same example and assume instead that z is given by $z = 20x_1 + 20x_2$. Again, z_{max} is obtained when the line $z = 20x_1 + 20x_2$ touches the region of feasible solution. However, in this case, this line coincides with one of the constraint lines, namely the line $x_1 + x_2 = 20$ (Figure 3-1). Any point on the portion PQ of this line is a solution to this problem. The solution is thus not unique.

Example 3.3: (unbounded solution)

Consider the example

$$\text{maximize } z = x_1 + x_2$$

subject to the conditions

$$\begin{aligned} x_1 - x_2 &\geq -1 \\ -x_1 + 3x_2 &\leq 5 \end{aligned}$$

$$x_1 \geq 0, x_2 \geq 0$$

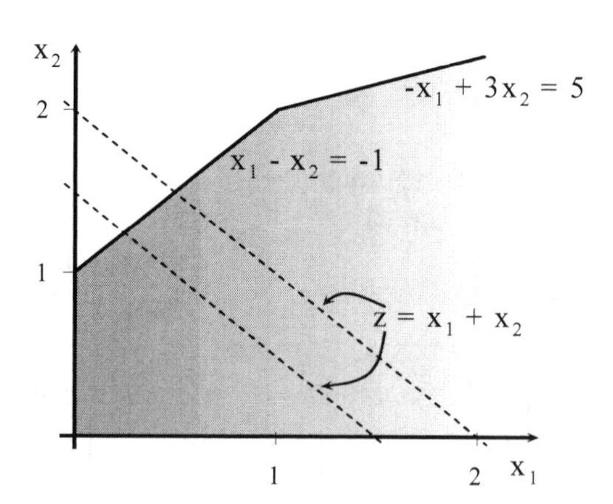

Figure 3-2: Unbounded Solution

For this example the region of feasible solution is given by the shaded area in Figure 3-2. The line representing z can be moved parallel to itself in the direction of increasing z and z can be made as large as one wishes. The problem thus has no solution as the solution is unbounded.

Example 3.4: (inconsistent constraints)

Consider the example

$$\text{maximize } z = x_1 - x_2$$

subject to the constraints

$$\begin{aligned} x_1 + 2x_2 &\leq 2 \\ x_1 + 2x_2 &\geq 3 \end{aligned}$$

$$x_1 \geq 0, x_2 \geq 0$$

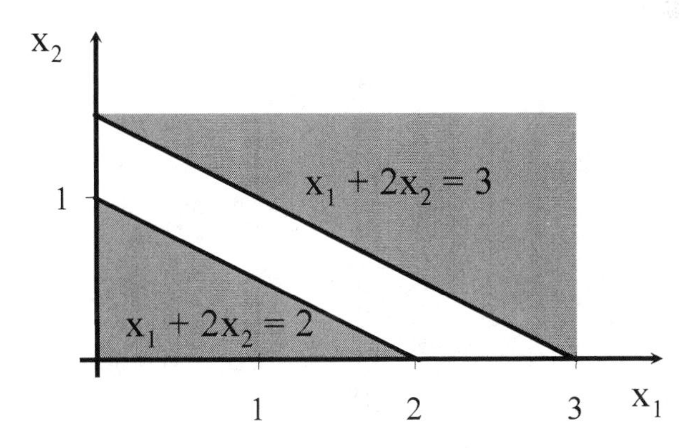

Figure 3-3: Inconsistent Constraints

In this example the constraints are inconsistent as there is no common feasibility region between them, as shown in Figure 3-3.

In Section 3.2, some notations and definitions are given. In Section 3.3, a well-known version of the simplex method is described and in Section 3.4 the simplex tableau is illustrated with a solved example. In Section 3.5, the two-phase method of the simplex algorithm is introduced.

In Section 3.6, the duality theory in linear programming is presented and in Section 3.7, degeneracy, which often occurs in linear programming, is described and its resolution discussed. In Section 3.8, the relationships between linear programming and the discrete linear L_1 and Chebyshev approximations are established. Finally, in Section 3.9, the stability of linear programs is considered.

3.2 Notations and definitions

Let us recall the linear programming problem formulated by (3.1.1-3). The following are well-known notations and definitions:

(a) The **objective function**: Function z of (3.1.1) is known as the objective function.

(b) The **prices**: Constants (c_i) in (3.1.1) are known as the prices associated with the variables (x_i).

(c) The **requirement vector**: Vector **b**, whose elements are $(b_1, \ldots, b_m)$ of (3.1.2), is often known as the requirement vector. It is preferable to have the b_i all positive. If one or more of the b_i is negative, the corresponding inequality might be multiplied by –1 and the inequality sign reversed.

(d) **Slack and surplus variables**: The simplex method described in Section 3.2, in effect, deals with constraints in the form of equalities, not in the form of inequalities as those of (3.1.4) for example. It is thus necessary to convert these inequalities into equalities. This is done by using the slack and surplus variables.

A slack variable: An inequality of the form

$$a_{i1}x_1 + a_{i2}x_2 + \ldots + a_{in}x_n \leq b_i$$

is transferred to the equality

$$a_{i1}x_1 + a_{i2}x_2 + \ldots + a_{in}x_n + x_{i+n} = b_i$$

by adding to the l.h.s. a new variable, say $x_{i+n} > 0$, known as a slack variable.

A surplus variable: Also, the inequality

$$a_{i1}x_1 + a_{i2}x_2 + \ldots + a_{in}x_n \geq b_i$$

is transferred to the equality

$$a_{i1}x_1 + a_{i2}x_2 + \ldots + a_{in}x_n - x_{i+n} = b_i$$

by subtracting from the l.h.s. a new variable, say $x_{i+n} \geq 0$, known as a surplus variable.

(e) **Artificial variables**: It is sometimes desirable, as in the two-phase method of Section 3.5, to add other variables to the problem. Such variables are neither slack nor surplus. They are known as artificial variables.

(f) A **hyper plane**: Anyone of the resulting equalities in (3.1.2) is known as a hyper plane. This is the generalization of a plane when it has more than three variables.

(g) A **solution**: Any set of variables $x_1, x_2, \ldots, x_n$ that satisfies the sets of constraints (3.1.2) is known as a solution to the given programming problem.

(h) A **basic solution**: Let us now assume that the m constraints (3.1.2) are converted, by using the slack and surplus variables, to m equations in the form $\mathbf{Ax} = \mathbf{b}$, where $\mathbf{A}$ is an m by N matrix, $m < N$ of rank m. Then a basic solution to this set is obtained by equating $(N - m)$ variables (x_j) to 0's and solving for the remaining m variables (x_i). These m variables are known as basic variables, or the basic solution.

(i) **Basis matrix**: The m linearly independent columns of matrix $\mathbf{A}$, in the system $\mathbf{Ax} = \mathbf{b}$, associated with the m basic variables form an m by m matrix, known as the basis matrix. If the basis matrix is denoted by $\mathbf{B}$, the basic solution $\mathbf{x}_B$ is given by

$$\mathbf{x}_B = \mathbf{B}^{-1}\mathbf{b}$$

and each of the other $(N - m)$ variables (x_j) is 0.

(j) A **basic feasible solution**: Any basic solution that also satisfies the non-negativity constraints (3.1.3) is called a basic feasible solution.

(k) A **vertex** in the region of feasible solution: It is easy to show that a basic feasible solution is a vector whose elements are the coordinates of a vertex (or a corner) in the region of feasible solution.

(l) **Degenerate solution**: If one or more basic variables x_i is 0, then the basic solution to $\mathbf{Ax} = \mathbf{b}$ is known as a degenerate solution. Often degeneracy in linear programming should be resolved. See Section 3.7.

(m) **Optimal basic feasible solution**: Any basic feasible solution that maximizes (or minimizes) the function z of (3.1.1) is called an optimal basic feasible solution.

3.3 The simplex algorithm

By adding the necessary slack variables and by subtracting the necessary surplus variables in the set of constraints (3.1.2), we get an underdetermined set of linear equations. Problem (3.1.1-3) might be reformulated as follows.

Maximize the objective function

$$z = c_1 x_1 + \ldots + c_n x_n + 0x_{n+1} + \ldots + 0x_N \tag{3.3.1}$$

subject to the constraints

$$\begin{aligned} \sum_{i=1}^{n} a_{pi} x_i &= b_p, \quad p = 1, 2, \ldots, w \\ \sum_{i=1}^{n} a_{ji} x_i + x_{n+j} &= b_j, \quad j = 1, 2, \ldots, u \\ \sum_{i=1}^{n} a_{ki} x_i - x_{n+u+k} &= b_k, \quad k = 1, 2, \ldots, v \end{aligned} \tag{3.3.2}$$

and

$$x_1, \ldots, x_n \geq 0, \quad x_{n+1}, \ldots, x_N \geq 0 \tag{3.3.3}$$

In writing (3.3.1-3), we have assumed that in (3.1.2) there are w equalities, u inequalities with ≤ 0 signs and v inequalities with ≥ 0 signs. Also in (3.3.1), $N = n + u + v$. Note that the coefficients (the prices) of the slack and surplus variables in (3.3.1) are 0's.

In vector-matrix notation, this formulation reduces to

$$\text{maximize } z = (\mathbf{c}, \mathbf{x}) \tag{3.3.4}$$

subject to the constraints

$$\mathbf{Ax} = \mathbf{b} \tag{3.3.5}$$

and

$$\mathbf{x} \geq \mathbf{0} \tag{3.3.6}$$

Vector **c** in (3.3.4) is vector **c** appearing in (3.3.1), namely

$$\mathbf{c} = (c_1, \ldots, c_n, 0, \ldots, 0)^T$$

and vector **b** in (3.3.5) is the requirement vector (b_i) of (3.3.2) which is assumed to have non-negative elements. Matrix **A** is an m by N matrix given by (assume that $u = 3$, $v = 1$ and $w = 1$)

$$(3.3.7) \qquad \mathbf{A} = \begin{bmatrix} a_{11} & a_{12} & a_{13} & a_{14} & \cdots & a_{1,n-1} & a_{1,n} & 1 & 0 & 0 & 0 \\ a_{21} & a_{22} & a_{23} & a_{24} & \cdots & a_{2,n-1} & a_{2,n} & 0 & 1 & 0 & 0 \\ a_{31} & a_{32} & a_{33} & a_{34} & \cdots & a_{3,n-1} & a_{3,n} & 0 & 0 & 1 & 0 \\ a_{41} & a_{42} & a_{43} & a_{44} & \cdots & a_{4,n-1} & a_{4,n} & 0 & 0 & 0 & -1 \\ a_{51} & a_{52} & a_{53} & a_{54} & \cdots & a_{5,n-1} & a_{5,n} & 0 & 0 & 0 & 0 \end{bmatrix}$$

Problem (3.3.4-6) is solved by the simplex algorithm, described next.

The simplex algorithm is an iterative one. It starts by calculating the coordinates of one of the vertices (corners) of the region of feasible solution. It then goes from one vertex to a neighboring one in such a way that at each step the objective function z increases (or decreases), until the optimal solution is reached. This process takes a finite number of steps, usually between m and 2m steps, where m is the number of constraints in (3.3.2).

If the programming problem has an unbounded solution or if it has an inconsistent set of constraints, the simplex algorithm will detect this in the course of the solution. The simplex method will also indicate whether the solution is unique. Another important merit is that the simplex algorithm detects any redundancy in the set of constraints (3.1.2). It removes such redundant constraints and proceeds to find the solution after discarding them. By redundant constraints, we mean that one or more equations in (3.3.5) are linearly dependent on other equations in (3.3.5).

Consider the linear programming problem given by (3.3.4-6). Let us further assume that $\text{rank}(\mathbf{A}|\mathbf{b}) = \text{rank}(\mathbf{A})$, where matrix $(\mathbf{A}|\mathbf{b})$ denotes the m by $(N+1)$ matrix with $\mathbf{b}$ as the $(N+1)^{\text{th}}$ column of matrix $\mathbf{A}$.

3.3.1 Initial basic feasible solution

As indicated, the simplex method solves the problem by first finding any basic feasible solution to the problem, i.e., by first finding m non-negative coordinates of a vertex of the region of feasible solution.

In the case where a slack variable exists in each constraint equation in (3.3.5), matrix **A** of (3.3.7) has the form

(3.3.8) $$\mathbf{A} = (\mathbf{R}|\mathbf{I})$$

where **I** is an m-unit matrix. In this case, the slack variables themselves form an initial basic feasible solution. Let us write $\mathbf{x} = (\mathbf{x}_0, \mathbf{x}_s)^T$, where $\mathbf{x}_0$ contains the original variables and $\mathbf{x}_s$ contains the m slack variables. Then it is obvious that by setting $\mathbf{x}_0 = \mathbf{0}$, we have

$$\mathbf{I}\mathbf{x}_s = \mathbf{b}$$

The slack variables $\mathbf{x}_s = \mathbf{b}$ and they form an initial basic feasible solution.

The parameters of interest in the simplex algorithm are

(i) Vectors $(\mathbf{y}_j)$, given by

$$\mathbf{y}_j = \mathbf{B}^{-1}\mathbf{a}_j = \mathbf{a}_j, \quad j = 1, 2, \ldots, n$$

since $\mathbf{B}^{-1} = \mathbf{I}$ here

(ii) Scalars $(z_j - c_j)$, known as the marginal costs, given by

$$z_j - c_j = (\mathbf{c}_B, \mathbf{y}_j) - c_j = -c_j, \quad j = 1, 2, \ldots, n$$

as the prices associated with the slack variables (elements of $\mathbf{c}_B$ here) are 0's. See (3.3.13) and (3.3.14).

Hence, for this initial basic feasible solution, no computation is needed to obtain the initial parameters of interest in the simplex algorithm.

3.3.2 Improving the initial basic feasible solution

Once an initial vertex is available, one iteration in the simplex algorithm is used to find a neighboring vertex associated with a larger value of z. This is done in a straightforward manner using what is known as the **simplex tableau.**

Let the columns of matrix **A** in (3.3.5) be denoted by $\mathbf{a}_1, \mathbf{a}_2, \ldots, \mathbf{a}_N$. Then equation (3.3.5) may be written as

(3.3.9) $$x_1\mathbf{a}_1 + x_2\mathbf{a}_2 + \ldots + x_m\mathbf{a}_m + \ldots + x_N\mathbf{a}_N = \mathbf{b}$$

Let the initial basic feasible solution be the m non-negative (x_i) denoted by $x_1^{(1)}, x_2^{(1)}, \ldots, x_m^{(1)}$. The superscripts on the x_i denote that the x_i are those of the first vertex. Let the columns $\mathbf{a}_i$ associated with the x_i be denoted by $\mathbf{a}_i^{(1)}$. These columns form the initial basis matrix $\mathbf{B}$.

The initial basic solution is given by

$$x_1^{(1)}\mathbf{a}_1^{(1)} + x_2^{(1)}\mathbf{a}_2^{(1)} + \ldots + x_m^{(1)}\mathbf{a}_m^{(1)} = \mathbf{b} \tag{3.3.10}$$

The initial value z of (3.3.4), denoted by $z^{(1)}$ is also given by

$$z^{(1)} = c_1^{(1)}x_1^{(1)} + c_2^{(1)}x_2^{(1)} + \ldots + c_m^{(1)}x_m^{(1)} \tag{3.3.11}$$

The columns in (3.3.10) are assumed to be linearly independent. Thus these columns (vectors) form a basis for an m-dimensional real space $\mathbf{E}_m$, and therefore any other m-dimensional real vector can be expressed as a linear combination of the m-vectors $\mathbf{a}_i^{(1)}$. In particular the (N – m) non-basic columns $\mathbf{a}_j$ in (3.3.9) may be expressed in terms of the basis columns $\mathbf{a}_i^{(1)}$ of (3.3.10) in the form

$$\mathbf{a}_j^{(1)} = y_{1j}^{(1)}\mathbf{a}_1^{(1)} + y_{2j}^{(1)}\mathbf{a}_2^{(1)} + \ldots + y_{mj}^{(1)}\mathbf{a}_m^{(1)} \tag{3.3.12}$$

$$j = m+1, m+2, \ldots, N$$

where the $y_{ij}^{(1)}$ are none other than the elements of vector $\mathbf{y}_j^{(1)}$ given by

$$\mathbf{y}_j^{(1)} = \mathbf{B}^{-1}\mathbf{a}_j^{(1)}, \quad j = m+1, m+2, \ldots, N \tag{3.3.13}$$

where $\mathbf{B}$ is the basis matrix whose columns are $(\mathbf{a}_1^{(1)}, \mathbf{a}_2^{(1)}, \ldots, \mathbf{a}_m^{(1)})$.

For j = m+1, m+2, …, N, let us denote the scalar product $z_j^{(1)}$ by

$$z_j^{(1)} = (\mathbf{c}_B, \mathbf{y}_j^{(1)})$$

or

$$z_j^{(1)} = c_1^{(1)}y_{1j}^{(1)} + c_2^{(1)}y_{2j}^{(1)} + \ldots + c_m^{(1)}y_{mj}^{(1)} \tag{3.3.14}$$

where the elements of $\mathbf{c}_B$ are those of (3.3.11), associated with the basic columns $(\mathbf{a}_1^{(1)}, \mathbf{a}_2^{(1)}, \ldots, \mathbf{a}_m^{(1)})$.

Our task now is to replace one of the basic columns $\mathbf{a}_i^{(1)}$ by a non-basic column $\mathbf{a}_j^{(1)}$ in such a way that the new basic variables $x_i^{(2)}$ will all be non-negative and also that the new value $z^{(2)}$ will be greater than $z^{(1)}$. This is done by manipulating equations (3.3.10-14).

Let us multiply (3.3.12) and (3.3.14) by a positive parameter θ and subtract the resulting equations from (3.3.10) and (3.3.11) respectively. We thus get

$$(x_1^{(1)} - \theta y_{1j}^{(1)})\mathbf{a}_1^{(1)} + (x_2^{(1)} - \theta y_{2j}^{(1)})\mathbf{a}_2^{(1)} + \ldots + \tag{3.3.15}$$

$$(x_m^{(1)} - \theta y_{mj}^{(1)})\mathbf{a}_m^{(1)} + \theta \mathbf{a}_j^{(1)} = \mathbf{b}$$

and

(3.3.16) $$(x_1^{(1)} - \theta y_{1j}^{(1)})c_1^{(1)} + \ldots + (x_m^{(1)} - \theta y_{mj}^{(1)})c_m^{(1)} + \theta c_j^{(1)}$$
$$= z^{(1)} + \theta(c_j^{(1)} - z_j^{(1)}), \quad j = m+1, m+2, \ldots, N$$

All the coefficients but one of $\mathbf{a}_k^{(1)}$ in (3.3.15) will form the new basic variables. Also, the r.h.s. of (3.3.16) will be the value $z^{(2)}$. Therefore, to achieve our goal, we search for an index j for which $(c_j^{(1)} - z_j^{(1)})$ in the r.h.s. of (3.3.16) is positive. Also, for this value of j, θ is chosen so that

(3.3.17) $\theta = \theta_{min} = \min_i(x_i^{(1)}/y_{ij}^{(1)}) = x_u^{(1)}/y_{uj}^{(1)}$ (say), $\quad y_{ij}^{(1)} > 0$

This choice of θ, reduces one of the coefficients in (3.3.15) to 0 and leaves the remaining m coefficients positive. Such coefficients are the coordinates of the new vertex, and are denoted by $x_1^{(2)}, x_2^{(2)}, \ldots, x_m^{(2)}$. The value $z^{(2)}$ is now greater than $z^{(1)}$ and is given by

(3.3.18) $$z^{(2)} = z^{(1)} + \theta_{min}(c_j^{(1)} - z_j^{(1)})$$

In general, there may be several values of j for which $(c_j^{(1)} - z_j^{(1)}) > 0$. It would be reasonable to choose θ for which $z^{(2)}$ in (3.3.18) yields the greatest increase over $z^{(1)}$. However, this requires some extra computational efforts.

Instead, one usually selects j for which $(c_j^{(1)} - z_j^{(1)})$ is the largest and thus the increase in $z^{(1)}$ will be

$$\delta z = (\min_i(x_i^{(1)}/y_{ij}^{(1)}))\max_j(c_j^{(1)} - z_j^{(1)}), \quad (c_j^{(1)} - z_j^{(1)}) > 0, \quad y_{ij}^{(1)} > 0$$

The coordinates of the new vertex are now the coefficients of the vectors $\mathbf{a}_i^{(1)}$ in (3.3.15), where θ is given by (3.3.17).

The new values of $(c_j^{(2)} - z_j^{(2)})$ are given by

$$(c_u^{(2)} - z_u^{(2)}) = 0$$
$$(c_j^{(2)} - z_j^{(2)}) = (c_j^{(1)} - z_j^{(1)}) - \theta\,(c_u^{(1)} - z_u^{(1)}), \quad j \neq u$$

where u is defined in (3.3.17).

This iteration is repeated several times until the optimal value z_{max} is obtained. The value z_{max} is achieved when $(c_j - z_j) \leq 0$ for all j.

The above calculation is greatly simplified if we arrange the parameters of interest in the form of a table called the simplex tableau.

3.4 The simplex tableau

Without loss of generality, let the basis matrix **B** consist of the first m columns of matrix **A** in (3.3.5) and let us pre-multiply (3.3.5) by $\mathbf{B}^{-1}$. We shall then get the equation (for m = 5)

$$\begin{bmatrix} 1 & 0 & 0 & 0 & 0 & y_{1,m+1}^{(1)} & \cdots & y_{1,N}^{(1)} \\ 0 & 1 & 0 & 0 & 0 & y_{2,m+1}^{(1)} & \cdots & y_{2,N}^{(1)} \\ 0 & 0 & 1 & 0 & 0 & y_{3,m+1}^{(1)} & \cdots & y_{3,N}^{(1)} \\ 0 & 0 & 0 & 1 & 0 & y_{4,m+1}^{(1)} & \cdots & y_{4,N}^{(1)} \\ 0 & 0 & 0 & 0 & 1 & y_{5,m+1}^{(1)} & \cdots & y_{5,N}^{(1)} \end{bmatrix} \begin{bmatrix} x_1^{(1)} \\ x_2^{(1)} \\ \cdot \\ \cdot \\ x_N^{(1)} \end{bmatrix} = \begin{bmatrix} b_1^{(1)} \\ b_2^{(1)} \\ b_3^{(1)} \\ b_4^{(1)} \\ b_5^{(1)} \end{bmatrix}$$

and $x_{m+1} = x_{m+2} = \ldots = x_N = 0$.

In this matrix equation, it is obvious that the $b_i^{(1)}$ are the basic variables. Also, for the non-basic vectors, $a_{ij}^{(1)} = y_{ij}^{(1)}$, from which the parameters $z_j^{(1)}$ may be calculated. In other words, this equation gives the information needed for the next iteration.

Once an iteration has been performed, it would be advantageous to obtain a new set of parameters from which the information needed for the next iteration is obtained. These concepts of the simplex tableau are best illustrated by an example.

Consider Example 3.1, which was solved geometrically in Section 3.1. By adding the slack variables to the inequalities (3.1.4), we get the formulation

(3.4.1a) $\quad$ maximize $z = 20x_1 + 30x_2 + 0x_3 + 0x_4 + 0x_5$

subject to the constraints

(3.4.1b)
$$\begin{aligned} x_1 + x_2 + x_3 \quad\quad\quad &= 20 \\ x_1 + 3x_2 \quad + x_4 \quad &= 50 \\ 2x_1 + x_2 \quad\quad + x_5 &= 30 \end{aligned}$$

(3.4.1c) $\quad x_1 \geq 0, x_2 \geq 0, x_3 \geq 0, x_4 \geq 0, x_5 \geq 0$

Tableau 3.4.1 is the simplex tableau for the first iteration.

Tableau 3.4.1

		$\mathbf{c}^T$	20	30	0	0	0	
$\mathbf{c}_B$	**B**	$\mathbf{b}_B$	$\mathbf{a}_1$	$\mathbf{a}_2$	$\mathbf{a}_3$	$\mathbf{a}_4$	$\mathbf{a}_5$	θ_i
0	$\mathbf{a}_3$	20	1	1	1	0	0	20
0	$\mathbf{a}_4$	50	1	(3)	0	1	0	(50/3)
0	$\mathbf{a}_5$	30	2	1	0	0	1	30
z = 0		$(z_i - c_i)$	–20	(–30)	0	0	0	

The second column from the left shows the vectors forming the current basis matrix **B**, namely $\mathbf{a}_3$, $\mathbf{a}_4$ and $\mathbf{a}_5$. The first column on the left shows the parameters $\mathbf{c}_B$ associated with these vectors, namely 0, 0 and 0. The third column from the left gives the initial basic feasible solution, i.e., the coordinates of the initial vertex, namely $x_3 = 20$, $x_4 = 50$ and $x_5 = 30$.

The inner product of the column $\mathbf{c}_B$ with each of the non-basic columns, namely $\mathbf{a}_1$ and $\mathbf{a}_2$, is obtained. The z_i, calculated as in (3.3.14) are subtracted from the top row of the tableau, and the result is written in the bottom row; the $(c_i - z_i)$ row.

However, in this book we adopt the opposite notation and record the values $(z_i - c_i)$ instead. Lastly, the value of z is entered in the last row, namely $z = (\mathbf{c}_B, \mathbf{b}_B) = 0$ in this tableau.

In Tableau 3.4.1, it is shown that the basic feasible solution is associated with the unit (vectors) $\mathbf{a}_3$, $\mathbf{a}_4$ and $\mathbf{a}_5$. The entries in columns $\mathbf{a}_1$ and $\mathbf{a}_2$ are simply $\mathbf{y}_1$ and $\mathbf{y}_2$.

It is expected that one iteration in the simplex algorithm is used to find a new feasible basic solution (a neighboring vertex) associated with a larger value of z. This is done as follows.

From the last row in Tableau 3.4.1, it is found that maximum $(c_j - z_j)$, or minimum $(z_j - c_j)$ is $(z_2 - c_2)$. Then if $\mathbf{a}_2$ replaces one of the basic columns $\mathbf{a}_3$, $\mathbf{a}_4$ or $\mathbf{a}_5$, the resulting value of z will be larger than the current one. This is done by first calculating the parameters θ_i, where θ_i are obtained by dividing the elements of column $\mathbf{b}_B$ by the corresponding elements of $\mathbf{a}_2$ which have the minimum $(z_j - c_j)$.

Since we are interested in the parameters θ_i that have positive values, and that the elements of $\mathbf{b}_B$ are non-negative, the elements of $\mathbf{a}_2$ used in the divisions are the positive elements, which they are in this case. The possible θ_i are recorded in the last column to the right of the tableau, from which it is found that θ_2 is the minimum of the θ_i.

Therefore, if $\mathbf{a}_2$ replaces $\mathbf{a}_4$ in the basis, the new basis matrix will have $\mathbf{a}_3$, $\mathbf{a}_2$ and $\mathbf{a}_5$ as its columns. The new basic solution will be feasible (all its elements are non-negative) and the value of z will increase.

The so-called **pivotal entry** in the tableau is now decided by the intersection of the key column (of minimum $(z_j - c_j)$) and the key row (of minimum θ_i). The pivotal entry is then used to establish the simplex tableau for the next iteration, which results in a larger value of z.

Step (1)

Divide all the entries in the key row by the pivotal entry. This step is equivalent to dividing one of the equations in a linear set of equations by a constant and it does not result in any change in the solution.

Step (2)

Reduce all the entries in the key column, except the pivotal entry itself, which is now unity, to zero. This is done by subtracting an appropriate multiplier of the altered key row from each of the other rows of the column $\mathbf{b}_B$ and the $\mathbf{a}_i$. This step is equivalent to one step of the Gauss-Jordan (elimination step) method, outlined in Section 4.6.

The calculated entries in the $(z_i - c_i)$ row of the simplex tableau, including the z entry, are obtained by extending step (2) to the $(z_i - c_i)$ row also. This is done by subtracting an appropriate multiplier of the altered key row from the $(z_i - c_i)$ row that reduces the $(z_j - c_j)$ entry of the key column to 0. The results are recorded in Tableau 3.4.2.

In Tableau 3.4.2, the key column is that of $\mathbf{a}_1$ and the key row is the first row, and thus the pivotal entry is determined. It is used to construct the next tableau and this is done by observing the two steps described above. The tableau for the next iteration is Tableau 3.4.3.

The end row of Tableau 3.4.3 contains non-negative entries and this indicates that no further improvement to the solution is possible and the programming problem is now solved. The optimal solution is $x_1 = 5$, $x_2 = 15$ and $z_{max} = 550$.

Tableau 3.4.2

		$\mathbf{c}^T$	20	30	0	0	0	
$\mathbf{c}_B$	$\mathbf{B}$	$\mathbf{b}_B$	$\mathbf{a}_1$	$\mathbf{a}_2$	$\mathbf{a}_3$	$\mathbf{a}_4$	$\mathbf{a}_5$	θ_i
0	$\mathbf{a}_3$	10/3	(2/3)	0	1	−1/3	0	(5)
30	$\mathbf{a}_2$	50/3	1/3	1	0	1/3	0	50
0	$\mathbf{a}_5$	40/3	5/3	0	0	−1/3	1	8
z = 500		$(z_i - c_i)$	(−10)	0	0	10	0	

Tableau 3.4.3

		$\mathbf{c}^T$	20	30	0	0	0
$\mathbf{c}_B$	$\mathbf{B}$	$\mathbf{b}_B$	$\mathbf{a}_1$	$\mathbf{a}_2$	$\mathbf{a}_3$	$\mathbf{a}_4$	$\mathbf{a}_5$
20	$\mathbf{a}_1$	5	1	0	3/2	−1/2	0
30	$\mathbf{a}_2$	15	0	1	−1/2	1/2	0
0	$\mathbf{a}_5$	5	0	0	−5/2	1/2	1
z = 550		$(z_i - c_i)$	0	0	15	5	0

It is easy to see that the increase in the value of z after each iteration is given by $[-\theta_{min}(z_i - c_i)_{min}]$. That is, if $z^{(1)}$ and $z^{(2)}$ are the values of z in tableaux k and (k + 1) respectively

$$z^{(2)} = z^{(1)} - \theta_{min}(z_i - c_i)_{min}$$

where θ_{min} and $(z_i - c_i)_{min}$ are those of tableau k.

In this example, if $z^{(1)}$ and $z^{(2)}$ are those of Tableaux 3.4.1 and 3.4.2 respectively

$$z^{(2)} = 0 + (50/3)\times 30 = 500$$

Also

$$z^{(3)} = z^{(2)} + (5)\times 10 = 550$$

where $z^{(3)}$ is that in Tableau 3.4.3.

In this example, we found that because of the slack variables, an identity matrix appears in matrix **A** and as a result, the initial calculation is an easy task. However, in many cases no identity matrix appears in **A**. This will be the case when some constraints have equality signs or $\geq$ signs. In such cases, the initial basic feasible solution is constructed by the so-called two-phase method, which is described in the next section.

3.5 The two-phase method

In this method, artificial variables are added first to every equality constraint and also to every constraint having a surplus variable, so that matrix **A** of (3.3.5) may have the form (3.3.8).

The problem is solved in two-phases. In phase 1, we try to derive all the artificial variables to 0. In phase 2, we try to maximize the actual objective function z, starting from a basic feasible solution that either contains no artificial vectors or that contains some artificial vectors at zero level; that is, whose associated $x_i = 0$.

3.5.1 Phase 1

In phase 1, we assign to each artificial variable a price $c_i = -1$. To all other variables, which include the original n variables, the slack and surplus variables, we assign the price $c_i = 0$. We then use the simplex method techniques to maximize the function

$$z = -1x_{1a} - 1x_{2a} - \ldots - 1x_{ka}$$

where the x_{ia} are the artificial variables.

Since in the simplex method any variable is not allowed to be negative, the x_{ia} are also non-negative and hence z would be non-positive. The maximum value of z will be 0. If there is no redundancy in the constraints, maximum $z = 0$ and no artificial variable will appear in the basis.

Furthermore, z may become 0 before the optimality criterion is satisfied. In this case, some artificial vectors appear in the basis at a zero level. This case may result from the presence of redundancy in the original constraints. In any case, if z is 0 before the optimality criterion is satisfied, we end phase 1 and start working on phase 2.

On the other hand, if maximum $z < 0$, then the artificial variables in the basis cannot be driven to 0 and the original problem has no feasible solution.

3.5.2 Phase 2

In phase 2, we assign the actual prices c_i to each legitimate variable and a price 0 to each slack and surplus variable. Also, we assign 0 to each artificial variable that may appear in the basis at zero level. The simplex method is then used to maximize the original function z. The first tableau in phase 2 is itself the last tableau in phase 1. The only difference is that the $(z_i - c_i)$ for the last tableau in phase 1 are altered to account for change in the prices.

If at the beginning of phase 2, one or more artificial variables appear at zero level, we ensure that such artificial variables remain at zero level at each iteration in phase 2. See the second example in Hadley ([7], pp. 154-158).

3.5.3 Detection of the exceptional cases

In Section 3.1, different exceptional cases of a linear programming problem were described. Such cases are detected easily from the simplex tableaux in the course of the solution of the problem. At the end of phase 1, we found that the following possibilities exist:

(a) The presence of redundancy in the original constraints.

(b) The possibility of inconsistent constraints.

(c) Feasible problem.
If no artificial vectors appear in the basis, then we have found an initial basic feasible solution, and thus the original problem is feasible. The original constraints are consistent and none of them is redundant.

However, in phase 2, the following may occur:

(d) An unbounded solution.
Suppose that in a simplex tableau in phase 2, it was found that for some non-basic variable j, $(z_j - c_j)$ is negative, but all the elements of the corresponding $\mathbf{y}_j$ are non-positive. Then the solution to this problem is unbounded. See Hadley ([7], pp. 93, 94).

(e) Non-unique optimal solution.
If the optimal solution is degenerate, i.e., one or more elements in the basic feasible solution is 0, then we may have a non-unique optimal solution.

3.6 Duality theory in linear programming

The dual problem

If a problem has a dual, the dual problem is usually given in terms of the same variables of the original problem, with the roles of certain variables being interchanged. The original problem, in this case, is known as the primal problem.

Consider the primal problem

(3.6.1a) $$\text{maximize } z = (\mathbf{c}, \mathbf{x})$$

subject to the constraints

(3.6.1b) $$\mathbf{Ax} \leq \mathbf{b}$$

and

(3.6.1c) $$\mathbf{x} \geq \mathbf{0}$$

The dual problem to (3.6.1) is

(3.6.2a) $$\text{minimize } Z = (\mathbf{b}, \mathbf{w})$$

subject to the constraints

(3.6.2b) $$\mathbf{A}^T\mathbf{w} \geq \mathbf{c}$$

and

(3.6.2c) $$\mathbf{w} \geq \mathbf{0}$$

Note in (3.6.1) and (3.6.2) that the roles of **b** and **c** are interchanged and the matrices of the constraints are the transpose of one another. Hence, if we consider Example 3.1, also given in (3.4.1), as a primal problem, its dual will be

(3.6.3a) $$\text{minimize } Z = 20w_1 + 50w_2 + 30w_3$$

subject to the constraints

(3.6.3b) $$\begin{aligned} w_1 + w_2 + 2w_3 &\geq 20 \\ w_1 + 3w_2 + w_3 &\geq 30 \end{aligned}$$

and

(3.6.3c) $$w_1 \geq 0, w_2 \geq 0 \text{ and } w_3 \geq 0$$

The solution of a dual problem is related in some defined way to the solution of its primal problem. We convert (3.6.3a) to a maximization problem and also subtract surplus variables in (3.6.3b). We get

(3.6.4a) $$\text{maximize } Z = -20w_1 - 50w_2 - 30w_3$$

subject to the constraints

(3.6.4b)
$$\begin{aligned} w_1 + w_2 + 2w_3 - w_4 \quad &= 20 \\ w_1 + 3w_2 + w_3 \quad - w_5 &= 30 \end{aligned}$$

and

(3.6.4c) $$w_1 \geq 0,\ w_2 \geq 0,\ w_3 \geq 0,\ w_4 \geq 0 \text{ and } w_5 \geq 0$$

We only write the initial and final tableaux for each of the primal and the dual and also omit the rows and the columns corresponding to the artificial variables.

The primal problem (3.4.1) is

Initial

	c^T	20	30	0	0	0
0	20	1	1	1	0	0
0	50	1	3	0	1	0
0	30	2	1	0	0	1
z = 0		–20	–30	0	0	0

Final

	c^T	20	30	0	0	0
20	5	1	0	3/2	–1/2	0
30	15	0	1	–1/2	1/2	0
0	5	0	0	–5/2	1/2	1
z = 550		0	0	15	5	0

The dual problem (3.6.4) is

Initial

	b^T	–20	–50	–30	0	0
0	20	1	1	2	–1	0
0	30	1	3	1	0	–1
Z = 0		20	50	30	0	0

Final

	b^T	–20	–50	–30	0	0
–20	15	1	0	5/2	–3/2	1/2
–50	5	0	1	–1/2	1/2	–1/2
Z = –550		0	0	5	5	15

Let us examine the final tableaux of the two problems. We notice that the basic variables for the dual, namely 15 and 5, appear in the last row of the primal. Also, the basic variables in the primal, namely 5, 15 and 5, appear in the last row of the primal.

Also, the columns of the non-basic vectors in the dual are found in the rows of the primal with negative signs.

It is sometimes easier to solve the dual problem rather than the primal. If a linear programming problem contains many constraints and only a few variables, then the dual to this problem contains only a few constraints and many variables. In this case, it is much easier to solve the dual problem, which has a smaller basis matrix. For this reason, for all the algorithms of the L_1 and Chebyshev solutions of overdetermined linear equations in this book, we found that solving the dual forms of the linear programming problems are easier than solving the primal forms.

Moreover, a knowledge of the properties of the dual problems also leads to a much better understanding of all aspects of linear programming theory. It led, for example, to the discovery of the dual simplex algorithm and to the primal dual algorithm. See for example, Hadley [7].

3.6.1 Fundamental properties of the dual problems

Theorem 3.1

The dual of the dual is the primal.

Theorem 3.2

If $\mathbf{x}$ is any feasible solution to (3.6.1) and $\mathbf{w}$ is any feasible solution to (3.6.2), then

$$(\mathbf{c}, \mathbf{x}) \leq (\mathbf{b}, \mathbf{w}), \quad \text{that is, } z \leq Z$$

Theorem 3.3

If $\mathbf{x}$ is a feasible solution to (3.6.1) and $\mathbf{w}$ is a feasible solution to (3.6.2) such that

$$(\mathbf{c}, \underline{\mathbf{x}}) = (\mathbf{b}, \underline{\mathbf{w}})$$

then $\underline{\mathbf{x}}$ is an optimal solution to (3.6.1) and $\underline{\mathbf{w}}$ is an optimal solution to (3.6.2).

Theorem 3.4

If one of the set of problems (3.6.1) and (3.6.2) has an optimal solution, then the other also has an optimal solution.

Theorem 3.5

If the maximum value of z is unbounded, then the dual has no feasible solution.

Note 3.1

The converse of Theorem 5.5 is not true; that is, if the dual has no feasible solution, this does not imply that maximum z is unbounded. Neither problem has a feasible solution.

3.6.2 Dual problems with mixed constraints

The primal problem (3.6.1) has all its constraints in (3.6.1b) with the same ≤ sign. Consequently all the constraints in the dual in (3.6.2b) have the ≥ signs. However, a primal problem may have both kinds of inequality signs and also may have some equality signs. We first observe that an ≥ sign in the primal is reversed by multiplying the whole inequality by –1.

We also observe that a constraint with an equality sign

$$\Sigma a_{ij}x_j = b_i$$

may be replaced by the two inequalities

(3.6.5a)
$$\Sigma a_{ij}x_j \leq b_i$$
$$\Sigma a_{ij}x_j \geq b_i$$

When the last inequality is multiplied by –1, it is converted to

(3.6.5b)
$$-\Sigma a_{ij}x_j \leq -b_i$$

In other words, a constraint with an equality sign is replaced by two constraints each having ≤ signs. Therefore, a programming problem with mixed constraints may always be converted to the constraints with ≤ signs.

Theorem 3.6

An equality constraint in the primal corresponds to an unrestricted (in sign) variable in the dual.

Proof:

An unrestricted (in sign) variable say w, may be written as the difference of two non-negative variables

(3.6.6) $\quad w = w^1 - w^2$, where $w^1 \geq 0$ and $w^2 \geq 0$

On the other hand, an equality constraint in the primal, may be replaced by the two inequalities (3.6.5a) and (3.6.5b), which yield two variables w^1 and w^2 in the dual. The dual column and the dual price corresponding to w^1 are the negative to those of w^2. Thus w^1 and w^2 may be replaced by one unrestricted variable w, as in (3.6.6).

We conclude that the column corresponding to an unrestricted variable is the one obtained had the original equation in the primal was not replaced by two inequalities.

3.6.3 The dual simplex algorithm

It is observed in the simplex algorithm that any basic feasible solution with all the $(z_i - c_i) \geq 0$ is an optimal solution. It is also observed that the z_i are completely independent of the requirement vector $\mathbf{b}_B$. These observations present an interesting alternative to the method of solution.

We may start the simplex tableau in the dual with a basic, but not feasible solution (not all the elements of $\mathbf{b}_B$ are non-negative) to the linear programming problem which has all the $(z_i - c_i) \geq 0$. If we then move from this basic solution to another by changing one basic vector at a time in such a way that we keep all the $(z_i - c_i) \geq 0$, an optimal solution would be obtained in a finite number of steps. This constitutes the idea of the dual simplex algorithm.

The dual simplex algorithm does not have the general applicability of the simplex algorithm because it is not always easy to start with all the $(z_i - c_i) \geq 0$. The algorithm is given this name since the criterion for changing the basis is that for the dual, not for the primal problem.

In the dual simplex algorithm, one first determines the vector that leaves the basis and then the vector that enters the basis. This is the reverse of what is done in the simplex algorithm. The criteria for changing the basis are as follows:

(a) For the vector to remove from the basis, choose

$$b_{Br} = \min_i(b_{Bi}), \quad b_{Bi} < 0$$

Column $\mathbf{a}_r$ is removed from the basis and x_{Br} is driven to 0.

(b) For the vector to enter the basis, choose

$$\theta = (z_k - c_k)/y_{rk} = \max_j((z_j - c_j)/y_{rj}), \quad y_{rj} < 0$$

The increase in z is $\underline{z} = b_{Br}(z_k - c_k)/y_{rk}$.

This algorithm is particularly useful in solving bounded linear programming problems, as it eliminates the necessity for the introduction of the artificial variables. A bounded linear programming problem has the elements of its vector **x** bounded between upper and lower bounds. Instead of $\mathbf{x} \geq \mathbf{0}$ in (3.6.1c), we have

$$d_i \geq x_i \geq 0, \quad i = 1, 2, ..., n$$

where the (d_i) are given constants.

A numerical example is solved in detail in Chapter 5; Example 5.1, for a bounded linear programming problem using the dual simplex algorithm.

3.7 Degeneracy in linear programming and its resolution

3.7.1 Degeneracy in the simplex method

It is noted that in the simplex method, if one or more of the elements of the basic solution $\mathbf{b}_B$ is 0, then the solution is known to be degenerate. Degeneracy may occur in the initial basic feasible solution, or it may occur in the course of the solution.

The initial basic feasible solution is degenerate if and only if one or more of the elements in the initial $\mathbf{b}_B$ column is 0. Degeneracy occurs in the course of the solution if there are two or more values of i which have the same

$$\theta_{min} = \min_i(x_i/y_{ij}), \quad y_{ij} > 0$$

In this case, in view of (3.3.17), two or more of the nonzero x_{Bi} of the current solution are reduced to zero for the next iteration, while only one of the x_i that is zero in the current solution becomes positive for the next iteration.

When degeneracy occurs, the value of θ_{min} for the iteration that follows will certainly be 0 and in view of (3.3.18), the new solution $z^{(2)} = z^{(1)}$; that is, the objective function does not increase. Moreover, as noted earlier, this degeneracy may persist for several successive iterations and there is a possibility of cycling; that is, a set of bases may be introduced again every few iterations. Hadley ([7], p. 190) presents an (artificial) example in which cycling occurred. In that example there were two equal θ_{min} and the wrong basis vector was removed. Hadley however states that no actual problem has ever cycled.

Resolution of degeneracy is done, in effect, by perturbing the elements of the $\mathbf{b}_B$ column as explained next. In the next two sections we deal with degeneracy in the simplex method. In the third section we mention two cases where degeneracy was resolved in the dual simplex algorithm.

3.7.2 Avoiding initial degeneracy in the simplex algorithm

Assume that k elements in the initial $\mathbf{b}_B$ column are 0's. Select a small number ε such that $k\varepsilon \ll 1$. Replace the first zero element in $\mathbf{b}_B$ by ε, the second zero element by 2ε, … and the kth zero element in $\mathbf{b}_B$ by $k\varepsilon$. At the end of the problem, if necessary let $\varepsilon \to 0$. This would resolve the problem of initial degeneracy.

3.7.3 Resolving degeneracy resulting from equal θ_{min}

We replace the set of variables x_i by another set x_i', as follows

$$\begin{aligned} x_1' &= x_1 - \varepsilon_1 \\ x_2' &= x_2 - \varepsilon_2 \\ \dots \quad & \dots \quad \dots \\ x_N' &= x_N - \varepsilon_N \end{aligned}$$

where $\varepsilon_1, \varepsilon_2, \dots, \varepsilon_N$ are very small positive numbers and that

$$\varepsilon_1 \neq \varepsilon_2 \neq \dots \neq \varepsilon_N$$

We might instead make this procedure systematic by taking $\varepsilon_1 = \varepsilon$, $\varepsilon_2 = \varepsilon^2, \dots, \varepsilon_N = \varepsilon^N$ and define the variables

$$x_j' = x_j - \varepsilon^j, \quad j = 1, 2, \dots, N$$

The ith constraint now becomes

$$b_i = a_{i1}x_1' + a_{i2}x_2' + \dots + a_{iN}x_N'$$

or

$$b_i = a_{i1}(x_1 - \varepsilon) + a_{i2}(x_2 - \varepsilon^2) + \dots + a_{iN}(x_N - \varepsilon^N)$$

Or in terms of the original variables, the new variables b_i' are

$$b_i' = b_i + a_{i1}\varepsilon + a_{i2}\varepsilon^2 + \ldots + a_{in}\varepsilon^n$$

This means that the perturbation, in effect, is to the $\mathbf{b}_B$ column and the degeneracy is resolved.

In practice, it is not necessary to follow this procedure literally. It is observed that the biggest change in the elements b_{Bi} is caused by the ε term in (3.7.1), as the higher orders of ε have negligible effects. We can thus predict how the tie in θ_{min} will be broken without actually calculating b_{Bi}'.

3.7.4 Resolving degeneracy in the dual simplex method

In the dual simplex algorithm, degeneracy occurs when one or more of the marginal costs is zero. A method for resolving this kind of degeneracy is described in Section 5.6. This method is also used in Section 21.3 for resolving the same kind of degeneracy.

3.8 Linear programming and linear approximation

Wagner [13] was the first to successfully formulate both the linear L_1 and the Chebyshev approximation problems as linear programming problems in both the primal and dual forms. Given, as usual, is the overdetermined system of linear equations

$$\mathbf{Ca} = \mathbf{f}$$

$\mathbf{C} = (c_{ij})$ is a given real n by m matrix of rank k, $k \leq m < n$ and $\mathbf{f} = (f_i)$ is a given real n-vector. The residual vector $\mathbf{r}$ is

$$\mathbf{r} = \mathbf{Ca} - \mathbf{f}$$

Its ith element, r_i is given by

$$r_i = \sum_{j=1}^{m} c_{ij}a_j - f_i, \quad i = 1, 2, \ldots, n$$

3.8.1 Linear programming and the L_1 approximation

The L_1 approximation problem is defined as to find the solution vector $\mathbf{a}$ for the equation $\mathbf{Ca} = \mathbf{f}$ such that the L_1 norm of the residual $\mathbf{r}$

be as small as possible. That is

$$\text{minimize } Z = \sum_{i=1}^{n} |r_i|$$

Since the residuals (r_i) are unrestricted in sign; that is, r_i may be >, = or < 0, i = 1, 2, …, n, we write

$$r_i = u_i - v_i$$

Hence

$$|r_i| = u_i + v_i$$

(u_i) and (v_i) are the elements of the n-vectors **u** and **v** respectively and that

$$u_i > 0 \text{ and } v_i > 0, \quad i = 1, 2, \ldots, n$$

The primal form of the linear programming problem for the L_1 approximation problem is now

(3.8.1a) $$\text{minimize } Z = \sum_{i=1}^{n} u_i + \sum_{i=1}^{n} v_i$$

subject to the constraints

$$\mathbf{r} = \mathbf{u} - \mathbf{v} = \mathbf{Ca} - \mathbf{f}$$

or

$$\mathbf{Ca} - \mathbf{u} + \mathbf{v} = \mathbf{f}$$

(3.8.1b) a_i unrestricted in sign, i = 1, 2, …, m

(3.8.1c) $$u_i \geq 0, v_i \geq 0, \quad i = 1, 2, \ldots, n$$

From Section 3.6, the dual form to problem (3.8.1) is given by

(3.8.2a) $$\text{maximize } z = \mathbf{f}^T\mathbf{w}$$

subject to

(3.8.2b) $$\mathbf{C}^T\mathbf{w} = \mathbf{0}$$

$$w_i \leq 1, \quad i = 1, 2, \ldots, n$$

$$w_i \geq -1, \quad i = 1, 2, \ldots, n$$

where $\mathbf{C}^T$ is the transpose of $\mathbf{C}$ and the n-vector $\mathbf{w} = (w_i)$. The last two sets of constraints reduce to the constraints

(3.8.2c) $$-1 \le w_i \le 1, \quad i = 1, 2, \ldots, n$$

Problem (3.8.2) is a linear programming problem with bounded variables (w_i). That is, by defining $b_i = w_i + 1$, $i = 1, 2, \ldots, n$, we get the following formulation of the problem

(3.8.3a) $$\text{maximize } z = \mathbf{f}^T(\mathbf{b} - \mathbf{e})$$

subject to the constraints

(3.8.3b) $$\mathbf{C}^T\mathbf{b} = \mathbf{C}^T\mathbf{e}$$

(3.8.3c) $$0 \le b_i \le 2, \quad i = 1, 2, \ldots, n$$

$\mathbf{e}$ is an n-vector of 1's and the elements of the vector $\mathbf{b} = (b_i)$ are bounded. Problem (3.8.3) is a linear programming problem with non-negative bounded variables [7]. In Chapter 5, this problem is solved by the dual simplex algorithm.

3.8.2 Linear programming and Chebyshev approximation

The Chebyshev approximation problem is defined as to find the solution vector $\mathbf{a}$ for the equation $\mathbf{Ca} = \mathbf{f}$ such that the L_∞ norm of the residual $\mathbf{r}$ be as small as possible. Let

$$z = \max_i |r_i|, \quad i = 1, 2, \ldots, n$$

Let $h \ge 0$, be the value $z = \max_i |r_i|$. Hence, since r_i is unrestricted in sign, $i = 1, 2, \ldots, n$, the primal form of the linear programming problem is

$$\text{minimize } h$$

subject to

$$r_i \le h$$
$$r_i \ge -h$$

The above two inequalities reduce to

$$-h \le \sum_{j=1}^{m} c_{ij}a_j - f_i \le h, \quad i = 1, 2, \ldots, n$$

In vector-matrix form, these inequalities become

$$\begin{aligned} \mathbf{Ca} + h\mathbf{e} &\geq \mathbf{f} \\ -\mathbf{Ca} + h\mathbf{e} &\geq -\mathbf{f} \end{aligned}$$

$h \geq 0$ and a_j, $j = 1, 2, \ldots, m$, unrestricted in sign.

From Section 3.6, the dual form of this formulation is

$$\text{maximize } z = [\mathbf{f}^T \ -\mathbf{f}^T]\mathbf{b} \tag{3.8.4a}$$

subject to

$$\begin{bmatrix} \mathbf{C}^T & -\mathbf{C}^T \\ \mathbf{e}^T & \mathbf{e}^T \end{bmatrix} \mathbf{b} = \mathbf{e}_{m+1} \tag{3.8.4b}$$

$$b_i > 0, \quad i = 1, 2, \ldots, 2n \tag{3.8.4c}$$

$\mathbf{e}_{m+1}$ is an $(m + 1)$-vector, which is the last column in an $(m + 1)$-unit matrix. The 2n-vector $\mathbf{b} = (b_i)$. In Chapter 10, problem (3.8.4) is solved by the simplex method.

Relevant linear programming formulations are also made for other problems in this book, such as the one-sided and the bounded L_1 and Chebyshev solutions of overdetermined linear equations and the formulations for the solutions of the underdetermined linear equations of Chapters 20-23.

3.9 Stability of the solution in linear programming

In practice, it is common to solve linear programming problems that have several hundred equations. Consequently, round-off error plays a significant role in the accuracy of the result.

The basis matrix for a simplex tableau differs from that of a preceding tableau in only one column. The inverse of the basis matrix is updated at each simplex step by the Gauss-Jordan method rather than by inverting a new matrix.

While this generally gives satisfactory results, it is susceptible to round-off error in two respects. First, if the basis matrix is continually updated, computational errors are propagated in the problem from step to step. Second, because the Gauss-Jordan method is applied without pivoting, inaccurate results are obtained when small pivots are met.

Wolfe [15] described some methods of round-off control. This include (1) conditioning of the linear programming problem and (2) precautions to observe in using the simplex algorithm.

Pierre [10] described in detail some scaling techniques applied to the linear programming problem before the simplex algorithm is applied. This process has the effect of conditioning the problem; that is, by minimizing the condition number of the basis matrix. In this respect, see also Noble ([9], Section 13.5).

Clasen [5] gave a criterion by which tolerances in linear programming problems are calculated. Any number that is smaller in absolute value than a specified tolerance is considered as round-off error and is replaced by 0.

Storoy [12] described a simple method for error control in the simplex algorithm. He also presented a method for improving the obtained solutions.

Bartels [1] discussed the stability of Gauss-Jordan elimination method by a round-off error analysis. He then implemented the simplex method based on the Hessenberg **LU** decomposition of the basis matrices.

In order to be able to select large pivots, Bartels and Golub [2] implemented the **LU** factorization method with row interchanges of the basis matrix. Additional accuracy is obtained by iteratively refining the optimal solution, using a technique due to Wilkinson [14].

Bartels, Golub and Saunders [3] described computational methods for updating the inverse of the basis matrix by both the **LU** decomposition method and by the orthogonalization method of Householder's transformation.

Bartels, Stoer and Zenger [4] used a triangular decomposition method for the basis matrix, which ensures numerical stability of the results for ill-conditioned problems. A variation of this technique has been used in the algorithm for calculating the L_1 approximation problem in Chapter 5 and another variation has been used in the algorithm for the restricted Chebyshev approximation problem in Chapter 13.

References

1. Bartels, R.H., A stabilization of the simplex method, *Numerische Mathematik*, 16(1971)414-434.
2. Bartels, R.H. and Golub, G.H., The simplex method of linear programming using LU decomposition, *Communications of ACM*, 12(1969)266-268.
3. Bartels, R.H., Golub, G.H. and Saunders, M.A., Numerical techniques in mathematical programming, *Nonlinear Programming*, Rosen, J.B., Mangasarian, O.L. and Ritter, K. (eds.), Academic Press, New York, pp. 123-176, 1970.
4. Bartels, R.H, Stoer, J. and Zenger, Ch., A realization of the simplex method based on triangular decomposition, *Handbook for Automatic Computation, Vol. II: Linear Algebra*, Wilkinson, J.H. and Reinsch, C. (eds.), Springer-Verlag, New York, pp. 152-190, 1971.
5. Clasen, R.J., Techniques for automatic tolerance control in linear programming, *Communications of ACM*, 9(1966)802-803.
6. Faigle, U., Kern, W. and Still, G., *Algorithmic Principles of Mathematical Programming*, Kluwer Academic Publishers, London, 2002.
7. Hadley, G., *Linear Programming*, Addison-Wesley, Reading, MA, 1962.
8. Ignizio, J.P. and Cavalier, T.M., *Linear Programming*, Prentice Hall, Englewood Cliffs, NJ, 1993.
9. Noble, B., *Applied Linear Algebra*, Prentice-Hall, Englewood Cliffs, NJ, 1969.
10. Pierre, D.A., *Optimization Theory with Applications*, John Wiley & Sons, New York, 1969.
11. Sierksma, G., *Linear and Integer Programming, Theory and Practice*, Second Edition, Marcel Dekker Inc., New York, 2002.
12. Storoy, S., Error control in the simplex technique, *BIT*, 7(1967)216-225.

13. Wagner, H.M., Linear programming techniques for regression analysis, *Journal of American Statistical Association*, 54(1959)206-212.
14. Wilkinson, J.H., *Rounding Errors in Algebraic Processes*, Prentice-Hall, Englewood Cliffs, NJ, 1963.
15. Wolfe, P., Error in the solution of linear programming problems, *Proceedings of a symposium conducted by the MRC and the University of Wisconsin*, Vol. 2, Rall, L.B., (ed.), John Wiley, New York, pp. 271-284, 1966.

Chapter 4

Efficient Solutions of Linear Equations

4.1 Introduction

This is another tutorial chapter. It deals with the solution of real non-singular systems of linear equations and inversion of matrices. This chapter is intended to be an introduction to Chapter 17, on the least squares problem and the pseudo-inversion of matrices.

It is required to solve the system of linear equations

$$\mathbf{Ax} = \mathbf{b}$$

$\mathbf{A} = (a_{ij})$ is an n by n real non-singular matrix, $\mathbf{b} = (b_i)$ an n-real vector and $\mathbf{x} = (x_j)$ is the solution n-vector. We assume that matrix $\mathbf{A}$ is of a reasonable size, and that it is not sparse.

We start in Section 4.2 with the familiar subject of vector and matrix norms and some relevant theorems. A norm for either a vector or a matrix gives an assessment to the size of the vector or the matrix.

In Section 4.3, elementary matrices, which are used to perform elementary operations on a matrix equation, are introduced. In Sections 4.4 and 4.5, two of the direct methods for solving linear equations, which are among the most efficient known methods are described. These are the Gauss **LU** factorization method with complete pivoting and the Householder's **QR** factorization method with pivoting. These two methods have been studied extensively by many authors. However, we adopt here the approach of Wilkinson [16]. The inverse of a square non-singular matrix is calculated.

In Section 4.6, a note on the Gauss-Jordan elimination method for a set of underdetermined system of linear equations is given. Gauss-Jordan method is the key elimination method for updating simplex tableaux in linear programming. We end this chapter with

Section 4.7, where a presentation of rounding error analysis for simple and extended arithmetic operations is given.

4.2 Vector and matrix norms and relevant theorems

Given an n-dimensional vector **x**, real or complex, it is useful in mathematical analysis to have a single non-negative number which gives an assessment of the size of **x**. This number is known as the norm of **x** and is denoted by $||\mathbf{x}||$. We already defined and used vector norms in Chapter 2. We elaborate here on this subject.

4.2.1 Vector norms

Let vector $\mathbf{x} = (x_1, x_2, \ldots, x_n)^T$. The norm of **x** plays the same role as the modulus in the case of a complex number. It satisfies

(i) $||\mathbf{x}|| > 0$, unless $\mathbf{x} = \mathbf{0}$; $||\mathbf{x}|| = 0$ implies $\mathbf{x} = \mathbf{0}$,

(ii) $||c\mathbf{x}|| = |c|\ ||\mathbf{x}||$, c is a complex scalar, and

(iii) $||\mathbf{x} + \mathbf{y}|| \leq ||\mathbf{x}|| + ||\mathbf{y}||$ (known as the **triangle inequality**) and as a result, $||\mathbf{x} - \mathbf{y}|| \geq ||\mathbf{x}|| - ||\mathbf{y}||$.

As described in Chapter 2, there are three vector norms in common use. They are derived from the p or the **Holder's norm**

(4.2.1) $$||\mathbf{x}||_p = \left[\sum_{i=1}^{n} |x_i|^p\right]^{1/p}, \quad 1 \leq p \leq \infty$$

Then for p = 1, the $\mathbf{L_1}$ **norm** of **x** is

(4.2.2) $$||\mathbf{x}||_1 = \sum_{i=1}^{n} |x_i|$$

For p = 2, the L_2 or the **Euclidean vector norm** of **x** is

(4.2.3) $$||\mathbf{x}||_2 = \mathrm{sqrt}\left[\sum_{i=1}^{n} |x_i|^2\right]$$

Again, for $p = \infty$, the L_∞ or the **Chebyshev vector norm** of **x** is

(4.2.4) $$||\mathbf{x}||_\infty = \max|x_i|, \quad i = 1, 2, \ldots, n$$

It is easy to show that these three norms are related by the following inequalities

$$(1/\text{sqrt}(n))||\mathbf{x}||_1 \le ||\mathbf{x}||_2 \le ||\mathbf{x}||_1, \quad (1/n)||\mathbf{x}||_1 \le ||\mathbf{x}||_\infty \le ||\mathbf{x}||_1$$

$$||\mathbf{x}||_\infty \le ||\mathbf{x}||_1 \le n\, ||\mathbf{x}||_\infty, \qquad ||\mathbf{x}||_\infty \le ||\mathbf{x}||_2 \le \text{sqrt}(n)||\mathbf{x}||_\infty$$

$$(1/\text{sqrt}(n))||\mathbf{x}||_2 \le ||\mathbf{x}||_\infty \le ||\mathbf{x}||_2, \qquad ||\mathbf{x}||_2 \le ||\mathbf{x}||_1 \le \text{sqrt}(n)||\mathbf{x}||_2$$

A useful inequality known as **Schwartz's inequality** is

$$|(\mathbf{x}, \mathbf{y})| \le ||\mathbf{x}||_2\, ||\mathbf{y}||_2$$

where $(\mathbf{x}, \mathbf{y})$ denotes the inner product of vectors $\mathbf{x}$ and $\mathbf{y}$.

4.2.2 Matrix norms

Similarly, given an n by n real or complex matrix $\mathbf{A} = (a_{ij})$, the norm of $\mathbf{A}$ satisfies the conditions

(i) $||\mathbf{A}|| > 0$, unless $\mathbf{A} = \mathbf{0}$; $||\mathbf{A}|| = 0$ implies $\mathbf{A} = \mathbf{0}$,

(ii) $||c\mathbf{A}|| = |c|\, ||\mathbf{A}||$; c is a complex scalar,

(iii) $||\mathbf{A} + \mathbf{B}|| \le ||\mathbf{A}|| + ||\mathbf{B}||$; ($\mathbf{A}$ and $\mathbf{B}$ are of the same size), and

(iv) $||\mathbf{AB}|| \le ||\mathbf{A}||\, ||\mathbf{B}||$.

There are matrix norms that are said to be **natural norms**, i.e., that are associated with (subordinate to or induced by) vector norms. Since for any vector norm, we expect that $||\mathbf{Ax}|| \le ||\mathbf{A}||\, ||\mathbf{x}||$, the subordinate matrix norm associated with a vector norm is defined by

$$||\mathbf{A}||_p = \max(||\mathbf{Ax}||_p / ||\mathbf{x}||_p), \quad \mathbf{x} \ne \mathbf{0}$$

or equally

$$||\mathbf{A}||_p = \max||\mathbf{Ax}||_p, \quad ||\mathbf{x}||_p = 1$$

4.2.3 Hermitian matrices and vectors

Definition 4.1

A square matrix $\mathbf{A}$ is known as a **Hermitian** matrix if $\mathbf{A}^H = \mathbf{A}$, where $\mathbf{A}^H$ is the complex conjugate transpose of $\mathbf{A}$. This class of matrices includes symmetric matrices when the elements of $\mathbf{A}$ are real. The following are Hermitian matrices

$$\begin{bmatrix} a & b \\ b & d \end{bmatrix} \quad \text{and} \quad \begin{bmatrix} a & b+ci \\ b-ci & d \end{bmatrix}$$

where a, b, c and d are real elements and i = sqrt(−1).

Similarly let $\mathbf{x}$ be a column vector whose elements are complex. Then $\mathbf{x}^H$ is a row vector whose elements are the complex conjugates of those of $\mathbf{x}$. It follows that

$$(\mathbf{x}^H)^H = \mathbf{x}, \ (\mathbf{A}^H)^H = \mathbf{A} \text{ and } (\mathbf{AB})^H = \mathbf{B}^H\mathbf{A}^H$$

and $\mathbf{x}^H\mathbf{x}$ is real and positive, $\mathbf{x} \neq \mathbf{0}$. In other words, Hermitian transposes have similar properties to those of ordinary transposes.

Theorem 4.1

Let $\mathbf{A}$ be a Hermitian matrix. Then

(i) $\mathbf{A}$ has real eigenvalues

(ii) $\mathbf{A}^H\mathbf{A}$ has real non-negative eigenvalues.

Proof:

Let λ be an eigenvalue of a Hermitian matrix $\mathbf{A}$ and $\mathbf{x}$ the corresponding eigenvector

$$\mathbf{Ax} = \lambda\mathbf{x}$$

Hence

$$\mathbf{x}^H\mathbf{Ax} = \lambda\mathbf{x}^H\mathbf{x}$$

Now $\mathbf{x}^H\mathbf{x}$ is real and positive, for $\mathbf{x} \neq \mathbf{0}$. Again, $\mathbf{x}^H\mathbf{Ax}$ is scalar and thus real, since

$$(\mathbf{x}^H\mathbf{Ax})^H = \mathbf{x}^H\mathbf{A}^H\mathbf{x} = \mathbf{x}^H\mathbf{Ax}$$

proving (i) that is λ real.

The proof of (ii) follows since

$$\mathbf{x}^H\mathbf{A}^H\mathbf{Ax} = (\mathbf{Ax}, \mathbf{Ax}) = \mathbf{x}^H\mathbf{A}^H\mathbf{Ax} \tag{4.2.5}$$

$$= (\mathbf{Ax})^H(\mathbf{Ax}) = \lambda^2\mathbf{x}^H\mathbf{x} = \sigma\mathbf{x}^H\mathbf{x}$$

Let the eigenvalues of $(\mathbf{A}^H\mathbf{A})$ be $\sigma_i(\mathbf{A}^H\mathbf{A})$. Then sqrt($\sigma_i(\mathbf{A}^H\mathbf{A})$) are known as the **singular values** of $\mathbf{A}$ and are denoted by $s_i(\mathbf{A})$, i.e.,

$$s_i^2(\mathbf{A}) = \text{sqrt}(\sigma_i(\mathbf{A}^H\mathbf{A})), \quad i = 1, 2, \ldots, n$$

Theorem 4.2

Let **A** be an n by n matrix. Corresponding to the given vector norms of Section 4.2.1, there are matrix norms $||\mathbf{A}||_p$, for p =1, 2 and ∞, that are subordinate norms:

(i) $||\mathbf{A}||_1$ = maximum absolute column sum of **A**

(ii) $||\mathbf{A}||_2 = \max_i s_i(\mathbf{A})$ (maximum singular value of **A**)

(iii) $||\mathbf{A}||_\infty$ = maximum absolute row sum of **A**.

Proof:

To prove (i), from (4.2.2)

$$\|\mathbf{Ax}\|_1 = \sum_{i=1}^{n}\left|\sum_{j=1}^{n} a_{ij}x_j\right| \le \sum_{i=1}^{n}\sum_{j=1}^{n} |a_{ij}||x_j|$$

and by re-arranging the summations

$$\|\mathbf{Ax}\|_1 \le \sum_{i=1}^{n} |a_{ij}| \sum_{j=1}^{n} |x_j| \le \left(\max_j \sum_{i=1}^{n} |a_{ij}|\right)\left(\sum_{j=1}^{n} |x_j|\right)$$

Since the summation of the far right = $||\mathbf{x}||_1$

$$\|\mathbf{A}\|_1 = \max(\|\mathbf{Ax}\|_1 / \|\mathbf{x}\|_1) \le \max_j \sum_{i=1}^{n} |a_{ij}|$$

Suppose that the maximum sum is for column k. Choose $x_k = 1$ and $x_i = 0$, for $i \neq k$. For this choice of **x**, equality is obtained.

To prove (ii), pre-multiply matrix **A** by $\mathbf{A}^H$. Then $(\mathbf{A}^H\mathbf{A})$, being Hermitian, has an orthogonal set of n eigenvectors $\{\mathbf{y}_i\}$, each associated with a non-negative eigenvalue (Theorem 4.1 (ii)). Let **x** be written as a linear combination of $(\mathbf{y}_i)$, i.e.,

$$\mathbf{x} = \sum_{j=1}^{n} c_j \mathbf{y}_j$$

where the c_j are constants. Then from (4.2.5)

$$||\mathbf{Ax}||_2^2 / ||\mathbf{x}||_2^2 = (\mathbf{x}^H\mathbf{A}^H\mathbf{Ax}) / (\mathbf{x}, \mathbf{x})$$

$$= \sum_{j=1}^{n} |c_j|^2 s_j^2(\mathbf{A}) / \sum_{j=1}^{n} |c_j|^2$$

or

$$||\mathbf{Ax}||_2 \,/\, ||\mathbf{x}||_2 \le s_{max}(\mathbf{A})$$

Moreover, by choosing $\mathbf{x} = \mathbf{y}_i$ associated with $\sigma_{max}(\mathbf{A}^H\mathbf{A})$, an equality is obtained.

To prove (iii), from (4.2.4)

$$\|\mathbf{A}\|_\infty = \max_i \left| \sum_{j=1}^{n} a_{ij}x_j \right| \le \max_i \sum_{j=1}^{n} |a_{ij}||x_j| \le \max_i \sum_{j=1}^{n} |a_{ij}| \max |x_j|$$

Hence

$$\|\mathbf{A}\|_\infty = \max \|\mathbf{Ax}\|_\infty / \|\mathbf{x}\|_\infty \le \max_i \sum_{j=1}^{n} |a_{ij}|$$

Again, assume that $i = k$ gives the maximum sum on the right. Construct a vector $\mathbf{x}$ with $x_j = 1$ if $a_{kj} > 0$ and $x_j = -1$ if $a_{kj} < 0$. For this $\mathbf{x}$, an equality is attained. The theorem is thus proved.

As a result of this theorem

$$||\mathbf{I}|| = 1$$

where $\mathbf{I}$ is an n-unit matrix and the norm is any natural norm.

4.2.4 Other matrix norms

There are other matrix norms [9, 11] that satisfy the four conditions of Section 4.2.2 and are defined by

$$\|\mathbf{A}\|_q = \left[\sum_{i,j} |a_{ij}|^q \right]^{1/q}, \quad 1 < q < 2$$

For the case $q = 2$, the norm is known as the **Schur** or the **Euclidean matrix norm**. It is easily calculated and hence often used

$$\|\mathbf{A}\|_E = \left[\sum_{i,j} |a_{ij}|^2\right]^{1/2}$$

This matrix norm is consistent with the Euclidean vector norm $\|\mathbf{x}\|_2$ but it is not subordinate to any vector norm, since for an n-unit matrix $\mathbf{I}$, $\|\mathbf{I}\|_E = \text{sqrt}(n)$.

4.2.5 Euclidean and the spectral matrix norms

An important property shared by both the Euclidean and spectral matrix norms is given by the following.

Theorem 4.3

The Euclidean matrix norm and the 2 (the spectral) matrix norm have invariant properties under unitary transformation. The same is true for the 2-vector norm. For unitary matrices **U** and **V**, we have

$$\|\mathbf{x}\|_2 = \|\mathbf{Ux}\|_2$$
$$\|\mathbf{A}\| = \|\mathbf{UAV}\| = \|\mathbf{UA}\| = \|\mathbf{AV}\|$$

(**U** is said to be unitary if $\mathbf{U}^H\mathbf{U} = \mathbf{U}\mathbf{U}^H = \mathbf{I}$; that is, $\mathbf{U}^H = \mathbf{U}^{-1}$. A real unitary matrix is called **orthonormal**).

Proof:

To prove the first part of the theorem, since **U** is unitary, $\|\mathbf{U}^H\mathbf{U}\|_2 = \|\mathbf{I}\|_2 = 1$

$$\|\mathbf{Ux}\|_2^2 = (\mathbf{Ux})^H(\mathbf{Ux}) = \mathbf{x}^H\mathbf{U}^H\mathbf{Ux} = \mathbf{x}^H\mathbf{x} = \|\mathbf{x}\|_2^2$$

To prove the second part of the theorem, we have for example

$$\|\mathbf{UA}\|_2^2 = \|(\mathbf{UA})^H\mathbf{UA}\|_2 = \|\mathbf{A}^H\mathbf{U}^H\mathbf{UA}\|_2 = \|\mathbf{A}^H\mathbf{A}\|_2 = \|\mathbf{A}\|_2^2$$

For the Euclidean matrix norm, again consider for example **UA**, then

$$\|\mathbf{UA}\|_E = \|\mathbf{A}\|_E$$

The proof follows from the fact that the Euclidean length of each column of **UA** equals the Euclidean length of that column and the theorem is proved.

4.2.6 Euclidean norm and the singular values

Definition 4.2

The trace of an n square matrix **A**, denoted by tr(**A**), is the sum of its diagonal elements. For two square matrices **A** and **B**, tr(**AB**) = tr(**BA**).

As a result:

(i) tr(**A**) = the sum of the eigenvalues of **A**.

(ii) The sum of the eigenvalues of **AB** = sum of the eigenvalues of **BA**. In fact, **AB** and **BA** have identical eigenvalues. See also Theorem 4.10.

(iii) $||\mathbf{A}||_E^2 = \mathrm{tr}(\mathbf{A}^H\mathbf{A})$.

Since $||\mathbf{A}||_E^2$ equals the trace of $\mathbf{A}^H\mathbf{A}$ which equals the sum of the eigenvalues of $(\mathbf{A}^H\mathbf{A})$, $\Sigma_i\ \sigma_i(\mathbf{A}^H\mathbf{A})$, and again, since $\sigma_i(\mathbf{A}^H\mathbf{A}) = s_i^2(\mathbf{A})$, we have

$$\|\mathbf{A}\|_E^2 = \sum_{i=1}^{n} s_i^2(\mathbf{A})$$

Then since $||\mathbf{A}||_2 = s_1(\mathbf{A})$, the largest singular value of **A**, we have the relations

(4.2.6) $$||\mathbf{A}||_2 \le ||\mathbf{A}||_E \le \mathrm{sqrt}(n)||\mathbf{A}||_2$$

Theorem 4.4

Let $||\mathbf{A}||_\alpha$ and $||\mathbf{A}||_\beta$ be any two matrix norms. Then there exist positive numbers a and b such that

$$a \le ||\mathbf{A}||_\alpha \,/\, ||\mathbf{A}||_\beta \le b$$

Assume that **A** is an n by n matrix. The following relations between matrix norms exist

$(1/\mathrm{sqrt}(n))||\mathbf{A}||_E \le ||\mathbf{A}||_2 \le ||\mathbf{A}||_E$, $(1/\mathrm{sqrt}(n))||\mathbf{A}||_E \le ||\mathbf{A}||_\infty \le \mathrm{sqrt}(n)||\mathbf{A}||_E$

$(1/\mathrm{sqrt}(n))||\mathbf{A}||_E \le ||\mathbf{A}||_1 \le \mathrm{sqrt}(n)||\mathbf{A}||_E$, $(1/\mathrm{sqrt}(n))||\mathbf{A}||_2 \le ||\mathbf{A}||_1 \le \mathrm{sqrt}(n)||\mathbf{A}||_2$

$(1/\mathrm{sqrt}(n))||\mathbf{A}||_2 \le ||\mathbf{A}||_\infty \le \mathrm{sqrt}(n)||\mathbf{A}||_2$, $(1/n)||\mathbf{A}||_\infty \le ||\mathbf{A}||_1 \le n||\mathbf{A}||_\infty$

Theorem 4.5 ([13], p. 133)

$$||\mathbf{AB}||_E \le ||\mathbf{A}||_E\ ||\mathbf{B}||_2 \text{ and } ||\mathbf{AB}||_E \le ||\mathbf{A}||_2\ ||\mathbf{B}||_E$$

Theorem 4.6

For any matrix norm, we have

$$\| \mathbf{I} \| \geq 1, \ \ \|A^{-1}\| \geq 1/\|\mathbf{A}\|, \ \ \|\mathbf{A}^k\| \leq \|\mathbf{A}\|^k$$

4.2.7 Eigenvalues and the singular values of the sum and the product of two matrices

Theorem 4.7: Wielandt-Hoffman theorem ([16], p. 104)

Let $\mathbf{C} = \mathbf{A} + \mathbf{B}$, where **A**, **B** and **C** are real symmetric n-matrices having eigenvalues α_i, β_i and γ_i respectively, arranged in a non-increasing order; that is, $\alpha_1 \geq \alpha_2 \geq \ldots \geq \alpha_n$, etc. Then

$$\sum_{i=1}^{n} (\gamma_i - \alpha_i)^2 \leq \|\mathbf{B}\|_E^2 = \sum_{i=1}^{n} \beta_i^2$$

This theorem relates the perturbations in the eigenvalues to the Euclidean norm of the perturbation matrix **B**.

Theorem 4.8

Let $\mathbf{C} = \mathbf{A} + \mathbf{B}$, where **A**, **B** and **C** are real symmetric n-matrices having singular values $s_i(\mathbf{A})$, $s_i(\mathbf{B})$ and $s_i(\mathbf{C})$ respectively, arranged in a non-decreasing order; $\alpha_1 \leq \alpha_2 \leq \ldots \leq \alpha_n$. Then

$$s_i(\mathbf{A}) - s_1(\mathbf{B}) \leq s_i(\mathbf{C}) \leq s_i(\mathbf{A}) + s_n(\mathbf{B}), \quad i = 1, 2, \ldots, n$$

In particular, since $\|\mathbf{B}\|_2 = s_n(\mathbf{B})$ and since from (4.2.6), $\|\mathbf{B}\|_2 \leq \|\mathbf{B}\|_E$

$$s_i(\mathbf{C}) \leq s_i(\mathbf{A}) + \|\mathbf{B}\|_E, \quad i = 1, 2, \ldots, n$$

Definitions 4.3

Given an n by m matrix **A**, then ([11], p. 25)

(i) A k(≤n) by s(≤m) submatrix of **A** is obtained from **A** by deleting certain rows and columns from **A**.

(ii) A principal submatrix of **A** is a submatrix, the diagonal elements of which are diagonal elements of **A**.

(iii) A leading submatrix of **A** is a submatrix (not necessary a square one) that lies in the upper left hand corner of **A**.

(vi) A leading principal submatrix of **A** is a square submatrix that lies in the upper left hand corner of **A**.

Theorem 4.9 ([11], p. 317)

Let **A** be a real symmetric n-matrix and **B** be a principal (n – 1)-submatrix of **A**. Let (α_i) be the eigenvalues of **A** and (β_i) be the eigenvalues of **B**, arranged in a non-decreasing order. Then

$$\alpha_1 \le \beta_1 \le \alpha_2 \le \beta_2 \ldots \le \alpha_{n-1} \le \beta_{n-1} \le \alpha_n$$

Theorem 4.10 ([16], p. 54)

(i) Let **A** and **B** be square real n-matrices. Then the eigenvalues of **AB** and of **BA** are identical,

(ii) Let **A** and **B** be an n by m and an m by n real martices respectively, $n \ge m$. Then **AB** and **BA** have m identical eigenvalues and the extra (n – m) eigenvalues of **AB** are 0's.

4.2.8 Accuracy of the solution of linear equations

Theorem 4.11

Assume that $\|d\mathbf{A}\| < 1$, where d**A** is a perturbation to matrix **A**. Then $(\mathbf{I} + d\mathbf{A})$ is non-singular and

$$\frac{1}{(1+\|d\mathbf{A}\|)} \le \left\|(\mathbf{I} + d\mathbf{A})^{-1}\right\| \le \frac{1}{(1-\|d\mathbf{A}\|)}$$

Proof:

Let us write

$$\mathbf{I} = (\mathbf{I} + d\mathbf{A})(\mathbf{I} + d\mathbf{A})^{-1} \tag{4.2.7}$$

Then by taking norms

$$1 \le \|(\mathbf{I} + d\mathbf{A})\|\ \|(\mathbf{I} + d\mathbf{A})^{-1}\| \le (1 + \|d\mathbf{A}\|)\|(\mathbf{I} + d\mathbf{A})^{-1}\|$$

which proves the left inequality of the theorem.

To prove the right inequality, write (4.2.7) in the form

$$(\mathbf{I} + d\mathbf{A})^{-1} = \mathbf{I} - d\mathbf{A}(\mathbf{I} + d\mathbf{A})^{-1}$$

By taking norms, we get

$$\|(\mathbf{I} + d\mathbf{A})^{-1}\| \le 1 + \|d\mathbf{A}(\mathbf{I} + d\mathbf{A})^{-1}\| \le 1 + \|d\mathbf{A}\|\ \|(\mathbf{I} + d\mathbf{A})^{-1}\|$$

By dividing by $\|(\mathbf{I} + d\mathbf{A})^{-1}\|$ and arranging the terms, we get the right inequality of the theorem.

Suppose that in a certain problem a small change $d\alpha$ in the data α causes a change dx in the result x. If it is possible to write

$$|dx/x| = K|d\alpha/\alpha|$$

then K is called the **condition number** for the relative change in x caused by a relative change in α. Let us consider the matrix equation

$$\mathbf{Ax} = \mathbf{b}$$

If vector **b** is perturbed by d**b** and matrix **A** is perturbed by d**A**, the solution has to be given by

$$(\mathbf{A} + d\mathbf{A})(\mathbf{x} + d\mathbf{x}) = (\mathbf{b} + d\mathbf{b}) \tag{4.2.8}$$

Theorem 4.12

Assume that **A** is a square non-singular matrix and that $||\mathbf{A}^{-1}||\ ||d\mathbf{A}|| < 1$. Then if $\mathbf{x} \neq \mathbf{0}$

$$\frac{\|d\mathbf{x}\|}{\|\mathbf{x}\|} \leq CK(\mathbf{A})\left[\frac{\|d\mathbf{A}\|}{\|\mathbf{A}\|} + \frac{\|d\mathbf{b}\|}{\|\mathbf{b}\|}\right]$$

where $C = 1 / [1 - K(\mathbf{A})||d\mathbf{A}||/||\mathbf{A}||]$ and $K(\mathbf{A})$ is the condition number of **A**. The spectral condition number of **A** is given by

$$K(\mathbf{A}) = ||\mathbf{A}||_2\ ||\mathbf{A}^{-1}||_2$$

See also Section 17.5.1.

Proof:

Subtract $\mathbf{Ax} = \mathbf{b}$, from (4.2.8)

$$(\mathbf{A} + d\mathbf{A})d\mathbf{x} = d\mathbf{b} - d\mathbf{A}\ \mathbf{x} \tag{4.2.9}$$

Then

$$d\mathbf{x} = (\mathbf{A} + d\mathbf{A})^{-1}(d\mathbf{b} - d\mathbf{A}\ \mathbf{x}) = (\mathbf{I} + \mathrm{A}^{-1}d\mathrm{A})^{-1}\mathbf{A}^{-1}(d\mathbf{b} - d\mathrm{A}\ \mathbf{x})$$

and by taking norms

$$||d\mathbf{x}|| \leq ||(\mathbf{I} + \mathbf{A}^{-1}d\mathbf{A})^{-1}||\ ||\mathbf{A}^{-1}||(||d\mathbf{b}|| + ||d\mathbf{A}||\ ||\mathbf{x}||)$$

By using the right inequality of Theorem 4.11, after replacing d**A** $\mathbf{A}^{-1}d\mathbf{A}$ and since $||\mathbf{A}^{-1}||\ ||d\mathbf{A}|| < 1$

$$\left\|(\mathbf{I}+\mathbf{A}^{-1}\mathrm{d}\mathbf{A})^{-1}\right\| \le \frac{1}{(1-\left\|\mathbf{A}^{-1}\mathrm{d}\mathbf{A}\right\|)} \le \frac{1}{(1-\left\|\mathbf{A}^{-1}\right\|\left\|\mathrm{d}\mathbf{A}\right\|)}$$

Since $||\mathbf{A}^{-1}||\ ||\mathrm{d}\mathbf{A}|| = \mathrm{K}(\mathbf{A})||\mathrm{d}\mathbf{A}||/||\mathbf{A}||$, we get

$$||(\mathbf{I}+\mathbf{A}^{-1}\mathrm{d}\mathbf{A})^{-1}|| \le \mathrm{C}$$

where C is defined above. Then by dividing by $||\mathbf{x}||$ and noting that $||\mathbf{b}|| \le ||\mathbf{A}||\ ||\mathbf{x}||$, the theorem is proved.

Again, since in the denominator of C, the quantity $\mathrm{K}(\mathbf{A})||\mathrm{d}\mathbf{A}||/||\mathbf{A}|| = ||\mathbf{A}^{-1}||\ ||\mathrm{d}\mathbf{A}||$, if $||\mathbf{A}^{-1}||\ ||\mathrm{d}\mathbf{A}|| << 1$, $\mathrm{C} \approx 1$ and we get

$$\frac{\|\mathrm{d}\mathbf{x}\|}{\|\mathbf{x}\|} \le \mathrm{K}(\mathbf{A})\left[\frac{\|\mathrm{d}\mathbf{A}\|}{\|\mathbf{A}\|}+\frac{\|\mathrm{d}\mathbf{b}\|}{\|\mathbf{b}\|}\right] \tag{4.2.10}$$

If $\mathrm{d}\mathbf{b} = \mathbf{0}$, we get from (4.2.9)

$$\mathrm{d}\mathbf{x} = -(\mathbf{I}+\mathbf{A}^{-1}\mathrm{d}\mathbf{A})^{-1}\mathbf{A}^{-1}\mathrm{d}\mathbf{A}\ \mathbf{x}$$

Then by taking norms and using the right inequality of Theorem 4.11, after replacing $\mathrm{d}\mathbf{A}$ by $\mathbf{A}^{-1}\mathrm{d}\mathbf{A}$

$$||\mathrm{d}\mathbf{x}|| \le ||(\mathbf{I}+\mathbf{A}^{-1}\mathrm{d}\mathbf{A})^{-1}||\ ||\mathbf{A}^{-1}\mathrm{d}\mathbf{A}||\ ||\mathbf{x}||$$

and

$$\frac{\|\mathrm{d}\mathbf{x}\|}{\|\mathbf{x}\|} \le \left[\frac{\left\|\mathbf{A}^{-1}\mathrm{d}\mathbf{A}\right\|}{1-\left\|\mathbf{A}^{-1}\mathrm{d}\mathbf{A}\right\|}\right]$$

This inequality is used in estimating the relative error in the solution vector $\mathbf{x}$, due to rounding error in some matrix computations, using backward error analysis. It is itself inequality (4.7.12) below.

Note 4.1

It is noted that if the set of equations $\mathbf{Ax} = \mathbf{b}$ is ill-conditioned, resulting in a large relative error $||\mathrm{d}\mathbf{x}||/||\mathbf{x}||$, we expect $\mathrm{K}(\mathbf{A})$ in (4.2.10) to be large. Yet, the converse is not necessarily true. A large $\mathrm{K}(\mathbf{A})$ does not necessarily imply that $\mathbf{Ax} = \mathbf{b}$ is ill-conditioned. The reason for this is that Theorem 4.12 gives an upper bound for the ratio $||\mathrm{d}\mathbf{x}||/||\mathbf{x}||$ and this upper bound may not be attained.

4.3 Elementary matrices

Elementary operations to a given matrix equation involving matrix **A** are mostly based on pre-multiplying and/or post-multiplying **A** progressively by one or more of the matrices known as elementary matrices. We describe here 3 of the elementary matrices.

(a) Diagonal matrix: A diagonal matrix $\mathbf{D} = \mathrm{diag}(\alpha, \beta, \gamma, \ldots)$ is an elementary matrix. Pre-multiplying matrix **A** by **D** results in multiplying each element of the first row of **A** by α and each element of the second row of **A** by β, …, etc. Post-multiplying **A** by **D** results in multiplying each element of the first column of **A** by α and each element of the second column of **A** by β, …, etc.

(b) **Permutation matrix**: A permutation matrix $\mathbf{P}_{ij}$ is an n-unit matrix **I**, except that its i and j columns are interchanged. Pre-multiplying **A** by $\mathbf{P}_{ij}$ results in the interchange of rows i and j of **A**. Post-multiplying **A** by $\mathbf{P}_{ij}$ results in the interchange of columns i and j of **A**. We note that $\mathbf{P}_{ij} = \mathbf{P}_{ij}^T = \mathbf{P}_{ij}^{-1}$ and $\mathbf{P}_{ij}\mathbf{P}_{ij} = \mathbf{I}$.

(c) A **matrix** $\mathbf{M}_i$, which is an n-unit matrix **I**, except for its column i which has the form (for the case n = 5 and i = 2)

$$(m_{12}, 1, m_{32}, m_{42}, m_{52})^T$$

is another elementary matrix. Pre-multiplying **A** by $\mathbf{M}_2$ results in a matrix **B**, $\mathbf{B} = \mathbf{M}_2\mathbf{A}$. The second row of **B** equals the second row of **A**. But the first row of **B** is the first row of **A** + m_{12} times the second row of **A**. The third row of **B** is the third row of **A** + m_{32} times the second row of **A**, …, etc. Post-multiplying **A** by $\mathbf{M}_2^T$, where the superscript T refers to the transpose, gives a similar result with the columns of **A** instead.

Note 4.2

It is noted that the elementary matrices described above are never stored in the computer. They are described here in order to elucidate the operations that equation $\mathbf{Ax} = \mathbf{b}$ undergoes.

In particular, the operations of the permutation matrices are recorded by index vectors. For example, an index vector $\mathbf{IR}(j) = (1, 2, 5, 4, 3)$ indicates that rows 3 and 5 of matrix **A** have been interchanged.

Note 4.3

Elementary operations on matrices are performed on matrices of any size, not only on square matrices (Chapter 17).

4.4 Gauss LU factorization with complete pivoting

Given a system of linear equations $\mathbf{Ax} = \mathbf{b}$, the Gauss elimination method or the Gauss **LU** factorization is described as follows. The method applies elementary operations successively to matrix **A** and vector **b**, which finally reduce matrix **A** to an upper triangular matrix **U** while vector **b** becomes vector $\mathbf{L}^{-1}\mathbf{b}$, where **L** is a unit lower triangular matrix (its diagonal elements are 1's).

Gauss **LU** factorization without pivoting is as follows. Let $\mathbf{A} = \mathbf{A}^{(1)}$ and $\mathbf{b} = \mathbf{b}^{(1)}$. Then

(i) Add multiples of the first row of equation $\mathbf{A}^{(1)}\mathbf{x} = \mathbf{b}^{(1)}$ to each of the second, third, …, n^{th} rows of equation $\mathbf{A}^{(1)}\mathbf{x} = \mathbf{b}^{(1)}$ so as to eliminate x_1 from such equations. This results in the equation $\mathbf{A}^{(2)}\mathbf{x} = \mathbf{b}^{(2)}$.

(ii) Add multiples of the second row of the new equation $\mathbf{A}^{(2)}\mathbf{x} = \mathbf{b}^{(2)}$ to each of the new third, forth, …, etc. rows of $\mathbf{A}^{(2)}\mathbf{x} = \mathbf{b}^{(2)}$, so as to eliminate x_2 from such equations. This results in the equation $\mathbf{A}^{(3)}\mathbf{x} = \mathbf{b}^{(3)}$.

(iii) Continue this process until an upper triangular set of equations is obtained as shown (for n = 4)

$$
\begin{aligned}
a_{11}^{(1)}x_1 + a_{12}^{(1)}x_2 + a_{13}^{(1)}x_3 + a_{14}^{(1)}x_4 &= b_1^{(1)} \\
a_{22}^{(2)}x_2 + a_{23}^{(2)}x_3 + a_{24}^{(2)}x_4 &= b_2^{(2)} \\
a_{33}^{(3)}x_3 + a_{34}^{(3)}x_4 &= b_3^{(3)} \\
a_{44}^{(4)}x_4 &= b_4^{(4)}
\end{aligned}
\tag{4.4.1}
$$

In (4.4.1) the $a_{1j}^{(1)}$ are the elements of the first row of $\mathbf{A}^{(1)}$ and $a_{2j}^{(2)}$ are the elements of the second row of $\mathbf{A}^{(2)}$ and so on. Again, $b_1^{(1)}$ is the first element of $\mathbf{b}^{(1)}$ and so on. The solution is obtained by back substitution in the triangular system (4.4.1) starting with x_4.

The first element $a_{11}^{(1)}$ of matrix $\mathbf{A}^{(1)}$ is the pivot element in process (i) above. Likewise the element $a_{22}^{(2)}$ of matrix $\mathbf{A}^{(2)}$ is the pivot element in process (ii) above and so on.

We now show that this method gives the factorization

$$\mathbf{A} = \mathbf{LU}$$

As indicated above, **U** is the upper triangular matrix appearing on the l.h.s. of (4.4.1) and $\mathbf{L}^{-1}\mathbf{b}$ is the vector appearing on the r.h.s. of (4.4.1).

4.4.1 Importance of pivoting

In the Gauss elimination method, complete pivoting is achieved by means of interchanging the rows of the system of equations $\mathbf{A}^{(i)}\mathbf{x} = \mathbf{b}^{(i)}$, and/or the columns of the matrices $\mathbf{A}^{(i)}$, where $i = 1, 2, \ldots, n-1$ such that the pivot element $a_{11}^{(1)}$ is the largest element in absolute value in the whole of matrix $\mathbf{A}^{(1)}$, while $a_{22}^{(2)}$ is the largest element in absolute value among the current $a_{ij}^{(2)}$ of $\mathbf{A}^{(2)}$, with $i, j \geq 2, \ldots$, and so on.

Partial pivoting is achieved by interchanging the columns of $\mathbf{A}^{(1)}, \mathbf{A}^{(2)}, \ldots$ such that $a_{11}^{(1)}$ is the largest element in absolute value in first row of $\mathbf{A}^{(1)}$, $a_{22}^{(2)}$ is the largest element in absolute value among the current $a_{2j}^{(2)}$ of $\mathbf{A}^{(2)}$, with $j \geq 2 \ldots$, and so on. By doing this, we avoid pivots that are small in absolute value. Otherwise, due to the finite precision of the computer word, inaccurate or even incorrect results are obtained.

Example 4.1

This is a well-known classical example [3] which illustrates the importance of pivoting in the Gauss elimination process. Assume that all the results are rounded to four decimal places. Let us solve the system of two equations

$$\begin{aligned} 0.0001x_1 + 1.00x_2 &= 1.00 \\ 1.00x_1 + 1.00x_2 &= 2.00 \end{aligned} \tag{4.4.2}$$

The true result rounded to 4 decimals is

$$x_1 = 10000/9999 = 1.0001 \text{ and } x_2 = 0.9999$$

Yet, Gauss elimination without partial pivoting (without exchanging columns) is done by using the first equation in (4.4.2) to eliminate x_1 from the second equation. That is by subtracting 10000 times the first equation from the second equation and rounding the result to 4 decimal places, which gives

$$\begin{aligned} 0.0001x_1 + 1.00x_2 &= 1.00 \\ -9999x_2 &= -9998 \end{aligned}$$

from which $x_2 = 1.0$ and $x_1 = 0.0$ (wrong result).

On the other hand, with pivoting (by first exchanging the two columns on the l.h.s. of (4.4.2)), and subtracting the first equation from the second, we get the triangular system

$$\begin{aligned} 1.00x_2 + 1.00x_1 &= 1.00 \\ (1.0 - 0.0001)x_1 &= 1.00 \end{aligned}$$

giving $x_1 = 1.00/0.9999 = 1.0001$ and $x_2 = 0.9999$ (correct result).

4.4.2 Using complete pivoting

Let $(\mathbf{A}|\mathbf{b})$ denote the n by (n + 1) matrix with vector **b** situated to the right side of matrix **A**.

The Gauss elimination process with complete pivoting consists of (n – 1) major steps in which matrix $(\mathbf{A}|\mathbf{b}) = (\mathbf{A}^{(1)}|\mathbf{b}^{(1)})$ is reduced successively to $(\mathbf{A}^{(2)}|\mathbf{b}^{(2)})$, $(\mathbf{A}^{(3)}|\mathbf{b}^{(3)})$, …, $(\mathbf{A}^{(n)}|\mathbf{b}^{(n)})$ as follows:

(i) Choose the element of $a_{ij}^{(1)}$ of matrix $\mathbf{A}^{(1)}$ of maximum absolute value among all the elements of $\mathbf{A}^{(1)}$, as the pivot. Suppose this element is $a_{uv}^{(1)}$. Interchange rows 1 and u in the complete n by (n + 1) matrix $(\mathbf{A}^{(1)}|\mathbf{b}^{(1)})$ and interchange columns 1 and v of $\mathbf{A}^{(1)}$. This may be done by pre-multiplying $(\mathbf{A}^{(1)}|\mathbf{b}^{(1)})$ by a permutation matrix $\mathbf{S}_1$ and post-multiplying $\mathbf{A}^{(1)}$ by another permutation matrix $\mathbf{P}_1$.

(ii) Calculate and record the (n – 1) multipliers m_{i1}

$$m_{i1} = a_{i1}^{(1)}/a_{11}^{(1)}, \quad i = 2, 3, \dots, n$$

where $a_{i1}^{(1)}$ and $a_{11}^{(1)}$ are the elements of the permuted matrix $\mathbf{A}^{(1)}$.

(iii) Calculate the matrix $(\mathbf{A}^{(2)}|\mathbf{b}^{(2)})$ as follows

$$\text{for } i = 2, \dots, n$$

$$\begin{aligned} a_{ij}^{(2)} &= a_{ij}^{(1)} - m_{i1}a_{1j}^{(1)}, \quad j = 2, 3, \dots, n \\ b_i^{(2)} &= b_i^{(1)} - m_{i1}b_1^{(1)} \end{aligned}$$

We remark that

(4.4.3) $\mathbf{S}_1 = \mathbf{S}_1^{\mathrm{T}} = \mathbf{S}_1^{-1}, \mathbf{P}_1 = \mathbf{P}_1^{\mathrm{T}} = \mathbf{P}_1^{-1}, \mathbf{S}_1\mathbf{S}_1 = \mathbf{I}, \ \mathbf{P}_1\mathbf{P}_1 = \mathbf{I}$

Then in vector-matrix notation, step (i) is equivalent to changing equation $\mathbf{A}^{(1)}\mathbf{x} = \mathbf{b}^{(1)}$ to

$$\mathbf{S}_1\mathbf{A}^{(1)}\mathbf{P}_1\mathbf{P}_1\mathbf{x} = \mathbf{S}_1\mathbf{b}^{(1)}$$

Or in terms of the permuted matrix $\underline{\mathbf{A}}^{(1)}$ and permuted vectors $\underline{\mathbf{x}}$ and $\underline{\mathbf{b}}^{(1)}$, where $\underline{\mathbf{A}}^{(1)} = \mathbf{S}_1\mathbf{A}^{(1)}\mathbf{P}_1$, $\underline{\mathbf{x}} = \mathbf{P}_1\mathbf{x}$ and $\underline{\mathbf{b}}^{(1)} = \mathbf{S}_1\mathbf{b}^{(1)}$

$$\underline{\mathbf{A}}^{(1)}\underline{\mathbf{x}} = \underline{\mathbf{b}}^{(1)}$$

Step (iii) is equivalent to pre-multiplying $(\underline{\mathbf{A}}^{(1)}|\underline{\mathbf{b}}^{(1)})$ by matrix $\mathbf{M}_1$ in order to obtain matrix $(\mathbf{A}^{(2)}|\mathbf{b}^{(2)})$. $\mathbf{M}_1$ is a unit matrix, except for its first column, which is given by

$$(1, -m_{21}, -m_{31}, \ldots, -m_{n1})^T$$

where $m_{21}, m_{31}, \ldots, m_{n1}$ are given in step (ii) above. Hence, we get

$$(\mathbf{A}^{(2)}|\mathbf{b}^{(2)}) = \mathbf{M}_1(\underline{\mathbf{A}}^{(1)}|\underline{\mathbf{b}}^{(1)})$$

Or in other words

$$\mathbf{A}^{(2)} = \mathbf{M}_1\mathbf{S}_1\mathbf{A}^{(1)}\mathbf{P}_1 \text{ and } \mathbf{b}^{(2)} = \mathbf{M}_1\mathbf{S}_1\mathbf{b}^{(1)}$$

The second major operation is similar. Determine the element of maximum absolute value among the current $a_{ij}^{(2)}$ with $i, j \geq 2$. Suppose this element is $a_{rs}^{(2)}$. Interchange rows 2 and r in the complete n by (n + 1) matrix $(\mathbf{A}^{(2)}|\mathbf{b}^{(2)})$ and columns 2 and s of $\mathbf{A}^{(2)}$. Calculate the (n – 2) multipliers

$$m_{i2} = a_{i2}^{(2)}/a_{22}^{(2)}, \quad i = 3, 4, \ldots, n$$

Again $a_{i2}^{(2)}$ and $a_{22}^{(2)}$ are the elements of the permuted matrix $\mathbf{A}^{(2)}$.

Calculate the elements of $(\mathbf{A}^{(3)}|\mathbf{b}^{(3)})$ in the same manner as before. This is equivalent to changing equation $\mathbf{A}^{(2)}\underline{\mathbf{x}} = \mathbf{b}^{(2)}$ to

$$\mathbf{S}_2\mathbf{A}^{(2)}\mathbf{P}_2\mathbf{P}_2\underline{\mathbf{x}} = \mathbf{S}_2\mathbf{b}^{(2)}$$

Then we pre-multiply it by $\mathbf{M}_2$, where $\mathbf{M}_2$ is a unit matrix, except for its second column, which is given by

$$(0, 1, -m_{32}, -m_{42}, \ldots, -m_{n2})^T$$

The subsequent operations follow easily until we finally get

$$\mathbf{U}\underline{\mathbf{x}} = \mathbf{A}^{(n)}\underline{\mathbf{x}} = \mathbf{b}^{(n)}$$

where

(4.4.4) $\quad \mathbf{M}_{(n-1)}\mathbf{S}_{(n-1)} \ldots \mathbf{M}_2\mathbf{S}_2\mathbf{M}_1\mathbf{S}_1\mathbf{A}\mathbf{P}_1\mathbf{P}_2 \ldots \mathbf{P}_{(n-1)} = \mathbf{U}$

By pre-multiplying (4.4.4) successively by $\mathbf{M}_{(n-1)}^{-1}$, $\mathbf{S}_{(n-1)}$, …, $\mathbf{M}_2^{-1}$, $\mathbf{S}_2$, $\mathbf{M}_1^{-1}$, $\mathbf{S}_1$, …, and making use of (4.4.3), we get

(4.4.5) $\mathbf{A}\mathbf{P}_1\mathbf{P}_2 \ldots \mathbf{P}_{(n-1)} = \mathbf{S}_1\mathbf{M}_1^{-1}\mathbf{S}_2\mathbf{M}_2^{-1} \ldots \mathbf{S}_{(n-1)}\mathbf{M}_{(n-1)}^{-1}\mathbf{U}$

It is not difficult to show that

(a) M_i^{-1} is simply M_i with the signs of the off-diagonal elements reversed, and

(b)
$$\mathbf{S}_1\mathbf{M}_1^{-1}\mathbf{S}_2\mathbf{M}_2^{-1} \ldots \mathbf{S}_{(n-1)}\mathbf{M}_{(n-1)}^{-1}$$
$$= \mathbf{S}_1\mathbf{S}_2 \ldots \mathbf{S}_{(n-1)}\mathbf{M}_1^{-1}\mathbf{M}_2^{-1} \ldots \mathbf{M}_{(n-1)}^{-1}$$

(c) Also, $\mathbf{M}_1^{-1}\mathbf{M}_2^{-1} \ldots \mathbf{M}_{(n-1)}^{-1} = \mathbf{L}$, a unit lower triangular matrix, whose diagonal elements are 1's and the sub-diagonal elements are the calculated m_{ij} elements. Hence

$$\mathbf{S}_1\mathbf{M}_1^{-1}\mathbf{S}_2\mathbf{M}_2^{-1} \ldots \mathbf{S}_{(n-1)}\mathbf{M}_{(n-1)}^{-1} = \mathbf{S}_1\mathbf{S}_2 \ldots \mathbf{S}_{(n-1)}\mathbf{L}$$

We may thus write (4.4.5) in the form

$$\mathbf{AP} = \mathbf{SLU}$$

where $\mathbf{S} = \mathbf{S}_1\mathbf{S}_2 \ldots \mathbf{S}_{(n-1)}$ and $\mathbf{P} = \mathbf{P}_1\mathbf{P}_2 \ldots \mathbf{P}_{(n-1)}$ and by pre-multiplying by **S** we get

(4.4.6) $$\underline{\mathbf{A}} = \mathbf{SAP} = \mathbf{LU}$$

As a result of pivoting, the **LU** factorization in (4.4.6) is not for matrix **A** but for matrix $\underline{\mathbf{A}} = \mathbf{SAP}$, which is the permuted matrix **A**. Equation $\mathbf{Ax} = \mathbf{b}$ thus reduces to

(4.4.7) $$\mathbf{LU}\underline{\mathbf{x}} = \underline{\mathbf{b}}$$

where

$$\underline{\mathbf{x}} = \mathbf{Px} \text{ and } \underline{\mathbf{b}} = \mathbf{Sb}$$

The solution of (4.4.7) is obtained in two steps, i.e., by solving the two triangular systems, $\mathbf{Ly} = \underline{\mathbf{b}}$ and $\mathbf{U}\underline{\mathbf{x}} = \mathbf{y}$, from which $\underline{\mathbf{x}}$ is obtained. Finally, the solution of $\mathbf{Ax} = \mathbf{b}$ is $\mathbf{x} = \mathbf{P}\underline{\mathbf{x}}$.

4.4.3 Pivoting and the rank of matrix A

We noted earlier that elementary operations may be performed on any matrix systems, not only on square non-singular systems. Besides the importance of pivoting illustrated by Example 4.1, an equally important purpose of pivoting is to determine the rank of matrix **A**.

Let us assume that complete pivoting has been used in deriving the triangular equations (4.4.1) from the given set of equations **Ax** = **b**. If rank(**A**) < n, we expect one or more rows of **A** to be linearly dependent on the other rows. However, after each step of the Gauss elimination method, such dependent rows will depend on one row less than in the previous step. Hence, if rank(A) = k < n, after step k in the Gauss elimination method, the set of equations (4.4.1) will instead have the form (for n = 4 and k = 2)

$$
\begin{array}{rrrrcl}
a_{11}^{(1)}x_1 + & a_{12}^{(1)}x_2 + & a_{13}^{(1)}x_3 + & a_{14}^{(1)}x_4 & = & b_1^{(1)} \\
& a_{22}^{(2)}x_2 + & a_{23}^{(2)}x_3 + & a_{24}^{(2)}x_4 & = & b_2^{(2)} \\
& & 0 & + 0 & = & b_3^{(3)} \\
& & & 0 & = & b_4^{(4)}
\end{array}
\tag{4.4.8}
$$

In (4.4.8), if $b_3^{(2)}$ and $b_4^{(2)}$ are 0's, then rank(**A**|**b**) = rank(**A**) = 2 and system (4.4.8), being an underdeterminerd system, has an infinte number of solutions But if $b_3^{(2)}$ and/or $b_4^{(2}$ is not 0, rank(**A**|**b**) > rank(**A**), and (4.4.8) is inconsistent and has no solution.

Theorem 4.13 ([10], Section 3.5)

Let **A** be an n by n matrix and **b** be an n-vector. Then:

(i) System **Ax** = **b** has a solution if and only if rank(**A**|**b**) = rank(**A**).

(ii) If rank(**A**|**b**) = rank(**A**) = n, the solution is unique.

(iii) If rank(**A**|**b**) = rank(**A**) < n, the solution is not unique.

(iv) If rank(**A**|**b**) > rank(**A**), system **Ax** = **b** is inconsistent and it has no solution.

Example 4.2

A simple example of inconsistent set of two equations is

$$
\begin{aligned}
4x_1 + 6x_2 &= 6 \\
2x_1 + 3x_2 &= 4
\end{aligned}
$$

A Gauss elimination step produces

$$
\begin{aligned}
4x_1 + 6x_2 &= 6 \\
0 &= 1
\end{aligned}
$$

The l.h.s. of the second equation is half the l.h.s. of the first equation. That is, $\text{rank}(\mathbf{A}) = 1$, while $\text{rank}(\mathbf{A}|\mathbf{b}) = 2$. The two equations represent two parallel lines that will never intersect. This system is inconsistent and has no solution.

The following is a special case of the previous theorem.

Theorem 4.14

Consider the solution of the homogeneous set of equations $\mathbf{Ax} = \mathbf{0}$ (i.e., $\mathbf{b} = 0$), where $\mathbf{A}$ is an n by n matrix. Then:

(i) If $\text{rank}(\mathbf{A}) = n$, the equations have the unique solution $\mathbf{x} = \mathbf{0}$,

(ii) If $\text{rank}(\mathbf{A}) = k < n$, the solution is not unique.

4.5 Orthogonal factorization methods

Orthogonal factorization methods are used in factorizing an n by n matrix $\mathbf{A}$ in the form of $\mathbf{A} = \mathbf{QR}$. They include Householder, Givens and the Gram-Schmidt methods [5, 6, 8]. We shall consider here the first method only and also assume that $\mathbf{A}$ and $\mathbf{b}$ are real. For Householder's method, we introduce the following elementary orthogonal matrices.

4.5.1 The elementary orthogonal matrix H

Let

$$\mathbf{H} = \mathbf{I} - 2\mathbf{w}\mathbf{w}^T \tag{4.5.1}$$

where $\mathbf{w}$ is an n-dimensional vector such that $(\mathbf{w}, \mathbf{w}) = \mathbf{w}^T\mathbf{w} = 1$. Then $\mathbf{H}$ is symmetric and also orthonormal; $\mathbf{H}^T\mathbf{H} = \text{I}$, since

$$\begin{aligned}\mathbf{H}^T\mathbf{H} &= (\mathbf{I} - 2\mathbf{w}\mathbf{w}^T)(\mathbf{I} - 2\mathbf{w}\mathbf{w}^T) \\ &= \mathbf{I} - 4\mathbf{w}\mathbf{w}^T + 4\mathbf{w}(\mathbf{w}, \mathbf{w})\mathbf{w}^T \\ &= \mathbf{I}\end{aligned}$$

$\mathbf{H}$ is known as an elementary orthogonal matrix.

4.5.2 Householder's QR factorization with pivoting

This method gives the factorization of a square matrix $\mathbf{A}$ into

$$(4.5.2) \qquad \mathbf{A} = \mathbf{QR}$$

$\mathbf{Q}$ is an orthonormal matrix, $\mathbf{Q}^T\mathbf{Q} = \mathbf{I}$, and $\mathbf{R}$ an upper triangular. The transformation of matrix $\mathbf{A}$ into an upper triangular matrix $\mathbf{R}$ is done by pre-multiplying $\mathbf{A}$ respectively by the $(n-1)$ elementary orthogonal matrices $\mathbf{H}^{(1)}, \mathbf{H}^{(2)}, \ldots, \mathbf{H}^{(n-1)}$.

Let $\mathbf{A} = \mathbf{A}^{(1)}$ and let $\mathbf{A}^{(2)}, \mathbf{A}^{(3)}, \ldots, \mathbf{A}^{(n)}$ be defined by

$$\mathbf{A}^{(k+1)} = \mathbf{H}^{(k)}\mathbf{A}^{(k)}, \quad k = 1, 2, \ldots, n-1$$

The transformation $\mathbf{A}^{(2)} = \mathbf{H}^{(1)}\mathbf{A}^{(1)}$ produces $(n-1)$ 0's below the first element in column 1 of $\mathbf{A}^{(2)}$. Likewise, the transformation $\mathbf{A}^{(3)} = \mathbf{H}^{(2)}\mathbf{A}^{(2)}$, produces $(n-2)$ 0's below the second element in column 2 of $\mathbf{A}^{(3)}$ and also leaves the 0's obtained in the previous step unchanged. The process continues until we obtain an upper triangular matrix $\mathbf{A}^{(n)}$. To illustrate this process, we consider the 4 by 4 matrix $\mathbf{A} = (a_{ij})$.

We wish to determine the elements of the vector $\mathbf{w}^{(1)} = (w_1^{(1)}, w_2^{(1)}, w_3^{(1)}, w_4^{(1)})^T$ such that $\mathbf{A}^{(2)} = \mathbf{H}^{(1)}\mathbf{A}^{(1)}$ has zero elements in the positions (2, 1), (3, 1) and (4, 1). We have from (4.5.1)

$$\mathbf{A}^{(2)} = \mathbf{A}^{(1)} - 2\mathbf{w}^{(1)}\mathbf{w}^{(1)T}\mathbf{A}^{(1)}$$

Let $\mathbf{w}^{(1)T}\mathbf{A}^{(1)} = (d_1, d_2, d_3, d_4)$. The elements in the first column of $\mathbf{A}^{(2)}$ are thus

$$(4.5.3) \quad (a_{11}^{(1)} - 2w_1^{(1)}d_1), (a_{21}^{(1)} - 2w_2^{(1)}d_1), (a_{31}^{(1)} - 2w_3^{(1)}d_1), \text{ and } (a_{41}^{(1)} - 2w_4^{(1)}d_1)$$

Each of the last 3 elements in (4.5.3) is equated to 0

$$(4.5.4) \qquad \begin{aligned} a_{21}^{(1)} - 2w_2^{(1)}d_1 &= 0 \\ a_{31}^{(1)} - 2w_3^{(1)}d_1 &= 0 \\ a_{41}^{(1)} - 2w_4^{(1)}d_1 &= 0 \end{aligned}$$

We also observe that since $\mathbf{H}^{(1)}$ is orthogonal, the sum of the squares of the elements of any column of $\mathbf{A}^{(1)}$ is invariant. Thus from (4.5.3) and (4.5.4), $(a_{11}^{(1)} - 2w_1^{(1)}d_1)^2 + 0 + 0 + 0 =$ the sum of the squares of the 4 elements of the first column of $\mathbf{A}^{(1)} = C^2$. Hence

$$(4.5.5) \qquad a_{11}^{(1)} - 2w_1^{(1)}d_1 = \pm C \text{ or } 2w_1^{(1)}d_1 = a_{11}^{(1)} \pm C$$

By squaring equation (4.5.5) as well as the three equations of (4.5.4) and adding the results, we get

(4.5.6) $$2d_1^2 = C^2 \pm 2Ca_{11}^{(1)}$$

To ensure numerical stability, the + or – sign in (4.5.6) is chosen according to whether $a_{11}^{(1)}$ is > 0 or < 0 respectively. The elements of $\mathbf{w}^{(1)}$ are thus obtained from (4.5.5) and (4.5.4) by substituting d_1 from (4.5.6), from which $\mathbf{H}^{(1)}$ is easily calculated.

To calculate the elements of $w^{(2)}$ of $\mathbf{H}^{(2)} = \mathbf{I} - 2\mathbf{w}^{(2)}\mathbf{w}^{(2)T}$, we proceed in the same manner. We observe that the first element of $\mathbf{w}^{(2)}$ is 0, i.e.,

$$\mathbf{w}^{(2)} = (0, w_2^{(2)}, w_3^{(2)}, w_4^{(2)})^T$$

This will ensure that the first row and also the 0's in the first column of $\mathbf{A}^{(2)}$ are left unaltered in the transformation $\mathbf{A}^{(3)} = \mathbf{H}^{(2)}\mathbf{A}^{(2)}$.

Likewise, in calculating $\mathbf{w}^{(3)}$, its first two elements are 0's. This also ensures that the first two rows as well as the 0's obtained in the first two columns of $\mathbf{A}^{(3)}$ are left unaltered in the subsequent transformation, and so on.

For computational purposes, to summarize and simplify the above calculation of $\mathbf{w}^{(k)}$ and $\mathbf{H}^{(k)}$ for k = 1, 2, …, n – 1, let

$$\mathbf{H}^{(k)} = \mathbf{I} - \beta_k\mathbf{w}^{(k)}\mathbf{w}^{(k)T}$$

For k = 1, 2, …, n – 1, $\mathbf{H}^{(k)}$ and $\mathbf{A}^{(k+1)}$ are generated as follows [1]

(4.5.7)
$$\begin{aligned}
\sigma_k &= \mathrm{sqrt}(\Sigma_i\, |a_{ik}^{(k)}|^2) \quad \text{(sum from i = k to n)} \\
\beta_k &= [\sigma_k(\sigma_k + |a_{kk}^{(k)}|)]^{-1} \\
w_i^{(k)} &= 0, \text{ for } i < k \\
w_k^{(k)} &= \mathrm{sgn}(a_{kk}^{(k)})(\sigma_k + |a_{kk}^{(k)}|) \\
w_i^{(k)} &= a_{ik}^{(k)}, \text{ for } i > k
\end{aligned}$$

$\mathbf{H}^{(k)}$ may thus be computed from (4.5.7) and consequently $\mathbf{A}^{(k+1)} = \mathbf{H}^{(k)}\mathbf{A}^{(k)}$ is computed.

4.5.3 Pivoting in Householder's method

At the beginning of the kth step, k = 1, 2, …, n – 1, we compute σ_j^2 from (4.5.7) for each value of j from k to n. If σ_p^2 is the maximum sum, interchange columns k and p of $\mathbf{A}^{(k)}$ in the full array. This is done by post-multiplying $\mathbf{A}^{(k)}$ by the permutation matrix $\mathbf{P}_k$, and in order not to change the structure of the equation $\mathbf{A}^{(k)}\mathbf{x} = \mathbf{b}^{(k)}$ we pre-multiply $\mathbf{x}$ by $\mathbf{P}_k$. That is

$$\mathbf{A}^{(k)}\mathbf{P}_k\mathbf{P}_k\mathbf{x} = \mathbf{b}^{(k)}$$

This equation is then pre-multiplied by $\mathbf{H}^{(k)}$. Hence, at the end of the $(n-1)^{th}$ step we get

$$\mathbf{A}^{(n)}\underline{\mathbf{x}} = \mathbf{b}^{(n)} \tag{4.5.8}$$

By denoting $\mathbf{R}$ as an upper triangular matrix

$$\mathbf{A}^{(n)} = \mathbf{HAP} = \mathbf{R},\ \ \underline{\mathbf{x}} = \mathbf{Px}\ \text{ and }\ \mathbf{b}^{(n)} = \mathbf{Hb}$$

Equation (4.5.8) is re-written as

$$\mathbf{HAP}\underline{\mathbf{x}} = \mathbf{R}\underline{\mathbf{x}} = \mathbf{Hb} \tag{4.5.9}$$

where $\mathbf{H} = \mathbf{H}^{(n-1)} \ldots \mathbf{H}^{(2)}\mathbf{H}^{(1)}$ and $\mathbf{P} = \mathbf{P}_1\mathbf{P}_2 \ldots \mathbf{P}_{n-1}$

Since $\mathbf{H}$ is orthonormal, $\mathbf{H}^{-1} = \mathbf{H}^T$, and by pre-multiplying (4.5.9) by $\mathbf{H}^T$ we get $\mathbf{QR}\underline{\mathbf{x}} = \mathbf{b}$, where $\mathbf{Q} = \mathbf{H}^T$ is an orthonormal matrix. The $\mathbf{QR}$ factorization is thus not for matrix $\mathbf{A}$ but for the permuted matrix $\underline{\mathbf{A}} = \mathbf{AP}$. The solution $\underline{\mathbf{x}}$ is obtained by solving the triangular system $\mathbf{R}\underline{\mathbf{x}} = \mathbf{Q}^T\mathbf{b}$, and finally $\mathbf{x}$ is obtained from $\mathbf{x} = \mathbf{P}\underline{\mathbf{x}}$.

4.5.4 Calculation of the matrix inverse A^{-1}

We may use either Gauss' or Householder's method in calculating $\mathbf{A}^{-1}$. In the equation $\mathbf{Ax} = \mathbf{b}$, we take successively $\mathbf{b} = \mathbf{e}_1, \mathbf{e}_2, \ldots, \mathbf{e}_n$, where $\mathbf{e}_i$ is the ith column of the n-unit matrix $\mathbf{I}$. The calculated solution $\mathbf{x}_i$ is the ith column of the inverse $\mathbf{A}^{-1}$.

Proof:

Let

$$\mathbf{Ax}_1 = \mathbf{e}_1,\ \ \mathbf{Ax}_2 = \mathbf{e}_2,\ \ \mathbf{Ax}_n = \mathbf{e}_n$$

Then since $[\mathbf{e}_1\ \mathbf{e}_2 \ldots \mathbf{e}_n] = \mathbf{I}$

$$[\mathbf{x}_1\ \mathbf{x}_2 \ldots \mathbf{x}_n] = \mathbf{A}^{-1}$$

4.6 Gauss-Jordan method

Gauss-Jordan elimination method is of importance in the solution of linear programming problems [3, 7]. This method proceeds as in the Gauss elimination method, described at the beginning of Section 4.4, but without pivoting. The only difference is that at the kth step, x_k is eliminated not only from all equations below the kth equation, but also from all equations above the kth equation. If we have a system of n equations in n unknowns, after step n, we get (for n = 4)

$$\begin{aligned} a_{11}^{(1)}x_1 &= b_1^{(1)} \\ a_{22}^{(2)}x_2 &= b_2^{(2)} \\ a_{33}^{(3)}x_3 &= b_3^{(3)} \\ a_{44}^{(4)}x_4 &= b_4^{(4)} \end{aligned}$$

Then the solution of the system is obtained by dividing equation i by the coefficient $a_{ii}^{(i)}$, i = 1, 2, …, n.

We are interested here in the solutions of an underdetermined system of linear equations by the Gauss-Jordan elimination method. Let us suppose that we have applied this method to a system of 4 equation in 6 unknown and that the system is of rank 4. Let us assume that we have also divided each equation i by the coefficient a_{ii}

$$\begin{aligned} x_1 \qquad\qquad\qquad + a_{15}x_5 + a_{16}x_6 &= b_1^{(1)} \\ x_2 \qquad\qquad + a_{25}x_5 + a_{26}x_6 &= b_2^{(2)} \\ x_3 \qquad + a_{35}x_5 + a_{36}x_6 &= b_3^{(3)} \\ x_4 + a_{45}x_5 + a_{46}x_6 &= b_4^{(4)} \end{aligned} \tag{4.6.1}$$

Hence, if we assign any prescribed values to x_5 and x_6, 0's say, x_1, x_2, x_3 and x_4 in (4.6.1) are known.

Let us assume now that we want to solve the given system not for x_1, x_2, x_3 and x_4 but for x_1, x_2, x_6 and x_4 instead. Rather than solving the original set of equations from the beginning, one iteration of the Gauss-Jordan method is enough to give us the required result. This is done in two steps, as follows:

(1) Divide the third equation (which contains x_3) by a_{36}, so that the coefficient of x_6 in this equation is unity,

(2) Subtract a suitable multiple of the obtained equation from each of the other three equations so that x_6 is eliminated from such equations. That is, apply a Gauss-Jordan elimination step to eliminate x_6 from all equations except the third one.

The new set of equations will now have the form

$$(4.6.2)\qquad \begin{aligned} x_1 \quad\; + a_{13}x_3 + \quad\;\; a_{15}x_5 \quad\quad &= b_1^{(1)} \\ x_2 + a_{23}x_3 + \quad\;\; a_{25}x_5 \quad\quad &= b_2^{(2)} \\ a_{33}x_3 + \quad\;\; a_{35}x_5 + x_6 &= b_3^{(3)} \\ a_{43}x_3 + x_4 + a_{45}x_5 \quad\quad &= b_4^{(4)} \end{aligned}$$

from which the solution for x_1, x_2, x_6 and x_4 is obtained. Obviously, the coefficients and the r.h.s. of (4.6.1) and of (4.6.2) are not the same. The method described above is that of changing basis in the simplex algorithm, which is used for solving linear programming problems.

Note 4.4

We observe in (4.6.1), by assigning 0's to x_5 and x_6, the values of x_1, x_2, x_3 and x_4 are themselves the r.h.s. of these equations, namely = $b_1^{(1)}$, $b_2^{(2)}$, $b_3^{(3)}$ and $b_4^{(4)}$ respectively. Hence, we name vector **b** as the basic solution.

We end this chapter with an analysis of rounding errors in arithmetic operations.

4.7 Rounding errors in arithmetic operations

4.7.1 Normalized floating-point representation

The general format of a normalized floating-point number consists of a sign bit, an exponent and a mantissa, as shown in Figure 4-1. Single- and double-precision numbers differ only in the number of bits allocated to their exponent and mantissa.

±	Exponent	Mantissa

Figure 4-1: General floating-point format

Most present day implementations use the IEEE (Institute of Electrical and Electronics Engineers) normalized floating-point representation. The single-precision type "float" in C occupies 32 bits (binary digits), consisting of a sign bit, an 8-bit exponent and a 23-bit mantissa. The double-precision type "double" occupies 64 bits, consisting of a sign bit, an 11-bit exponent and a 52-bit mantissa. There is an implied leading "1." in the mantissa, so s.p and d.p. mantissas are actually 24 and 53 bits long respectively (even though the most-significant bit is not stored in memory) resulting in $1 \leq \text{mantissa} < 2$.

We will simplify our examination of round-off error in floating-point arithmetic operations by eliminating the implied leading bit in the mantissa. Instead, the mantissa is normalized by constraining the most-significant bit to be nonzero. This format was typical of pre-IEEE floating-point implementations.

We can generalize the normalized floating-point representation to any base, such as binary, octal, decimal, hexadecimal, etc. We denote the base by $\beta = 2, 8, 10, 16$, etc. [2, 3, 15]. Let t1 be the number of digits in a s.p. mantissa and t2 be the number of digits in a d.p. mantissa. A normalized s.p. floating-point number x is interpreted as

$$x = \pm\beta^b(0.d_1d_2 \ldots d_{t1})$$

where b is the exponent, the "." is the decimal point and $d_1, d_2, \ldots, d_{t1}$ are the digits of the mantissa. Because $d_1 \neq 0$, we get

$$(1/\beta) \leq 0.d_1d_2\ldots d_{t1} < 1 \qquad (4.7.1)$$

so $(½) \leq 0.d_1d_2\ldots d_{t1} < 1$ (binary), $(1/10) \leq 0.d_1d_2\ldots d_{t1} < 1$ (decimal), $(1/16) \leq 0.d_1d_2\ldots d_{t1} < 1$ (hexadecimal), etc. The value of x is thus calculated by

$$x = \pm\beta^b\left(\frac{d_1}{\beta} + \frac{d_2}{\beta^2} + \quad \ldots \quad + \frac{d_{t1}}{\beta^{t1}}\right)$$

For example, the decimal numbers 4983 and 0.004983 are expressed in normalized floating-point representation as $10^4(0.4983)$ and $10^{-2}(.4983)$ respectively.

Due to the finite length of floating-point representation, numbers are rounded-off or truncated after each arithmetic operation. In this chapter, round-off error is calculated for simple floating-point arithmetic operations. Bounds for round-off errors are then obtained for extended simple operations and for simple matrix calculations. Backward round-off error analysis is introduced by an example.

4.7.2 Overflow and underflow in arithmetic operations

In an IEEE-format s.p. number, the exponent b has the range $-127 \leq b \leq 128$, the stored value of which is offset by 127. This means that a s.p. number has a range of the order of $2^{-127} \leq x \leq 2^{128}$; specifically, $1.175494351 \times 10^{-38} \leq x \leq 3.402823466 \times 10^{38}$. A d.p. number has a range of the order of $2^{-1023} \leq x \leq 2^{1024}$; specifically, $2.2250738585072014 \times 10^{-308} \leq x \leq 1.7976931348623158 \times 10^{308}$.

Beyond the bounds of the given floating-point representation, overflow or underflow occurs.

4.7.3 Arithmetic operations in a d.p. accumulator

Floating-point processors perform all simple arithmetic operations in at-least double-precision (often higher than d.p.), regardless of whether the result is stored in single- or double-precision.

Consider the following simplified examples in which we assume a s.p. representation supporting 4 decimal places of precision, a d.p. of 8 decimal places and a d.p. accumulator also of 8 decimal places:

(a) **Addition (subtraction) of two s.p. floating-point numbers.** Let x and y each be a 4-digit floating-point decimal number and let us calculate $z = (x + y)$. Before the arithmetic operation, the smaller number is given an exponent equal to that of the larger number. In the following three examples, we show how round-off error can be affected by the numerical signs and the relative magnitudes of the two numbers.

(1a) Consider $10^{-6}(0.1015) - 10^{-7}(0.9852)$:

$$\begin{array}{r} 10^{-6}(0.1\ 0\ 1\ 5\ 0\ 0\ 0\ 0) \\ -10^{-6}(0.0\ 9\ 8\ 5\ 3\ 0\ 0\ 0) \\ \hline 10^{-6}(0.0\ 0\ 2\ 9\ 7\ 0\ 0\ 0) \end{array}$$

We denote the difference $\underline{z}$, normalized in d.p. and then stored in single-precision as

$$\underline{z} = \mathrm{fl}(x - y) = 10^{-8}(0.2970)$$

where fl denotes the floating point calculation in s.p. In this example, we see that there is no round-off error.

(1b) Consider the same numbers as above, but in an addition operation rather than a subtraction:

$$\begin{array}{r} 10^{-6}(0.1\ 0\ 1\ 5\ 0\ 0\ 0\ 0) \\ +10^{-6}(0.0\ 9\ 8\ 5\ 3\ 0\ 0\ 0) \\ \hline 10^{-6}(0.2\ 0\ 0\ 0\ 3\ 0\ 0\ 0) \end{array}$$

The rounded sum is stored in single-precision as

$$\underline{z} = \text{fl}(x + y) = 10^{-6}(0.2000)$$

giving a round-off error of $10^{-10}(0.3000)$, which was realized by simply changing the sign of the arithmetic operation.

(2) Consider $10^4(0.8314) + 10^1(0.5241)$:
There is a difference of 3 between the exponents of the two numbers. Hence, the second number is shifted to the right 3 places before the addition takes place

$$\begin{array}{r} 10^{4}(0.8\ 3\ 1\ 4\ 0\ 0\ 0\ 0) \\ +10^{4}(0.0\ 0\ 0\ 5\ 2\ 4\ 1\ 0) \\ \hline 10^{4}(0.8\ 3\ 1\ 9\ 2\ 4\ 1\ 0) \end{array}$$

The rounded sum $\underline{z} = \text{fl}(x + y) = 10^4(0.8319)$, giving a round-off error of $10^{-4}(0.2410)$.

(3) Consider $10^{-6}(0.3145) + 10^4(0.6758)$:
The difference in exponent between the two numbers is 10, (i.e., > the 8 decimal places in the accumulator), and thus the computed sum $\underline{z} = \text{fl}(x + y) = 10^4\ (0.6758) = y$, giving a round-off error equal to the whole of the smaller number.

(b) **Multiplication (division) of two s.p. floating-point numbers.**
Let x and y each be a 4-digit floating-point decimal number and let us calculate $z = xy$. Let us consider the following two examples:

(1) Consider two numbers of similar magnitude:

$$10^{-4}(0.1714) \times 10^{-3}(0.1213) = 10^{-8}(0.20790820)$$

The computed result $\underline{z} = \text{fl}(xy) = 10^{-8}(0.2079)$, giving a round-off error of $10^{-13}(0.8200)$.

(2) Consider two numbers of dissimilar magnitude:

$$10^{-4}(0.8204) \times 10^{6}(0.3325) = 10^{2}(0.27278300)$$

The product is then rounded to $\underline{z} = fl(xy) = 10^{2}(0.2728)$. The least-significant digit is rounded up from 7 to 8, giving a round-off error of $10^{-2}(0.1700)$, which is much greater than the smaller number.

Theorem 4.15

Let us take two normalized floating-point s.p. numbers x and y and let 'op' denote any of the arithmetic operations +, –, × or /. Let (x'op'y) be computed in d.p. before rounding it back to s.p. Let the calculated result be denoted by $\underline{z}$. Then

$$\underline{z} = fl(x\text{'op'}y) = (x\text{'op'}y)(1 + \varepsilon)$$
$$|\varepsilon| \le (1/2)\beta^{(1-t1)}$$

Proof:

Let $\underline{z}$ before and after rounding be respectively

$$\beta^{b}(0.d_1d_2\ldots d_{t1}d_{t1+1}\ldots) \text{ and } \beta^{b}(0.d_1d_2\ldots \underline{d}_{t1})$$

where $\underline{d}_{t1}$ is the rounded digit. Hence, the error resulting from the rounding operation is

$$\text{error} = \beta^{b}|0.d_1d_2\ldots d_{t1}d_{t1+1}\ldots - 0.d_1d_2\ldots \underline{d}_{t1}| \le \beta^{b}[(1/2)\beta^{-t1}]$$

The relative error (R.E.) = error/true value, is given by

$$\text{R.E.} \le \beta^{b}[(1/2)\beta^{-t1)}]/\beta^{b}(0.d_1d_2\ldots d_{t1}d_{t1+1}\ldots)$$

Yet, we know from (4.7.1) that $(1/\beta) \le (0.d_1d_2\ldots d_{t1}d_{t1+1}\ldots)$ and thus

$$\text{R.E.} \le [(1/2)\beta^{-t1)}]/(1/\beta) \le (1/2)\beta^{(1-t1)}$$

and the theorem is proved.

This theorem tells us that

$$|\varepsilon| \le 2^{-t1} \quad \text{(binary)}$$
$$|\varepsilon| \le (1/2)10^{(1-t1)} \quad \text{(decimal)}$$
$$|\varepsilon| \le (1/2)16^{(1-t1)} \quad \text{(hexadecimal)}$$

Truncating of the result

In some computers, the result $\underline{z}$ in the previous theorem is truncated to s.p. t1 digits rather than rounded. In this case, we have

Theorem 4.16

(4.7.2) $\underline{z}(\text{truncated}) = z(1+\varepsilon)$, where $|\varepsilon| \le \beta^{(1-t1)}$

It is also possible to prove the following alternative to the above two theorems.

Theorem 4.17

Let 'op' denote any of the arithmetic operations +, –, × or /. Then

$$\underline{z} = \text{fl}(x\text{'op'}y) = (x\text{'op'}y)/(1+\varepsilon)$$

where

$$|\varepsilon| \le (1/2)\beta^{(1-t1)} \quad \text{(rounded operation)}$$
$$|\varepsilon| \le \beta^{(1-t1)} \quad \text{(truncated operation)}$$

4.7.4 Computation of the square root of a s.p. number

The bound for the error made in obtaining the square root depends on the algorithm used to extract the root of the number. However, it will always be assumed that [15]

$$\text{fl}[\text{sqrt}(x)] = \text{sqrt}(x)(1+\varepsilon), \text{ where } |\varepsilon| \le 1.00001(1/2)\beta^{(1-t1)}$$

4.7.5 Arithmetic operations in a s.p. accumulator

Let us see what happens when arithmetic operations are instead performed in a s.p. accumulator. Addition and subtraction operations are more affected than multiplication or division. In the former, either none, one or two rounding operations are performed for the single arithmetic operation. Let us illustrate the worst case by the example

$$10^3(0.9741) + 10^2(0.4936)$$

When a d.p. accumulator is used, the result is computed as follows

$$10^3(0.97410000) + 10^3(0.04936000) = 10^4(0.1023)$$

Yet, in s.p. we have

$$10^3(0.9741) + 10^3(0.0494) = 10^3(1.0235)$$

and the normalized result is $10^4(0.1024)$; that is, the smaller number is rounded up and the final result is also rounded up. In this case, let the error in these two rounding operations be respectively ξ and η. Let y be the smaller number, which is rounded off first; then ξ and η may be given by

$$|\xi| \le \beta^b(1/2)\beta^{(-1-t1)} \text{ and } |\eta| \le \beta^b(1/2)\beta^{-t1}$$

Thus we have

$$|\xi + \eta| \le |\xi| + |\eta| \le (1 + (1/\beta))\beta^b(1/2)\beta^{-t1}$$

Theorem 4.18

For a single-precision accumulator

$$\underline{z} = \mathrm{fl}(x + y) = (x + y)(1 + \delta)$$
$$|\delta| \le (1/2)(1 + (1/\beta))\beta^{1-t1}$$

Hence, $|\delta| \le (1.5)2^{-t1}$ (binary), $|\delta| \le (0.55)10^{1-t1}$ (decimal) and for hexadecimal $|\delta| \le (1/2)(1 + (1/16))16^{1-t1}$.

For multiplication or division, the error depends on how the operation is performed for the given floating-point processor. In any case, it is unlikely that an error greater than 1 in the least significant digit of the final result will occur, so the result of the previous theorem will still be valid for the multiplication and division case.

4.7.6 Arithmetic operations with two d.p. numbers

In case there is no quadruple word for two d.p. numbers in which simple operations take place, we get

$$\mathrm{fl}_2(x\text{'op'}y) = (x\text{'op'}y)(1 + \delta), \quad |\delta| \le (1/2)(1 + (1/\beta))\beta^{1-t2}$$

where t2 = the number of digits in the mantissa in a d.p. word.

From now on, ε_1 and ε_2 denote the round-off error in a s.p. operation performed first in a d.p. accumulator, and the d.p. operation performed in a d.p. accumulator respectively, where

$$(4.7.3) \qquad \varepsilon_1 = (1/2)\beta^{(1-t1)} \text{ and } \varepsilon_2 = (1/2)(1 + (1/\beta))\beta^{(1-t2)}$$

Let us now consider extended simple arithmetic operations of s.p. numbers.

4.7.7 Extended simple s.p. operations in a d.p. accumulator

Consider the following operations:

(a) **Extended multiplication**:

Let us calculate w = xyz. This is done in two steps, namely by calculating fl(xy) then fl(fl(xy)z). Thus

$$\underline{w} = fl((xy)(1 + \delta_1)z)$$

$$= (xyz)(1 + \delta_1)(1 + \delta_2), \quad |\delta_1|, |\delta_2| \le \varepsilon_1$$

where ε_1 is defined in (4.7.3). In general, by assuming $|\delta_i| \le \varepsilon_1$

$$\underline{w} = fl(x_1\, x_2 \ldots x_n)$$

$$= (x_1\, x_2 \ldots x_n)(1 + \delta_1)(1 + \delta_2)\ldots(1 + \delta_{n-1})$$

(b) **Extended addition (summation)**:

Consider w = (x + y + z). In the same manner, this is done in two steps, by calculating fl(x+y) then fl(fl(x + y) + z). We get

$$\underline{w} = fl((x + y)(1 + \delta_1) + z)$$

$$= ((x + y)(1 + \delta_1) + z)(1 + \delta_2)$$

$$= (x + y)(1 + \delta_1)(1 + \delta_2) + z(1 + \delta_2)$$

In general

$$\underline{w} = fl(x_1 + x_2 + \ldots + x_n)$$

$$= (x_1 + x_2)(1 + \delta_1)(1 + \delta_2)\ldots(1 + \delta_{n-1})$$

$$+ x_3(1 + \delta_2)\ldots(1 + \delta_{n-1}) + \ldots + x_{n-1}(1 + \delta_{n-2})(1 + \delta_{n-1})$$

$$+ x_n(1 + \delta_{n-1})$$

$|\delta_i| \le \varepsilon_1$

(c) **Inner products**:

Let us compute $w = (x_1y_1 + x_2y_2)$

$$\underline{w} = (x_1y_1(1 + \delta_1) + x_2y_2(1 + \delta_2))(1 + \delta_3), \quad |\delta_i| \le \varepsilon_1$$

This result may also be extended.

Because the product of the factors $(1 + \delta_i)$ appear so often, it is useful for practical purposes to obtain some rounding error bounds for such products.

Lemma 4.1

Assume that $|\delta_i| \le \varepsilon_1$, i = 1, 2, …, n, and that $n\varepsilon_1 < 0.01$. Then

(4.7.4) $$1 - n\varepsilon_1 \le \Pi_1{}^n(1 + \delta_i) \le 1 + 1.01n\varepsilon_1$$

Proof:

From the assumption that $|\delta_i| < \varepsilon_1$

$$(1 - \varepsilon_1)^n \le \Pi_1{}^n(1 + \delta_i) \le (1 + \varepsilon_1)^n$$

It is easy to establish that

$$1 - n\varepsilon_1 \le (1 - \varepsilon_1)^n$$

Also, assuming that $n\varepsilon_1 < 0.01$, it is easy to show that

$$(1 + \varepsilon_1)^n < 1 + 1.01n\varepsilon_1$$

and by applying (4.7.5a, b) to (4.7.4), the lemma is proved.

By using (4.7.4), we get the following bounds for the extended multiplication, extended summation and inner products respectively.

(a) $\underline{w} = fl(x_1\, x_2 \ldots x_n) = (x_1\, x_2 \ldots x_n)(1 + \tau)$, where

(4.7.6) $$1 - (n - 1)\varepsilon_1 \le 1 + \tau \le 1 + 1.01(n - 1)\varepsilon_1$$

(b) $\underline{w} = fl(x_1 + x_2 + \ldots + x_n)$
$= fl(x_1 + x_2)(1 + \tau_2) + x_3(1 + \tau_3) + \ldots + x_n(1 + \tau_n)$, where

(4.7.7) $1 - (n + 1 - r)\varepsilon_1 \le (1 + \tau_r) \le 1 + 1.01(n + 1 - r)\varepsilon_1,\ r = 2, 3, \ldots, n$

(c) $\underline{w} = fl(\mathbf{x}, \mathbf{y}) = (x_1y_1)(1+\tau_1) + x_2y_2(1+\tau_2) + \ldots + x_n\, y_n\, (1+\tau_n)$

where

$$1 - n\varepsilon_1 \le (1 + \tau_1) \le 1 + 1.01n\varepsilon_1$$

$$1 - (n + 2 - r)\varepsilon_1 \le (1 + \tau_r) \le 1 + 1.01(n + 2 - r)\varepsilon_1, \quad r = 2, 3, \ldots, n$$

Remark 4.1

The error in the extended-product case depends on the order in which the numbers are multiplied. Yet, the upper bound of the error as given in (4.7.6) is independent of such order. Also, the relative error (R.E.) is appreciably small and from (4.7.6) is

$$|\text{R.E.}| < 1.01(n - 1)\varepsilon_1$$

Remark 4.2

In the summation operation, both the error and the upper bound in (4.7.7) depend on the order in which the numbers are added. The upper bound of this error will be smaller if the numbers were added in order of increasing absolute magnitude. That is because the largest factor $(1+\tau_i)$ will be associated with the smallest x_i. However, in this process there is no guarantee that the relative error will be small.

$$|R.E.| \leq \left|\sum_{i=1}^{n} x_i\tau_i\right| / \left|\sum_{i=1}^{n} x_i\right|$$

It may happen that the positive and negative terms in the summation of the denominator cancel each other in such a way that the summation is very small. It may also happen that the terms in the numerator will support each other. In such a case the relative error will be very high. The same situation may occur in the inner product case. This phenomenon of cancellation is very crucial and it magnifies the harmful effect of the round-off error.

4.7.8 Alternative expressions for summations and inner-product operations

(a) The expression for summations may take the form

$$\begin{aligned} \underline{w} &= fl(x_1 + x_2 + \dots + x_n) \\ &= (x_1 + x_2 + \dots + x_n) + e_1 \end{aligned}$$

where

(4.7.8) $|e_1| \leq 1.01\varepsilon_1[(n-1)|x_1 + x_2| + (n-2)|x_3| + \dots + 2|x_{n-1}| + |x_n|]$

(b) Also, for the inner product operation, we may have

$$\underline{w} = fl(\mathbf{x}, \mathbf{y}) = (\mathbf{x}, \mathbf{y}) + e_2, \text{ where}$$

(4.7.9) $|e_2| \leq 1.01\varepsilon_1[n|x_1|\,|y_1| + n|x_2|\,|y_2| + (n-1)|x_3|\,|y_3| + \dots + 2|x_n|\,|y_n|]$

4.7.9 More conservative error bounds

(a) The summation case, from (4.7.8)

$$\begin{aligned} \underline{w} &= fl(x_1 + x_2 + \dots + x_n) \\ &= (x_1 + x_2 + \dots + x_n) + e_3, \text{ where} \end{aligned}$$

$$|e_3| \le 1.01(n-1)\varepsilon_1(|x_1| + |x_2| + \ldots + |x_n|)$$

(b) The inner product case, from (4.7.9)

$$\underline{w} = fl(\mathbf{x}, \mathbf{y}) = (\mathbf{x}, \mathbf{y}) + e_4, \text{ where}$$

$$|e_4| \le 1.01 n\varepsilon_1(|\mathbf{x}|, |\mathbf{y}|)$$

where $|\mathbf{x}|$ and $|\mathbf{y}|$ are the two vectors whose elements are respectively the absolute elements of $\mathbf{x}$ and $\mathbf{y}$.

4.7.10 D.p. summations and inner-product operations

In most computers there are facilities for accumulating the sum of several terms in a d.p. accumulator for each partial sum and the whole terms are added. The final result is then rounded to s.p.

(a) The summation operation.
First we obtain an expression similar to (4.7.7), namely

$$fl_2(x_1 + x_2 + \ldots + x_n)$$
$$= (x_1 + x_2)(1 + \eta_2) + x_3(1 + \eta_3) + \ldots + x_n(1 + \eta_n)$$

and

$$1 - (n + 1 - r)\varepsilon_2$$
$$\le (1 + \eta_r) \le 1 + 1.01(n + 1 - r)\varepsilon_2, \quad r = 2, 3, \ldots, n$$

and from (4.7.3)

$$\varepsilon_2 = (1/2)(1 + (1 + (1/\beta)))\beta^{(1-t2)}$$

Then when this result is rounded to s.p. we get

$$\underline{w} = fl\left[fl_2\left(\sum_{i=1}^{n} x_i\right)\right]$$

$$= [(x_1 + x_2)(1 + \eta_2) + x_3(1 + \eta_3)$$
$$+ \ldots + x_n(1 + \eta_n)](1 + \varepsilon), \quad |\varepsilon| \le \varepsilon_1$$

From this result we write

$$\underline{w} = (x_1 + x_2 + \ldots + x_n) + E_1, \text{ where}$$

$$|E_1| \le \varepsilon_1 \left| \sum_{i=1}^{n} x_i \right| + 1.01(n-1)\varepsilon_2 \sum_{i=1}^{n} |x_i|$$

(b) The inner product case.
In the same manner as above

$$\underline{w} = \mathrm{fl}(\mathrm{fl}_2(\mathbf{x}, \mathbf{y})) = (\mathbf{x}, \mathbf{y}) + E_2,$$

where

$$|E_2| \le \varepsilon_1 |(\mathbf{x}, \mathbf{y})| + 1.01 n \varepsilon_2 (|\mathbf{x}|, |\mathbf{y}|)$$

We observe that the second term on the right for both $|E_1|$ and $|E_2|$ is a second order term, compared with the first term.

4.7.11 Rounding error in matrix computation

The obtained results so far are now applied to a simple matrix computation. Let $\mathbf{A} = (a_{ij})$ be an n by n matrix and let $\mathbf{x}$ be an n-vector. Let us compute the following:

(i) $\mathbf{y} = \mathbf{Ax}$. The computed i^{th} element of $\mathbf{y}$ is

$$y_i = \mathrm{fl}\left[\sum_{j=1}^{n} a_{ij} x_j\right] = \sum_{j=1}^{n} a_{ij} x_j + E_i$$

From (4.7.9)

$$|E_i| \le 1.01\varepsilon_1 [n|a_{i1}|\,|x_1| + n|a_{i2}|\,|x_2| + (n-1)|a_{i3}|\,|x_3| + \dots + 2|a_{in}|\,|x_n|]$$

We may write

$$\underline{\mathbf{y}} = \mathbf{Ax} + \mathbf{E}_1$$

$$|\mathbf{E}_1| \le 1.01\, |\mathbf{A}|\, \mathbf{D}\, |\mathbf{x}|\, \varepsilon_1$$

and $\mathbf{D}$ is a diagonal matrix given by

$$\mathbf{D} = \mathrm{diag}(n, n, n-1, \dots, 2)$$

Again, $|\mathbf{A}|$ denotes the matrix whose elements are the absolute elements of matrix $\mathbf{A}$. Thus by taking the Euclidean norms of the above inequality, we get the following bound

$$\|\mathbf{E}_1\|_2 \leq 1.01\ n\varepsilon_1 \|\mathbf{A}\|_E \|\mathbf{x}\|_2$$

(ii) $\mathbf{C} = \mathbf{AB}$, where each of $\mathbf{A}$, $\mathbf{B}$ and $\mathbf{C}$ is an n by n real matrix. In the same manner we write down

$$\underline{\mathbf{C}} = \text{fl}(\mathbf{AB}) = \mathbf{AB} + \mathbf{E}_2$$

where

$$\|\mathbf{E}_2\|_E \leq 1.01\ n\varepsilon_1 \|\mathbf{A}\|_E \|\mathbf{B}\|_E$$

See for example Wilkinson ([15], p. 83).

However, if d.p. accumulations of inner products are made, the results for (i) and (ii) are

(i) $\underline{\mathbf{y}} = \text{fl}(\text{fl}_2(\mathbf{Ax})) = \mathbf{Ax} + \mathbf{E}_3$, where

$$\|\mathbf{E}_3\|_2 \leq \varepsilon_1 \|\mathbf{Ax}\|_2 + 1.01 n\varepsilon_2 \|\mathbf{A}\|_E \|\mathbf{x}\|_2$$

(ii) $\underline{\mathbf{C}} = \text{fl}(\text{fl}_2(\mathbf{AB})) = \mathbf{AB} + \mathbf{E}_4$, where

$$\|\mathbf{E}_4\|_E \leq \varepsilon_1 \|\mathbf{AB}\|_E + 1.01 n\varepsilon_2 \|\mathbf{A}\|_E \|\mathbf{B}\|_2$$

4.7.12 Forward and backward round-off error analysis

Let us now assume that we are calculating the value of

$$\mathbf{x} = g(a_1, a_2, \ldots, a_n)$$

where the a_i are given. As a result of the round-off error, the computed value of $\mathbf{x}$, namely $\underline{\mathbf{x}}$ will differ from the exact value $\mathbf{x}$ by $d\mathbf{x} = \underline{\mathbf{x}} - \mathbf{x}$.

In the forward error analysis, we attempt to obtain some bound on $d\mathbf{x}$. In the backward error analysis, we do not concern ourselves with the value of $d\mathbf{x}$. Instead, we try to show that the computed value $\underline{\mathbf{x}}$ is exactly equal to

$$\underline{\mathbf{x}} = g(a_1 + da_1, a_2 + da_2, \ldots, a_n + da_n)$$

for some values of $da_1, da_2, \ldots, da_n$, and we calculate the bounds for the da_i. The analysis has to be continued after this step to obtain some bounds for $\|d\mathbf{x}\|$, usually by using some perturbation technique [3, 14].

In practice, it is found that the backward round-off error analysis is much simpler than the forward analysis, in particular in the floating-point computation. We give here a simple example to illustrate the idea of the backward error analysis. Suppose that we are computing the elements x_i from the lower triangular set of equations

$$(4.7.10) \qquad \begin{array}{lllll} a_{11}x_1 & & & & = b_1 \\ a_{21}x_1 & + a_{22}x_2 & & & = b_2 \\ a_{31}x_1 & + a_{32}x_2 & + a_{33}x_3 & & = b_3 \\ a_{41}x_1 & + a_{42}x_2 & + a_{43}x_3 & + a_{44}x_4 & = b_4 \end{array}$$

The variables $x_1, x_2, \ldots,$ are computed in succession

$$\underline{x}_1 = fl(b_1/a_{11})$$

$$\underline{x}_i = fl[(-a_{i1}x_1 - a_{i2}x_2 \ldots + b_i)/a_{ii}], \ i = 2, 3, 4$$

From Theorem 4.17

$$\underline{x}_1 = b_1/[a_{11}(1 + \delta_{11})]$$

and

$$\underline{x}_i = fl[(-a_{i1}(1 + \delta_{i1})\underline{x}_1 - \ldots - a_{i,\,i-1}(1 + \delta_{i,\,i-1})\underline{x}_{i-1} + b_i)] \\ / [a_{ii}(1 + \delta_{ii})(1 + \varepsilon_{ii})], \quad i = 2, 3, 4$$

where

$$|\delta_{ii}|, |\varepsilon_{ii}| \le \varepsilon_1, \quad i = 1, 2, 3, 4, \qquad |\delta_{i1}| \le 1.01(i-1)\varepsilon_1, \quad i = 2, 3, 4$$

and
$$|\delta_{ij}| \le 1.01(i + 1 - j)\varepsilon_1, \quad i = 2, 3, 4, \quad j < i$$

Hence, we get

$$\begin{array}{lll} a_{11}(1 + \delta_{11})\underline{x}_1 & & = b_1 \\ a_{21}(1 + \delta_{21})x_1 & + a_{22}(1 + \delta_{22})(1 + \varepsilon_{22})x_2 & = b_{2,} \text{ etc.} \end{array}$$

or in matrix form

$$(\mathbf{A} + d\mathbf{A})\underline{\mathbf{x}} = \mathbf{b}$$

where

$$|d\mathbf{A}| \le 1.01\varepsilon_1 \begin{bmatrix} |a_{11}| & 0 & 0 & 0 \\ |a_{21}| & 2|a_{22}| & 0 & 0 \\ 2|a_{31}| & 2|a_{32}| & 2|a_{33}| & 0 \\ 3|a_{41}| & 3|a_{42}| & 2|a_{43}| & 2|a_{44}| \end{bmatrix}$$

Hence, by taking the L_∞ norm of $d\mathbf{A}$, (maximum absolute row sum)

$$||d\mathbf{A}||_\infty \le (1/2) \times 1.01 \times 4 \times 5\varepsilon_1 \times \max_{ij}|a_{ij}|$$

If instead, **A** is n by n (not 4 by 4), the error bound would be

(4.7.11) $\|\mathbf{dA}\|_\infty \le (1/2)1.01n(n+1)\varepsilon_1 \times \max_{ij}|a_{ij}|$

In other words, the computed solution to (4.7.10) is the exact solution of $(\mathbf{A} + \mathrm{d}\mathbf{A})\underline{\mathbf{x}} = \mathbf{b}$, where $\|\mathrm{d}\mathbf{A}\|_\infty$ is given above and the error $\mathrm{d}\mathbf{x} = (\underline{\mathbf{x}} - \mathbf{x})$ is calculated from Theorem 4.12 by taking $\mathrm{d}\mathbf{b} = \mathbf{0}$ and assuming $\|\mathbf{A}^{-1}\mathrm{d}\mathbf{A}\| < 1$

(4.7.12) $$\frac{\|\mathrm{d}\mathbf{x}\|}{\|\mathbf{x}\|} \le \left[\frac{\|\mathbf{A}^{-1}\mathrm{d}\mathbf{A}\|}{1 - \|\mathbf{A}^{-1}\mathrm{d}\mathbf{A}\|}\right]$$

4.7.13 Statistical error bounds and concluding remarks

We conclude that the upper bounds of round-off error may be reached only in very special cases. The round-off error in a single arithmetic operation lies between $-\varepsilon_1$ and ε_1. To realize the worst-case estimated upper bound in the error of an accumulation of simple arithmetic operations, every single round-off error would have to be at either extreme $-\varepsilon_1$ or ε_1 and the data of the problem would have to be arranged in such a form that the errors all accumulate in the same direction. In the above, the obtained error bound for $\|\mathrm{d}\mathbf{A}\|_\infty$ is given by (4.7.11). In practice, due to the statistical distribution of the round-off errors the factor n(n + 1) in (4.7.11) is more likely to be n.

We note that the first detailed round-off error analysis in fixed point arithmetic operations (not discussed in this section) was given by von Neumann and Goldstine [12]. The corresponding analysis in floating-point operations was first given by Wilkinson [14]. The backward round-off error analysis was first introduced by Givens [4].

The pioneering contribution to the subject of round-off error analysis in algebraic processes in both fixed and floating-point computation is due to Wilkinson. Most of his published and unpublished work is summarized in his two books [15, 16].

References

1. Businger, P. and Golub, G., Linear least squares solution by Householder transformation, *Numerische Mathematik*, 7(1965)269-276.

2. Fisher, M.E., *Introductory Numerical Methods with the NAG Software Library*, The University of Western Australia, Crawley, 1988.
3. Forsythe, G.E. and Moler, C.B., *Computer Solution of Linear Algebraic Systems*, Prentice-Hall, Englewood Cliffs, NJ, 1967.
4. Givens, J.W., Numerical computation of the characteristic values of a real matrix, *Oak Ridge National Laboratory*, ORNL-1574, 1954.
5. Givens, J.W., Computation of plane unitary rotations transforming a general matrix to triangular form, *Journal of SIAM*, 6(1958)26-50.
6. Golub, G.H. and Van Loan, C.F., *Matrix Computation*, Third Edition, The Johns Hopkins University Press, Baltimore, 1996.
7. Hadley, G., *Linear Programming*, Addison-Wesley, Reading, MA, 1962.
8. Householder, A.S., Unitary triangularization of a non-symmetric matrix, *Journal of ACM*, 5(1958)339-342.
9. Lancaster, P., *Theory of Matrices*, Academic Press, New York, 1969.
10. Noble, B., *Applied Linear Algebra*, Prentice-Hall, Englewood Cliffs, NJ, 1969.
11. Stewart, G.W., *Introduction to Matrix Computations*, Academic Press, New York, 1973.
12. von Neumann, J. and Goldstine, H.H., Numerical inverting of matrices of high order, *Bulletine of American Mathematical Society*, 53(1947)1021-1099.
13. Westlake, J.R., *A Handbook of Numerical Matrix Inversion and Solution of Linear Equations*, John Wiley & Sons, New York, 1968.
14. Wilkinson, J.H., Error analysis of floating-point computation, *Numerische Mathematik*, 2(1960)319-340.
15. Wilkinson, J.H., *Rounding Errors in Algebraic Processes*, Prentice-Hall, Englewood Cliffs, NJ, 1963.
16. Wilkinson, J.H., *The Algebraic Eigenvalue Problem*, Clarendon Press, Oxford, 1965.

PART 2

The L_1 Approximation

Chapter 5

Linear L_1 Approximation

5.1 Introduction

Given is the overdetermined system of linear equations

$$\mathbf{Ca} = \mathbf{f} \tag{5.1.1}$$

where $\mathbf{C} = (c_{ij})$ is a real n by m matrix of rank k, $k \leq m \leq n$ and $\mathbf{f} = (f_i)$ is a real n-vector. In general, this is an inconsistent linear system and has no exact solution; only an approximate one. The residual vector for the system $\mathbf{Ca} = \mathbf{f}$, denoted by $\mathbf{r}$ is

$$\mathbf{r} = \mathbf{Ca} - \mathbf{f}$$

In this chapter, we consider the linear L_1 solution of $\mathbf{Ca} = \mathbf{f}$, which requires that the L_1 norm of the errors (the residuals) in the approximation problem be as small as possible.

Since the early seventies, there has been an increasing interest in the linear L_1 approximation. According to Bloomfield and Steiger ([14], pp. 33, 34), it has been given many names, such as Discrete L_1 approximation [1, 2, 8], L_1 solution of overdetermined linear equations [1, 3, 4], Linear discrete L_1 norm problem [6], L_1 norm minimization [22], Least absolute deviations (LAD) [14], Minimum sum of absolute errors (MSAE) [17], L^1–Approximation [18], Least absolute value (LAV) [23], L_1 linear regression [24], Least absolute value regression [28] and others.

The robustness of the L_1 solution against odd points or outliers (one or more inaccurate element in vector $\mathbf{f}$) was illustrated in Figure 2-1. In this context, Rosen et al. [22] give necessary and sufficient conditions that correct solution is obtained when there are some errors not only in vector $\mathbf{f}$, but also in the coefficient matrix $\mathbf{C}$.

The L_1 solution of system $\mathbf{Ca} = \mathbf{f}$, or the **discrete L_1 approximation**, is the m-vector $\mathbf{a} = (a_i)$ that minimizes the L_1 norm of the residuals; that is

$$\text{(5.1.2)} \qquad \text{minimize } Z = \sum_{i=1}^{n} |r_i|$$

where r_i is the ith element of vector $\mathbf{r}$ and is given by

$$\text{(5.1.3)} \qquad r_i = \sum_{j=1}^{m} c_{ij}a_j - f_i, \quad i = 1, 2, \ldots, n$$

Abdelmalek [1] showed that the discrete linear L_1 approximation is equivalent to the solution of the overdetermined linear equation $\mathbf{Ca} = \mathbf{f}$ in the L_1 norm, as explained in Chapter 2. For the case when matrix $\mathbf{C}$ satisfies the Haar condition, Usow [25] treated the discrete L_1 approximation by solving a geometric problem equivalent to solving $\mathbf{Ca} = \mathbf{f}$ in the L_1 norm. Abdelmalek [1] generalized Usow's algorithm for the non-Haar case and showed that his algorithm is completely equivalent to a dual simplex method applied to a linear programming problem with non-negative bounded variables. However, one iteration of Usow's algorithm is equivalent to one or more iterations in the latter.

The most widely used methods for solving the L_1 approximation employ linear programming techniques, in the primal or the dual form. Wagner [26] was the first to successfully formulate the L_1 approximation problem (5.1.1-3) to a linear programming one in both the primal and dual forms.

Barrodale and Roberts [8, 9] used a modification of the simplex method of linear programming to solve the primal problem. Their routine is able to skip certain intermediate simplex iterations. Matrix $\mathbf{C}$ of (5.1.1) need not be of full rank, minimum computer storage is required and an initial basic feasible solution is easily computed.

To advance the starting basis in the linear programming solution for the L_1 approximation problem, Sklar and Armstrong [23] solved the equation $\mathbf{Ca} = \mathbf{f}$ first in the least squares (LS) sense. They computed and ordered the absolute LS residuals. Then the initial basis for the L_1 problem corresponds to the columns related to the m smallest absolute LS residuals. This is because the L_1 approximation is supposed to interpolate at least m of the given data points, m being the rank of matrix **C**. Their method requires solving the problem in the LS sense first and assumes that the coefficient matrix **C** is of full rank m. See Section 5.1.1.

Bloomfield and Steiger [13] presented an algorithm identical to that of Barrodale and Roberts [8, 9] except that they differ in the start up of the algorithm. They assume that rank(**C**) = m and use the characterization theorem outlined in Section 5.1.1. They consider all m combinations with zero residuals and use an exchange method that is a variation of the simplex method.

Among other methods that use linear programming is that of Narula and Wellington [17]. They developed one algorithm that solves both the L_1 approximation problem, calling it the **minimum sum of absolute errors (MSAE) regression** and the Chebyshev approximation problem, calling it **minimization of the maximum absolute error (MMAE) regression**. They maintain that the algorithm for MSAE regression is that of Barrodale and Roberts [8] and the algorithm for MMAE regression retains the basic elements of Barrodale and Phillips [7].

Following the analysis of Usow's method, Abdelmalek [2, 3, 4] next solved the linear programming problem in the dual form, in such a way that certain intermediate simplex iterations are skipped. Numerical results [3] on a large number of test cases show that this method is comparable to that of Barrodale and Roberts [7, 8]. This is not surprising since the two algorithms are identical as shown by Armstrong and Godfrey [6]. However, the two methods differ in obtaining the initial basic solution and the way a vector leaves and enters the basis. Another difference is that in Abdelmalek [3, 4], for the purpose of numerical stability for ill-conditioned problems, a triangular decomposition to the basis matrix is used. Nevertheless, a program without triangular decomposition of the basis, which is faster, is also included here.

Robers and Ben-Israel [20] and Robers and Robers [21] described an interval programming technique to solve the bounded variables dual linear programming problem (5.2.3) below. No intermediate simplex iterations are skipped in their method.

Techniques for solving the L_1 approximation problem other than linear programming, include that of Bartels et al. [10] who used a projected gradient method for choosing a descent direction to minimize a piecewise differential function. Wesolowsky [28] used a technique closely related to that of Bartels et al. [10] with different rules of descent. Bartels and Conn [11] used a descent method, which is a variant of the simplex method.

In the following, we describe the solution using the dual form of the linear programming formulation of the L_1 approximation problem. It reduces to a programming problem with non-negative bounded variables. No artificial variables are needed in the algorithm and as we noted earlier, it deals with rank deficient cases. Also, minimum computer storage is used and an initial basic feasible solution is easily computed. We are able to skip certain intermediate simplex iterations and hence, the algorithm is an efficient one. In Section 3.8.1, we outlined the formulation of the L_1 problem as a linear programming problem. For the sake of completeness, we formulate this problem again here.

5.1.1 Characterization of the L_1 solution

For characterization theorems of the L_1 optimal solution, see for example, Madsen et al. [16], Pinkus [18], Powell [19] and Watson [27]. The most important characterization is the following. For practical purposes, for an optimal solution, if matrix **C** in (5.1.1) has rank k, $k \leq m$, then at least k of the residuals are 0's. This characterization (theorem) is a direct result of the solution of the dual linear programming problem of the L_1 approximation. See Section 5.7. However, simplex methods calculate such optimal solutions, but not every optimal solution has this characterization ([24], p. 60).

5.2 Linear programming formulation of the problem

Since the residuals (r_i) in (5.1.3) are unrestricted in sign; that is, r_i may be >, =, or < 0, i = 1, 2, …, n, we write

(5.2.1a) $$r_i = u_i - v_i$$

Hence

(5.2.1b) $$|r_i| = u_i + v_i$$

where u_i and v_i are such that

(5.2.1c) $$u_i \geq 0 \text{ and } v_i \geq 0, \quad i = 1, 2, \ldots, n$$

When $r_i > 0$, we have $u_i > 0$ and $v_i = 0$ and when $r_i < 0$, we have $v_i > 0$ and $u_i = 0$. Obviously, when $r_i = 0$, $u_i = v_i = 0$. Hence, from (5.2.1b), (5.1.2) becomes

$$\text{minimize } Z = \sum_{i=1}^{n} u_i + \sum_{i=1}^{n} v_i$$

In vector-matrix notation, let vector $\mathbf{u} = (u_j)$ and vector $\mathbf{v} = (v_j)$. Hence, the residual vector r is

$$\mathbf{r} = \mathbf{u} - \mathbf{v} = \mathbf{Ca} - \mathbf{f}$$

The primal form of the linear programming problem is

(5.2.2a) $$\text{minimize } Z = \mathbf{e}^T\mathbf{u} + \mathbf{e}^T\mathbf{v}$$

subject to the constraints

(5.2.2b) $$\begin{bmatrix} \mathbf{C} & -\mathbf{I}_n & \mathbf{I}_n \end{bmatrix} \begin{bmatrix} \mathbf{a} \\ \mathbf{u} \\ \mathbf{v} \end{bmatrix} = \mathbf{f}$$

(5.2.2c) $$a_i \text{ unrestricted in sign}, \quad i = 1, 2, \ldots, m$$

(5.2.2d) $$u_i \geq 0, v_i \geq 0, \quad i = 1, 2, \ldots, n$$

In (5.2.2a), $\mathbf{e}$ is an n-vector, each element of which equals 1, and the superscript T refers to the transpose. In (5.2.2b), $\mathbf{I}_n$ is an n-unit matrix.

The dual form to problem (5.2.2) is

(5.2.3a) $$\text{maximize } z = \mathbf{f}^T\mathbf{w}$$

subject to

(5.2.3b) $$\mathbf{C}^T\mathbf{w} = \mathbf{0}$$

(5.2.3c) $$\begin{aligned} \mathbf{w} &\leq \mathbf{e} \\ \mathbf{w} &\geq -\mathbf{e} \end{aligned}$$

where the n-vector $\mathbf{w} = (w_i)$. The last two sets of constraints reduce to the following constraints

$$-1 \leq w_i \leq 1, \quad i = 1, 2, \ldots, n$$

Then by defining $b_i = w_i + 1$, $i = 1, 2, \ldots, n$, we get the following formulation of the problem.

(5.2.4a) $$\text{maximize } z = \mathbf{f}^T(\mathbf{b} - \mathbf{e})$$

subject to the constraints

$$\mathbf{C}^T\mathbf{b} = \mathbf{C}^T\mathbf{e}$$

or

(5.2.4b) $$\mathbf{C}^T\mathbf{b} = \sum_{i=1}^{n} \mathbf{C}_i^T$$

(5.2.4c) $$0 \leq b_i \leq 2, \quad i = 1, 2, \ldots, n$$

$\mathbf{C}_i^T$ is the ith column of matrix $\mathbf{C}^T$ and $\mathbf{b} = (b_i)$. We see that (5.2.4) is a linear programming problem with non-negative bounded variables [15]. We solve here this dual form (5.2.4).

5.3 Description of the algorithm

In (5.2.4c), for an optimum solution, the lower bounds of parameter b_i, $i = 1, 2, \ldots, n$, are 0's and their upper bounds are 2. A dual simplex algorithm for solving this problem is described here. Without loss of generality, assume that matrix $\mathbf{C}$ is of rank m.

Theorem 5.1

A necessary and sufficient condition for a nonzero program for system (5.2.4) to be optimal is that (n – m) elements of vector **b**, each has the value 0 (lower bound) or 2 (upper bound), and that the other m elements of **b** are basic variables [15].

Let a basis indicator set for vector **b** be the index set $\mathbf{I(b)} \subset \{1, 2, \ldots, n\}$ with the property that the variables $\{b_i \mid i \in \mathbf{I(b)}\}$ are basic variables. Let also the index sets $\mathbf{L(b)}$ and $\mathbf{U(b)}$ be indicators for the non-basic variables $\{b_i \mid i \notin \mathbf{I(b)}\}$ that are respectively at their lower bounds (= 0) and at their upper bounds (= 2). Let **B** denote the basis matrix and the basic solution be $\mathbf{b}_B = (b_{Bj})$, $j = 1, 2, \ldots, m$.

As usual, the simplex tableau is formed by calculating the non-basic vectors $\mathbf{y}_i$ and their marginal costs $(z_i - f_i)$, $i = 1, 2, \ldots, n$. Then

$$\mathbf{y}_i = \mathbf{B}^{-1}\mathbf{C}_i^T \tag{5.3.1}$$

and

$$z_i - f_i = \mathbf{f}_B{}^T\mathbf{y}_i - f_i \tag{5.3.2}$$

The elements of $\mathbf{f}_B = (f_i)$, $i \in \mathbf{I}(\mathbf{b})$.

Since some of the non-basic variables may be at their upper bound (= 2), from (5.2.4b), the basic solution

$$\mathbf{b}_B = \mathbf{B}^{-1}\left[\sum_{i=1}^{n} \mathbf{C}_i^T - 2 \sum_{i \subset \mathbf{U}(\mathbf{b})} \mathbf{C}_i^T\right] \tag{5.3.3}$$

By denoting the first term on the right by $\mathbf{b}_{B0}$ and by (5.3.1)

$$\mathbf{b}_B = \mathbf{b}_{B0} - 2 \sum_{i \subset \mathbf{U}(\mathbf{b})} \mathbf{y}_i \tag{5.3.4}$$

Theorem 5.2

A basic feasible solution is maximal, if the marginal costs for the non-basic variables $(z_i - f_i)$, $i \notin \mathbf{I}(\mathbf{b})$, satisfy the relations [15]

$$z_i - f_i \geq 0, \quad i \in \mathbf{L}(\mathbf{b})$$

$$z_i - f_i \leq 0, \quad i \in \mathbf{U}(\mathbf{b})$$

The solution of this dual problem is summarized as follows. A non-basic column may replace a basic column, may go from a zero bound to an upper bound or may go from an upper bound to zero. The optimal solution is characterized by Theorem 5.1.

Theorem 5.3

At any stage of the computation, the residuals (r_i) of (5.1.3) are themselves the marginal costs $(z_i - f_i)$ for the same reference

$$z_i - f_i = r_i, \quad i = 1, 2, \ldots, n$$

Proof:

For any $i \notin \mathbf{I}(\mathbf{b})$, from (5.3.1), (5.3.2) is given by

$$z_i - f_i = \mathbf{f_B}^T\mathbf{B}^{-1}\mathbf{C}_i^T - f_i = \mathbf{C}_i\mathbf{B}^{-T}\mathbf{f_B} - f_i$$

where $\mathbf{C}_i$ is the ith row of $\mathbf{C}$. Also, the elements of $\mathbf{f_B} = (f_j)$, $j \in \mathbf{I}(\mathbf{b})$. Thus $\mathbf{B}^{-T}\mathbf{f_B} = (a_1, \ldots, a_m)^T$, and hence

$$z_i - f_i = \sum_{j=1}^{m} c_{ij}a_j - f_i = r_i, \quad i = 1, 2, \ldots, n$$

For $i \in \mathbf{I}(\mathbf{b})$; that is, for the reference equations $(z_i - f_i) = 0$, the residual $r_i = 0$, which proves the theorem.

As a result, the objective function z of (5.1.2) is given by

$$z = \sum_{i=1}^{n} |z_i - f_i|$$

Theorem 5.4

The solution vector of the L_1 approximation problem is given by

$$\mathbf{a}^T = \mathbf{f_B}^T\mathbf{B}^{-1} \tag{5.3.5}$$

where $\mathbf{B}^{-1}$ is the inverse of the basis matrix for the optimum solution of the programming problem and $\mathbf{f_B}$ is associated with the optimal solution.

The proof follows also from the fact that $\mathbf{B}^{-T}\mathbf{f_B} = (a_1, \ldots, a_m)^T$.

5.4 The dual simplex method

We start solving (5.2.4) by choosing any m linearly independent columns of $\mathbf{C}^T$ to form the basis matrix $\mathbf{B}$. The simplex tableau is then formed by calculating vectors $(\mathbf{y}_i)$ from (5.3.1) and the marginal costs $(z_i - f_i)$ from (5.3.2), for $i = 1, 2, \ldots, n$.

The following steps constitute none other than a dual simplex algorithm for the non-negative bounded variables (b_i) as described, for example, in Hadley ([15], pp. 387-394). The choice of the vector that leaves the basis is made first. Then the vector that enters the basis is determined. At first, any non-basic variable b_i is given the lower bound 0.

(1) Scan the marginal costs. For all marginal costs $(z_i - f_i) < 0$, let their b_i take the value 2 (upper bound). Indicate that by putting a mark "x" above the corresponding columns in the simplex tableau. Then the $\mathbf{b}_B$ is calculated from (5.3.4).

(2) Scan the elements of the basic solution $\mathbf{b}_B$. If all the elements b_{Bs}, s = 1, 2, …, m, are bounded, i.e., $0 \le b_{Bs} \le 2$, an optimal solution is reached. Otherwise go to step (3).

(3) The first element b_{Bi} that is < 0 or > 2 is considered. The corresponding column in the basis is to be replaced by a non-basic column according to one of the steps (3.1-4) below. The scanning proceeds from element b_{Bi+1} and back again from b_{B1}.

Let at any iteration, $\mathbf{C}_j^T$ be associated with b_{Bi}, and let $\mathbf{C}_r^T$ replace $\mathbf{C}_j^T$ in the basis. The following steps are employed.

Case (1)

If $b_{Bi} < 0$, $\mathbf{C}_r^T$ is determined from

$$\theta_r = \max(\theta_1, \theta_2) < 0$$

where

$$\theta_1 = (z_r - f_r)/y_{ir} = \max_s((z_s - f_s)/y_{is}), \quad y_{is} < 0, \quad s \in \mathbf{L}(\mathbf{b})$$

and

$$\theta_2 = (z_r - f_r)/y_{ir} = \max_s((z_s - f_s)/y_{is}), \quad y_{is} > 0, \quad s \in \mathbf{U}(\mathbf{b})$$

(3.1) If $\theta_r = \theta_1$, transfer the simplex tableau and go to step (2) above.

(3.2) If $\theta_r = \theta_2$, transfer the simplex tableau, add 2 (upper bound) to the new b_{Bi}. Remove the mark "x" from column r since b_r is no longer at its upper bound. Go to step (2).

Case (2)

If $b_{Bi} > 2$, $\mathbf{C}_r^T$ is determined from

$$\tau_r = \max(\tau_1, \tau_2) > 0$$

where

$$\tau_1 = (z_r - f_r)/y_{ir} = \min_s((z_s - f_s)/y_{is}), \quad y_{is} > 0, \quad s \in \mathbf{L}(\mathbf{b})$$

and

$\tau_2 = (z_r - f_r)/y_{ir} = \min_s((z_s - f_s)/y_{is}), \quad y_{is} < 0, \quad s \in \mathbf{U}(\mathbf{b})$

(3.3) If $\tau_r = \tau_1$, transfer the simplex tableau. Mark column $\mathbf{C}_j^T$ with "x" to indicate that b_j is now at its upper bound (= 2). Subtract $2\mathbf{y}_j$ from $\mathbf{b}_B$ and go to step (2).

(3.4) If $\tau_r = \tau_2$, transfer the simplex tableau as usual. Remove the mark "x" from column $\mathbf{C}_r^T$ and place an "x" on column $\mathbf{C}_j^T$. Add 2 to the new b_{Bi} and subtract $2\mathbf{y}_j$ from $\mathbf{b}_B$. Go to step (2).

In view of (5.3.4), steps (3.1-4) are analyzed as follows. When a non-basic column $\mathbf{C}_r^T$ at its upper bound enters the basis, as in steps (3.2) and (3.4), it is no more at its upper bound and according to (5.3.4), we add $2\mathbf{y}_r$ to $\mathbf{b}_B$, or in effect, we add 2 to b_{Bi}. Also, when a basic column $\mathbf{C}_j^T$ leaves the basis to its upper bound, as in steps (3.3) and (3.4), we subtract $2\mathbf{y}_j$ from $\mathbf{b}_B$. To illustrate the steps in this algorithm, so far, consider this example.

Example 5.1

Solve the following system of equations in the L_1 norm

$$
\begin{aligned}
a_1 + a_2 &= 3 \\
a_1 - a_2 &= 1 \\
a_1 + 2a_2 &= 7 \\
2a_1 + 4a_2 &= 11.1 \\
3a_1 + a_2 &= 7.2
\end{aligned} \tag{5.4.1}
$$

In the initial data, column $\mathbf{b}_{B0}$ is the sum of the 5 columns of $\mathbf{C}^T$. Matrix $\mathbf{B}^{-1}$ is the inverse of the matrix formed by the first 2 columns of $\mathbf{C}^T$.

Initial Data

$\mathbf{B}^{-1}$		$\mathbf{f}^T$ / $\mathbf{b}_{B0}$	3 / $\mathbf{C}_1^T$	1 / $\mathbf{C}_2^T$	7 / $\mathbf{C}_3^T$	11.1 / $\mathbf{C}_4^T$	7.2 / $\mathbf{C}_5^T$
0.5	0.5	8	1	1	1	2	3
0.5	–0.5	7	1	–1	2	4	1

In the initial data $\mathbf{B}^{-1}$ is not calculated yet, and in view of (5.3.3), vector $\mathbf{b}_{B0}$ is the algebraic sum of the five columns of $\mathbf{C}_i^T$.

Tableau 5.4.1 is formed by multiplying the initial data by $\mathbf{B}^{-1}$.

However, we actually obtain Tableau 5.4.1 by pivoting over the first element in row 1 of the initial data and apply a Gauss-Jordan elimination step, as explained in Chapter 4. Then we pivot over the second element in row 2 in the obtained tableau and apply a Gauss-Jordan elimination step. We then calculate the marginal costs $(z_i - f_i)$, $i \notin \mathbf{I}(\mathbf{b})$.

We find in Tableau 5.4.1 that $(z_i - f_i) < 0$, for $i = 3, 4$ and 5. Hence, the corresponding b_i is given the upper bound 2, and this is identified by marks "x" on these columns. Vector $\mathbf{b}_B$ is modified by subtracting $2(\mathbf{y}_3 + \mathbf{y}_4 + \mathbf{y}_5)$ from it.

In Tableau 5.4.1, the elements of column $\mathbf{b}_B$ are –5.5 and 1.5. The first element violates the inequalities $0 \le \mathbf{b}_{B1} \le 2$ of (5.2.4c). Column $\mathbf{C}_1^T$ is the column in the basis that is associated with this element. Hence, we replace this column in the basis.

Tableau 5.4.1

					x	x	x
	$\mathbf{f}^T$		3	1	7	11.1	7.2
$\mathbf{b}_B$	$\mathbf{B}$		$\mathbf{C}_1^T$	$\mathbf{C}_2^T$	$\mathbf{C}_3^T$	$\mathbf{C}_4^T$	$\mathbf{C}_5^T$
7.5–2(1.5+3+2) = –5.5	$\mathbf{C}_1^T$		1	0	1.5	3	(2)
0.5–2(–0.5–1+1) = 1.5	$\mathbf{C}_2^T$		0	1	–0.5	–1	1
z = 6.3	$z_i - f_i$		0	0	–3	–3.1	–0.2
		θ_r			–2	–1.03	–0.1

Tableau 5.4.2

					x	x	
	$\mathbf{f}^T$		3	1	7	11.1	7.2
$\mathbf{b}_B$	$\mathbf{B}$		$\mathbf{C}_1^T$	$\mathbf{C}_2^T$	$\mathbf{C}_3^T$	$\mathbf{C}_4^T$	$\mathbf{C}_5^T$
–2.75+2 = –0.75	$\mathbf{C}_5^T$		0.5	0	0.75	1.5	1
4.25	$\mathbf{C}_2^T$		–0.5	1	–1.25	(–2.5)	0
z = 5.75	$z_i - f_i$		0.1	0	–2.85	–2.8	0
		τ_r			2.28	1.12	

The vector that enters the basis is calculated from steps (3.1-4) above via calculating the ratios θ_r or τ_r. In Tableau 5.4.1 the parameters θ_r represent the ratios of the elements of $(z_i - f_i)$ over the corresponding elements of the first row in that tableau.

For this Tableau we apply step (3.2) and the pivot element is shown between brackets; that is, column $\mathbf{C}_5^T$ will replace column $\mathbf{C}_1^T$ in the basis. We calculate Tableau 5.4.2, by applying a Gauss-Jordan step and remove the mark "x" from the top of column $\mathbf{C}_5^T$.

For Tableau 5.4.2, both elements of $\mathbf{b}_B$ violate the inequalities (5.2.4c), so we have the choice of replacing $\mathbf{C}_5^T$ or $\mathbf{C}_2^T$ in the basis. We shall replace $\mathbf{C}_2^T$ by $\mathbf{C}_4^T$ by observing step (3.4) above. Again, in Tableau 5.4.2, the parameters τ_r represent the ratios of the elements of $(z_i - f_i)$ over the corresponding elements of the second row in that tableau.

We get Tableau 5.4.3, in which both elements of $\mathbf{b}_B$ satisfy the inequalities (5.2.4c). Hence, Tableau 5.4.3 gives the optimal solution for Example 5.1.

Tableau 5.4.3

			x	x		
	$\mathbf{f}^T$	3	1	7	11.1	7.2
$\mathbf{b}_B$	$\mathbf{B}$	$\mathbf{C}_1^T$	$\mathbf{C}_2^T$	$\mathbf{C}_3^T$	$\mathbf{C}_4^T$	$\mathbf{C}_5^T$
1.8 $-2(0.6) = 0.6$	$\mathbf{C}_5^T$	0.2	0.6	0	0	1
$-1.7 +2-2(-0.4) = 1.1$	$\mathbf{C}_4^T$	0.2	-0.4	0.5	1	0
$z = 3.23$	$z_i - f_i$	0.66	-1.12	-1.45	0	0

Note also that the elements of the last row in Tableau 5.4.3 are the residuals of system (5.4.1). From Theorem 5.3, the sum of their absolute vales $= 0.66 + 1.12 + 1.45 + 0 + 0 = 3.23 = z$. From (5.3.5) or by solving the fifth and fourth equations of (5.4.1) of Example 5.1, we get the solution of the problem; $\mathbf{a} = (a_1, a_2)^T = (1.77, 1.89)^T$.

5.5 Modification to the algorithm

We note in the above section that the process of adding $2\mathbf{y}_r$ and/or subtracting $2\mathbf{y}_j$ vectors is done after the simplex tableau has been changed by applying a Gauss-Jordan elimination step.

The efficiency of the algorithm is improved by skipping certain intermediate Gauss-Jordan iterations [2]. This is done with the following modifications:

(1) Reverse the order of the two processes. The process of adding the $2\mathbf{y}_r$ and/or subtracting the $2\mathbf{y}_j$ vectors from $\mathbf{b}_B$ is done first. Then it is followed by the Gauss-Jordan step.

(2) More importantly, the changing of the tableau may be postponed until some non-basic columns each enter then leave the basis. This is because the corresponding b_{Bi} does not yet satisfy the bounds $0 \le b_{Bi} \le 2$ of (5.2.4c). This process continues until the last non-basic column for which b_{Bi} satisfies this inequality is found. This ensures that in replacing a basic column, the maximum decrease in the objective function z has been achieved.

Let us assume that we start the procedure of the previous paragraph from a basic solution given by a simplex tableau, say tableau 1. Let us also assume that the $(k - 1)$ non-basic columns have each entered, then left the basis for which b_{Bi} violated the inequalities (5.2.4c). Assume also that a kth column entered the basis without leaving, because b_{Bi} satisfied the inequalities (5.2.4c). Hence, in this method, we have skipped calculating the $(k - 1)$ intermediate tableaux 2, 3, …, k that correspond to the $(k - 1)$ non-basic columns that each entered, then left the basis. Only tableau $k + 1$ is calculated.

All necessary data needed for pursuing this procedure is contained in the ith row of tableau 1 and the marginal costs of this tableau. We need to know the (b_{Bi}) and the pivot elements (y_{ir}) for the $(k - 1)$ intermediate iterations. We also note that each intermediate tableau is obtained by pivoting over an element y_{ir} in row i of the previous (intermediate) tableau. As a result, the ith element of the basic solution and the pivot elements for intermediate tableaux $t = 2, 3, \ldots, k$, are $(b_{Bi}/y_{i,t-1})$ and $(y_{it}/y_{i,t-1})$ respectively.

We also know that the $(k + 1)^{th}$ tableau is the tableau that we would have obtained, had we changed tableau 1 once, by pivoting over y_{ik}.

Let

(5.5.1a) $$\underline{b}'_{Bi} = b_{Bi} \text{ and } \underline{y}_{i,t} = y_{i,t}, \quad t = 1$$

(5.5.1b) $$\underline{b}'_{Bi} = b'_{Bi}/y_{i,t-1} \text{ and } \underline{y}_{i,t} = y_{i,t}/y_{i,t-1}, \quad t = 2, 3, \ldots, k$$

where $\underline{b}'_{Bi}$ and $\underline{y}_{i,t}$ are respectively the ith element of $\mathbf{b}'_B$ and the pivot element in tableaux t = 1, 2, …, k. Also, $\mathbf{b}'_B$ is $\mathbf{b}_B$ with vector(s) $2\mathbf{y}_r$ added to it and/or vector(s) $2\mathbf{y}_j$ subtracted from it. Again, let us start an iteration from a basic solution given by, say, tableau 1. This algorithm modifies the algorithm of Section 5.4 as follows.

Starting step

Scan b_{Bk} for k = 1, 2, …, m. Consider the smallest of parameters $b_{Bk} < 0$ and $(2 - b_{Bk})$. Let $\mathbf{C}_r^T$ replaces $\mathbf{C}_j^T$ and each time calculate $\underline{b}'_{Bi}$ from (5.5.1). This is done until the calculated $\underline{b}'_{Bi}$ satisfies $0 \le \underline{b}'_{Bi} \le 2$. Change the simplex tableau and re-scan the elements of $\mathbf{b}_B$. This is repeated until all the elements b_{Bk} satisfy $0 \le b_{Bk} \le 2$. Steps (3.1-4) of Section 5.4 are now modified as follows.

Case (1')

If $b_{Bi} < 0$, the sequence of the non-basic columns $\mathbf{C}_r^T$ that enter the basis and may leave the basis, corresponds to the parameters $\theta_r = (z_r - f_r)/y_{ir} < 0$, $r \notin \mathbf{I}(\mathbf{b})$, starting from the algebraically larger ratio.

(3.1') If $\underline{y}_{ir} < 0$, $\mathbf{b}'_B$ is not changed. Go to step (3.5').

(3.2') If $\underline{y}_{ir} > 0$, add $2\mathbf{y}_r$ to $\mathbf{b}'_B$. Remove the "x" from $\mathbf{C}_r^T$ indicating that b_r is no more at its upper bound. Go to step (3.5').

Case (2')

If $b_{Bi} > 2$, the sequence of the non-basic columns $\mathbf{C}_r^T$ that enter the basis and may leave the basis corresponds to the parameters $\tau_r = (z_r - f_r)/y_{ir} > 0$, $r \notin \mathbf{I}(\mathbf{b})$, starting from the smallest ratio.

(3.3') If $\underline{y}_{ir} > 0$, subtract $2\mathbf{y}_j$ from $\mathbf{b}'_B$ and place a mark "x" over $\mathbf{C}_j^T$ indicating that it is now at its upper bound. Go to step (3.5').

(3.4') If $\underline{y}_{ir} < 0$, add $2\mathbf{y}_r$ and subtract $2\mathbf{y}_j$ from $\mathbf{b}'_B$. Remove the "x" from $\mathbf{C}_r^T$ and place an "x" over $\mathbf{C}_j^T$. Go to step (3.5').

(3.5') Calculate $\underline{b}'_{Bi}$ from (5.5.1a, b). If the resulting $\underline{b}'_{Bi}$ still violates $0 \le \underline{b}'_{Bi} \le 2$, go to either case (1') or case (2'), according to whether $\underline{b}'_{Bi} < 0$ or > 2 respectively. If the resulting $\underline{b}'_{Bi}$ satisfies $0 \le \underline{b}'_{Bi} \le 2$, change the tableau. Go to the starting step.

5.6 Occurrence of degeneracy

In the dual simplex method degeneracy arises when there exist one or more non-basic column r, for which the ratio $(z_r - f_r)/y_{ir} = 0$. It should be decided whether it is $\tau_r = 0$ or $\theta_r = 0$. This is resolved as follows. If $b_r = 0$, we replace $(z_r - f_r)$ by a small number δ, and if $b_r = 2$, we replace $(z_r - f_r)$ by $-\delta$. We then calculate $(z_r - f_r)/y_{ir}$, and it will definitely be either > or < 0, and the degeneracy is resolved.

The small number δ represents the round-off error in floating-point arithmetic operations, denoted in our software by PREC, and set to 10^{-6} and 10^{-16} for single- and double-precision computation respectively (see Sections 2.4 and 4.7.1).

Example 5.2

Solve the following system of equations in the L_1 norm.

$$\begin{aligned}
-2a_1 & & &= 6 \\
8a_1 &+ 9a_2 &&= 6 \\
-8a_1 & &&= 24 \\
21a_1 &+ 18a_2 &&= 3 \\
12a_1 &- 9a_2 &&= -6 \\
-32a_1 &- 13.5a_2 &&= -9
\end{aligned} \tag{5.6.1}$$

For simplicity of presentation, we are not using pivoting in obtaining part 1 of the solution. As indicated above, in part 1 of this problem, columns 1 and 2 form the initial basis matrix.

Initial Data

f^T	6	6	24	3	−6	−9
$\Sigma_i C_i^T$	C_1^T	C_2^T	C_3^T	C_4^T	C_5^T	C_6^T
−1	−2	8	−8	21	12	−32
4.5	0	9	0	18	−9	−13.5

Tableau 5.6.2 (part 1)

f^T	6	6	24	3	−6	−9
b_{B0}	C_1^T	C_2^T	C_3^T	C_4^T	C_5^T	C_6^T
2.5	1	0	4	−2.5	−10	10
0.5	0	1	0	2	−1	−1.5

Tableau 5.6.1 (not shown) is obtained by pivoting over the first nonzero element in the first row of the initial data and applying a Gauss-Jordan step. Tableau 5.6.2 is obtained by pivoting over the first nonzero element in the second row in Tableau 5.6.1 and applying a Gauss-Jordan step.

Tableau 5.6.3 is Tableau 5.6.2, with the marginal costs $(z_i - f_i)$, $i = 1, 2, \ldots, 6$, added to it, and with $\mathbf{b}_B$ calculated.

We have an initial degeneracy, as $\tau_r = (z_i - f_i)/y_{ir} = 0$, for $r = 3$. Since $b_3 = 0$, we replace the 0 value of $(z_3 - f_3)$ by $+\delta$ and $(z_3 - f_3)/y_{13} > 0$. From $\tau_r = (z_i - f_i)/y_{ir}$ in Tableau 5.6.3, the sequence of the columns that enter the basis and may then leave the basis is $\mathbf{C}_3^T, \mathbf{C}_4^T, \mathbf{C}_5^T, \mathbf{C}_6^T$. The pivot $\underline{y}_{13} = 4 > 0$, and according to step (3.3'), $2\mathbf{y}_1$ is subtracted from $\mathbf{b}_B$, and a mark "x" is placed above $\mathbf{C}_1^T$. The new $\underline{b}'_{B1} = (27.5 - 2)/4 = 25.5/4 > 2$, and we still have case (2') of the algorithm. Now $\mathbf{C}_3^T$ leaves the basis and $\mathbf{C}_4^T$ enters the basis. From (5.5.1b), the pivot $\underline{y}_{14} = -2.5/4 < 0$. Hence, from step (3.4'), $2\mathbf{y}_3$ is subtracted from and $2\mathbf{y}_4$ is added to the previously calculated $\mathbf{b'}_B$. A mark "x" is placed over $\mathbf{C}_3^T$ and the "x" over $\mathbf{C}_4^T$ is removed. The new $\underline{b}'_{B1} = (25.5 - 8 - 5)/(-2.5)$, which $= 12.5/(-2.5) < 0$, and we now have case (1') of the algorithm.

Tableau 5.6.3 (part 2)

	$\mathbf{f}^T$	x		x			
		6	6	24	3	−6	−9
$\mathbf{b}_B$		$\mathbf{C}_1^T$	$\mathbf{C}_2^T$	$\mathbf{C}_3^T$	$\mathbf{C}_4^T$	$\mathbf{C}_5^T$	$\mathbf{C}_6^T$
2.5+5+20 = 27.5		1	0	4	−2.5	−10	10
0.5–4+2 = −1.5		0	1	0	2	−1	−1.5
z = 126	$z_i - f_i$	0	0	0	−6	−60	60
	τ_r			δ/4	2.4	6	6

As $\mathbf{C}_4^T$ leaves the basis, either $\mathbf{C}_5^T$ or $\mathbf{C}_6^T$ enters the basis as they both have the same ratio τ_r. Let $\mathbf{C}_5^T$ enter the basis. The pivot $\underline{y}_{15} = -10/(-2.5) > 0$, and from step (3.2') of case (1') we add $2\mathbf{y}_5$ to the previously obtained $\mathbf{b'}_B$ and remove the "x" from $\mathbf{C}_5^T$. The obtained $\underline{b}'_{B1} = (12.5 - 20)/(-10) = 0.75$, which is between 0 and 2. We thus change the tableau and get Tableau 5.6.4. That ends this iteration.

Tableau 5.6.4

	$\mathbf{f}^T$	x 6	6	x 24	3	–6	–9
$\mathbf{b}_B$		C_1^T	C_2^T	C_3^T	C_4^T	C_5^T	C_6^T
0.75		–0.1	0	–0.4	0.25	1	–1
1.25		–0.1	1	–0.4	2.25	0	–2.5
z = 39	$z_i - f_i$	–6	0	–24	9	0	0

In Tableau 5.6.4, both the elements of $\mathbf{b}_B$ are between 0 and 2 and Tableau 5.6.4 is optimal. Hence, this example requires 3 iterations, 2 in part 1 and 1 in part 2.

The solution vector **a** is calculated from the fifth and second equations of system (5.6.1). These two equations correspond to the final basis (in Tableau 5.6.4)

$$a_1 = 0 \text{ and } a_2 = 2/3$$

From Theorem 5.3, the residuals are the marginal costs

$$\mathbf{r} = (-6, 0, -24, 9, 0, 0)^T$$

and z = 39.

5.7 A significant property of the L_1 approximation

Assume that the n by m matrix **C** of **Ca** = **f** is of rank m, m < n. Then the best L_1 approximation of system **Ca** = **f** has m equations with zero residuals r_i. This property is a direct consequence of using the dual form of the linear programming for this problem, since the m equations in **Ca** = **f** associated with the basis matrix have zero marginal costs, $(z_i - f_i)$, i.e., zero residuals r_i.

This means that if n discrete points in the x-y plane are being approximated by a plane curve in the L_1 norm, the curve will pass through m of the discrete points. This is known as the **interpolation property** of the L_1 approximation, described in Section 2.3.1.

If the rank of matrix **C** is k, k ≤ m < n, there exists an L_1 solution to system **Ca** = **f** such that there are k equations in **Ca** = **f** for which the residuals are 0's. However, as noted in Section 5.1.1, Spath ([24], p. 60) argues that while optimal solutions calculated by simplex methods produce this property, not every optimal solution has this characterization.

For the purpose of numerical stability for ill conditioned systems $\mathbf{Ca} = \mathbf{f}$, we employ a triangular decomposition to the basis matrix. This is described in the next section.

5.8 Triangular decomposition of the basis matrix

In our modification to part 1 of the algorithm (Section 5.5), we apply m Gauss-Jordan elimination steps with complete pivoting.

In part 2, a triangular decomposition method for the basis matrix is used [3, 4]. This method is a variation of the method of Bartels et al. [12]. Without loss of generality, let rank($\mathbf{C}$) = m.

Let at the end of part 1, the basis be denoted by $\mathbf{I}_0(\mathbf{b})$, the inverse of the basis matrix be $\mathbf{B}_0^{-1}$, the basic solution be $\mathbf{b}_{B0}$, and the columns of the simplex tableau be $\mathbf{Y}_0$. In other words, the programming problem (5.2.4) is described by the 5-tuple

$$\{\mathbf{I}_0(\mathbf{b}), \mathbf{I}, \mathbf{B}_0^{-1}, \mathbf{b}_{B0}, \mathbf{Y}_0\}$$

where an m-unit matrix $\mathbf{I}$ is added.

For part 2, the inverse of the basis matrix as well as the other parameters in the 5-tuple are updated. The inverse of the basis matrix $\mathbf{B}^{-1}$ may be updated as

$$\mathbf{B}^{-1} = \mathbf{E}\mathbf{B}_0^{-1}$$

$\mathbf{E}$ is a nonsingular m-square matrix and $\mathbf{E}^{-1}$ is decomposed into

$$\mathbf{L}\mathbf{E}^{-1} = \mathbf{P}$$

$\mathbf{L}$ is a nonsingular m-square matrix and $\mathbf{P}$ is a nonsingular upper triangular m-matrix. Now, $\mathbf{P}$ and $\mathbf{B}_0^{-1}$ are stored and updated. For each iteration in part 2, the above 5-tuple change to

$$\{\mathbf{I}(\mathbf{b}), \mathbf{P}, \mathbf{G}^{-1}, \mathbf{V}_B, \mathbf{X}\} \tag{5.8.1}$$

In (5.8.1)

$$\mathbf{G}^{-1} = \mathbf{L}\mathbf{B}_0^{-1}, \ \mathbf{V}_B = \mathbf{L}\mathbf{b}_{B0}, \ \mathbf{X} = \mathbf{L}\mathbf{Y}_0 \tag{5.8.1a}$$

$$\mathbf{B}^{-1} = \mathbf{P}^{-1}\mathbf{G}^{-1} \tag{5.8.2a}$$

The basic solution is given by

(5.8.2b) $$\mathbf{b}_B = \mathbf{P}^{-1}\left[\mathbf{V}_B - 2\sum_j \mathbf{x}_j + 2\sum_r \mathbf{x}_r\right] = \mathbf{P}^{-1}\underline{\mathbf{V}}_B$$

The summation over j is for vectors that go from zero bound to an upper bound, and the summation over r are for vectors that go from upper bound to zero bound. Vector $\mathbf{b}_B$ is calculated from

$$\mathbf{P}\mathbf{b}_B = \underline{\mathbf{V}}_B$$

by back substitution.

Vectors $(\mathbf{y}_i)$ in the simplex tableau are given by

(5.8.2c) $$\mathbf{y}_i = \mathbf{P}^{-1}\mathbf{x}_i, \quad i = 1, 2, \ldots, n$$

The marginal costs which are the residuals are, as usual, given by

(5.8.2d) $$r_i = \mathbf{f}_B{}^T\mathbf{y}_i - f_i, \quad i = 1, 2, \ldots, n$$

where $\mathbf{f}_B$ is associated with the basis matrix.

In part 2, the non-basic column r, that enters the basis, is determined from

(5.8.3) $$\alpha_r = \pm\min_s |r_s/y_{is}|, \quad s \notin \mathbf{I}(\mathbf{b})$$

where α_r denotes τ_r or θ_r of Section 5.4.

If $|\alpha_r| \leq 1$, the residuals in (5.8.2d) are updated as

$$r'_j = r_j - \alpha_r y_{ij}, \quad j \notin \mathbf{I}(\mathbf{b})$$

However, if $|\alpha_r| > 1$, the r'_j are calculated from (5.8.2c) and (5.8.2d); that is, $r'_j = \mathbf{u}^T\mathbf{x}_j - f_j$, where $\mathbf{u}$ is the solution of $\mathbf{P}^T\mathbf{u} = \mathbf{f}_B$.

For the elements y_{is}, $s \notin \mathbf{I}(\mathbf{b})$, from (5.8.2c), $y_{is} = \mathbf{w}^T\mathbf{x}_s$, where $\mathbf{w}^T$ is row i of $\mathbf{P}^{-1}$and $\mathbf{P}^T\mathbf{w} = \mathbf{e}_i$, where $\mathbf{e}_i$ is column i of an m-unit matrix.

Let now

$$\{\mathbf{I}'(\mathbf{b}), \mathbf{P}', \mathbf{G}'^{-1}, \mathbf{V}'_B, \mathbf{X}'\}$$

be the update of (5.8.1) for which $r \notin \mathbf{I}(\mathbf{b})$ replaces $s \in \mathbf{I}(\mathbf{b})$. Then

$$\mathbf{I}'(\mathbf{b}) = \{i, \ldots, i_{s-1}, i_{s+1}, \ldots, i_m, r\}$$

and

$$\underline{\mathbf{P}} = \{\mathbf{p}_1, \ldots, \mathbf{p}_{s-1}, \mathbf{p}_{s+1}, \ldots, \mathbf{p}_m, \mathbf{x}_r\}$$

$\underline{\mathbf{P}}$ is an upper Hessenberg matrix. Its (m – s) nonzero sub-diagonal elements are annihilated by (m – s) Gauss elimination steps with row permutation.

Finally, the solution of the L_1 problem is given by Theorem 5.4

$$\mathbf{a}^T = \mathbf{f}_B{}^T\mathbf{B}^{-1} \tag{5.8.4}$$

where $\mathbf{B}^{-1}$ is the inverse of the basis matrix given by (5.8.2a), for the optimum solution of the programming problem.

5.9 Arithmetic operations count

In the majority of algorithms, the number of additions/ subtractions in the arithmetic operations is comparable to the number of multiplications/divisions (m/d). Yet, the CPU time of an addition or a subtraction is very small, compared with the CPU time taken to execute a multiplication or a division. Hence, in this section we shall only count the number of (m/d) for this algorithm.

For part 1 of this algorithm, each iteration requires an average of $(nm + 0.5m^2)$ m/d. This is to apply a Gauss-Jordan elimination step to the simplex tableau and to calculate the inverse of the basis $\mathbf{B}_0{}^{-1}$. At the end of part 1, from (5.8.2d), the residuals (r_i), $i \notin \mathbf{I}(\mathbf{b})$, are calculated, which requires $m(n - m)$ m/d. Then $\mathbf{b}_{B0}$ is calculated, which requires $m{\times}n$ additions and subtractions, which are small, compared with other m/d, and may be neglected.

For part 2, from (5.8.2b), $\mathbf{b}_B$ is calculated and it requires $0.5m^2$ m/d. If the residuals (r_i), $i \notin \mathbf{I}(\mathbf{b})$, are calculated from (5.8.3), $(n - m)$ m/d are required. Otherwise, approximately $(mn - 0.5m^2)$ m/d are required to calculate vector $\mathbf{u}$ and the inner product $\mathbf{u}^T\mathbf{x}_i$, for $i \notin \mathbf{I}(\mathbf{b})$.

The $(n - m)$ elements y_{is}, $s \notin \mathbf{I}(\mathbf{b})$, need an average of 0.5mn m/d to calculate vector $\mathbf{w}$ and the inner product $\mathbf{w}^T\mathbf{x}_s$. The parameter α_r requires $(n - m)$ divisions. For more detail, see [3]. We come to the following conclusion.

Each iteration in part 2 needs an average of about $mn + 1.17m^2$ m/d if $|\alpha_r| \le 1$ and about $2mn + 0.67m^2$ m/d if $|\alpha_r| > 1$. Numerical experience shows that for most test examples conducted in [3], most of the α_r satisfy $|\alpha_r| < 1$. Very few examples have $|\alpha_r| > 1$ for the first few iterations; the $|\alpha_r|$ then become ≤ 1 for the remaining iterations. Hence, in practice, the arithmetic operations count per iteration in part 2 is comparable to that of part 1. At last, the solution of the L_1

problem as given by (5.8.4), where matrix $\mathbf{B}^{-1}$ and vector $\mathbf{f}_B$ are associated with the optimal solution needs $1.5m^2$ m/d.

Our final conclusion is that in part 1 of the algorithm the arithmetic operations count per iteration is $[m(n-m)]$ m/d. In part 2, the arithmetic operations count per iteration is on the average of approximately $[mn + 1.17m^2]$ m/d if $|\alpha_r| \le 1$ and $[2mn + 0.67m^2]$ if $|\alpha_r| > 1$. That means that the arithmetic operations count per iteration in either part 1 or part 2 of the algorithm is a linear function of n and m^2. It is also known that in linear programming problems, the number of iterations is proportional to m.

A number of examples were solved in [3] on the IBM 360/67 computer and their CPU seconds were recorded (Table 1 in [3], p. 226). For problem 4 in that table, matrix $\mathbf{C}(n\times m)$ has the values n = 23, 53, 103 and 203, for m = 6. In problem 5, matrix $\mathbf{C}(n\times m)$ has the values n = 21, 51, 101 and 202, for m = 11.

Table 5.1 (problem 4 in [3])

Example	$\mathbf{C}(n\times m)$	Iterations	CPU seconds	CPU/iter	(CPU/iter)/n in 10^{-3}sec.
1	23×6	9	0.08	0.0089	0.386
2	53×6	10	0.21	0.0210	0.396
3	103×6	12	0.48	0.0400	0.388
4	203×6	11	1.03	0.0936	0.461

Table 5.2 (problem 5 in [3])

Example	$\mathbf{C}(n\times m)$	Iterations	CPU seconds	CPU/iter	(CPU/iter)/n in 10^{-3}sec.
1	21×11	18	0.24	0.0133	0.633
2	51×11	27	0.88	0.0326	0.639
3	101×11	29	1.71	0.0590	0.584
4	201×11	36	5.12	0.1422	0.708

In Tables 5.1 and 5.2, we show that the CPU times/iteration for these two problems are approximately proportional to the parameter n.

In other words, (CPU times/iteration)/n is almost the same for the 4 cases in each problem.

5.10 Numerical results and comments

The functions LA_L1() and LA_Lone() both implement this algorithm. LA_Lone() is the same as LA_L1(), except that LA_Lone() does not use a triangular decomposition of the basis matrix. LA_Lone() implements the algorithm described in [2] and LA_L1() implements the algorithm [3] given in FORTRAN IV in [4]. LA_Lone() is faster than LA_L1(), and it is used by other programs in this book. DR_L1() and DR_Lone() both test the following examples.

Example 5.3

This example is the same as that solved in the L_1 norm in ([2], pp. 848-849). The obtained results are z = 90, the residual vector r = $(-5.6, -4.0, 48.0, -22.4, 0, 0, 10)^T$ and $\mathbf{a} = (-0.2, 0.4, 0.0)^T$.

LA_Lone() and LA_L1() yield the same values of z and **r** as obtained in [2], in 3 iterations. However, they compute a different solution vector $\mathbf{a} = (0, 0.6, -0.2)^T$, indicating that it is not unique.

Example 5.4

For $0 \leq x \leq 1$, approximate in the L_1 norm the function $f(x) = \exp(x)$, for $x \leq (0.5)$, and $f(x) = \exp(0.5)$, for $x > 0.5$. The approximating function is

$$a_1 + a_2 \sin(x) + a_3 \cos(x) + a_4 \sin(2x) + a_5 \cos(2x) + a_6 \sin(3x) \\ + a_7 \cos(3x) + a_8 \sin(4x) + a_9 \cos(4x) + a_{10} \sin(5x) + a_{11} \cos(5x)$$

This problem is the same as problem 2 of example 5 in [3]. The coordinates of x are taken at 0.02 intervals, i.e., (0.0, 0.02, 0.04, …, 1.0). Matrix **C** in **Ca** = **f** is a 51 by 11 matrix and is ill-conditioned.

When the solution is computed in s.p, both LA_L1() and LA_Lone() give rank(**C**) = 8 and z = 0.3242. However, the elements of vector **a** agree only to 2 decimal places; $\mathbf{a} = (-7.97, 0, 0, 7.40, 18.72, 0, -19.64, -3.83, 15.03, 0.39, -5.15)^T$. This indicates that for ill conditioned problems in single-precision (s.p.), LA_L1() gives more accurate results.

In double-precision (d.p.), LA_L1() and LA_Lone() give identical results (**a** and **r**), in 27 iterations; rank(**C**) = 11 and z = 0.1564. It is observed, however, that although the value of z in d.p. is about half the value of z in s.p., the values of the elements of **a** in d.p. are very large, compared with those of **a** in s.p., indicating that matrix **C** for this problem is ill-conditioned.

Example 5.5

This is the curve fitting Example 2.1. We have the set of 8 points in the x-y plane: (1, 2), (2, 2.5), (3, 2), (4, 6.5), (5, 3.5), (6, 4.5), (7, 6), (8, 7), with the fourth point as an outlier. It is required to approximate this discrete set in the L_1 norm by the vertical parabola $y = a_1 + a_2x + a_3x^2$, where a_1, a_2 and a_3 are unknowns.

In this example, the equation $\mathbf{Ca} = \mathbf{f}$ is given by (2.2.2). The result, computed by both LA_L1() and LA_Lone(), is $z = 4.857$ and $\mathbf{r} = (0, -0.429, 0.357, -3.643, 0.071, 0, -0.357, 0)^T$. The solution vector $\mathbf{a} = (2.143, -0.25, 0.107)^T$ is unique. The fitting parabola is given by the solid curve in Figure 2-1.

This example is solved again where the given 8 points are approximated by the curve $y = a_1 + a_2x + a_3x^2 + a_4x^2$, where a_1, a_2, a_3 and a_4 are to be calculated. Matrix **C** is an 8 by 4 matrix, where the first 3 columns are the same as in equation (2.2.2) and the fourth column is the same as the third column; that is, matrix **C** is rank deficient. The results **r** and z are the same as in the previous paragraph but the solution vector $\mathbf{a} = (2.143, -0.25, 0.107, 0.0)^T$ and **a** is not unique due to rank deficiency of matrix **C**. We shall be using this example again in other chapters of this book.

Finally, we note that in our experience, cycling (as a result of the occurrence of degeneracy) never occurred, and no failure of our programs has been recorded. We also observe that our algorithm compares favorably with that of Barrodale and Roberts [8, 9]. See the comments later.

The techniques used in this algorithm, i.e., solving a linear programming problem with non-negative bounded variables has proved to be valuable in solving other problems such as the **L_1 solution of overdetermined systems of linear equations with bounded variables** of Chapter 7 and the **bounded and L_1 bounded solutions of underdetermined systems of linear equations** of Chapter 21.

A large number of examples have been tested by the current

method [3, 4] and compared with those of Barrodale and Roberts (BR) [9]. A typical result is that the total CPU time for the current method is about 15% higher than (BR). Spath ([24], pp. 74, 82) argues that this is partially true. We also note that when we used partial pivoting in part 1 of our algorithm LA_L1() instead of complete pivoting, our method was slightly faster. As noted earlier, it was concluded by Armstrong and Godfrey [6] that the two algorithms, ours and that of Barrodale and Roberts (BR) [9], are identical except for the starting basis.

Bloomfield and Steiger [14] presented a unified treatment of the role of the L_1 approximation in many areas, such as statistics and numerical analysis. They also discussed and described different algorithms for solving the L_1 approximation problem. In their book ([14], pp. 219-236) they compared the CPU times and the iteration counts between three algorithms. These are of Barrodale and Roberts (BR) [9], Bartels et al. (BCS) [10] and Bloomfield and Steiger (BS) [13]. The comparison was over a variety of problems, some deterministic and some random.

The coefficient matrices have row values n ranging from 50 to 600 and some from 100 to 900, in increments and column values $k = 3, 4, \ldots, 8$. A characteristic difference was observed between the (BR) and the other two. For all k, as n increases, the CPU times increase almost proportional to n^2 while for the other two the CPU increase linearly. A comparison was also made between the (BCS) and the (BS) algorithms on some other data. The CPU times for the (BS) algorithm was between one half to one third of that of the (BCS).

Spath [24] collected data (matrix **C** and vector **f**) for 42 examples that were tested on a number of routines after converting them to FORTRAN 77 on the IBM PC AT 102. He compared them with respect to computer storage, CPU time and accuracy of results. The routines were those of Robers and Robers (RR) [21] (believed to be the oldest published routine and needing the most computer storage) Barrodale and Roberts (BR) [8], Armstrong et al. (AFK) [5], ours (NA) [4], Bloomfield and Steiger (BS) [13] and Bartels and Conn (BC) [11]. The last algorithm is for constrained L_1 approximation problem, and was modified for the case when the constraints do not exist. Spath displayed the results of the last 5 methods ([24], pp. 79-82).

He found that the algorithm of (BS) needs two arrays of size nm. However, unlike other methods, matrix **C** is not overwritten. The (BR) method needs the smallest storage and the fewest number of lines of code. He also proved that the (AFK) method was the fastest of all available algorithms. The (BS) method is the fastest of all algorithms only if n is at least two orders of magnitude larger than m. Due to the number of lines of code executed by the (BC) method, it is slower than the (BS) and (BR) methods. For realistic sizes of n and m, (BR) does best after (AFK) and (BS).

Spath's chose the method of Armstrong et al. [5] as the best, followed by Abdelmalek [4] and Bloomfield and Steiger [13], in comparison with Barrodale and Roberts [9]. Moreover, among the algorithms tested by Spath, only the methods of Barrodale and Roberts [8, 9] and of Abdelmalek [3, 4] can deal with the case when matrix **C** is rank deficient, i.e., has rank(**C**) $\leq$ m. Spath concluded, however, that the results of the different algorithms are sensitive to the value of the tolerance parameter EPS.

References

1. Abdelmalek, N.N., On the discrete L_1 approximation and L_1 solutions of overdetermined linear equations, *Journal of Approximation Theory*, 11(1974)38-53.
2. Abdelmalek, N.N., An efficient method for the discrete linear L_1 approximation problem, *Mathematics of Computation*, 29(1975) 844-850.
3. Abdelmalek, N.N., L_1 solution of overdetermined systems of linear equations, *ACM Transactions on Mathematical Software*, 6(1980)220-227.
4. Abdelmalek, N.N., Algorithm 551: A FORTRAN subroutine for the L_1 solution of overdetermined systems of linear equations [F4], *ACM Transactions on Mathematical Software*, 6(1980)228-230.
5. Armstrong, R.D., Frome, E.L. and Kung, D.S., Algorithm 79-01: A revised simplex algorithm for the absolute deviation curve fitting problem, *Communications on Statistics-Simulation and Computation*, B8(1979)175-190.

6. Armstrong, R.D. and Godfrey, J., Two linear programming algorithms for the linear discrete L_1 norm problem, *Mathematics of Computation*, 33(1979)289-300.
7. Barrodale, I. and Phillips, C., An improved algorithm for discrete Chebyshev linear approximation, *Proceedings of the Fourth Manitoba Conference on Numerical Mathematics*, Hartnell, B.L. and Williams, H.C. (eds.), Winnipeg, Manitoba, Canada, pp. 177-190, 1975.
8. Barrodale, I. and Roberts, F.D.K., An improved algorithm for discrete l_1 approximation, *SIAM Journal on Numerical Analysis*, 10(1973)839-848.
9. Barrodale, I. and Roberts, F.D.K., Algorithm 478, Solution of an overdetermined system of equations in the l_1 norm [F4], *Communications of ACM*, 17(1974)319-320.
10. Bartels, R.H., Conn, A.R. and Sinclair, J.W., Minimization techniques for piecewise differentiable functions: The l_1 solution of an overdetermined linear system, *SIAM Journal on Numerical Analysis*, 15(1978)224-241.
11. Bartels, R.H. and Conn, A.R., Algorithm 563: A program for linear constrained discrete l_1 problems, *ACM Transactions on Mathematical Software*, 6(1980)609-614.
12. Bartels, R.H, Stoer, J. and Zenger, Ch., A realization of the simplex method based on triangular decomposition, *Handbook for Automatic Computation, Vol. II: Linear Algebra*, Wilkinson, J.H. and Reinsch, C. (eds.), Springer-Verlag, New York, pp. 152-190, 1971.
13. Bloomfield, P. and Steiger, W.L., Least absolute deviations curve-fitting, *SIAM Journal on Scientific and Statistical Computing*, 1(1980)290-301.
14. Bloomfield, P. and Steiger, W.L., *Least Absolute Deviations, Theory, Applications, and Algorithms*, Birkhauser, Boston, 1983.
15. Hadley, G., *Linear Programming*, Addison-Wesley, Reading, MA, 1962.
16. Madsen, K., Nielsen, H.B. and Pinar, M.C., New characterizations of l_1 solutions to overdetermined systems of linear equations, *Operations Research Letters*, 16(1994)159-166.

17. Narula, S.C. and Wellington, J.F., An efficient algorithm for the MSAE and the MMAE regression problems, *SIAM Journal on Scientific and Statistical Computing*, 9(1988)717-727.
18. Pinkus, A.M., *On L^1-Approximation*, Cambridge University Press, London, 1989.
19. Powell, M.J.D., *Approximation Theory and Methods*, Cambridge University Press, London, 1981.
20. Robers, P.D. and Ben-Israel, A., An interval programming algorithm for discrete linear L_1 approximation problems, *Journal of Approximation Theory*, 2(1969)323-336.
21. Robers, P.D. and Robers, S.S., Algorithm 458: Discrete linear L_1 approximation by interval linear programming, *Communications of ACM*, 16(1973)629-631.
22. Rosen, J.B., Park, H., Glick, J. and Zhang, L., Accurate solution to overdetermined linear equations with errors using L_1 norm minimization, *Computational Optimization and Applications*, 17(2000)329-341.
23. Sklar, M.G. and Armstrong, R.D., Least absolute value and Chebyshev estimation utilizing least squares results, *Mathematical Programming*, 24(1982)346-352.
24. Spath, H., *Mathematical Algorithms for Linear Regression*, Academic Press, English Edition, London, 1991.
25. Usow, K.H., On L_1 approximation. II. Computation for discrete functions and discretization effects, *SIAM Journal on Numerical Analysis*, 4(1967)233-244.
26. Wagner, H.M., Linear programming techniques for regression analysis, *Journal of American Statistical Association*, 54(1959) 206-212.
27. Watson, G.A., *Approximation Theory and Numerical Methods*, John Wiley & Sons, New York, 1980.
28. Wesolowsky, G.O., A new descent algorithm for the least absolute value regression problem, *Communications in Statistics – Simulation and Computation*, B10(1981)479-491.

5.11 DR_L1

```
/*-----------------------------------------------------------------------
DR_L1
-------------------------------------------------------------------------
This program is a driver for the function LA_L1(), which solves an
overdetermined system of linear equations in the L1 (L-One) norm.
It uses a dual simplex method and a triangular factorization of the
basis matrix.

The overdetermined system has the form

                        c*a = f

"c" is a given real n by m matrix of rank k, k <= m <= n.
"f" is a given real n vector.
"a" is the solution m vector.

This program contains 3 examples whose results appear in the
text.
-----------------------------------------------------------------------*/

#include "DR_Defs.h"
#include "LA_Prototypes.h"

#define Na          7
#define Ma          3
#define Nb          51
#define Mb          11
#define Nc          8
#define Mc          4

void DR_L1 (void)
{
    /*-----------------------------------------
    Constant matrices/vectors
    -----------------------------------------*/
    static tNumber_R c1init[Na][Ma] =
    {
        {  -2.0,    0.0,   -2.0 },
        {   8.0,    9.0,   17.0 },
        {  36.0,   18.0,   54.0 },
        {  -8.0,    0.0,   -8.0 },
        {  21.0,   18.0,   39.0 },
```

```
        { 12.0,  -9.0,   3.0 },
        { -32.0, -13.5, -45.5 }
    };

    static tNumber_R f1[Na+1] =
    {   NIL,
        6.0, 6.0, -48.0, 24.0, 3.0, -6.0, -9.0
    };

    static tNumber_R c3init[Nc][Mc] =
    {
        { 1.0, 1.0,  1.0,  1.0 },
        { 1.0, 2.0,  4.0,  4.0 },
        { 1.0, 3.0,  9.0,  9.0 },
        { 1.0, 4.0, 16.0, 16.0 },
        { 1.0, 5.0, 25.0, 25.0 },
        { 1.0, 6.0, 36.0, 36.0 },
        { 1.0, 7.0, 49.0, 49.0 },
        { 1.0, 8.0, 64.0, 64.0 }
    };

    static tNumber_R f3[Nc+1] =
    {   NIL,
        2.0, 2.5, 2.0, 6.5, 3.5, 4.5, 6.0, 7.0
    };

    /*-----------------------------------------
    Variable matrices/vectors
    -----------------------------------------*/
    tMatrix_R   ct      = alloc_Matrix_R (MM_COLS, NN_ROWS);
    tVector_R   f       = alloc_Vector_R (NN_ROWS);
    tVector_R   r       = alloc_Vector_R (NN_ROWS);
    tVector_R   a       = alloc_Vector_R (MM_COLS);

    tMatrix_R   c1      = init_Matrix_R (&(c1init[0][0]), Na, Ma);
    tMatrix_R   c3      = init_Matrix_R (&(c3init[0][0]), Nc, Mc);

    int         m, n, irank, iter;
    int         i, j, Iexmpl;
    tNumber_R   dx, x1, x2, x3, x4, x5, g, g1, z;

    eLaRc       rc = LaRcOk;

    prn_dr_bnr ("DR_L1, L1 Solution of an Overdetermined System "
                " of Linear Equations");
```

```
for (Iexmpl = 1; Iexmpl <= 3; Iexmpl++)
{
    switch (Iexmpl)
    {
        case 1:
            n = Na;
            m = Ma;
            for (i = 1; i <= n; i++)
            {
                f[i] = f1[i];
                for (j = 1; j <= m; j++) ct[j][i] = c1[i][j];
            }
            break;

        case 2:
            n = Nb;
            m = Mb;
            dx = 0.02;
            g1 = exp (0.5);
            for (i = 1; i <= n; i++)
            {
                x1 = dx * (tNumber_R)(i-1);
                x2 = x1 + x1;
                x3 = x1 + x2;
                x4 = x2 + x2;
                x5 = x2 + x3;
                ct[1][i]  = 1.0;
                ct[2][i]  = sin (x1);
                ct[3][i]  = cos (x1);
                ct[4][i]  = sin (x2);
                ct[5][i]  = cos (x2);
                ct[6][i]  = sin (x3);
                ct[7][i]  = cos (x3);
                ct[8][i]  = sin (x4);
                ct[9][i]  = cos (x4);
                ct[10][i] = sin (x5);
                ct[11][i] = cos (x5);
                g = exp (x1);
                f[i] = g1;
                if (g < g1) f[i] = g;
            }
            break;

        case 3:
```

```
            n = Nc;
            m = Mc;
            for (i = 1; i <= n; i++)
            {
                f[i] = f3[i];
                for (j = 1; j <= m; j++) ct[j][i] = c3[i][j];
            }
            break;

        default:
            break;
    }

    prn_algo_bnr ("L1");

    prn_example_delim();
    PRN ("Example #%d: Size of Matrix \"c\", %d by %d\n",
         Iexmpl, n, m);
    prn_example_delim();
    PRN ("L1 Solution of an Overdetermined System\n");
    prn_example_delim();
    PRN ("r.h.s. Vector \"f\"\n");
    prn_Vector_R (f, n);
    PRN ("Transpose of Coefficient Matrix, \"ct\"\n");
    prn_Matrix_R (ct, m, n);

    rc = LA_L1 (m, n, ct, f, &irank, &iter, r, a, &z);

    if (rc >= LaRcOk)
    {
        PRN ("\n");
        PRN ("Results of the L1 Approximation\n");
        PRN ("L1 solution vector \"a\"\n");
        prn_Vector_R (a, m);
        PRN ("L1 residual vector \"r\"\n");
        prn_Vector_R (r, n);
        PRN ("L1 norm \"z\" = %8.4f\n", z);
        PRN ("Rank of matrix \"c\" = %d, No. of Iterations "
             "= %d\n", irank, iter);
    }

    prn_la_rc (rc);
}

free_Matrix_R (ct, MM_COLS);
```

```
    free_Vector_R (f);
    free_Vector_R (r);
    free_Vector_R (a);

    uninit_Matrix_R (c1);
    uninit_Matrix_R (c3);
}
```

5.12 LA_L1

```
/*-----------------------------------------------------------------
LA_L1
-------------------------------------------------------------------
This program solves an overdetermined system of linear equations
in the L1 (L-One) norm.  It uses a dual simplex method to the dual
linear programming formulation of the problem.  In this program
certain intermediate simplex iterations are skipped.

For the purpose of numerical stability, this program uses a
triangular decomposition to the basis matrix.

The system of linear equations has the form

                    c*a = f

"c" is a given real n by m matrix of rank k, k <= m <= n.
"f" is a given real n vector.

The problem is to calculate the elements of the real m vector
"a" that gives the minimum L1 residual norm z.

        z = |r[1]| + |r[2]| + ... + |r[n]|

where r[i] is the ith residual and is given by

 r[i] = c[i][1]*a[1] + c[i][2]*a[2] + ... + c[i][m]*a[m] - f[i],
                      i = 1, 2, ..., n

Inputs
m       Number of columns of matrix "c" in the system c*a = f.
n       Number of rows of matrix "c" in the system c*a = f.
mmm     An integer = (m* (m+3))/2 - 1.
ct      A real m by n matrix containing the transpose of matrix "c"
        of the system c*a = f.
f       A real n vector containing the r.h.s. of the system c*a = f.

Local Variables
ginv    A real m square matrix that contains the inverse of the
        basis matrix in the linear programming problem.
vb      A real m vector containing the basic solution in the linear
        programming problem.
ic      An integer m vector containing the indices of the columns of
```

```
        "ct" that form the columns of the basis matrix.
ir      An integer m vector containing the row indices of of matrix
        "ct".
ib      A sign n vector.  Its elements have the values +1 or -1.
        ib[j] = +1 indicates that column j of matrix "ct" is in
        the basis or is at its lower bound 0.  ib[j] = -1
        indicates that column j is at its upper bound 2.
p       A real (((m+1)*((m+1)+3))/2-1) vector whose first
        ((irank*(irank+3))/2)-1 elements contain the
        ((irank*(irank+1))/2) elements of an upper triangular matrix
        + extra (irank-1) working locations.  See the comments in
        LA_l1_pslv().
bp      A real m vector whose first "irank" elements are the r.h.s.
        of the triangular equation p*xp = bp.
xp      A real m vector whose first "irank" elements are the solution
        of the triangular equation p*xp = bp or the triangular
        equation p(transpose)*xp = bp.

Outputs
irank   The calculated rank of matrix "c".
iter    Number of iterations or the number of times the simplex
        tableau is changed by a Gauss-Jordon elimination step.
a       A real m vector containing the L1 solution of the system
        c*a = f.
r       A real n vector containing the residual vector
        r = (c*a - f).
z       The minimum L1 norm of the residual vector "r".

Returns one of
        LaRcSolutionUnique
        LaRcSolutionProbNotUnique
        LaRcSolutionDefNotUniqueRD
        LaRcNoFeasibleSolution
        LaRcErrBounds
        LaRcErrNullPtr
        LaRcErrAlloc
--------------------------------------------------------------------*/

#include "LA_Prototypes.h"

eLaRc LA_L1 (int m, int n, tMatrix_R ct, tVector_R f, int *pIrank,
    int *pIter, tVector_R r, tVector_R a, tNumber_R *pZ)
{
    tVector_R   t       = alloc_Vector_R (n);
    tVector_R   uf      = alloc_Vector_R (m);
```

```
tVector_R   bp      = alloc_Vector_R (m);
tVector_R   xp      = alloc_Vector_R (m);
tMatrix_R   ginv    = alloc_Matrix_R (m, m);
tVector_R   vb      = alloc_Vector_R (m);
tVector_I   ib      = alloc_Vector_I (n);
tVector_I   ic      = alloc_Vector_I (n);
tVector_I   ir      = alloc_Vector_I (m);
tVector_R   alfa    = alloc_Vector_R (n);
tVector_R   p       = alloc_Vector_R
                      (((m + 1) * ((m + 1) + 3)) / 2 - 1);

int         iout = 0, jin = 0, icj = 0, iciout = 0, icjin = 0,
            ivo = 0, itest = 0;
int         i = 0, j = 0, kk = 0, ijk = 0;
int         nmm = 0, irank1 = 0, irnkm1 = 0;
tNumber_R   tpeps = 0.0;
tNumber_R   alpha = 0.0, pivot = 0.0, pivotn = 0.0, pivoto = 0.0,
            xb = 0.0, bxb = 0.0;

/* Validation of the data before executing the algorithm */
eLaRc       rc = LaRcSolutionUnique;
VALIDATE_BOUNDS ((0 < m) && (m <= n) && !((n == 1) && (m == 1)));
VALIDATE_PTRS   (ct && f && pIrank && pIter && r && a && pZ);
VALIDATE_ALLOC  (t && uf && bp && xp && ginv && vb && ib && ic &&
                 ir && alfa && p);

/* Initialization */
tpeps = 2.0 + EPS;
*pIrank = m;
nmm = (m * (m + 3))/2;
*pIter = 0;
for (j = 1; j <= n; j++)
{
    ib[j] = 1;
    ic[j] = j;
}
for (j = 1; j <= m; j++)
{
    ir[j] = j;
    vb[j] = 0.0;
    a[j] = 0.0;
    for (i = 1; i <= m; i++)
    {
        ginv[i][j] = 0.0;
    }
```

```
        ginv[j][j] = 1.0;
    }

    iout = 0;
    /* Part 1 of the algorithm */
    LA_l1_part_1 (m, n, ct, ic, ir, ginv, pIrank, pIter);

    /* Part 2 of the algorithm */
    irank1 = *pIrank + 1;
    irnkm1 = *pIrank - 1;

    /* Initial residuals and basic solution */
    LA_l1_part_2 (n, ct, f, ic, ib, uf, vb, pIrank, r);

    /* Initializing the triiangular matrix */
    LA_l1_triang_matrix (m, p, pIrank);

    for (ijk = 1; ijk <= n; ijk++)
    {
        ivo = 0;

        LA_l1_pslv (1, pIrank, p, vb, xp);

        /* Determine the vector that leaves the basis */
        LA_l1_vleav (&ivo, &iout, &xb, xp, pIrank);
        if (ivo == 0)
        {
            LA_l1_pslv (2, pIrank, p, uf, vb);

            /* Calculate the results */
            rc = LA_l1_res (pIrank, m, n, r, ginv, vb, ir, xp, a,
                            pZ);
            GOTO_CLEANUP_RC (rc);
        }

        iciout = ic[iout];
        t[iciout] = 1.0;
        bxb = xb;
        for (i = 1; i <= *pIrank; i++)
        {
            bp[i] = 0.0;
        }
        bp[iout] = 1.0;
        LA_l1_pslv (2, pIrank, p, bp, xp);
```

```
/* Determine the alfa ratios */
LA_l1_alfa (iout, n, ct, ic, ib, xp, t, alfa, pIrank, r);

pivoto = 1.0;
itest = 0;

for (kk = 1; kk <= n; kk++)
{
    /* Determine the vector that enters the basis */
    LA_l1_vent (ivo, &jin, &itest, n, ic, alfa, pIrank);

    /* No feasible solution has been found */
    if (itest != 1)
    {
        GOTO_CLEANUP_RC (LaRcNoFeasibleSolution);
    }

    icjin = ic[jin];
    pivot = t[icjin];
    alpha = alfa[icjin];
    pivotn = pivot/pivoto;

    /* Skipping simplex iteration */
    LA_l1_skip_iters (iciout, icjin, xb, &bxb, pivotn, ct,
                      ib, t, vb, pIrank);

    xb = bxb/pivot;
    if (xb < -EPS || xb > tpeps) itest = 0;
    alfa[icjin] = 0.0;
    if (itest == 1) break;
    pivoto = pivot;
    iciout = icjin;
}
r[icjin] = 0.0;
*pIter = *pIter + 1;

/* Update vector p */
LA_l1_update_p (iout, jin, ct, ic, p, pIrank);

if (iout != *pIrank)
{
    for (i = iout; i <= irnkm1; i++)
    {
        /* Update (ginv, vb, ct) */
        LA_l1_update_ginv (i, n, ct, ir, p, ginv, vb,
```

```
                                        pIrank);
            }
        }

        for (j = 1; j <= *pIrank; j++)
        {
            icj = ic[j];
            uf[j] = f[icj];
        }
        LA_l1_calcul_r (alpha, n, ct, f, ic, ib, uf, xp, t, p,
                        pIrank, r);
    }

CLEANUP:

    free_Vector_R (t);
    free_Vector_R (uf);
    free_Vector_R (bp);
    free_Vector_R (xp);
    free_Matrix_R (ginv, m);
    free_Vector_R (vb);
    free_Vector_I (ib);
    free_Vector_I (ic);
    free_Vector_I (ir);
    free_Vector_R (alfa);
    free_Vector_R (p);

    return rc;
}

/*-----------------------------------------------------------------------
Square non-singular system
-------------------------------------------------------------------------
This function solves the square real non-singular system of linear
equations
                          p*xp = vb

or the square real non-singular system of linear equations

                      p(transpose)*xp = vb

"p"  is an upper triangular matrix.
"vb" is the right hand side vector.
"xp" is the solution vector.
```

```
Inputs
id      An integer specifying the action to be performed.
        If id = 1 the equation "p*xp=vb" is solved.
        if id = any integer other than 1, the equation
        "p(transpose)*xp=vb" is solved.
k       The number of equations of the given system.
p       An (((m+1)*((m+1)+3))/2-1) vector whose first (k+1) elements
        contain the first k elements of row 1 of the upper triangular
        matrix plus an extra location to the right.  Its next k
        elements contain the (k-1) elements of row 2 of the upper
        triangular matrix plus an extra location to the right,
        ..., etc.
vb      An m vector whose first k elements contain the r.h.s. vector
        of the given system.

Outputs
xp      A real m vector whose first k elements contain the solution
        to the given system.
-----------------------------------------------------------------*/
void LA_l1_pslv (int id, int *pIrank, tVector_R p, tVector_R vb,
    tVector_R xp)
{
    int         i, l, ll, j, jj, kk, kd, km1, kkd, kkm1;
    tNumber_R   s;

    /* Solution of the upper triangular system */
    if (id == 1)
    {
        l = (*pIrank-1) + (*pIrank * (*pIrank+1))/2;
        xp[*pIrank] = vb[*pIrank]/p[l];
        if (*pIrank != 1)
        {
            kd = 3;
            km1 = *pIrank - 1;
            for (i = 1; i <= km1; i++)
            {
                j = *pIrank - i;
                l = l - kd;
                kd = kd + 1;
                s = vb[j];
                ll = l;
                jj = j;
                for (kk = 1; kk <= i; kk++)
                {
                    ll = ll + 1;
```

```
                jj = jj + 1;
                s = s - p[ll] * (xp[jj]);
            }
            xp[j] = s/p[l];
        }
    }
}

/* Solution of the lower triangular system */
if (id != 1)
{
    xp[1] = vb[1]/p[1];
    if (*pIrank != 1)
    {
        l = 1;
        kd = *pIrank + 1;
        for (i = 2; i <= *pIrank; i++)
        {
            l = l + kd;
            kd = kd - 1;
            s = vb[i];
            kk = i;
            kkm1 = i - 1;
            kkd = *pIrank;
            for (j = 1; j <= kkm1; j++)
            {
                s = s - p[kk] * xp[j];
                kk = kk + kkd;
                kkd = kkd - 1;
            }
            xp[i] = s/p[l];
        }
    }
}
}

/*-------------------------------------------------------------------
A Gauss-Jordan elimination step in LA_L1()
-------------------------------------------------------------------*/
void LA_l1_gauss_jordn (int iout, int jin, int lj, int *pIrank,
    int n, tMatrix_R ct, tMatrix_R ginv, tVector_I ic)
{
    int         i, j;
    tNumber_R   pivot, d;
```

```
    pivot = ct[iout][jin];
    for (j = 1; j <= n; j++)
    {
        ct[iout][j] = ct[iout][j]/pivot;
    }
    for (j = 1; j <= iout; j++)
    {
        ginv[iout][j] = ginv[iout][j]/pivot;
    }
    for (i = 1; i <= *pIrank; i++)
    {
        if (i != iout)
        {
            d = ct[i][jin];
            for (j = 1; j <= n; j++)
            {
                ct[i][j] = ct[i][j] - d * ct[iout][j];
            }
            for (j = 1; j <= iout; j++)
            {
                ginv[i][j] = ginv[i][j] - d * ginv[iout][j];
            }
        }
    }

    /* Swap two elements of vector "ic" */
    swap_elems_Vector_I (ic, iout, lj);
}

/*-------------------------------------------------------------------
Calculate the results of LA_L1()
-------------------------------------------------------------------*/
eLaRc LA_l1_res (int *pIrank, int m, int n, tVector_R r,
    tMatrix_R ginv, tVector_R vb, tVector_I ir, tVector_R xp,
    tVector_R a, tNumber_R *pZ)
{
    int         i, j, k, ij;
    tNumber_R   d, e, s;

    for (i = 1; i <= *pIrank; i++)
    {
        s = 0.0;
        for (k = 1; k <= *pIrank; k++)
        {
            s = s + vb[k] * ginv[k][i];
```

```
        }
        ij = ir[i];
        a[ij] = s;
    }
    s = 0.0;
    for (j = 1; j <= n; j++)
    {
        s = s + fabs (r[j]);
    }
    *pZ = s;

    if (*pIrank < m)
        return LaRcSolutionDefNotUniqueRD;

    e = 2.0 - EPS;
    for (i = 1; i <= m; i++)
    {
        d = xp[i];
        if (d < -EPS || d > e)
            return LaRcSolutionProbNotUnique;
    }

    return LaRcSolutionUnique;
}

/*-------------------------------------------------------------------
Part 1 of LA_L1()
-------------------------------------------------------------------*/
void LA_l1_part_1 (int m, int n, tMatrix_R ct, tVector_I ic,
    tVector_I ir, tMatrix_R ginv, int *pIrank, int *pIter)
{
    int         i, j, k, lj = 0, li = 0;
    int         iout, jin = 0, icj;
    tNumber_R   d, g, piv;

    for (iout = 1; iout <= m; iout++)
    {
        if (iout <= *pIrank)
        {
            piv = 0.0;
            for (j = iout; j <= n; j++)
            {
                icj = ic[j];
                for (i = iout; i <= *pIrank; i++)
                {
```

```
          d = ct[i][icj];
          if (d < 0.0) d = -d;
          if (d > piv)
          {
              li = i;
              jin = icj;
              lj = j;
              piv = d;
          }
      }
  }

  /* Detection of rank deficiency of matrix "c"*/
  if (piv < EPS)
  {
      *pIrank = iout - 1;
  }

  if (piv > EPS)
  {
      if (li != iout)
      {
          for (j = 1; j <= n; j++)
          {
              g = ct[iout][j];
              ct[iout][j] = ct[li][j];
              ct[li][j] = g;
          }

          /* Swap two elements of vector "ir" */
          swap_elems_Vector_I (ir, iout, li);

          /* Swap two rows of matrix "ginv" */
          if (iout != 1)
          {
              k = iout - 1;
              for (j = 1; j <= k; j++)
              {
                  d = ginv[li][j];
                  ginv[li][j] = ginv[iout][j];
                  ginv[iout][j] = d;
              }
          }
      }
```

```
                LA_l1_gauss_jordn (iout, jin, lj, pIrank, n, ct,
                                   ginv, ic);
                *pIter = *pIter + 1;
            }
        }
    }
}

/*-----------------------------------------------------------------
Part 2 of LA_L1()
-----------------------------------------------------------------*/
void LA_l1_part_2 (int n, tMatrix_R ct, tVector_R f, tVector_I ic,
    tVector_I ib, tVector_R uf, tVector_R vb, int *pIrank,
    tVector_R r)
{
    int         i, j, icj, irank1;
    tNumber_R   s, sa;

    /* Calculating initial residuals r[] */
    irank1 = *pIrank + 1;
    for (j = 1; j <= *pIrank; j++)
    {
        icj = ic[j];
        r[icj] = 0.0;
        uf[j] = f[icj];
    }
    for (j = irank1; j <= n; j++)
    {
        icj = ic[j];
        s = -f[icj];
        for (i = 1; i <= *pIrank; i++)
        {
            s = s + uf[i] * ct[i][icj];
        }
        r[icj] = s;
        if (s < 0.0) ib[icj] = -1;
    }

    /* Calculating the initial basic solution vb[] */
    for (i = 1; i <= *pIrank; i++)
    {
        s = 0.0;
        for (j = 1; j <= n; j++)
        {
            sa = ct[i][j];
```

```
            if (ib[j] == -1) sa = -sa;
            s = s + sa;
        }
        vb[i] = s;
    }
}

/*-----------------------------------------------------------------------
Initializing the triangular matrix in LA_L1()
-----------------------------------------------------------------------*/
void LA_l1_triang_matrix (int m, tVector_R p, int *pIrank)
{
    int         i, k, kd, nmm;

    nmm = (m * (m + 3))/2;
    for (i = 1; i <= nmm; i++)
    {
        p[i] = 0.0;
    }
    k = 1;
    kd = *pIrank + 1;
    for (i = 1; i <= *pIrank; i++)
    {
        p[k] = 1.0;
        k = k + kd;
        kd = kd - 1;
    }
}

/*-----------------------------------------------------------------------
Determine the vector that leaves the basis in LA_L1()
-----------------------------------------------------------------------*/
void LA_l1_vleav (int *pIvo, int *pIout, tNumber_R *pXb,
    tVector_R xp, int *pIrank)
{
    int         i;
    tNumber_R   d, e, g, tpeps;

    tpeps = 2.0 + EPS;
    g = 1.0;
    for (i = 1; i <= *pIrank; i++)
    {
        e = xp[i];
        if (e > tpeps || e < -EPS)
        {
```

```
            if (e > tpeps)
            {
                d = 2.0 - e;
                if (d < g)
                {
                    g = d;
                    *pIvo = 1;
                    *pIout = i;
                    *pXb = e;
                }
            }
            if (e < -EPS)
            {
                d = e;
                if (d < g)
                {
                    g = d;
                    *pIvo = -1;
                    *pIout = i;
                    *pXb = e;
                }
            }
        }
    }
}

/*----------------------------------------------------------------------
Determine the alfa ratios for LA_L1()
----------------------------------------------------------------------*/
void LA_l1_alfa (int iout, int n, tMatrix_R ct, tVector_I ic,
    tVector_I ib, tVector_R xp, tVector_R t, tVector_R alfa,
    int *pIrank, tVector_R r)
{
    int         i, j, icj, irank1;
    tNumber_R   d, e, s;
    tNumber_R   alfmx;

    irank1 = *pIrank + 1;
    alfmx = 0.0;
    for (j = irank1; j <= n; j++)
    {
        icj = ic[j];
        alfa[icj] = 0.0;
        s = 0.0;
        for (i = iout; i <= *pIrank; i++)
```

```
        {
            s = s + xp[i] * (ct[i][icj]);
        }
        e = s;
        t[icj] = e;
        if (fabs (e) > EPS)
        {
            d = r[icj];
            if (fabs (d) < PREC) d = PREC * (ib[icj]);
            alfa[icj] = d/e;
        }
    }

}

/*-----------------------------------------------------------------
Determine the vector that enters the basis for LA_L1()
-----------------------------------------------------------------*/
void LA_l1_vent (int ivo, int *pJin, int *pItest, int n,
    tVector_I ic, tVector_R alfa, int *pIrank)
{
    int         j, icj, irank1;
    tNumber_R   d, e, alfmn, alfmx;

    irank1 = *pIrank + 1;
    alfmx = 1.0/EPS;
    alfmn = -alfmx;
    for (j = irank1; j <= n; j++)
    {
        icj = ic[j];
        e = alfa[icj];
        d = e * ivo;
        if (d > 0.0)
        {
            if (ivo == -1)
            {
                if (e > alfmn)
                {
                    alfmn = e;
                    *pJin = j;
                    *pItest = 1;
                }
            }
            if (ivo == 1)
            {
```

```
                if (e < alfmx)
                {
                    alfmx = e;
                    *pJin = j;
                    *pItest = 1;
                }
            }
        }
    }
}

/*-------------------------------------------------------------------
Skipping simplex iterations in LA_L1()
-------------------------------------------------------------------*/
void LA_l1_skip_iters (int iciout, int icjin, tNumber_R xb,
    tNumber_R *pBxb, tNumber_R pivotn, tMatrix_R ct, tVector_I ib,
    tVector_R t, tVector_R vb, int *pIrank)
{
    int         i;
    tNumber_R   tpeps;

    tpeps = 2.0 + EPS;

    if (xb < -EPS)
    {
        if (pivotn > 0.0)
        {
            for (i = 1; i <= *pIrank; i++)
            {
                vb[i] = vb[i] + ct[i][icjin] + ct[i][icjin];
            }
            *pBxb = *pBxb + t[icjin] + t[icjin];
            ib[icjin] = 1;
        }
    }

    if (xb > tpeps)
    {
        for (i = 1; i <= *pIrank; i++)
        {
            vb[i] = vb[i] - ct[i][iciout] - ct[i][iciout];
        }
        *pBxb = *pBxb - t[iciout] - t[iciout];
        ib[iciout] = -1;
```

```
        if (pivotn < 0.0)
        {
            for (i = 1; i <= *pIrank; i++)
            {
                vb[i] = vb[i] + ct[i][icjin] + ct[i][icjin];
            }
            *pBxb = *pBxb + t[icjin] + t[icjin];
            ib[icjin] = 1;
        }
    }
}

/*-------------------------------------------------------------------
Update vector p for LA_L1()
-------------------------------------------------------------------*/
void LA_l1_update_p (int iout, int jin, tMatrix_R ct, tVector_I ic,
    tVector_R p, int *pIrank)
{
    int         i, j, k, k1, kd;
    int         icjin, irnkm1;

    icjin = ic[jin];
    irnkm1 = *pIrank - 1;

    if (iout != *pIrank)
    {
        for (j = iout; j <= irnkm1; j++)
        {
            k = j;
            k1 = k + 1;
            kd = *pIrank;

            /* Swap two elements of vector "ic" */
            swap_elems_Vector_I (ic, k, k1);

            for (i = 1; i <= k1; i++)
            {
                p[k] = p[k + 1];
                k = k + kd;
                kd = kd - 1;
            }
        }

        /* Swap elements "irank" and "jin" of vector "ic" */
        swap_elems_Vector_I (ic, jin, *pIrank);
```

```
        k = *pIrank;
        kd = *pIrank;
        for (i = 1; i <= *pIrank; i++)
        {
            p[k] = ct[i][icjin];
            k = k + kd;
            kd = kd - 1;
        }
    }
}

/*--------------------------------------------------------------------
Update matrix ginv in LA_L1()
--------------------------------------------------------------------*/
void LA_l1_update_ginv (int i, int n, tMatrix_R ct, tVector_I ir,
    tVector_R p, tMatrix_R ginv, tVector_R vb, int *pIrank)
{
    int          j, k, l;
    int          i1, ii, kd, kk, kl, irank1;

    tNumber_R    d, e, g;

    irank1 = *pIrank + 1;
    ii = i;
    i1 = i + 1;
    k = 0;
    kd = irank1;
    for (j = 1; j <= ii; j++)
    {
        k = k + kd;
        kd = kd - 1;
    }
    kk = k;
    kl = k - kd;
    l = kl;
    g = p[k];
    d = p[l];
    if (g < 0.0) g = - g;
    if (d < 0.0) d = - d;
    if (g > d)
    {
        for (j = ii; j <= *pIrank; j++)
        {
            /* Swap two elements of vector "p" */
```

```
            swap_elems_Vector_R (p, k, l);

            k = k + 1;
            l = l + 1;
        }
        /* Swap two elements of vector "ir" */
        swap_elems_Vector_I (ir, i, i1);

        /* Swap two rows of matrix "ct" */
        swap_rows_Matrix_R (ct, i, i1);

        /* Swap two rows of matrix "ginv" */
        swap_rows_Matrix_R (ginv, i, i1);

        /* Swap two columns of matrix "ginv" */
        for (j = 1; j <= *pIrank; j++)
        {
            e = ginv[j][i];
            ginv[j][i] = ginv[j][i1];
            ginv[j][i1] = e;
        }

        /* Swap two elements in real vector "vb" */
        swap_elems_Vector_R (vb, i, i1);
    }
    e = p[kk]/p[kl];
    k = kk;
    l = kl;
    for (j = ii; j <= *pIrank; j++)
    {
        p[k] = p[k] - e * p[l];
        k = k + 1;
        l = l + 1;
    }
    for (j = 1; j <= n; j++)
    {
        ct[i1][j] = ct[i1][j] - e * (ct[i][j]);
    }
    for (j = 1; j <= *pIrank; j++)
    {
        ginv[i1][j] = ginv[i1][j] - e * ginv[i][j];
    }
    vb[i1] = vb[i1] - e * vb[i];
}
```

```
/*-----------------------------------------------------------------
Calculate the residual vector "r" for LA_L1()
-----------------------------------------------------------------*/
void LA_l1_calcul_r (tNumber_R alpha, int n, tMatrix_R ct,
    tVector_R f, tVector_I ic, tVector_I ib, tVector_R uf,
    tVector_R xp, tVector_R t, tVector_R p, int *pIrank,
    tVector_R r)
{
    int         i, j, icj, irank1;
    tNumber_R   d, s;

    irank1 = *pIrank + 1;
    if (fabs (alpha) <= 1.0)
    {
        for (j = irank1; j <= n; j++)
        {
            icj = ic[j];
            r[icj] = r[icj] - alpha * t[icj];
        }
    }
    if (fabs (alpha) > 1.0)
    {
        LA_l1_pslv (2, pIrank, p, uf, xp);
        for (j = irank1; j <= n; j++)
        {
            icj = ic[j];
            s = -f[icj];
            for (i = 1; i <= *pIrank; i++)
            {
                s = s + xp[i] * (ct[i][icj]);
            }
            r[icj] = s;
        }
    }
    for (j = irank1; j <= n; j++)
    {
        icj = ic[j];
        d = r[icj] * ib[icj];
        if (d < 0.0) r[icj] = 0.0;
    }
}
```

5.13 DR_Lone

```
/*---------------------------------------------------------------------
DR_Lone
-----------------------------------------------------------------------
This program is a driver for the function LA_Lone(), which solves an
overdetermined system of linear equations in the L1 (L-One) norm,
using a dual simplex method.

The overdetermined system has the form

                        c*a = f

"c" is a given real n by m matrix of rank k, k <= m <= n.
"f" is a given real n vector.
"a" is the solution m vector.

This program contains the 3 examples whose results appear in the
text.
---------------------------------------------------------------------*/

#include "DR_Defs.h"
#include "LA_Prototypes.h"

#define Na          7
#define Ma          3
#define Nb          51
#define Mb          11
#define Nc          8
#define Mc          4

void DR_Lone (void)
{
    /*-----------------------------------------
      Constant matrices/vectors
    -----------------------------------------*/
    static tNumber_R c1init[Na][Ma] =
    {
        {  -2.0,   0.0,  -2.0 },
        {   8.0,   9.0,  17.0 },
        {  36.0,  18.0,  54.0 },
        {  -8.0,   0.0,  -8.0 },
        {  21.0,  18.0,  39.0 },
        {  12.0,  -9.0,   3.0 },
```

```
    { -32.0, -13.5, -45.5 }
};

static tNumber_R f1[Na+1] =
{   NIL,
    6.0, 6.0, -48.0, 24.0, 3.0, -6.0, -9.0
};

static tNumber_R c3init[Nc][Mc] =
{
    { 1.0, 1.0,  1.0,  1.0 },
    { 1.0, 2.0,  4.0,  4.0 },
    { 1.0, 3.0,  9.0,  9.0 },
    { 1.0, 4.0, 16.0, 16.0 },
    { 1.0, 5.0, 25.0, 25.0 },
    { 1.0, 6.0, 36.0, 36.0 },
    { 1.0, 7.0, 49.0, 49.0 },
    { 1.0, 8.0, 64.0, 64.0 }
};

static tNumber_R f3[Nc+1] =
{   NIL,
    2.0, 2.5, 2.0, 6.5, 3.5, 4.5, 6.0, 7.0
};

/*-----------------------------------------
  Variable matrices/vectors
-----------------------------------------*/
tMatrix_R   ct      = alloc_Matrix_R (MM_COLS, NN_ROWS);
tVector_R   f       = alloc_Vector_R (NN_ROWS);
tVector_R   r       = alloc_Vector_R (NN_ROWS);
tVector_R   a       = alloc_Vector_R (MM_COLS);

tMatrix_R   c1      = init_Matrix_R (&(c1init[0][0]), Na, Ma);
tMatrix_R   c3      = init_Matrix_R (&(c3init[0][0]), Nc, Mc);

int         irank, iter;
int         i, j, m, n, Iexmpl;

tNumber_R   dx, x1, x2, x3, x4, x5, g, g1, z;

eLaRc       rc = LaRcOk;

prn_dr_bnr ("DR_Lone, L-One Solution of an Overdetermined "
            "System of Linear Equations");
```

```
z = 0.0;
for (Iexmpl = 1; Iexmpl <= 3; Iexmpl++)
{
    switch (Iexmpl)
    {
    case 1:
        n = Na;
        m = Ma;
        for (i = 1; i <= n; i++)
        {
            f[i] = f1[i];
            for (j = 1; j <= m; j++)
                ct[j][i] = c1[i][j];
        }
        break;

    case 2:
        n = Nb;
        m = Mb;
        dx = 0.02;
        g1 = exp (0.5);
        for (i = 1; i <= n; i++)
        {
            x1 = dx * (tNumber_R)(i-1);
            x2 = x1 + x1;
            x3 = x1 + x2;
            x4 = x2 + x2;
            x5 = x2 + x3;
            ct[1][i]  = 1.0;
            ct[2][i]  = sin (x1);
            ct[3][i]  = cos (x1);
            ct[4][i]  = sin (x2);
            ct[5][i]  = cos (x2);
            ct[6][i]  = sin (x3);
            ct[7][i]  = cos (x3);
            ct[8][i]  = sin (x4);
            ct[9][i]  = cos (x4);
            ct[10][i] = sin (x5);
            ct[11][i] = cos (x5);
            g = exp (x1);
            f[i] = g1;
            if (g < g1) f[i] = g;
        }
        break;
```

```
    case 3:
        n = Nc;
        m = Mc;
        for (i = 1; i <= n; i++)
        {
            f[i] = f3[i];
            for (j = 1; j <= m; j++)
            {
                ct[j][i] = c3[i][j];
            }
        }
        break;

    default:
        break;
    }

    prn_algo_bnr ("Lone");

    prn_example_delim();
    PRN ("Example #%d: Size of matrix \"c\", %d by %d\n",
         Iexmpl, n, m);
    prn_example_delim();
    PRN ("L-One Solution of an Overdetermined System"
         " of Linear Equations\n");
    prn_example_delim();
    PRN ("r.h.s. Vector \"f\"\n");
    prn_Vector_R (f, n);
    PRN ("Transpose of Coefficient Matrix, \"ct\"\n");
    prn_Matrix_R (ct, m, n);

    rc = LA_Lone (m, n, ct, f, &irank, &iter, r, a, &z);

    if (rc >= LaRcOk)
    {
        PRN ("\n");
        PRN ("Results of the L-One Approximation\n");
        PRN ("L-One solution vector, \"a\"\n");
        prn_Vector_R (a, m);
        PRN ("L-One residual vector \"r\"\n");
        prn_Vector_R (r, n);
        PRN ("L-One norm \"z\" = %8.4f\n", z);
        PRN ("Rank of matrix \"c\" = %d, No. of Iterations ="
             " %d\n", irank, iter);
```

```
        }

        prn_la_rc (rc);
    }

    free_Matrix_R (ct, MM_COLS);
    free_Vector_R (f);
    free_Vector_R (r);
    free_Vector_R (a);

    uninit_Matrix_R (c1);
    uninit_Matrix_R (c3);
}
```

5.14 LA_Lone

```
/*--------------------------------------------------------------------
LA_Lone
----------------------------------------------------------------------
This program solves an overdetermined system of linear equations
in the L1 (L-One) norm.  It uses a dual simplex method to the dual
linear programming formulation of the problem.  In this program
certain intermediate simplex iterations are skipped.

The system of linear equations has the form

                      c*a = f

"c" is a given real n by m matrix of rank k, k <= m <= n.
"f" is a given real n vector.

The problem is to calculate the elements of the real m vector
"a" that gives the minimum L1 residual norm z.

          .    z = |r[1]| + |r[2]| + ... + |r[n]|

where r[i] is the ith residual and is given by

 r[i] = c[i][1]*a[1] + c[i][2]*a[2] + ... + c[i][m]*a[m] - f[i],
                         i = 1, 2, ..., n

Inputs
m       Number of columns of matrix "c" in the system c*a = f.
n       Number of rows of matrix "c" in the system c*a = f.
ct      A real m by n matrix containing the transpose of matrix "c"
        of the system c*a = f.
f       A real n vector containing the r.h.s. of the system c*a = f.

Local Variables
binv    A real m square matrix containing the inverse of the basis
        matrix in the linear programming problem.
bv      A real m vector containing the basic solution in the linear
        programming problem.
th      A real n vector containing the ratios

                         th[j] = r[j]/ct[iout][j]

        "iout" corresponds to the basic vector that leaves the
```

```
        basis.
icbas   An integer m vector containing the indices of the columns
        of "ct" that form the columns of the basis matrix.
irbas   An integer m vector containing the indices of the rows of
        "ct".
ibound  An n sign vector.  Its elements have the values +1 or -1.
        ibound[j] = +1 indicates that column j of matrix "ct" is in
        the basis or is at its lower bound 0.  ibound[j] = -1
        indicates that column j is at its upper bound 2.

Outputs
irank   The calculated rank of matrix "c".
iter    Number of iterations, or the number of times the simplex
        tableau is changed by a Gauss-Jordon elimination step.
a       A real m vector containing the L1 solution of the system
        c*a = f.
r       A real n vector containing the residual vector
        r = (c*a - f).
z       The minimum L1 norm of the residual vector "r".

Returns one of
        LaRcSolutionUnique
        LaRcSolutionProbNotUnique
        LaRcSolutionDefNotUniqueRD
        LaRcNoFeasibleSolution
        LaRcErrBounds
        LaRcErrNullPtr
        LaRcErrAlloc
-----------------------------------------------------------------------*/

#include "LA_Prototypes.h"

eLaRc LA_Lone (int m, int n, tMatrix_R ct, tVector_R f, int *pIrank,
    int *pIter, tVector_R r, tVector_R a, tNumber_R *pZ)
{
    tVector_I   icbas   = alloc_Vector_I (m);
    tVector_I   irbas   = alloc_Vector_I (m);
    tVector_R   th      = alloc_Vector_R (n);
    tMatrix_R   binv    = alloc_Matrix_R (m, m);
    tVector_R   bv      = alloc_Vector_R (m);
    tVector_I   ibound  = alloc_Vector_I (n);

    int         iout = 0, jin = 0, jout = 0, ivo = 0;
    int         i = 0, j = 0, ij = 0, kk = 0, itest = 0, ipart = 0;
    tNumber_R   tpeps = 0.0, xb = 0.0;
```

```
tNumber_R   pivot = 0.0, pivotn = 0.0, pivoto = 0.0;

/* Validation of the data before executing the algorithm */
eLaRc        rc = LaRcSolutionUnique;
VALIDATE_BOUNDS ((0 < m) && (m <= n) && !((n == 1) && (m == 1)));
VALIDATE_PTRS   (ct && f && pIrank && pIter && r && a && pZ);
VALIDATE_ALLOC  (icbas && irbas && th && binv && bv && ibound);

/* Initialization */
xb = 0.0;
tpeps = 2.0 + EPS;
*pIrank = m;
*pIter = 0;
*pZ = 0.0;
ipart = 1;
for (j = 1; j <= n; j++)
{
    r[j] = 0.0;
    ibound[j] = 1;
}

for (j = 1; j <= m; j++)
{
    a[j] = 0.0;
    bv[j] = 0.0;
    irbas[j] = j;
    icbas[j] = 0;
    for (i = 1; i <= m; i++) binv[i][j] = 0.0;
    binv[j][j] = 1.0;
}

/* Part 1 of the algorithm */
LA_lone_part_1 (ipart, n, ct, icbas, irbas, binv, bv, ibound,
                pIrank, pIter, r);

/* Part 2 of the algorithm */
/* Calculating the initial residuals (marginal costs)
    and the initial basic solution */
ipart = 2;
LA_lone_part_2 (n, ct, f, icbas, bv, ibound, pIrank, r);

for (kk = 1; kk <= n; kk++)
{
    ivo = 0;
```

```
    /* Determine the vector that leaves the basis */
    LA_lone_vleav (&ivo, &iout, pIrank, &xb, bv);

    /* Calculate the results */
    if (ivo == 0)
    {
        rc = LA_lone_res (m, n, f, icbas, irbas, binv, bv,
                          pIrank, r, a, pZ);
        GOTO_CLEANUP_RC (rc);
    }
    jout = icbas[iout];

    /* Calculation of the possible parameters th[j] */
    LA_lone_th (iout, n, ct, icbas, ibound, th, pIrank, r);

    pivoto = 1.0;
    itest = 0;

    for (ij = 1; ij <= n; ij++)
    {
        /* Determine the vector that enters the basis */
        LA_lone_vent (ivo, &itest, &jin, n, th);

        /* Solution is not feasible */
        if (itest != 1)
        {
            GOTO_CLEANUP_RC (LaRcNoFeasibleSolution);
        }
        pivot = ct[iout][jin];
        pivotn = pivot/pivoto;

        if (xb < -EPS)
        {
            if (pivotn > 0.0)
            {
                for (i = 1; i <= *pIrank; i++)
                {
                    bv[i] = bv[i] + ct[i][jin] + ct[i][jin];
                }
            }
            ibound[jin] = 1;
        }
        if (xb > tpeps)
        {
            for (i = 1; i <= *pIrank; i++)
```

```
                {
                    bv[i] = bv[i] - ct[i][jout] - ct[i][jout];
                }
                ibound[jout] = -1;
                if (pivotn < 0.0)
                {
                    for (i = 1; i <= *pIrank; i++)
                    {
                        bv[i] = bv[i] + ct[i][jin] + ct[i][jin];
                    }
                    ibound[jin] = 1;
                }
            }
            xb = bv[iout]/pivot;
            if (xb < -EPS || xb > tpeps) itest = 0;
            th[jin] = 0.0;
            icbas[iout] = jin;
            if (itest == 1) break;
            pivoto = pivot;
            jout = jin;
        }
        /* A Gauss-Jordan elimination step */
        LA_lone_gauss_jordn (ipart, iout, jin, n, ct, icbas, binv,
                             bv, ibound, pIrank, r);
        *pIter = *pIter + 1;
    }

CLEANUP:

    free_Vector_I (icbas);
    free_Vector_I (irbas);
    free_Vector_R (th);
    free_Matrix_R (binv, m);
    free_Vector_R (bv);
    free_Vector_I (ibound);

    return rc;
}

/*-------------------------------------------------------------------
Part 1 of LA_Lone()
-------------------------------------------------------------------*/
void LA_lone_part_1 (int ipart, int n, tMatrix_R ct, tVector_I icbas,
    tVector_I irbas, tMatrix_R binv, tVector_R bv, tVector_I ibound,
    int *pIrank, int *pIter, tVector_R r)
```

```
{
    int             i, j, li = 0, lj;
    int             iout, jin = 0;

    tNumber_R       d, piv;

    for (iout = 1; iout <= *pIrank; iout++)
    {
        if (iout <= *pIrank)
        {
            piv = 0.0;
            for (j = 1; j <= n; j++)
            {
                for (i = iout; i <= *pIrank; i++)
                {
                    d = ct[i][j];
                    if (d < 0.0) d = -d;
                    if (d > piv)
                    {
                        li = i;
                        jin = j;
                        piv = d;
                    }
                }
            }

            /* Detection of rank deficiency of matrix "ct" */
            if (piv < EPS)
            {
                *pIrank = iout - 1;
                ipart = 2;
            }
            if (ipart == 2) break;
            if (li != iout)
            {
                /* Swap two elements of vector "irbas" */
                swap_elems_Vector_I (irbas, li, iout);

                /* Swap two rows of matrix "ct" */
                swap_rows_Matrix_R (ct, li, iout);

                if (iout != 1)
                {
                    lj = iout - 1;
                    for (j = 1; j <= lj; j++)
```

```
                    {
                        d = binv[li][j];
                        binv[li][j] = binv[iout][j];
                        binv[iout][j] = d;
                    }
                }
            }
            if (ipart == 2) break;
            /* A Gauss-Jordan elimination step */
            LA_lone_gauss_jordn (ipart, iout, jin, n, ct, icbas,
                          binv, bv, ibound, pIrank, r);
            *pIter = *pIter + 1;
        }
    }
}

/*-----------------------------------------------------------------------
Part 2 of LA_Lone()
-----------------------------------------------------------------------*/
void LA_lone_part_2 (int n, tMatrix_R ct, tVector_R f,
    tVector_I icbas, tVector_R bv, tVector_I ibound, int *pIrank,
    tVector_R r)
{
    int             i, j, k, ic;

    tNumber_R       s, sa;

    for (j = 1; j <= n; j++)
    {
        ic = 0;
        for (i = 1; i <= *pIrank; i++) if (j == icbas[i]) ic = 1;
        if (ic == 0)
        {
            s = - f[j];
            for (i = 1; i <= *pIrank; i++)
            {
                k = icbas[i];
                s = s + f[k]*ct[i][j];
            }
            r[j] = s;
            if (s  <= 0.0) ibound[j] = - 1;
        }
    }
    for (i = 1; i <= *pIrank; i++)
    {
```

```
        s = 0.0;
        for (j = 1; j <= n; j++)
        {
            sa = ct[i][j];
            if (ibound[j] == -1) sa = - sa;
            s = s + sa;
        }
        bv[i] = s;
    }
}

/*-----------------------------------------------------------------
Calculate the "th" ratios in LA_lone()
-----------------------------------------------------------------*/
void LA_lone_th (int iout, int n, tMatrix_R ct, tVector_I icbas,
    tVector_I ibound, tVector_R th, int *pIrank, tVector_R r)
{
    int             i, j, ic;
    tNumber_R       d, e, gg, thmax;

    thmax = 0.0;

    /* Calculation of the possible parameters th[j] */
    for (j = 1; j <= n; j++)
    {
        th[j] = 0.0;
        ic = 0;
        for (i = 1; i <= *pIrank; i++)
        {
            if (j == icbas[i]) ic = 1;
        }
        if (ic == 0)
        {
            e = ct[iout][j];
            if (fabs (e) > EPS)
            {
                d = r[j];
                if (fabs (d) < PREC) d = PREC*ibound[j];
                th[j] = d/e;
                gg = th[j];
                if (gg <0.0) gg = - gg;
                if (gg > thmax) thmax = gg;
            }
        }
    }
```

```
}

/*-------------------------------------------------------------------
Determine the vector that enters the basis in LA_Lone()
--------------------------------------------------------------------*/
void LA_lone_vent (int ivo, int *pItest, int *pJin, int n,
    tVector_R th)
{
    int             j, ij;

    tNumber_R       d, e, thmax, thmin;

    for (ij = 1; ij <= n; ij++)
    {
        thmax = 1.0/ (EPS*EPS);
        thmin = -thmax;
        for (j = 1; j <= n; j++)
        {
            e = th[j];
            d = e * ivo;
            if (d > 0.0)
            {
                if (ivo == -1)
                {
                    if (e > thmin)
                    {
                        thmin = e;
                        *pJin = j;
                        *pItest = 1;
                    }
                }
                if (ivo == 1)
                {
                    if (e < thmax)
                    {
                        thmax = e;
                        *pJin = j;
                        *pItest = 1;
                    }
                }
            }
        }
    }
}
```

```
/*-------------------------------------------------------------------
Determine the vector that leaves the basis in LA_Lone()
-------------------------------------------------------------------*/
void LA_lone_vleav (int *pIvo, int *pIout, int *pIrank,
    tNumber_R *pXb, tVector_R bv)
{
    int             i;
    tNumber_R       d, e, g, tpeps;

    tpeps = 2.0 + EPS;
    g = 1.0;
    for (i = 1; i <= *pIrank; i++)
    {
        e = bv[i];
        if (e > tpeps || e < -EPS)
        {
            if (e > tpeps)
            {
                d = 2.0 - e;
                if (d < g)
                {
                    g = d;
                    *pIvo = 1;
                    *pIout = i;
                    *pXb = e;
                }
            }
            if (e < -EPS)
            {
                d = e;
                if (d < g)
                {
                    g = d;
                    *pIvo = -1;
                    *pIout = i;
                    *pXb = e;
                }
            }
        }
    }
}

/*-------------------------------------------------------------------
A Gauss-Jordan elimination step in LA_Lone()
-------------------------------------------------------------------*/
```

```
void LA_lone_gauss_jordn (int ipart, int iout, int jin, int n,
    tMatrix_R ct, tVector_I icbas, tMatrix_R binv, tVector_R bv,
    tVector_I ibound, int *pIrank, tVector_R r)
{
    int i, j, ic;
    tNumber_R  pivot, e, d;

    pivot = ct[iout][jin];

    icbas[iout] = jin;
    for (j = 1; j <= n; j++)
    {
        ct[iout][j] = ct[iout][j]/pivot;
    }
    for (j = 1; j <= *pIrank; j++)
    {
        binv[iout][j] = binv[iout][j]/pivot;
    }
    if (ipart != 1) bv[iout] = bv[iout]/pivot;
    for (i = 1; i <= *pIrank; i++)
    {
        if (i != iout)
        {
            e = ct[i][jin];
            for (j = 1; j <= n; j++)
            {
                ct[i][j] = ct[i][j] - e*ct[iout][j];
            }
            for (j = 1; j <= *pIrank; j++)
            {
                binv[i][j] = binv[i][j] - e*binv[iout][j];
            }
            if (ipart != 1)
            {
                bv[i] = bv[i] - e*bv[iout];
            }
        }
    }
    if (ipart != 1)
    {
        e = r[jin];
        for (j = 1; j <= n; j++) r[j] = r[j] - e*ct[iout][j];
        for (j = 1; j <= n; j++)
        {
            ic = 0;
```

```
            for (i = 1; i <= *pIrank; i++)
            {
                if (j == icbas[i]) ic = 1;
                if (ic == 0)
                {
                    d = r[j]*ibound[j];
                    if (d < 0.0) r[j] = 0.0;
                }
            }
        }
    }
}

/*-----------------------------------------------------------------------
Calculate the results of LA_Lone()
-----------------------------------------------------------------------*/
eLaRc LA_lone_res (int m, int n, tVector_R f, tVector_I icbas,
    tVector_I irbas, tMatrix_R binv, tVector_R bv, int *pIrank,
    tVector_R r, tVector_R a, tNumber_R *pZ)
{
    int             i, j, k;
    tNumber_R       s, sa;

    for (j = 1; j <= *pIrank; j++)
    {
        s = 0.0;
        for (i = 1; i <= *pIrank; i++)
        {
            k = icbas[i];
            s = s + f[k] * (binv[i][j]);
        }
        k = irbas[j];
        a[k] = s;
    }

    s = 0.0;
    for (j = 1; j <= n; j++)
    {
        sa = r[j];
        if (sa < 0.0) sa = - sa;
        s = s + sa;
    }
    *pZ = s;

    if (*pIrank < m)
```

```
        return LaRcSolutionDefNotUniqueRD;

    for (i = 1; i <= m; i++)
    {
        if (bv[i] <= EPS || bv[i] >= 2.0 - EPS)
            return LaRcSolutionProbNotUnique;
    }

    return LaRcSolutionUnique;
}
```

Chapter 6

One-Sided L_1 Approximation

6.1 Introduction

In the previous chapter, an algorithm for obtaining the L_1 solution of an overdetermined system of linear equations is given. That L_1 solution is a double-sided one, meaning that some of the elements of the residual vector have values ≥ 0 and others have values < 0. In other words, for the discrete linear L_1 approximation, some of the points lie above or on the approximating surface (curve) and some lie below the approximating surface. Hence, the approximation is the **ordinary** or **double-sided L_1 approximation**.

In this chapter, we present the **linear one-sided L_1 approximation problem**. In this approximation, all the given discrete points lie either above or on, or below and on the approximating surface. When all the discrete points lie above or on the approximating surface, this is known as the **one-sided L_1 approximation from above**. When all the discrete points lie below or on the approximating surface, the approximation is known as the **one-sided L_1 approximation from below**. In two dimensional case, this is illustrated by Figure 6-1.

We shall consider the problem of the one-sided L_1 approximation from below. However, the analysis and presentation of the problem from above are almost identical. The described algorithm is manipulated slightly so that it can be applied to the latter case as well.

The problem is presented here as a linear programming problem and we pursue the analysis of the dual form of the linear programming presentation. The algorithm is much simpler than the algorithm for the ordinary L_1 approximation in Chapter 5.

We should note that there are problems that have (ordinary) L_1

approximation but do not have one-sided L_1 approximation from above and/or one-sided L_1 approximation from below. See the numerical examples in Section 6.5. See also the practical example, Example 16.2.

Consider the overdetermined system of linear equations

$$\mathbf{Ca} = \mathbf{f} \tag{6.1.1a}$$

$\mathbf{C} = (c_{ij})$, is a given real n by m matrix of rank k, $k \leq m \leq n$ and $\mathbf{f} = (f_i)$ is a given real n-vector. The (ordinary) L_1 solution of system $\mathbf{Ca} = \mathbf{f}$ is the m-vector $\mathbf{a} = (a_j)$ that minimizes the L_1 norm z of the residuals

$$z = \sum_{i=1}^{n} |r_i| \tag{6.1.1b}$$

where r_i is the ith residual and is given by

$$r_i = \sum_{j=1}^{m} c_{ij}a_j - f_i, \quad i = 1, 2, \ldots, n \tag{6.1.1c}$$

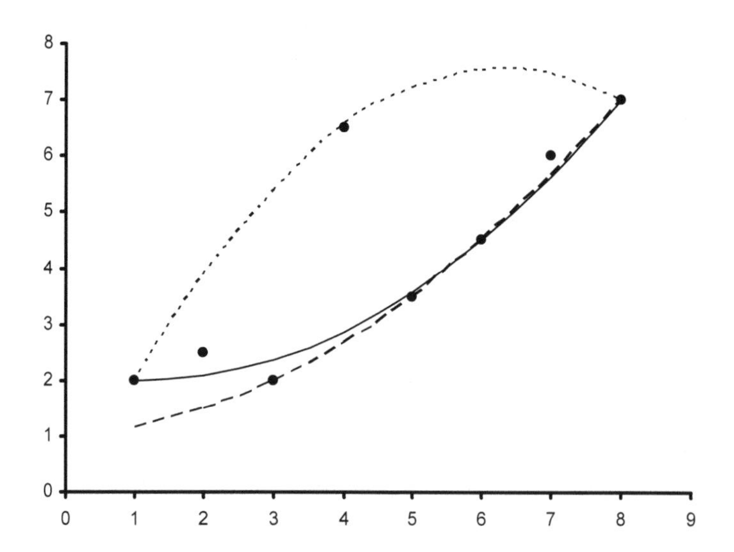

Figure 6-1: Curve fitting with vertical parabolas of a set of 8 points using L_1 approximation and one-sided L_1 approximations

This figure gives curve fitting with vertical parabolas of the set of

8 points shown in Figure 2-1. The solid curve is the ordinary L_1 approximation. The dashed curve is the one-sided L_1 approximation from above and the dotted curve is the one-sided L_1 approximation from below.

The special case when the system of equation $\mathbf{Ca} = \mathbf{f}$, is consistent, i.e., the residual $\mathbf{r} = \mathbf{Ca} - \mathbf{f} = \mathbf{0}$, is not of interest here. We thus assume that $\mathbf{r} \neq \mathbf{0}$.

When the elements of the residual vector $\mathbf{r}$ satisfy the additional conditions

(6.1.1d) $$r_i \leq 0, \quad i = 1, 2, \ldots, n$$

or in effect

(6.1.1e) $$\mathbf{Ca} \leq \mathbf{f}$$

we have the one-sided L_1 solution from above of system $\mathbf{Ca} = \mathbf{f}$; that is, for any equation i, i = 1, ..., n, in $\mathbf{Ca} = \mathbf{f}$, the observed value f_i is greater than (or equal to) the calculated value $(c_{i1}a_1 + c_{i2}a_2 + \cdots + c_{im}a_m)$.

If the inequalities (6.1.1d) are reversed, i.e.,

$$r_i \geq 0, \quad i = 1, 2, \ldots, n$$

we have the one-sided L_1 solution from below.

As indicated above, we formulate here the problem of the one-sided L_1 solution from below as a linear programming one. We use the dual formulation of the problem as we did for the ordinary L_1 approximation in the previous chapter. However, we use here the simplex method, not the dual simplex method that we used in the previous chapter. In this algorithm no conditions are imposed on the coefficient matrix. It may be a rank deficient one. An initial basic solution is obtained with a small effort. The described algorithm applies as well to the one-sided L_1 solution from above.

In Section 6.2, the problem is presented as a special case of a general constrained L_1 approximation problem. In Section 6.3, the linear programming formulation of the problem is given. In Section 6.4, the algorithm is described and a numerical example is solved. A note on the linear one-sided L_1 solution from above is also given and the interpolating properties of the one-sided L_1 approximation is described. In Section 6.5, numerical results are presented and compared with other techniques for solving the same problem.

6.1.1 Applications of the algorithm

One-sided L_p approximations have applications to the numerical solution of operator equations, to ordinary differential equations and to integral equations. See Watson [23, 24].

The one-sided approximation in the L_1 norm is applied to the degree reduction of interval polynomial of the so-called **Bézier curves** in computer aided design. See Deng et al. [13]. The one-sided L_1 and the one-sided Chebyshev solutions of overdetermined systems are also applied to the solution of overdetermined linear inequalities [2]. The latter is a basic problem in pattern classification. See Tou and Gonzalez ([22], pp. 40-41, 48-49) and also Chapter 16.

6.1.2 Characterization and uniqueness

For the characterization and uniqueness of the best one-sided L_1 approximation of a continuous function, see Pinkus [21] and for harmonic functions see Armitage et al. [3]. For the uniqueness of the best one-sided L_1 approximation of continuous differentiable functions see Babenko and Glushko [5] and also Lenze [19]

6.2 A special problem of a general constrained one

A number of authors developed algorithms for general constrained L_1 approximation problems. By manipulating the constraints, each problem reduces to a one-sided L_1 approximation problem. In other words, their algorithms are general purpose algorithms, while ours is a special purpose one.

Using our notation, let $\mathbf{C}$ and $\mathbf{E}$ be matrices of appropriate dimensions and let $\mathbf{f}$ be the vector associated with $\mathbf{C}$, and $\mathbf{e}_a$ and $\mathbf{e}_b$ be two vectors associated with $\mathbf{E}$ respectively. Armstrong and Hultz (AH) [4] seek a solution vector $\mathbf{a}$ that satisfies the problem

(6.2.1a) $$\text{minimize } \|\mathbf{Ca} - \mathbf{f}\|_1$$

subject to

(6.2.1b) $$\mathbf{e}_a \leq \mathbf{Ea} \leq \mathbf{e}_b$$

If in (6.2.1b) we take $\mathbf{E} = \mathbf{C}$, $\mathbf{e}_a = \mathbf{f}$, and $\mathbf{e}_b$ is a very large vector, we get the one-sided L_1 approximation from below.

Each of Barrodale and Roberts (BR) [7, 9], Bartels and Conn (BC) [10, 11] and Dax (DA) [12] also proposed to solve a class of problems that includes the problem given by Armstrong and Hultz (AH) [4] as a special case. They minimize the L_1 norm of the residual subject to a mixture of linear equality and inequality constraints. Using our notation, let **C**, **G** and **D** be matrices of appropriate dimensions and let **f**, **g** and **d** be vectors associated with **C**, **G** and **D** respectively. Then they seek a solution vector **a** that satisfies

(6.2.2a) $$\text{minimize } \|\mathbf{Ca} - \mathbf{f}\|_1$$

subject to

(6.2.2b) $$\mathbf{Ga} = \mathbf{g} \text{ and } \mathbf{Da} \geq \mathbf{d}$$

They allow for the possibility that some but not all of the arrays (**G**, **g**) and (**D**, **d**) be vacuous. Barrodale and Roberts [7, 9] use a primal linear programming technique that is an extension of their method for the L_1 approximation without linear constraints [6]. Bartels and Conn [10, 11], however, use a penalty-function method, which solves a constrained optimization problem, while Dax [12] uses a steepest descent search direction method by first solving a linear least squares sub-problem. Hence, if in (6.2.2b) we take $\mathbf{G} = \mathbf{0}$, $\mathbf{g} = \mathbf{0}$, $\mathbf{D} = \mathbf{C}$ and $\mathbf{d} = \mathbf{f}$, problem (6.2.2) reduces to the one-sided L_1 problem from below. We shall comment on these methods in Section 6.5.

6.3 Linear programming formulation of the problem

In linear programming terminology [18], problem (6.1.1) is formulated as follows. See Lewis [20] and Watson [23, 24]. Since the residual vector **r** for $\mathbf{Ca} = \mathbf{f}$ is given by $\mathbf{r} = \mathbf{Ca} - \mathbf{f}$, and since all the elements of **r**, $r_i \geq 0$, $i = 1, 2, \ldots, n$, in vector-matrix notation, we have

$$\text{minimize } Z = \mathbf{e}^T(\mathbf{Ca} - \mathbf{f})$$

where **e** is an n-vector, each element of which is 1. Now since $\mathbf{e}^T\mathbf{f}$ is just a constant, this reduces to

(6.3.1a) $$\text{minimize } Z = \mathbf{e}^T(\mathbf{Ca})$$

subject to $\mathbf{Ca} - \mathbf{f} > \mathbf{0}$, which is

(6.3.1b) $$\mathbf{Ca} \geq \mathbf{f}$$

and

(6.3.1c) a_j unrestricted in sign, $j = 1, 2, \ldots, m$

It is easier to deal with the dual of problem (6.3.1), namely

(6.3.2a) $$\text{maximize } z = \mathbf{f}^T\mathbf{b}$$

subject to $\mathbf{C}^T\mathbf{b} = \mathbf{C}^T\mathbf{e}$, which is

(6.3.2b) $$\mathbf{C}^T\mathbf{b} = \sum_{i=1}^{n} \mathbf{C}_j^T$$

and

(6.3.2c) $b_i \geq 0, \quad i = 1, 2, \ldots, n$

In (6.3.2b) $\mathbf{C}_j^T$ is the jth column of matrix $\mathbf{C}^T$ and the r.h.s. of (6.3.2b) is the sum of the columns of $\mathbf{C}^T$.

6.4 Description of the algorithm

This problem may be solved by the two-phase method of linear programming, as described in Section 3.5. However, we should start with vector $\mathbf{b}_B$ whose elements are non-negative. This is easily done because of the simple structure of the problem, as explained next.

6.4.1 Obtaining an initial basic feasible solution

We note that the main body (the matrix of constraints) in the initial data in the programming problem is matrix $\mathbf{C}^T$ and from (6.3.2b), the basic solution vector $\mathbf{b}_B$ is given by

$$\mathbf{b}_B = \sum_{i=1}^{n} \mathbf{C}_j^T$$

The elements of vector $\mathbf{b}_B$ may or may not be non-negative, i.e., one or more of its elements may be < 0. Let the element $\mathbf{b}_{Bi} < 0$. If we now multiply $\mathbf{b}_{Bi}$ and the whole of row i in the initial data by -1, this amounts to multiplying column i of matrix $\mathbf{C}$ in the system $\mathbf{Ca} = \mathbf{f}$ by -1. Then the calculated element a_i of the solution vector $\mathbf{a}$ will not be a_i but $-a_i$. An index i would then be stored in an index vector and, at the end of the program, element a_i would be multiplied by –1.

At the end of phase 1, we have an initial basic feasible solution $\mathbf{b}_B$. If rank($\mathbf{C}$) = k < m, then only k Gauss-Jordan steps are needed. We then calculate the marginal costs $(z_j - f_j)$

$$z_j - f_j = \mathbf{f}_B{}^T\mathbf{y}_j - f_j, \quad j = 1, \ldots, n$$

Phase 2 of the algorithm is the ordinary simplex method. If at any iteration, a pivot element is not found, or the ratio b_{Bi}/y_{ij} is < EPS, the problem has no solution and the program terminates.

Lemma 6.1 (Theorem 5.3)

At any stage of the computation, the residuals (r_i) of (6.1.1c) are themselves the marginal costs $(z_i - f_i)$ for the same reference

$$z_i - f_i = r_i, \quad i = 1, 2, \ldots, n$$

As a result, for the optimum solution, the objective function z = the sum of the marginal costs, being all ≥ 0

$$z = \sum_{i=1}^{n} (z_i - f_i)$$

See Theorem 5.3.

Lemma 6.2

The solution vector of the one-sided L_1 approximation problem is given by

$$\mathbf{a}^T = \mathbf{f}_B{}^T\mathbf{B}^{-1} \tag{6.4.1}$$

where $\mathbf{B}^{-1}$ is the inverse of the basis matrix for the optimum solution and $\mathbf{f}_B$ is associated with the optimal solution. See Theorem 5.4.

Example 6.1

Obtain the L_1 solution from below of the following system

$$\begin{aligned} -a_1 - a_2 &= 2 \\ a_1 + 3a_2 &= 1 \\ a_1 + 2a_2 &= 1 \\ a_2 &= -3 \\ a_3 &= 0 \end{aligned} \tag{6.4.2}$$

In the tableau for the Initial Data, the left hand side is the algebraic sum of the columns of matrix $\mathbf{C}^T$. Tableau 6.4.1 represents the end of part 1. It is obtained by applying 3 Gauss-Jordan steps to the Initial Data. The basic solution is feasible; all elements of $\mathbf{b}_B$ are ≥ 0.

Initial Data

	$\mathbf{f}^T$	2	1	1	–3	0
$\Sigma_i \mathbf{C}_i^T$		$\mathbf{C}_1^T$	$\mathbf{C}_2^T$	$\mathbf{C}_3^T$	$\mathbf{C}_4^T$	$\mathbf{C}_5^T$
1		–1	1	1	0	0
5		–1	3	2	1	0
1		0	0	0	0	1

Tableau 6.4.1 (end of part 1)

	$\mathbf{f}^T$	2	1	1	–3	0
$\mathbf{f}_B^T$	$\mathbf{b}_B$	$\mathbf{C}_1^T$	$\mathbf{C}_2^T$	$\mathbf{C}_3^T$	$\mathbf{C}_4^T$	$\mathbf{C}_5^T$
2	1	1	0	–1/2	1/2	0
1	2	0	1	(1/2)	1/2	0
0	1	0	0	0	0	1
		0	0	–3/2	9/2	0

Tableau 6.4.2 (part 2)

	$\mathbf{f}^T$	2	1	1	–3	0
$\mathbf{f}_B^T$	$\mathbf{b}_B$	$\mathbf{C}_1^T$	$\mathbf{C}_2^T$	$\mathbf{C}_3^T$	$\mathbf{C}_4^T$	$\mathbf{C}_5^T$
2	3	1	1	0	1	0
1	4	0	2	1	1	0
0	1	0	0	0	0	1
z = 9		0	3	0	6	0

It took one Gauss-Jordan iteration to obtain the optimum solution of the problem. The residuals of the problem are given as the marginal costs (Lemma 6.1) in the last tableau and are $\mathbf{r} = (0, 0, 3, 6, 0)^T$ from which $z = 9$. From (6.4.1) or by solving the first, the third and the fifth equations in the set (6.4.2), we get the solution vector $\mathbf{a} = (-5, 3, 0)^T$.

6.4.2 One-sided L_1 solution from above

For the linear one-sided L_1 solution from above, as indicated earlier, the inequalities for the residuals are given by

$$r_i \leq 0, \quad i = 1, 2, \ldots, n$$

The algorithm described here for one-sided L_1 solution from below may be applied as well to the one-sided L_1 solution from above as follows. We multiply each element of matrix **C** and each element of vector **f** in **Ca** = **f** by –1. We get the equation $\underline{\mathbf{C}}\mathbf{a} = \underline{\mathbf{f}}$, where $\underline{\mathbf{C}} = -\mathbf{C}$ and $\underline{\mathbf{f}} = -\mathbf{f}$. We then apply the current algorithm to the equation $\underline{\mathbf{C}}\mathbf{a} = \underline{\mathbf{f}}$.

Then the elements $\underline{r_i}$ of the residual vector $\underline{\mathbf{r}} = \underline{\mathbf{C}}\mathbf{a} - \underline{\mathbf{f}}$, satisfy

$$\underline{r_i} \geq 0, \quad i = 1, 2, \ldots, n$$

which implies that for the given equation **Ca** = **f**

$$r_i \leq 0, \quad i = 1, 2, \ldots, n$$

meaning the one-sided solution from above.

The obtained solution vector **a** would be that for the L_1 solution from above, and the elements of the obtained residual vector **r** are to be multiplied by –1.

6.4.3 The interpolation property

A certain property is shared between the (ordinary) L_1 solution [1] and the one-sided L_1 solutions of overdetermined systems of linear equations. Let rank(**C**) = $k \leq m$. Then at least k equations of **Ca** = **f**, each has zero residual r_i. This property is a direct result of using the dual form of the linear programming formulation for both problems.

The reason is that k equations in **Ca** = **f**, associated with the basis matrix have zero marginal costs (residuals). This also means the following. If n discrete points in the x-y plane are being approximated by a plane curve, in the one-sided L_1 sense, the curve will pass through at least k of the discrete points. See Figure 6-1.

6.5 Numerical results and comments

LA_Loneside() implements this algorithm. DR_Loneside() tests 8 examples in which the data is taken from [14, 15, 16, 17].

Table 6.1 shows the results of 3 of the examples, computed in single-precision. For comparison purposes, the results for the (ordinary or the two-sided) L_1 solution for each example are included. The L_1 solution is calculated by LA_Lone() of Chapter 5.

Table 6.1

		One-sided L_1 from above		L_1 solution		One-sided L_1 from below	
Example	**C**(n×m)	Iter	z	Iter	z	Iter	z
1	4 × 2	no solution		2	18.4	no solution	
2	10× 5	7	10.00	6	10.00	no solution	
3	25×10	20	0.1408	21	0.0878	19	0.139

For each example, the number of iterations and the optimum L_1 norms z for the 3 cases are shown. In this table, "no solution" indicates that the problem has no one-sided L_1 solution. The results indicate that the number of iterations for the one-sided L_1 approximation are comparable to those for the ordinary L_1 approximation.

The current algorithm is a special purpose one for solving the one-sided L_1 approximation problem. However, the algorithms mentioned in Section 6.2, those of Armstrong and Hultz (AH) [4], Barrodale and Roberts (BR) [7-9], Bartels and Conn (BC) [10, 11] and Dax (DA) [12], are general purpose algorithms that may by used to solve the current problem.

Each of those algorithms necessitates that matrix **C** and vector **f** be stored twice in computer memory, once for (6.2.1a) and once for (6.2.1b), or once for (6.2.2a) and once for (6.2.2b), replacing **D** and **d** respectively. That would nearly double the number of arithmetic operations in their algorithms. Our algorithm, as a special purpose one, would thus be more efficient. The same observations are made at the end of Chapter 11 for the one-sided Chebyshev approximation problem.

References

1. Abdelmalek, N.N., On the discrete L_1 approximation and L_1 solutions of overdetermined linear equations, *Journal of Approximation Theory*, 11(1974)38-53.
2. Abdelmalek, N.N., Linear one-sided approximation algorithms for the solution of overdetermined systems of linear inequalities, *International Journal of Systems Science*, 15(1984)1-8.
3. Armitage, D.H., Gardiner, S.J., Haussmann, W. and Rogge, L., Best one-sided L^1 – approximation by harmonic functions, *Manuscripta Mathematica*, 96(1998)181-194.
4. Armstrong, R.D. and Hultz, J.W., An algorithm for a restricted discrete approximation problem in the L_1 norm, *SIAM Journal on Numerical Analysis*, 14(1977)555-565.
5. Babenko, V.F. and Glushko, V.N., On the uniqueness of elements of the best approximation and the best one-sided approximation in the space L_1, *Ukrainian Mathematical Journal*, 46(1994)503-513.
6. Barrodale, I. and Roberts, F.D.K., An improved algorithm for discrete l_1 approximation, *SIAM Journal on Numerical Analysis*, 10(1973)839-848.
7. Barrodale, I. and Roberts, F.D.K., Algorithms for restricted least absolute value estimation, *Communications on Statistics - Simulation and Computation*, B6(1977)353-363.
8. Barrodale, I. and Roberts, F.D.K., An efficient algorithm for discrete l_1 linear approximation with linear constraints, *SIAM Journal on Numerical Analysis*, 15(1978)603-611.
9. Barrodale, I. and Roberts, F.D.K., Algorithm 552: Solution of the constrained L_1 linear approximation problem, *ACM Transactions on Mathematical Software*, 6(1980)231-235.
10. Bartels, R.H. and Conn, A.R., Linearly constrained discrete l_1 problems, *ACM Transactions on Mathematical Software*, 6(1980)594-608.
11. Bartels, R.H. and Conn, A.R., Algorithm 563: A program for linearly constrained discrete l_1 problems, *ACM Transactions on Mathematical Software*, 6(1980)609-614.

12. Dax, A., The l_1 solution of linear equations subject to linear constraints, *SIAM Journal on Scientific and Statistical Computation*, 10(1989)328-340.
13. Deng, J., Feng, Y. and Chen, F., Best one-sided approximation of polynomials under L_1 norm, *Journal of Computational and Applied Mathematics*, 144(2002)161-174.
14. Duris, C.S., An exchange method for solving Haar and non-Haar overdetermined linear equations in the sense of Chebyshev, *Proceedings of Summer ACM Computer Conference*, (1968)61-65.
15. Duris, C.S. and Sreedharan, V.P., Chebyshev and l_1-solutions of linear equations using least squares solutions, *SIAM Journal on Numerical Analysis*, 5(1968)491-505.
16. Easton, M.C., A fixed point method for Tchebycheff solution of inconsistent linear equations, *Journal of Institute of Mathematics and Applications*, 12(1973)137-159.
17. Goldstein, A.A., Levine, N. and Hereshoff, J.B., On the best and least q^{th} approximation of an overdetermined system of linear equations, *Journal of ACM*, 4(1957)341-347.
18. Hadley, G., *Linear Programming*, Addison-Wesley, Reading, MA, 1962.
19. Lenze, B., Uniqueness in best one-sided L_1 – approximation by algebraic polynomials on unbounded intervals, *Journal of Approximation Theory*, 57(1989)169-177.
20. Lewis, J.T., Computation of best one-sided L_1 approximation, Mathematics of Computation, 24(1970)529-536.
21. Pinkus, A.M., *On L^1-Approximation*, Cambridge University Press, London, 1989.
22. Tou, J.T. and Gonzalez, R.C., *Pattern Recognition Principles*, Addison-Wesley, Reading, MA, 1974.
23. Watson, G.A., The calculation of best linear one-sided L_p approximations, *Mathematics of Computation*, 27(1973)607-620.
24. Watson, G.A., One-sided approximation and operator equations, *Journal of Institute of Mathematics and Applications*, 12(1973)197-208.

6.6 DR_Loneside

```
/*-----------------------------------------------------------------------
DR_Loneside
-------------------------------------------------------------------------
This program is a driver for the function LA_Loneside(), which
calculates the one-sided L-One solution from above or from below of
an overdetermined system of linear equations.

The overdetermined system has the form

                          c*a = f

"c" is a given real n by m matrix of rank k, k <= m <= n.
"f" is a given real n vector.
"a" is the solution m vector.

This driver contains 8 examples from which the results of examples
1, 6 and 7 are given in the text.  All the example are solved twice;
once for the one-sided L-One approximation from above and once for
the one-sided L-One approximation from below.
-----------------------------------------------------------------------*/

#include "DR_Defs.h"
#include "LA_Prototypes.h"

#define N1 4
#define M1 2
#define N2 5
#define M2 3
#define N3 6
#define M3 3
#define N4 7
#define M4 3
#define N5 8
#define M5 4
#define N6 10
#define M6 5
#define N7 25
#define M7 10
#define N8 8
#define M8 4

void DR_Loneside (void)
```

```
{
    /*----------------------------------------
      Constant matrices/vectors
    ----------------------------------------*/
    static tNumber_R c1init[N1][M1] =
    {
        { 0.0, -2.0 },
        { 0.0, -4.0 },
        { 1.0, 10.0 },
        {-1.0, 15.0 }
    };

    static tNumber_R c2init[N2][M2] =
    {
        { 1.0,  2.0, 0.0 },
        {-1.0, -1.0, 0.0 },
        { 1.0,  3.0, 0.0 },
        { 0.0,  1.0, 0.0 },
        { 0.0,  0.0, 1.0 }
    };

    static tNumber_R c3init[N3][M3] =
    {
        { 0.0, -1.0,  0.0 },
        { 1.0,  3.0, -4.0 },
        { 1.0,  0.0,  0.0 },
        { 0.0,  0.0,  1.0 },
        {-1.0,  1.0,  2.0 },
        { 1.0,  1.0,  1.0 }
    };

    static tNumber_R c4init[N4][M4] =
    {
        { 1.0, 0.0,  1.0 },
        { 1.0, 2.0,  2.0 },
        { 1.0, 2.0,  0.0 },
        { 1.0, 1.0,  0.0 },
        { 1.0, 0.0, -1.0 },
        { 1.0, 0.0,  0.0 },
        { 1.0, 1.0,  1.0 }
    };

    static tNumber_R c5init[N5][M5] =
    {
        { 1.0, -3.0,  9.0, -27.0 },
```

```
    { 1.0, -2.0,  4.0,  -8.0 },
    { 1.0, -1.0,  1.0,  -1.0 },
    { 1.0,  0.0,  0.0,   0.0 },
    { 1.0,  1.0,  1.0,   1.0 },
    { 1.0,  2.0,  4.0,   8.0 },
    { 1.0,  3.0,  9.0,  27.0 },
    { 1.0,  4.0, 16.0,  64.0 }
};

static tNumber_R c6init[N6][M6] =
{
    { 1.0, 0.0,  0.0,  0.0,  0.0 },
    { 0.0, 1.0,  0.0,  0.0,  0.0 },
    { 0.0, 0.0,  1.0,  0.0,  0.0 },
    { 0.0, 0.0,  0.0,  1.0,  0.0 },
    { 0.0, 0.0,  0.0,  0.0,  1.0 },
    { 1.0, 1.0,  1.0,  1.0,  1.0 },
    { 0.0, 1.0,  1.0,  1.0,  1.0 },
    {-1.0, 0.0, -1.0, -1.0, -1.0 },
    { 1.0, 1.0,  0.0,  1.0,  1.0 },
    { 1.0, 1.0,  1.0,  0.0,  1.0 }
};

static tNumber_R c8init[N8][M8] =
{
    { 1.0, 1.0,   1.0,  1.0 },
    { 1.0, 2.0,   4.0,  4.0 },
    { 1.0, 3.0,   9.0,  9.0 },
    { 1.0, 4.0,  16.0, 16.0 },
    { 1.0, 5.0,  25.0, 25.0 },
    { 1.0, 6.0,  36.0, 36.0 },
    { 1.0, 7.0,  49.0, 49.0 },
    { 1.0, 8.0,  64.0, 64.0 }
};

static tNumber_R f1[N1+1] =
{   NIL,
    -12.0, 6.0, 0.0, 5.0
};

static tNumber_R f2[N2+1] =
{   NIL,
    1.0, 2.0, 1.0, -3.0, 0.0
};
```

```
static tNumber_R f3[N3+1] =
{   NIL,
    1.0, 2.0, 3.0, 2.0, 2.0, 4.0
};

static tNumber_R f4[N4+1] =
{   NIL,
    0.0, -2.0, 1.0, -1.0, 5.0, 7.0, 0.0
};

static tNumber_R f5[N5+1] =
{   NIL,
    3.0, -3.0, -2.0, 0.0, 7.0, -1.0, 5.0, 2.0
};

static tNumber_R f6[N6+1] =
{   NIL,
    1.0, -1.0, 0.0, -1.0, 1.0, 0.0, 2.0, 3.0, -3.0, -2.0
};

static tNumber_R f7[N7+1] =
{   NIL,
    0.0872673, 0.0872794, 0.0873029, 0.0873315, 0.0873576,
    0.3491184, 0.3498802, 0.3513824, 0.3532572, 0.3550109,
    0.6111334, 0.6150641, 0.6230824, 0.6336395, 0.6441493,
    0.8733883, 0.8841621, 0.9071868, 0.9400757, 0.9766021,
    1.135895,  1.157550,  1.206257,  1.283258,  1.384432
};

static tNumber_R f8[N8+1] =
{   NIL,
    2.0, 2.5, 2.0, 6.5, 3.5, 4.5, 6.0, 7.0
};

/*----------------------------------------
  Variable matrices/vectors
----------------------------------------*/
tMatrix_R   ct      = alloc_Matrix_R (MM_COLS, NN_ROWS);
tVector_R   f       = alloc_Vector_R (NN_ROWS);
tVector_R   r       = alloc_Vector_R (NN_ROWS);
tVector_R   a       = alloc_Vector_R (MM_COLS);
tMatrix_R   c7      = alloc_Matrix_R (N7, M7);

tMatrix_R   c1      = init_Matrix_R (&(c1init[0][0]), N1, M1);
tMatrix_R   c2      = init_Matrix_R (&(c2init[0][0]), N2, M2);
```

```
tMatrix_R   c3      = init_Matrix_R (&(c3init[0][0]), N3, M3);
tMatrix_R   c4      = init_Matrix_R (&(c4init[0][0]), N4, M4);
tMatrix_R   c5      = init_Matrix_R (&(c5init[0][0]), N5, M5);
tMatrix_R   c6      = init_Matrix_R (&(c6init[0][0]), N6, M6);
tMatrix_R   c8      = init_Matrix_R (&(c8init[0][0]), N8, M8);

int         irank, iter, iside, kase;
int         i, j, k, m, n, Iexmpl;
tNumber_R   d, dd, ddd, e, ee, eee, z;

eLaRc       rc = LaRcOk;

for (j = 1; j <= 5; j++)
{
    d = 0.15* (j-3);
    dd = d*d;
    ddd = d*dd;
    for (i = 1; i <= 5; i++)
    {
        e = 0.15* (i-3);
        ee = e*e;
        eee = e*ee;
        k = 5* (j-1) + i;
        c7[k][1] = 1.0;
        c7[k][2] = d;
        c7[k][3] = e;
        c7[k][4] = dd;
        c7[k][5] = ee;
        c7[k][6] = e*d;
        c7[k][7] = ddd;
        c7[k][8] = eee;
        c7[k][9] = dd*e;
        c7[k][10] = ee*d;
    }
}

prn_dr_bnr ("DR_Loneside, One-Sided L-One Solutions of an "
            "Overdetermined System of Linear Equations");

for (kase = 1; kase <= 2; kase++)
{
    if (kase == 1) iside = 1;
    if (kase == 2) iside = 0;
    for (Iexmpl = 1; Iexmpl <= 8; Iexmpl++)
    {
```

```
switch (Iexmpl)
{
    case 1:
        n = N1;
        m = M1;
        for (i = 1; i <= n; i++)
        {
            f[i] = f1[i];
            for (j = 1; j <= m; j++) ct[j][i] = c1[i][j];
        }
        break;

    case 2:
        n = N2;
        m = M2;
        for (i = 1; i <= n; i++)
        {
            f[i] = f2[i];
            for (j = 1; j <= m; j++) ct[j][i] = c2[i][j];
        }
        break;

    case 3:
        n = N3;
        m = M3;
        for (i = 1; i <= n; i++)
        {
            f[i] = f3[i];
            for (j = 1; j <= m; j++) ct[j][i] = c3[i][j];
        }
        break;

    case 4:
        n = N4;
        m = M4;
        for (i = 1; i <= n; i++)
        {
            f[i] = f4[i];
            for (j = 1; j <= m; j++) ct[j][i] = c4[i][j];
        }
        break;

    case 5:
        n = N5;
        m = M5;
```

```
            for (i = 1; i <= n; i++)
            {
                f[i] = f5[i];
                for (j = 1; j <= m; j++) ct[j][i] = c5[i][j];
            }
            break;

        case 6:
            n = N6;
            m = M6;
            for (i = 1; i <= n; i++)
            {
                f[i] = f6[i];
                for (j = 1; j <= m; j++) ct[j][i] = c6[i][j];
            }
            break;

        case 7:
            n = N7;
            m = M7;
            for (i = 1; i <= n; i++)
            {
                f[i] = f7[i];
                for (j = 1; j <= m; j++) ct[j][i] = c7[i][j];
            }
            break;

        case 8:
            n = N8;
            m = M8;
            for (i = 1; i <= n; i++)
            {
                f[i] = f8[i];
                for (j = 1; j <= m; j++) ct[j][i] = c8[i][j];
            }

            break;

        default:
            break;
    }

    prn_algo_bnr ("Loneside");
    prn_example_delim();
    PRN ("Example #%d: Size of matrix \"c\", %d by %d\n",
```

```
                 Iexmpl, n, m);
            prn_example_delim();
            if (iside == 1)
                PRN ("One-sided L-One Solution from Above\n");
            else
                PRN ("One-sided L-One Solution from Below\n");
            prn_example_delim();
            PRN ("r.h.s. Vector \"f\"\n");
            prn_Vector_R (f, n);
            PRN ("Transpose of Coefficient Matrix, \"ct\"\n");
            prn_Matrix_R (ct, m, n);

            rc = LA_Loneside (iside, m, n, ct, f, &irank, &iter, r,
                              a, &z);

            if (rc >= LaRcOk)
            {
                PRN ("\n");
                PRN ("Results of the One-sided L1 Approximation\n");
                PRN ("One-sided L-One solution vector, \"a\"\n");
                prn_Vector_R (a, m);
                PRN ("One-sided L-One residual vector, \"r\"\n");
                prn_Vector_R (r, n);
                PRN ("One-sided L-One norm \"z\" = %8.4f\n", z);
                PRN ("Rank of Matrix \"c\" = %d,"
                     " Number of Iterations = %d\n", irank, iter);
            }

            prn_la_rc (rc);
        }
    }
    free_Matrix_R (ct, MM_COLS);
    free_Vector_R (f);
    free_Vector_R (r);
    free_Vector_R (a);
    free_Matrix_R (c7, N7);

    uninit_Matrix_R (c1);
    uninit_Matrix_R (c2);
    uninit_Matrix_R (c3);
    uninit_Matrix_R (c4);
    uninit_Matrix_R (c5);
    uninit_Matrix_R (c6);
    uninit_Matrix_R (c8);
}
```

6.7 LA_Loneside

```
/*-----------------------------------------------------------------------
LA_Loneside
-------------------------------------------------------------------------
This program calculates the one-sided L-One solution from above or
from below of an overdetermined system of linear equations. It uses
a modified simplex method to the linear programming formulation of
the problem.

The system of linear equations has the form

                       c*a = f

"c" is a given real n by m matrix of rank k <= m <= n.
"f" is a given real n vector.

The problem is to calculate the elements of the real m vector
"a" that gives the minimum L1 residual norm z.

              z = |r[1]| + |r[2]| + ... + |r[n]|

where r[i] is the ith residual and is given by

 r[i] = c[i][1]*a[1] + c[i][2]*a[2] + ... + c[i][m]*a[m] - f[i],
                         i = 1, 2, ..., n

subject to the conditions

    r[i] => 0, for the one-sided L-One solution from below
or
    r[i] =< 0, for the one-sided L-One solution from above.

Inputs
iside   An integer specifying the action to be performed.
        If iside = 1, the one-sided L-One solution from above is
        calculated.
        If iside != 1, the one-sided L-One solution from below is
        calculated.
m       Number of columns of matrix "c" in the system c*a = f.
n       Number of rows of matrix "c" in the system c*a = f.
ct      A real m by n matrix containing the transpose of matrix "c"
        of the system c*a = f.
f       A real n vector containing the r.h.s. of the system c*a = f.
```

```
Local Variables
binv    A real m square matrix containing the inverse of the basis
        matrix in the linear programming problem.
bv      A real m vector containing the basic solution in the linear
        programming problem.
icbas   An integer m vector containing the indices of the columns
        of "ct" that form the columns of the basis matrix.
irbas   An integer m  vector containing the indices of the rows
        of "ct".

Outputs
irank   The calculated rank of matrix "c".
iter    Number of iterations, or the number of times the simplex
        tableau is changed by a Gauss-Jordon elimination step.
a       A real m vector containing the one-sided L-One solution of
        the system c*a = f.
r       A real n vector containing the one-sided L-One residual
        vector r = (c*a - f).
z       The optimum one-sided L-One norm of the problem.

Returns one of
        LaRcSolutionUnique
        LaRcSolutionProbNotUnique
        LaRcSolutionDefNotUniqueRD
        LaRcNoFeasibleSolution
        LaRcErrBounds
        LaRcErrNullPtr
        LaRcErrAlloc
-------------------------------------------------------------------*/

#include "LA_Prototypes.h"

eLaRc LA_Loneside (int iside, int m, int n, tMatrix_R ct,
    tVector_R f, int *pIrank, int *pIter, tVector_R r, tVector_R a,
    tNumber_R *pZ)
{
    tMatrix_R   binv    = alloc_Matrix_R (m, m);
    tVector_R   bv      = alloc_Vector_R (m);
    tVector_I   icbas   = alloc_Vector_I (m);
    tVector_I   irbas   = alloc_Vector_I (m);

    int         i = 0, j = 0, kk = 0;
    int         iout = 0, jin = 0, ivo = 0, itest = 0;
```

```
/* Validation of data before executing the algorithm */
eLaRc       rc = LaRcSolutionUnique;
VALIDATE_BOUNDS ((0 < m) && (m <= n) && !((n == 1) && (m == 1)));
VALIDATE_PTRS   (ct && f && pIrank && pIter && r && a && pZ);
VALIDATE_ALLOC  (binv && bv && icbas && irbas);

/* Initialization */
*pIrank = m;
*pIter = 0;
for (j = 1; j <= m; j++)
{
    a[j] = 0.0;
    icbas[j] = 0;
    irbas[j] = j;
    for (i = 1; i <= m; i++)
    {
        binv[i][j] = 0.0;
    }
    binv[j][j] = 1.0;
}
for (j = 1; j <= n; j++)
{
    r[j] = 0.0;
}

/* One-sided L-One solution from above */
if (iside == 1)
{
    for (j = 1; j <= n; j++)
    {
        f[j] = -f[j];
        for (i = 1; i <= m; i++) ct[i][j] = - ct[i][j];
    }
}

/* Calculate the initial basic solution */
LA_loneside_basic_sol (m, n, ct, irbas, bv);

/* Determine the rank of matrix "ct" */
LA_loneside_part_1 (m, n, ct, icbas, irbas, binv, bv, pIrank,
                    pIter, r);

/* Part 2 of the algorithm */
/* Calculate the marginal costs */
LA_loneside_marg_costs (m, n, ct, f, icbas, r);
```

```
    for (kk = 1; kk <= n*n; kk++)
    {
        /* Determine the vector that enters the basis */
        LA_loneside_vent (&ivo, &jin, m, n, icbas, r);

        if (ivo == 0)
        {
            /* Calculate the results */
            rc = LA_loneside_res (iside, m, n, f, icbas, irbas, binv,
                                  bv, pIrank, r, a, pZ);
            GOTO_CLEANUP_RC (rc);
        }
        itest = 0;
        /* Determine the vector that leaves the basis */
        LA_loneside_vleav (jin, &iout, &itest, m, ct, bv);

        if (itest != 1)
        {
            GOTO_CLEANUP_RC (LaRcNoFeasibleSolution);
        }

        /* A Gauss-Jordan elimination step */
        LA_loneside_gauss_jordn (iout, jin, m, n, ct, icbas, binv,
                                 bv, r);
        *pIter = *pIter + 1;
    }

CLEANUP:

    free_Matrix_R (binv, m);
    free_Vector_R (bv);
    free_Vector_I (icbas);
    free_Vector_I (irbas);

    return rc;
}

/*-------------------------------------------------------------------
Determine the rank of matrix "c" in LA_Loneside()
-------------------------------------------------------------------*/
void LA_loneside_part_1 (int m, int n, tMatrix_R ct,
    tVector_I icbas, tVector_I irbas, tMatrix_R binv, tVector_R  bv,
    int *pIrank, int *pIter, tVector_R r)
{
```

```
int             i, j, k, li = 0, jin = 0, iout;
tNumber_R       d, g, piv;

for (iout = 1; iout <= m; iout++)
{
    if (iout <= *pIrank)
    {
        piv = 0.0;
        for (j = 1; j <= n; j++)
        {
            for (i = iout; i <= *pIrank; i++)
            {
                d = ct[i][j];
                if (d < 0.0) d = -d;
                if (d > piv)
                {
                    li = i;
                    jin = j;
                    piv = d;
                }
            }
        }

        /* Detection of rank deficiency */
        if (piv > 0.0)
        {
            if (li != iout)
            {
                /* Swap of two elements of vector "irbas" */
                swap_elems_Vector_I (irbas, li, iout);

                /* Swap of two elements of vector "bv" */
                swap_elems_Vector_R (bv, li, iout);

                /* Swap of two rows of matrix "ct" */
                swap_rows_Matrix_R (ct, li, iout);

                /* Swap parts of two rows of matrix "binv" */
                if (iout != 1)
                {
                    k = iout - 1;
                    for (j = 1; j <= k; j++)
                    {
                        g = binv[li][j];
                        binv[li][j] = binv[iout][j];
```

```
                        binv[iout][j] = g;
                    }
                }
            }

            LA_loneside_gauss_jordn (iout, jin, m, n, ct, icbas,
                                     binv, bv, r);
            *pIter = *pIter + 1;
        }
        if (piv < EPS)
        {
            /* Solution is not unique */
            *pIrank = iout - 1;
        }
    }
  }
}

/*-----------------------------------------------------------------------
Calculation of the initial marginal costs in LA_Loneside()
-----------------------------------------------------------------------*/
void LA_loneside_marg_costs (int m, int n, tMatrix_R ct, tVector_R f,
    tVector_I icbas, tVector_R r)
{
    int             i, j, k, ibc;
    tNumber_R       s;

    for (j = 1; j <= n; j++)
    {
        r[j] = 0.0;
        ibc = 0;
        for (i = 1; i <= m; i++)
        {
            if (j == icbas[i]) ibc = 1;
        }
        if (ibc == 0)
        {
            s = - f[j];
            for (i = 1; i <= m; i++)
            {
                k = icbas[i];
                s = s + f[k] * ct[i][j];
            }
            r[j] = s;
        }
```

```
    }
}

/*-----------------------------------------------------------------
Calculation of the initial basic solution in LA_Loneside()
-----------------------------------------------------------------*/
void LA_loneside_basic_sol (int m, int n, tMatrix_R ct,
    tVector_I irbas, tVector_R  bv)
{
    int             i, j;
    tNumber_R       s;

    for (i = 1; i <= m; i++)
    {
        s = 0.0;
        for (j = 1; j <= n; j++)
        {
            s = s + ct[i][j];
        }
        bv[i] = s;
    }
    for (i = 1; i <= m; i++)
    {
        if (bv[i] < -EPS)
        {
            bv[i] = -bv[i];
            irbas[i] = -i;
            for (j = 1; j <= n; j++)
            {
                ct[i][j] = -ct[i][j];
            }
        }
    }
}

/*-----------------------------------------------------------------
Determine the vector that enters the basis in LA_Loneside()
-----------------------------------------------------------------*/
void LA_loneside_vent (int *pIvo, int *pJin, int m, int n,
    tVector_I icbas, tVector_R r)
{
    int             i, j, ic;
    tNumber_R       d, g;

    *pIvo = 0;
```

```
    g = 1.0/ (EPS*EPS);
    for (j = 1; j <= n; j++)
    {
        ic = 0;
        for (i = 1; i <= m; i++)
        {
            if (j == icbas[i]) ic = 1;
        }
        if (ic == 0)
        {
            d = r[j];
            if (d < 0.0)
            {
                if (d < g)
                {
                    *pIvo = 1;
                    g = d;
                    *pJin = j;
                }
            }
        }
    }
}

/*-----------------------------------------------------------------------
Determine the vector that leaves the basis in LA_Loneside()
-----------------------------------------------------------------------*/
void LA_loneside_vleav (int jin, int *pIout, int *pItest, int m,
    tMatrix_R ct, tVector_R  bv)
{
    int             i;
    tNumber_R       d, g, thmax;

    thmax = 1.0/ (EPS*EPS);
    for (i = 1; i <= m; i++)
    {
        d = ct[i][jin];
        if (d > EPS)
        {
            g = bv[i]/d;
            if (g <= thmax)
            {
                thmax = g;
                *pIout = i;
                *pItest = 1;
```

```
            }
        }
    }
}

/*-----------------------------------------------------------------
A Gauss-Jordan elimination step in LA_Loneside()
-----------------------------------------------------------------*/
void LA_loneside_gauss_jordn (int iout, int jin, int m, int n,
    tMatrix_R ct, tVector_I icbas, tMatrix_R binv,
    tVector_R bv, tVector_R r)
{
    int             i, j;
    tNumber_R       pivot, d;

    pivot = ct[iout][jin];
    for (j = 1; j <= n; j++) ct[iout][j] = ct[iout][j]/pivot;
    for (j = 1; j <= m; j++) binv[iout][j] = binv[iout][j]/pivot;
    bv[iout] = bv[iout]/pivot;
    for (i = 1; i <= m; i++)
    {
        if (i != iout)
        {
            d = ct[i][jin];
            for (j = 1; j <= n; j++)
            {
                ct[i][j] = ct[i][j] - d * (ct[iout][j]);
            }
            for (j = 1; j <= m; j++)
            {
                binv[i][j] = binv[i][j] - d * (binv[iout][j]);
            }
            bv[i] = bv[i] - d * (bv[iout]);
        }
    }
    icbas[iout] = jin;
    d = r[jin];
    for (j = 1; j <= n; j++) r[j] = r[j] - d * (ct[iout][j]);
}

/*-----------------------------------------------------------------
Calculate the results of LA_Loneside()
-----------------------------------------------------------------*/
eLaRc LA_loneside_res (int iside, int m, int n, tVector_R f,
    tVector_I icbas, tVector_I irbas, tMatrix_R binv, tVector_R bv,
```

```
    int *pIrank, tVector_R r, tVector_R a, tNumber_R *pZ)
{
    int             i, j, k;
    tNumber_R       s;

    for (j = 1; j <= *pIrank; j++)
    {
        s = 0.0;
        for (i = 1; i <= *pIrank; i++)
        {
            k = icbas[i];
            s = s + f[k]*binv[i][j];
        }
        k = irbas[j];
        if (k < 0) k = -k;
        a[k] = s;
        if (irbas[j] < 0) a[k] = -s;
    }

    s = 0.0;
    for (j = 1; j <= n; j++)
    {
        s = s + r[j];
    }

    *pZ = s;
    if (iside == 1)
    {
        for (j = 1; j <= n; j++) r[j] = -r[j];
    }

    if (*pIrank < m)
        return LaRcSolutionDefNotUniqueRD;

    for (i = 1; i <= m; i++)
    {
        if (bv[i] < EPS)
            return LaRcSolutionProbNotUnique;
    }

    return LaRcSolutionUnique;
}
```

Chapter 7

L_1 Approximation with Bounded Variables

7.1 Introduction

In Chapter 5, an algorithm for calculating the L_1 solution of overdetermined systems of linear equations is given. In this solution the L_1 norm of the residual vector, is as small as possible. In Chapter 6, the one-sided L_1 solution of overdetermined linear equations was presented, in which the L_1 norm of the residual vector is as small as possible and such that all the elements of the residual vector are either non-positive or non-negative.

In this chapter we present yet another kind of the linear L_1 approximation, known as L_1 approximation with bounded variables [5]. Here, the additional constraints are on the elements of the solution vector, not on the elements of the residual vector as in the previous chapter. The elements of the solution vector are to be bounded between –1 and 1.

For example, given the (x, y) data of Figure 2-1, the solid curve in Figure 7-1, is for the ordinary L_1 approximation, and is given by $y = 2.143 - 0.25x + 0.107x^2$. The elements of the solution vector are (2.143, –0.25, 0.107) and they are not all bounded between –1 and 1.

Yet for the same data, for the L_1 approximation with bounded variables between –1 and 1, the approximating curve is given by $y = 1 + 0.083x + 0.083x^2$, where the elements of the solution vector are between –1 and 1. For comparison purposes, the two approximating curves are shown in Figure 7-1.

In this chapter, the problem is solved by linear programming techniques, where initial basic solution is obtained with very little computational effort. Minimum computer storage is required and no conditions are imposed on the coefficient matrix. It could be a rank deficient one.

Consider the overdetermined system of linear equations

$$\mathbf{Ca} = \mathbf{f}$$

$\mathbf{C} = (c_{ij})$ is a given real n by m matrix of rank k, $k \le m \le n$, and $f = (f_i)$ is a given real n-vector. The L_1 solution of system $\mathbf{Ca} = \mathbf{f}$ is the m-vector $\mathbf{a} = (a_i)$ that minimizes the L_1 norm z of the residuals

(7.1.1) $$z = \sum_{i=1}^{n} |r_i|$$

where r_i is the ith element of the residual vector

$$\mathbf{r} = \mathbf{Ca} - \mathbf{f}$$

When the elements of the solution vector $\mathbf{a}$ are required to satisfy the additional conditions

(7.1.2) $$-1 \le a_i \le 1, \quad i = 1, 2, \ldots, m$$

we have the problem of calculating the L_1 solution of system $\mathbf{Ca} = \mathbf{f}$ with bounded variables between –1 and 1.

If instead of the constraints (7.1.2), we require the elements of vector $\mathbf{a}$ to satisfy the constraints

(7.1.3) $$c_i \le a_i \le d_i, \quad i = 1, 2, \ldots, m$$

where vectors $\mathbf{c} = (c_i)$ and $\mathbf{d} = (d_i)$ are given m-vectors, by substituting variables, these constraints reduce to the constraints (7.1.2) in the new variables. Let

(7.1.4) $$a_i = 0.5[(d_i - c_i)z_i + (d_i + c_i)], \quad i = 1, 2, \ldots, m$$

That is

$$z_i = [2a_i - (d_i + c_i)]/(d_i - c_i), \quad i = 1, 2, \ldots, m$$

Hence, when $a_i = d_i$, $z_i = 1$ and when $a_i = c_i$, $z_i = -1$. In vector-matrix form, (7.1.4) is given by

(7.1.5) $$\mathbf{a} = \mathbf{Gz} + \mathbf{g}$$

$\mathbf{G}$ is a diagonal m-matrix whose i^{th} diagonal element is $0.5(d_i - c_i)$ and $\mathbf{g}$ is an m-vector whose i^{th} element is $0.5(d_i + c_i)$. By substituting (7.1.5) into $\mathbf{Ca} = \mathbf{f}$, one gets

$$\mathbf{Dz} = \mathbf{h}$$

where

$$\mathbf{D} = \mathbf{CG} \text{ and } \mathbf{h} = \mathbf{f} - \mathbf{Cg}$$

and the elements of the new solution vector **z** are bounded between –1 and 1

$$-1 \leq z_i \leq 1, \quad i = 1, 2, \ldots, m$$

Once the solution vector **z** is obtained, the solution vector **a** of the given system **Ca** = **f** is calculated by substituting **z** into (7.1.5).

The linear L_1 approximation with non-negative parameters may be formulated as an L_1 approximation with bounded variables, as explained in Section 7.1.1.

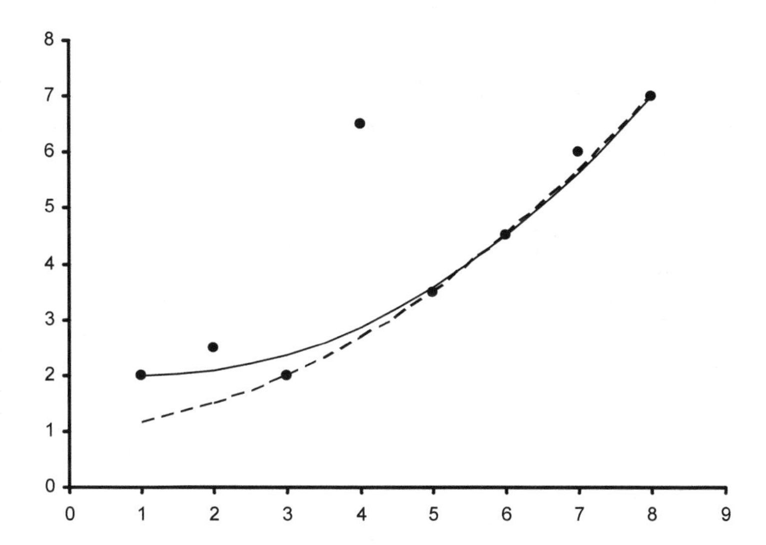

Figure 7-1: Curve fitting with vertical parabolas of a set of 8 points using L_1 approximation and L_1 approximation with bounded variables between –1 and 1

The solid curve is the L_1 approximation and the dashed curve is the L_1 approximation with bounded variables.

In Section 7.2, the bounded L_1 approximation problem would be treated as a special case of some general purpose algorithms. We shall comment on these algorithms in Section 7.5. In Section 7.3, the linear programming formulation of the problem is presented and necessary lemmas are given. In Section 7.4, the algorithm is described, and in Section 7.5, numerical results and comments are given.

7.1.1 Linear L_1 approximation with non-negative parameters (NNL1)

If in (7.1.3) we take $(c_i) = 0$ and $(d_i) = \text{Big}$, i = 1, 2, …, m, where Big is a large number, we get the non-negative linear L_1 (**NNL1**) approximation. Spath ([13], p. 250) suggested to take $\text{Big} = 100 \times \max|f_i|$, i = 1, 2, …, n, where (f_i) are the elements of vector **f** in the system $\mathbf{Ca} = \mathbf{f}$. See Chapter 12.1.1 for the non-negative L-infinity approximation (**NNLI**).

7.2 A special problem of a general constrained one

As in Chapter 6, this problem would be treated as a special case of one of the four general purpose algorithms. These are of Armstrong and Hultz [6], Barrodale and Roberts [7, 8], Bartels and Conn [9, 10] and Dax [11].

Using our notation, let **C** and **E** be matrices of appropriate dimensions and let **f** be the vector associated with **C**, and $\mathbf{e}_1$ and $\mathbf{e}_2$ be two vectors associated with **E** respectively. Using a special purpose primal linear programming method, Armstrong and Hultz (AH) [6] seek a solution vector **a** that satisfies the problem

$$\text{minimize } \|\mathbf{Ca} - \mathbf{f}\|_1 \tag{7.2.1a}$$

subject to

$$\mathbf{e}_1 \leq \mathbf{Ea} \leq \mathbf{e}_2 \tag{7.2.1b}$$

Hence, if we take in (7.2.1b), matrix $\mathbf{E} = \mathbf{I}_m$, an m-unit matrix, and take $\mathbf{e}_1 = -\mathbf{e}_m$ and $\mathbf{e}_2 = \mathbf{e}_m$, where each element of $\mathbf{e}_m$ is 1, (7.2.1a, b) reduce to the problem of bounded L_1 approximation between –1 and 1.

Once more, consider equations (6.2.2a-b) using our notation. Let **C**, **G** and **D** be matrices of appropriate dimensions and let **f**, **g** and **d** be vectors associated with **C**, **G** and **D** respectively. Each of Barrodale and Roberts (BR) [7, 8], Bartels and Conn (BC) [9, 10] and Dax (DA) [11] seek a vector **a** that satisfies

(7.2.2a) $$\text{minimize } \|\mathbf{Ca} - \mathbf{f}\|_1$$

subject to

(7.2.2b) $$\mathbf{Ga} = \mathbf{g} \text{ and } \mathbf{Da} \geq \mathbf{d}$$

They allow the possibility that some but not all of the arrays ($\mathbf{G}$, $\mathbf{g}$) and ($\mathbf{D}$, $\mathbf{d}$) be vacuous.

Hence, if in (7.2.2b), we take $\mathbf{G} = \mathbf{0}$, $\mathbf{g} = \mathbf{0}$, $\mathbf{D} = [\mathbf{I}_m, -\mathbf{I}_m]^T$ and $\mathbf{d} = [-\mathbf{e}_m, -\mathbf{e}_m]^T$, (7.2.2a, b) reduce to the problem of the L_1 approximation with bounded variables. As indicated earlier, we shall comment on these methods in Section 7.5.

In the following, we describe an algorithm for the L_1 approximation with bounded variables between –1 and 1, using linear programming techniques. The algorithm is an extension of the algorithm for the ordinary L_1 solution of overdetermined linear equations. In the linear programming formulation, the initial basic feasible solution is obtained with little computational effort, and intermediate simplex iterations are skipped [2, 3].

7.3 Linear programming formulation of the problem

Since the elements of the residual vector $\mathbf{r} = (r_i)$ are unrestricted in sign, we may write the residual vector as

$$\mathbf{r} = \mathbf{r}_1 - \mathbf{r}_2$$

where vectors $\mathbf{r}_1, \mathbf{r}_2 \geq \mathbf{0}$. It is understood that when $r_i > 0$, $r_{1i} > 0$ and $r_{2i} = 0$ and when $r_i < 0$, $r_{1i} = 0$ and $r_{2i} > 0$. When $r_i = 0$, $r_{1i} = r_{2i} = 0$, where r_{1i} and r_{2i} are the ith elements of vectors $\mathbf{r}_1$ and $\mathbf{r}_2$ respectively.

This problem may be reduced to a linear programming problem in the primal form as follows.

$$\text{minimize } Z = \mathbf{e}_n{}^T\mathbf{r}_1 + \mathbf{e}_n{}^T\mathbf{r}_2$$

subject to

$$\mathbf{Ca} - \mathbf{I}_n\mathbf{r}_1 + \mathbf{I}_n\mathbf{r}_2 = \mathbf{f}$$

$$-\mathbf{I}_m\mathbf{a} \geq -\mathbf{e}_m$$
$$\mathbf{I}_m\mathbf{a} \geq -\mathbf{e}_m$$

$$a_j,\ j = 1, 2, \ldots, m, \text{ unrestricted in sign}, \quad \mathbf{r}_1, \mathbf{r}_2 \geq 0$$

Here, $\mathbf{e}_n$ and $\mathbf{e}_m$ are n and m-vectors respectively, each element of which is 1. Also, $\mathbf{I}_n$ and $\mathbf{I}_m$ are n- and m-unit matrices respectively.

It is more beneficial to work with the dual form of this primal problem, namely (Chapter 3)

$$\text{maximize } z = \mathbf{f}^T\mathbf{w} - \mathbf{e}_m{}^T\mathbf{u} - \mathbf{e}_m{}^T\mathbf{v}$$

subject to

$$\mathbf{C}^T\mathbf{w} - \mathbf{I}_m\mathbf{u} + \mathbf{I}_m\mathbf{v} = \mathbf{0}$$

$$\mathbf{w} \leq \mathbf{e}_n$$
$$\mathbf{w} \geq -\mathbf{e}_n$$

$$u_i, v_i \geq 0, \quad i = 1, 2, \ldots, m$$

The last two vector inequalities reduce to

$$-1 \leq w_i \leq 1, \quad i = 1, 2, \ldots, n$$

where **w**, **u** and **v** are respectively n-, m- and m-vectors for the dual linear programming problem. As in [1, 2], by letting $b_i = w_i + 1$, $i = 1, 2, \ldots, n$, this problem reduces to (in vector-matrix form)

$$\text{maximize } z = \begin{bmatrix} \mathbf{f}^T & -\mathbf{e}_m{}^T & -\mathbf{e}_m{}^T \end{bmatrix} \begin{bmatrix} \mathbf{b} \\ \mathbf{u} \\ \mathbf{v} \end{bmatrix} - \mathbf{f}^T\mathbf{e}_n$$

or since $\mathbf{f}^T\mathbf{e}_n$ is just a constant term, the above reduces to

$$\text{(7.3.1a)} \qquad \text{maximize } z = \begin{bmatrix} \mathbf{f}^T & -\mathbf{e}_m{}^T & -\mathbf{e}_m{}^T \end{bmatrix} \begin{bmatrix} \mathbf{b} \\ \mathbf{u} \\ \mathbf{v} \end{bmatrix}$$

subject to

$$\text{(7.3.1b)} \qquad \begin{bmatrix} \mathbf{C}^T & -\mathbf{I}_m & \mathbf{I}_m \end{bmatrix} \begin{bmatrix} \mathbf{b} \\ \mathbf{u} \\ \mathbf{v} \end{bmatrix} = \mathbf{C}^T\mathbf{e}_n$$

$$\text{(7.3.1c)} \qquad 0 \leq b_i \leq 2, \quad i = 1, 2, \ldots, n$$

$$\text{(7.3.1d)} \qquad u_i, v_i \geq 0, \quad i = 1, 2, \ldots, m$$

Let in (7.3.1a)

(7.3.2a) $$\mathbf{g} = [\mathbf{f}^T \ -\mathbf{e}_m^T \ -\mathbf{e}_m^T]$$

and in (7.3.1b)

(7.3.2b) $$\mathbf{D} = [\mathbf{C}^T \ -\mathbf{I}_m \ \mathbf{I}_m]$$

The r.h.s. of (7.3.1b) is none other that the sum of the columns of matrix $\mathbf{C}^T$. Compare the r.h.s. of (7.3.1b) with those of (5.2.4b) and of (6.3.2b).

Problem (7.3.1) is solved by using a dual simplex algorithm with non-negative bounded variables [1-4, 12]. Since matrix **D** in (7.3.2b) is of dimension m by (n + 2m), a simplex tableau for m constraints in (n + 2m) variables is constructed for problem (7.3.1). Note also that the matrix of constraints **D** in (7.3.2b) is of full rank; rank(**D**) = m. though matrix **C** may be rank deficient.

The basis matrix, denoted by **B** has m linearly independent columns in the simplex tableau. Vectors $\mathbf{y}_j$ are given by

(7.3.3a) $$\mathbf{y}_j = \mathbf{B}^{-1}\mathbf{D}_j, \quad j = 1, 2, \ldots, n+2m$$

where $\mathbf{D}_j$ is the jth column of matrix **D** in (7.3.2b).

Let $\mathbf{b}_B$ denote the initial basic solution. Since in (7.3.1c), variables b_i, i = 1, 2, …, n, are bounded between 0 and 2, as in [2]. In this problem, $\mathbf{b}_B$ is given by

(7.3.3b) $$\mathbf{b}_B = \sum_{i \varepsilon \mathbf{I}(\mathbf{b})} \mathbf{y}_i + \sum_{i \varepsilon \mathbf{L}(\mathbf{b})} \mathbf{y}_i - \sum_{i \varepsilon \mathbf{U}(\mathbf{b})} \mathbf{y}_i$$

The summations are respectively over the basic variables b_i, the non-basic variables b_i at their lower bound (= 0) and the non-basic variables b_i at their upper bound (= 2). The technique used here is very similar to that used for the (ordinary) linear L_1 approximation problem of Chapter 5.

The marginal costs, denoted by $(z_j - g_j)$, are given by

(7.3.3c) $$z_j - g_j = \mathbf{g}_B{}^T\mathbf{y}_j - g_j, \quad j = 1, 2, \ldots, n+2m$$

where the elements of the m-vector $\mathbf{g}_B$ are associated with the basic variables and vector **g** is defined by (7.3.2a).

7.3.1 Properties of the matrix of constraints

We will make use of the fact that there is a kind of asymmetry in parts of the matrix of constraints **D**. There exist the two matrices $-\mathbf{I}_m$ and $\mathbf{I}_m$ as part of **D** in (7.3.2b). We take the m-unit matrix $\mathbf{I}_m$ as the initial basis matrix **B**.

Definition

Let i and j, $(n + 1) \leq i, j \leq (n + 2m)$, be the indices of any two columns in matrix **D** such that $|i - j| = m$. We define columns i and j as two corresponding columns. Consider the following lemmas.

Lemma 7.1

Any two corresponding columns should not appear together in any basis matrix. Otherwise, the basis matrix would be singular.

Lemma 7.2

At any stage of the computation, the corresponding columns of the simplex tableau are related. If one column is known, the other is easily derived. The same is true about their marginal costs. From (7.3.3a) and (7.3.3c) respectively we get

(7.3.4a) $\quad \mathbf{y}_i + \mathbf{y}_{i+m} = 0, \quad i = n+1, n+2, \ldots, n+m$

(7.3.4 b) $\quad (z_i - g_i) + (z_{i+m} - g_{i+m}) = 2, \quad i = n+1, n+2, \ldots, n+m$

These two lemmas indicate that only m columns of the last 2m columns in matrix **D** need to be stored by the program, and no corresponding columns exist in these m columns. Hence, from the matrix of constraints **D**, we need to store m constraints in only (n + m) variables, since the **y** vectors of the other m variables and their marginal costs would be known from (7.3.4a, b).

Lemma 7.3

At any stage of the computation, the last m columns in the simplex m by (n + 2m) tableau are themselves the m columns of the inverse of the basis matrix; $\mathbf{B}^{-1}$.

Lemma 7.4 (Lemma 2 in [1])

The residuals r_i, $i = 1, \ldots, n$, of (7.2.1c) are themselves the marginal costs for the first n columns in the simplex tableau

$$r_i = (z_i - g_i), \quad i = 1, 2, \ldots, n$$

Hence, the objective function z of (7.1.1) is given by

$$z = \sum_{i=1}^{n} |r_i| = \sum_{i=1}^{n} |z_i - g_i| \tag{7.3.5}$$

Lemma 7.5

At any stage of the computation, the solution of the L_1 problem of the bounded variables is given by

$$\mathbf{a}^T = \mathbf{g}_B{}^T \mathbf{B}^{-1}$$

See Lemmas 5.4 and 6.2.

The steps taken to solve problem (7.3.1) are almost identical to those used in solving (5.2.4). However, we make use here of the asymmetry that exists in the matrix of constraints **D**, as indicated above. This is explained in the following section.

7.4 Description of the algorithm

Lemma 7.3 indicates that the last m columns in the simplex tableau are themselves the m columns of matrix $\mathbf{B}^{-1}$. Since matrix $\mathbf{B}^{-1}$ is always available, and from the discussion following (7.3.4a, b), in the initial tableau, we only need to store the first n columns of matrix **D** of (7.3.2b). We call the initial tableau and the following tableaux, the **condensed tableaux**. We consider the columns of matrix $\mathbf{B}^{-1}$ (and their corresponding columns) together with their marginal costs as part of the simplex tableau.

We need an (n + m)-index indicator vector whose elements are +1 or −1. For the first n elements of this indicator, if b_j, $1 \leq j \leq n$, of (7.3.1c), is a basic variable, or if it is at its lower bound (= 0), index j has the value +1. If b_j is at its upper bound (= 2), index j has value −1. For the remaining m elements of this index vector, if column j of matrix $\mathbf{B}^{-1}$, $1 \leq j \leq m$, has its marginal cost stored, the index (n + j) has the value +1. Else, it has the value −1.

The algorithm is in 2 parts. In part 1, the last m columns of matrix **D**; columns (n + m + 1), …, (n + 2m), form an m-unit matrix $\mathbf{I}_m$ and they form the inverse of the initial basis matrix, $\mathbf{B}^{-1}$. This does not require changing the simplex tableau being columns of an m-unit matrix.

In part 2, one calculates the basic solution $\mathbf{b}_B$ from (7.3.3b) and the marginal costs $(z_i - g_i)$, $i = 1, 2, \ldots, n$, from (7.3.3c) for the condensed tableau. We also calculate the marginal costs for $\mathbf{B}^{-1}$. From (7.3.1a) and from the above discussion, vector $(-\mathbf{e}_m)$ would be the price vector for the columns of $\mathbf{B}^{-1}$.

The algorithm proceeds almost exactly as in part 2 in Chapter 5, for the ordinary L_1 solution, where intermediate simplex iterations are skipped. The only difference between the algorithm of Chapter 5 and this one is as follows. From (7.3.1c), if b_j, $1 \le j \le n$, is in the basis for the optimum solution, b_j has to be bounded; i.e., $0 \le b_j \le 2$. While, from (7.3.1d), if u_j or v_j, $1 \le j \le m$, is in the basis, u_j or v_j should only satisfy the non-negativity conditions; i.e., $u_j, v_j \ge 0$.

From Lemma 7.4, the residuals r_i, $i = 1, 2, \ldots, n$, are themselves the marginal costs for the first n columns in the simplex tableau. The bounded L_1 error norm z is calculated from (7.3.5). The optimal solution vector **a** is calculated from Lemma 7.5.

7.5 Numerical results and comments

LA_Lonebv() is an extension to LA_Lone() of Chapter 5, where again certain intermediate simplex iterations are skipped. This algorithm has been tested with many examples, including examples with rank deficient matrices **C**. No failures were encountered.

DR_Lonebv() tests the 8 examples that were solved in Chapter 6 for the one-sided L_1 approximation problem.

Table 7.1

		L_1 Solution		L_1 solution with bounded variables	
Example	**C**(n×m)	Iterations	z	Iterations	z
1	4 × 2	2	18.40	1	20.2
2	10× 5	6	10.00	2	11.00
3	25×10	21	0.0878	12	3.548

The results satisfy the inequalities (7.1.2), namely, $-1 \le a_i \le 1$. Table 7.1 shows the results for 3 of the 8 examples.

For each example, the number of iterations and the optimum norm for the L_1 solution with bounded variables are shown. For comparison purposes, the results for the ordinary L_1 approximation problem are also given. We observe that the number of iterations for this algorithm are, in general, smaller than those for the corresponding ordinary L_1 case.

Analogous to the comments at the end of Chapter 6, our algorithm in this chapter is a special purpose algorithm compared with the algorithms of Armstrong and Hultz (AH) [6], Barrodale and Roberts (BR) [7, 8], Bartels and Conn (BC) [9, 10] and Dax (DA) [11] mentioned in Section 7.2. Their algorithms are general purpose ones. They need more computer storage than ours and as a result, more computation. This point was explained in detail at the end of Chapter 6.

References

1. Abdelmalek, N.N., On the discrete linear L_1 approximation and L_1 solutions of overdetermined linear equations, *Journal of Approximation Theory*, 11(1974) 38-53.
2. Abdelmalek, N.N., An efficient method for the discrete linear L_1 approximation problem, *Mathematics of Computation*, 29(1975) 844-850.
3. Abdelmalek, N.N., L_1 solution of overdetermined systems of linear equations, *ACM Transactions on Mathematical Software*, 6(1980) 220-227.
4. Abdelmalek, N.N., Algorithm 551: A FORTRAN subroutine for the L_1 solution of overdetermined systems of linear equations [F4], *ACM Transactions on Mathematical Software*, 6(1980)228-230.
5. Abdelmalek, N.N., Chebyshev and L_1 solutions of overdetermined systems of linear equations with bounded variables, *Numerical Functional Analysis and Optimization*, 8(1985-86)399-418.

6. Armstrong, R.D. and Hultz, J.W., An algorithm for a restricted discrete approximation in the L_1 norm, *SIAM Journal on Numerical Analysis*, 14(1977)555-565.
7. Barrodale, I. and Roberts, F.D.K., An efficient algorithm for discrete l_1 linear approximation with linear constraints, *SIAM Journal on Numerical Analysis*, 15(1978)603-611.
8. Barrodale, I. and Roberts, F.D.K., Algorithm 552: Solution of the constrained L_1 linear approximation problem, *ACM Transactions on Mathematical Software*, 6(1980)231-235.
9. Bartels, R.H. and Conn, A.R., Linearly constrained discrete l_1 problems, *ACM Transactions on Mathematical Software*, 6(1980)594-608.
10. Bartels, R.H. and Conn, A.R., Algorithm 563: A program for linearly constrained discrete l_1 problems, *ACM Transactions on Mathematical Software*, 6(1980)609-614.
11. Dax, A., The l_1 solution of linear equations subject to linear constraints, *SIAM Journal on Scientific and Statistical Computation*, 10(1989)328-340.
12. Hadley, G., *Linear Programming*, Addison-Wesley, Reading, MA, 1962.
13. Spath, H., *Mathematical Algorithms for Linear Regression*, Academic Press, English Edition, London, 1991.

7.6 DR_Lonebv

```
/*--------------------------------------------------------------------
DR_Lonebv
----------------------------------------------------------------------
This program is a driver for the function LA_Lonebv(), which solves
an overdetermined system of linear equations in the L-One norm
subject to the constraints that each element of the solution vector
"a" is bounded between -1 and 1;

              -1 <= a[j] <= 1,   j = 1, 2, ..., m

The overdetermined system has the form

                        c*a = f

"c" is a given real n by m matrix of rank k, k <= m <= n.
"f" is a given real n vector.
"a" is the solution m vector.

This driver contains the 8 examples whose results are given in the
text.
--------------------------------------------------------------------*/

#include "DR_Defs.h"
#include "LA_Prototypes.h"

#define NN_MM_ROWS    (NN_ROWS + MM_COLS)
#define N1            4
#define M1            2
#define N2            5
#define M2            3
#define N3            6
#define M3            3
#define N4            7
#define M4            3
#define N5            8
#define M5            4
#define N6            10
#define M6            5
#define N7            25
#define M7            10
#define N8            8
#define M8            4
```

```
void DR_Lonebv (void)
{
    /*----------------------------------------
      Constant matrices/vectors
    ----------------------------------------*/
    static tNumber_R c1init[N1][M1] =
    {
        { 0.0, -2.0 },
        { 0.0, -4.0 },
        { 1.0, 10.0 },
        {-1.0, 15.0 }
    };

    static tNumber_R c2init[N2][M2] =
    {
        { 1.0,  2.0, 0.0 },
        {-1.0, -1.0, 0.0 },
        { 1.0,  3.0, 0.0 },
        { 0.0,  1.0, 0.0 },
        { 0.0,  0.0, 1.0 }
    };

    static tNumber_R c3init[N3][M3] =
    {
        { 0.0, -1.0,  0.0 },
        { 1.0,  3.0, -4.0 },
        { 1.0,  0.0,  0.0 },
        { 0.0,  0.0,  1.0 },
        {-1.0,  1.0,  2.0 },
        { 1.0,  1.0,  1.0 }
    };

    static tNumber_R c4init[N4][M4] =
    {
        { 1.0, 0.0,  1.0 },
        { 1.0, 2.0,  2.0 },
        { 1.0, 2.0,  0.0 },
        { 1.0, 1.0,  0.0 },
        { 1.0, 0.0, -1.0 },
        { 1.0, 0.0,  0.0 },
        { 1.0, 1.0,  1.0 }
    };

    static tNumber_R c5init[N5][M5] =
```

```
{
    { 1.0, -3.0,  9.0, -27.0 },
    { 1.0, -2.0,  4.0,  -8.0 },
    { 1.0, -1.0,  1.0,  -1.0 },
    { 1.0,  0.0,  0.0,   0.0 },
    { 1.0,  1.0,  1.0,   1.0 },
    { 1.0,  2.0,  4.0,   8.0 },
    { 1.0,  3.0,  9.0,  27.0 },
    { 1.0,  4.0, 16.0,  64.0 }
};

static tNumber_R c6init[N6][M6] =
{
    { 1.0, 0.0,  0.0,  0.0,  0.0 },
    { 0.0, 1.0,  0.0,  0.0,  0.0 },
    { 0.0, 0.0,  1.0,  0.0,  0.0 },
    { 0.0, 0.0,  0.0,  1.0,  0.0 },
    { 0.0, 0.0,  0.0,  0.0,  1.0 },
    { 1.0, 1.0,  1.0,  1.0,  1.0 },
    { 0.0, 1.0,  1.0,  1.0,  1.0 },
    {-1.0, 0.0, -1.0, -1.0, -1.0 },
    { 1.0, 1.0,  0.0,  1.0,  1.0 },
    { 1.0, 1.0,  1.0,  0.0,  1.0 }
};

static tNumber_R c8init[N8][M8] =
{
    { 1.0, 1.0,   1.0,  1.0 },
    { 1.0, 2.0,   4.0,  4.0 },
    { 1.0, 3.0,   9.0,  9.0 },
    { 1.0, 4.0,  16.0, 16.0 },
    { 1.0, 5.0,  25.0, 25.0 },
    { 1.0, 6.0,  36.0, 36.0 },
    { 1.0, 7.0,  49.0, 49.0 },
    { 1.0, 8.0,  64.0, 64.0 }
};

static tNumber_R f1[N1+1] =
{   NIL,
    -12.0, 6.0, 0.0, 5.0
};

static tNumber_R f2[N2+1] =
{   NIL,
    1.0, 2.0, 1.0, -3.0, 0.0
```

```
};

static tNumber_R f3[N3+1] =
{   NIL,
    1.0, 2.0, 3.0, 2.0, 2.0, 4.0
};

static tNumber_R f4[N4+1] =
{   NIL,
    0.0, -2.0, 1.0, -1.0, 5.0, 7.0, 0.0
};

static tNumber_R f5[N5+1] =
{   NIL,
    3.0, -3.0, -2.0, 0.0, 7.0, -1.0, 5.0, 2.0
};

static tNumber_R f6[N6+1] =
{   NIL,
    1.0, -1.0, 0.0, -1.0, 1.0, 0.0, 2.0, 3.0, -3.0, -2.0
};

static tNumber_R f7[N7+1] =
{   NIL,
    0.0872673, 0.0872794, 0.0873029, 0.0873315, 0.0873576,
    0.3491184, 0.3498802, 0.3513824, 0.3532572, 0.3550109,
    0.6111334, 0.6150641, 0.6230824, 0.6336395, 0.6441493,
    0.8733883, 0.8841621, 0.9071868, 0.9400757, 0.9766021,
    1.135895,  1.157550,  1.206257,  1.283258,  1.384432
};

static tNumber_R f8[N8+1] =
{   NIL,
    2.0, 2.5, 2.0, 6.5, 3.5, 4.5, 6.0, 7.0
};

/*-----------------------------------------
  Variable matrices/vectors
-----------------------------------------*/
tMatrix_R   ct      = alloc_Matrix_R (MM_COLS, NN_ROWS);
tVector_R   f       = alloc_Vector_R (NN_ROWS);
tVector_R   rbv     = alloc_Vector_R (NN_MM_ROWS);
tVector_R   a       = alloc_Vector_R (MM_COLS);
tMatrix_R   binv    = alloc_Matrix_R (MM_COLS, MM_COLS);
tVector_R   bv      = alloc_Vector_R (MM_COLS);
```

```
tVector_R   thbv    = alloc_Vector_R (NN_MM_ROWS);
tVector_I   icbas   = alloc_Vector_I (MM_COLS);
tVector_I   ibbv    = alloc_Vector_I (NN_MM_ROWS);
tMatrix_R   c7      = alloc_Matrix_R (N7, M7);

tMatrix_R   c1      = init_Matrix_R (&(c1init[0][0]), N1, M1);
tMatrix_R   c2      = init_Matrix_R (&(c2init[0][0]), N2, M2);
tMatrix_R   c3      = init_Matrix_R (&(c3init[0][0]), N3, M3);
tMatrix_R   c4      = init_Matrix_R (&(c4init[0][0]), N4, M4);
tMatrix_R   c5      = init_Matrix_R (&(c5init[0][0]), N5, M5);
tMatrix_R   c6      = init_Matrix_R (&(c6init[0][0]), N6, M6);
tMatrix_R   c8      = init_Matrix_R (&(c8init[0][0]), N8, M8);

int         iter;
int         i, j, k, m, n, Iexmpl;
tNumber_R   d, dd, ddd, e, ee, eee, z;

eLaRc       rc = LaRcOk;

prn_dr_bnr ("DR_Lonebv, Bounded L-One Solution of an "
            "Overdetermined System of Linear Equations");

z = 0.0;

for (j = 1; j <= 5; j++)
{
    d = 0.15* (j-3);
    dd = d*d;
    ddd = d*dd;
    for (i = 1; i <= 5; i++)
    {
        e = 0.15* (i-3);
        ee = e*e;
        eee = e*ee;
        k = 5* (j-1) + i;
        c7[k][1] = 1.0;
        c7[k][2] = d;
        c7[k][3] = e;
        c7[k][4] = dd;
        c7[k][5] = ee;
        c7[k][6] = e*d;
        c7[k][7] = ddd;
        c7[k][8] = eee;
        c7[k][9] = dd*e;
        c7[k][10] = ee*d;
```

```
        }
    }

    for (Iexmpl = 1; Iexmpl <= 8; Iexmpl++)
    {
        switch (Iexmpl)
        {
        case 1:
            n = N1;
            m = M1;
            for (i = 1; i <= n; i++)
            {
                f[i] = f1[i];
                for (j = 1; j <= m; j++) ct[j][i] = c1[i][j];
            }
            break;
        case 2:
            n = N2;
            m = M2;
            for (i = 1; i <= n; i++)
            {
                f[i] = f2[i];
                for (j = 1; j <= m; j++) ct[j][i] = c2[i][j];
            }
            break;
        case 3:
            n = N3;
            m = M3;
            for (i = 1; i <= n; i++)
            {
                f[i] = f3[i];
                for (j = 1; j <= m; j++) ct[j][i] = c3[i][j];
            }
            break;
        case 4:
            n = N4;
            m = M4;
            for (i = 1; i <= n; i++)
            {
                f[i] = f4[i];
                for (j = 1; j <= m; j++) ct[j][i] = c4[i][j];
            }
            break;
        case 5:
            n = N5;
```

```
        m = M5;
        for (i = 1; i <= n; i++)
        {
            f[i] = f5[i];
            for (j = 1; j <= m; j++) ct[j][i] = c5[i][j];
        }
        break;
    case 6:
        n = N6;
        m = M6;
        for (i = 1; i <= n; i++)
        {
            f[i] = f6[i];
            for (j = 1; j <= m; j++) ct[j][i] = c6[i][j];
        }
        break;
    case 7:
        n = N7;
        m = M7;
        for (i = 1; i <= n; i++)
        {
            f[i] = f7[i];
            for (j = 1; j <= m; j++) ct[j][i] = c7[i][j];
        }
        break;
    case 8:
        n = N8;
        m = M8;
        for (i = 1; i <= n; i++)
        {
            f[i] = f8[i];
            for (j = 1; j <= m; j++) ct[j][i] = c8[i][j];
        }
        break;
    default:
        break;
    }

    prn_algo_bnr ("Lonebv");
    prn_example_delim();
    PRN ("Example #%d: Size of matrix \"c\" %d by %d\n",
         Iexmpl, n, m);
    prn_example_delim();
    PRN ("Bounded L-One Solution of an Overdetermined "
         "Equations\n");
```

```
        prn_example_delim();
        PRN ("r.h.s. Vector \"f\"\n");
        prn_Vector_R (f, n);
        PRN ("Transpose of Coefficient Matrix, \"ct\"\n");
        prn_Matrix_R (ct, m, n);

        rc = LA_Lonebv (m, n, ct, f, icbas, binv, bv, ibbv, thbv,
                        &iter, rbv, a, &z);

        if (rc >= LaRcOk)

        {
            PRN ("\n");
            PRN ("Results of the Bounded L-One Solution\n");
            PRN ("Bounded L-One solution vector \"a\"\n");
            prn_Vector_R (a, m);
            PRN ("Bounded L-One residual vector \"r\"\n");
            prn_Vector_R (rbv, n);
            PRN ("Bounded L-One norm \"z\" = %8.4f\n", z);
            PRN ("No. of Iterations = %d\n", iter);
        }

        prn_la_rc (rc);
    }

    free_Matrix_R (ct, MM_COLS);
    free_Vector_R (f);
    free_Vector_R (rbv);
    free_Vector_R (a);
    free_Matrix_R (binv, MM_COLS);
    free_Vector_R (bv);
    free_Vector_R (thbv);
    free_Vector_I (icbas);
    free_Vector_I (ibbv);
    free_Matrix_R (c7, N7);

    uninit_Matrix_R (c1);
    uninit_Matrix_R (c2);
    uninit_Matrix_R (c3);
    uninit_Matrix_R (c4);
    uninit_Matrix_R (c5);
    uninit_Matrix_R (c6);
    uninit_Matrix_R (c8);
}
```

7.7 LA_Lonebv

```
/*-----------------------------------------------------------------
LA_Lonebv
-------------------------------------------------------------------
This program calculates the L-One solution of an overdetermined
system of linear equations subject to the conditions that the
elements of the solution vector be bounded between -1 and +1.

This program uses a modified simplex method to the linear
programming formulation of the problem.  In this method certain
intermediate simplex iterations are skipped.

The system of linear equations has the form

                          c*a = f

"c" is a given real n by m matrix of rank k, k <= m <= n.
"f" is a given real n vector.

The problem is to calculate the elements of the solution vector
"a" that minimizes the L-One norm z

          z = |rbv[1]| + |rbv[2]| + ... + |rbv[n]|

subject to the constraints

             -1 <= a[j] <= 1,   j = 1, 2, ..., m

rbv[i] is the ith residual and is given by

 rbv[i] = c[i][1]*a[1] + c[i][2]*a[2] + ... + c[i][m]*a[m] - f[i],
                          i = 1, 2, ..., n

Inputs
m       Number of columns of matrix "c" of the system c*a = f.
n       Number of rows of matrix "c" of the system c*a = f.
ct      A real m by n matrix containing the transpose of matrix "c"
        of the system c*a = f.
f       A real n vector containing the r.h.s. of the system c*a = f.

Other Parameters
binv    A real m square matrix containing the inverse of the basis
        matrix in the linear programming problem.
```

```
bv      A real m vector containing the basic solution in the linear
        programming problem.
thbv    An (n + m) vector containing the ratios

                    thbv[j] = rbv[j]/ct[iout][j]

        "iout" corresponds to the basic vector that leaves the
        basis.
ibbv    A sign (n + m) vector.  Its first n elements have the
        values 1 or -1.
        ibbv[j] = 1 indicates that column j of matrix "ct" is in
        the basis or at its lower bound 0.
        ibbv[j] = -1 indicates that column "j" is at its upper
        bound 2.
icbas   An integer m vector containing the indices of the columns
        of "ct" forming the basis matrix.

Outputs
iter    Number of iterations, or the number of times the simplex
        tableau is changed by a Gauss-Jordan step.
a       A real m vector containing the bounded L-One solution of
        the system c*a = f.
rbv     An (n + m) vector.  Its first n elements are the bounded
        L-One residual vector r = c*a - f.
z       The minimum bounded L-One norm of the residuals "rbv".

Returns one of
        LaRcSolutionFound
        LaRcNoFeasibleSolution
        LaRcErrBounds
        LaRcErrNullPtr
-------------------------------------------------------------------*/

#include "LA_Prototypes.h"

eLaRc LA_Lonebv (int m, int n, tMatrix_R ct, tVector_R f,
    tVector_I icbas, tMatrix_R binv, tVector_R bv, tVector_I ibbv,
    tVector_R thbv, int *pIter, tVector_R rbv, tVector_R a,
    tNumber_R *pZ)
{
    int          ij = 0, kk = 0, n1 = 0, nm = 0;
    int          iout = 0, jout = 0, jin = 0, ivo = 0, itest = 0;
    tNumber_R    pivot = 0.0, pivoto = 0.0, tpeps = 0.0, xb = 0.0;

    /* Validation of the data before executing the algorithm */
```

```
eLaRc       rc = LaRcSolutionFound;
VALIDATE_BOUNDS ((0 < m) && (m <= n) && !((n == 1) && (m == 1)));
VALIDATE_PTRS   (ct && f && icbas && binv && bv && ibbv && thbv
                 && pIter && rbv && a && pZ);

nm = n + m;
n1 = n + 1;
tpeps = 2.0 + EPS;
*pIter = 0;

/* Part 1 of the algorithm. Obtain a feasible basic solution */
LA_lonebv_part_1 (m, n, ct, f, icbas, binv, bv, ibbv, rbv);

/* Part 2 of the algorithm */
for (kk = 1; kk <= n*n; kk++)
{
    ivo = 0;
    xb = 0.0;

    /* Determine the vector that leaves the basis */
    LA_lonebv_vleav (&ivo, &iout, &xb, m, n, icbas, bv);

    /* Calculate the results */
    if (ivo == 0)
    {
        LA_lonebv_res (m, n, f, icbas, binv, rbv, a, pZ);
        GOTO_CLEANUP_RC (LaRcSolutionFound);
    }

    jout = icbas[iout];

    /* Calculate the possible ratios thbv[j] */
    LA_lonebv_thbv (ivo, iout, m, n, ct, icbas, binv, ibbv, thbv,
                    rbv);

    pivoto = 1.0;
    itest = 0;

    for (ij = 1; ij <= n*n; ij++)
    {
        /* Determine the vector that enters the basis */
        LA_lonebv_vent (ivo, &jin, &itest, m, n, thbv);

        /* Solution is not feasible */
        if (itest != 1)
```

```
            {
                GOTO_CLEANUP_RC (LaRcNoFeasibleSolution);
            }

            /* Update the solution vector "bv" */
            LA_lonebv_update_bv (iout, jout, jin, &pivot, pivoto,
                                 &xb, m, n, ct, binv, bv, ibbv);

            if ((jin > n && xb > -EPS) || (xb >= -EPS && xb<= tpeps))
            {
                thbv[jin] = 0.0;
            }
            else
            {
                itest = 0;
                thbv[jin] = 0.0;
            }
            icbas[iout] = jin;
            if (itest == 1) break;
            pivoto = pivot;
            jout = jin;
        }

        /* A Gauss-Jordan elimination step */
        LA_lonebv_gauss_jordn (jin, iout, m, n, ct, icbas, binv, bv,
                               ibbv, rbv);
        *pIter = *pIter + 1;
    }

CLEANUP:

    return rc;
}

/*-------------------------------------------------------------------
Part 1; Initialization for LA_Lonebv()
-------------------------------------------------------------------*/
void LA_lonebv_part_1 (int m, int n, tMatrix_R ct, tVector_R f,
    tVector_I icbas, tMatrix_R binv, tVector_R bv, tVector_I ibbv,
    tVector_R rbv)
{
    int         i, j, k;
    tNumber_R   s, sa;

    /* Part 1 of the algorithm. Obtain a feasible basic solution */
```

```
    for (j = 1; j <= m; j++)
    {
        k = n + j;
        icbas[j] = k;
        ibbv[k] = -1;
        rbv[k] = 0.0;
        for (i = 1; i <= m; i++) binv[i][j] = 0.0;
        binv[j][j] = 1.0;
    }

    /* Calculate the residuals (marginal costs) */
    for (j = 1; j <= n; j++)
    {
        ibbv[j] = 1;
        s = - f[j];
        for (i = 1; i <= m; i++) s = s - ct[i][j];
        rbv[j] = s;
        if (s < -EPS) ibbv[j] = -1;
    }

    /* Calculate the initial basic solution */
    for (i = 1; i <= m; i++)
    {
        s = 0.0;
        for (j = 1; j <= n; j++)
        {
            sa = ct[i][j];
            if (ibbv[j] == -1) sa = -sa;
            s = s + sa;
        }
        bv[i] = s;
    }
}

/*-------------------------------------------------------------------
Determine the vector that leaves the basis in LA_Lonebv()
-------------------------------------------------------------------*/
void LA_lonebv_vleav (int *pIvo, int *pIout, tNumber_R *pXb, int m,
    int n, tVector_I icbas, tVector_R bv)
{
    int         i, k;
    tNumber_R   e, d, g, tpeps;

    tpeps = 2.0 + EPS;
    g = 1.0;
```

```
    for (i = 1; i <= m; i++)
    {
        e = bv[i];
        k = icbas[i];
        if (e < -EPS)
        {
            d = e;
            if (d < g)
            {
                *pIvo = -1;
                g = d;
                *pIout = i;
                *pXb = e;
            }
        }
        if (k <= n && e > tpeps)
        {
            d = 2.0 - e;
            if (d < g)
            {
                *pIvo = 1;
                g = d;
                *pIout = i;
                *pXb = e;
            }
        }
    }
}

/*------------------------------------------------------------------
Calculate the parameters of vector "thbv" in LA_Lonebv()
------------------------------------------------------------------*/
void LA_lonebv_thbv (int ivo, int iout, int m, int n, tMatrix_R ct,
    tVector_I icbas, tMatrix_R binv, tVector_I ibbv, tVector_R thbv,
    tVector_R rbv)
{
    int         i, j, ii, i0 = 0, k, nm;
    tNumber_R   e = 0, d, gg, thmax;

    nm = n + m;

    thmax = 0.0;
    for (j = 1; j <= nm; j++)
    {
        thbv[j] = 0.0;
```

```
    ii = 0;
    for (i = 1; i <= m; i++)
    {
        if (j == icbas[i])
        {
            ii = 1;
            i0 = i;
        }
    }
    if (ii == 1)
    {
        if (j > n && i0 == iout)
        {
            thbv[j] = -2.0;
            rbv[j] = 2.0;
            ibbv[j] = -ibbv[j];
            gg = thbv[j];
            if (gg < 0.0) gg = -gg;
            if (gg > thmax) thmax = gg;
        }
    }
    else if (ii == 0)
    {
        if (j <= n) e = ct[iout][j];
        if (j > n)
        {
            k = j - n;
            d = -1.0;
            if (ibbv[j] == -1) d = -d;
            e = d * binv[iout][k];
            if ((ivo == -1 && e > EPS) || (ivo == 1 && e < -EPS))
            {
                e = -e;
                rbv[j] = 2.0 - rbv[j];
                ibbv[j] = -ibbv[j];
            }
        }

        if (fabs (e) > EPS)
        {
            d = rbv[j];
            if (fabs (d) < PREC) d = PREC * ibbv[j];
            if (fabs (d) < PREC && j > n) d = PREC;
            thbv[j] = d/e;
        }
```

```
        }
    }
}

/*-----------------------------------------------------------------------
Determine the vector that enters the basis in LA_Lonebv()
-----------------------------------------------------------------------*/
void LA_lonebv_vent (int ivo, int *pJin, int *pItest, int m, int n,
    tVector_R thbv)
{
    int         j, nm;
    tNumber_R   e, d, thmax, thmin;

    nm = n + m;
    thmax = 1.0/EPS;
    thmin = -thmax;
    for (j = 1; j <= nm; j++)
    {
        e = thbv[j];
        d = e * ivo;
        if (d > 0.0)
        {
            if (ivo == -1)
            {
                if (e > thmin)
                {
                    thmin = e;
                    *pJin = j;
                    *pItest = 1;
                }
            }
            if (ivo == 1)
            {
                if (e < thmax)
                {
                    thmax = e;
                    *pJin = j;
                    *pItest = 1;
                }
            }
        }
    }
}

/*-----------------------------------------------------------------------
```

```
Update the basic solution bv in LA_Lonebv()
-----------------------------------------------------------------------*/
void LA_lonebv_update_bv (int iout, int jout, int jin,
    tNumber_R *pPivot, tNumber_R pivoto, tNumber_R *pXb, int m,
    int n, tMatrix_R ct, tMatrix_R binv, tVector_R bv,
    tVector_I ibbv)
{
    int         i, kin, nm;
    tNumber_R   e, pivotn, tpeps;
    nm = n + m;

    tpeps = 2.0 + EPS;
    if (jin > n)
    {
        kin = jin - n;
        e = -1.0;
        if (ibbv[jin] == -1) e = -e;
        *pPivot = e * (binv[iout][kin]);
    }
    if (jin <= n) *pPivot = ct[iout][jin];

    pivotn = *pPivot/pivoto;
    if ((jin > n && *pXb > -EPS) ||  *pXb > tpeps)
    {
        if (jout <= n)
        {
            for (i = 1; i <= m; i++)
            {
                bv[i] = bv[i] - ct[i][jout] - ct[i][jout];
            }
            ibbv[jout] = -1;
            if (pivotn <= 0.0)
            {
                for (i = 1; i <= m; i++)
                {
                    bv[i] = bv[i] + ct[i][jin] + ct[i][jin];
                }
                ibbv[jin] = 1;
            }
        }
    }
    else if (pivotn > 0.0)
    {
        for (i = 1; i <= m; i++)
        {
```

```
            bv[i] = bv[i] + ct[i][jin] + ct[i][jin];
        }
        ibbv[jin] = 1;
    }
    *pXb = bv[iout]/ (*pPivot);
}

/*-----------------------------------------------------------------------
A Gauss-Jordan elimination step in LA_Lonebv()
-----------------------------------------------------------------------*/
void LA_lonebv_gauss_jordn (int jin, int iout, int m, int n,
    tMatrix_R ct, tVector_I icbas, tMatrix_R binv, tVector_R bv,
    tVector_I ibbv, tVector_R rbv)
{
    tNumber_R   pivot, e = 0, d;
    int         i, ic = 0, j, k, kin = 0, n1, nm;

    nm = n + m;
    n1 = n + 1;
    pivot = ct[iout][jin];
    if (jin > n)
    {
        kin = jin - n;
        e = -1.0;
        if (ibbv[jin] == -1) e = -e;
        pivot = e * binv[iout][kin];
    }
    for (j = 1; j <= n; j++) ct[iout][j] = ct[iout][j]/pivot;
    for (j = 1; j <= m; j++) binv[iout][j] = binv[iout][j]/pivot;
    bv[iout] = bv[iout]/pivot;
    for (i = 1; i <= m; i++)
    {
        if (i != iout)
        {
            d = ct[i][jin];
            if (jin > n) d = e * binv[i][kin];
            for (j = 1; j <= n; j++)
            {
                ct[i][j] = ct[i][j] - d * ct[iout][j];
            }
            for (j = 1; j <= m; j++)
            {
                binv[i][j] = binv[i][j] - d * binv[iout][j];
            }
            bv[i] = bv[i] - d * bv[iout];
```

```
        }
    }
    d = rbv[jin];
    for (j = 1; j <= n; j++)
    {
        rbv[j] = rbv[j] - d * ct[iout][j];
    }
    for (j = n1; j <= nm; j++)
    {
        e = -1.0;
        if (ibbv[j] == -1) e = -e;
        k = j - n;
        rbv[j] = rbv[j] - e * d * binv[iout][k];
    }
    for (j = 1; j <= n; j++)
    {
        for (i = 1; i <= m; i++)
        {
            ic = 0;
            if (j == icbas[i]) ic = 1;
        }
        if (ic == 0)
        {
            d = rbv[j] * ibbv[j];
            if (d >= 0.0) break;
            rbv[j] = 0.0;
        }
    }
}

/*-------------------------------------------------------------------
Calculate the results of LA_Lonebv()
------------------------------------------------------------------*/
void LA_lonebv_res (int m, int n, tVector_R f, tVector_I icbas,
    tMatrix_R binv, tVector_R rbv, tVector_R a, tNumber_R *pZ)
{
    tNumber_R   e, s;
    int         i, j, k;

    for (j = 1; j <= m; j++)
    {
        s = 0.0;
        for (i = 1; i <= m; i++)
        {
            k = icbas[i];
```

```
            e = -1.0;
            if (k <= n) e = f[k];
            s = s + e * binv[i][j];
        }
        a[j] = s;
    }
    s = 0.0;
    for (j = 1; j <= n; j++) s = s + fabs (rbv[j]);
    *pZ = s;
}
```

Chapter 8

L_1 Polygonal Approximation of Plane Curves

8.1 Introduction

Polygonal approximation of a given plane curve is done by first digitizing the given curve into discrete points and then approximating the digitized curve by a polygon of connected straight lines. The points at which the lines join are usually, but not necessarily, a subset of the digitized points of the curve. There are certain particulars about polygonal approximation that are summarized in the following sections.

8.1.1 Two basic issues

There are two basic issues associated with piecewise approximation in general, including polygonal approximation:

(1) Find a polygonal approximation of the given curve such that the measured residual or error norm between the discrete points of the curve and the straight line in each segment does not exceed a specified tolerance ε.

(2) Given a specified number of segments, find the polygonal approximation of the given curve such that the error norms in all the segments are nearly equal. This is known as the near balances norm approximation.

There might be a third issue, which is:

(3) Find a polygonal approximation of the given curve such that the error norms in all the segments are nearly equal and that each does not exceed a specified tolerance ε. This outcome could be achieved by assuming a number of segments n, and calculating the polygonal approximation for near balanced

norms. If the calculated norm is $> \varepsilon$, increase n by 1 at a time and the process is repeated until the obtained norm is $\leq \varepsilon$. If the calculated norm is $< \varepsilon$, the process is reversed by decreasing n by 1, and repeated until the calculated norm starts to be $> \varepsilon$.

8.1.2 Approaches for polygonal approximation

In the literature, a large number of polygonal approximation methods have been proposed. The various approaches of these algorithms can be classified into categories such as:

(a) Sequential or scan-along approach: This approach is utilized by Kurozumi and Davis [18], Sklansky and Gonzalez [35] and by Wall and Danielsson [39] as well as the approach of our algorithm [2]. In the sequential approach, one starts at a point P_a, and scans sequentially along the following points of the digitized curve. The scan stops at point P_b, when the norm of the errors between the intermediate points and the line joining P_a and P_b begins to exceed a specified tolerance ε. The method is repeated by starting at point P_b and so on.

(b) Split approach: This approach has been used by Ramer [25] and by Duda and Hart [9]. In this approach, the digitized points of a curve segment is approximated by a straight line connecting the first and the last points of the segment. If the calculated error norm is larger than the specified value ε, the curve segment is broken into two segments at the curve point most distant from the straight line segment and the new error norm is calculated. This process is repeated until each curve segment is approximated by a straight line connecting its end points and the error norm for each segment is $\leq$ the specified value ε.

(c) Split or merge approach: This is demonstrated by Pavlidis [21] and by Ansari and Delp [4]. In this approach, one starts from an initial segmentation of the given digitized curve and calculates the error norms between the given points and the straight lines. The segmentation is then split or merged by adjusting segment end points until the error norms are driven under the pre-specified bound.

(d) Dominant point or significant vertices detection approach: This is explained by Teh and Chin [37], Ansari and Delp [4], and by Ray and Ray [27]. See also Sarkar [33] for polygonal approximation of chain-coded curves. This method extracts the most significant points in the curve before approximating it. These are the peak points, positive maximum and negative minimum curvature points. The method then partitions the curves between the dominant points.

Hamann and Chen [13] presented a method for selecting a subset of points from the finite set of curve points. The selected subset is based on assigning weights to all the initial data points. Only the most significant points are used in the piecewise approximation. Sato [34] presented a **point choice function** that relates the original points and the chosen points. Sarkar et al. [32] presented a genetic algorithm based approach to locate a specified number of significant points.

Johnson and Vogt [16] presented a geometric method for polygonal approximation of convex arcs. The method tends to produce many points where the curvature is high and few points where the curvature is low. Pei and Lin [22] introduced a technique for detecting the dominant points on a digital curve by a scale-space filtering with a Gaussian kernel.

(e) K-means based approach: This method was elaborated upon by Yin [41]. In this approach, the digitized points are first partitioned into k connected arbitrary clusters and their principal axes are computed. The perpendicular distances from the points in the cluster to the principal axis of the cluster are calculated. Points are reassigned to a neighboring cluster if they have shorter distances to the principal axis of the neighboring cluster. This process is repeated until the points of each cluster have the shortest distances to the principal axis of their own cluster.

(f) Genetic algorithms approach, particularly for closed digitized curves: Such algorithms are described by Yin [42], Huang and Sun [15], Sun and Huang [36] and by Ho and Chen [14]. A disadvantage of all the methods outlined in (a) to (e) above is that the final polygonal approximation results depend on the selection of the initial points and the arbitrary initial solution.

See for example, Kolesnikov and Franti [17]. Genetic algorithms are supposed to rectify this disadvantage. Genetic algorithms can be viewed as stochastic search algorithms based on natural genetic systems, by stimulating the biological model of evolution. These algorithms simulate the survival of the fittest elements, which reproduce and then compete with each other. That produces an optimal or near optimal solution in a search space for optimization problems. See for example, Golden [11] and Michalewicz [20].

Yin [42] presented three genetic algorithms for polygonal approximation of digital curves. The first one minimizes the number of sides of the polygon such that the approximation error norms, each does not exceed a specified value. The second one minimizes the error norm for a given number of sides. The third one determines the approximating polygon automatically without any given condition.

8.1.3 Other unique approaches

There are other unique approaches to polygonal approximations of plane curves.

(a) Lopes et al. [19] considered the problem of computing a polygonal approximation for a plane curve presented implicitly by a given function. A robust and adaptive polygonal approximation was proposed. By robust they mean that the algorithm captures both the topology and the geometry of the curve. By adaptive, they mean that the polygonal approximation is adapted to the geometry of the curve; having longer edges for the flat parts of the curve where the curvature is low, and shorter edges where the curvature is high.

(b) Badi'i and Peikari [6] presented a linear functional approximation of planar curves based on adaptive segmentation procedure. Their method alleviates the need for specifying the number of segments and minimizing the error norm.

(c) Rannou and Gregor [26] presented a polygonal approximation algorithm for closed contours with a unique feature. Their algorithm creates polygons whose edges are all of the same length. Then a one-dimensional description of the given shape is possible by using the interior angles between the approximating lines.

8.1.4 Criteria by which error norm is chosen

Polygonal approximation may also be classified according to the criterion by which the error norm ε between the straight line and the digitized points of a segment is chosen.

(1) The mean square (the L_2) error norm was used by Cantoni [7], Pavlidis [21], Salotti [31] and Sarkar et al. [32].

(2) The uniform or the Chebyshev norm was used by Ramer [25], Tomek [38], Williams [40], Kurozumi and Davis [18], Dunham [10], Sklansky and Gonzalez [35] and Zhu and Seneviratne [43].

(3) The L_1 error norm was used by a number of authors. Johnson and Vogt [16] used a geometric method for polygonal approximation of a planar convex curve. Their method minimizes the area between the curve and the polygonal arc. This method is a kind of an L_1 piecewise approximation method. Ray and Ray [28] presented a technique that uses the L_1 error norm criterion. Wall and Danielsson [39] also used the area deviation for each line segment, which is itself the L_1 error norm criterion. We shall comment in Section 8.5 on these 3 algorithms that also use the L_1 error norm criterion.

(4) The city block metric criterion was used by Pikaz and Dinstein [24] to define the maximal distance of a point (x, y) between the line and the curve as $|x| + |y|$. That is, the sum of the sides of the triangle whose diagonal is the perpendicular distance between the point (x, y) and the approximating line.

8.1.5 Direction of error measure

One more issue by which the errors between the digitized points and the straight line approximating them are measured is considered:

(a) The error may be measured in the direction perpendicular to the approximating line, known as the Euclidean distance, such as in the case of Pavlidis [21], Sarkar et al. [32], Ramer [25],

Williams [40], Kurozumi and Davis [18], Sklansky and Gonzalez [35], Dunham [10] and Zhu and Seneviratne [43], or

(b) The error may be measured in the direction along the y-axis, such as in the case of Cantoni [7], Tomek [38] and our technique.

8.1.6 Comparison and stability of polygonal approximations

Arkin et al. [5] developed a method for comparing two polygons, one is stored as a model for a particular object and the other as an approximate one of the object. The method is based on the L_2 distance between certain functions of the two polygons.

Rosin [30] developed several measures to assess the stability of polygonal approximation algorithms under perturbation of the given data and changes of the algorithms scale parameters.

We present here an algorithm for obtaining a polygonal approximation in the L_1 norm of a plane curve of an arbitrary shape [2]. In the presence of spikes and wild points in a given set of discrete points, the L_1 approximation is usually recommended over other norms to approximate this set of points. See Section 2.3, Rice and White [29] and Abdelmalek [3].

In the current algorithm, the L_1 (error) norm in any segment is not to exceed a pre-assigned value. The errors between the discrete points and the approximating lines are measured along the y-axis direction. We shall discuss this point in Section 8.5. The given curve may be an open curve, as in Figure 8-1 or a closed one, as in Figure 8-2.

The given curve is first digitized into a number of discrete points, not necessarily at equal distances. The algorithm is then applied to the discrete points. The vertices of the polygon are a subset of the points of the given digitized curve. In other words, the approximating straight line to any segment of the given curve passes through the two end points of the segment. The algorithm uses linear programming techniques [12], making it more efficient than other recorded methods.

In Section 8.2, the L_1 approximation problem is outlined. In Section 8.3, the algorithm is described. In Section 8.4, the linear programming techniques are implemented. In Section 8.5, numerical results and also comments on related methods to ours for calculating the polygonal approximation of plane curves in the L_1 norm are given.

8.1.7 Applications of the algorithm

Piecewise approximation of plane curves, including polygonal approximation, represent the data in waveforms and the boundaries of digital images. This facilitates feature extraction, shape recognition [8] and pattern classification of the given waveforms or images [21]. Besides, polygonal representation of a digital curve reduces the amount of data needed to process the given shape.

Polygonal approximation is also used in map simplification in cartography, pre-processing in electrocardiography and in electro-encephalography, transient analysis of speech signals, and other applications. For this see the references in Perez and Vidal [23].

8.2 The L_1 approximation problem

Let us be given a two-dimensional open or closed curve. Let this curve be digitized and let the digitized set be ordered at K consecutive points. The first and the last points coincide for a closed curve. The digitized points may not necessarily be equally spaced. Let n be the number of pieces (segments) in the approximation and let (z_j), $j = 1, 2, \ldots, n$, be the L_1 residual (error) norms for the n pieces. The number of pieces n is not known beforehand.

Consider any segment j, $1 \leq j \leq n$ of the given curve. Let this segment consist of N digitized points, with coordinates $(x_i, f(x_i))$, $i = 1, 2, \ldots, N$, $N \geq 2$. Let the N points be approximated by the straight line

$$\mathbf{y} = a_1 + a_2\mathbf{x} \qquad (8.2.1)$$

that minimizes the L_1 norm z_j (or briefly z), of the residuals r_i

$$z = \sum_{i=1}^{N} |r_i| \qquad (8.2.2a)$$

where

$$r_i = a_1 + a_2 x_i - f_i, \quad i = 1, 2, \ldots, N \qquad (8.2.2b)$$

and f_i denotes $f(x_i)$.

It is known that this problem reduces to the problem of obtaining the L_1 solution of an overdetermined system of linear equations. See Chapter 2. In vector-matrix notation, the linear system is

$$\mathbf{Ca} = \mathbf{f} \tag{8.2.3}$$

C is the coefficient matrix in (8.2.3), vector $\mathbf{a} = (a_1, a_2)^T$ and vector $\mathbf{f} = (f_1, f_2, \ldots, f_N)^T$. Or we have

$$\begin{bmatrix} 1 & x_1 \\ 1 & x_2 \\ \cdot & \cdot \\ \cdot & \cdot \\ \cdot & \cdot \\ 1 & x_N \end{bmatrix} \begin{bmatrix} a_1 \\ a_2 \end{bmatrix} = \begin{bmatrix} f_1 \\ f_2 \\ \cdot \\ \cdot \\ \cdot \\ f_N \end{bmatrix}$$

The L_1 solution to system (8.2.3) is the real vector **a** that minimizes the norm z given by (8.2.2a). As indicated above, the approximating straight line that approximates segment j passes through the end points; points 1 and N of segment j.

8.3 Description of the algorithm

Let the piecewise approximation for any segment j, $1 \leq j \leq n$, be such that its residual or error norm $z_j \leq \varepsilon$, a pre-assigned value. This algorithm employs the following steps:

(1) Take j = 1 (the first segment).

(2) Take N = 2, which is the minimum initial number of points for segment j to be approximated by the straight line (8.2.1). Calculate the solution vector $(a_1, a_2)^T$ for the line that passes through the 2 points. The norm z = 0, since it corresponds to a perfect fit to the 2 points. Go to step (3). In case points 1 and N lie on a vertical or a nearly vertical line, a_2 (the slope of the line) would be a very large number and the computation may break down. In this case, increase the number of points N by 1, as many times as necessary, until the end points 1 and N are not on a vertical or a nearly vertical line.

(3) Add the point $(N + 1)$ of the curve to segment j and calculate the new solution (a_1, a_2) for the straight line that passes through the points 1 and $(N + 1)$. Calculate z from (8.2.2a). Take $N = N + 1$.

(4) If $z < \varepsilon$, go to step (3). If $z > \varepsilon$, take $N = N - 1$, which is the final number of points for segment j, with the solution (a_1, a_2) as the required L_1 solution for segment j. Take $j = j + 1$ and go to step (2) for segment $(j + 1)$.

Since this algorithm is for a polygonal approximation, the last end point of segment j is the first end point of segment $(j + 1)$.

8.4 Linear programming technique

Implementing linear programming techniques enables us to calculate, in an efficient way, the norm z each time we add a new point to the N points of segment j. Let $\mathbf{C}^T$ form the initial data of a linear programming problem and let **f** be the cost vector to the initial data. **C** and **f** are those of equation (8.2.3).

For $N \geq 2$, let a 2 by 2 unit matrix $\mathbf{I}_2$ be the initial basis matrix **B**. Let **B** be an extension to the initial data, given by Tableau 8.4.1. Hence initially, $\mathbf{B}^{-1} = \mathbf{I}_2$. As the tableau is updated, $\mathbf{B}^{-1}$ is also updated.

Tableau 8.4.1 (Initial Data)

f_1	f_2	f_3	...	f_N
1	1	1	...	1
x_1	x_2	x_3	...	x_N

We update Tableau 8.4.1 by applying two Gauss-Jordan elimination steps, which reduce columns 1 and N of $\mathbf{C}^T$ to the first and second columns of $\mathbf{I}_2$ respectively. We get Tableau 8.4.2.

If points 1 and N lie on a vertical or near vertical line, the pivot element for the second elimination step in Tableau 8.4.1 would be 0, or a very small number, and the computation would break down (Lemma 8.3). In this case, as indicated earlier, we increase the number of points N by 1, more than once if necessary, until the pivot element for the added point is greater than a specified tolerance.

Tableau 8.4.2

f_1	f_2	f_3	...	f_N
1	y_{12}	y_{13}	...	0
0	y_{22}	y_{23}	...	1
0	c_2-f_2	c_3-f_3		0

In Tableau 8.4.2, vectors $(\mathbf{y}_i)$, i = 1, 2, …, N, are given by

$$\mathbf{y}_i = \mathbf{B}^{-1}\begin{bmatrix}1\\x_i\end{bmatrix}, \quad i = 2, 3, \ldots, N-1 \tag{8.4.1}$$

The marginal costs in Tableau 8.4.2 are given by

$$c_i - f_i = f_1 y_{1i} + f_N y_{2i} - f_i, \quad i = 2, 3, \ldots, N \tag{8.4.2}$$

Lemma 8.1

The marginal costs denoted here by $(c_i - f_i)$, i = 1, 2, …, N, in any tableau are themselves the residuals, given by (8.2.2b)

$$c_i - f_i = r_i, \quad i = 1, 2, \ldots, N$$

and since $r_1 = r_N = 0$, the L_1 norm z of (8.2.2a) is

$$z = \sum_{i=2}^{N-1} |c_i - f_i| \tag{8.4.3}$$

See Theorem 5.3 and also [1].

Lemma 8.2

The solution vector **a** of the interpolating straight line is given by

$$(a_1, a_2) = (f_1, f_N)\,\mathbf{B}^{-1} \tag{8.4.4}$$

See Theorem 5.4 and [1].

Lemma 8.3

If a new added point (N + 1) lies on a vertical line with point 1 of the segment at hand, then the two points would have the same x-coordinate, and we would have $\mathbf{y}_{N+1} = (1, 0)^T$. The next elimination step breaks down, since the pivot element is 0.

In view of this lemma, the users of the software should ensure that no two points of the digitized curve have the same x-coordinate.

8.4.1 The algorithm using linear programming

The following steps are employed:

(1) Take j = 1 (the first segment).

(2) Take N = 2. Form Tableau 8.4.1 and Tableau 8.4.2 for the two points. For N = 2, z = 0. Go to step (3).

(3) Add the point (N + 1) of the curve to segment j. From (8.4.1) and (8.4.2) calculate the vector $\mathbf{y}_{N+1}$ from (8.4.1) and the marginal cost $(c_{N+1} - f_{N+1})$ from (8.4.2). Append $\mathbf{y}_{N+1}$, f_{N+1} and $(c_{N+1} - f_{N+1})$ to the right of Tableau 8.4.2. Apply an elimination step, which reduces vector $\mathbf{y}_{N+1}$ to the second column of the unit matrix $\mathbf{I}_2$ and also reduces $(c_{N+1} - f_{N+1})$ to 0. Take N = N + 1. Go to step (4).

(4) Calculate the norm z from (8.4.3). If $z < \varepsilon$, the pre-assigned norm, go to step (3). If $z > \varepsilon$, take N = N – 1, which is the final number of points for segment j. From (8.4.4) calculate the vector $(a_1, a_2)^T$ as the L_1 solution for segment j. Take j = j + 1 and go to step (2) for segment (j + 1).

The solution vector **a** is calculated only once; i.e., at the end of step (4).

8.5 Numerical results and comments

LA_L1pol() calculates the number of segments in the polygonal approximation in the L_1 error norm, the end points and the parameters (a_1 and a_2) of each straight line. It also calculates the residuals (errors) at all the points of the digitized curve. From the end points of the n segments or from the parameters of the straight lines (a_1, a_2), we draw the polygonal approximation.

DR_L1pol() tests 4 examples, 2 of which are presented here. The first example is with unequal x-intervals. The second example is a closed figure and its data is not equally spaced.

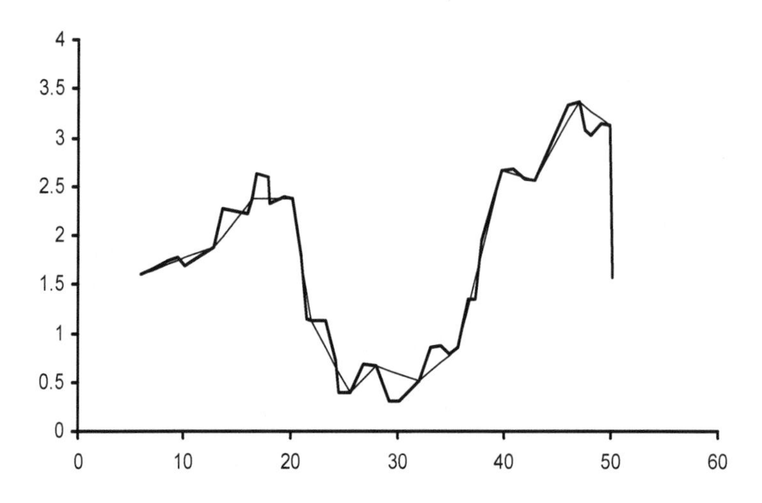

Figure 8-1: L_1 polygonal approximation for a waveform. The L_1 norm in any segment is ≤ 0.6

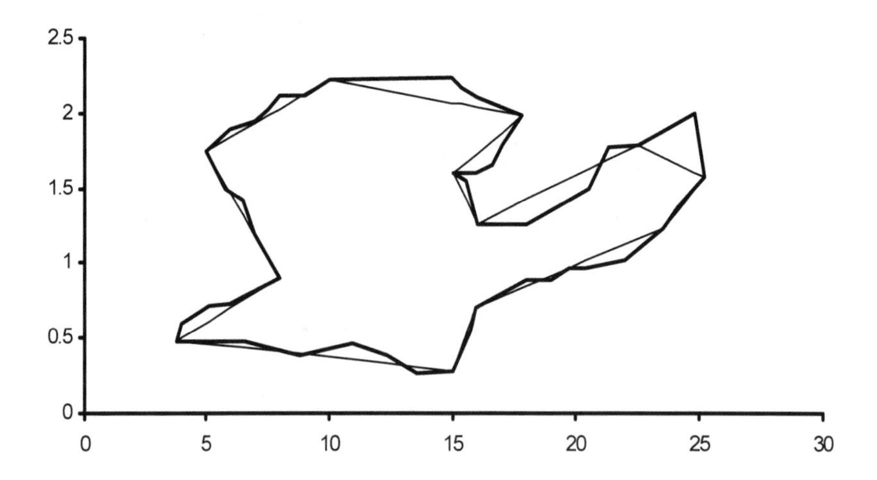

Figure 8-2: L_1 polygonal approximation for a contour. The L_1 norm in any segment is ≤ 0.4

In calculating the polygonal approximation of plane curves in the Chebyshev norm, Dunham [10], Kurozumi and Davis [18], Ramer [25], Williams [40] and Zhu and Seneviratne [43] defined the errors as the Euclidean distances between the given points and the approximating line, i.e., perpendicular to the approximating line in each segment.

This requirement is not needed in the L_1 polygonal approximation such as in our algorithm. The reason is that the L_1 error norm for a digitized curve segment is approximately proportional to the area between the segment and the approximating straight line. This area is approximately the same when the residuals are measured along the y-axis or when the error is measured perpendicular to the approximating straight line using the straight line as the base, as in Ray and Ray [28] and in Wall and Danielsson [39].

We now comment on 3 of the algorithms that deal with the same problem as ours. These are the algorithms of Ray and Ray [28], Wall and Danielsson [39] and Johnson and Vogt [16].

The number of arithmetic operations in the method of Ray and Ray [28] including the calculation of square roots, is much higher than ours. The method of Wall and Danielsson [39] encounters some difficulties when the curvature of the given curve changes sign.

Once more, Johnson and Vogt [16] use a geometric method that is a generalization of a method that minimizes the area between the curve and the polygonal arc. This method is a kind of an L_1 piecewise approximation method. In contrast to our algorithm, in their method the given curve has to be a convex one.

References

1. Abdelmalek, N.N., On the discrete linear L_1 approximation and L_1 solutions of overdetermined linear equations, *Journal of Approximation Theory*, 11(1974)38-53.
2. Abdelmalek, N.N., Polygonal approximation of planar curves in the L_1 norm, *International Journal of Systems Science*, 17(1986)1601-1608.
3. Abdelmalek, N.N., Noise filtering in digital images and approximation theory, *Pattern Recognition*, 19(1986)417-424.

4. Ansari, N. and Delp, E.J., On detecting dominant points, *Pattern Recognition*, 24(1991)441-451.
5. Arkin, E.M., Chew, L.P., Huttenlocher, D.P., Kedem, K. and Mitchell, J.S.B., An efficient computable metric for comparing polygonal shapes, *IEEE Transactions on Pattern Analysis and Machine Intelligence*, 13(1991)209-215.
6. Badi'i, F. and Peikari, B., Functional approximation of planar curves via adaptive segmentation, *International Journal of Systems Science*, 13(1982)667-674.
7. Cantoni, A., Optimal curve fitting with piecewise linear functions, *IEEE Transactions on Computers*, 20(1971)59-67.
8. Davis, L.S., Shape matching using relaxation techniques, *IEEE Transactions on Pattern Analysis and Machine Intelligence*, 1(1979)60-72.
9. Duda, R.O. and Hart, P.E., *Pattern Classification and Scene Analysis*, John Wiley & Sons, New York, 1973.
10. Dunham, J.G., Optimum uniform piecewise linear approximation of planar curves, *IEEE Transactions on Pattern Analysis and Machine Intelligence*, 8(1986)67-75.
11. Golden, D.E., *Genetic Algorithms in Search, Optimization, and Machine Learning*, Addison-Wesley, Reading, MA, 1989.
12. Hadley, G., *Linear Programming*, Addison-Wesley, Reading, MA, 1962.
13. Hamann, B. and Chen, J-L., Data point selection for piecewise linear curve approximation, *Computer Aided Geometric Design*, 11(1994)289-301.
14. Ho, S-Y. and Chen, Y-C., An efficient evolutionary algorithm for accurate polygonal approximation, *Pattern Recognition*, 34(2001)2305-2317.
15. Huang, S-C. and Sun, Y-N., Polygonal approximation using genetic algorithms, *Pattern Recognition*, 32(1999)1409-1420.
16. Johnson, H.H. and Vogt, A., A geometric method for approximating convex arcs, *SIAM Journal on Applied Mathematics*, 38(1980)317-325.
17. Kolesnikov, A. and Franti, P., Polygonal Approximation of Closed Contours, *Lecture Notes in Computer Science*, 2749(2003)778-785.

18. Kurozumi, Y. and Davis, W.A., Polygonal approximation by the minimax method, *Computer Graphics and Image Processing*, 19(1982)248-264.
19. Lopes, H., Oliveira, J.B. and de Figueiredo, L.H., Robust adaptive polygonal approximation of implicit curves, *Computers and Graphics*, 26(2002)841-852.
20. Michalewicz, Z., *Genetic Algorithms + Data Structures = Evolution Programs*, Second Extended Edition, Springer-Verlag, New York, 1992.
21. Pavlidis, T., Polygonal approximations by Newton's method, *IEEE Transactions on Computers*, 26(1977)800-807.
22. Pei, S-C. and Lin, C-N., The detection of dominant points on digital curves by scale-space filtering, *Pattern Recognition*, 25(1992)1307-1314.
23. Perez, J-C. and Vidal, E., Optimum polygonal approximation of digitized curves, *Pattern Recognition Letters*, 15(1994)743-750.
24. Pikaz, A. and Dinstein, I.H., Optimal polygonal approximation of digital curves, *Pattern Recognition*, 28(1995)373-379.
25. Ramer, U., An iterative procedure for the polygonal approximation of plane curves, *Computer Graphics and Image Processing*, 1(1972)244-256.
26. Rannou, F. and Gregor, J., Equilateral polygon approximation of closed contours, *Pattern Recognition*, 29(1996)1105-1115.
27. Ray, B.K. and Ray, K.S., An algorithm for detection of dominant points and polygonal approximation of digitized curves, *Pattern Recognition Letters*, 13(1992)849-856.
28. Ray, B.K. and Ray, K.S., Determination of optimal polygon from digital curve using L_1 norm, *Pattern Recognition*, 26(1993)505-509.
29. Rice, J.R. and White, J.S., Norms for smoothing and estimation, *SIAM Review*, 6(1964)243-256.
30. Rosin, P.L., Assessing the behavior of polygonal approximation algorithms, *Pattern Recognition*, 36(2003)505-518.
31. Salotti, M., Optimal polygonal approximation of digitized curves using the sum of square deviations criterion, *Pattern Recognition*, 35(2002)435-443.

32. Sarkar, B., Singh, L.K. and Sarkar, D., A genetic algorithm-based approach for detection of significant vertices for polygonal approximation of digital curves, *International Journal of Image and Graphics*, 4(2004)223-239.
33. Sarkar, D., A simple algorithm for detection of significant vertices for polygonal approximation of chain-coded curves, *Pattern Recognition Letters*, 14(1993)959-964.
34. Sato, Y., Piecewise linear approximation of plane curves by perimeter optimization, *Pattern Recognition*, 25(1992)1535-1543.
35. Sklansky, J. and Gonzalez, V., Fast polygonal approximation of digitized curves, *Pattern Recognition*, 12(1980)327-331.
36. Sun, Y-N. and Huang, S-C., Genetic algorithms for error-bounded polygonal approximation, *International Journal of Pattern Recognition and Artificial Intelligence*, 14(2000)297-314.
37. Teh, C-H. and Chin, R.T., On the detection of dominant points on digital curves, *IEEE Transactions on Pattern Analysis and Machine Intelligence*, 11(1989)859-872.
38. Tomek, I., Two algorithms for piecewise-linear continuous approximation of functions of one variable, *IEEE Transactions on Computers*, 23(1974)445-448.
39. Wall, K. and Danielsson, P-E., A fast sequential method for polygonal approximation of digitized curves, *Computer Vision, Graphics, and Image Processing,* 28(1984)220-227.
40. Williams, C.M., An efficient algorithm for the piecewise linear approximation of planar curves, *Computer Graphics and Image Processing*, 8(1978)286-293.
41. Yin, P-Y., Algorithms for straight line fitting using k-means, *Pattern Recognition Letters*, 19(1998)31-41.
42. Yin, P-Y., Genetic algorithms for polygonal approximation of digital curves, *International Journal of Pattern Recognition and Artificial Intelligence*, 13(1999)1061-1082.
43. Zhu, Y. and Seneviratne, L.D., Optimal polygonal approximation of digitized curves, *IEEE Proceedings on Vision, Image and Signal Processing*, 144(1997)8-14.

8.6 DR_L1pol

```
/*--------------------------------------------------------------------
DR_L1pol
----------------------------------------------------------------------
This is a driver for the function LA_L1pol(), which calculates the
polygonal straight line approximation of a given data point set {x,y}
that results from discretizing a given plane curve y = f(x).

The approximation by LA_L1pol() is such that the approximating
straight line interpolates the end points of the segment at hand
and also such that the L-One residual norm for any segment not
to exceed a pre-assigned tolerance denoted by "enorm".

From an n data points (x,y) we form the overdetermined system of
linear equations

                        c*a = f

"c" is a real n by 2 matrix. Each element of the first column of "c"
is 1.  The second column of "c" contains the x-coordinates of the
set {x,y} arranged in a sequential order.

"f" is a real n vector whose elements are the y coordinates of
the given data set {x,y}.

"a" is the solution 2 vector (for each segment there is a different
solution vector "a").

This driver contains 4 test examples.  The results of 2 of them are
given in the text.
--------------------------------------------------------------------*/

#include "DR_Defs.h"
#include "LA_Prototypes.h"

#define Ncp       28
#define Ndp       44
#define Nep       49
#define Ngp       64

void DR_L1pol (void)
{
    /*----------------------------------------
```

```
  Constant matrices/vectors
----------------------------------------*/
static tNumber_R fc[Ncp+1] =
{   NIL,
     8.0, 11.0, 13.0, 11.2,  9.1, 10.8, 14.8, 16.0, 15.1, 14.0,
    14.7, 15.8, 16.8, 15.6, 13.0, 14.3, 13.8, 10.6,  9.3,  9.6,
    10.8, 11.2,  9.0,  7.0,  5.8,  6.8,  4.8,  3.9
};

static tNumber_R xd[Ndp+1] =
{   NIL,
     6.0,  7.0,  8.5,  9.5, 10.2, 12.9, 13.6, 15.9, 16.4, 16.8,
    17.8, 18.0, 19.3, 20.1, 21.0, 21.5, 21.9, 23.2, 24.2, 24.5,
    25.5, 26.8, 28.0, 29.2, 30.2, 32.0, 33.1, 34.0, 34.9, 35.7,
    36.6, 37.3, 37.8, 39.7, 40.8, 41.9, 42.8, 46.0, 47.0, 47.6,
    48.1, 49.0, 49.9, 50.2
};

static tNumber_R fd[Ndp+1] =
{   NIL,
    1.6,  1.65, 1.75, 1.78, 1.7,  1.88, 2.28,  2.23, 2.38, 2.63,
    2.6,  2.32, 2.4,  2.38, 1.8,  1.15, 1.12,  1.13, 0.72, 0.4,
    0.39, 0.68, 0.67, 0.31, 0.31, 0.51, 0.86,  0.87, 0.79, 0.85,
    1.35, 1.35, 1.95, 2.67, 2.68, 2.58, 2.57,  3.33, 3.36, 3.08,
    3.03, 3.15, 3.12, 1.58
};

static tNumber_R xe[Nep+1] =
{   NIL,
     3.8,  6.6,  8.8, 10.9, 12.3, 13.5, 15.0, 15.6, 15.7, 15.8,
    15.9, 18.0, 19.0, 19.7, 20.4, 22.0, 23.5, 24.1, 25.2, 24.8,
    22.5, 21.3, 20.5, 18.0, 16.0, 15.5, 15.0, 15.9, 16.6, 17.0,
    17.8, 16.0, 15.3, 14.9, 10.0,  9.0,  8.0,  7.5,  7.0,  6.0,
     5.0,  5.8,  6.5,  7.0,  8.0,  6.0,  5.1,  4.0,  3.8
};

static tNumber_R fe[Nep+1] =
{   NIL,
    0.48, 0.48, 0.39, 0.46, 0.38, 0.27, 0.28, 0.5,  0.55, 0.6,
    0.7,  0.89, 0.89, 0.96, 0.96, 1.02, 1.23, 1.38, 1.57, 2.0,
   1.79,  1.77, 1.5,  1.26, 1.26, 1.55, 1.6,  1.6,  1.65, 1.78,
   1.98,  2.1,  2.17, 2.23, 2.22, 2.11, 2.11, 2.02, 1.95, 1.89,
   1.74,  1.5,  1.41, 1.19, 0.9,  0.73, 0.72, 0.6,  0.48
};
```

```
static tNumber_R xg[Ngp+1] =
{   NIL,
     2.8,  3.9,  5.6,  6.7,  8.2,  9.5, 11.3, 12.4, 12.4, 14.1,
    14.1, 15.0, 19.0, 19.8, 21.0, 20.9, 25.2, 28.0, 32.5, 31.6,
    32.5, 32.1, 32.6, 31.5, 32.0, 30.1, 26.1, 24.2, 23.1, 21.7,
    19.9, 19.0, 18.7, 19.0, 18.1, 18.1, 18.7, 18.9, 18.1, 18.1,
    17.6, 17.4, 16.9, 16.8, 17.0, 17.1, 16.5, 15.9, 16.0, 16.3,
    15.2, 13.3, 13.0, 12.5, 12.1, 11.5, 10.5, 10.0,  9.5,  9.0,
     8.1,  8.5,  5.5,  2.8
};

static tNumber_R fg[Ngp+1] =
{   NIL,
    1.18, 0.9,  0.8,  0.6,  0.51, 0.6,  0.52, 0.6,  0.73, 0.8,
    0.98, 0.92, 1.58, 1.58, 1.89, 1.95, 2.3,  2.32, 1.92, 1.8,
    1.8,  1.7,  1.52, 1.3,  1.02, 0.82, 0.5,  0.48, 0.52, 0.5,
    0.6,  0.66, 0.8,  0.88, 0.88, 0.99, 0.99, 1.15, 1.13, 1.38,
    1.38, 1.46, 1.46, 1.58, 1.58, 1.73, 1.7,  1.8,  1.9,  1.98,
    2.1,  2.08, 2.09, 2.09, 2.03, 2.1,  2.0,  2.05, 1.92, 1.9,
    1.75, 1.6,  1.1,  1.18
};

/*-----------------------------------------
  Variable matrices/vectors
-----------------------------------------*/
tMatrix_R   ct      = alloc_Matrix_R (2, NN_ROWS);
tMatrix_R   ctn     = alloc_Matrix_R (2, NN_ROWS);
tMatrix_R   binv    = alloc_Matrix_R (2, 2);
tMatrix_R   ap      = alloc_Matrix_R (KK_PIECES, 2);
tVector_R   f       = alloc_Vector_R (NN_ROWS);
tVector_R   r       = alloc_Vector_R (NN_ROWS);
tVector_R   rp      = alloc_Vector_R (NN_ROWS);
tVector_R   zp      = alloc_Vector_R (KK_PIECES);
tVector_R   a       = alloc_Vector_R (2);
tVector_R   v       = alloc_Vector_R (2);
tVector_I   icbas   = alloc_Vector_I (2);
tVector_I   ixl     = alloc_Vector_I (KK_PIECES);

int         m, n, npiece, Iexmpl;
int         j;

tNumber_R   enorm;

eLaRc       rc = LaRcOk;
```

```
prn_dr_bnr ("DR_L1pol, Polygonal L1 Approximation of a Plane"
            " Curve");

m = 2;
for (Iexmpl = 1; Iexmpl <= 4; Iexmpl++)
{
    switch (Iexmpl)
    {
        case 1:
            enorm = 1.8;
            n = Ncp;
            for (j = 1; j <= n; j++)
            {
                f[j] = fc[j];
                ct[1][j] = 1.0;
                ct[2][j] = j;
            }
            break;

        case 2:
            enorm = 0.6;
            n = Ndp;
            for (j = 1; j <= n; j++)
            {
                f[j] = fd[j];
                ct[1][j] = 1.0;
                ct[2][j] = xd[j];
            }
            break;

        case 3:
            enorm = 0.48;
            n = Nep;
            for (j = 1; j <= n; j++)
            {
                f[j] = fe[j];
                ct[1][j] = 1.0;
                ct[2][j] = xe[j];
            }
            break;

        case 4:
            enorm = 0.8;
            n = Ngp;
            for (j = 1; j <= n; j++)
```

```
            {
                f[j] = fg[j];
                ct[1][j] = 1.0;
                ct[2][j] = xg[j];
            }
            break;

        default:
            break;
    }

    prn_algo_bnr ("L1pol");

    prn_example_delim();
    PRN ("Example #%d: Size of Matrix \"c\", %d by %d\n",
         Iexmpl, n, m);
    prn_example_delim();
    PRN ("Polygonal L1 Approximation of a Plane Curve\n");
    prn_example_delim();
    PRN ("r.h.s. Vector \"f\"\n");
    prn_Vector_R (f, n);
    PRN ("Transpose of Coefficient Matrix, \"ct\"\n");
    prn_Matrix_R (ct, m, n);

    rc = LA_L1pol (m, n, enorm, ct, f, ctn, binv, icbas, r, a,
                   ixl, v, rp, ap, zp, &npiece);

    if (rc >= LaRcOk)
    {
        PRN ("\n");
        PRN ("Results of Polygonal L1 Approximation\n");
        PRN ("Pre-assigned tolerance 'enorm' = %8.4f\n", enorm);
        PRN ("Number of lines (pieces) = %d\n", npiece);
        PRN ("Starting point of each line\n");
        prn_Vector_I (ixl, npiece);
        PRN ("Solution Vector \"a\" for each line\n");
        prn_Matrix_R (ap, npiece, 2);
        PRN ("Residual Vector \"r\" for the n points\n");
        prn_Vector_R (rp, n);
        PRN ("L-One norms of the residuals for the "
             " \"npiece\" lines\n");
        prn_Vector_R (zp, npiece);
    }

    prn_la_rc (rc);
```

```
    }

    free_Matrix_R (ct, 2);
    free_Matrix_R (ctn, 2);
    free_Matrix_R (binv, 2);
    free_Matrix_R (ap, KK_PIECES);
    free_Vector_R (f);
    free_Vector_R (r);
    free_Vector_R (rp);
    free_Vector_R (zp);
    free_Vector_R (a);
    free_Vector_R (v);
    free_Vector_I (icbas);
    free_Vector_I (ixl);
}
```

8.7 LA_L1pol

```
/*-----------------------------------------------------------------------
LA_L1pol
-------------------------------------------------------------------------
This program calculates straight line polygon to a discrete
point set {x,y}.  The approximation is such that the L-One error
norm for any line not to exceed a pre-assigned tolerance denoted
by "enorm".  The number of lines in the polygon is not known before
hand. The polygon may be an open or a closed one.

Given is a set of points {x,y} resulting from digitizing a given
plane curve of the form y = f(x). The points may not be equally
spaced. We form the overdetermined system of linear equations

                              c*a = f

"c" is a real n by 2 matrix.  Each element of its first column
is 1 and its second column contains the x-coordinates of the points
(x,y) in a sequential order.
"f" is a real n vector whose elements are the y coordinates of
the points (x,y).

Inputs
m       An integer = 2 which is the number of unknowns in the
        straight line y = a[1] + a[2]*x.
n       The number of points of the set {x,y}.
ct      A real 2 by n matrix containing the transpose of matrix "c"
        of system c*a = f. "ct" is not destroyed in the computation.
f       A real n vector containing the r.h.s. of the system c*a = f.
        This vector also is not destroyed in the computation.
enorm   A given real parameter; a pre-assigned tolerance such that
        the L-One residual norm for any segment is <= enorm.

Outputs
npiece  The number of straight lines of the polygonal approximation.
ixl     An integer "npiece" vector containing the indices of the
        first elements of the "npiece" segments.
        For example, if ixl = (1,5,12,22,...), then the first
        segment contains points 1 to 5, the second segment contains
        points 5 to 12, and so on.
ap      A real "npiece" by 2 matrix.  The 2 elements of its first
        row contain the coefficients of the first line. The 2
        elements of the second row contain the coefficients of
```

```
        the second line and so on.
rp      A real n vector containing the residual values at the n
        points of the set {x,y}.
zp      A real "npiece" vector containing the npiece optimum values
        of the L-One norms for the "npiece" segments.

Returns one of
        LaRcSolutionFound
        LaRcErrBounds
        LaRcErrNullPtr
--------------------------------------------------------------------*/
#include "LA_Prototypes.h"

eLaRc LA_L1pol (int m, int n, tNumber_R enorm, tMatrix_R ct,
    tVector_R f, tMatrix_R ctn, tMatrix_R binv, tVector_I icbas,
    tVector_R r, tVector_R a, tVector_I ixl, tVector_R v,
    tVector_R rp, tMatrix_R ap, tVector_R zp, int *pNpiece)
{
    int         is = 0, ie = 0, nu = 0, je = 0;
    int         i = 0, j = 0, k = 0, ij = 0, ijk = 0, kase = 0;
    tNumber_R   z = 0.0, piv = 0.0;

    /* Validation of the data before executing the algorithm */
    eLaRc       rc = LaRcSolutionFound;
    VALIDATE_BOUNDS ((m == 2) && (m <= n) && !((n == 1) && (m == 1)));
    VALIDATE_PTRS   (ct && f && ctn && binv && icbas && r && a && ixl
                     && v && rp && ap && zp && pNpiece);

    /* Initialization */
    *pNpiece = 1;

    /* "is" means i(start) and  "ie" means i(end) */
    is = 1;
    for (ijk = 1; ijk <= n; ijk++)
    {
        /* kase == 0: The start of the algorithm */
        je = 0;
        kase = 0;
        z = 0.0,
            zp[*pNpiece] = 0.0;
        ie = is + m - 1;
        for (j = is; j <= ie; j++)
        {
            rp[j] = 0.0;
        }
```

```
for (i = 1; i <= m; i++)
{
    ap[*pNpiece][i] = 0.0;
}
if (ie > n) ie = n;
ixl[*pNpiece] = is;
nu = ie - is + 1;
if (nu < m)
{
    GOTO_CLEANUP_RC (LaRcSolutionFound);
}

if (kase == 0)
{
    for (j = is; j <= ie; j++)
    {
        for (i = 1; i <=m; i++)
        {
            ctn[i][j] = ct[i][j];
        }
    }
}

for (ij = 1; ij <= n; ij++)
{
    LA_l1pol_residuals (m, is, ie, ctn, f, icbas, binv,
                        r, a, &z, kase);

    if (z > enorm)
    {
        is = ie - 1;
        *pNpiece = *pNpiece + 1;
        break;
    }
    zp[*pNpiece] = z;
    for (j = 1; j <= m; j++)
    {
        ap[*pNpiece][j] = a[j];
    }
    for (j = is; j <= ie; j++)
    {
        rp[j] = r[j];
    }
    kase = -1;
    je = ie + 1;
```

```
            if (je > n)
            {
                GOTO_CLEANUP_RC (LaRcSolutionFound);
            }

            /* Check if the first and last points lie on a vertical
               or near vertical line */
            LA_l1pol_vertic_line (je, m, ct, ctn, f, r, binv, a, v,
                                  &piv);

            for (k = 1; k <= n; k++)
            {
                if (fabs (piv) > EPS) break;
                if (fabs (piv) <= EPS)
                {
                    je = ie + 1;
                    if (je > n)
                    {
                        GOTO_CLEANUP_RC (LaRcSolutionFound);
                    }
                    LA_l1pol_vertic_line (je, m, ct, ctn, f, r, binv,
                                          a, v, &piv);
                }
            }
            ie = je;
        }
    }

CLEANUP:

    return rc;
}

/*-------------------------------------------------------------------
This  function solves  the first and the last equations of an
overdetermined system.  It uses a linear programming method, such
that if an extra equation is added to the system, it solves the first
and the new last equation. It does that in one simplex step.
It also calculates the L-One norm "z" of the residual vector of the
segment at hand from the simplex tableau.
-------------------------------------------------------------------*/
void LA_l1pol_residuals (int m, int n1, int n2, tMatrix_R ct,
    tVector_R f, tVector_I icbas, tMatrix_R binv, tVector_R r,
    tVector_R a, tNumber_R *pZ, int kase)
```

```
{
    int             i, j, k, jin, ibc, iout;
    tNumber_R       e, s, z, pivot;

    /* kase = 0 indicates the start of a new segment */
    if (kase == 0)
    {
        for (j = 1; j <= m; j++)
        {
            a[j] = 0.0;
            icbas[j] = 0;
            for (i = 1; i <= m; i++)
            {
                binv[i][j] = 0.0;
            }
            binv[j][j] = 1.0;
        }
        z = 0.0;
        iout = 1;
        jin = n1;

        /* A Gauss-Jordan elimination step */
        LA_l1pol_gauss_jordn (n1, n2, m, iout, jin, ct, binv, icbas);
    }

    /* kase != 0  The segment at hand is enlarged by adding to its
       end an extra point */

    iout = 2;
    jin = n2;
    pivot = ct[iout][jin];

    /* Detection of a singular basis matrix */
    if (fabs (pivot) >= EPS)
    {
        LA_l1pol_gauss_jordn (n1, n2, m, iout, jin, ct, binv, icbas);
        if (kase != 0)
        {
            e = r[jin];
            for (j = n1; j <= n2; j++)
            {
                r[j] = r[j] - e * (ct[iout][j]);
            }
        }
    }
```

```
    else
        return;

    /* Calculate the residuals (marginal costs) */
    for (j = n1; j <= n2; j++)
    {
        r[j] = 0.0;
        ibc = 0;
        for (i = 1; i <= m; i++)
        {
            if (j == icbas[i]) ibc = 1;
        }
        if (ibc == 0)
        {
            s = -f[j];
            for (i = 1; i <= m; i++)
            {
                k = icbas[i];
                s = s + f[k] * ct[i][j];
            }
            r[j] = s;
        }
    }

    LA_l1pol_gauss_jordn (n1, n2, m, iout, jin, ct, binv, icbas);
    LA_l1pol_res (n1, n2, m, f, r, a, binv, icbas, pZ);
}

/*-------------------------------------------------------------------
A Gauss-Jordan elimination step in LA_L1pol()
-------------------------------------------------------------------*/
void LA_l1pol_gauss_jordn (int n1, int n2, int m, int iout, int jin,
    tMatrix_R ct, tMatrix_R binv, tVector_I icbas)
{
    int             i, j;
    tNumber_R       e, pivot;

    pivot = ct[iout][jin];
    icbas[iout] = jin;
    for (j = n1; j <= n2; j++)
    {
        ct[iout][j] = ct[iout][j]/pivot;
    }
    for (j = 1; j <= m; j++)
    {
```

```
        binv[iout][j] = binv[iout][j]/pivot;
    }
    for (i = 1; i <= m; i++)
    {
        if (i != iout)
        {
            e = ct[i][jin];
            for (j = n1; j <= n2; j++)
            {
                ct[i][j] = ct[i][j] - e * (ct[iout][j]);
            }
            for (j = 1; j <= m; j++)
            {
                binv[i][j] = binv[i][j] - e * (binv[iout][j]);
            }
        }
    }
}

/*-----------------------------------------------------------------
Calculate the results of LA_L1pol()
-----------------------------------------------------------------*/
void LA_l1pol_res (int n1, int n2, int m, tVector_R f, tVector_R r,
    tVector_R a, tMatrix_R binv, tVector_I icbas, tNumber_R *pZ)
{
    int             i, j, k;
    tNumber_R       s;

    for (j = 1; j <= m; j++)
    {
        s = 0.0;
        for (i = 1; i <= m; i++)
        {
            k = icbas[i];
            s = s + f[k] * binv[i][j];
        }
        a[j] = s;
    }

    s = 0.0;
    for (j = n1; j <= n2; j++)
    {
        s = s + fabs (r[j]);
    }
    *pZ = s;
```

```
}

/*-------------------------------------------------------------------
Test if the 2 end points lie on a vertical or near vertical line in
LA_L1pol()
-------------------------------------------------------------------*/
void LA_l1pol_vertic_line (int je, int m, tMatrix_R ct,
    tMatrix_R ctn, tVector_R  f, tVector_R r, tMatrix_R binv,
    tVector_R a, tVector_R v, tNumber_R *pPiv)
{
    int             i, k;
    tNumber_R       s;

    for (i = 1; i <= m; i++)
    {
        v[i] = ct[i][je];
    }

    /* Calculating the marginal costs (the residuals) */
    for (i = 1; i <= m; i++)
    {
        s = 0.0;
        for (k = 1; k <= m; k++)
        {
            s = s + binv[i][k]*v[k];
        }
        ctn[i][je] = s;
    }
    s = -f[je];
    for (i = 1; i <= m; i++)
    {
        s = s + a[i] * ct[i][je];
    }
    r[je] = s;
    *pPiv = ctn[2][je];
}
```

Chapter 9

Piecewise L_1 Approximation of Plane Curves

9.1 Introduction

Piecewise linear approximation of a plane curve implies that the plane curve is divided into segments, or pieces, and each piece is approximated by a simple curve. Unlike polygonal approximation, where the approximating lines are joined to form a polygon, the approximating curves in piecewise approximations are not joined to one another. In this respect, Pavlidis [8] argues that the continuity requirement at the joints does not appear to be essential for some applications, and even desirable for other applications.

Discontinuity is desirable in the detection of feature selection for pattern recognition and picture processing. It is also desirable in detecting jumps in waveforms. Besides that, calculating the approximations for separate segments reduces the computation complexity and time.

In the majority of the published works on piecewise linear approximation, the approximating curves are either constants (horizontal lines) or straight lines. However, the approximations could be by any kind of polynomials.

In the presence of spikes in the given curve, the approximation in the L_1 norm is usually recommended over other norms. This is because the approximating curve in the L_1 norm almost totally ignores the presence of the spikes, as explained in Section 2.3.

The problem of piecewise approximation in the L_1 norm has not received much attention. The obvious reason is that the segmentation requires the use of algorithms for solving the L_1 approximation problem that were not available to many authors. On the other hand, piecewise approximations in the Chebyshev and the L_2 norms received considerable attention. See the references in Chapters 15 and 18 respectively. We note here that Sklar and Armstrong [10] described a procedure for piecewise linear approximation for L_p norm curve fitting, where $1 < p < 2$.

In this chapter, two algorithms for the piecewise linear L_1 approximation of plane curves are presented [4] for the two cases:

(a) when the L_1 norm in any segment is not to exceed a pre-assigned value, and

(b) when the number of segments is given and a **balanced** (equal) L_1 norm solution for all the segments is required.

The problem is solved by first digitizing the given curve. Then either algorithm is applied to the discrete points. This work is analogous to that presented in Chapters 15 and 18 for the Chebyshev [3] and the L_2 [5] norms respectively.

In Section 9.2, the characteristics of the piecewise approximation are presented. In Section 9.3, the discrete linear L_1 approximation problem is outlined. In Section 9.4, the two piecewise linear L_1 approximation algorithms are described. In Section 9.5, numerical results and comments are given.

9.1.1 Applications of piecewise approximation

As indicated in the previous chapter, piecewise approximation of plane curves has several applications. It is used in the area of image processing such as image segmentation, image compression and segmentation/reconstruction of range images. It is also used in feature extraction, noise filtering, speech recognition and numerous other fields. See the references cited in Section 8.1.1.

9.2 Characteristics of the piecewise approximation

Piecewise approximation possesses certain characteristics that were laid down by Lawson [7]. These characteristic properties are for segmented rational Chebyshev approximation. Yet, they apply as well for linear approximation and for norms other than the Chebyshev norm. See also Pavlidis and Maika [9].

Let us be given the plane curve $y = f(x)$ and let it be defined on the interval $[a, b]$. Let this curve be digitized at the K points $(x_i, f(x_i))$,

$i = 1, 2, \ldots, K$, where $x_1 = a$ and $x_K = b$. Let n be the number of pieces (segments) in the approximation and let (z_j), $j = 1, 2, \ldots, n$, be the L_1 residual or error norms for the n pieces. For the first algorithm, the number of segments is not known beforehand.

Assume that f(x) is continuous and satisfies Lipschitz condition on [a, b]. Hence, the error norm z_j, $1 \leq j \leq n$, has the following characteristics.

Characteristic 1

(a) z_j is a continuous function of the end points of the segment,
(b) z_j is non-increasing in the segment left end point, and
(c) z_j is non-decreasing in the segment right end point.

Characteristic (b) means that if a point is deleted from the left end of segment j, z_j, the error norm for segment j, will not increase, but is likely to decease or remain unchanged. Characteristic (c) means that if a point is added to the right end of segment j, z_j will not decrease, but is likely to increase or remain unchanged.

Characteristic 2

If a piecewise approximation is calculated for the curve f(x), (not for the digitized points of the curve), and if $z_1 = z_2 = \ldots = z_n$, then the solution is optimal. Such a solution always exists and is known as a **balanced** (equal) error norm solution. However, in the case of the approximation of the discrete points of the digitized curve, a balanced error norm solution may not exist. One attempts to obtain a near-balanced piecewise approximation instead.

Lawson [7] suggested an iterative technique for obtaining such a balanced error norm solution. The technique performs well on smooth data, but it might fail (cycle) if at some intermediate step, a segment has a zero-error norm. This happened when segmentation by straight lines were used. Pavlidis [8] was able to rectify this point. We should also note that this algorithm does not result in a unique near equal error norm solution. The final result depends on the initial segmentation of the given digitized curve. We have met this situation in our experimentation. Pavlidis [8] encountered the same situation in his experimentation.

As indicated above, the problem is solved by first digitizing the given curve into discrete points. Then either algorithm is applied to the discrete data. The two algorithms use the discrete linear L_1 approximation function LA_Lone() of Chapter 5 [2].

The first algorithm has the option of calculating connected or disconnected piecewise linear L_1 approximations, while the second algorithm calculates disconnected piecewise L_1 approximations only. In the connected piecewise approximation, the x-coordinate of the right end point of segment j is the x-coordinate of the starting (left) point of segment (j + 1). In the disconnected piecewise approximation, the x-coordinate of the adjacent point to the right end point of segment j is the x-coordinate of the starting left point of segment (j + 1). See Figures 9.1-3 below.

9.3 The discrete linear L_1 approximation problem

Consider any segment j, $1 \leq j \leq n$, of the given curve f(x). Let this segment consist of N digitized points with coordinates $(x_i, f(x_i))$, i = 1, 2, …, N. Let these N discrete points be approximated by

$$\mathbf{L}(\mathbf{a}, \mathbf{x}) = \sum_{i=1}^{M} a_i \phi_i(\mathbf{x})$$

which minimizes the L_1 norm z_j (or briefly z), of the residuals $r(x_i)$

$$z = \sum_{i=1}^{N} |r(x_i)|$$

where

$$r(x_i) = \mathbf{L}(\mathbf{a}, x_i) - f(x_i), \quad i = 1, 2, \ldots, N \tag{9.3.1}$$

The functions $(\phi_i(x))$, i = 1, 2, …, M, $M \leq N$, are given real linearly independent approximating functions and $\mathbf{a} = (a_i)$ is an M real vector to be calculated.

This problem reduces to the problem of obtaining the L_1 solution of the overdetermined system of equations [1] (Section 2.2)

$$\mathbf{Ca} = \mathbf{f}$$

$\mathbf{C}$ is an N by M matrix given by $\mathbf{C} = (c_{ij}) = (\phi_j(x_i))$, i = 1, 2, …, N, j = 1, 2, …, M. and $\mathbf{f}$ is the N-vector $\mathbf{f} = (f_i) = (f(x_i))$. The L_1 solution to system **Ca** = **f** is the real M-vector **a** that minimizes the L_1 norm

$$z = \sum_{i=1}^{N} |r_i|$$

where r_i is the ith residual given by (9.3.1), or by

$$r_i = r(x_i) = \sum_{j=1}^{M} c_{ij}a_j - f_i, \quad i = 1, 2, \ldots, N$$

This problem has been efficiently solved as a linear programming problem in the dual form [2, 6] (Chapter 5). The implementation of the algorithm for the discrete linear L_1 approximation to this problem is described in the next section.

9.4 Description of the algorithms

9.4.1 Piecewise linear L_1 approximation with pre-assigned tolerance

Let the piecewise approximation for any segment j, $1 \leq j \leq n$, be such that the L_1 error norm z_j, be $z_j \leq \varepsilon$, where ε is a pre-assigned parameter. The number of segments n is not known yet. For this algorithm, we use the sequential or scan-along approach. See Section 8.1.2. It is described in the following steps:

(1) Take j = 1 (first segment).

(2) Starting from the first point in segment j in the digitized curve, take a number of points N = M, which is the minimum initial number of points for segment j. M is the number of terms in the approximating function **L(a, x)**. Use the function **L(a, x)** to calculate matrix **C** and vector **f**, as in **Ca** = **f**. From **C**, **f**, and the parameters M and N calculate the L_1 approximation using LA_Lone() of Chapter 5. This corresponds to a perfect fit, with $z_j = 0$, assuming that the N by M matrix **C** is of rank M. In fact, since we assumed that the functions $(\phi_i(x))$, i = 1, …, M, are linearly independent, matrix **C** in **Ca** = **f** would be of rank M. Go to step (3).

(3) Add point $(N + 1)$ to the right of the N points of segment j. Call the L_1 approximation function LA_Lone() and calculate the L_1 norm for the $(N + 1)$ points. Go to step (4).

(4) If the value of the norm z_j is $\leq \varepsilon$, where ε is the pre-assigned value, go to step (3). If $z_j > \varepsilon$, take $N = N - 1$, which is the final number of points for segment j, with the corresponding z_j as the L_1 solution for segment j. Take $j = j + 1$ and go to step (2) for segment $(j + 1)$.

If the number of points in the last segment is $\leq M$, we get a perfect fit for this segment with $z_n = 0$.

9.4.2 Piecewise linear approximation with near-balanced L_1 norms

Given is the number of segments n. It is required that the calculated L_1 residual norms for the n segments be equal; i.e., $z_1 = z_2 = \ldots = z_n$. As indicated earlier, it may be possible to fulfill this requirement for the piecewise approximation of the given curve itself. However, in the case of the approximation of the discrete points of the digitized curve, a balanced residual norm solution may not exist, so a near-balanced solution is obtained instead.

The initial indices of the first point in each segment are calculated. This is done by dividing the total number of the digitized points of the given curve into n approximate numbers. The steps of this algorithm are similar to those of Pavlidis [8]:

(1) Call the L_1 approximation function LA_Lone() to calculate the L_1 approximation for the n segments, where n is specified. Let their optimum L_1 residual norms be $(z_1, \ldots, z_n)$.

(2) Set iflag(j) = 1, for j = 1, 2, …, n. iflag(j) is an index indicator for segment j such that if both iflag(j) = 0 and iflag(j + 1) = 0, we skip step (3) for segments j and (j + 1). This is indicated shortly.

(3) For j = 1, 3, 5, …, call LA_Lone() to calculate the L_1 residual norms z_j and z_{j+1}. Compare the norms z_j and z_{j+1}. If $z_j < z_{j+1}$, go to step (3.1). Otherwise if $z_j > z_{j+1}$, go to step (4.1).

(3.1) If $z_j < z_{j+1}$, add the left end point of segment $(j + 1)$ to the right end of segment j. Call LA_Lone() to calculate the L_1 approximation for the enlarged segment j and for the shortened segment $(j + 1)$. Call the new L_1 residual norms z_j^{new} and z_{j+1}^{new}. Go to step (3.2).

(3.2) If $|z_j^{new} - z_{j+1}^{new}| < |z_j - z_{j+1}|$, set $z_j = z_j^{new}$ and $z_{j+1} = z_{j+1}^{new}$. Also set iflag(j) = 0 and iflag(j + 1) = 0.
Repeat step (3) for j = 2, 4, 6, …
If iflag(j) = 1, $1 \le j \le n$, repeat step (3), otherwise terminate. If in repeating step (3), iflag = 0 for any two adjacent segments j and (j + 1), skip step (3) for segments j and (j + 1).

(4.1) If $z_j > z_{j+1}$, delete the right end point of segment j by adding it to the left end of segment (j + 1). Go to step (4.2).

(4.2) This step uses equivalent logic to step (3.2).

9.5 Numerical results and comments

Each of the functions LA_L1pw1() and LA_L1pw2() calculate the number of segments in the piecewise approximation (for LA_L1pw2(), n is given), the starting points of the n segments, the coefficients of the approximating curves for the n segments, the residuals at each point of the digitized curve and finally the L_1 residual norms for the n segments.

LA_L1pw1() computes for the case when the L_1 residual norm in any segment is not to exceed a pre-assigned value ε. LA_L1pw2() computes for the case when the number of segments is given and a near-balanced L_1 error norm solution is required. Both of these functions use LA_Lone() of Chapter 5.

LA_L1pw1() has the option of calculating connected or disconnected piecewise L_1 approximations, while LA_L1pw2() can only calculate disconnected piecewise L_1 approximations. This was explained at the end of Section 9.2. DR_L1pw1() and DR_L1pw2() test several examples.

In order to compare the results of these two algorithms with those of the algorithms for the Chebyshev norm [2] and for the L_2 norm [3], we chose the same curve. This curve is digitized with equal x-intervals into 28 points, and the data points are fitted with vertical parabolas. Each is of the form $y = a_1 + a_2x + a_3x^2$. The results of LA_L1pw1() are shown in Figures 9-1 and 9-2. The results of LA_L1pw2() are shown in Figure 9-3, where the number of segments was set to n = 4.

The L_1 norms of Figures 9-1 and 9-2 are (5.940, 4.543, 5.480, 4.414) and n = 4, and (5.940, 4.820, 5.213, 3.038, 1.367) and n = 5, respectively. The L_1 norms of Figure 9-3 are (5.940, 4.543, 5.480, 4.414), for n = 4.

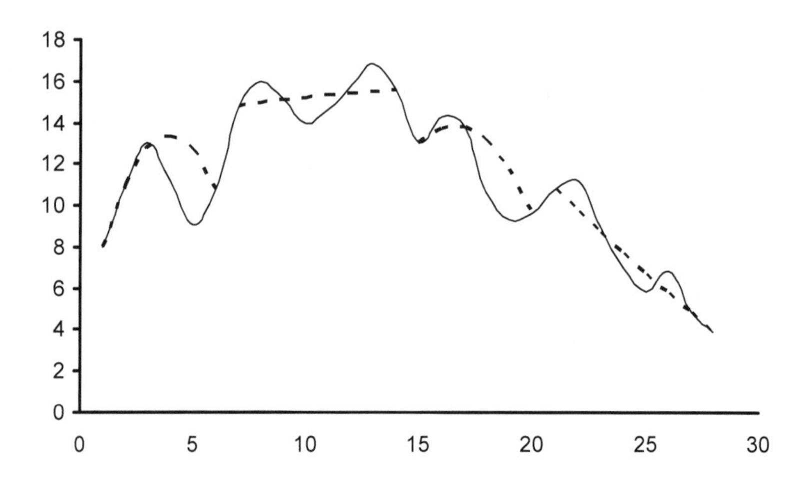

Figure 9-1: Disconnected linear L_1 piecewise approximation with vertical parabolas. The L_1 residual norm in any segment ≤ 6.2

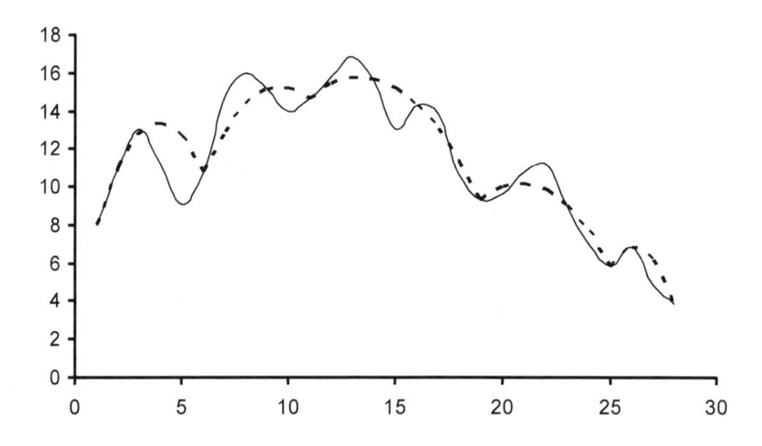

Figure 9-2: Connected linear L_1 piecewise approximation with vertical parabolas. The L_1 residual norm in any segment ≤ 6.2

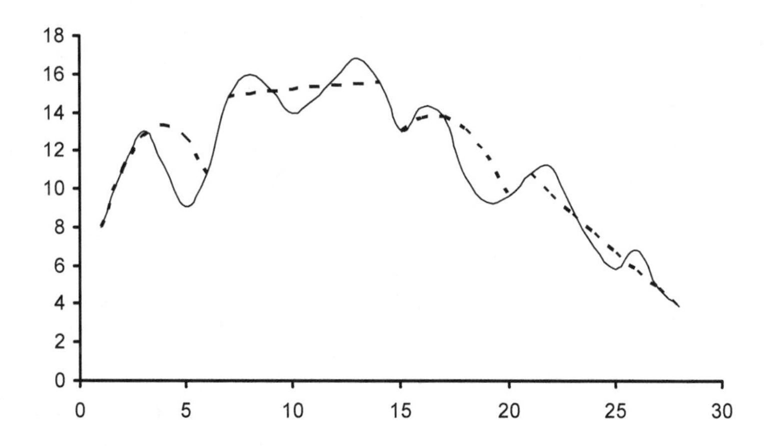

Figure 9-3: Near-balanced residual norm solution. Disconnected linear L_1 piecewise approximation with vertical parabolas. Number of segments = 4

We observe that LA_L1pw2() did not produce a balanced norm solution (equal L_1 norms for all the segments) in Figure 9-3. Also, by mere chance in this example, Figures 9-1 and 9-3 give identical results.

We should note that for each segment, the error between the digitized points and the approximating curve is measured along the y-axis direction. That is, not in the direction perpendicular to the approximating curve.

The requirement that the error between the digitized points and the approximate curve be measured in a direction perpendicular to the approximating curve is not needed in the L_1 piecewise approximation, such as in our algorithm. The reason is that the L_1 error norm for a digitized curve segment is approximately proportional to the area between the digitized points and the approximating curve. This area is approximately the same when the residuals (errors) are measured in the y-direction or when measured perpendicular to the approximating curve. This same point was made in the previous chapter for polygonal approximation of plane curves in the L_1 norm.

In a previous version of our algorithms [4], we utilized parametric linear programming techniques [6] and made use of the interpolating property of the L_1 approximation (Chapters 2 and 5). In these techniques, information from a previous simplex iteration in step (3) of Section 9.4.1 or Section 9.4.2 are stored and used in a current iteration. This increases the code complexity but reduces the number of iterations.

References

1. Abdelmalek, N.N., On the discrete linear L_1 approximation and L_1 solutions of overdetermined linear equations, *Journal of Approximation Theory*, 11(1974)38-53.
2. Abdelmalek, N.N., An efficient method for the discrete linear L_1 approximation problem, *Mathematics of Computation* 29(1975)844-850.
3. Abdelmalek, N.N., Piecewise linear Chebyshev approximation of planar curves, *International Journal of Systems Science*, 14(1983)425-435.
4. Abdelmalek, N.N., Piecewise linear L_1 approximation of planar curves, *International Journal of Systems Science*, 16(1985)447-455.
5. Abdelmalek, N.N., Piecewise linear least-squares approximation of planar curves, *International Journal of Systems Science*, 21(1990)1393-1403.
6. Hadley, G., *Linear Programming*, Addison-Wesley, Reading, MA, 1962.
7. Lawson, C.L., Characteristic properties of the segmented rational minimax approximation problem, *Numerische Mathematik*, 6(1964)293-301.
8. Pavlidis, T., Waveform segmentation through functional approximation, *IEEE Transactions on Computers*, 22(1973)689-697.
9. Pavlidis, T. and Maika, A.P., Uniform piecewise polynomial approximation with variable joints, *Journal of Approximation Theory*, 12(1974)61-69.
10. Sklar, M.G. and Armstrong, R.D., A piecewise linear approximation procedure for Lp norm curve fitting, *Journal of Statistical Computation and Simulation*, 52(1995)323-335.

9.6 DR_L1pw1

```
/*-----------------------------------------------------------------------
DR_L1pw1
-------------------------------------------------------------------------
This is a driver for the function LA_L1pw1(), which calculates a
linear piecewise L-One approximation of a given data point set {x,y}
that results from discretizing a given plane curve y = f(x).
The points of the set might not be equally spaced.

The approximation by LA_L1pw1() is such that the L-One residual
norm for any segment is not to exceed a pre-assigned tolerance
denoted by "enorm".

LA_L1pw1() calculates the connected or the disconnected linear
piecewise L-One approximation, according to the value of an integer
parameter "konect" set by the user.
See comments in program LA_L1pw1().

From the approximating curve we form the overdetermined system of
linear equations

                            c*a = f

"c" is a real n by m matrix of rank k, k <= m < n.
n is the number of digitized points of the given plane curve.

m is the number of terms in the approximating curves.  If for
example, the approximating curves are vertical parabolas of
the form
                      y = a1 + a2*x + a3*x*x
then m = 3.

"f" is a real n vector whose elements are the y coordinates of
the data set {x,y}.

"a" is the solution m vector.  There are different "a" solution
vectors for the different segments.

This driver contains 1 test example. A given curve is digitized
into 28 points at equal x intervals.  The points are piecewise
approximated by vertical parabolas of the form

                      y = a1 + a2*x + a3*x*x
```

```
The results for the disconnected and of the connected piecewise
L-One approximation are given in the text.
-------------------------------------------------------------------*/

#include "DR_Defs.h"
#include "LA_Prototypes.h"

#define Nc          28

void DR_L1pw1 (void)
{
    /*-----------------------------------------
      Constant matrices/vectors
    -----------------------------------------*/
    static tNumber_R fc[Nc+1] =
    {   NIL,
         8.0, 11.0, 13.0, 11.2,  9.1, 10.8, 14.8, 16.0, 15.1, 14.0,
        14.7, 15.8, 16.8, 15.6, 13.0, 14.3, 13.8, 10.6,  9.3,  9.6,
        10.8, 11.2,  9.0,  7.0,  5.8,  6.8,  4.8,  3.9
    };

    /*-----------------------------------------
      Variable matrices/vectors
    -----------------------------------------*/
    tMatrix_R   ct      = alloc_Matrix_R (MM_COLS, NN_ROWS);
    tMatrix_R   rpl     = alloc_Matrix_R (KK_PIECES, NN_ROWS);
    tMatrix_R   ap      = alloc_Matrix_R (KK_PIECES, MM_COLS);
    tVector_R   f       = alloc_Vector_R (NN_ROWS);
    tVector_R   zp      = alloc_Vector_R (KK_PIECES);
    tVector_I   irankp  = alloc_Vector_I (KK_PIECES);
    tVector_I   ixl     = alloc_Vector_I (KK_PIECES);

    int         j, k, m, n;
    int         konect, npiece;
    tNumber_R   enorm;
    eLaRc       prnRc;

    eLaRc       rc = LaRcOk;

    prn_dr_bnr ("DR_L1pw1, L1 Piecewise Approximation of a Plane "
                "with Pre-assigned Norm");

    konect = 0;
    for (k = 1; k <= 2; k++)
```

```
{
    switch (k)
    {
    case 1:
        enorm = 6.2;
        n = Nc;
        m = 3;
        for (j = 1; j <= n; j++)
        {
            f[j] = fc[j];
            ct[1][j] = 1.0;
            ct[2][j] = j;
            ct[3][j] = j*j;
        }
        break;
    default:
        break;
    }

    prn_algo_bnr ("L1pw1");
    if (k == 1) konect = 0;
    if (k == 2) konect = 1;
    prn_example_delim();
    PRN ("konect = %d: Size of matrix \"c\" %d by %d\n",
          konect, n, m);
    PRN ("L1 Piecewise Approximation with Pre-assigned Norm\n");
    PRN ("Pre-assigned Norm \"enorm\" = %8.4f\n", enorm);
    prn_example_delim();
    if (konect == 1)
        PRN ("Connected L1 Piecewise Approximation\n");
    else
        PRN ("Disconnected L1 Piecewise Approximation\n");
    prn_example_delim();
    PRN ("r.h.s. Vector \"f\"\n");
    prn_Vector_R (f, n);
    PRN ("Transpose of Coefficient Matrix, \"ct\"\n");
    prn_Matrix_R (ct, m, n);

    rc = LA_L1pw1 (m, n, enorm, konect, ct, f, ixl, irankp, rp1,
                   ap, zp, &npiece);

    if (rc >= LaRcOk)
    {
        PRN ("\n");
        PRN ("Results of the L1 Piecewise Approximation\n");
```

```
            PRN ("Calculated number of segments (pieces) = %d\n",
                 npiece);
            PRN ("Staring points of the \"npiece\" segments\n");
            prn_Vector_I (ixl, npiece);
            PRN ("Coefficients of the approximating curves\n");
            prn_Matrix_R (ap, npiece, m);
            PRN ("Residual vectors for the \"npiece\" segments\n");
            prnRc = LA_pw1_prn_rp1 (konect, npiece, n, ixl, rp1);
            PRN ("L1 residual norms for the \"npiece\" segments\n");
            prn_Vector_R (zp, npiece);
            if (prnRc < LaRcOk)
            {
                PRN ("Error printing PW1 results\n");
            }
        }

        prn_la_rc (rc);
    }

    free_Matrix_R (ct, MM_COLS);
    free_Matrix_R (rp1, KK_PIECES);
    free_Matrix_R (ap, KK_PIECES);
    free_Vector_R (f);
    free_Vector_R (zp);
    free_Vector_I (irankp);
    free_Vector_I (ixl);
}
```

9.7 LA_L1pw1

```
/*--------------------------------------------------------------------
LA_L1pw1
--------------------------------------------------------------------
This program calculates a linear piecewise L1 (L-One) approximation
to a discrete point set {x,y}.  The approximation is such that the
L1 residual (error) norm for any segment is not to exceed a given
tolerance "enorm".  The number of segments (pieces) is not known
before hand.

Given is a set of points {x,y}.  From the approximating functions
of the piecewise approximation, one forms the overdetermined system
of linear equations
                          c*a = f

This program uses LA_Lone() for obtaining the L-One solution of
overdetermined system of linear equations.

LA_L1pw1() has the option of calculating connected or disconnected
piecewise L1 approximation, according to the value of an integer
parameter "konect".

In the connected piecewise approximation the x-coordinate of the
end point of segment j say, is the x-coordinate of the starting
point of segment (j+1).  In the disconnected piecewise
approximation, the x-coordinate of the adjacent point to the end
point of segment j is the x-coordinate of the starting point of
segment (j+1).  See the comments on "ixl" below.

Inputs
m       Number of terms in the approximating functions.
n       Number of points to be piecewise approximated.
ct      An m by n real matrix containing the transpose of matrix
        "c" of the system c*a = f.
        Matrix "ct" is not destroyed in the computation.
f       An real n vector containing the r.h.s. of the system
        c*a = f. This vector contains the y-coordinates of the
        given point set.  This vector is not destroyed in the
        computation.
enorm   A real pre-assigned parameter, such that the L-One residual
        norm for any segment is <= enorm.
konect  An integer specifying the action to be performed.
        If konect = 1, the program calculates the connected L-One
```

```
        piecewise approximation.
        If konect != 1, the program calculates the disconnected
        L-One piecewise approximation.

Outputs
npiece  Obtained number of segments or pieces of the approximation.
ixl     An integer "npiece" vector containing the indices of the
        first elements of the "npiece" segments.
        For example, if ixl = (1,5,12,22,...), and if konect = 1,
        then the first segment contains points 1 to 5, the second
        segment contains points 5 to 12, the third segment contains
        points 12 to 22 and so on.
        Again if ixl = (1,5,12,22,...), and if konect !=1, then the
        first segment contains points 1 to 4, the second segment
        contains points 5 to 11, the third segment contains points 12
        to 21 ..., etc.
ap      A real "npiece" by m  matrix.  Its first row contains the
        the coefficients of the approximating curve for the first
        segment.  The second row contains the coefficients of the
        approximating curve for the second segment and so on.
        If any row j is all zeros, this indicates that vector "a" of
        segment j is not calculated as the number of points of
        segment j is <= m and there is a perfect fit by the
        approximating curve for segment j.
rp1     A real "npiece" by n matrix.  Its first row contains the
        residuals for the points of the first segment.  Its second
        row contains the residuals for the points of the second
        segment, and so on.
zp      A real "npiece" vector containing the "npiece" optimum
        values of the L-One residual norms for the "npiece" segments.
        If zp[j] == 0.0, it indicates that there is a perfect fit for
        segment j.

Returns one of
        LaRcSolutionFound
        LaRcErrBounds
        LaRcErrNullPtr
        LaRcErrAlloc
-------------------------------------------------------------------*/

#include "LA_Prototypes.h"

eLaRc LA_L1pw1 (int m, int n, tNumber_R enorm, int konect,
    tMatrix_R ct, tVector_R f, tVector_I ixl, tVector_I irankp,
    tMatrix_R rp1, tMatrix_R ap, tVector_R zp, int *pNpiece)
```

```
{
    tMatrix_R   ctp     = alloc_Matrix_R (m, n);
    tVector_R   fp      = alloc_Vector_R (n);
    tVector_R   r       = alloc_Vector_R (n);
    tVector_R   a       = alloc_Vector_R (m);

    int         i = 0, j = 0, je = 0, ji = 0, jj = 0, is = 0, ie = 0,
                nu = 0, irank = 0;
    int         ijk = 0, iter = 0;
    tNumber_R   z = 0.0;

    /* Validation of the data before executing the algorithm */
    eLaRc       rc = LaRcSolutionFound;
    VALIDATE_BOUNDS ((0 < m) && (m <= n) && !((n == 1) && (m == 1))
                    && (0.0 < enorm));
    VALIDATE_PTRS   (ct && f && ixl && irankp && rp1 && ap && zp &&
                    pNpiece);
    VALIDATE_ALLOC  (ctp && fp && r && a);

    *pNpiece = 1;

    /* "is" means i(start) for the segment at hand */
    is = 1;

    for (ijk = 1; ijk <= n; ijk++)
    {
        /* Initializing the data for "npiece" */
        ie = is + m - 1;
        LA_pw1_init (pNpiece, is, ie, m, rp1, ap, zp);

        irank = m;
        z = 0;

        for (j = is; j <= ie; j++)
        {
            ji = j - is + 1;
            fp[ji] = f[j];
            for (i = 1; i <= m; i++)
            {
                ctp[i][ji] = ct[i][j];
            }
        }

        for (jj = 1; jj <= n; jj++)
        {
```

```
            nu = ie - is + 1;

            rc = LA_Lone (m, nu, ctp, fp, &irank, &iter, r, a, &z);
            if (rc < LaRcOk)
            {
                GOTO_CLEANUP_RC (rc);
            }

            if (z > enorm + EPS)
            {
                break;
            }
            else
            {
                LA_pwl_map (m, nu, r, a, z, rp1, ap, zp, pNpiece);

                ixl[*pNpiece] = is;
                je = ie + 1;

                if (je > n)
                {
                    GOTO_CLEANUP_RC (LaRcSolutionFound);
                }

                ie = je;
                nu = ie - is + 1;

                for (j = is; j <= ie; j++)
                {
                    ji = j - is + 1;
                    fp[ji] = f[j];
                    for (i = 1; i <= m; i++)
                    {
                        ctp[i][ji] = ct[i][j];
                    }
                }
            }
        }

        is = ie;
        if (konect == 1) is = ie - 1;
        *pNpiece = *pNpiece + 1;
        ixl[*pNpiece] = is;
    }
```

```
CLEANUP:

    free_Matrix_R (ctp, m);
    free_Vector_R (fp);
    free_Vector_R (r);
    free_Vector_R (a);

    return rc;
}
```

9.8 DR_L1pw2

```
/*--------------------------------------------------------------------
DR_L1pw2
--------------------------------------------------------------------
This is a driver for the function LA_L1pw2() which calculates the
"near balanced" piecewise linear L-One approximation of a given
data point set (x,y) resulting from the discretization of a plane
curve y = f(x).

Given is an integer number "npiece" which is the number of segments
(pieces) in the approximation.

The approximation by LA_L1pw2() is such that the L-One residual norms
for all segments are nearly equal, hence the name "near balanced"
piecewise approximation.

From the approximating curves we form the overdetermined system of
linear equations

                              c*a = f

"c" is a real n by m matrix of rank k, k <= m < n.
n is the number of digitized points of the given plane curve.
m is the number of terms in the approximating curves.  If for
example, the piecewise approximating curves are vertical parabolas
of the form
                      y = a1 + a2*x + a3*x*x
then m = 3.

"f" is a real n vector whose elements are the y coordinates of
the data set {x,y}.

"a" is the solution m vector.  There are different "a" solution
vectors for the different segments.

This driver contains 1 test example.
A given curve is digitized into 28 points at equal x intervals.  The
points are piecewise approximated by vertical parabolas of the form

                      y = a1 + a2*x + a3*x*x

The results for piecewise L-One approximation are given in the text.
--------------------------------------------------------------------*/
```

```
#include "DR_Defs.h"
#include "LA_Prototypes.h"

#define Nc          28

void DR_L1pw2 (void)
{
    /*----------------------------------------
      Constant matrices/vectors
    ----------------------------------------*/
    static tNumber_R fc[Nc+1] =
    {   NIL,
         8.0, 11.0, 13.0, 11.2,  9.1, 10.8, 14.8, 16.0, 15.1, 14.0,
        14.7, 15.8, 16.8, 15.6, 13.0, 14.3, 13.8, 10.6,  9.3,  9.6,
        10.8, 11.2,  9.0,  7.0,  5.8,  6.8,  4.8,  3.9
    };

    /*----------------------------------------
      Variable matrices/vectors
    ----------------------------------------*/
    tMatrix_R   ct      = alloc_Matrix_R (MM_COLS, NN_ROWS);
    tVector_R   rp2     = alloc_Vector_R (NN_ROWS);
    tMatrix_R   ap      = alloc_Matrix_R (KK_PIECES, MM_COLS);
    tVector_R   f       = alloc_Vector_R (NN_ROWS);
    tVector_R   zp      = alloc_Vector_R (KK_PIECES);
    tVector_I   ixl     = alloc_Vector_I (KK_PIECES);

    int         j, m, n;
    int         Iexmpl, npiece;
    eLaRc       prnRc;

    eLaRc       rc = LaRcOk;

    prn_dr_bnr ("DR_L1pw2, L1 Piecewise Approximation of a Plane "
                "Curve with Near Equal Residual Norms");

    for (Iexmpl = 1; Iexmpl <= 1; Iexmpl++)
    {
        switch (Iexmpl)
        {
        case 1:
            npiece = 4;
            n = Nc;
            m = 3;
```

```
        for (j = 1; j <= n; j++)
        {
            f[j] = fc[j];
            ct[1][j] = 1.0;
            ct[2][j] = j;
            ct[3][j] = j*j;
        }
        break;
    default:
        break;
    }

    prn_algo_bnr ("L1pw2");

    prn_example_delim();
    PRN ("Size of matrix \"c\" %d by %d\n", n, m);
    prn_example_delim();
    PRN ("L1 Piecewise Approximation with Near Equal Norms\n");
    PRN ("Given number of segments (pieces) = %d\n", npiece);
    prn_example_delim();
    PRN ("r.h.s. Vector \"f\"\n");
    prn_Vector_R (f, n);
    PRN ("Transpose of Coefficient Matrix, \"ct\"\n");
    prn_Matrix_R (ct, m, n);

    rc = LA_L1pw2 (m, n, npiece, ct, f, ap, rp2, zp, ixl);

    if (rc >= LaRcOk)
    {
        PRN ("\n");
        PRN ("Results of the L1 Piecewise Approximation\n");
        PRN ("Starting points of the \"npiece\" segments\n");
        prn_Vector_I (ixl, npiece);
        PRN ("Coefficients of the \"npiece\" approximating "
             " curves\n");
        prn_Matrix_R (ap, npiece, m);
        PRN ("Residuals at the given points\n");
        prnRc = LA_pw2_prn_rp2 (npiece, n, ixl, rp2);
        PRN ("L1 residual norms for the \"npiece\" segments\n");
        prn_Vector_R (zp, npiece);

        if (prnRc < LaRcOk)
        {
            PRN ("Error printing PW2 results:  ");
        }
```

```
        }

        prn_la_rc (rc);
    }

    free_Matrix_R (ct, MM_COLS);
    free_Vector_R (rp2);
    free_Matrix_R (ap, KK_PIECES);
    free_Vector_R (f);
    free_Vector_R (zp);
    free_Vector_I (ixl);
}
```

9.9 LA_L1pw2

```
/*------------------------------------------------------------------
LA_L1pw2
--------------------------------------------------------------------
This program calculates the "near balanced" piecewise linear
L1 (L-One) approximation  of a given data point set {x,y} resulting
from the discretization of a plane curve y = f(x).

Given is an integer number "npiece" which is the number of segments
in the approximation.

The approximation by LA_L1pw2() is such that the L1 residual norms
for all segments are nearly equal, hence the name "near balanced"
piecewise approximation.

From the approximating functions (curves) one forms the
overdetermined system of linear equations

                        c*a = f

Inputs
npiece  Given umber of segments (pieces) of the approximation.
m       Number of terms in the approximating curves.
n       Number of points to be piecewise approximated.
ct      A real m by n matrix containing the transpose of matrix "c"
        of the system c*a = f.  This matrix is not destroyed in the
        computation.
f       A real n vector containing the r.h.s. of the system c*a = f.
        This vector contains the y-coordinates of the given point
        set. This vector is not destroyed in the computation.

Outputs
ixl     An integer "npiece' vector containing the indices of the
        first elements of the "npiece" segments.
        For example, if ixl = (1,5,12,22,...), then the first
        segment contains points 1 to 4, the second segment  contains
        points 5 to 11, the third segment contains points 12
        to 21 ..., etc.
ap      A real "npiece" by m matrix.  Its first row contains the
        coefficients of the approximating curve for the first
        segment. The second row contains the coefficients of the
        approximating curve for the second segment and so on.
rp2     A real n vector containing the residual values of the n
```

```
        points of the given set {x,y}.
zp      A real "npiece" vector containing the optimum L-One
        "npiece" residual norms for the "npiece" segments.

Returns one of
        LaRcSolutionFound
        LaRcErrBounds
        LaRcErrNullPtr
        LaRcErrAlloc
--------------------------------------------------------------------*/

#include "LA_Prototypes.h"

eLaRc LA_L1pw2 (int m, int n, int npiece, tMatrix_R ct, tVector_R f,
    tMatrix_R ap, tVector_R rp2, tVector_R zp, tVector_I ixl)
{
    tVector_R   al      = alloc_Vector_R (m);
    tVector_R   rl      = alloc_Vector_R (n);
    tVector_R   bv      = alloc_Vector_R (m);
    tMatrix_R   binv    = alloc_Matrix_R (m, m);
    tVector_R   th      = alloc_Vector_R (n);
    tVector_I   icbas   = alloc_Vector_I (m);
    tVector_I   irbas   = alloc_Vector_I (m);
    tVector_I   ibound  = alloc_Vector_I (n);
    tVector_R   ar      = alloc_Vector_R (m);
    tVector_R   rr      = alloc_Vector_R (n);
    tMatrix_R   ctp     = alloc_Matrix_R (m, n);
    tVector_R   fp      = alloc_Vector_R (n);
    tVector_I   iflag   = alloc_Vector_I (npiece);

    int         i = 0, j = 0, k = 0, is = 0, ie = 0, ji = 0, nu = 0,
                ibc = 0, kl = 0, klp1 = 0;
    int         iend = 0, iter = 0, irank = 0, kase = 0;
    int         ijk = 0, isl = 0, isr = 0, isrn = 0, iel = 0,
                ieln = 0, ier = 0;
    int         ipcp1 = 0, istart = 0;
    tNumber_R   zl = 0.0, zln = 0.0, zrn = 0.0;
    tNumber_R   con = 0.0, conn = 0.0;

    /* Validation of the data before executing the algorithm */
    eLaRc       rc = LaRcSolutionFound;
    VALIDATE_BOUNDS ((0 < m) && (m <= n) && !((n == 1) && (m == 1))
                     && (1 < npiece));
    VALIDATE_PTRS   (ct && f && ap && rp2 && zp && ixl);
    VALIDATE_ALLOC  (al && rl && bv && binv && th && icbas && irbas
```

```
                  && ibound && ar && rr && ctp && fp && iflag);

    kase = 0;
    for (k = 1; k <= npiece; k++)
    {
        iflag[k] = 1;

        /* Initializing L1pw2 */
        LA_pw2_init (k, npiece, m, n, &is, &ie, ct, f, ctp, fp, ixl);

        /* Calculating the L1 approximations of the "npiece"
           segments */
        nu = ie - is + 1;

        rc = LA_Lone (m, nu, ctp, fp, &irank, &iter, rl, al, &zl);
        if (rc < LaRcOk)
        {
            GOTO_CLEANUP_RC (rc);
        }

        zp[k] = zl;

        /* Mapping initial data for the "npiece" segments */
        for (i = 1; i <= m; i++)
        {
            ap[k][i] = al[i];
        }
        for (j = is; j <= ie; j++)
        {
            ji = j - is + 1;
            rp2[j] = rl[ji];
        }
    }

    /*---Process of balancing the L-One norms----*/
    istart = 1;
    iend = npiece - 1;
    ipcp1 = npiece + 1;
    ixl[ipcp1] = n + 1;

    for (ijk = 1; ijk < n*n; ijk++)
    {
        for (kl = istart; kl <= iend; kl=kl+2)
        {
            klp1 = kl + 1;
```

```
        con = fabs (zp[klp1] - zp[kl]);
        isl = ixl[kl];
        isr = ixl[klp1];
        iel = isr - 1;
        if (kl < iend) ier = ixl[kl + 2] - 1;
        if (kl == iend) ier = n;

        /* The case where : -----z[i]<z[i+1] */
        if (zp[kl] < zp[klp1])
        {
            ieln = iel + 1;
            isrn = ieln + 1;
            nu = ieln - isl + 1;
            for (j = isl; j <= ieln; j++)
            {
                ji = j - isl + 1;
                fp[ji] = f[j];
                for (i = 1; i <= m; i++)
                {
                    ctp[i][ji] = ct[i][j];
                }
            }

            rc = LA_Lone (m, nu, ctp, fp, &irank, &iter, rl, al,
                          &zln);
            if (rc < LaRcOk)
            {
                GOTO_CLEANUP_RC (rc);
            }

            nu = ier - isrn + 1;
            for (j = isrn; j <= ier; j++)
            {
                ji = j - isrn + 1;
                fp[ji] = f[j];
                for (i = 1; i <= m; i++)
                {
                    ctp[i][ji] = ct[i][j];
                }
            }

            rc = LA_Lone (m, nu, ctp, fp, &irank, &iter, rr, ar,
                          &zrn);
            if (rc < LaRcOk)
            {
```

```
            GOTO_CLEANUP_RC (rc);
        }

        conn = fabs (zrn - zln);
        iflag[kl] = 1;
        iflag[klp1] = 1;
        if (conn >   con)
        {
            iflag[kl] = 0;
            iflag[klp1] = 0;
            continue;
        }
    }
    /* The case where : -----z[i]>z[i+1] */
    else if (zp[kl] > zp[klp1])
    {
        isrn = isr - 1;
        ieln = isrn - 1;
        nu = ieln - isl + 1;
        for (j = isl; j <= ieln; j++)
        {
            ji = j - isl + 1;
            fp[ji] = f[j];
            for (i = 1; i <= m; i++)
            {
                ctp[i][ji] = ct[i][j];
            }
        }

        rc = LA_Lone (m, nu, ctp, fp, &irank, &iter, rl, al,
                      &zln);
        if (rc < LaRcOk)
        {
            GOTO_CLEANUP_RC (rc);
        }

        nu = ier - isrn + 1;
        for (j = isrn; j <= ier; j++)
        {
            ji = j - isrn + 1;
            fp[ji] = f[j];
            for (i = 1; i <= m; i++)
            {
                ctp[i][ji] = ct[i][j];
            }
```

```
            }

            rc = LA_Lone (m, nu, ctp, fp, &irank, &iter, rr, ar,
                          &zrn);
            if (rc < LaRcOk)
            {
                GOTO_CLEANUP_RC (rc);
            }

            conn = fabs (zrn - zln);
            iflag[kl] = 1;
            iflag[klp1] = 1;
            if (conn > con)
            {
                iflag[kl] = 0;
                iflag[klp1] = 0;
                continue;
            }
        }
        for (j = isl; j <= ieln; j++)
        {
            ji = j - isl + 1;
            rp2[j] = rl[ji];
        }
        for (j = isrn; j <= ier; j++)
        {
            ji = j - isrn + 1;
            rp2[j] = rr[ji];
        }
        for (i = 1; i <= m; i++)
        {
            ap[kl][i] = al[i];
            ap[klp1][i] = ar[i];
        }
        zp[kl] = zln;
        zp[klp1] = zrn;
        ixl[klp1] = isrn;
        isr = isrn;
        iel = isr - 1;
    }

    is = 2;
    if (istart == 2) is = 1;
    istart = is;
    ibc = 0;
```

```
        for (j = 1; j <= npiece; j++)
        {
            if (iflag[j] != 0) ibc = 1;
        }
        if (ibc == 0)
        {
            GOTO_CLEANUP_RC (LaRcSolutionFound);
        }
    }

CLEANUP:

    free_Vector_R (al);
    free_Vector_R (rl);
    free_Vector_R (bv);
    free_Matrix_R (binv, m);
    free_Vector_R (th);

    free_Vector_I (icbas);
    free_Vector_I (irbas);
    free_Vector_I (ibound);

    free_Vector_R (ar);
    free_Vector_R (rr);

    free_Matrix_R (ctp, m);
    free_Vector_R (fp);
    free_Vector_I (iflag);

    return rc;
}
```

PART 3

The Chebyshev Approximation

Chapter 10

Linear Chebyshev Approximation

10.1 Introduction

In this chapter and the following four chapters, 5 algorithms for the discrete linear Chebyshev (L_∞ or minimax) approximations are described. In this chapter, we consider the first of these algorithms, the (ordinary) linear Chebyshev approximation, which requires that the Chebyshev or L_∞ norm of the residuals (errors) in the approximation be as small as possible [1, 2].

In real life it is important to know the largest error between the approximation of a given function and the true value of the function itself. Hence, it is called the **minimax** approximation.

Consider the overdetermined system of linear equations

$$\mathbf{Ca} = \mathbf{f} \tag{10.1.1}$$

$\mathbf{C} = (c_{ij})$ is a given real n by m matrix of rank k, $k \leq m < n$, and $\mathbf{f} = (f_i)$ is a given real n-vector. The Chebyshev solution of system (10.1.1) is the m-vector $\mathbf{a} = (a_j)$ that minimizes the Chebyshev or the L_∞ norm z

$$z = \max_i |r_i| \tag{10.1.2}$$

where r_i is the ith residual in (10.1.1), given by

$$r_i = \sum_{j=1}^{m} c_{ij}a_j - f_i, \quad i = 1, 2, \ldots, n \tag{10.1.3}$$

The first algorithm for solving the Chebyshev approximation problem, we are aware of, was by Stiefel [19] for the case when matrix $\mathbf{C}$ in the system $\mathbf{Ca} = \mathbf{f}$ satisfies the Haar condition, (every m by m sub-matrix of $\mathbf{C}$ is nonsingular). This algorithm is known as the exchange algorithm. It is also known by Cheney ([10], p. 45) as the ascent method. Later, both Stiefel [20] and Osborne and Watson [15] showed that the exchange algorithm is exactly equivalent to a simplex algorithm of a linear programming problem.

Barrodale and Phillips [6, 7] implemented a simplex algorithm for obtaining the Chebyshev solution of **Ca** = **f** using the dual form of the linear programming formulation of the problem.

Among the other methods that solve the linear Chebyshev approximation is that of Bartels and Golub [8, 9]. They presented a numerically stable version of Siefel exchange algorithm [20] and used an **LU** decomposition technique for the basis matrix together with an iterative improvement. In their method matrix **C** has to be of full rank.

Narula and Wellington [14] developed one algorithm that solves both the L_1 approximation problem (minimum sum of absolute errors (**MSAE**) regression) and the Chebyshev approximation problem (minimization of the maximum absolute error (**MMAE**) regression). Their algorithm exploits the similarities between the two problems.

Armstrong and Sklar [5] used the revised simplex method with multiple pivoting to solve the dual form of the linear programming problem. They also used an **LU** decomposition technique and in their method matrix **C** is a full rank matrix.

To advance the starting basis in the linear programming solution for the Chebyshev approximation, Sklar and Armstrong [17] solved **Ca** = **f** first in the L_2 (least squares) sense. Then using the characterization theorem (Section 10.1.1), the initial basis in the linear programming problem corresponds to the (m + 1) largest residuals in absolute value of the L_2 solution. This method would be advantageous when the least squares estimate is already available.

Coleman and Li [11] presented a new iterative quadratically convergent algorithm for solving the linear Chebyshev approximation problem. In their method the required number of iterations to solve the problem is insensitive to the problem size and being quadratically convergent, a solution can be obtained with high accuracy. However, in their method, at each iteration, a weighted least squares problem is solved. They showed that their method converges in fewer steps than that of Barrodale and Phillips [7].

Pinar and Elhedhli [16] presented an algorithm based on the application of quadratic penalty functions to a primal linear programming formulation of the Chebyshev approximation problem. Comparisons were made between the Barrodale and Phillips [7] and the predictor-corrector primal-dual interior point algorithms.

We obtain here the Chebyshev solution of the overdetermined system of linear equations $\mathbf{Ca} = \mathbf{f}$ using the dual form of the linear programming formulation of the problem. Our method differs in significant points from that of Barrodale and Phillips [6] as explained in Section 10.5. However, like [6], our algorithm needs minimum computer storage and imposes no conditions on the coefficient matrix, such as the Haar condition or the full rank condition.

The algorithm presented here is in 3 parts. In part 1, a numerically stable initial basic solution is obtained. The initial basic solution may be feasible or not. Also, if $\mathbf{C}$ is not of full rank, the rank of matrix $\mathbf{C}$ is determined.

In part 2, if the initial basic solution is not feasible, it is made feasible by applying a number of Gauss Jordan elimination steps. To end part 2, the marginal costs as well as the objective function z are calculated. The objective function z may not be > 0. If $z < 0$, it is made positive with a small effort.

Part 3 consists of a modified simplex algorithm that makes use of the kind of asymmetry of the simplex tableau.

In Section 10.1.1, the characterization of the Chebyshev solution is outlined. In Section 3.8.2, we outlined the formulation of the Chebyshev problem as a linear programming problem. For the sake of completeness, we present this formulation again in Section 10.2. In Section 10.3, the algorithm is described and a numerical example is solved in detail. In Section 10.4, a significant property of the Chebyshev approximation is described. In Section 10.5, numerical results and comments are given.

10.1.1 Characterization of the Chebyshev solution

For the characterization theorems of the residuals for an optimum solution, see for example, Appa and Smith [3] and Watson [22]. The most important characteristic is described as follows. If matrix $\mathbf{C}$ in $\mathbf{Ca} = \mathbf{f}$ has rank k, $k \leq m$, then there are at least $(k + 1)$ residuals having values $\pm z$, where z is the optimum norm solution. See Sections 2.3.1 and 10.4.

10.2 Linear programming formulation of the problem

The Chebyshev solution problem of system $\mathbf{Ca} = \mathbf{f}$ was first formulated as a linear programming problem in both the primal and the dual formats by Wagner [21]. Let $h \geq 0$, be the value $z = \max_i|r_i|$ in (10.1.2). The primal form of the linear programming problem is

$$\text{minimize } h$$

subject to

$$r_i \leq h \text{ and } r_i \geq -h, \quad i = 1, 2, \ldots, n$$

From (10.1.3) the above two inequalities reduce to

$$-h \leq \sum_{j=1}^{m} c_{ij}a_j - f_i \leq h, \quad i = 1, 2, \ldots, n$$

In vector-matrix form, they become

$$\begin{aligned} \mathbf{Ca} + h\mathbf{e} &\geq \mathbf{f} \\ -\mathbf{Ca} + h\mathbf{e} &\geq -\mathbf{f} \end{aligned}$$

$$h \geq 0 \text{ and } a_j, j = 1, 2, \ldots, m, \text{ unrestricted in sign}$$

$\mathbf{e}$ is an n-vector, each element of which is 1. That is

$$\text{minimize } h$$

subject to

$$\begin{bmatrix} \mathbf{C} & \mathbf{e} \\ -\mathbf{C} & \mathbf{e} \end{bmatrix} \begin{bmatrix} \mathbf{a} \\ h \end{bmatrix} \geq \begin{bmatrix} \mathbf{f} \\ -\mathbf{f} \end{bmatrix} \tag{10.2.1}$$

$$h \geq 0 \text{ and } a_j, j = 1, 2, \ldots, m, \text{ unrestricted in sign}$$

It is much easier to solve the dual form of this formulation

$$\text{maximize } z = [\mathbf{f}^T \; -\mathbf{f}^T]\mathbf{b} \tag{10.2.2a}$$

subject to

$$\begin{bmatrix} \mathbf{C}^T & -\mathbf{C}^T \\ \mathbf{e}^T & \mathbf{e}^T \end{bmatrix} \mathbf{b} = \mathbf{e}_{m+1} \tag{10.2.2b}$$

(10.2.2c) $$b_i \geq 0, \quad i = 1, 2, \ldots, 2n$$

where $\mathbf{e}_{m+1}$ is an (m + 1)-vector, which is the last column in an (m + 1)-unit matrix, $\mathbf{e}^T$ is an n-row vector whose elements are 1's, and the 2n-vector $\mathbf{b} = (b_i)$.

For simplicity of presentation, we write (10.2.2b) in the form

(10.2.2d) $$\mathbf{Db} = \mathbf{e}_{m+1}$$

and let the columns of matrix **D** be denoted by $\mathbf{D}_j$, j = 1, 2, …, 2n.

To simplify the analysis, assume that rank(**C**) = m and thus rank(**D**) = (m + 1). We construct a simplex tableau for problem (10.2.2), which is for (m + 1) constraints in 2n variables. We call this the **large tableau**, because it has 2n variables. This is to distinguish it from the condensed tableaux of (m + 1) constraints in only n variables, described later.

Let **B** denote the basis matrix for problem (10.2.2). It would be of rank (m + 1). Let also $(\mathbf{y}_j)$ be the columns forming the simplex tableau, $(z_j - f_j)$ the marginal costs for column $(\mathbf{y}_j)$, $\mathbf{b}_B$ the basic solution and z the objective function. Let also the elements of the 2n-vector $[\mathbf{f}^T \; -\mathbf{f}^T]$ of (10.2.2a), associated with the basic variables be the (m + 1)-vector $\mathbf{f}_B$. Then as usual, we have for j = 1, 2, …, 2n.

(10.2.3a) $$\mathbf{y}_j = \mathbf{B}^{-1}\mathbf{D}_j$$
(10.2.3b) $$(z_j - f_j) = \mathbf{f}_B{}^T\mathbf{y}_j - f_j$$
(10.2.3c) $$\mathbf{b}_B = \mathbf{B}^{-1}\mathbf{e}_{m+1}$$
(10.2.3d) $$z = \mathbf{f}_B{}^T\mathbf{b}_B$$

10.2.1 Property of the matrix of constraints

Note the kind of asymmetry in the right and left halves of matrix **D** in (10.2.2d). This kind of asymmetry enables us to use a condensed simplex tableau for this problem, as explained in Section 10.3. From (10.2.2b), we note that for j = 1, 2, …, n

(10.2.4) $$\mathbf{D}_j = \begin{bmatrix} \mathbf{C}_j{}^T \\ 1 \end{bmatrix} \quad \text{and} \quad \mathbf{D}_{j+n} = \begin{bmatrix} -\mathbf{C}_j{}^T \\ 1 \end{bmatrix}$$

where $\mathbf{C}_j{}^T$ is column j of $\mathbf{C}^T$.

We have $\mathbf{C}_j^T$ as part of vector $\mathbf{D}_j$, and $-\mathbf{C}_j^T$ as part of vector $\mathbf{D}_{j+n}$, for j = 1, 2, …, n. Hence, from (10.2.4)

(10.2.5) $$\mathbf{D}_j + \mathbf{D}_{j+n} = 2\mathbf{e}_{m+1}, \quad 1 \le j \le n$$

Definition

Because there is a kind of asymmetry between the two halves of the matrix of constraints **D** in (10.2.2b), we define any column j, $1 \le j \le n$, and the column (j + n) in this matrix as two corresponding columns.

Lemma 10.4 and 10.5 below give useful relations between any two corresponding columns in the simplex tableau and between their marginal costs. Thus we use a condensed simplex tableau having (m + 1) constraints in only n variables.

Lemma 10.1

At any stage of the computation, the basic solution vector $\mathbf{b}_B$ equals the $(m + 1)^{th}$ column of $\mathbf{B}^{-1}$.

This follows directly from (10.2.3c), since $\mathbf{e}_{m+1}$ is the $(m + 1)^{th}$ column of an (m + 1)-unit matrix.

Lemma 10.2

The optimum solution **a** and z for problem (10.2.2) is given by

$$\mathbf{B}^T \begin{bmatrix} \mathbf{a} \\ z \end{bmatrix} = \mathbf{f}_B$$

or in effect

$$(\mathbf{a}^T\ z) = \mathbf{f}_B^T\mathbf{B}^{-1}$$

Proof:

To prove the first equation, we observe that the (m + 1) linearly independent rows of matrix $\mathbf{B}^T$ are themselves the (m + 1) rows from the l.h.s. matrix of (10.2.1), and that the (m + 1) elements of $\mathbf{f}_B$ are the corresponding elements from the r.h.s. of (10.2.1). We also observe that when the optimum solution is reached, h = z and the inequalities in (10.2.1) become equalities.

By taking the transpose of the first equation, the second equation follows.

Lemma 10.3

At any stage of the computation, the algebraic sum of the elements of the basic solution $\mathbf{b}_B$ equals 1.

Proof:

Consider the last equation in the system (10.2.2b), which is

$$(\mathbf{e}^T, \mathbf{e}^T)\mathbf{b} = 1$$

Since the elements of the vector $(\mathbf{e}^T, \mathbf{e}^T)$ are 1's, this inner product equals the sum of the elements of vector $\mathbf{b}$. Then since the non-basic elements of $\mathbf{b}$ are all 0's, the lemma is proved.

This lemma is useful in checking the simplex tableau in the process of the solution of the problem. See Tableaux 10.3.1-5 below.

Lemma 10.4

Any two corresponding columns, i and j, $|i - j| = n$, where $1 \leq i, j \leq 2n$, should not appear together in any basis.

See for example, Lemma 4.3 in [15].

Lemma 10.5

(10.2.6a) $$\mathbf{y}_i + \mathbf{y}_j = 2\mathbf{b}_B$$

and

(10.2.6b) $$(z_i - f_i) + (z_j - f_j) = 2z$$

Proof:

By multiplying (10.2.5) by $\mathbf{B}^{-1}$, from (10.2.3a, c), (10.2.6a) is proved. Also, from (10.2.3a-d) and that $f_i = -f_j$, (10.2.6b) is proved.

Let us assume that we have obtained an initial basic solution that is not feasible; that is, one or more elements of the basic solution $\mathbf{b}_B$ is negative. Consider the following lemma.

Lemma 10.6

Let there be one or more elements of the basic solution $\mathbf{b}_B$ that is < 0. Let b_{Bs}, $1 \leq s \leq (m + 1)$, be one of those elements and let i be the basic column associated with b_{Bs}. Let j be the corresponding column to the basic column i, i.e., $|i - j| = n$. Then if column i is replaced by column j in the basis, the new element $b_{Bs}^{(n)}$ will be > 0. Moreover, the elements of the new basic solution $b_{Bk}^{(n)}$ $k \neq s$, each will keep the sign of its old value.

Proof:

Since we assume that column i is the basic column associated with b_{Bs}

$$\mathbf{y}_i = (0, \ldots, 1, \ldots, 0)^T \tag{10.2.7}$$

where the 1 is in the s position. If now column j replaces column i in the basis, and we apply a Gauss-Jordan step to the simplex tableau, we get the following. The pivot to change the tableau is y_{sj}. Then after the Gauss-Jordan step we get

$$b_{Bs}^{(n)} = b_{Bs}/y_{sj} \tag{10.2.8a}$$

$$b_{Bk}^{(n)} = b_{Bk} - (y_{kj}/y_{sj})b_{Bs}, \quad k \neq s \tag{10.2.8b}$$

From (10.2.6a) and (10.2.7), $y_{sj} = 2b_{Bs} - 1$ and $y_{kj} = 2b_{Bk}$, $k \neq s$. Then by substituting the values of y_{sj} and y_{kj} into (10.2.8b), it is easy to show that (10.2.8a, b) become

$$b_{Bs}^{(n)} = b_{Bs}/y_{sj} \tag{10.2.9a}$$

$$b_{Bk}^{(n)} = -b_{Bk}/y_{sj}, \quad k \neq s \tag{10.2.9b}$$

As b_{Bs} is assumed negative, $y_{sj} = 2b_{Bs} - 1$ is also negative. Thus in (10.2.8a), $b_{Bs}^{(n)} > 0$. In (10.2.9b), the parameters $b_{Bk}^{(n)}$, $k \neq s$, each has the same sign as its old value b_{Bk}. The lemma is thus proved.

Lemma 10.7

(a) Consider any basic solution, feasible or not. Then there corresponds two basis matrices denoted by $\mathbf{B}_{(1)}$ and $\mathbf{B}_{(2)}$, each giving the same basic solution. Every column in one has its corresponding column in the other basis, arranged in the same order.

(b) The corresponding two values of z are equal in magnitude but opposite in sign.

Proof:

Let $\mathbf{B}_{(1)}$ be given in partitioned form as $\mathbf{B}_{(1)} = (\mathbf{H}/\mathbf{h})$, where $\mathbf{H}$ is an m by (m + 1) matrix and $\mathbf{h}$ is an (m + 1)-row vector, then from (10.2.2b), and since no two corresponding columns appear together in any basis, $\mathbf{B}_{(2)} = (-\mathbf{H}/\mathbf{h})$. Hence, if $\mathbf{B}_{(1)}^{-1} = (\mathbf{G} \mid \mathbf{g})$, where $\mathbf{G}$ is an (m + 1) by m matrix and **g** an (m + 1)-vector, then $\mathbf{B}_{(2)}^{-1} = (-\mathbf{G} \mid \mathbf{g})$.

For $\mathbf{B}_{(1)}$, from (10.2.3c), the basic solution is $\mathbf{b}_{B(1)} = (\mathbf{G}|\mathbf{g})\mathbf{e}_{m+1} = \mathbf{g}$. For $\mathbf{B}_{(2)}$, the basic solution is $\mathbf{b}_{B(2)} = (-\mathbf{G}|\mathbf{g})\mathbf{e}_{m+1} = \mathbf{g}$, which proves part (a) of the lemma.

To prove part (b) of the lemma, if $\mathbf{f}_{B(1)}$ is associated with $\mathbf{B}_{(1)}$, then from (10.2.3b), $-\mathbf{f}_{B(1)}$ would be associated with $\mathbf{B}_{(2)}$. For $\mathbf{B}_{(1)}$, $z_{(1)} = \mathbf{f}_{B(1)}{}^T\mathbf{g}$, and for $\mathbf{B}_{(2)}$, $z_{(2)} = -\mathbf{f}_{B(1)}{}^T\mathbf{g} = -z_{(1)}$ which proves (b) of the lemma.

Lemma 10.8

Let us use (10.2.3a, b) to construct two simplex tableaux $\mathbf{T}_{(1)}$ and $\mathbf{T}_{(2)}$ that correspond respectively to the bases $\mathbf{B}_{(1)}$ and $\mathbf{B}_{(2)}$ defined in Lemma 10.7. Let also j be the corresponding column to column i, where $1 \le i, j \le 2n$. Then we have the following:

(1) $\mathbf{y}_i$ in $\mathbf{T}_{(1)}$ equals $\mathbf{y}_j$ in $\mathbf{T}_{(2)}$, and
(2) the marginal cost $(z_i - f_i)$ in $\mathbf{T}_{(1)} = -(z_j - f_j)$ in $\mathbf{T}_{(2)}$.

Proof:

For $\mathbf{T}_{(1)}$, $\mathbf{B}_{(1)}^{-1} = (\mathbf{G} \mid \mathbf{g})$, and for $\mathbf{T}_{(2)}$, $\mathbf{B}_{(2)}^{-1} = (-\mathbf{G} \mid \mathbf{g})$, from (10.2.4) and (10.2.3a)

$$\mathbf{y}_i = (\mathbf{G}\mathbf{C}_i^T + \mathbf{g})$$

and for $\mathbf{T}_{(2)}$

$$\mathbf{y}_j = (\mathbf{G}\mathbf{C}_i^T + \mathbf{g}) = \mathbf{y}_i$$

which proves the first part of the lemma.

The second part is established by (10.2.3b), the above equation and the facts that $\mathbf{f}_{B(2)} = -\mathbf{f}_{B(1)}$ and $f_j = -f_i$.

10.3 Description of the algorithm

We construct a condensed simplex tableau for problem (10.2.2), for (m + 1) constraints in only n variables. An n-index indicator vector whose elements are +1 or −1 is needed. If column i, $1 \le i \le n$, is in the condensed tableau, index i of this vector has the value +1. If the corresponding column to column i is in the condensed tableau, index i has a value −1. The algorithm is in 3 parts.

In part 1, we obtain an initial basic solution, feasible or not. This is simply done by performing a finite number of Gauss-Jordan elimination steps to the initial data. We use partial pivoting, where the pivot is the largest element in absolute value in row i of tableau (i – 1), i = 1, 2, …, m+1. Tableau 0 denotes the initial data.

If rank(**C**) = k < m, a row (or more) of $\mathbf{C}^T$ is linearly dependent on one or more of the preceding rows, a zero row in the tableau is obtained and is discarded from the following tableaux. An exception from this rule is the last row in the tableau. The simplex tableau becomes for (k + 1) constraints in n variables.

In part 2 of the algorithm, we examine the elements of the basic solution $\mathbf{b}_B$. If one or more elements in $\mathbf{b}_B$ is < 0, the initial basic solution is not feasible. For each basic column i associated with a negative element of $\mathbf{b}_B$, we apply Lemma 10.6. We make use of (10.2.6a) and replace the column $\mathbf{y}_i$ by its corresponding column $\mathbf{y}_j$. We reverse the sign of the element in vector $\mathbf{f}_B$ associated with this column. According to Lemma 10.6, by applying a Gauss-Jordan step after each replacement, the new basic solution will be feasible. The calculation of the marginal costs $(z_i - f_i)$ follows from (10.2.3b), and the objective function z from (10.2.3d).

If z < 0, we use Lemmas 10.7 and 10.8, and replace the columns of the basis matrix by its corresponding columns. We mainly keep the simplex tableau unchanged, except for the f_i and $\mathbf{f}_B$ values, the marginal costs and z. Such parameters have their signs reversed. We now have a basic feasible solution and z > 0. This ends part 2 of the algorithm.

Part 3 is the ordinary simplex algorithm. The difference is in the choice of the non-basic column that enters the basis. It has the most negative marginal cost among the non-basic columns in the condensed tableau and their corresponding columns. Relation (10.2.6b) is used for calculating the marginal costs of the corresponding columns. Consider the following numerical example.

Example 10.1

This example is, in effect, Example 5.1 solved in Chapter 5 for the L_1 approximation. Matrix **C** is a 5 by 2 matrix of rank 2. Obtain the Chebyshev solution for the system

$$
\begin{aligned}
a_1 + a_2 &= -3 \\
a_1 - a_2 &= -1 \\
a_1 + 2a_2 &= -7 \\
2a_1 + 4a_2 &= -11.1 \\
3a_1 + a_2 &= -7.2
\end{aligned}
\tag{10.3.1}
$$

The Initial Data and the condensed tableaux are shown for this example. In the initial data and in Tableaux 10.3.1 and 10.3.2, the pivot is chosen as the largest element in absolute value in row i of tableau (i – 1). The pivot element in each tableau is bracketed.

Initial Data

	$\mathbf{f}^T$	–3	–1	–7	–11.1	–7.2
B	$\mathbf{b}_B$	$\mathbf{D}_1$	$\mathbf{D}_2$	$\mathbf{D}_3$	$\mathbf{D}_4$	$\mathbf{D}_5$
	0	1	1	1	2	(3)
	0	1	–1	2	4	1
	1	1	1	1	1	1

Tableau 10.3.1 (part 1)

		$\mathbf{f}^T$	–3	–1	–7	–11.1	–7.2
$\mathbf{f}_B$	**B**	$\mathbf{b}_B$	$\mathbf{D}_1$	$\mathbf{D}_2$	$\mathbf{D}_3$	$\mathbf{D}_4$	$\mathbf{D}_5$
–7.2	$\mathbf{D}_5$	0	1/3	1/3	1/3	2/3	1
		0	2/3	–4/3	5/3	(10/3)	0
		1	2/3	2/3	2/3	1/3	0

Tableau 10.3.2

		$\mathbf{f}^T$	–3	–1	–7	–11.1	–7.2
$\mathbf{f}_B$	**B**	$\mathbf{b}_B$	$\mathbf{D}_1$	$\mathbf{D}_2$	$\mathbf{D}_3$	$\mathbf{D}_4$	$\mathbf{D}_5$
–7.2	$\mathbf{D}_5$	0	0.2	0.6	0	0	1
–11.1	$\mathbf{D}_4$	0	0.2	–0.4	0.5	1	0
		1	0.6	(0.8)	0.5	0	0

Tableau 10.3.3 gives an initial basic solution $\mathbf{b}_B$ that is not feasible

since the first element of $\mathbf{b}_B$ is < 0. We make use of Lemma 10.6. Hence, $\mathbf{y}_5$, which is associated with this element, is replaced by its corresponding vector $\mathbf{y}_{10}$ and f_5 is replaced by f_{10}, which $= -f_5$. From (10.2.6a), $\mathbf{y}_{10} = 2\mathbf{b}_B - \mathbf{y}_5$, or $\mathbf{y}_{10} = (-5/2, 1, 5/2)^T$. A Gauss-Jordan step is performed giving tableau 10.3.4. In this tableau the marginal costs and z are also calculated. Though the initial basic solution $\mathbf{b}_B$ is feasible, yet $z < 0$. We use Lemma 10.7 and get Tableau 10.3.4*, where $z > 0$.

Tableau 10.3.3

		$\mathbf{f}^T$	−3	−1	−7	−11.1	−7.2
$\mathbf{f}_B$	**B**	$\mathbf{b}_B$	$\mathbf{D}_1$	$\mathbf{D}_2$	$\mathbf{D}_3$	$\mathbf{D}_4$	$\mathbf{D}_5$
−7.2	$\mathbf{D}_5$	−3/4	−1/4	0	−3/8	0	1
−11.1	$\mathbf{D}_4$	1/2	1/2	0	3/4	1	0
−1	$\mathbf{D}_2$	5/4	3/4	1	5/8	0	0

Tableau 10.3.4 (part 2)

		$\mathbf{f}^T$	−3	−1	−7	−11.1	−7.2
$\mathbf{f}_B$	**B**	$\mathbf{b}_B$	$\mathbf{D}_1$	$\mathbf{D}_2$	$\mathbf{D}_3$	$\mathbf{D}_4$	$\mathbf{D}_{10}$
7.2	$\mathbf{D}_{10}$	0.3	0.1	0	3/20	0	1
−11.1	$\mathbf{D}_4$	0.2	0.4	0	3/5	1	0
−1	$\mathbf{D}_2$	0.5	0.5	1	1/4	0	0
	$z = -0.56$		−1.22	0	1.17	0	0

Tableau 10.3.4*

		$\mathbf{f}^T$	−3	−1	−7	−11.1	−7.2
$\mathbf{f}_B$	**B**	$\mathbf{b}_B$	$\mathbf{D}_6$	$\mathbf{D}_7$	$\mathbf{D}_8$	$\mathbf{D}_9$	$\mathbf{D}_5$
−7.2	$\mathbf{D}_5$	0.3	0.1	0	3/20	0	1
11.1	$\mathbf{D}_9$	0.2	0.4	0	(3/5)	1	0
1	$\mathbf{D}_7$	0.5	0.5	1	1/4	0	0
	$z = 0.56$		1.22	0	−1.17	0	0

In tableau 10.3.4* and its corresponding part, $(z_8 - f_8)$ is the most negative marginal cost, and $\mathbf{D}_8$ replaces $\mathbf{D}_9$ in the basis, which gives tableau 10.3.5.

In each of Tableau 10.3.5 and its corresponding part, we search for the most negative marginal cost. We find that $(z_6 - f_6)$ is the most negative marginal cost, which from (10.2.6b), $= 2z - (z_1 - f_1) = -0.1$. Hence, in Tableau 10.3.5, $\mathbf{D}_1$ replaces $\mathbf{D}_6$ and a Gauss-Jordan step is performed, giving tableau 10.3.6, which gives the optimum Chebyshev norm with $z = 1$. Note that in all the tableaux, the algebraic sum of the elements of $\mathbf{b}_B$ equals 1 (Lemma 10.3).

Tableau 10.3.5 (part 3)

		$\mathbf{f}^T$	−3	−1	−7	−11.1	−7.2
$\mathbf{f}_B$	**B**	$\mathbf{b}_B$	$\mathbf{D}_6$	$\mathbf{D}_7$	$\mathbf{D}_8$	$\mathbf{D}_9$	$\mathbf{D}_5$
−7.2	$\mathbf{D}_5$	1/4	0	0	0	−1/4	1
7	$\mathbf{D}_8$	1/3	2/3	0	1	5/3	0
1	$\mathbf{D}_7$	0.42	1/3	1	0	−0.42	0
	$z = 0.95$		2	0	0	1.95	0

Tableau 10.3.6

		$\mathbf{f}^T$	−3	−1	−7	−11.1	−7.2
$\mathbf{f}_B$	**B**	$\mathbf{b}_B$	$\mathbf{D}_1$	$\mathbf{D}_7$	$\mathbf{D}_8$	$\mathbf{D}_9$	$\mathbf{D}_5$
−3	$\mathbf{D}_1$	1/2	1	0	0	−1/2	2
7	$\mathbf{D}_8$	1/3	0	0	1	5/3	0
1	$\mathbf{D}_7$	1/6	0	1	0	−1/6	−1
	$z = 1$		0	0	0	1.9	0.2

The Chebyshev solution of system (10.3.1) is obtained by solving the equations that correspond to the basis in tableau 10.3.6, namely $\mathbf{D}_1$, $\mathbf{D}_8$ and $\mathbf{D}_7$. We observe that $\mathbf{D}_8$ and $\mathbf{D}_7$ are the corresponding columns of $\mathbf{D}_3$ and $\mathbf{D}_2$ respectively; that is, from Lemma 10.2

$$\begin{aligned} a_1 + a_2 + z &= -3 \\ -a_1 - 2a_2 + z &= 7 \\ -a_1 + a_2 + z &= 1 \end{aligned}$$

or

$$\begin{aligned} a_1 + a_2 + 3 &= -z \\ -a_1 - 2a_2 - 7 &= -z \\ -a_1 + a_2 - 1 &= -z \end{aligned} \tag{10.3.2}$$

By comparing the above 3 equations with their counterparts in the set (10.3.1), we find that the residuals in these 3 equations are $\pm z = \pm 1$. By replacing z with 1.0 (from Tableau 10.3.6), we get

$$z = 1, \ a_1 = -2, \ a_2 = -2$$

However, in the software of this chapter, **a** and z are calculated from the second equation of Lemma 10.2.

Nevertheless, the set (10.3.2) is known as the **reference equation set** for this example, as explained in the following section.

10.4 A significant property of the Chebyshev approximation

Assume that matrix **C** of **Ca** = **f** in (10.1.1) is of rank m. It was noted that the basis matrix **B** is of rank (m + 1); that is, (m + 1) linearly independent columns in the simplex tableau for problem (10.2.2) form the basis matrix **B**. These columns correspond to (m + 1) linearly independent equations in system (10.2.1) (with the inequality sign made an equality sign), known as the **reference equation set**. As noted in the previous section, the equation in Lemma 10.2, which is equation (10.3.2), is a reference equation set.

A significant property of the Chebyshev approximation is stated as follows. The residuals r_i of the (m + 1) equations in the reference equation set each has the value $\pm z$, arranged in any order, where z is the optimum Chebyshev norm. The residuals of all other equations of the set **Ca** = **f**, in absolute value are $\leq z$. If rank(**C**) = k < m, then the reference equation set consists of (k + 1) equations and the residuals of (k + 1) equations each has the value $\pm z$.

10.4.1 The equioscillation property of the Chebyshev norm

If n (> m) discrete points in the x-y plane are being approximated by a plane curve in the Chebyshev norm, the residuals for (m + 1) of these points will be ±z in an oscillating order, where z is the optimum Chebyshev norm. A residual $r_i = +z$, is followed by residual $r_{i+1} = -z$, followed by $r_{i+2} = -z$ and so on, where i, i+1, i+2, …, are consecutive points that correspond to the reference set. The residuals of all other points in the discrete set in absolute value ≤ z.

The above is illustrated in Figure 2-1, where z = 1.797 and the optimum values $r_3 = +1.797$, $r_4 = -1.797$, $r_5 = +1.797$, $r_8 = -1.797$ and the reference equation set involves equations 3, 4, 5 and 8 of system describe by equation (2.2.2). Hence, this property is known as the equioscillation property of the Chebyshev norm.

10.5 Numerical results and comments

LA_Linf() implements this algorithm [2]. DR_Linf() tests the same 8 examples that were solved in Chapter 6 for the one-sided L_1 approximation problem. Table 10.5.1 shows the results of 3 of the examples, computed in single-precision. These 3 results are typical of the other test cases.

Table 10.5.1

Example	**C**(n×m)	Iterations	z	Unique
1	4 × 2	5	10.0	no
2	10× 5	9	1.7778	yes
3	25×10	23	0.0071	no

For each example, the number of iterations and the optimum Chebyshev norm z are shown.

The Chebyshev solution is not unique when matrix **C** is a rank deficient matrix. Also, when the optimum basic feasible solution $\mathbf{b}_B$ is degenerate, i.e., $\mathbf{b}_B$ has one or more zero elements, the Chebyshev solution is most probably not unique.

The algorithm of Barrodale and Phillips [6, 7] is the nearest to ours. Yet, it differs from ours in significant points. They introduced (m + 2) artificial variables in the linear programming problem, which were used as slack variables. In our algorithm no artificial variables are needed. In part 1 of our algorithm it is possible to obtain a numerically stable basic solution, which is always desirable in any iterative procedure. Also, we do not need to calculate the marginal costs and objective function z until the end of part 2 of the algorithm. In our algorithm, if rank(**C**) = k < m, the simplex tableaux would correspond to (k + 1) constraints in n variables and the amount of computation is considerably reduced. On the other hand, in this case in [6], some of the slack variables will persist in all the tableaux.

Spath [18] collected data (matrix **C** and vector **f**) for 42 examples. He tested these examples using different routines, after converting them to FORTRAN 77. He compared the routines with respect to computer storage, computation time and accuracy of results. Among these algorithms are those of Bartels and Golub [9], Barrodale and Phillips [7], ours [2] and Armstrong and Kung [4].

Spath concluded that the method of Armstrong and Kung [4] contains some errors, which yielded incorrect results for some test cases. According to Grant and Hopkins [12], if working correctly, it is faster than that of Barrodale and Phillips [7] only for small values of m.

The total CPU time on the IBM PC AT 102, for the 42 examples were 34, 46, 56 and 73 seconds respectively for the methods of Barrodale and Phillips [7], ours [2], Bartels and Golub [9] and Armstrong and Kung [4]. For pseudo-randomly generated data, the method of Barrodale and Phillips [7] took longer CPU time than the other 3 methods in 3 out of 4 cases (values of m and n), and nearly the same value for the 4th case. Finally, only the methods of Barrodale and Phillips [7] and ours [2] accommodate rank deficient coefficient matrix **C**. Spath also stated that the results are sensitive to the value of the computer tolerance parameter EPS.

References

1. Abdelmalek, N.N., Chebyshev solution of overdetermined systems of linear equations, *BIT*, 15(1975)117-129.

2. Abdelmalek, N.N., A computer program for the Chebyshev solution of overdetermined systems of linear equations, *International Journal for Numerical Methods in Engineering*, 10(1976)1197-1202.
3. Appa, G. and Smith, C., On L_1 and Chebyshev estimation, *Mathematical Programming*, 5(1973)73-87.
4. Armstrong, R.D. and Kung, D.S., Algorithm AS 135: Min-Max estimates for a linear multiple regression problem, *Applied Statistics*, 28(1979)93-100.
5. Armstrong, R.D. and Sklar, M.G., A linear programming algorithm for curve fitting in the L_∞ norm, *Numerical Functional Analysis and Optimization*, 2(1980)187-218.
6. Barrodale, I. and Phillips, C., An improved algorithm for discrete Chebyshev linear approximation, *Proceedings of the Fourth Manitoba conference on Numerical Mathematics*, Hartnell, B.L. and Williams, H.C. (eds.), Winnipeg, Manitoba, Canada, pp. 177-190, 1975.
7. Barrodale, I. and Phillips, C., Algorithm 495: Solution of an overdetermined system of linear equations in the Chebyshev norm, *ACM Transactions on Mathematical Software*, 1(1975)264-270.
8. Bartels, R.H. and Golub, G.H., Stable numerical methods for obtaining the Chebyshev solution of an overdetermined system of equations, *Communications of ACM*, 11(1968)401-406.
9. Bartels, R.H. and Golub, G.H., Algorithm 328: Chebyshev solution to an overdetermined linear system, *Communications of ACM*, 11(1968)428-430.
10. Cheney, E.W., *Introduction to Approximation Theory*, McGraw-Hill, New York, 1966.
11. Coleman, T.F. and Li, Y., A global and quadratically convergent method for linear L_∞ problems, *SIAM Journal on Numerical Analysis*, 29(1992)1166-1186.
12. Grant, P.M. and Hopkins, T.R., A remark on algorithm AS 135: Min-Max estimates for linear multiple regression problems, *Applied Statistics*, 32(1983)345-347.
13. Hadley, G., *Linear Programming*, Addison-Wesley, Reading, MA, 1962.

14. Narula, S.C. and Wellington, J.F., An efficient algorithm for the MSAE and the MMAE regression problems, *SIAM Journal on Scientific and Statistical Computing*, 9(1988)717-727.
15. Osborne, M.R. and Watson, G.A., On the best linear Chebyshev approximation, *Computer Journal*, 10(1967)172-177.
16. Pinar, M.C., and Elhedhli, S., A penalty continuation method for the l_∞ solution of overdetermined linear systems, *BIT – Numerical Mathematics*, 38(1998)127-150.
17. Sklar, M.G. and Armstrong, R.D., Least absolute value and Chebyshev estimation utilizing least squares results, *Mathematical Programming*, 24(1982)346-352.
18. Spath, H., *Mathematical Algorithms for Linear Regression*, Academic Press, English Edition, London, 1991.
19. Stiefel, E., Uber diskrete und lineare Tschebyscheff-approximation, *Numerische Mathematik*, 1(1959)1-28.
20. Stiefel, E., Note on Jordan elimination, linear programming and Tchebycheff approximation. *Numerische Mathematik*, 2(1960)1-17.
21. Wagner, H.M., Linear programming techniques for regression analysis, *Journal of American Statistical Association*, 54(1959)206-212.
22. Watson, G.A., *Approximation Theory and Numerical Methods*, John Wiley & Sons, New York, 1980.

10.6 DR_Linf

```
/*-----------------------------------------------------------------------
DR_Linf
-------------------------------------------------------------------------
This is  a driver for the function LA_Linf(), which solves an
overdetermined system of linear equations in the L-infinity or the
Chebyshev norm.

The overdetermined system has the form

                          c*a = f

"c" is a given real n by m matrix of rank k, k <= m < n.
"f" is a given real n vector.
"a" is the solution m vector.

This driver contains the 8 examples whose results are given in the
text.
-----------------------------------------------------------------------*/

#include "DR_Defs.h"
#include "LA_Prototypes.h"

#define N1 4
#define M1 2
#define N2 5
#define M2 3
#define N3 6
#define M3 3
#define N4 7
#define M4 3
#define N5 8
#define M5 4
#define N6 10
#define M6 5
#define N7 25
#define M7 10
#define N8 8
#define M8 4

void DR_Linf (void)
{
    /*------------------------------------------
```

```
  Constant matrices/vectors
----------------------------------------*/
static tNumber_R c1init[N1][M1] =
{
    { 0.0, -2.0 },
    { 0.0, -4.0 },
    { 1.0, 10.0 },
    {-1.0, 15.0 }
};

static tNumber_R c2init[N2][M2] =
{
    { 1.0,  2.0, 0.0 },
    {-1.0, -1.0, 0.0 },
    { 1.0,  3.0, 0.0 },
    { 0.0,  1.0, 0.0 },
    { 0.0,  0.0, 1.0 }
};

static tNumber_R c3init[N3][M3] =
{
    { 0.0, -1.0,  0.0 },
    { 1.0,  3.0, -4.0 },
    { 1.0,  0.0,  0.0 },
    { 0.0,  0.0,  1.0 },
    {-1.0,  1.0,  2.0 },
    { 1.0,  1.0,  1.0 }
};

static tNumber_R c4init[N4][M4] =
{
    { 1.0, 0.0,  1.0 },
    { 1.0, 2.0,  2.0 },
    { 1.0, 2.0,  0.0 },
    { 1.0, 1.0,  0.0 },
    { 1.0, 0.0, -1.0 },
    { 1.0, 0.0,  0.0 },
    { 1.0, 1.0,  1.0 }
};

static tNumber_R c5init[N5][M5] =
{
    { 1.0, -3.0,  9.0, -27.0 },
    { 1.0, -2.0,  4.0,  -8.0 },
    { 1.0, -1.0,  1.0,  -1.0 },
```

```
        { 1.0,  0.0,  0.0,   0.0 },
        { 1.0,  1.0,  1.0,   1.0 },
        { 1.0,  2.0,  4.0,   8.0 },
        { 1.0,  3.0,  9.0,  27.0 },
        { 1.0,  4.0, 16.0,  64.0 }
    };

    static tNumber_R c6init[N6][M6] =
    {
        { 1.0, 0.0,  0.0,  0.0,  0.0 },
        { 0.0, 1.0,  0.0,  0.0,  0.0 },
        { 0.0, 0.0,  1.0,  0.0,  0.0 },
        { 0.0, 0.0,  0.0,  1.0,  0.0 },
        { 0.0, 0.0,  0.0,  0.0,  1.0 },
        { 1.0, 1.0,  1.0,  1.0,  1.0 },
        { 0.0, 1.0,  1.0,  1.0,  1.0 },
        {-1.0, 0.0, -1.0, -1.0, -1.0 },
        { 1.0, 1.0,  0.0,  1.0,  1.0 },
        { 1.0, 1.0,  1.0,  0.0,  1.0 }
    };

    static tNumber_R c8init[N8][M8] =
    {
        { 1.0, 1.0,   1.0,  1.0 },
        { 1.0, 2.0,   4.0,  4.0 },
        { 1.0, 3.0,   9.0,  9.0 },
        { 1.0, 4.0,  16.0, 16.0 },
        { 1.0, 5.0,  25.0, 25.0 },
        { 1.0, 6.0,  36.0, 36.0 },
        { 1.0, 7.0,  49.0, 49.0 },
        { 1.0, 8.0,  64.0, 64.0 }
    };

    static tNumber_R f1[N1+1] =
    {   NIL,
        -12.0, 6.0, 0.0, 5.0
    };

    static tNumber_R f2[N2+1] =
    {   NIL,
        1.0, 2.0, 1.0, -3.0, 0.0
    };

    static tNumber_R f3[N3+1] =
    {   NIL,
```

```
    1.0, 2.0, 3.0, 2.0, 2.0, 4.0
};

static tNumber_R f4[N4+1] =
{   NIL,
    0.0, -2.0, 1.0, -1.0, 5.0, 7.0, 0.0
};

static tNumber_R f5[N5+1] =
{   NIL,
    3.0, -3.0, -2.0, 0.0, 7.0, -1.0, 5.0, 2.0
};

static tNumber_R f6[N6+1] =
{   NIL,
    1.0, -1.0, 0.0, -1.0, 1.0, 0.0, 2.0, 3.0, -3.0, -2.0
};

static tNumber_R f7[N7+1] =
{   NIL,
    0.0872673, 0.0872794, 0.0873029, 0.0873315, 0.0873576,
    0.3491184, 0.3498802, 0.3513824, 0.3532572, 0.3550109,
    0.6111334, 0.6150641, 0.6230824, 0.6336395, 0.6441493,
    0.8733883, 0.8841621, 0.9071868, 0.9400757, 0.9766021,
    1.135895,  1.157550,  1.206257,  1.283258,  1.384432
};

static tNumber_R f8[N8+1] =
{   NIL,
    2.0, 2.5, 2.0, 6.5, 3.5, 4.5, 6.0, 7.0
};

/*-----------------------------------------
  Variable matrices/vectors
-----------------------------------------*/
tMatrix_R   ct      = alloc_Matrix_R (MMc_COLS, NN_ROWS);
tVector_R   f       = alloc_Vector_R (NN_ROWS);
tVector_R   r       = alloc_Vector_R (NN_ROWS);
tVector_R   a       = alloc_Vector_R (MMc_COLS);
tMatrix_R   binv    = alloc_Matrix_R (MMc_COLS, MMc_COLS);
tVector_R   bv      = alloc_Vector_R (MMc_COLS);
tVector_I   ibound  = alloc_Vector_I (NN_ROWS);
tVector_I   icbas   = alloc_Vector_I (MMc_COLS);
tVector_I   irbas   = alloc_Vector_I (MMc_COLS);
tMatrix_R   c7      = alloc_Matrix_R (N7, M7 + 1);
```

```
tMatrix_R   c1      = init_Matrix_R (&(c1init[0][0]), N1, M1);
tMatrix_R   c2      = init_Matrix_R (&(c2init[0][0]), N2, M2);
tMatrix_R   c3      = init_Matrix_R (&(c3init[0][0]), N3, M3);
tMatrix_R   c4      = init_Matrix_R (&(c4init[0][0]), N4, M4);
tMatrix_R   c5      = init_Matrix_R (&(c5init[0][0]), N5, M5);
tMatrix_R   c6      = init_Matrix_R (&(c6init[0][0]), N6, M6);
tMatrix_R   c8      = init_Matrix_R (&(c8init[0][0]), N8, M8);

int         iter, irank;
int         i, j, k, m, n, Iexmpl;
tNumber_R   d, dd, ddd, e, ee, eee, z;

eLaRc       rc = LaRcOk;

for (j = 1; j <= 5; j++)
{
    d = 0.15* (j-3);
    dd = d*d;
    ddd = d*dd;
    for (i = 1; i <= 5; i++)
    {
        e = 0.15* (i-3);
        ee = e*e;
        eee = e*ee;
        k = 5* (j-1) + i;
        c7[k][1] = 1.0;
        c7[k][2] = d;
        c7[k][3] = e;
        c7[k][4] = dd;
        c7[k][5] = ee;
        c7[k][6] = e*d;
        c7[k][7] = ddd;
        c7[k][8] = eee;
        c7[k][9] = dd*e;
        c7[k][10] = ee*d;
    }
}

prn_dr_bnr ("DR_Linf, Chebyshev Solution of an Overdetermined "
            "System of Linear Equations");

for (Iexmpl = 1; Iexmpl <= 8; Iexmpl++)
{
    switch (Iexmpl)
```

```
    {
        case 1:
            n = N1;
            m = M1;
            for (i = 1; i <= n; i++)
            {
                f[i] = f1[i];
                for (j = 1; j <= m; j++) ct[j][i] = c1[i][j];
            }
            break;

        case 2:
            n = N2;
            m = M2;
            for (i = 1; i <= n; i++)
            {
                f[i] = f2[i];
                for (j = 1; j <= m; j++) ct[j][i] = c2[i][j];
            }
            break;

        case 3:
            n = N3;
            m = M3;
            for (i = 1; i <= n; i++)
            {
                f[i] = f3[i];
                for (j = 1; j <= m; j++) ct[j][i] = c3[i][j];
            }
            break;

        case 4:
            n = N4;
            m = M4;
            for (i = 1; i <= n; i++)
            {
                f[i] = f4[i];
                for (j = 1; j <= m; j++) ct[j][i] = c4[i][j];
            }
            break;

        case 5:
            n = N5;
            m = M5;
```

```
            for (i = 1; i <= n; i++)
            {
                f[i] = f5[i];
                for (j = 1; j <= m; j++) ct[j][i] = c5[i][j];
            }
            break;

        case 6:
            n = N6;
            m = M6;
            for (i = 1; i <= n; i++)
            {
                f[i] = f6[i];
                for (j = 1; j <= m; j++) ct[j][i] = c6[i][j];
            }
            break;

        case 7:
            n = N7;
            m = M7;
            for (i = 1; i <= n; i++)
            {
                f[i] = f7[i];
                for (j = 1; j <= m; j++) ct[j][i] = c7[i][j];
            }
            break;

        case 8:
            n = N8;
            m = M8;
            for (i = 1; i <= n; i++)
            {
                f[i] = f8[i];
                for (j = 1; j <= m; j++) ct[j][i] = c8[i][j];
            }
            break;

        default:
            break;
    }

    prn_algo_bnr ("Linf");

    prn_example_delim();
    PRN ("Example #%d: Size of matrix \"c\" %d by %d\n",
```

```
                Iexmpl, n, m);

        prn_example_delim();
        PRN ("Chebyshev Solution of an Overdetermined System\n");

        prn_example_delim();
        PRN ("r.h.s. Vector \"f\"\n");
        prn_Vector_R (f, n);

        PRN ("Transpose of Coefficient Matrix, \"ct\"\n");
        prn_Matrix_R (ct, m, n);

        rc = LA_Linf (m, n, ct, f, &irank, &iter, r, a, &z);

        if (rc >= LaRcOk)
        {
            PRN ("\n");
            PRN ("Results of the Chebyshev Approximation\n");
            PRN ("Chebyshev solution vector, \"a\"\n");
            prn_Vector_R (a, m);

            PRN ("Chebyshev residual vector \"r\"\n");
            prn_Vector_R (r, n);

            PRN ("Chebyshev norm \"z\" = %8.4f\n", z);
            PRN ("Rank of matrix \"c\" = %d, No. of Iterations "
                "= %d\n", irank, iter);
        }

        prn_la_rc (rc);
    }

    free_Matrix_R (ct, MMc_COLS);
    free_Vector_R (f);
    free_Vector_R (r);
    free_Vector_R (a);
    free_Matrix_R (binv, MMc_COLS);
    free_Vector_R (bv);
    free_Vector_I (ibound);
    free_Vector_I (icbas);
    free_Vector_I (irbas);
    free_Matrix_R (c7, N7);

    uninit_Matrix_R (c1);
    uninit_Matrix_R (c2);
```

```
    uninit_Matrix_R (c3);
    uninit_Matrix_R (c4);
    uninit_Matrix_R (c5);
    uninit_Matrix_R (c6);
    uninit_Matrix_R (c8);
}
```

10.7 LA_Linf

```
/*-----------------------------------------------------------------------
LA_Linf
/*-----------------------------------------------------------------------
This program solves an overdetermined system of linear equations in
the Chebyshev (L-infinity) norm.  It uses a modified simplex method
to the linear programming formulation of the problem.

The system of linear equations has the form

                          c*a = f

"c" is a given real n by m matrix of rank k <= m < n.
"f" is a given real n vector.

The problem is to calculate the elements of the m vector "a" that
gives the minimum Chebyshev norm z.

                z = max|r[i]|,   i=1,2,...,n

where r[i] is the ith residual and is given by

 r[i] = c[i][1]*a[1] + c[i][2]*a[2] + ... + c[i][m]*a[m] - f[i],
                         i = 1, 2, . . ., n

Inputs
m       Number of columns of matrix "c" in the system c*a = f.
n       Number of rows of matrix "c" in the system c*a = f.
ct      A real (m+1) by n matrix. Its first m rows and its n
        columns contain the transpose of matrix "c" of the system
        c*a = f. Its (m+1)th row will be filled with ones by the
        program.
f       A real n vector containing the r.h.s. of the system c*a = f.

Local Variables
binv    A real (m+1) square matrix containing the inverse of the
        basis matrix in the linear programming problem.
bv      A real (m+1) vector containing the basic solution in the
        linear programming problem.
icbas   An integer (m+1) vector containing the indices of the
        columns of matrix "ct" forming the basis matrix.
irbas   A integer (m+1) vector containing the indices of the rows
        of matrix "ct".
```

```
Outputs
irank   Calculated rank of matrix "c".
iter    Number of iterations, or the number of times the simplex
        tableau is changed by a Gauss-Jordon elimination step..
a       A real (m + 1) vector. Its first m elements are the
        Chebyshev solution of the system  c*a = f.
r       A real n vector containing the residual vector
        r = (c*a - f).
z       The minimum Chebyshev norm of the residual vector "r".

Returns one of
        LaRcSolutionUnique
        LaRcSolutionProbNotUnique
        LaRcSolutionDefNotUniqueRD
        LaRcNoFeasibleSolution
        LaRcErrBounds
        LaRcErrNullPtr
        LaRcErrAlloc
-------------------------------------------------------------------*/

#include "LA_Prototypes.h"

eLaRc LA_Linf (int m, int n, tMatrix_R ct, tVector_R f, int *pIrank,
    int *pIter, tVector_R r, tVector_R a, tNumber_R *pZ)
{
    tVector_I   icbas   = alloc_Vector_I (m + 1);
    tVector_I   irbas   = alloc_Vector_I (m + 1);
    tVector_R   th      = alloc_Vector_R (n);
    tMatrix_R   binv    = alloc_Matrix_R (m + 1, m + 1);
    tVector_R   bv      = alloc_Vector_R (m + 1);
    tVector_I   ibound  = alloc_Vector_I (n);

    int         i = 0, ij = 0, j = 0, kl = 0, m1 = 0;
    int         itest = 0, iout = 0, jin = 0, ivo = 0;
    tNumber_R   d = 0.0;

    /* Validation of data before executing the algorithm */
    eLaRc       rc = LaRcSolutionUnique;
    VALIDATE_BOUNDS ((0 < m) && (m < n));
    VALIDATE_PTRS   (ct && f && pIrank && pIter && r && a && pZ);
    VALIDATE_ALLOC  (icbas && irbas && th && binv && bv && ibound);

    /* Part 1 of the algorithm */
    m1 = m + 1;
```

```
kl = 1;
*pIter = 0;
*pIrank = m;
*pZ = 0.0;

/* Initializing the data */
LA_linf_init (m, n, ct, icbas, irbas, binv, ibound, r, a);

iout = 0;

/* Part 1 of the algorithm.
   Detecting the rank of matrix "c" */
LA_linf_part_1 (&kl, m, n, ct, f, icbas, irbas, binv, bv, ibound,
                pIrank, pIter);

/* Part 2 of the algorithm.
   Obtaining a basic feasible solution */
LA_linf_part_2 (kl, m, n, ct, f, icbas, binv, bv, ibound, pIter);

/* Part 3 of the algorithm.
   Calculating the initial marginal costs and the norm z */
LA_linf_part_3 (kl, m, n, ct, f, icbas, binv, bv, ibound, r, pZ);

for (ij = 1; ij <= n*n; ij++)
{
    ivo = 0;
    /* Determine the vector that enters the basis */
    LA_linf_vent (&ivo, &jin, kl, m, n, icbas, r, pZ);

    if (ivo == 0)
    {
        /* Calculate the results */
        rc = LA_linf_res (kl, m, n, f, icbas, irbas, binv, bv,
                          ibound, r, a, *pIrank, pZ);
        GOTO_CLEANUP_RC (rc);
    }
    if (ivo != 1)
    {
        for (i = kl; i <= m1; i++)
        {
            ct[i][jin] = bv[i] + bv[i] - ct[i][jin];
        }
        r[jin] = *pZ + *pZ - r[jin];
        f[jin] = -f[jin];
        ibound[jin] = -ibound[jin];
```

```
        }
        itest = 0;

        /* Determine the vector that leaves the basis */
        LA_linf_vleav (&itest, &iout, jin, kl, m, ct, bv);

        /* No feasible solution is available */
        if (itest != 1)
        {
            GOTO_CLEANUP_RC (LaRcNoFeasibleSolution);
        }

        /* A Gauss-Jordan elimination step */
        LA_linf_gauss_jordn (iout, jin, kl, m, n, ct, icbas, binv,
                             bv, pIter);
        d = r[jin];
        for (j = 1; j <= n; j++)
        {
            r[j] = r[j] - d * (ct[iout][j]);
        }
        *pZ = *pZ - d * (bv[iout]);
    }

CLEANUP:

    free_Vector_I (icbas);
    free_Vector_I (irbas);
    free_Vector_R (th);
    free_Matrix_R (binv, m + 1);
    free_Vector_R (bv);
    free_Vector_I (ibound);

    return rc;
}

/*----------------------------------------------------------------------
Initializing the data of LA_Linf()
----------------------------------------------------------------------*/
void LA_linf_init (int m, int n, tMatrix_R ct, tVector_I icbas,
    tVector_I irbas, tMatrix_R binv, tVector_I ibound, tVector_R r,
    tVector_R a)
{
    int         i, j, m1;

    m1 = m + 1;
```

```
    for (j = 1; j <= m1; j++)
    {
        a[j] = 0.0;
        icbas[j] = 0;
        irbas[j] = j;
        for (i = 1; i <= m1; i++)
        {
            binv[i][j] = 0.0;
        }
        binv[j][j] = 1.0;
    }
    for (j = 1; j <= n; j++)
    {
        ct[m1][j] = 1.0;
        r[j] = 0.0;
        ibound[j] = 1;
    }
}

/*-----------------------------------------------------------------------
Part 1 of the algorithm LA_Linf()
-----------------------------------------------------------------------*/
void LA_linf_part_1 (int *pKl, int m, int n, tMatrix_R ct,
    tVector_R f, tVector_I icbas, tVector_I irbas, tMatrix_R binv,
    tVector_R bv, tVector_I ibound, int *pIrank, int *pIter)
{
    int         j, iout, jin = 0;
    tNumber_R    d, piv,

        m1 = m + 1;
    for (iout = 1; iout <= m1; iout++)
    {
        piv = 0.0;
        for (j = 1; j <= n; j++)
        {
            d = ct[iout][j];
            if (d < 0.0) d = -d;
            if (d > piv)
            {
                jin = j;
                piv = d;
            }
        }
        if (piv > EPS)
        {
```

```
            /* A Gauss-Jordan elimination step */
            LA_linf_gauss_jordn (iout, jin, *pKl, m, n, ct, icbas,
                                 binv, bv, pIter);
        }

        /* Detect the rank of matrix "c" */
        LA_linf_detect_rank (pKl, iout, jin, m, n, piv, ct, f, icbas,
                             irbas, binv, bv, ibound, pIrank, pIter);
    }
}

/*--------------------------------------------------------------------
Detection of rank deficiency of matrix "c" in LA_Linf()
--------------------------------------------------------------------*/
void LA_linf_detect_rank (int *pKl, int iout, int jin, int m, int n,
    tNumber_R piv, tMatrix_R ct, tVector_R f, tVector_I icbas,
    tVector_I irbas, tMatrix_R binv, tVector_R bv, tVector_I ibound,
    int *pIrank, int *pIter)
{
    int         i, j, k, m1, icb;
    m1 = m + 1;
    if ((piv < EPS) && iout < m1)
    {
        swap_rows_Matrix_R (ct, *pKl, iout);
        k = irbas[iout];
        irbas[iout] = irbas[*pKl];
        irbas[*pKl] = 0;
        for (j = *pKl; j <= m; j ++)
        {
            binv[iout][j] = binv[*pKl][j];
            binv[*pKl][j] = 0.0;
        }
        icbas[iout] = icbas[*pKl];
        icbas[*pKl] = 0;
        for (i = *pKl; i <= m1; i++)
        {
            binv[i][iout] = binv[i][*pKl];
            binv[i][*pKl] = 0.0;
        }
        *pIrank = *pIrank - 1;
        *pKl = *pKl + 1;
    }
    if ((piv < EPS) && iout == m1)
    {
        for (j = 1; j <= n; j++)
```

```
        {
            icb = 0;
            for (i = *pKl; i <= m; i++)
            {
                if (j == icbas[i]) icb = 1;
            }
            if (icb == 0)
            {
                jin = j;
                break;
            }
        }
        f[jin] = -f[jin];
        ibound[jin] = -ibound[jin];
        for (i = *pKl; i <= m; i++)
        {
            ct[i][jin] = -ct[i][jin];
        }
        ct[m1][jin] = 2.0 - ct[m1][jin];
        /* A Gauss-Jordan elimination step */
        LA_linf_gauss_jordn (iout, jin, *pKl, m, n, ct, icbas, binv,
                             bv, pIter);
    }
}

/*-------------------------------------------------------------------
Part 2 of the algorithm.
Obtaining a basic feasible solution in LA_Linf()
-------------------------------------------------------------------*/
void LA_linf_part_2 (int kl, int m, int n, tMatrix_R ct, tVector_R f,
    tVector_I icbas, tMatrix_R binv, tVector_R bv, tVector_I ibound,
    int *pIter)
{
    int         i, k, m1;
    int         iout, jin;

    jin = 0;
    iout = 0;
    m1 = m + 1;

    for (i = kl; i <= m; i++)
    {
        if (bv[i] < 0.0)
        {
            iout = i;
```

```
            jin = icbas[i];
            f[jin] = -f[jin];
            ibound[jin] = -ibound[jin];
            for (k = kl; k <= m1; k++)
            {
                ct[k][jin] = bv[k] + bv[k] - ct[k][jin];
            }
            /* A Gauss-Jordan elimination step */
            LA_linf_gauss_jordn (iout, jin, kl, m, n, ct, icbas,
                binv, bv, pIter);
        }
    }
}

/*-------------------------------------------------------------------
Part 3 of the algorithm.
Calculating the initial marginal costs and the norm z in LA_Linf()
-------------------------------------------------------------------*/
void LA_linf_part_3 (int kl, int m, int n, tMatrix_R ct, tVector_R f,
    tVector_I icbas, tMatrix_R binv, tVector_R bv,
    tVector_I ibound, tVector_R r, tNumber_R *pZ)
{
    int         i, j, k, m1, icb;
    tNumber_R   s;

    m1 = m + 1;
    for (j = 1;  j <= n; j++)
    {
        r[j] = 0.0;
        icb = 0;
        for (i = kl; i <= m1; i++)
        {
            if (j == icbas[i]) icb = 1;
        }
        if (icb == 0)
        {
            s = -f[j];
            for (i = kl; i <= m1; i++)
            {
                k = icbas[i];
                s = s + ct[i][j] * (f[k]);
            }
            r[j] = s;
        }
    }
```

```
    s = 0.0;
    for (i = kl; i <= m1; i++)
    {
        k = icbas[i];
        s = s + bv[i] * (f[k]);
    }
    *pZ = s;
    if (*pZ < 0.0)
    {
        for (j = 1; j <= n; j++)
        {
            f[j] = -f[j];
            ibound[j] = -ibound[j];
            r[j] = -r[j];
        }
        *pZ = -*pZ;
        for (j = kl; j <= m; j++)
        {
            for (i = kl; i <= m1; i++)
            {
                binv[i][j] = -binv[i][j];
            }
        }
    }
}

/*-----------------------------------------------------------------
Determine the vector that enters the basis in LA_Linf()
-----------------------------------------------------------------*/
void LA_linf_vent (int *pIvo, int *pJin, int kl, int m, int n,
    tVector_I icbas, tVector_R r, tNumber_R *pZ)
{
    int         i, j, m1, icb;
    tNumber_R   d, e, g, tz;

    m1 = m + 1;
    g = 1.0/ (EPS*EPS);
    tz = *pZ + *pZ + EPS;
    for (j = 1; j <= n; j++)
    {
        icb = 0;
        for (i = kl; i <= m1; i++)
        {
            if (j == icbas[i]) icb = 1;
        }
```

```
        if (icb == 0)
        {
            d = r[j];
            if (d < -EPS)
            {
                e = d;
                if (e  < g)
                {
                    g = e;
                    *pJin = j;
                    *pIvo = 1;
                }
            }
            else if (d >= tz)
            {
                e = tz - d;
                if (e < g)
                {
                    g = e;
                    *pJin = j;
                    *pIvo = -1;
                }
            }
        }
    }
}

/*-------------------------------------------------------------------
Determine the vector that leaves the basis in LA_Linf()
-------------------------------------------------------------------*/
void LA_linf_vleav (int *pItest, int *pIout, int jin, int kl, int m,
    tMatrix_R ct, tVector_R bv)
{
    int         i, m1;
    tNumber_R   d, g, thmax;

    m1 = m + 1;
    thmax = 1.0/ (EPS*EPS);
    for (i = kl; i <= m1; i++)
    {
        d = ct[i][jin];
        if (d > EPS)
        {
            g = bv[i]/d;
            if (g <= thmax)
```

```
            {
                thmax = g;
                *pIout = i;
                *pItest = 1;
            }
        }
    }
}

/*-----------------------------------------------------------------------
A Gauss-Jordan elimination step in LA_Linf()
-----------------------------------------------------------------------*/
void LA_linf_gauss_jordn (int iout, int jin, int kl, int m, int n,
    tMatrix_R ct, tVector_I icbas, tMatrix_R binv, tVector_R bv,
    int *pIter)
{
    int         i, j, l, m1;
    tNumber_R   d, pivot;

    m1 = m + 1;
    pivot = ct[iout][jin];

    for (j = 1; j <= n; j++)
    {
        ct[iout][j] = ct[iout][j]/pivot;
    }
    l = m1;
    /*ko = kl + *pIter;
    if (ko < m1) l = ko; */
    for (j = kl; j <= m1; j++)
    {
        binv[iout][j] = binv[iout][j]/pivot;
    }
    for (i = kl; i <= m1; i++)
    {
        if (i != iout)
        {
            d = ct[i][jin];
            for (j = 1; j <= n; j++)
            {
                ct[i][j] = ct[i][j] - d * (ct[iout][j]);
            }
            for (j =kl; j <= m1; j++)
            {
                binv[i][j] = binv[i][j] - d * (binv[iout][j]);
```

```
            }
        }
    }
    for (i = kl; i <= m1; i++)
    {
        bv[i] = binv[i][m1];
    }
    *pIter = *pIter + 1;
    icbas[iout] = jin;
}

/*-----------------------------------------------------------------------
Calculate the results of LA_Linf()
-----------------------------------------------------------------------*/
eLaRc LA_linf_res (int kl, int m, int n,  tVector_R f,
    tVector_I icbas, tVector_I irbas, tMatrix_R binv, tVector_R bv,
    tVector_I ibound, tVector_R r, tVector_R a, int irank,
    tNumber_R *pZ)
{
    int         i, j, k, m1;
    tNumber_R   s, d;

    m1 = m + 1;
    for (j = kl; j <= m1; j++)
    {
        s = 0.0;
        for (i = kl; i <= m1; i++)
        {
            k = icbas[i];
            s = s + f[k]*binv[i][j];
        }

        k = irbas[j];
        a[k] = s;
    }

    for (j = 1; j <= n; j++)
    {
        d = r[j] - *pZ;
        if (ibound[j] == -1) d = -d;
        r[j] = d;
    }

    if (irank < m)
        return LaRcSolutionDefNotUniqueRD;
```

```
    for (i = 1; i <= m1; i++)
    {
        if (bv[i] < EPS)
            return LaRcSolutionProbNotUnique;
    }

    return LaRcSolutionUnique;
}
```

Chapter 11

One-Sided Chebyshev Approximation

11.1 Introduction

In the previous chapter, an algorithm for obtaining the linear (ordinary) Chebyshev approximation is given [1]. The linear Chebyshev approximation is a double-sided one, meaning that some of the elements of the residual vector have values ≥ 0 and the others have values < 0. In other words, for the discrete linear Chebyshev approximation, some of the discrete points are above or on, and some are below the approximating curve (surface). Hence, the approximation is the ordinary or the double-sided one.

In this chapter, we study the linear one-sided Chebyshev approximation problem. In this problem, all the residuals are either non-positive or non-negative. In other words, for the Chebyshev approximation of discrete points, all the discrete points are either above or on the approximating curve (surface), or below or on the approximating curve. In the former case, when all the discrete points are above or on the approximating curve, we have the one-sided Chebyshev approximation from above. The one-sided Chebyshev approximation from below is described in a similar manner. This is illustrated by Figure 11-1, which is analogous to Figure 6-1 for the linear one-sided L_1 approximations.

Consider the overdetermined system of linear equations

$$\mathbf{Ca} = \mathbf{f}$$

$\mathbf{C} = (c_{ij})$, is a given real n by m matrix of rank k, $k \leq m < n$ and $\mathbf{f} = (f_i)$ is a given real n-vector. The L_∞ or the Chebyshev solution of system $\mathbf{Ca} = \mathbf{f}$ is the m-vector $\mathbf{a} = (a_i)$ that minimizes the Chebyshev norm z of the residuals

(11.1.1) $$z = \max|r_i|, \quad i = 1, 2, \ldots, n$$

where r_i is the ith residual and is given by

(11.1.2) $$r_i = f_i - \sum_{j=1}^{m} c_{ij}a_j, \quad i = 1, 2, \ldots, n$$

The special case when the system of equation $\mathbf{Ca} = \mathbf{f}$ is consistent, i.e., the residual $\mathbf{r} = \mathbf{f} - \mathbf{Ca} = \mathbf{0}$, is not of interest here. We thus assume that $\mathbf{r} \neq \mathbf{0}$. Note that the residuals in (11.1.2) are chosen to be the negative of the residuals in Chapter 10. This is to facilitate the analysis in this chapter. As noted in Chapter 2, such choice of the definition of the residuals is arbitrary.

When the solution vector satisfies the additional conditions

(11.1.3) $$r_i \geq 0, \quad i = 1, 2, \ldots, n$$

which in vector-matrix form is

(11.1.3a) $$\mathbf{f} \geq \mathbf{Ca}$$

we have the one-sided Chebyshev solution from above of system $\mathbf{Ca} = \mathbf{f}$; that is, from (11.1.3a), for any equation i in $\mathbf{Ca} = \mathbf{f}$, $i = 1, 2, \ldots, n$, the observed value f_i of vector $\mathbf{f}$ is $\geq$ the calculated element i of $\mathbf{Ca}$. If the inequalities in (11.1.3) are reversed, we have the problem of the one-sided Chebyshev solution from below.

We also note that there are problems that have an (ordinary) Chebyshev approximation but do not have one-sided Chebyshev approximation from above and/or from below. See the numerical examples in Section 11.5.

We consider here the problem of the one-sided Chebyshev approximation from above. However, the analysis and presentation of the problem from below are almost identical and the described algorithm applies to it as well. The problem is presented here as a linear programming problem and we pursue the analysis of the dual form of the linear programming presentation. In this algorithm, minimum computer storage is required, no artificial variables are needed and no conditions are imposed on the coefficient matrix.

In Section 11.2, the problem is presented as a special case of general constrained Chebyshev approximation problems. We shall comment on these algorithms in Section 11.5. In Section 11.3, the linear programming formulation of the problem is given and necessary lemmas are derived. In Section 11.4, the algorithm is described and a numerical example is solved. A note on the linear one-sided Chebyshev solution from below is also presented. In Section 11.5, numerical results are given. Simple relationships between the one-sided and the ordinary Chebyshev approximations are observed. Comparisons with other methods for solving the same problem are also given.

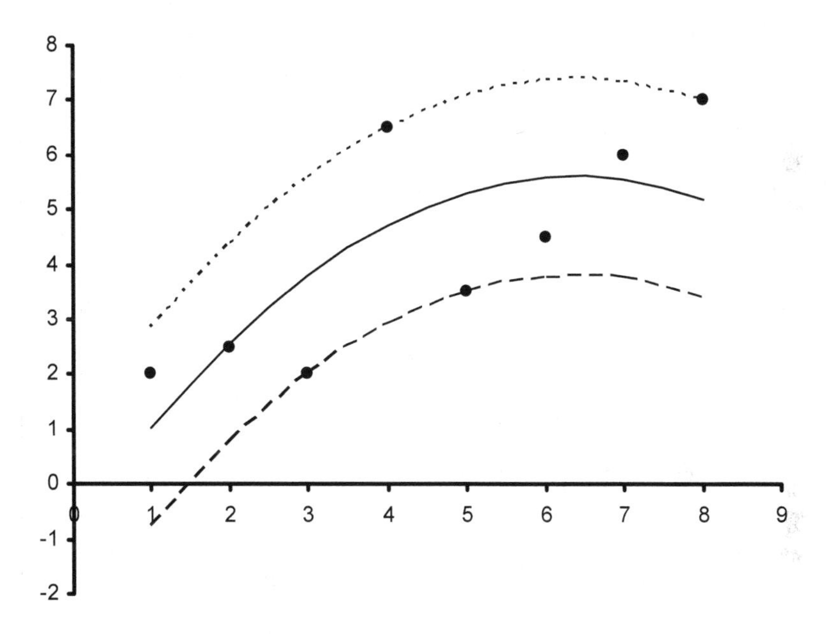

Figure 11-1: Curve fitting with vertical parabolas of a set of 8 points using Chebyshev and one-sided Chebyshev approximations

This figure shows curve fitting with vertical parabolas of the set of 8 points shown in Figure 2-1. The solid curve is the ordinary Chebyshev approximation. The dashed curve is the one-sided Chebyshev approximation from above and the dotted curve is the one-sided Chebyshev approximation from below.

11.1.1 Applications of the algorithm

As indicated in Chapter 6, the discrete linear one-sided approximation problems arise in the discretization of the continuous one-sided approximation problems. The latter problems arise, for example, in the solution of operator equations [13, 14]. They are also applied to the solution of overdetermined linear inequalities. The latter is a basic problem in pattern classification ([12], pp. 40-41, 48-49). See Chapter 16.

11.2 A special problem of a general constrained one

There exist algorithms for constrained Chebyshev approximation problems. By manipulating the constraints, the problem reduces to a one-sided Chebyshev approximation one. This was the case with the one-sided L_1 approximation problem presented in Section 6.2.

Using our notation, we start with Dax [6] who presented an algorithm for solving the one-sided Chebyshev approximation from below, namely

$$\text{minimize } ||\mathbf{Ca} - \mathbf{f}||_\infty$$

subject to

$$\mathbf{Ca} \geq \mathbf{f}$$

In his method, a search direction is obtained via the solution of the least squares problem, which means that a least squares solution should be available first.

Joe and Bartels [8], using a penalty linear programming approach, presented an algorithm to solve the problem

$$\text{minimize } ||\mathbf{Ca} - \mathbf{f}||_\infty \tag{11.2.1a}$$

subject to

$$\mathbf{Ga} = \mathbf{g} \text{ and } \mathbf{Ea} \geq \mathbf{e} \tag{11.2.1b}$$

If in (11.2.2b) we take $\mathbf{G} = \mathbf{0}$, $\mathbf{g} = \mathbf{0}$ and take $\mathbf{E} = \mathbf{C}$ and $\mathbf{e} = \mathbf{f}$, (11.2.1b) reduces to

$$\mathbf{Ca} \geq \mathbf{f}$$

and we get the one-sided Chebyshev solution from below.

Sklar [11] presented a linear programming algorithm with multiple pivoting and an **LU** decomposition technique to solve the problem

$$\text{minimize } ||\mathbf{Ca} - \mathbf{f}||_\infty \tag{11.2.2a}$$

subject to

(11.2.2b) $$\mathbf{e_a} \leq \mathbf{Ea} \leq \mathbf{e_b}$$

His algorithm is a generalization of that of Armstrong and Sklar [3]. If we take $\mathbf{E} = \mathbf{C}$, $\mathbf{e_a} = -\infty$ and $\mathbf{e_b} = \mathbf{f}$, we get the one-sided Chebyshev solution from above. By $-\infty$, we mean a large negative number, such as say -10^6.

Roberts and Barrodale [10] presented an algorithm to solve the problem

(11.2.3a) $$\text{minimize } \|\mathbf{Ca} - \mathbf{f}\|_\infty$$

subject to

(11.2.3b) $$\mathbf{Ga} = \mathbf{g} \text{ and } \mathbf{e_a} \leq \mathbf{Ea} \leq \mathbf{e_b}$$

Their algorithm is the extension of the algorithm of Barrodale and Phillips for the ordinary Chebyshev solution [4, 5]. If in (11.2.3b) we take $\mathbf{G} = \mathbf{0}$, $\mathbf{g} = \mathbf{0}$, problem (11.2.3) reduces to problem (11.2.2).

11.3 Linear programming formulation of the problem

Watson [14, 15] reduced the problem of the one-sided Chebyshev approximation from above to a linear programming problem as follows. Let in (11.1.1), $h = \max|r_i|$, $i = 1, 2, \ldots, n$. Then from (11.1.2) and (11.1.3), in vector-matrix form, the primal linear programming formulation of this problem is

$$\text{minimize } h$$

subject to

$$\mathbf{0} \leq \mathbf{f} - \mathbf{Ca} \leq h\mathbf{e}$$

$\mathbf{e}$ is an n-vector each element of which equals 1 and $\mathbf{0}$ is an n-zero vector. The above two inequalities reduce to

$$\mathbf{Ca} + h\mathbf{e} \geq \mathbf{f}$$

$$-\mathbf{Ca} \geq -\mathbf{f}$$

This problem may be stated as

$$\text{minimize } Z = \mathbf{e}_{m+1}^T \begin{bmatrix} \mathbf{a} \\ h \end{bmatrix}$$

subject to

$$\begin{bmatrix} \mathbf{C} & \mathbf{e} \\ -\mathbf{C} & \mathbf{0} \end{bmatrix} \begin{bmatrix} \mathbf{a} \\ h \end{bmatrix} \geq \begin{bmatrix} \mathbf{f} \\ -\mathbf{f} \end{bmatrix}$$

where $\mathbf{e}_{m+1}$ is an $(m + 1)$-vector that is the $(m + 1)^{th}$ column of an $(m + 1)$-unit matrix.

It is more efficient to go to the dual of this problem [7, 15]; i.e.,

(11.3.1a) $$\text{maximize } z = [\mathbf{f}^T \ \ -\mathbf{f}^T]\mathbf{b}$$

subject to the constraints

(11.3.1b) $$\begin{bmatrix} \mathbf{C}^T & -\mathbf{C}^T \\ \mathbf{e}^T & \mathbf{0}^T \end{bmatrix} \mathbf{b} = \mathbf{e}_{m+1}$$

and $\mathbf{b}$ is a 2n-vector

(11.3.1c) $$b_i \geq 0, \quad i = 1, 2, \ldots, 2n$$

Let us write (11.3.1b) as

(11.3.1d) $$\mathbf{D}\mathbf{b} = \mathbf{e}_{m+1}$$

where $\mathbf{D}$ is the matrix on the l.h.s. of (11.3.1b).

As usual, for problem (11.3.1), a simplex tableau for $(m + 1)$ constraints in 2n variables is constructed. This is denoted by the large tableau, in order to distinguish it from the condensed tableaux, which we shall use in this algorithm.

Without loss of generality, we assume that rank($\mathbf{C}$) = m. Let the basis matrix at any stage of the computation be denoted by $\mathbf{B}$, which is an $(m + 1)$ non-singular square matrix. For any column j in the simplex tableau, the vector $\mathbf{y}_j$ and its marginal cost are given by

(11.3.2a) $$\mathbf{y}_j = \mathbf{B}^{-1}\mathbf{D}_j, \quad j = 1, 2, \ldots, 2n$$

(11.3.2b) $$(z_j - f_j) = \mathbf{f}_B{}^T\mathbf{y}_j - f_j, \quad j = 1, 2, \ldots, 2n$$

The elements of $\mathbf{f}_B$ are those associated with the basic variables.

The basic solution, denoted by $\mathbf{b}_B$ is given by

(11.3.2c) $$\mathbf{b}_B = \mathbf{B}^{-1}\mathbf{e}_{m+1}$$

and the objective function by

(11.3.2d) $$z = \mathbf{f}_B{}^T\mathbf{b}_B$$

11.3.1 Properties of the matrix of constraints

The analysis of the linear programming problem is initiated by the kind of asymmetry in the matrix of constraints $\mathbf{D}$ in (11.3.1d). Let us denote the jth column matrix $\mathbf{D}$ by $\mathbf{D}_j$ and the jth column of matrix $\mathbf{C}^T$ by $\mathbf{C}_j{}^T$. Then from (11.3.1b), for $j = 1, 2, \ldots, n$

(11.3.3) $$\mathbf{D}_j = \begin{bmatrix} \mathbf{C}_j{}^T \\ 1 \end{bmatrix} \quad \text{and} \quad \mathbf{D}_{j+n} = \begin{bmatrix} -\mathbf{C}_j{}^T \\ 0 \end{bmatrix}$$

We have $\mathbf{C}_j{}^T$ as part of vector $\mathbf{D}_j$ and $-\mathbf{C}_j{}^T$ as part of vector $\mathbf{D}_{j+n}$, for $j = 1, 2, \ldots, n$. See also Section 10.2.1. This kind of asymmetry in the matrix of constraints $\mathbf{D}$, enables us to use a condensed simplex tableau of $(m + 1)$ constraints in only n variables. From (11.3.3)

(11.3.4) $$\mathbf{D}_j + \mathbf{D}_{j+n} = \mathbf{e}_{m+1}$$

Definition

As in Section 10.2, we define any two columns j and $(j + n)$, $1 \leq j \leq n$ in the matrix of constraints $\mathbf{D}$ in (11.3.1d) as two corresponding columns.

The following 8 lemmas correspond respectively to the 8 lemmas in Chapter 10.

Lemma 11.1

At any stage of the computation, the $(m + 1)^{th}$ column of $\mathbf{B}^{-1}$ is the basic solution vector $\mathbf{b}_B$. This follows directly from (11.3.2c).

Lemma 11.2

The optimum solution vector $\mathbf{a}$ and z for problem (11.3.1) are given by

$$\mathbf{B}^T \begin{bmatrix} \mathbf{a} \\ z \end{bmatrix} = \mathbf{f}_B$$

or

$$(\mathbf{a}^T\ z) = \mathbf{f}_B{}^T\mathbf{B}^{-1}$$

Lemma 11.3

Assume that $\mathbf{b}_B$ is a basic feasible solution. Then the algebraic sum of the elements of $\mathbf{b}_B \geq 1$.

Proof:

Consider the last equation in (11.3.1b), which is

$$(\mathbf{e}^T\ \mathbf{0}^T)\mathbf{b} = 1$$

The elements of the vector $(\mathbf{e}^T\ \mathbf{0}^T)$ are 1's and 0's. Also, as $\mathbf{b}_B$ is assumed feasible, each element of $\mathbf{b}_B$ is ≥ 0. Since the non-basic elements of vector $\mathbf{b}$ are all 0's, the algebraic sum of the elements of $\mathbf{b}_B \geq (\mathbf{e}^T\ \mathbf{0}^T)\mathbf{b} = 1$ and the lemma is proved.

Lemma 11.4

Any two corresponding columns i and j, $|i - j| = n$, $1 \leq i, j < 2n$, should not appear together in any basis matrix.

For the proof see Lemma 3 in [15]

Lemma 11.5

Let $\mathbf{b}_B$ be any basic solution, and z be the corresponding objective function for the programming problem (11.3.1). Let also columns j and $(j + n)$, $1 \leq j \leq n$, be two corresponding columns in the matrix of constraints in (11.3.1b). Then at any stage of the computation

(11.3.5a) $$\mathbf{y}_j + \mathbf{y}_{j+n} = \mathbf{b}_B, \quad 1 \leq j \leq n$$

and a

(11.3.5b) $$(z_j - f_j) + (z_{j+n} - f_{j+n}) = z, \quad 1 \leq j \leq n$$

Proof:

The proof of this lemma follows the proof of Lemma 10.5. That is, by multiplying (11.3.4) by $\mathbf{B}^{-1}$, from (11.3.2a) and Lemma 11.1, we get (11.3.5a). From (11.3.5a), (11.3.2a, b), Lemma 11.1, and that $f_{j+n} = -f_j$, (11.3.5b) is obtained.

Lemma 11.6

Assume that we have obtained an initial basic solution that is not feasible; there exist one or more elements of the basic solution $\mathbf{b}_B$ that is < 0. Let $\mathbf{b}_{Bs}$, $1 \le s \le (m + 1)$, be one of these elements. Let i be the basic column associated with $\mathbf{b}_{Bs}$. Let j be the corresponding column to the basic column i; $|i - j| = n$. Then if column i is replaced by column j in the basis, and a Gauss-Jordan step is performed on the simplex tableau, the new element $\mathbf{b}_{Bs}$ of the basic solution will be > 0. Moreover, the elements of the new basic solution $\mathbf{b}_{Bk}$, $k \ne s$, $k = 1, 2, \ldots, m+1$, each will have the sign of its old value.

The proof here follows the steps of the proof of Lemma 10.6.

Lemma 11.7 (Lemma 5 in [2])

Let a basis matrix $\mathbf{B}_{(1)}$ be formed by choosing (m + 1) linearly independent columns out of the 2n columns in the matrix of constraints in (11.3.1b) such that no two corresponding columns appear together in $\mathbf{B}_{(1)}$. Let another basis matrix $\mathbf{B}_{(2)}$ be formed, the columns of which are the corresponding columns of $\mathbf{B}_{(1)}$ arranged in the same order and assume that $\mathbf{B}_{(2)}$ is non-singular. Let $\mathbf{b}_{B(1)}$ and $z_{(1)}$ be respectively the basic solution and the objective function associated with $\mathbf{B}_{(1)}$ and $\mathbf{b}_{B(2)}$ and $z_{(2)}$ be the respective parameters associated with $\mathbf{B}_{(2)}$. Then

$$\mathbf{b}_{B(2)} = c\mathbf{b}_{B(1)}$$

and

$$z_{(2)} = -cz_{(1)}$$

where c is a constant given by

$$c = 1/(s_{m+1} - 1) \qquad (11.3.6)$$

where s_{m+1} is the algebraic sum of the elements of $\mathbf{b}_{B(1)}$. If $\mathbf{b}_{B(1)}$ is feasible, from Lemma 11.3, $s_{m+1} > 1$ and then $c > 0$.

Lemma 11.8 (Lemma 6 in [2])

Consider the two bases $\mathbf{B}_{(1)}$ and $\mathbf{B}_{(2)}$ of the previous lemma. Let

$(\mathbf{B}_{(1)})_i$ and $(\mathbf{B}_{(2)})_i$ be respectively column i of $\mathbf{B}_{(1)}$ and of $\mathbf{B}_{(2)}$. Let also column i of $\mathbf{B}_{(1)}^{-1}$ and of $\mathbf{B}_{(2)}^{-1}$ be respectively $(\mathbf{B}_{(1)}^{-1})_i$ and $(\mathbf{B}_{(2)}^{-1})_i$.

Let us construct two simplex tableaux $\mathbf{T}_{(1)}$ and $\mathbf{T}_{(2)}$ for $\mathbf{B}_{(1)}$ and $\mathbf{B}_{(2)}$ respectively, where the columns of $\mathbf{T}_{(2)}$ are the corresponding columns of $\mathbf{T}_{(1)}$ arranged in the same order. Then

$$(\mathbf{B}_{(2)}^{-1})_i = -(\mathbf{B}_{(1)}^{-1})_i + cs_i\mathbf{b}_{B(1)}$$

$$\mathbf{y}_i \text{ in } T_{(2)} = \mathbf{y}_i \text{ in } \mathbf{T}_{(1)} + (1 - c_i)c\mathbf{b}_{B(1)}$$

where s_i is the algebraic sum of the elements of column $(\mathbf{B}_{(1)}^{-1})_i$, c_i is the algebraic sum of the elements of $\mathbf{y}_i$ in $T_{(1)}$ and c is given by (11.3.6).

Lemma 11.9

Unlike the ordinary linear Chebyshev approximation problem (Chapter 10), there are problems whose one-sided Chebyshev solution(s) do not exist. The corresponding linear programming problem (11.3.1) would have an unbounded solution.

See Lemma 7 in [2], the numerical results in Section 11.5 and also Example 16.2.

11.4 Description of the algorithm

The algorithm is similar to that of the ordinary Chebyshev approximation of Chapter 10. It is again in three parts. In part 1, a numerically stable initial basic solution, feasible or not, is obtained. Also, the rank of matrix **C** is determined. In part 2, if the initial basic solution is not feasible, it is made feasible. Then the objective function z and the marginal costs are calculated. Part 3 is a slightly modified simplex algorithm.

We start by constructing a condensed simplex tableau for (m + 1) constraints in only n variables, namely, the first n variables. An n-index indicator vector whose elements are +1 or –1 is needed. If column j, $1 \leq j \leq n$, is in the condensed tableau, index j of this vector has the value +1. If column (j + n), $1 \leq j \leq n$, is in the condensed tableau, index j of this vector has the value –1.

An initial basic solution, feasible or not is obtained. This is done by simply applying a finite number of Gauss-Jordan elimination steps to the initial data. We use partial pivoting. The pivot is the largest element in absolute value in row i of tableau (i – 1), i = 1, 2, …, m+1. Tableau 0 is the initial data.

If one of the first m rows of matrix $\mathbf{C}^T$ is linearly dependent on one or more of the preceding rows, i.e., rank($\mathbf{C}$) = k < m, during the elimination processes in part 1 of the algorithm, we get a zero row in the tableau. Such row(s) are discarded from the following tableaux.

From (11.3.1b), the continuation of any of these rows in the large tableau will also be a row consisting of zero elements. In this case, the simplex tableau is for (k + 1) constraints in n variables. An exception from this rule is the last row in the tableau, as this row cannot be linearly dependent on any of the preceding rows. If the last row in the condensed tableau consists of all zero elements, the continuation of this row in the large tableau will not contain all zero elements.

For the $(k + 1)^{th}$ iteration in part 1, we chose the pivot as the largest positive element in the last row of the tableau and the extension of this row (in the large tableau). This ends part 1 of the algorithm.

If one or more elements of the obtained basic solution $\mathbf{b}_B$ is < 0, the initial basic solution is not feasible. For each basic column j associated with a negative element in the basic solution, we utilize Lemma 11.6. Column $\mathbf{y}_j$ in the simplex tableau is replaced by its corresponding column. Then by applying a Gauss-Jordan step after each replacement, the new basic solution will be feasible.

The objective function z is then calculated from (11.3.2c, d). Then if z < 0, we calculate the algebraic sum of the elements of $\mathbf{b}_B$. If this algebraic sum > 1, we construct the simplex tableau of the corresponding columns of the current tableau. We also construct the inverse of the corresponding basis matrix. This is done in the manner given by Lemma 11.7 and Lemma 11.8. Hence, z becomes > 0. The marginal costs $(z_i - f_i)$ are then calculated from (11.3.5b).

If z < 0 and the algebraic sum of the elements of $\mathbf{b}_B = 1$, $\mathbf{B}_{(2)}^{-1}$ would be singular, and we proceed to calculate the marginal costs and go to part 3 of the algorithm. In this case, in part 3, z is bound to increase algebraically and will eventually change to a positive quantity and if the solution exists, z will eventually increase to its optimum value. This ends part 2 of the algorithm, as we now have an initial basic feasible solution and z ≥ 0.

Part 3 is the ordinary simplex algorithm. The only difference is in the choice of the non-basic column that enters the basis. The column to enter the basis is that which has the most negative marginal cost among the non-basic columns in the current condensed tableau and their corresponding columns. Relation (11.3.5b) is used for calculating the marginal costs of the corresponding columns. The above steps are illustrated by the following detailed numerical example.

Example 11.1

Obtain the one-sided Chebyshev solution from above of the following system.

$$\begin{aligned} a_1 + 2a_2 &= -1 \\ -a_1 - a_2 &= -2 \\ a_1 + 3a_2 &= -1 \\ a_2 &= -4 \end{aligned} \quad (11.4.1)$$

The initial data are shown in the condensed tableau. The pivot is chosen as the largest element in absolute value in row i of tableau (i – 1). The pivot element in each tableau is bracketed.

Initial Data

$\mathbf{f}^T$	–1	–2	–1	–4
b_B	$\mathbf{D}_1$	$\mathbf{D}_2$	$\mathbf{D}_3$	$\mathbf{D}_4$
0	(1)	–1	1	0
0	2	–1	3	1
1	1	1	1	1

Tableau 11.4.1 (part 1)

		$\mathbf{f}^T$	–1	–2	–1	–4
f_B		b_B	$\mathbf{D}_1$	$\mathbf{D}_2$	$\mathbf{D}_3$	$\mathbf{D}_4$
–1	$\mathbf{D}_1$	0	1	–1	1	0
		0	0	(1)	1	1
		1	0	2	0	1

In Tableau 11.4.2 and its corresponding part, the pivot is chosen as the largest positive element, which is y_{37}. For this reason we write down Tableau 11.4.2*, where $\mathbf{y}_7$ replaces $\mathbf{y}_3$ in the basis.

Tableau 11.4.2

		$\mathbf{f}^T$	−1	−2	−1	−4
$\mathbf{f}_B$	**B**	$\mathbf{b}_B$	$\mathbf{D}_1$	$\mathbf{D}_2$	$\mathbf{D}_3$	$\mathbf{D}_4$
−1	$\mathbf{D}_1$	0	1	0	2	1
−2	$\mathbf{D}_2$	0	0	1	1	1
		1	0	0	−2	−1

Tableau 11.4.2*

		$\mathbf{f}^T$	−1	−2	1	−4
$\mathbf{f}_B$	**B**	$\mathbf{b}_B$	$\mathbf{D}_1$	$\mathbf{D}_2$	$\mathbf{D}_7$	$\mathbf{D}_4$
−1	$\mathbf{D}_1$	0	1	0	−2	1
−2	$\mathbf{D}_2$	0	0	1	−1	1
		1	0	0	(3)	−1

Tableau 11.4.3 gives an initial basic feasible solution but $z < 0$. Meanwhile, the algebraic sum of the elements of $\mathbf{b}_B = 1.33 > 1$. Hence, we make use of Lemmas 11.7 and 11.8 and replace Tableau 11.4.3 by Tableau 11.4.3*. In this Tableau, $z > 0$ and we calculate the marginal costs from (11.3.5b).

Tableau 11.4.3

		$\mathbf{f}^T$	−1	−2	1	−4
$\mathbf{f}_B$	**B**	$\mathbf{b}_B$	$\mathbf{D}_1$	$\mathbf{D}_2$	$\mathbf{D}_7$	$\mathbf{D}_4$
−1	$\mathbf{D}_1$	2/3	1	0	0	1/3
−2	$\mathbf{D}_2$	1/3	0	1	0	2/3
1	$\mathbf{D}_7$	1/3	0	0	1	−1/3

$z = -1$

Tableau 11.4.3* (part 2)

		$\mathbf{f}^T$	1	2	−1	4
f_B	B	b_B	$\mathbf{D}_5$	$\mathbf{D}_6$	$\mathbf{D}_3$	$\mathbf{D}_8$
1	$\mathbf{D}_5$	2	1	0	0	1
2	$\mathbf{D}_6$	1	0	1	0	(1)
−1	$\mathbf{D}_3$	1	0	0	1	0
	z = 3		0	0	0	−1

Tableau 11.4.4 (part 3)

		$\mathbf{f}^T$	1	2	−1	4
f_B	B	b_B	$\mathbf{D}_5$	$\mathbf{D}_6$	$\mathbf{D}_3$	$\mathbf{D}_8$
1	$\mathbf{D}_5$	1	1	−1	0	0
4	$\mathbf{D}_8$	1	0	1	0	1
−1	$\mathbf{D}_3$	1	0	0	1	0
	z = 4		0	1	0	0

In Tableau 11.4.3*, $(z_8 - f_8)$ is the most negative marginal cost and thus $\mathbf{D}_8$ replaces $\mathbf{D}_6$ in the basis. Tableau 11.4.4 gives the optimum one-sided Chebyshev solution from above with $z = 4$. The optimum solution vector **a** of system (11.4.1) is given by the solution of the equations in (11.4.1) associated with the basic vectors, namely $\mathbf{D}_5$, $\mathbf{D}_8$ and $\mathbf{D}_3$ in Tableau 11.4.4. That is

$$\begin{aligned} -a_1 - 2a_2 \quad &= 1 \\ - a_2 \quad &= 4 \\ a_1 + 3a_2 + z &= -1 \end{aligned}$$

which gives $a_1 = 7$, $a_2 = -4$ and $z = 4$. The residual $\mathbf{r} = (0, 1, 4, 0)^T$.

However, since the initial data (matrix **C** and vector **f**) are destroyed in the computation, vector **a** and z are calculated from the second equation in Lemma 11.2.

11.4.1 One-sided Chebyshev solution from below

For the one-sided Chebyshev solution from below, the problem would be

(11.4.2a) $$\text{minimize } \|\mathbf{Ca} - \mathbf{f}\|_\infty$$

subject to

(11.4.2.b) $$\mathbf{Ca} \geq \mathbf{f}$$

The formulation of the dual form of the linear programming problem is straightforward and is given by

$$\text{maximize } z = [\mathbf{f}^T \ -\mathbf{f}^T]\mathbf{b}$$

subject to the constraints

$$\begin{bmatrix} \mathbf{C}^T & -\mathbf{C}^T \\ \mathbf{0}^T & \mathbf{e}^T \end{bmatrix} \mathbf{b} = \mathbf{e}_{m+1}$$

and

$$b_i > 0, \quad i = 1, 2, \ldots, 2n$$

Our algorithm for one-sided Chebyshev from above can also be applied to the one-sided Chebyshev from below problem. The function LA_Linfside() calculates either the one-sided Chebyshev solution from above or from below, depending on an indicator specified by the user.

For the solution from below, we follow the steps of Section 6.4.2. We multiply both matrix $\mathbf{C}$ and vector $\mathbf{f}$ by -1. The inequality (11.4.2b) reduces to

$$[-\mathbf{C}]\mathbf{a} \leq [-\mathbf{f}]$$

and the problem is solved as a one-sided Chebyshev approximation from above. The solution vector $\mathbf{a}$ would be for the one-sided from below. However, the residual vector $\mathbf{r}$, would be multiplied by -1.

11.5 Numerical results and comments

LA_Linfside() implements this algorithm. DR_Linfside() tests the same 8 examples that were solved in Chapter 6 for the one-sided L_1 approximation problem. The driver tests these examples twice; once for the one-sided Chebyshev solution from above and once for the one-sided Chebyshev solution from below.

Table 11.5.1 shows the results of 3 of the examples, computed in single-precision. For each example, the number of iterations and the optimum norm for the one-sided Chebyshev solution from above, the ordinary Chebyshev solution, and the one-sided Chebyshev solution from below are shown. In this table, "no solution" indicates that the programming problem has an unbounded solution.

The results show that the number of iterations for the one-sided Chebyshev solution from above (or from below), if either exists, is of the same order of that of the (ordinary) Chebyshev solution of the given problems.

Table 11.5.1

		One-sided from above		Chebyshev solution		One-sided from below	
Example	**C**(n×m)	Iter	z	Iter	z	Iter	z
1	4 × 2	no solution		5	10.0	no solution	
2	10× 5	12	3.00	9	1.7778	no solution	
3	25×10	23	0.0142	23	0.0071	23	0.0142

Example 1 has no one-sided Chebyshev solution from above nor from below. Example 2 has one-sided Chebyshev solution from above but not from below. The third example has both one-sided Chebyshev solutions. For Example 3, consider the next section.

11.5.1 Simple relationships between the Chebyshev and one-sided Chebyshev approximations

It is observed in example 3 that the value of z for either the one-sided Chebyshev approximation from above or from below, is exactly twice that of the ordinary Chebyshev approximation.

This relationship exists when one of the approximating functions is a constant term. In this case, the one-sided Chebyshev approximations are just shifted ordinary Chebyshev approximations and the optimum value of z for the one-sided Chebyshev approximation is twice the optimum value of z for the ordinary Chebyshev approximation.

In the example of Figure 11-1, the approximating functions for the one-sided Chebyshev approximation from above, the Chebyshev approximation and the one-sided Chebyshev approximation from below are $(-2.6 + 2x - 0.156x^2)$, $(-0.8 + 2x - 0.156x^2)$ and $(1 + 2x - 0.156x^2)$ respectively. The three approximating functions differ only in the constant terms. The difference between the constant terms in the second and the first functions = the difference between the constant terms in the third and the second functions = the Chebyshev deviation 1.8.

This and other relationships between the one-sided and the ordinary approximations for certain classes of approximating functions, for certain types of norms, have been addressed by Phillips [9].

We now comment on the algorithms of Dax [6], Joe and Bartels [8], Sklar [11] and Roberts and Barrodale [10] mentioned in Section 11.2. Dax's algorithm is more involved than ours, as it requires the least squares solution of the system $\mathbf{Ca} = \mathbf{f}$. In the other three algorithms, one would store matrix **C** twice; once for (11.2.1a), (11.2.2a) or (11.2.3a) and once for (11.2.1b), (11.2.2b) or (11.2.3b), where **E** is replaced by **C**. The same can be said about storing vector **f** twice. Hence, their algorithms need more computer storage than ours and nearly double the number of arithmetic operations.

We claim that for this problem, a special purpose program such as ours would be more efficient than a general purpose one such as those of [8], [11] and [10]. See similar comments at the end of Chapter 6, for the one-sided L_1 approximation problem.

References

1. Abdelmalek, N.N., Chebyshev solution of overdetermined systems of linear equations, *BIT*, 15(1975)117-129.
2. Abdelmalek, N.N., The discrete linear one-sided Chebyshev approximation, *Journal of Institute of Mathematics and Applications*, 18(1976)361-370.
3. Armstrong, R.D. and Sklar, M.G., A linear programming algorithm for curve fitting in the L_∞ norm, *Numerical Functional Analysis and Optimization*, 2(1980)187-218.

4. Barrodale, I. and Phillips, C., An improved algorithm for discrete Chebyshev linear approximation, *Proceedings of the Fourth Manitoba Conference on Numerical Mathematics*, Hartnell, B.L. and Williams, H.C. (eds.), Winnipeg, Manitoba, Canada, pp. 177-190, 1975.
5. Barrodale, I. and Phillips, C., Algorithm 495: Solution of an overdetermined system of linear equations in the Chebyshev norm, *ACM Transactions on Mathematical Software*, 1(1975)264-270.
6. Dax, A., The minimax solution of linear equations subject to linear constraints, *IMA Journal of Numerical Analysis*, 9(1989)95-109.
7. Hadley, G., *Linear Programming*, Addison-Wesley, Reading, MA, 1962.
8. Joe, B. and Bartels, R., An exact penalty method for constrained, discrete, linear l_∞ data fitting, *SIAM Journal on Scientific and Statistical Computation*, 4(1983)76-84.
9. Phillips, D.L., A note on best one-sided approximations, *Communications of ACM*, 14(1971)598-600.
10. Roberts, F.D.K. and Barrodale, I., An algorithm for discrete Chebyshev linear approximation with linear constraints, *International Journal for Numerical Methods in Engineering*, 15(1980)797-807.
11. Sklar, M.G., L_∞ norm estimation with linear restrictions on the parameters, *Numerical Functional Analysis and Optimization*, 3(1981)53-68.
12. Tou, J.T. and Gonzalez, R.C., *Pattern Recognition Principles*, Addison-Wesley, Reading, MA, 1974.
13. Watson, G.A., One-sided approximation and operator equations, *Journal of Institute of Mathematics and Applications*, 12(1973)197-208.
14. Watson, G.A., The calculation of best linear one-sided L_p approximations, *Mathematics of Computation*, 27(1973)607-620.
15. Watson, G.A., On the best linear one-sided Chebyshev approximation, *Journal of Approximation Theory*, 7(1973)48-58.

11.6 DR_Linfside

```
/*-----------------------------------------------------------------
DR_Linfside
-------------------------------------------------------------------
This program is a driver for the function LA_Linfside(), which
calculates the one-sided Chebyshev solution from above or from below
of an overdetermined system of linear equations.

The overdetermined system has the form

                              c*a = f

"c" is a given real n by m matrix of rank k, k <= m < n.
"f" is a given real n vector.
"a" is the solution m vector.

This driver contains 8 examples from which the results of examples
1, 6 and 7 are given in the text.  All the examples are solved twice;
once for the one-sided Chebyshev approximation from above and once
for the one-sided Chebyshev approximation from below.
-----------------------------------------------------------------*/

#include "DR_Defs.h"
#include "LA_Prototypes.h"

#define N1          4
#define M1          2
#define N2          5
#define M2          3
#define N3          6
#define M3          3
#define N4          7
#define M4          3
#define N5          8
#define M5          4
#define N6          10
#define M6          5
#define N7          25
#define M7          10
#define N8          8
#define M8          4

void DR_Linfside (void)
```

```
{
    /*----------------------------------------
      Constant matrices/vectors
    ----------------------------------------*/
    static tNumber_R c1init[N1][M1] =
    {
        { 0.0, -2.0 },
        { 0.0, -4.0 },
        { 1.0, 10.0 },
        {-1.0, 15.0 }
    };

    static tNumber_R c2init[N2][M2] =
    {
        { 1.0,  2.0, 0.0 },
        {-1.0, -1.0, 0.0 },
        { 1.0,  3.0, 0.0 },
        { 0.0,  1.0, 0.0 },
        { 0.0,  0.0, 1.0 }
    };

    static tNumber_R c3init[N3][M3] =
    {
        { 0.0, -1.0,  0.0 },
        { 1.0,  3.0, -4.0 },
        { 1.0,  0.0,  0.0 },
        { 0.0,  0.0,  1.0 },
        {-1.0,  1.0,  2.0 },
        { 1.0,  1.0,  1.0 }
    };

    static tNumber_R c4init[N4][M4] =
    {
        { 1.0, 0.0,  1.0 },
        { 1.0, 2.0,  2.0 },
        { 1.0, 2.0,  0.0 },
        { 1.0, 1.0,  0.0 },
        { 1.0, 0.0, -1.0 },
        { 1.0, 0.0,  0.0 },
        { 1.0, 1.0,  1.0 }
    };

    static tNumber_R c5init[N5][M5] =
    {
        { 1.0, -3.0,  9.0, -27.0 },
```

```
    { 1.0, -2.0,  4.0,  -8.0 },
    { 1.0, -1.0,  1.0,  -1.0 },
    { 1.0,  0.0,  0.0,   0.0 },
    { 1.0,  1.0,  1.0,   1.0 },
    { 1.0,  2.0,  4.0,   8.0 },
    { 1.0,  3.0,  9.0,  27.0 },
    { 1.0,  4.0, 16.0,  64.0 }
};

static tNumber_R c6init[N6][M6] =
{
    { 1.0, 0.0,  0.0,  0.0,  0.0 },
    { 0.0, 1.0,  0.0,  0.0,  0.0 },
    { 0.0, 0.0,  1.0,  0.0,  0.0 },
    { 0.0, 0.0,  0.0,  1.0,  0.0 },
    { 0.0, 0.0,  0.0,  0.0,  1.0 },
    { 1.0, 1.0,  1.0,  1.0,  1.0 },
    { 0.0, 1.0,  1.0,  1.0,  1.0 },
    {-1.0, 0.0, -1.0, -1.0, -1.0 },
    { 1.0, 1.0,  0.0,  1.0,  1.0 },
    { 1.0, 1.0,  1.0,  0.0,  1.0 }
};

static tNumber_R c8init[N8][M8] =
{
    { 1.0, 1.0,   1.0,  1.0 },
    { 1.0, 2.0,   4.0,  4.0 },
    { 1.0, 3.0,   9.0,  9.0 },
    { 1.0, 4.0,  16.0, 16.0 },
    { 1.0, 5.0,  25.0, 25.0 },
    { 1.0, 6.0,  36.0, 36.0 },
    { 1.0, 7.0,  49.0, 49.0 },
    { 1.0, 8.0,  64.0, 64.0 }
};

static tNumber_R f1[N1+1] =
{   NIL,
    -12.0, 6.0, 0.0, 5.0
};

static tNumber_R f2[N2+1] =
{   NIL,
    1.0, 2.0, 1.0, -3.0, 0.0
};
```

```
static tNumber_R f3[N3+1] =
{   NIL,
    1.0, 2.0, 3.0, 2.0, 2.0, 4.0
};

static tNumber_R f4[N4+1] =
{   NIL,
    0.0, -2.0, 1.0, -1.0, 5.0, 7.0, 0.0
};

static tNumber_R f5[N5+1] =
{   NIL,
    3.0, -3.0, -2.0, 0.0, 7.0, -1.0, 5.0, 2.0
};

static tNumber_R f6[N6+1] =
{   NIL,
    1.0, -1.0, 0.0, -1.0, 1.0, 0.0, 2.0, 3.0, -3.0, -2.0
};

static tNumber_R f7[N7+1] =
{   NIL,
    0.0872673, 0.0872794, 0.0873029, 0.0873315, 0.0873576,
    0.3491184, 0.3498802, 0.3513824, 0.3532572, 0.3550109,
    0.6111334, 0.6150641, 0.6230824, 0.6336395, 0.6441493,
    0.8733883, 0.8841621, 0.9071868, 0.9400757, 0.9766021,
    1.135895,  1.157550,  1.206257,  1.283258,  1.384432
};

static tNumber_R f8[N8+1] =
{   NIL,
    2.0, 2.5, 2.0, 6.5, 3.5, 4.5, 6.0, 7.0
};

/*----------------------------------------
  Variable matrices/vectors
----------------------------------------*/
tMatrix_R   ct      = alloc_Matrix_R (MMc_COLS, NN_ROWS);
tVector_R   f       = alloc_Vector_R (NN_ROWS);
tVector_R   r       = alloc_Vector_R (NN_ROWS);
tVector_R   a       = alloc_Vector_R (MMc_COLS);
tMatrix_R   c7      = alloc_Matrix_R (N7, M7);

tMatrix_R   c1      = init_Matrix_R (&(c1init[0][0]), N1, M1);
tMatrix_R   c2      = init_Matrix_R (&(c2init[0][0]), N2, M2);
```

```
tMatrix_R   c3      = init_Matrix_R (&(c3init[0][0]), N3, M3);
tMatrix_R   c4      = init_Matrix_R (&(c4init[0][0]), N4, M4);
tMatrix_R   c5      = init_Matrix_R (&(c5init[0][0]), N5, M5);
tMatrix_R   c6      = init_Matrix_R (&(c6init[0][0]), N6, M6);
tMatrix_R   c8      = init_Matrix_R (&(c8init[0][0]), N8, M8);

int         irank, iter, iside, kase;
int         i, j, k, m, n, Iexmpl;
tNumber_R   d, dd, ddd, e, ee, eee, z;

eLaRc       rc = LaRcOk;

iside = 1;

for (j = 1; j <= 5; j++)
{
    d = 0.15* (j-3);
    dd = d*d;
    ddd = d*dd;
    for (i = 1; i <= 5; i++)
    {
        e = 0.15* (i-3);
        ee = e*e;
        eee = e*ee;
        k = 5* (j-1) + i;
        c7[k][1] = 1.0;
        c7[k][2] = d;
        c7[k][3] = e;
        c7[k][4] = dd;
        c7[k][5] = ee;
        c7[k][6] = e*d;
        c7[k][7] = ddd;
        c7[k][8] = eee;
        c7[k][9] = dd*e;
        c7[k][10] = ee*d;
    }
}
prn_dr_bnr ("DR_Linfside, One-Sided Chebyshev Solutions of an "
            "Overdetermined System of Linear Equations");

for (kase = 1; kase <= 2; kase++)
{
    iside = 2 - kase;
    for (Iexmpl = 1; Iexmpl <= 8; Iexmpl++)
    {
```

```
switch (Iexmpl)
{
    case 1:
        n = N1;
        m = M1;
        for (i = 1; i <= n; i++)
        {
            f[i] = f1[i];
            for (j = 1; j <= m; j++) ct[j][i] = c1[i][j];
        }
        break;

    case 2:
        n = N2;
        m = M2;
        for (i = 1; i <= n; i++)
        {
            f[i] = f2[i];
            for (j = 1; j <= m; j++) ct[j][i] = c2[i][j];
        }
        break;

    case 3:
        n = N3;
        m = M3;
        for (i = 1; i <= n; i++)
        {
            f[i] = f3[i];
            for (j = 1; j <= m; j++) ct[j][i] = c3[i][j];
        }
        break;

    case 4:
        n = N4;
        m = M4;
        for (i = 1; i <= n; i++)
        {
            f[i] = f4[i];
            for (j = 1; j <= m; j++) ct[j][i] = c4[i][j];
        }
        break;

    case 5:
        n = N5;
        m = M5;
```

```
            for (i = 1; i <= n; i++)
            {
                f[i] = f5[i];
                for (j = 1; j <= m; j++) ct[j][i] = c5[i][j];
            }
            break;

        case 6:
            n = N6;
            m = M6;
            for (i = 1; i <= n; i++)
            {
                f[i] = f6[i];
                for (j = 1; j <= m; j++) ct[j][i] = c6[i][j];
            }
            break;

        case 7:
            n = N7;
            m = M7;
            for (i = 1; i <= n; i++)
            {
                f[i] = f7[i];
                for (j = 1; j <= m; j++) ct[j][i] = c7[i][j];
            }
            break;

        case 8:
            n = N8;
            m = M8;
            for (i = 1; i <= n; i++)
            {
                f[i] = f8[i];
                for (j = 1; j <= m; j++) ct[j][i] = c8[i][j];
            }

            break;

        default:
            break;
    }

    prn_algo_bnr ("Linfside");
    prn_example_delim();
    PRN ("Example #%d: Size of matrix \"c\", %d by %d\n",
```

```
                Iexmpl, n, m);
            prn_example_delim();
            if (iside == 1)
                PRN ("One-sided Chebyshev Solution from above\n");
            else
                PRN ("One-sided Chebyshev Solution from below\n");
          . prn_example_delim();
            PRN ("r.h.s. Vector \"f\"\n");
            prn_Vector_R (f, n);
            PRN ("Transpose of Coefficient Matrix, \"ct\"\n");
            prn_Matrix_R (ct, m, n);

            rc = LA_Linfside (iside, m, n, ct, f, &irank, &iter, r,
                              a, &z);

            if (rc >= LaRcOk)
            {
                PRN ("\n");
                PRN ("Results of the "
                     "One-sided Chebyshev Approximation\n");
                PRN ("One-sided Chebyshev solution vector, \"a\"\n");
                prn_Vector_R (a, m);
                PRN ("One-sided Chebyshev residual vector \"r\"\n");
                prn_Vector_R (r, n);
                PRN ("One-sided Chebyshev norm \"z\" = %8.4f\n", z);
                PRN ("Rank of matrix \"c\" = %d, No. of Iterations ="
                     " %d\n", irank, iter);
            }

            prn_la_rc (rc);
        }
    }
    free_Matrix_R (ct, MMc_COLS);
    free_Vector_R (f);
    free_Vector_R (r);
    free_Vector_R (a);
    free_Matrix_R (c7, N7);
    uninit_Matrix_R (c1);
    uninit_Matrix_R (c2);
    uninit_Matrix_R (c3);
    uninit_Matrix_R (c4);
    uninit_Matrix_R (c5);
    uninit_Matrix_R (c6);
    uninit_Matrix_R (c8);
}
```

11.7 LA_Linfside

```
/*-----------------------------------------------------------------------
LA_Linfside
-------------------------------------------------------------------------
This program calculates the one-sided Chebyshev solution from above
or from below of an overdetermined system of linear equations.  It
uses a modified simplex method to the linear programming formulation
of the problem.

The system of linear equations has the form

                              c*a = f

"c" is a given real n by m matrix of rank k, k <= m < n.
"f" is a given real n vector.

The problem is to calculate the elements of the m vector
"a" that gives the minimum Chebyshev residual norm z.

                  z=max|r[i]|,   i = 1, 2, ..., n

where r[i] is the ith residual and is given by

 r[i] = f[i] - (c[i][1]*a[1] + c[i][2]*a[2] + ... + c[i][m]*a[m]),
                        i = 1, 2, ..., n

subject to the conditions

    r[i] => 0, for the one-sided L-One solution from above.
or
    r[i] =< 0,  for the one-sided L-One solution from below.

[Note: For LA_Linfside vector r = f - ca, while for
       LA_Loneside, r = ca - f].

Inputs
iside   An integer specifying the action to be performed.
        If iside = 1, the one-sided Chebyshev solution from above is
        calculated.
        If iside != 1, the one-sided Chebyshev solution from below is
        calculated.
m       Number of columns of matrix "c" in the system c*a = f.
n       Number of rows of matrix "c" in the system c*a = f.
ct      A real (m+1) by n matrix. Its first m rows and its n
        columns contain the transpose of matrix "c" of the system
        c*a = f. Its (m+1)th row will be filled with ones by the
        program.
```

```
f       A real n vector containing the r.h.s. of the system c*a = f.

Local Variables
binv    A real (m + 1) square matrix containing the inverse of the
        basis matrix in the linear programming problem.
bv      A real (m + 1) vector containing the basic solution in the
        linear programming problem.
icbas   An integer (m + 1) vector containing the indices of the
        columns of matrix "ct" forming the basis matrix.
irbas   An integer (m + 1) vector containing the indices of the
        rows of matrix "ct".

Outputs
irank   Calculated rank of matrix "c".
iter    Number of iterations, or the number of times the simplex
        tableau is changed by a Gauss-Jordan step.
a       A real (m + 1) vector. Its first m elements are the
        one-sided Chebyshev solution to the system c*a = f.
r       A real n vector containing the one-sided Chebyshev residual
        vector r = (f - c*a).
z       The optimum one-sided Chebyshev norm of the problem.

Returns one of
        LaRcSolutionUnique
        LaRcSolutionProbNotUnique
        LaRcSolutionDefNotUniqueRD
        LaRcNoFeasibleSolution
        LaRcErrBounds
        LaRcErrNullPtr
        LaRcErrAlloc
-------------------------------------------------------------------*/

#include "LA_Prototypes.h"

eLaRc LA_Linfside (int iside, int m, int n, tMatrix_R ct,
    tVector_R f, int *pIrank, int *pIter, tVector_R r, tVector_R a,
    tNumber_R *pZ)
{
    tMatrix_R   binv    = alloc_Matrix_R (m + 1, m + 1);
    tVector_R   bv      = alloc_Vector_R (m + 1);
    tVector_I   icbas   = alloc_Vector_I (m + 1);
    tVector_I   irbas   = alloc_Vector_I (m + 1);
    tVector_I   ibound  = alloc_Vector_I (n);

    int         i = 0, j = 0, kl = 0, ml = 0;
    int         ijk = 0, iout = 0, jin = 0, ivo = 0, itest = 0;
    tNumber_R   d = 0.0, piv = 0.0;

    /* Validation of the data before executing the algorithm */
```

```
eLaRc       rc = LaRcSolutionUnique;
VALIDATE_BOUNDS ((0 < m) && (m < n));
VALIDATE_PTRS   (ct && f && pIrank && pIter && r && a && pZ);
VALIDATE_ALLOC  (binv && bv && icbas && irbas && ibound);

*pIrank = m;
*pIter = 0;
m1 = m + 1;
kl = 1;
/* Initializing the data */
LA_linfside_init (iside, m, n, ct, irbas, binv, ibound, a);

iout = 0;

/* Part 1 of the algorithm */
for (iout = 1; iout <= m1; iout++)
{
    piv = 0.0;
    for (j = 1; j <= n; j++)
    {
        d = fabs (ct[iout][j]);
        if (d > piv)
        {
            jin = j;
            piv = d;
        }
    }

    if (piv > EPS)
    {
        /* A Gauss-Jordan elimination step, */
        LA_linfside_gauss_jordn (iout, jin, kl, m, n, ct, icbas,
                                 binv, bv);
        *pIter = *pIter + 1;
    }

    /* Detection of rank deficiency of matrix "c" */
    LA_linfside_detect_rank (piv, iout, jin, &kl, m, n, ct, f,
                             icbas, irbas, binv, bv, ibound,
                             pIrank, pIter);
}

/* Part 2; obtaining a basic feasible solution */
LA_linfside_part_2 (kl, m, n, ct, f, icbas, binv, bv, ibound,
                    pIter);

/* Calculating the initial residual vector r and norm z */
LA_linfside_resid_norm (kl, m, n, ct, f, icbas, binv, bv, ibound,
                        r, pZ);
```

```
/* Part 3 of the algorithm */
for (ijk = 1; ijk <= n*n; ijk++)
{
    ivo = 0;

    /* Determine the vector that enters the basis */
    LA_linfside_vent (&ivo, &jin, kl, m, n, icbas, r, pZ);

    /* Calculate the results */
    if (ivo == 0)
    {
        rc = LA_linfside_res (iside, m, n, kl, iout, f, icbas,
                              irbas, binv, bv, ibound, r, a,
                              *pIrank, pZ);
        GOTO_CLEANUP_RC (rc);
    }
    if (ivo != 1)
    {
        for (i = kl; i <= m1; i++)
        {
            ct[i][jin] = bv[i] - ct[i][jin];
        }
        r[jin] = *pZ - r[jin];
        f[jin] = -f[jin];
        ibound[jin] = -ibound[jin];
    }
    itest = 0;

    /* Determine the vector that leaves the basis */
    LA_linfside_vleav (&itest, &iout, jin, kl, m, ct, bv);

    if (itest != 1)
    {
        GOTO_CLEANUP_RC (LaRcNoFeasibleSolution);
    }

    /* A Gauss-Jordan elimination step */
    LA_linfside_gauss_jordn (iout, jin, kl, m, n, ct, icbas,
                             binv, bv);
    *pIter = *pIter + 1;
    d = r[jin];
    for (j = 1; j <= n; j++)
    {
        r[j] = r[j] - d * (ct[iout][j]);
    }
    *pZ = *pZ - d * (bv[iout]);
}
```

```
CLEANUP:

    free_Matrix_R (binv, m + 1);
    free_Vector_R (bv);
    free_Vector_I (icbas);
    free_Vector_I (irbas);
    free_Vector_I (ibound);

    return rc;
}

/*-----------------------------------------------------------------
Initializing the program data for LA_Linfside()
-----------------------------------------------------------------*/
void LA_linfside_init (int iside, int m, int n, tMatrix_R ct,
    tVector_I irbas, tMatrix_R binv, tVector_I ibound, tVector_R a)
{
    int             i, j, m1;
    tNumber_R       e;

    m1 = m + 1;
    for (i = 1; i <= m1; i++)
    {
        a[i] = 0.0;
        irbas[i] = i;
    }
    for (j = 1; j <= m1; j++)
    {
        for (i = 1; i <=m1; i++)
        {
            binv[i][j] = 0.0;
        }
        binv[j][j] = 1.0;
    }
    e = 1.0;
    if (iside != 1) e = 0.0;
    for (j = 1; j <= n; j++)
    {
        ct[m1][j] = e;
        ibound[j] = 1;
    }
}

/*-----------------------------------------------------------------
Detection of the rank of matrix "c" in LA_Linfside()
-----------------------------------------------------------------*/
void LA_linfside_detect_rank (tNumber_R piv, int iout, int jin,
    int *pKl, int m, int n, tMatrix_R ct, tVector_R f,
    tVector_I icbas, tVector_I irbas, tMatrix_R binv, tVector_R bv,
```

```
    tVector_I ibound, int *pIrank, int *pIter)
{
    int             i, j, k, m1, icb;

    m1 = m + 1;

    if (piv < EPS && iout < m1)
    {
        swap_rows_Matrix_R (ct, *pKl, iout);

        k = irbas[iout];
        irbas[iout] = irbas[*pKl];
        irbas[*pKl] = 0;
        for (j = *pKl; j <= m; j++)
        {
            binv[iout][j] = binv[*pKl][j];
            binv[*pKl][j] = 0.0;
        }
        bv[*pKl] = 0.0;
        icbas[iout] = icbas[*pKl];
        icbas[*pKl] = 0;
        for (i = *pKl; i <= m1; i++)
        {
            binv[i][iout] = binv[i][*pKl];
            binv[i][*pKl] = 0.0;
        }
        *pIrank = *pIrank - 1;
        *pKl = *pKl + 1;
    }
    if (piv < EPS && iout == m1)
    {
        for (j = 1; j <= n; j++)
        {
            icb = 0;
            for (i = *pKl; i <= m; i++)
            {
                if (j == icbas[i]) icb = 1;
            }
            if (icb == 0)
            {
                jin = j;
                break;
            }
        }
        f[jin] = -f[jin];
        ibound[jin] = -ibound[jin];
        for (i = *pKl; i <= m; i++)
        {
            ct[i][jin] = -ct[i][jin];
```

```
        }
        ct[m1][jin] = 1.0 - ct[m1][jin];
        /* A Gauss-Jordan elimination step, */
        LA_linfside_gauss_jordn (iout, jin, *pKl, m, n, ct, icbas,
                                 binv, bv);
        *pIter = *pIter + 1;
    }
}

/*-----------------------------------------------------------------
Obtaining a basic feasible solution of the linear programming
problem in LA_Linfside()
-----------------------------------------------------------------*/
void LA_linfside_part_2 (int kl, int m, int n, tMatrix_R ct,
    tVector_R f, tVector_I icbas, tMatrix_R binv, tVector_R bv,
    tVector_I ibound, int *pIter)
{
    int             i, k, m1;
    int             iout, jin;

    m1 = m + 1;
    for (i = kl; i <= m1; i++)
    {
        if (bv[i] < 0.0)
        {
            iout = i;
            jin = icbas[i];
            f[jin] = -f[jin];
            ibound[jin] = -ibound[jin];
            for (k = kl; k <= m1; k++)
            {
                ct[k][jin] = bv[k] - ct[k][jin];
            }

            /* A Gauss-Jordan elimination step, */
            LA_linfside_gauss_jordn (iout, jin, kl, m, n, ct, icbas,
                                     binv, bv);
            *pIter = *pIter + 1;
        }
    }
}

/*-----------------------------------------------------------------
Calculating initial residual vector r and norm z in LA_Linfside()
-----------------------------------------------------------------*/
void LA_linfside_resid_norm (int kl, int m, int n, tMatrix_R ct,
    tVector_R f, tVector_I icbas, tMatrix_R binv, tVector_R bv,
    tVector_I ibound, tVector_R r, tNumber_R *pZ)
{
```

```
int             i, j, k, m1, icb;
tNumber_R       d, g, s;

m1 = m + 1;

s = 0.0;
for (i = kl; i <= m1; i++)
{
    k = icbas[i];
    s = s + bv[i] * (f[k]);
}
*pZ = s;
if (*pZ < 0.0)
{
    s = 0.0;
    for (i = kl; i <= m1; i++)
    {
        s = s + bv[i];
    }
    d = s - 1.0;
    if (fabs (d) > EPS)
    {
        g = 1.0/d;
        for (i = kl; i <= m1; i++)
        {
            bv[i] = g * (bv[i]);
        }
        *pZ = - g * (*pZ);
        for (j = 1; j <= n; j++)
        {
            f[j] = -f[j];
            ibound[j] = -ibound[j];
            icb = 0;
            for (i = kl; i <= m1; i++)
            {
                if (j == icbas[i]) icb = 1;
            }
            if (icb == 0)
            {
                s = 0.0;
                for (i = kl; i <= m1; i++)
                {
                    s = s + ct[i][j];
                }
                d = 1.0 - s;
                for (i = kl; i <= m1; i++)
                {
                    ct[i][j] = ct[i][j] + d * (bv[i]);
                }
```

```
                }
            }
            for (j = kl; j <= m; j++)
            {
                s = 0.0;
                for (i = kl; i <= m1; i++)
                {
                    s = s + binv[i][j];
                }
                d = s;
                for (i = kl; i <= m1; i++)
                {
                    binv[i][j] = -binv[i][j] + d * (bv[i]);
                }
            }
            for (i = kl; i <= m1; i++)
            {
                binv[i][m1] = bv[i];
            }
        }
    }
    for (j = 1; j <= n; j++)
    {
        r[j] = 0.0;
        icb = 0;
        for (i = kl; i <= m1; i++)
        {
            if (j == icbas[i]) icb = 1;
        }
        if (icb == 0)
        {
            s = -f[j];
            for (i = kl; i <= m1; i++)
            {
                k = icbas[i];
                s = s + ct[i][j] * (f[k]);
            }
            r[j] = s;
        }
    }
}

/*-----------------------------------------------------------------
Determine the vector that leaves the basis in LA_Linfside()
-----------------------------------------------------------------*/
void LA_linfside_vleav (int *pItest, int *pIout, int jin, int kl,
    int m, tMatrix_R ct, tVector_R bv)
{
    int             i, m1;
```

```
    tNumber_R       d, g, thmax;

    m1 = m + 1;
    thmax = 1.0/ (EPS * EPS);
    for (i = kl; i <= m1; i++)
    {
        d = ct[i][jin];
        if (d > EPS)
        {
            g = bv[i]/d;
            if (g < thmax)
            {
                thmax = g;
                *pIout = i;
                *pItest = 1;
            }
        }
    }
}

/*-------------------------------------------------------------------
Determine the vector that enters the basis in LA_Linfside()
-------------------------------------------------------------------*/
void LA_linfside_vent (int *pIvo, int *pJin, int kl, int m, int n,
    tVector_I icbas, tVector_R r, tNumber_R *pZ)
{
    int             i, j, m1, icb;
    tNumber_R       d, e, g, tz;

    m1 = m + 1;
    g = 1.0/ (EPS*EPS);
    tz = *pZ + EPS;
    for (j = 1; j <= n; j++)
    {
        icb = 0;
        for (i = kl; i <= m1; i++)
        {
            if (j == icbas[i]) icb = 1;
        }
        if (icb == 0)
        {
            d = r[j];
            if (d < -EPS)
            {
                e = d;
                if (e < g)
                {
                    g = e;
                    *pJin = j;
```

```
                    *pIvo = 1;
                }
            }
            else if (d >= tz)
            {
                e = tz - d;
                if (e < g)
                {
                    g = e;
                    *pJin = j;
                    *pIvo = -1;
                }
            }
        }
    }
}

/*---------------------------------------------------------------------
A Gauss-Jordan elimination step for LA_Linfside()
---------------------------------------------------------------------*/
void LA_linfside_gauss_jordn (int iout, int jin, int kl, int m,
    int n, tMatrix_R ct, tVector_I icbas, tMatrix_R binv,
    tVector_R bv)
{
    int             i, j, k, m1;
    tNumber_R       d, pivot;

    m1 = m + 1;
    pivot = ct[iout][jin];
    for (j = 1; j <= n; j++)
    {
        ct[iout][j] = ct[iout][j]/pivot;
    }
    k = m1;

    /*
    kj = kl + *pIter;
    if (kj < m1) k = kj;
    for (j = kl; j <= k; j++)
    */
    for (j = kl; j <= m1; j++)
    {
        binv[iout][j] = binv[iout][j]/pivot;
    }
    for (i = kl; i <= m1; i++)
    {
        if (i != iout)
        {
            d = ct[i][jin];
```

```
            for (j = 1; j <= n; j++)
            {
                ct[i][j] = ct[i][j] - d * (ct[iout][j]);
            }
            for (j = kl; j <= k; j++)
            {
                binv[i][j] = binv[i][j] - d * (binv[iout][j]);
            }
        }
    }
    for (i = kl; i <= m1; i++)
    {
        bv[i] = binv[i][m1];
    }
    icbas[iout] = jin;
}

/*-----------------------------------------------------------------------
Calculate the results of LA_Linfside()
-----------------------------------------------------------------------*/
eLaRc LA_linfside_res (int iside, int m, int n, int kl, int iout,
    tVector_R f, tVector_I icbas, tVector_I irbas, tMatrix_R binv,
    tVector_R bv, tVector_I ibound, tVector_R r, tVector_R a,
    int irank, tNumber_R *pZ)
{
    int             i, j, k, m1;
    tNumber_R       d, g, s, piv, pivot;

    m1 = m + 1;
    piv = 0.0;
    for (i = kl; i <= m1; i++)
    {
        d = bv[i];
        if (d > piv)
        {
            piv = d;
            iout = i;
        }
    }
    if (iout != m1)
    {
        for (j = kl; j <= m1; j++)
        {
            d = binv[iout][j];
            binv[iout][j] = binv[m1][j];
            binv[m1][j] = d;
        }
        k = icbas[iout];
        icbas[iout] = icbas[m1];
```

```
        icbas[m1] = k;
    }
    pivot = binv[m1][m1];
    for (j = kl; j <= m1; j++)
    {
        binv[m1][j] = binv[m1][j]/pivot;
    }
    for (i = kl; i <= m; i++)
    {
        d = binv[i][m1];
        for (j = kl; j <= m1; j++)
        {
            binv[i][j] = binv[i][j] - d * (binv[m1][j]);
        }
    }
    for (j = kl; j <= m; j++)
    {
        k = icbas[j];
        g = f[k];
        if (iside == 1 && ibound[k] == 1) g = g - *pZ;
        if (iside != 1 && ibound[k] == -1) g = g - *pZ;
        binv[j][m1] = g;
    }
    for (j = kl; j <= m; j++)
    {
        s = 0.0;
        for (i = kl; i <= m; i++)
        {
            s = s + binv[i][m1] * (binv[i][j]);
        }
        k = irbas[j];
        a[k] = s;
    }

    /* Calculation of the one-sided residuals */
    for (j = 1; j <= n; j++)
    {
        d = r[j];
        if (iside == 1)
        {
            if (ibound[j] == 1) d = *pZ - d;
            r[j] = d;
        }
        else if (iside != 1)
        {
            if (ibound[j] == -1) d = *pZ - d;
            r[j] = - d;
        }
    }
```

```
    if (irank < m)
        return LaRcSolutionDefNotUniqueRD;

    if (kl > 1)
        return LaRcSolutionProbNotUnique;

    for (i = 1; i <= m1; i++)
    {
        if (bv[i] < EPS)
            return LaRcSolutionProbNotUnique;
    }

    return LaRcSolutionUnique;
}
```

Chapter 12

Chebyshev Approximation with Bounded Variables

12.1 Introduction

In Chapter 10, an algorithm for calculating the Chebyshev solution of overdetermined systems of linear equations is given. In that algorithm, the Chebyshev norm of the residual vector, is as small as possible. In Chapter 11, an algorithm for the one-sided Chebyshev solution of overdetermined linear equations is presented. In that algorithm, the Chebyshev solution is subject to the additional constraints that all the elements of the residual vector are either non-positive or non-negative.

In this chapter, we present the Chebyshev approximation with bounded variables [3], where the additional constraints are on the elements of the solution vector, not on the elements of the residual vector. Typically, the elements of the solution vector are to be bounded between –1 and 1. This chapter is analogous to Chapter 7 for the linear bounded L_1 approximation.

The solid curve in Figure 12-1 is for the ordinary Chebyshev approximation, for the 8 points of Figure 2-1, and is given by $y = -0.8 + 2x - 0.156x^2$. The elements of the solution vector for this approximating curve are (–0.8, 2, –0.156). They are not all bounded between –1 and 1. For the same 8 points, the Chebyshev approximation with bounded variables between –1 and 1 is given by $y = 1 + x - 0.024x^2$. The dotted curve in Figure 12-1 shows this approximation.

In this chapter, linear programming techniques [7] are used to solve this problem, where minimum computer storage is required and no conditions are imposed on the coefficient matrix. The coefficient matrix could be a rank deficient one.

Consider the overdetermined system of linear equations

$$\mathbf{Ca} = \mathbf{f} \tag{12.1.1}$$

$\mathbf{C} = (c_{ij})$ is a given real n by m matrix of rank k, $k \leq m < n$, and $\mathbf{f} = (f_i)$ is a given real n-vector. The Chebyshev solution to system $\mathbf{Ca} = \mathbf{f}$ is the m-vector $\mathbf{a} = (a_j)$ that minimizes the Chebyshev norm z of the residuals

$$z = \max|r_i|, \quad i = 1, 2, \ldots, n$$

where r_i is the ith residual and is given by

$$r_i = \sum_{j=1}^{m} c_{ij}a_j - f_i, \quad i = 1, 2, \ldots, n \tag{12.1.2}$$

When the elements of the solution vector **a** is required to satisfy the additional conditions

$$-1 \leq a_i \leq 1, \quad i = 1, 2, \ldots, m \tag{12.1.3}$$

we have the problem of calculating the Chebyshev solution of (12.1.1) with bounded variables between –1 and 1.

Figure 12-1 shows curve fitting with vertical parabolas for the same 8 points shown in Figure 2-1. The solid curve is the ordinary Chebyshev approximation and the dotted curve is the Chebyshev approximation with bounded variables between –1 and 1.

If, instead of the constraints (12.1.3), we require the elements of vector **a** to satisfy the constraints

$$c_i \leq a_i \leq d_i, \quad i = 1, 2, \ldots, m \tag{12.1.4}$$

where (c_i) and (d_i) are elements of given m-vectors **c** and **d**, by substituting variables, (12.1.4) reduces to the constraints (12.1.3) in the new variables. This has been given in Section 7.1.

The Chebyshev approximation with non-negative parameters may be formulated as a Chebyshev approximation with bounded variables as well. See Section 12.1.1.

In Section 12.2, the Chebyshev approximation with bounded variables is formulated as a special problem of some general constrained problems. We shall comment on these methods in Section 12.5. In Section 12.3, the linear programming formulation of the problem is presented. In Section 12.4, the algorithm is described and in Section 12.5, numerical results and comments are given.

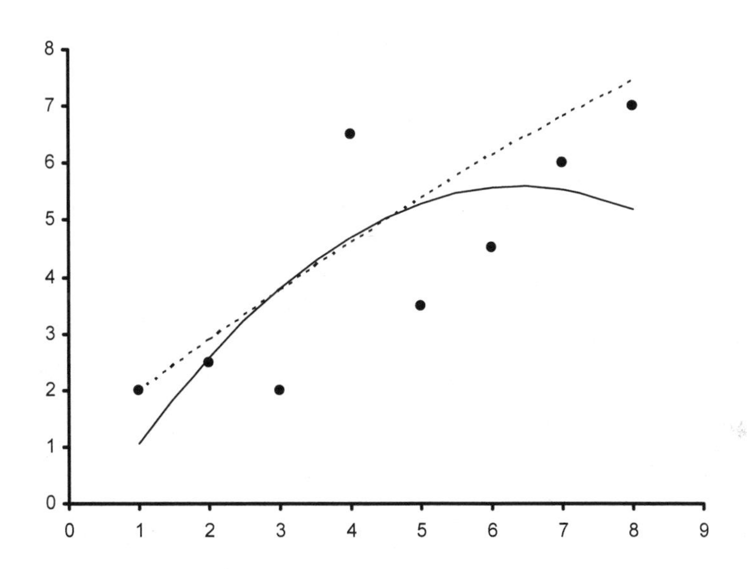

Figure 12-1: Curve fitting a set of 8 points using Chebyshev approximation and Chebyshev approximation with bounded variables between –1 and 1

12.1.1 Linear Chebyshev approximation with non-negative parameters (NNLI)

If in (12.1.4) we take $(c_i) = 0$ and $(d_i) = \text{Big}$, for $i = 1, 2, \ldots, m$, where Big is a large number, we have the linear L_∞ approximation with non-negative parameters, or non-negative L-infinity (**NNLI**). See Section 7.1.1 for the L_1 approximation with non-negative parameters, or non-negative L_1 approximation (**NNL1**).

12.2 A special problem of a general constrained one

Sklar [11] presented an algorithm to solve the problem (expressed here in our notation)

$$\text{minimize } ||\mathbf{Ca} - \mathbf{f}||_\infty$$

subject to

$$\mathbf{e}_a \le \mathbf{Ea} \le \mathbf{e}_b$$

His algorithm is a generalization of that of Armstrong and Sklar [4].

Also, Roberts and Barrodale [10] presented an algorithm to solve the problem

$$\text{minimize } \|\mathbf{Ca} - \mathbf{f}\|_\infty$$

subject to

$$\mathbf{Ga} = \mathbf{g} \text{ and } \mathbf{e}_a \le \mathbf{Ea} \le \mathbf{e}_b$$

Their algorithm is the extension of the algorithm of Barrodale and Phillips for the ordinary Chebyshev solution [5, 6].

In the above, **C**, **G** and **E** are matrices of appropriate dimensions, **f** is the vector associated with **C**, **g** is the vector associated with **G** and $\mathbf{e}_a$ and $\mathbf{e}_b$ are two vectors associated with **E**. Also, not all of the arrays (**G**, **g**) and (**E**, $\mathbf{e}_a$, $\mathbf{e}_b$) are vacuous.

If we take $\mathbf{G} = \mathbf{0}$, $\mathbf{g} = \mathbf{0}$ and take $\mathbf{E} = \mathbf{I}_m$ and $\mathbf{e}_a = -\mathbf{e}_m$ and $\mathbf{e}_b = \mathbf{e}_m$, where $\mathbf{I}_m$ and $\mathbf{e}_m$ are respectively an m-unit matrix and an m-vector, each element of which equals 1, we get the Chebyshev solution with bounded variables between 1 and –1. As mentioned earlier, we shall comment on these algorithms in Section 12.5.

12.3 Linear programming formulation of the problem

The problem given by (12.1.1-3) may be reduced to a linear programming problem as follows.

Let $h \ge 0$, be $h = \max_i|r_i|$ in (12.1.2). Hence, the primal form of the programming problem is

$$\text{minimize } h$$

subject to

$$-h \le \sum_{j=1}^{m} c_{ij}a_j - f_i \le h, \quad i = 1, 2, \ldots, n$$

and

$$-1 \le a_i \le 1, \quad i = 1, 2, \ldots, m$$

In vector-matrix notation, the above inequalities become

$$\begin{aligned} \mathbf{Ca} + h\mathbf{e}_n &\geq \mathbf{f} \\ -\mathbf{Ca} + h\mathbf{e}_n &\geq -\mathbf{f} \end{aligned}$$

$$\begin{aligned} -\mathbf{a} &\geq -\mathbf{e}_m \\ \mathbf{a} &\geq -\mathbf{e}_m \end{aligned}$$

where

$$a_j, j = 1, 2, \ldots, m, \text{ unrestricted in sign,} \quad h \geq 0$$

$\mathbf{e}_n$ and $\mathbf{e}_m$ are an n- and m-vectors, each element of which is 1.

This formulation may be written more conveniently as

$$\text{minimize } Z = \mathbf{d}_{m+1}^T \begin{bmatrix} \mathbf{a} \\ h \end{bmatrix}$$

subject to

$$\begin{bmatrix} \mathbf{C} & \mathbf{e}_n \\ -\mathbf{C} & \mathbf{e}_n \\ -\mathbf{I}_m & \mathbf{0} \\ \mathbf{I}_m & \mathbf{0} \end{bmatrix} \begin{bmatrix} \mathbf{a} \\ h \end{bmatrix} \geq \begin{bmatrix} \mathbf{f} \\ -\mathbf{f} \\ -\mathbf{e}_m \\ -\mathbf{e}_m \end{bmatrix}$$

$$a_j, j = 1, 2, \ldots, m, \text{ unrestricted in sign,} \quad h \geq 0$$

Here, $\mathbf{d}_{m+1}$ is an $(m + 1)$-vector whose first m elements are 0's and the $(m + 1)^{th}$ element is 1; that is, $\mathbf{d}_{m+1}$ is the $(m + 1)^{th}$ column of an $(m + 1)$-unit matrix. Here, $\mathbf{d}_{m+1}$ is the same as $\mathbf{e}_{m+1}$ in Chapters 10 and 11. Again, $\mathbf{I}_m$ is an m-unit matrix and $\mathbf{0}$ is an m-zero vector.

By rearranging the terms in the primal formulation, the dual form is (Section 3.6)

(12.3.1a) $$\text{maximize } z = [\mathbf{f}^T \ -\mathbf{e}_m^T \ -\mathbf{f}^T \ -\mathbf{e}_m^T]\mathbf{b}$$

subject to

(12.3.1b) $$\begin{bmatrix} \mathbf{C}^T & -\mathbf{I}_m & -\mathbf{C}^T & \mathbf{I}_m \\ \mathbf{e}_n^T & \mathbf{0}^T & \mathbf{e}_n^T & \mathbf{0}^T \end{bmatrix} \mathbf{b} = \mathbf{d}_{m+1}$$

(12.3.1c) $$b_i \geq 0, \quad i = 1, 2, \ldots, 2(n + m)$$

Let us write (12.3.1b) in the form

(12.3.1d) $$\mathbf{Db} = \mathbf{d}_{m+1}$$

where **D** is the coefficient matrix on the l.h.s. of (12.3.1b). **D** is an (m + 1) by 2(n + m) matrix and is of rank (m + 1), despite the fact that matrix **C** may be rank deficient.

As usual, let **B** denote the non-singular basis matrix for the linear programming problem (12.3.1). Since matrix **D** is of rank (m + 1), **B** is an (m + 1) square matrix. Let also ($\mathbf{y}_i$) be the columns forming the simplex tableau, ($z_j - f_j$) be the marginal costs for column ($\mathbf{y}_j$), $\mathbf{b}_B$ be the basic solution, and z be the objective function.

Let also the elements of the 2(n + m)-vector $\mathbf{g}^T = [\mathbf{f}^T \; -\mathbf{e}_m^T \; -\mathbf{f}^T \; -\mathbf{e}_m^T]$ of (12.3.1a) associated with the basic variables be the (m + 1)-vector $\mathbf{g}_B$. Then for j = 1, 2, …, 2(n + m), we have

(12.3.2a) $$\mathbf{y}_j = \mathbf{B}^{-1}\mathbf{D}_j$$

(12.3.2b) $$(z_j - g_j) = \mathbf{g}_B^T\mathbf{y}_j - g_j$$

(12.3.2c) $$\mathbf{b}_B = \mathbf{B}^{-1}\mathbf{d}_{m+1}$$

(12.3.2d) $$z = \mathbf{g}_B^T\mathbf{b}_B$$

where $\mathbf{D}_j$ is the jth column of matrix **D**.

12.3.1 Properties of the matrix of constraints

Consider the two halves of matrix **D**, in (12.3.1b, d), each is of (n + m) columns. While matrix $\mathbf{C}^T$ exists in the first n columns of the left half of **D**, matrix $-\mathbf{C}^T$ exists in the first n columns of the right half. The last m columns of the left half are the negatives of the last m columns of the right half. We also make use of the fact the a unit matrix exists in the last m columns of each of the two halves of **D**.

This kind of asymmetry and the existence of unit matrices in **D** will enable us to use a simplex tableau for this problem of only (m + 1) constraints in n variables, instead of a simplex tableau of (m + 1) constraints in 2(n + m) variables. This will be apparent from the coming analysis.

From the definition of matrix **D** in (12.3.1d), we have for j = 1, 2, …, n

(12.3.3a) $$\mathbf{D}_j = \begin{bmatrix} \mathbf{C}_j^T \\ 1 \end{bmatrix} \quad \text{and} \quad \mathbf{D}_{j+n+m} = \begin{bmatrix} -\mathbf{C}_j^T \\ 1 \end{bmatrix}$$

where again, $\mathbf{D}_j$ and $\mathbf{C}_j^T$ are respectively the jth column of **D** and the jth column of $\mathbf{C}^T$. Then we have

(12.3.3b) $$\mathbf{D}_j + \mathbf{D}_{j+n+m} = 2\mathbf{d}_{m+1}, \quad j = 1, \ldots, n$$

and also we have

(12.3.3c) $$\mathbf{D}_j + \mathbf{D}_{j+n+m} = \mathbf{0}, \quad j = n+1, \ldots, n+m$$

Definition

Let i and j, $1 \le i, j \le 2(n + m)$, be the indices of any two columns in matrix **D** such that $|i - j| = (n + m)$. We define columns i and j as two corresponding columns.

Consider the following lemmas.

Lemma 12.1

At any stage of the computation, the $(m + 1)^{th}$ column of matrix $\mathbf{B}^{-1}$ is the basic solution vector $\mathbf{b}_B$.

This follows directly from (12.3.2c) since $\mathbf{d}_{m+1}$ is the $(m + 1)^{th}$ column of an $(m + 1)$-unit matrix.

Lemma 12.2

At any stage of the computation, the solution vector **a** and the objective function z, are given by Lemma 10.2

(12.3.4a) $$\mathbf{B}^T \begin{bmatrix} \mathbf{a} \\ z \end{bmatrix} = \mathbf{g}_B$$

or

(12.3.4b) $$(\mathbf{a}^T\ z) = \mathbf{g}_B{}^T \mathbf{B}^{-1}$$

Lemma 12.3

Any two corresponding columns should not appear together in any basis matrix.

Lemma 12.4

For any two corresponding columns in the simplex tableau we have

$$\begin{aligned}
&\mathbf{y}_i + \mathbf{y}_{i+n+m} = 2\mathbf{b}_B, && i = 1, \ldots, n\\
&\mathbf{y}_i + \mathbf{y}_{i+n+m} = \mathbf{0}, && i = n+1, \ldots, n+m\\
&(z_i - g_i) + (z_{i+n+m} - g_{i+n+m}) = 2z, && i = 1, 2, \ldots, n\\
&(z_i - g_i) + (z_{i+n+m} - g_{i+n+m}) = 2, && i = n+1, \ldots, n+m
\end{aligned}$$

Proof:

By multiplying (12.3.3b, c) by $\mathbf{B}^{-1}$ and using (12.3.2a, c) the first two equations are proved. From (12.3.2a-c) the last two equations are proved.

Lemma 12.5

The residuals (r_i), $i = 1, 2, \ldots, n$, are given by one of the following relations

$$\begin{aligned}
r_i &= (z_i - g_i) - z, && i = 1, 2, \ldots, n\\
r_i &= z - (z_i - g_i), && i = n+m+1, \ldots, 2n+m
\end{aligned}$$

It is sufficient to prove the first relation. Let $1 \le j \le n$. Then since $\mathbf{C}_j^T$ is the jth column of $\mathbf{C}^T$, i.e., the jth row of $\mathbf{C}$, we write (12.1.2) as $r_j = \mathbf{a}^T\mathbf{C}_j^T - g_j$, or from the definition of $\mathbf{D}_j^T$ in (12.3.3a), $1 \le j \le n$, $r_j = (\mathbf{a}^T\ z)\mathbf{D}_j^T - g_j - z$.

By substituting (12.3.4b), (12.3.2a) and (12.3.2d) in succession in the last expression of r_j, we get $r_j = \mathbf{g}_B{}^T\mathbf{B}^{-1}\mathbf{D}_j^T - g_j - z = (z_i - g_i) - z$. This proves the first relation in the Lemma. The second relation is proved in the same manner. See also Lemma 13.4 and [2].

Lemma 12.6

The last m columns in the (m + 1) by 2(n + m) simplex tableau are the first m columns of the inverse of the basis matrix; $\mathbf{B}^{-1}$. Initially, they are the first m columns of an (m + 1)-unit matrix.

12.4 Description of the algorithm

We only need to store in the initial simplex tableau the first n columns of matrix **D** of (12.3.1b, d). We call this the **condensed tableau**. From Lemma 12.4, if any column in the simplex tableau and its marginal cost are known, the corresponding column and its marginal cost are easily derived.

From Lemma 12.6, the first m columns of the (m + 1) matrix $\mathbf{B}^{-1}$ are available. We consider the first m columns of matrix $\mathbf{B}^{-1}$ (and their corresponding columns) as part of the simplex tableau.

The algorithm is in 3 parts. In part 1, the simplex tableau is changed once when any one (usually the first column) of the first n columns of matrix $\mathbf{D}$ may form the $(m + 1)^{th}$ column of the basis matrix $\mathbf{B}$. After changing the simplex tableau once, the initial basic solution $\mathbf{b}_B$ is now available. From Lemma 12.1, $\mathbf{b}_B$ is itself the $(m + 1)^{th}$ column of $\mathbf{B}^{-1}$. We remark that for this Gauss-Jordan step, each of the pivot element and the $(m + 1)^{th}$ element of $\mathbf{b}_B$ is 1; that is, the $(m + 1)^{th}$ element of $\mathbf{b}_B$ is ≥ 0.

An (n + m)-index indicator vector whose elements are +1 or –1 is needed. If column j, $1 \leq j \leq n$, is in the condensed tableau, index j of this vector has the value +1. If the corresponding column to column j is in the condensed tableau, index j has a value –1. Similarly, if column j of matrix $\mathbf{B}^{-1}$, $1 \leq j \leq m$, has its marginal cost stored, the index (n + j) has the value +1. Otherwise, it has the value –1.

In part 2, if one or more elements i, $1 \leq i \leq m$, of the basic solution $\mathbf{b}_B$ is negative, the initial basic solution is not feasible. For each basic column j, $(2n + m + 1) \leq j \leq 2(n + m)$, associated with a negative element i in the basic solution, we make use of the second equation of Lemma 12.4 and replace column j by its corresponding column. Again, this does not require changing the simplex tableau, except that row i, which corresponds to the negative element of $\mathbf{b}_B$, is multiplied by –1. Row i of $\mathbf{B}^{-1}$ and the ith element of $\mathbf{b}_B$ are also multiplied by –1.

At the end of part 2, the initial objective function z is calculated from (12.3.2d) and z may have a positive or a negative value. If z is negative in part 3 of the algorithm, z increases monotonically until it acquires its optimum (positive) value.

Part 3 is almost identical to part 3 of the algorithm for the ordinary Chebyshev solution [1] described in Chapter 10. The only difference is the following. At the beginning of part 3, we calculate the marginal costs not only for the columns of the condensed tableau, but also for the first m columns of matrix $\mathbf{B}^{-1}$. From (12.3.1a) and the above discussion, the prices for the first m columns of $\mathbf{B}^{-1}$ would be the elements of the vector ($-\mathbf{e}_m$).

As in Chapter 10, the non-basic column to enter the basis is that which has the most negative marginal cost among:

(a) the non-basic columns in the current condensed tableau,
(b) the first m columns of matrix $\mathbf{B}^{-1}$, and
(c) the corresponding columns of (a) and (b).

The optimum residual vector **r** is calculated from Lemma 12.5 and the optimum ($\mathbf{a}^T z$) values are obtained from Lemma 12.2.

12.5 Numerical results and comments

LA_Linfbv() implements this algorithm. DR_Linfbv() tests 8 examples, including some with rank deficient matrices.

All of the results satisfy the inequalities (12.1.3). Table 12.5.1 shows the results of 3 of the examples.

Table 12.5.1

		Chebyshev Solution		Chebyshev solution with bounded variables	
Example	**C**(n×m)	Iterations	z	Iterations	z
1	4 × 2	5	10.0	5	10.0
2	10× 5	9	1.778	6	1.778
3	25×10	22	0.0071	19	0.2329

For each example, the number of iterations and the optimum norm for the Chebyshev solution with bounded variables are shown. For comparison purposes, the results of the ordinary Chebyshev approximation problem are also given.

The number of iterations for this algorithm are, in general, smaller than those for the corresponding ordinary Chebyshev case.

We now comment on other algorithms that solve this problem. Using an extension of the well-known exchange algorithm, Powell [9] presented an algorithm for obtaining the Chebyshev solution of system $\mathbf{Ca} = \mathbf{f}$ subject to the constraints (12.1.3). FORTRAN IV code for Powell's method was given by Madsen and Powell [8]. Excluding comments, the FORTRAN implementation consists of 289 lines of code, as compared with the FORTRAN IV implementation of our algorithm, which consists of 165 lines of code.

As noted in Section 12.2, both Sklar [11] and Roberts and Barrodale [10] presented general constrained Chebyshev algorithms that would solve the bounded Chebyshev problem as a special case.

We observe that our algorithm requires less computer storage and, as a special purpose algorithm, would be more efficient than a general purpose algorithm, such as [9] and [11]. It is also noted by Roberts and Barrodale ([10], p. 798) that Powell's algorithm [8, 9] is more efficient than theirs.

References

1. Abdelmalek, N.N., Chebyshev solution of overdetermined systems of linear equations, *BIT*, 15(1975)117-129.
2. Abdelmalek, N.N., The discrete linear restricted Chebyshev approximation, *BIT*, 17(1977)249-261.
3. Abdelmalek, N.N., Chebyshev and L_1 solutions of overdetermined systems of linear equations with bounded variables, *Numerical Functional Analysis and Optimization*, 8(1985-86)399-418.
4. Armstrong, R.D. and Sklar, M.G., A linear programming algorithm for curve fitting in the L_∞ norm, *Numerical Functional Analysis and Optimization*, 2(1980)187-218.
5. Barrodale, I. and Phillips, C., An improved algorithm for discrete Chebyshev linear approximation, *Proceedings of the Fourth Manitoba Conference on Numerical Mathematics*, Hartnell, B.L. and Williams, H.C. (eds.), Winnipeg, Manitoba, Canada, pp. 177-190, 1975.
6. Barrodale, I. and Phillips, C., Algorithm 495: Solution of an overdetermined system of linear equations in the Chebyshev norm, *ACM Transactions on Mathematical Software*, 1(1975)264-270.
7. Hadley, G., *Linear Programming*, Addison-Wesley, Reading, MA, 1962.
8. Madsen, K. and Powell, M.J.D., A FORTRAN subroutine that calculates the minimax solution of linear equations subject to

bounds on the variables, *United Kingdom Atomic Energy Research Establishment*, AERE-R7954, February 1975.

9. Powell, M.J.D., The minimax solution of linear equations subject to bounds on the variables, *Proceedings of the Fourth Manitoba Conference on Numerical Mathematics*, Hartnell, B.L. and Williams, H.C. (eds.), Winnipeg, Manitoba, Canada, pp. 53-107, 1975.
10. Roberts, F.D.K. and Barrodale, I., An algorithm for discrete Chebyshev linear approximation with linear constraints, *International Journal for Numerical Methods in Engineering*, 15(1980)797-807.
11. Sklar, M.G., L_∞ norm estimation with linear restrictions on the parameters, *Numerical Functional Analysis and Optimization*, 3(1981)53-68.

12.6 DR_Linfbv

```
/*--------------------------------------------------------------------
DR_Linfbv
----------------------------------------------------------------------
This program is a driver for the function LA_Linfbv(), which solves
an overdetermined system of linear equations in the Chebyshev norm
subject to the constraints that each element of the solution vector
is bounded between -1 and 1;

               -1 <= a[j] <= 1,    j = 1, 2, ..., m

The overdetermined system has the form

                         c*a = f

"c" is a given real n by m matrix of rank k, k <= m < n.
"f" is a given real n vector.
"a" is the solution m vector.

This driver contains the 8 examples whose results are given in the
text.
--------------------------------------------------------------------*/

#include "DR_Defs.h"
#include "LA_Prototypes.h"

#define NNMMc_ROWS  (NN_ROWS + MMc_COLS)
#define N1          4
#define M1          2
#define N2          5
#define M2          3
#define N3          6
#define M3          3
#define N4          7
#define M4          3
#define N5          8
#define M5          4
#define N6          10
#define M6          5
#define N7          25
#define M7          10
#define N8          8
#define M8          4
```

```
void DR_Linfbv (void)
{
    /*---------------------------------------
      Constant matrices/vectors
    ---------------------------------------*/
    static tNumber_R c1init[N1][M1] =
    {
        { 0.0, -2.0 },
        { 0.0, -4.0 },
        { 1.0, 10.0 },
        {-1.0, 15.0 }
    };

    static tNumber_R c2init[N2][M2] =
    {
        { 1.0,  2.0, 0.0 },
        {-1.0, -1.0, 0.0 },
        { 1.0,  3.0, 0.0 },
        { 0.0,  1.0, 0.0 },
        { 0.0,  0.0, 1.0 }
    };

    static tNumber_R c3init[N3][M3] =
    {
        { 0.0, -1.0,  0.0 },
        { 1.0,  3.0, -4.0 },
        { 1.0,  0.0,  0.0 },
        { 0.0,  0.0,  1.0 },
        {-1.0,  1.0,  2.0 },
        { 1.0,  1.0,  1.0 }
    };

    static tNumber_R c4init[N4][M4] =
    {
        { 1.0, 0.0,  1.0 },
        { 1.0, 2.0,  2.0 },
        { 1.0, 2.0,  0.0 },
        { 1.0, 1.0,  0.0 },
        { 1.0, 0.0, -1.0 },
        { 1.0, 0.0,  0.0 },
        { 1.0, 1.0,  1.0 }
    };

    static tNumber_R c5init[N5][M5] =
```

```
{
    { 1.0, -3.0,  9.0, -27.0 },
    { 1.0, -2.0,  4.0,  -8.0 },
    { 1.0, -1.0,  1.0,  -1.0 },
    { 1.0,  0.0,  0.0,   0.0 },
    { 1.0,  1.0,  1.0,   1.0 },
    { 1.0,  2.0,  4.0,   8.0 },
    { 1.0,  3.0,  9.0,  27.0 },
    { 1.0,  4.0, 16.0,  64.0 }
};

static tNumber_R c6init[N6][M6] =
{
    { 1.0, 0.0,  0.0,  0.0,  0.0 },
    { 0.0, 1.0,  0.0,  0.0,  0.0 },
    { 0.0, 0.0,  1.0,  0.0,  0.0 },
    { 0.0, 0.0,  0.0,  1.0,  0.0 },
    { 0.0, 0.0,  0.0,  0.0,  1.0 },
    { 1.0, 1.0,  1.0,  1.0,  1.0 },
    { 0.0, 1.0,  1.0,  1.0,  1.0 },
    {-1.0, 0.0, -1.0, -1.0, -1.0 },
    { 1.0, 1.0,  0.0,  1.0,  1.0 },
    { 1.0, 1.0,  1.0,  0.0,  1.0 }
};

static tNumber_R c8init[N8][M8] =
{
    { 1.0, 1.0,   1.0,  1.0 },
    { 1.0, 2.0,   4.0,  4.0 },
    { 1.0, 3.0,   9.0,  9.0 },
    { 1.0, 4.0,  16.0, 16.0 },
    { 1.0, 5.0,  25.0, 25.0 },
    { 1.0, 6.0,  36.0, 36.0 },
    { 1.0, 7.0,  49.0, 49.0 },
    { 1.0, 8.0,  64.0, 64.0 }
};

static tNumber_R f1[N1+1] =
{   NIL,
    -12.0, 6.0, 0.0, 5.0
};

static tNumber_R f2[N2+1] =
{   NIL,
    1.0, 2.0, 1.0, -3.0, 0.0
```

```
};

static tNumber_R f3[N3+1] =
{   NIL,
    1.0, 2.0, 3.0, 2.0, 2.0, 4.0
};

static tNumber_R f4[N4+1] =
{   NIL,
    0.0, -2.0, 1.0, -1.0, 5.0, 7.0, 0.0
};

static tNumber_R f5[N5+1] =
{   NIL,
    3.0, -3.0, -2.0, 0.0, 7.0, -1.0, 5.0, 2.0 };

static tNumber_R f6[N6+1] =
{   NIL,
    1.0, -1.0, 0.0, -1.0, 1.0, 0.0, 2.0, 3.0, -3.0, -2.0
};

static tNumber_R f7[N7+1] =
{   NIL,
    0.0872673, 0.0872794, 0.0873029, 0.0873315, 0.0873576,
    0.3491184, 0.3498802, 0.3513824, 0.3532572, 0.3550109,
    0.6111334, 0.6150641, 0.6230824, 0.6336395, 0.6441493,
    0.8733883, 0.8841621, 0.9071868, 0.9400757, 0.9766021,
    1.135895,  1.157550,  1.206257,  1.283258,  1.384432
};

static tNumber_R f8[N8+1] =
{   NIL,
    2.0, 2.5, 2.0, 6.5, 3.5, 4.5, 6.0, 7.0
};

/*----------------------------------------
  Variable matrices/vectors
----------------------------------------*/
tMatrix_R   ct      = alloc_Matrix_R (MMc_COLS, NN_ROWS);
tVector_R   f       = alloc_Vector_R (NN_ROWS);
tVector_R   r       = alloc_Vector_R (NNMMc_ROWS);
tVector_R   a       = alloc_Vector_R (MMc_COLS);
tMatrix_R   binv    = alloc_Matrix_R (MMc_COLS, MMc_COLS);
tVector_R   bv      = alloc_Vector_R (MMc_COLS);
tVector_I   ibound  = alloc_Vector_I (NNMMc_ROWS);
```

```
tVector_I   icbas   = alloc_Vector_I (MMc_COLS);
tVector_I   irbas   = alloc_Vector_I (MMc_COLS);
tMatrix_R   c7      = alloc_Matrix_R (N7, M7);

tMatrix_R   c1      = init_Matrix_R (&(c1init[0][0]), N1, M1);
tMatrix_R   c2      = init_Matrix_R (&(c2init[0][0]), N2, M2);
tMatrix_R   c3      = init_Matrix_R (&(c3init[0][0]), N3, M3);
tMatrix_R   c4      = init_Matrix_R (&(c4init[0][0]), N4, M4);
tMatrix_R   c5      = init_Matrix_R (&(c5init[0][0]), N5, M5);
tMatrix_R   c6      = init_Matrix_R (&(c6init[0][0]), N6, M6);
tMatrix_R   c8      = init_Matrix_R (&(c8init[0][0]), N8, M8);

int         iter, i, j, k, m, n, Iexmpl;
tNumber_R   d, dd, ddd, e, ee, eee, z;

eLaRc       rc = LaRcOk;

for (j = 1; j <= 5; j++)
{
    d = 0.15* (j-3);
    dd = d*d;
    ddd = d*dd;
    for (i = 1; i <= 5; i++)
    {
        e = 0.15* (i-3);
        ee = e*e;
        eee = e*ee;
        k = 5* (j-1) + i;
        c7[k][1] = 1.0;
        c7[k][2] = d;
        c7[k][3] = e;
        c7[k][4] = dd;
        c7[k][5] = ee;
        c7[k][6] = e*d;
        c7[k][7] = ddd;
        c7[k][8] = eee;
        c7[k][9] = dd*e;
        c7[k][10] = ee*d;
    }
}

prn_dr_bnr ("DR_Linfbv, Bounded Chebyshev Solution of an "
            "Overdetermined System of Linear Equations");
```

```
for (Iexmpl = 1; Iexmpl <= 8; Iexmpl++)
{
    switch (Iexmpl)
    {
        case 1:
            n = N1;
            m = M1;
            for (i = 1; i <= n; i++)
            {
                f[i] = f1[i];
                for (j = 1; j <= m; j++) ct[j][i] = c1[i][j];
            }
            break;

        case 2:
            n = N2;
            m = M2;
            for (i = 1; i <= n; i++)
            {
                f[i] = f2[i];
                for (j = 1; j <= m; j++) ct[j][i] = c2[i][j];
            }
            break;

        case 3:
            n = N3;
            m = M3;
            for (i = 1; i <= n; i++)
            {
                f[i] = f3[i];
                for (j = 1; j <= m; j++) ct[j][i] = c3[i][j];
            }
            break;

        case 4:
            n = N4;
            m = M4;
            for (i = 1; i <= n; i++)
            {
                f[i] = f4[i];
                for (j = 1; j <= m; j++) ct[j][i] = c4[i][j];
            }
            break;

        case 5:
```

```
            n = N5;
            m = M5;
            for (i = 1; i <= n; i++)
            {
                f[i] = f5[i];
                for (j = 1; j <= m; j++) ct[j][i] = c5[i][j];
            }
            break;
        case 6:
            n = N6;
            m = M6;
            for (i = 1; i <= n; i++)
            {
                f[i] = f6[i];
                for (j = 1; j <= m; j++) ct[j][i] = c6[i][j];
            }
            break;

        case 7:
            n = N7;
            m = M7;
            for (i = 1; i <= n; i++)
            {
                f[i] = f7[i];
                for (j = 1; j <= m; j++) ct[j][i] = c7[i][j];
            }
            break;

        case 8:
            n = N8;
            m = M8;
            for (i = 1; i <= n; i++)
            {
                f[i] = f8[i];
                for (j = 1; j <= m; j++) ct[j][i] = c8[i][j];
            }
            break;

        default:
            break;
    }

    prn_algo_bnr ("Linfbv");

    prn_example_delim();
```

```
        PRN ("Example #%d: Size of matrix \"c\" %d by %d\n",
             Iexmpl, n, m);
        prn_example_delim();
        PRN ("Bounded Chebyshev Solution "
             "of an Overdetermined System\n");
        prn_example_delim();
        PRN ("r.h.s. Vector \"f\"\n");
        prn_Vector_R (f, n);
        PRN ("Transpose of Coefficient Matrix, \"ct\"\n");
        prn_Matrix_R (ct, m, n);

        rc = LA_Linfbv (m, n, ct, f, icbas, irbas, binv, bv, ibound,
                        &iter, r, a, &z);

        if (rc >= LaRcOk)
        {
            PRN ("\n");
            PRN ("Results of the Bounded Chebyshev Solution\n");
            PRN ("Bounded Chebyshev solution vector, \"a\"\n");
            prn_Vector_R (a, m);
            PRN ("Bounded Chebyshev residual vector \"r\"\n");
            prn_Vector_R (r, n);
            PRN ("Bounded Chebyshev norm \"z\" = %8.4f\n", z);
            PRN ("No. of Iterations = %d\n", iter);
        }

        prn_la_rc (rc);
    }

    free_Matrix_R (ct, MMc_COLS);
    free_Vector_R (f);
    free_Vector_R (r);
    free_Vector_R (a);
    free_Matrix_R (binv, MMc_COLS);
    free_Vector_R (bv);
    free_Vector_I (ibound);
    free_Vector_I (icbas);
    free_Vector_I (irbas);
    free_Matrix_R (c7, N7);

    uninit_Matrix_R (c1);
    uninit_Matrix_R (c2);
    uninit_Matrix_R (c3);
    uninit_Matrix_R (c4);
    uninit_Matrix_R (c5);
```

```
    uninit_Matrix_R (c6);
    uninit_Matrix_R (c8);
}
```

12.7 LA_Linfbv

```
/*-----------------------------------------------------------------------
LA_Linfbv
-------------------------------------------------------------------------
This program calculates the Chebyshev solution of an overdetermined
system of linear equations subject to the conditions that the
elements of the solution vector be bounded between -1 and +1.

The system of linear equations has the form

                        c*a = f

"c" is a given real n by m matrix of rank k <= m < n.
"f" is a given real n vector.

The problem is to calculate the solution vector "a" that gives
the minimum Chebyshev norm z,

                z = max|r[i]|,   i = 1, 2, ..., n

subject to the constraints that the elements of vector "a" are
bounded between -1 and 1. That is

              -1 <= a[j] <= 1,   j = 1, 2, ..., m

r[i] is the ith residual and is given by

 r[i] = c[i][1]*a[1] + c[i][2]*a[2] + ... + c[i][m]*a[m] - f[i],
                        i = 1, 2, ..., n

Inputs
m       Number of columns of matrix "c" of the system c*a = f.
n       Number of rows of matrix "c" of the system c*a = f.
ct      A real (m+1) by n matrix. Its first m rows and its n
        columns contain the transpose of matrix "c" of the system
        c*a = f. Its (m+1)th row will be filled with ones by the
        program.
f       A real n vector containing the r.h.s. of the system c*a = f.

Other Parameters
binv    An (m+1) real square matrix containing the inverse of the
        basis matrix in the linear programming problem.
bv      An (m+1) real vector containing the basic  solution in the
```

```
        linear programming problem.
icbas   An (m+1) integer vector containing the indices of the
        columns of "ct" forming the basis.
irbas   An (m+1) integer vector containing the indices of the rows
        of matrix "ct".

Outputs
iter    Number of iterations, or the number of times the simplex
        tableau is changed by a Gauss-Jordan step.
a       An (m + 1) vector whose first m elements contain the bounded
        Chebyshev solution vector "a" to the system c*a = f.
r       An (n + m + 1) vector whose first n elements constitute the
        residual vector r = (c*a - f).
z       The minimum bounded Chebyshev norm of the residual vector r.

Returns one of
        LaRcSolutionFound
        LaRcNoFeasibleSolution
        LaRcErrBounds
        LaRcErrNullPtr
---------------------------------------------------------------------*/

#include "LA_Prototypes.h"

eLaRc LA_Linfbv (int m, int n, tMatrix_R ct, tVector_R f,
    tVector_I icbas, tVector_I irbas, tMatrix_R binv, tVector_R bv,
    tVector_I ibound, int *pIter, tVector_R r, tVector_R a,
    tNumber_R *pZ)
{
    int         i = 0, j = 0, ijk = 0, k = 0, m1 = 0, n1 = 0, nm = 0;
    int         iout = 0, jin = 0, ivo = 0, itest = 0;
    tNumber_R   d = 0.0, e = 0.0;

    /* Validation of the data before executing the algorithm */
    eLaRc       rc = LaRcSolutionFound;
    VALIDATE_BOUNDS ((0 < m) && (m < n));
    VALIDATE_PTRS   (ct && f && icbas && irbas && binv && bv &&
                     ibound && pIter && r && a && pZ);

    /* Initialization */
    m1 = m + 1;
    n1 = n + 1;
    nm = n + m;
    *pIter = 0;
```

```
/* Part 1 of the algorithm
Initializing program data */
LA_linfbv_init (m, n, ct, icbas, irbas, binv, ibound, r, a);

iout = m1;
jin = 1;
r[1] = 0.0;

/* A Gauss-Jordan elimination step */
LA_linfbv_gauss_jordn (m, n, iout, jin, ct, icbas, binv, bv,
                       ibound);
*pIter = *pIter + 1;

/* Part 2 of the algorithm.
Obtaining a feasible basic solution */
LA_linfbv_part_2 (m, n, ct, binv, bv, ibound);

/* Calculating the initial residual vector "r" and
norm "z" */
LA_linfbv_resid_norm (m, n, ct, f, icbas, bv, r, pZ);

/* Part 3 of the algorithm */
for (ijk = 1; ijk < n*n; ijk++)
{
    ivo = 0;
    /* Determine the vector that enters the basis */
    LA_linfbv_vent (&ivo, &jin, m, n, icbas, r, pZ);

    if (ivo == 0)
    {
        /* Calculate the results */
        LA_linfbv_res (m, n, f, icbas, irbas, binv, ibound, r, a,
                       pZ);
        GOTO_CLEANUP_RC (LaRcSolutionFound);
    }

    if (ivo == -1)
    {
        ibound[jin] = -ibound[jin];
        if (jin <= n)
        {
            for (i = 1; i <= m1; i++)
            {
                ct[i][jin] = bv[i] + bv[i] - ct[i][jin];
            }
```

```
                r[jin] = *pZ + *pZ - r[jin];
                f[jin] = -f[jin];
            }
            else if (jin > n)
            {
                r[jin] = 2.0 - r[jin];
            }
        }
        itest = 0;

        /* Determine the vector that leaves the basis */
        LA_linfbv_vleav (jin, &iout, &itest, m, n, ct, binv, bv,
                         ibound);

        /* Solution is not feasible */
        if (itest != 1)
        {
            GOTO_CLEANUP_RC (LaRcNoFeasibleSolution);
        }

        /* A Gauss-Jordan elimination step */
        LA_linfbv_gauss_jordn (m, n, iout, jin, ct, icbas, binv, bv,
                               ibound);
        *pIter = *pIter + 1;

        d = r[jin];
        for (j = 1; j <= n; j++)
        {
            r[j] = r[j] - d * (ct[iout][j]);
        }
        for (j = n1; j <= nm; j++)
        {
            e = -1.0;
            if (ibound[j] == -1) e = -e;
            k = j - n;
            r[j] = r[j] - e * (d) * (binv[iout][k]);
        }
        *pZ = *pZ - d * (bv[iout]);
    }

CLEANUP:

    return rc;
}
```

```
/*--------------------------------------------------------------------
A Gauss-Jordan elimination step in LA_Linfbv()
--------------------------------------------------------------------*/
void LA_linfbv_gauss_jordn (int m, int n, int iout, int jin,
    tMatrix_R ct, tVector_I icbas, tMatrix_R binv, tVector_R bv,
    tVector_I ibound)
{
    int             i, j, kin = 0, m1;
    tNumber_R       d, e = 0, pivot = 0;

    m1 = m + 1;
    if (jin <= n)
    {
        pivot = ct[iout][jin];
    }
    else if (jin > n)
    {
        kin = jin - n;
        e = -1.0;
        if (ibound[jin] == -1) e = 1.0;
        pivot = e * binv[iout][kin];
    }

    for (j = 1; j <= n; j++)
    {
        ct[iout][j] = ct[iout][j]/pivot;
    }
    for (j = 1; j <= m1; j++)
    {
        binv[iout][j] = binv[iout][j]/pivot;
    }
    for (i = 1; i <= m1; i++)
    {
        if (i != iout)
        {
            d = ct[i][jin];
            if (jin > n) d = e * (binv[i][kin]);
            for (j = 1; j <= n; j++)
            {
                ct[i][j] = ct[i][j] - d * (ct[iout][j]);
            }
            for (j = 1; j <= m1; j++)
            {
                binv[i][j] = binv[i][j] - d * (binv[iout][j]);
            }
```

```
        }
    }
    for (i = 1; i <= m1; i++)
    {
        bv[i] = binv[i][m1];
    }
    icbas[iout] = jin;
}

/*-----------------------------------------------------------------------
Calculate the results of LA_Linfbv()
-----------------------------------------------------------------------*/
void LA_linfbv_res (int m, int n, tVector_R f, tVector_I icbas,
    tVector_I irbas, tMatrix_R binv, tVector_I ibound,
    tVector_R r, tVector_R a, tNumber_R *pZ)
{
    int             i, j, k, m1;
    tNumber_R       d, e, s;

    m1 = m + 1;
    for (j = 1; j <= m1; j++)
    {
        s = 0.0;
        for (i = 1; i <= m1; i++)
        {
            k = icbas[i];
            e = -1.0;
            if (k <= n) e = f[k];
            s = s + e * (binv[i][j]);
        }
        k = irbas[j];
        a[k] = s;
    }
    for (j = 1; j <= n; j++)
    {
        d = r[j] - *pZ;
        if (ibound[j] == -1) d = -d;
        r[j] = d;
    }
}

/*-----------------------------------------------------------------------
Initializing program data of LA_Linfbv()
-----------------------------------------------------------------------*/
void LA_linfbv_init (int m, int n, tMatrix_R ct, tVector_I icbas,
```

```
    tVector_I irbas, tMatrix_R binv, tVector_I ibound, tVector_R r,
    tVector_R a)
{
    int             i, j, k, m1;

    m1 = m + 1;
    for (j = 1; j <= m; j++)
    {
        a[j] = -1.0;
        k = n + j;
        icbas[j] = k;
        ibound[k] = -1;
        r[k] = 0.0;
    }
    for (j = 1; j <= m1; j++)
    {
        irbas[j] = j;
        for (i = 1; i <= m1; i++)
        {
            binv[i][j] = 0.0;
        }
        binv[j][j] = 1.0;
    }
    for (j = 1; j <= n; j++)
    {
        ct[m1][j] = 1.0;
        ibound[j] = 1;
    }
}

/*-----------------------------------------------------------------
Obtaining a feasible basic solution of LA_Linfbv()
-----------------------------------------------------------------*/
void LA_linfbv_part_2 (int m, int n, tMatrix_R ct, tMatrix_R binv,
    tVector_R bv, tVector_I ibound)
{
    int             i, j, k, m1;

    m1 = m + 1;
    for (i = 1; i <= m; i++)
    {
        if (bv[i] < 0.0)
        {
            k = n + i;
            ibound[k] = -ibound[k];
```

```
            for (j = 1; j <= n; j++)
            {
                ct[i][j] = -ct[i][j];
            }
            for (j = 1; j <= m1; j++)
            {
                binv[i][j] = - binv[i][j];
            }
            bv[i] = - bv[i];
        }
    }
}

/*--------------------------------------------------------------------
Calculating the initial residual vector "r" and norm "z" in
LA_Linfbv()
--------------------------------------------------------------------*/
void LA_linfbv_resid_norm (int m, int n, tMatrix_R ct, tVector_R f,
    tVector_I icbas, tVector_R bv, tVector_R r, tNumber_R *pZ)
{
    int             i, j, m1, icb;
    tNumber_R       s;

    m1 = m + 1;
    for (j = 1;  j <= n; j++)
    {
        icb = 0;
        for (i = 1; i <= m1; i++)
        {
            if (j == icbas[i]) icb = 1;
        }
        if (icb == 0)
        {
            s = - f[j] + f[1] * (ct[m1][j]);
            for (i = 1; i <= m; i++)
            {
                s = s - ct[i][j];
            }
            r[j] = s;
        }
    }
    s = f[1] * (bv[m1]);
    for (i = 1;  i <= m; i++)
    {
        s = s - bv[i];
```

```
    }
    *pZ = s;
}

/*-----------------------------------------------------------------------
Determine the vector that enters the basis in LA_Linfbv()
-----------------------------------------------------------------------*/
void LA_linfbv_vent (int *pIvo, int *pJin, int m, int n,
    tVector_I icbas, tVector_R r, tNumber_R *pZ)
{
    int             i, j, m1, nm, icb;
    tNumber_R       d, e, g, tz;

    g = 1.0/ (EPS*EPS);
    m1 = m + 1;
    nm = n + m;
    for (j = 1; j <= nm; j++)
    {
        icb = 0;
        for (i = 1; i <= m1; i++)
        {
            if (j == icbas[i]) icb = 1;
        }
        if (icb == 0)
        {
            tz = *pZ + *pZ + EPS;
            if (j > n) tz = 2.0 + EPS;
            d = r[j];
            if (d < 0.0)
            {
                e = d;
                if (e < g)
                {
                    g = e;
                    *pJin = j;
                    *pIvo = 1;
                }
            }
            else if (d >= tz)
            {
                e = tz - d;
                if (e < g)
                {
                    g = e;
                    *pJin = j;
```

```
                    *pIvo = -1;
                }
            }
        }
    }
}

/*-----------------------------------------------------------------
Determine the vector that leaves the basis in LA_Linfbv()
-----------------------------------------------------------------*/
void LA_linfbv_vleav (int jin, int *pIout, int *pItest, int m, int n,
    tMatrix_R ct, tMatrix_R binv, tVector_R bv, tVector_I ibound)
{
    int             i, m1, kin;
    tNumber_R       d, e, g, thmax;

    m1 = m + 1;
    thmax = 1.0/ (EPS*EPS);
    if (jin <= n)
    {
        for (i = 1; i <= m1; i++)
        {
            d = ct[i][jin];
            if (d > EPS)
            {
                g  = bv[i]/d;
                if (g <= thmax)
                {
                    thmax = g;
                    *pIout = i;
                    *pItest = 1;
                }
            }
        }
    }
    else if (jin > n)
    {
        e = -1.0;
        if (ibound[jin] == -1) e = 1.0;
        for (i = 1; i <= m1; i++)
        {
            kin = jin - n;
            d = e * (binv[i][kin]);
            if (d > EPS)
            {
```

```
                g = bv[i]/d;
                if (g <= thmax)
                {
                    thmax = g;
                    *pIout = i;
                    *pItest = 1;
                }
            }
        }
    }
}
```

Chapter 13

Restricted Chebyshev Approximation

13.1 Introduction

Consider the overdetermined system of linear equations

$$\mathbf{Ca} = \mathbf{f}$$

$\mathbf{C} = (c_{ij})$ is a given real n by m matrix of rank k, $k \leq m < n$ and $\mathbf{f} = (f_i)$ is a given real n-vector. The Chebyshev solution of this system is the m-vector $\mathbf{a} = (a_i)$ that minimizes the Chebyshev norm z of the residual vector $\mathbf{r}$

$$z = \max|r_i|, \quad i = 1, 2, \ldots, n \tag{13.1.1}$$

where r_i is the ith residual (error) and is given by

$$r_i = f_i - \sum_{j=1}^{m} c_{ij}a_j, \quad i = 1, 2, \ldots, n \tag{13.1.2}$$

Note that r_i here is $-r_i$ in (10.1.3).

In the last three chapters, algorithms are presented for three different kinds of the linear Chebyshev approximations. In Chapter 10, the Chebyshev solution of system $\mathbf{Ca} = \mathbf{f}$, has the Chebyshev norm of vector $\mathbf{r}$ as small as possible. In Chapter 11, the one-sided Chebyshev solution of $\mathbf{Ca} = \mathbf{f}$ requires the additional constraints that the elements of the residual vector $\mathbf{r}$ be either non-positive or non-negative. In Chapter 12, for the bounded Chebyshev approximation, the additional constraints are that the elements of the Chebyshev solution vector $\mathbf{a}$, each be bounded between –1 and 1. In this chapter, an algorithm for yet another kind of the linear Chebyshev approximation, known as the **restricted** Chebyshev approximation, is presented. In this algorithm, the additional constraint is that the left hand side of **Ca** = **f** be bounded between 2 given n-vectors, namely

(13.1.3) $$\mathbf{l} \leq \mathbf{Ca} \leq \mathbf{u}$$

or

(13.1.3a) $$l_i \leq \sum_{i=1}^{m} c_{ij}a_j \leq u_i, \quad i = 1, 2, \ldots, n$$

where $\mathbf{l} = (l_i)$ and $\mathbf{u} = (u_i)$. That is, the elements of **Ca** are restricted between upper and lower ranges.

In deriving existence and characterization theorems to this problem, an extra condition to (13.1.3) is often imposed [17, 18, 22], namely

(13.1.4) $$\mathbf{l} \leq \mathbf{f} \leq \mathbf{u}$$

If this condition is not met, a solution for the problem may not exist. In our work, this condition is not imposed. However, if a solution does not exist, it could be because this condition is not met or because the linear programming problem has an unbounded solution. In either case, the computation is terminated.

Usually the assumption that matrix **C** satisfies the Haar condition is required [12, 18, 23]. This condition is not required in our work. Matrix **C** may even be a rank deficient matrix.

In this chapter, a numerically stable linear programming algorithm for calculating the restricted Chebyshev solution of overdetermined systems of linear equations is described. Minimum computer storage is required and no conditions are imposed on the coefficient matrix **C**.

The algorithm consists of two main parts. In part 1, a simplex algorithm is described. This part is analogous to that of obtaining the Chebyshev solution of system **Ca** = **f** given in Chapter 10. However, for the purpose of numerical stability, part 2 constitutes a triangular decomposition of the basis matrix.

The ordinary Chebyshev solution [1], the one-sided Chebyshev solutions [2] and the Chebyshev approximation by non-negative functions [17] are obtained as special cases in this algorithm. In Section 13.2, we state that this problem may be solved as a special case by algorithms of other constrained Chebyshev approximation problems. We shall comment on these algorithms in Section 13.7.

In Section 13.3, the linear programming formulation of the problem is given, together with necessary lemmas. In Section 13.4, the new algorithm is described and in Section 13.5, the triangular decomposition of the basis matrix is presented. In Section 13.6, arithmetic operations count for the algorithm are presented and in Section 13.7, numerical results and comments are given.

13.1.1 The semi-infinite programming problem

A interesting, related problem to the restricted Chebyshev approximation problem is the **semi-infinite programming** (SIP) problem, in which a given function f(x) is approximated in the Chebyshev norm over a set **X** that contains infinite elements.

The Chebyshev approximation by the function **Ca** is subject to the constraints (13.1.3) together with the additional constraints

$$b_i \le a_i \le c_i, \quad i = 1, 2, \ldots, m$$

where $\mathbf{b} = (b_i)$ and $\mathbf{c} = (c_i)$ are given m-vectors. That is, the SIP is a restricted problem with bounded variables as well.

The SIP problems occur in technical applications such as the study of propagation of water and air pollution [13, 14, 16].

13.1.2 Special cases

The following are special cases of the restricted Chebyshev approximation problem:

(a) If for all i, we take $l_i = -\infty$ and $u_i = \infty$, we get the ordinary Chebyshev solution [1] to system the **Ca** = **f** (Chapter 10).

(b) If for all i, we take $l_i = -\infty$ and $u_i = f_i$, we get the one-sided Chebyshev solution from above [2] (Chapter 11) to **Ca** = **f**. The one-sided Chebyshev solution from below is obtained in a similar way.

(c) If for all i, we take $l_i = 0$ and $u_i = \infty$, we have the problem of calculating the Chebyshev approximation by a **non-negative linear function** [7]. By ∞, we mean a large number, such as $100\times\max|f_i|$, i = 1, 2, …, n, where (f_i) are the elements of vector **f** in the system **Ca** = **f**.

(d) Arbitrary choices of l_i and u_i may also be made. Figure 13-1 shows curve fittings with vertical parabolas for the set of 8 points of Figure 2-1, as calculated by this algorithm.

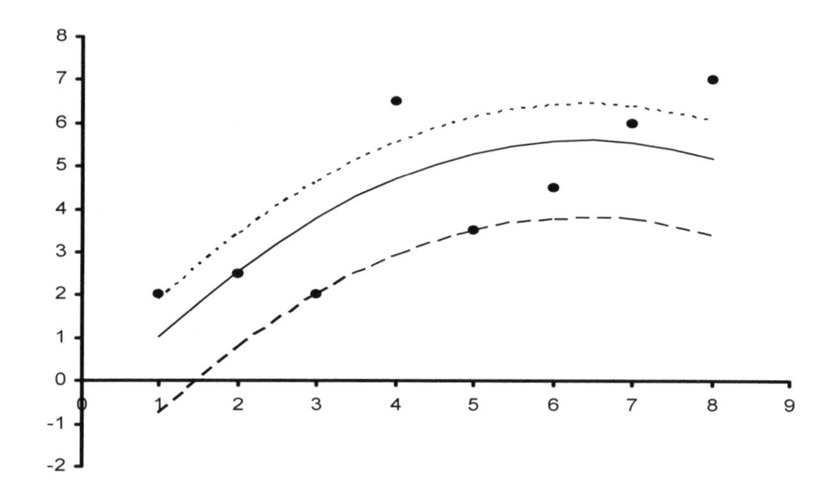

Figure 13-1: Curve fitting a set of 8 points using restricted Chebyshev approximations with arbitrary ranges

The solid curve is the (ordinary) Chebyshev approximation. The dashed curve is the one-sided Chebyshev approximation from above. The dotted curve is the Chebyshev approximation with arbitrary lower and upper ranges; $l_i = f_i - 0.5z_{C.S.}$ and $u_i = a$, very large number, where $z_{C.S.}$ is the optimum Chebyshev norm.

13.1.3 Applications of the restricted Chebyshev algorithm

The restricted Chebyshev approximation problem arises in several engineering applications. See the references in [12, 18]. See also [23, 24].

13.2 A special problem of general constrained algorithms

As in the previous two chapters, some authors developed constrained Chebyshev approximation algorithms that would solve this problem as a special case.

Using our notation, Sklar [21] presented an algorithm to solve the problem

$$\text{(13.2.1a)} \qquad \text{minimize } \|\mathbf{Ca} - \mathbf{f}\|_\infty$$

subject to

$$\text{(13.2.1b)} \qquad \mathbf{e}_a \le \mathbf{Ea} \le \mathbf{e}_b$$

His algorithm is an extension of that of Armstrong and Sklar [6].

Also, Roberts and Barrodale [20] presented the algorithm

$$\text{(13.2.2a)} \qquad \text{minimize } \|\mathbf{Ca} - \mathbf{f}\|_\infty$$

subject to

$$\text{(13.2.2b)} \qquad \mathbf{Ga} = \mathbf{g} \text{ and } \mathbf{e}_a \le \mathbf{Ea} \le \mathbf{e}_b$$

Their algorithm is an extension of the algorithm of Barrodale and Phillips for the ordinary Chebyshev solution [7, 8].

In the above **C**, **G** and **E** are matrices of appropriate dimensions, **f** is the vector associated with **C**, **g** is the vector associated with **G** and $\mathbf{e}_a$ and $\mathbf{e}_b$ are two vectors associated with **E**. Also, not all of the arrays (**G**, **g**) and (**E**, $\mathbf{e}_a$, $\mathbf{e}_b$) are vacuous. If in (13.2.1b) we take $\mathbf{E} = \mathbf{C}$ and $\mathbf{e}_a = \mathbf{l}$ and $\mathbf{e}_b = \mathbf{u}$, we get the restricted Chebyshev approximation problem. Also, if in (13.2.2b) we take $\mathbf{G} = \mathbf{0}$, $\mathbf{g} = \mathbf{0}$, we get problem (13.2.1), which reduces to the restricted Chebyshev approximation problem.

13.3 Linear programming formulation of the problem

This problem is formulated as a linear programming problem as follows. Let in (13.1.1), $h = \max_i |r_i|$, $h \ge 0$. The primal form is [24]

$$\text{minimize } h$$

subject to

$$-h \le f_i - \sum_{j=1}^{m} c_{ij} a_j \le h, \quad i = 1, 2, \ldots, n$$

and

$$l_i \le \sum_{j=1}^{m} c_{ij} a_j \le u_i, \quad i = 1, 2, \ldots, n$$

In vector-matrix notation, the above inequalities become

$$\begin{aligned} \mathbf{Ca} + h\mathbf{e} &\geq \mathbf{f} \\ -\mathbf{Ca} + h\mathbf{e} &\geq -\mathbf{f} \\ \mathbf{Ca} \quad &\geq \mathbf{l} \\ -\mathbf{Ca} \quad &\geq -\mathbf{u} \end{aligned}$$

$h \geq 0$ and a_j, $j = 1, 2, \ldots, m$, unrestricted in sign

This may be written more conveniently as

$$\text{minimize } Z = \mathbf{e}_{m+1}{}^T \begin{bmatrix} \mathbf{a} \\ h \end{bmatrix}$$

subject to

$$\begin{bmatrix} \mathbf{C} & \mathbf{e} \\ -\mathbf{C} & \mathbf{e} \\ \mathbf{C} & \mathbf{0} \\ -\mathbf{C} & \mathbf{0} \end{bmatrix} \begin{bmatrix} \mathbf{a} \\ h \end{bmatrix} \geq \begin{bmatrix} \mathbf{f} \\ -\mathbf{f} \\ \mathbf{l} \\ -\mathbf{u} \end{bmatrix}$$

$h \geq 0$ and a_j, $j = 1, 2, \ldots, m$, unrestricted in sign

where $\mathbf{e}_{m+1}$ is an $(m + 1)$-vector that is the $(m + 1)^{th}$ column of an $(m + 1)$-unit matrix. Also, $\mathbf{e}$ is an n-vector, each element of which equals 1, and the $\mathbf{0}$ is an n-zero vector. The two n-vectors $\mathbf{l} = (l_i)$ and $\mathbf{u} = (u_i)$.

It is more efficient to solve the dual of this problem, which is

(13.3.1a) $\quad$ $\text{maximize } z = [\mathbf{f}^T \ -\mathbf{f}^T \ \mathbf{l}^T \ -\mathbf{u}^T]\, \mathbf{b} = \mathbf{g}^T\mathbf{b}$

subject to

(13.3.1b) $$\begin{bmatrix} \mathbf{C}^T & -\mathbf{C}^T & \mathbf{C}^T & -\mathbf{C}^T \\ \mathbf{e}^T & \mathbf{e}^T & \mathbf{0}^T & \mathbf{0}^T \end{bmatrix} \mathbf{b} = \mathbf{e}_{m+1}$$

(13.3.1c) $\quad$ $b_i \geq 0, \quad i = 1, 2, \ldots, 4n$

Let (13.3.1b) be written in the form

(13.3.1d) $\quad$ $\mathbf{Db} = \mathbf{e}_{m+1}$

D is the (m + 1) by 4n coefficient matrix on the l.h.s. of (13.3.1b), and is the matrix of constraints in this linear programming problem.

To simplify the analysis, assume that rank(**C**) = m. Let for the linear programming problem (13.3.1), the basis matrix **B** be an (m + 1) square nonsingular matrix. In the simplex tableau, vectors ($\mathbf{y}_j$) are calculated, as usual, as

(13.3.2a) $$\mathbf{y}_j = \mathbf{B}^{-1}\mathbf{D}_j, \quad j = 1, 2, \ldots, 4n$$

where $\mathbf{D}_j$ is the jth column of matrix **D**.

The basic solution $\mathbf{b}_B$ is given by

(13.3.2b) $$\mathbf{b}_B = \mathbf{B}^{-1}\mathbf{e}_{m+1}$$

Let the (m + 1)-vector $\mathbf{g}_B$ be the vector whose elements are the prices for $\mathbf{b}_B$. Then for the marginal costs, denoted by $(z_j - g_j)$, we have

(13.3.2c) $$(z_j - g_j) = \mathbf{g}_B{}^T\mathbf{y}_j - g_j, \quad j = 1, 2, \ldots, 4n$$

The objective function z is

(13.3.2d) $$z = \mathbf{g}_B{}^T\mathbf{b}_B$$

Vector **g** is defined in (13.3.1a).

13.3.1 Properties of the matrix of constraints

As we encountered in the previous 3 chapters, we observe asymmetries in the matrix of constraints **D**. The description of the algorithm is prompted by such kind of asymmetries.

Consider the two halves of matrix **D**, in (13.3.1b, d), each is of 2n columns. While matrix $\mathbf{C}^T$ exists in the first n columns of the left half of **D**, matrix $-\mathbf{C}^T$ exists in the second n columns of this half. The first n columns of the right half are the negative of the last n columns of the right half.

From (13.3.1b), the following relationships exist between any column i, $1 \leq i \leq n$ and the columns i + n, i + 2n and i + 3n of matrix **D** in (13.3.1d).

(13.3.3a) $$\mathbf{D}_i + \mathbf{D}_{i+n} = 2\mathbf{e}_{m+1}, \ \mathbf{D}_i - \mathbf{D}_{i+2n} = \mathbf{e}_{m+1}, \ \mathbf{D}_i + \mathbf{D}_{i+3n} = \mathbf{e}_{m+1}$$

and

(13.3.3b) $$\mathbf{D}_{i+n} + \mathbf{D}_{i+2n} = \mathbf{e}_{m+1}, \ \mathbf{D}_{i+n} - \mathbf{D}_{i+3n} = \mathbf{e}_{m+1}, \ \mathbf{D}_{i+2n} + \mathbf{D}_{i+3n} = \mathbf{0}$$

Definition

Let i and j, $1 \le i, j \le 4n$, be two columns in the matrix of constraints **D** such that $|i - j| = n$ or $2n$ or $3n$. We define columns i and j as two corresponding columns.

From Lemmas 13.3, 13.4 and 13.5 below, we find that in the simplex algorithm, we need tableaux of (m + 1) constraints in only n variables (not 4n variables). We call such tableaux the **condensed tableaux**.

Lemma 13.1

From (13.3.2b), $\mathbf{b}_B$ is the $(m + 1)^{th}$ column of $\mathbf{B}^{-1}$ (Lemma 10.1).

Lemma 13.2

The solution vector **a** and objective function z at any stage of the computation are given by Lemma 10.2

$$(\mathbf{a}^T\ z) = \mathbf{g}_B{}^T\mathbf{B}^{-1}$$

Lemma 13.3

For any two corresponding columns, the following relations between the marginal costs are obtained:

(1) For $1 \le i \le n$

$$(13.3.4)\quad \begin{aligned}
(z_i - g_i) + (z_{j+n} - g_{j+n}) &= 2z \\
(z_i - g_i) - (z_{i+2n} - g_{i+2n}) &= z - (f_i - l_i) \\
(z_i - g_i) - (z_{i+3n} - g_{i+3n}) &= z + (u_i - f_i) \\
(z_{i+n} - g_{i+n}) + (z_{i+2n} - g_{i+2n}) &= z + (f_i - l_i) \\
(z_{i+n} - g_{i+n}) - (z_{i+3n} - g_{i+3n}) &= z - (u_i - f_i) \\
(z_{i+2n} - g_{i+2n}) + (z_{i+3n} - g_{i+3n}) &= (u_i - l_i)
\end{aligned}$$

(2) For $1 \le i \le n$

$$(13.3.5)\quad \mathbf{y}_i + \mathbf{y}_{i+n} = 2\mathbf{b}_B,\ \mathbf{y}_i - \mathbf{y}_{i+2n} = \mathbf{b}_B,\ \mathbf{y}_i + \mathbf{y}_{i+3n} = \mathbf{b}_B, \ldots$$

Relations (13.3.4) are established from (13.3.2a-d) since in (13.3.1a), for $1 \le i \le n$, $g_i = f_i$, $g_{i+n} = -f_i$, $g_{i+2n} = l_i$ and $g_{i+3n} = -u_i$. Also, relations (13.3.5) are proved from (13.3.3a, b) and (13.3.2a, b).

Lemma 13.4

Let z be the objective function and $(z_i - g_i)$, i = 1, 2, …, 4n, be the marginal costs for a basic solution for problem (13.3.1). Then the residuals r_i, $1 \leq i \leq n$ of (13.1.2) are given by any one of the relations

$$r_i = z - (z_i - g_i)$$
$$r_i = (z_{i+n} - g_{i+n}) - z$$
$$r_i = (f_i - l_i) - (z_{i+2n} - g_{i+2n})$$
$$r_i = (z_{i+3n} - g_{i+3n}) + (f_i - u_i)$$

The proof of the first relation follows the proof of Lemma 12.5. The proof of the other 3 relations are established from the first 3 relations in (13.3.4).

Lemma 13.5 (Lemma 3 in [3])

Any two corresponding columns in the simplex tableau can not appear together in any basis.

Lemma 13.6 (Lemma 4 in [3])

Let us have a basic feasible solution to problem (13.3.2). Then a non-basic vector i may replace a corresponding basic column j in the basis only if $|i - j| = 2n$.

Lemma 13.7

Not every problem has a restricted Chebyshev solution. As noted earlier, this is due to one of two cases:

(a) An optimal solution can not be reached, which may occur if the condition (13.1.4) is not met, or

(b) The corresponding linear programming problem (13.3.2) would have an unbounded solution.

13.4 Description of the algorithm

As in Chapter 10, the algorithm for solving problem (13.3.1) is in 3 parts. In part 1, an initial basic solution, feasible or not, is obtained. We note that the left half of matrix **D** in (13.3.1d) is the same as matrix **D** in equation (10.2.2b) and the right hand side of (13.3.1d) is itself the right hand side of (10.2.2b). Hence, part 1 here is made identical to part 1 in Chapter 10.

We should remark that in our work, the simplex tableau is not calculated explicitly, as in Chapter 10. This means that required vectors and elements of the tableau are calculated when they are needed. The given data of the problem, namely **C** and **f** of **Ca** = **f** are not destroyed (changed) in the computation, except perhaps for a –ve sign multiplication to some of the equations in $\mathbf{Ca} = \mathbf{f}$.

As indicated, we calculate the (condensed) tableaux of $(m + 1)$ constraints in only n variables (not in 4n variables). A 4n-index indicator vector whose elements are 1, 2, 3 or 4 is needed. If column i, $1 \leq i \leq n$, is in the condensed tableau, index i has the value 1. If a column j, corresponding to column i, $(n + 1) \leq j \leq 2n$, is in the condensed tableau, index j has a value 2, and so on.

In part 2, the basic solution is made feasible and an objective function $z \geq 0$ is calculated, as explained in detail in Chapter 10. However, for part 2 here, we have developed a simple technique [5] by which $\mathbf{B}^{-1}$, obtained at the end of part 1, is modified in one step by a matrix of rank 1. That is, no simplex steps are used in part 2. At the end of part 2, the marginal costs for the n stored columns are calculated.

Part 3 is the ordinary simplex method in which the column to enter the basis is that which has the most negative marginal cost among the non-basic columns. This is done in accordance with Lemma 13.6. For calculating the parameters of the corresponding columns, relations (13.3.4) are used. The results of the problem, **a** and z, are calculated from Lemma 13.2.

In part 3 also, a triangular decomposition method for the basis matrix is used. The details are briefly given in the next section.

13.5 Triangular decomposition of the basis matrix

A variation of the triangular decomposition method of Bartels et al. [10] is described here. Assume without loss of generality that rank($\mathbf{C}$) = m. Let $\mathbf{B}_0$ denote the basis matrix at the end of part 2 and $\mathbf{B}$ denote the basis matrix in part 3. Then we may write

$$\mathbf{B}^{-1} = \mathbf{E}\mathbf{B}_0^{-1} \tag{13.5.1}$$

where $\mathbf{E}$ is a nonsingular $(m + 1)$ square matrix that is the product of Frobenius matrices ([15], p. 48). Let $\mathbf{E}^{-1}$ be decomposed into

$$\mathbf{L}\mathbf{E}^{-1} = \mathbf{P} \tag{13.5.2}$$

where $\mathbf{L}$ is an $(m + 1)$ nonsingular square matrix and $\mathbf{P}$ is an $(m + 1)$ nonsingular upper triangular matrix. This is analogous to the decomposition (1.3.3) in [10]. Here, $\mathbf{P}$ and $\mathbf{B}_0^{-1}$ are stored and updated. This is done as follows. Let, in our algorithm, at the end of part 2, problem (13.3.2) be described by the 3-tuple

$$(13.5.3) \qquad \{\mathbf{I}_0(\mathrm{b}), \mathbf{I}, \mathbf{B}_0^{-1}\}$$

where $\mathbf{I}_0(\mathrm{b})$ represents the basis at the end of part 2, $\mathbf{I}$ is an (m + 1)-unit matrix and $\mathbf{B}_0^{-1}$ is the inverse of the basis matrix at the end of part 2. In each iteration in part 3, the above 3-tuple is updated to

$$(13.5.4) \qquad \{\mathbf{I(b)}, \mathbf{P}, \mathbf{G}^{-1}\}$$

In (13.5.4), $\mathbf{G}^{-1}$ would be given by $\mathbf{G}^{-1} = \mathbf{LB}_0^{-1}$.

The relation between the main parameters in Section 13.3 and the parameters in (13.5.1) are

$$(13.5.5a) \qquad \mathbf{B}^{-1} = \mathbf{P}^{-1}\mathbf{G}^{-1}$$

$$(13.5.5b) \qquad \mathbf{y}_j = \mathbf{P}^{-1}\mathbf{G}^{-1}\mathbf{D}_j, \quad j = 1, 2, \ldots, 4n$$

The updates are given in detail in [3].

13.6 Arithmetic operations count

As noted in Section 5.8, we shall only count the number of multiplications/divisions (m/d) per iteration for the algorithm.

In part 1, matrix $\mathbf{B}^{-1}$ is constructed from the Frobenius matrices $\mathbf{E}_i$, i = 1, 2, …, (m + 1), where $\mathbf{B}^{-1} = E_{m+1} \ldots E_2E_1$. See for example, Hadley ([15], pp. 48-50). For i = 1, we construct column 1 of $\mathbf{E}_1$. This step needs (m + 1) m/d. For i = 2, row 2 of the first tableau is calculated using $\mathbf{E}_1$ and rows 1 and 2 of matrix $\mathbf{D}$ of (13.3.1d), and $\mathbf{E}_2\mathbf{E}_1$. This step needs (n – 1) + (m + 1) + 2(m + 1) = (n – 1) + 3(m + 1) m/d for calculating respectively row 2 of the first tableau (minus the pivot element), the columns of the pivot element and the first 2 columns of $\mathbf{E}_{21}$. The remaining steps of part 1 follow in the same way.

Thus part 1 requires about $0.67(m + 1)^3 + 0.5n(m + 1)^2$ m/d. These operation counts are slightly less than those of part 1 in [1] obtained by applying the Gauss-Jordan elimination steps to the simplex tableau. Part 2 of our algorithm requires (m + 1)(m + 2) m/d.

For part 3, the calculation of vectors $\mathbf{y}_r$ needs about $1.5(m + 1)^2$ m/d; that is, to calculate vectors $\mathbf{x}_r = \mathbf{G}^{-1}\mathbf{D}_r$ and $\mathbf{Py}_r = \mathbf{x}_r$ by backward substitution. The marginal costs for the (n – m – 1) nonbasic columns of matrix **D** are calculated from Lemma 13.4, where the residuals are calculated from (13.1.2). Vector ($\mathbf{a}^T z$) is calculated by calculating vector **w** from $\mathbf{P}^T\mathbf{w} = \mathbf{g}_B$ by forward substitution and ($\mathbf{a}^T z$) = $\mathbf{w}^T\mathbf{G}^{-1}$. These operations need respectively about $0.5(m + 1)^2 + (m + 1)^2 = 1.5(m + 1)^2$ m/d. The m/d counts for the other parameters in part 3 follow easily. Part 3 meeds an average of $n(m+1) + 3.17(m + 1)^2$ m/d per iteration.

The arithmetic operations count/iteration is a linear function of n and m^3 in part 1, and of n and m^2 in part 3. We have reached a similar conclusion about the operations count in the L_1 approximation case (Chapter 5).

The above result compares favorably with those of related algorithms of Bartels et al. [9] and of Cline [11]. Also, numerical experience reported in [9] indicates that direct methods, in general, converge in a far greater number of iterations, compared with iterative methods such as ours.

Among the solved examples in table 1 in [3], 4 of the examples were solved by Bartels et al. [9] for the ordinary Chebyshev case. The reported number of iterations for their methods are very high, compared with the number of iterations of ours. The method of Cline [11] is very similar to that of Bartels et al. [9].

13.7 Numerical results and comments

LA_Restch() implements this algorithm. DR_Restch() has 40 test cases.

Let $\underline{z}$ be a large positive number, say $\underline{z} = 100\times\max|f_i|$. Then for each example, for i = 1, 2, …, n:

(a) $l_i = -\underline{z}$ and $u_i = \underline{z}$, the (ordinary) Chebyshev solution results calculated in Chapter 10 are obtained.

(b) $l_i = f_i - 0.5z_{C.S.}$ and $u_i = \underline{z}$, where $z_{C.S.}$ is the Chebyshev norm calculated from case (a).

(c) $l_i = -\underline{z}$ and $u_i = f_i$, the results for the one-sided Chebyshev solutions from above in Chapter 11 are obtained.

(d) $l_i = c_1$ and $u_i = c_2$, where $c_1 = \min(f_i)$ and $c_2 = \max(f_i)$, i = 1, 2, …, n.

(e) $l_i = 0$ and $u_i = \underline{z}$, (this is the **approximation by non-negative functions**).

Table 13.7.1 shows the results of 8 of the test cases, computed in single-precision.

Table 13.7.1

		Chebyshev Solution (a)		(b)		One-sided from above (c)		(d)	
Ex	**C**(n×m)	Iter	z	Iter	z	Iter	z	Iter	z
1	4 × 2	5	10.0	no solution		no solution		8	11.04
2	5 × 3	6	2.25	10	3.375	7	4.5	6	2.25
3	6 × 3	5	1.556	no solution		5	1.556	no solution	
4	7 × 3	7	2.625	10	3.938	8	5.25	8	3.0
5	8 × 4	7	3.786	10	5.679	8	7.571	7	3.7857
6	10× 5	9	1.778	13	4.889	13	3.0	9	1.778
7	25×10	21	0.0071	28	0.0106	28	0.0142	21	0.008
8	8 × 3	7	1.797	9	2.695	12	3.594	8	1.853

For each example, the optimum norms and the number of iterations for cases (a) to (d) are shown. In this table, "no solution" indicates that no feasible solution was obtained for these cases.

Cases (a), (b) and (c) of the results of example 8 are displayed in Figure 13-1. For all the examples, the number of iterations needed for the solution is comparable to the number of iterations required for obtaining the Chebyshev solution for the same example.

It is observed that for examples 2, 4, 7 and 8, the norms z for case (b) are $1.5z_{C.S.}$, and the norms z for case (c) are $2z_{C.S.}$ These relationships with $z_{C.S}$ were proved by Phillips [19] for certain approximations, including the one-sided Chebyshev approximations. See section 11.5.

As in Chapters 10, 11 and 12, our algorithm does not need any artificial variables and no conditions are imposed on the coefficient matrix **C**. Rank(**C**) is determined in part 1 of the algorithm and if rank(**C**) = k < m, the amount of calculation is considerably reduced.

The triangular decomposition method adopted here makes use of the numerically stable calculation of parts 1 and 2.

As indicated at the end of Section 13.6, the arithmetic operation counts for this algorithm [3, 4] compare favorably with related algorithms, such as those of Bartels et al. [9] and Cline [11].

Iterative refinement of the solutions of the two triangular systems occurring in the calculation in part 3 may be implemented if the problem is very badly conditioned. Again, the given data of the problem, namely **C** and **f** of **Ca** = **f**, are not changed, except perhaps for a –ve sign multiplication applied to some of the equations in **Ca** = **f**. Hence, iterative refinement to the final solution of the problem may be used.

Finally, our algorithm is a general one; that is, it calculates the restricted Chebyshev solution, including the ordinary and the one-sided Chebyshev solutions, as well as the approximation by non-negative functions, as special cases.

We now comment on the algorithms of Sklar [21] and Roberts and Barrodale [20] discussed in Section 13.2. As noted at the end of Chapter 11, in these algorithms one would store matrix **C** twice; once for (13.2.5a) and once for (13.2.5b), or once for (13.2.6a) and once for (13.2.6b), where **E** is replaced by **C**. Hence, their algorithms when compared with ours, need more computer storage and require nearly double the number of arithmetic operations. We claim that for this problem, a special purpose program such as ours would be more efficient than a general purpose one such as [21] and [20].

References

1. Abdelmalek, N.N., Chebyshev solution of overdetermined systems of linear equations, *BIT*, 15(1975)117-129.
2. Abdelmalek, N.N., The discrete linear one-sided Chebyshev approximation, *Journal of Institute of Mathematics and Applications*, 18(1976)361-370.
3. Abdelmalek, N.N., The discrete linear restricted Chebyshev approximation, *BIT*, 17(1977)249-261.
4. Abdelmalek, N.N., Computer program for the discrete linear restricted Chebyshev approximation, *Journal of Computational and Applied Mathematics*, 7(1981)141-150.

5. Abdelmalek, N.N., An exchange algorithm for the Chebyshev solution of overdetermined systems of linear equations, unpublished work.
6. Armstrong, R.D. and Sklar, M.G., A linear programming algorithm for curve fitting in the L_∞ norm, *Numerical Functional Analysis and Optimization*, 2(1980)187-218.
7. Barrodale, I. and Phillips, C., An improved algorithm for discrete Chebyshev linear approximation, *Proceedings of the Fourth Manitoba Conference on Numerical Mathematics*, Hartnell, B.L. and Williams, H.C. (eds.), Winnipeg, Manitoba, Canada, pp. 177-190, 1975.
8. Barrodale, I. and Phillips, C., Algorithm 495: Solution of an overdetermined system of linear equations in the Chebyshev norm, *ACM Transactions on Mathematical Software*, 1(1975)264-270.
9. Bartels, R.H, Conn, A.R. and Charalambous, C., Minimization techniques for piecewise differentiable functions: The l_∞ solution to overdetermined linear system, The Johns Hopkins University, Baltimore, MD, Technical report no. 247, May 1976.
10. Bartels, R.H, Stoer, J. and Zenger, Ch., A realization of the simplex method based on triangular decomposition, *Handbook for Automatic Computation, Vol. II: Linear Algebra*, Wilkinson, J.H. and Reinsch, C. (eds.), Springer-Verlag, New York, pp. 152-190, 1971.
11. Cline, A.K., A descent method for the uniform solution to overdetermined systems of linear equations, *SIAM Journal on Numerical Analysis*, 13(1976)293-309.
12. Gimlin, D.R., Cavin, III R.K. and Budge, Jr., M.C., A multiple exchange algorithm for calculation of best restricted approximations, *SIAM Journal on Numerical Analysis*, 11(1974)219-231.
13. Glashoff, K. and Gustafson, S.A., Numerical treatment of a parabolic boundary-value control problem, *Journal of Optimization Theory and Applications*, 19(1976)645-663.
14. Gustafson, S.A. and Kortanek, K.O., Numerical treatment of a class of semi-infinite programming problems, *Naval Research Logistics Quarterly*, 20(1973)477-504.

15. Hadley, G., *Linear Programming*, Addison-Wesley, Reading, MA, 1962.
16. Hettich, R., A Newton method for nonlinear Chebyshev approximation, *Lecture Notes in Mathematics No. 556*, Dold, A. and Eckmann, B. (eds.), Springer-Verlag, Berlin, pp. 222-236 (1976).
17. Jones, R.C. and Karlovitz, L.A., Iterative construction of constrained Chebyshev approximation of continuous functions, *SIAM Journal on Numerical Analysis*, 5(1968)574-585.
18. Lewis, J.T., Restricted range approximation and its application to digital filter design, *Mathematics of Computation*, 29(1975)522-539.
19. Phillips, D.L., A note on best one-sided approximations, *Communications of ACM*, 14(1971)598-600.
20. Roberts, F.D.K. and Barrodale, I., An algorithm for discrete Chebyshev linear approximation with linear constraints, *International Journal for Numerical Methods in Engineering*, 15(1980)797-807.
21. Sklar, M.G., L_∞ norm estimation with linear restriction on the parameters, *Numerical Functional Analysis and Optimization*, 3(1981)53-68.
22. Taylor, G.D., Approximation by functions having restricted ranges III, *Journal of Mathematical Analysis and Applications*, 27(1969)241-248.
23. Taylor, G.D. and Winter, M.J., Calculation of best restricted approximations, *SIAM Journal on Numerical Analysis*, 7(1970)248-255.
24. Watson, G.A., The calculation of best restricted approximations, *SIAM Journal on Numerical Analysis*, 11(1974)693-699.

13.8 DR_Restch

```
/*-----------------------------------------------------------------
DR_Restch
-------------------------------------------------------------------
This program is a driver for the function LA_Restch(), which
calculates the restricted Chebyshev solution of an overdetermined
systems of linear equations.  It uses a modified simplex method
with a triangular factorization of the basis matrix.

The overdetermined system has the form

                        c*a = f

"c" is a given real n by m matrix of rank k, k <= m < n.
"f" is a given real n vector.

The problem is to calculate the elements of the m vector "a" that
gives the minimum Chebyshev norm z.

            z=max|r[i]|,   i = 1, 2, ..., n

where r[i] is the ith residual and is given by
    r[i] = f[i] - (c[i][1]*a[1] + c[i][2]*a[2] + ... + c[i][m]*a[m])

subject to the conditions

    el[i] <= c[i][1]*a[1] + ... + c[i][m]*a[m] <= ue[i]
                                  i = 1, 2, ...,n

where el[i] and ue[i] are given

Also the following conditions should be satisfied

              el[i]<= f[i] <= ue[i],  i=1,...,n

This driver contains the 8 examples whose results are given in the
text.  Let "big" denote a large real number, say 1000 and "z" be
the Chebyshev solution of the given problem.

Then all the examples are solved for the 4 cases:

1)  el[i] = - big and ue[i] = big, i = 1, 2, ..., n
    The output will be the ordinary Chebyshev solution.
```

```
2)  el[i] = f[i] - 0.5*z and ue[i] = big, i = 1, 2, ..., n
    The output is the solution for arbitrary lower and upper ranges.

3)  el[i] = -big and ue[i] = f[i], i = 1, 2, ..., n
    The output is the one-sided Chebyshev solution from above.

4)  el[i] = c1 and ue[i] = c2
    where c1 is min{f[i]} and c2 = max{f[i]}; i = 1, 2, ..., n.
--------------------------------------------------------------------*/

#include "DR_Defs.h"
#include "LA_Prototypes.h"

#define N1 4
#define M1 2
#define N2 5
#define M2 3
#define N3 6
#define M3 3
#define N4 7
#define M4 3
#define N5 8
#define M5 4
#define N6 10
#define M6 5
#define N7 25
#define M7 10
#define N8 8
#define M8 4

void DR_Restch (void)
{
    /*--------------------------------------
      Constant matrices/vectors
    --------------------------------------*/
    static tNumber_R c1init[N1][M1] =
    {
        { 0.0, -2.0 },
        { 0.0, -4.0 },
        { 1.0, 10.0 },
        {-1.0, 15.0 }
    };

    static tNumber_R c2init[N2][M2] =
```

```
{
    { 1.0,  2.0, 0.0 },
    {-1.0, -1.0, 0.0 },
    { 1.0,  3.0, 0.0 },
    { 0.0,  1.0, 0.0 },
    { 0.0,  0.0, 1.0 }
};

static tNumber_R c3init[N3][M3] =
{
    { 0.0, -1.0,  0.0 },
    { 1.0,  3.0, -4.0 },
    { 1.0,  0.0,  0.0 },
    { 0.0,  0.0,  1.0 },
    {-1.0,  1.0,  2.0 },
    { 1.0,  1.0,  1.0 }
};

static tNumber_R c4init[N4][M4] =
{
    { 1.0, 0.0,  1.0 },
    { 1.0, 2.0,  2.0 },
    { 1.0, 2.0,  0.0 },
    { 1.0, 1.0,  0.0 },
    { 1.0, 0.0, -1.0 },
    { 1.0, 0.0,  0.0 },
    { 1.0, 1.0,  1.0 }
};

static tNumber_R c5init[N5][M5] =
{
    { 1.0, -3.0,  9.0, -27.0 },
    { 1.0, -2.0,  4.0,  -8.0 },
    { 1.0, -1.0,  1.0,  -1.0 },
    { 1.0,  0.0,  0.0,   0.0 },
    { 1.0,  1.0,  1.0,   1.0 },
    { 1.0,  2.0,  4.0,   8.0 },
    { 1.0,  3.0,  9.0,  27.0 },
    { 1.0,  4.0, 16.0,  64.0 }
};

static tNumber_R c6init[N6][M6] =
{
    { 1.0, 0.0,  0.0,  0.0,  0.0 },
    { 0.0, 1.0,  0.0,  0.0,  0.0 },
```

```
    { 0.0, 0.0,  1.0,  0.0,  0.0 },
    { 0.0, 0.0,  0.0,  1.0,  0.0 },
    { 0.0, 0.0,  0.0,  0.0,  1.0 },
    { 1.0, 1.0,  1.0,  1.0,  1.0 },
    { 0.0, 1.0,  1.0,  1.0,  1.0 },
    {-1.0, 0.0, -1.0, -1.0, -1.0 },
    { 1.0, 1.0,  0.0,  1.0,  1.0 },
    { 1.0, 1.0,  1.0,  0.0,  1.0 }
};

static tNumber_R c8init[N8][M8] =
{
    { 1.0, 1.0,   1.0,  1.0 },
    { 1.0, 2.0,   4.0,  4.0 },
    { 1.0, 3.0,   9.0,  9.0 },
    { 1.0, 4.0,  16.0, 16.0 },
    { 1.0, 5.0,  25.0, 25.0 },
    { 1.0, 6.0,  36.0, 36.0 },
    { 1.0, 7.0,  49.0, 49.0 },
    { 1.0, 8.0,  64.0, 64.0 }
};

static tNumber_R f1[N1+1] =
{   NIL,
    -12.0, 6.0, 0.0, 5.0
};

static tNumber_R f2[N2+1] =
{   NIL,
    1.0, 2.0, 1.0, -3.0, 0.0
};

static tNumber_R f3[N3+1] =
{   NIL,
    1.0, 2.0, 3.0, 2.0, 2.0, 4.0
};

static tNumber_R f4[N4+1] =
{   NIL,
    0.0, -2.0, 1.0, -1.0, 5.0, 7.0, 0.0
};

static tNumber_R f5[N5+1] =
{   NIL,
    3.0, -3.0, -2.0, 0.0, 7.0, -1.0, 5.0, 2.0
```

```
};

static tNumber_R f6[N6+1] =
{   NIL,
    1.0, -1.0, 0.0, -1.0, 1.0, 0.0, 2.0, 3.0, -3.0, -2.0
};

static tNumber_R f7[N7+1] =
{   NIL,
    0.0872673, 0.0872794, 0.0873029, 0.0873315, 0.0873576,
    0.3491184, 0.3498802, 0.3513824, 0.3532572, 0.3550109,
    0.6111334, 0.6150641, 0.6230824, 0.6336395, 0.6441493,
    0.8733883, 0.8841621, 0.9071868, 0.9400757, 0.9766021,
    1.135895,  1.157550,  1.206257,  1.283258,  1.384432
};

static tNumber_R f8[N8+1] =
{   NIL,
    2.0, 2.5, 2.0, 6.5, 3.5, 4.5, 6.0, 7.0
};

/* The elements of vector zz are the optimum Chebyshev norms for
   the 8 examples */
static tNumber_R zz[9] =
{   NIL,
    10.0, 2.25, 1.5556, 2.625, 3.786, 1.7778, 0.0071, 1.9
};

/*-----------------------------------------
  Variable matrices/vectors
-----------------------------------------*/
tMatrix_R   ct      = alloc_Matrix_R (MMc_COLS, NN_ROWS);
tVector_R   f       = alloc_Vector_R (NN_ROWS);
tVector_R   r       = alloc_Vector_R (NN_ROWS);
tVector_R   a       = alloc_Vector_R (MMc_COLS);
tVector_R   el      = alloc_Vector_R (NN_ROWS);
tVector_R   ue      = alloc_Vector_R (NN_ROWS);
tMatrix_R   c7      = alloc_Matrix_R (N7, M7);

tMatrix_R   c1      = init_Matrix_R (&(c1init[0][0]), N1, M1);
tMatrix_R   c2      = init_Matrix_R (&(c2init[0][0]), N2, M2);
tMatrix_R   c3      = init_Matrix_R (&(c3init[0][0]), N3, M3);
tMatrix_R   c4      = init_Matrix_R (&(c4init[0][0]), N4, M4);
tMatrix_R   c5      = init_Matrix_R (&(c5init[0][0]), N5, M5);
tMatrix_R   c6      = init_Matrix_R (&(c6init[0][0]), N6, M6);
```

```
tMatrix_R   c8      = init_Matrix_R (&(c8init[0][0]), N8, M8);

int         irank, iter, kase;
int         i, j, k, m, n, Iexmpl;
tNumber_R   cc1, cc2, d, dd, ddd, e, ee, eee, big, z;

eLaRc       rc = LaRcOk;

big = 1.0/ (EPS * EPS);
for (j = 1; j <= 5; j++)
{
    d = 0.15* (j-3);
    dd = d*d;
    ddd = d*dd;
    for (i = 1; i <= 5; i++)
    {
        e = 0.15* (i-3);
        ee = e*e;
        eee = e*ee;
        k = 5* (j-1) + i;
        c7[k][1] = 1.0;
        c7[k][2] = d;
        c7[k][3] = e;
        c7[k][4] = dd;
        c7[k][5] = ee;
        c7[k][6] = e*d;
        c7[k][7] = ddd;
        c7[k][8] = eee;
        c7[k][9] = dd*e;
        c7[k][10] = ee*d;
    }
}

prn_dr_bnr ("DR_Restch, Restricted Chebyshev Solution of an "
            "Overdetermined System of Linear Equations");

for (kase = 1; kase <= 4; kase++)
{
    for (Iexmpl = 1; Iexmpl <= 8; Iexmpl++)
    {
        switch (Iexmpl)
        {
            case 1:
                n = N1;
                m = M1;
```

```
        z = zz[1];
        for (i = 1; i <= n; i++)
        {
            f[i] = f1[i];
            for (j = 1; j <= m; j++) ct[j][i] = c1[i][j];
        }
        break;

    case 2:
        n = N2;
        m = M2;
        z = zz[2];
        for (i = 1; i <= n; i++)
        {
            f[i] = f2[i];
            for (j = 1; j <= m; j++) ct[j][i] = c2[i][j];
        }
        break;

    case 3:
        n = N3;
        m = M3;
        z = zz[3];
        for (i = 1; i <= n; i++)
        {
            f[i] = f3[i];
            for (j = 1; j <= m; j++) ct[j][i] = c3[i][j];
        }
        break;

    case 4:
        n = N4;
        m = M4;
        z = zz[4];
        for (i = 1; i <= n; i++)
        {
            f[i] = f4[i];
            for (j = 1; j <= m; j++) ct[j][i] = c4[i][j];
        }
        break;

    case 5:
        n = N5;
        m = M5;
        z = zz[5];
```

```
            for (i = 1; i <= n; i++)
            {
                f[i] = f5[i];
                for (j = 1; j <= m; j++) ct[j][i] = c5[i][j];
            }
            break;

        case 6:
            n = N6;
            m = M6;
            z = zz[6];
            for (i = 1; i <= n; i++)
            {
                f[i] = f6[i];
                for (j = 1; j <= m; j++) ct[j][i] = c6[i][j];
            }
            break;

        case 7:
            n = N7;
            m = M7;
            z = zz[7];
            for (i = 1; i <= n; i++)
            {
                f[i] = f7[i];
                for (j = 1; j <= m; j++) ct[j][i] = c7[i][j];
            }
            break;

        case 8:
            n = N8;
            m = M8;
            z = zz[8];
            for (i = 1; i <= n; i++)
            {
                f[i] = f8[i];
                for (j = 1; j <= m; j++) ct[j][i] = c8[i][j];
            }

            break;

        default:
            break;
    }
    if (kase == 1)
```

```
{
    for (i = 1; i <= n; i++)
    {
        el[i] = - big;
        ue[i] = big;
    }
}
if (kase == 2)
{
    for (i = 1; i <= n; i++)
    {
        el[i] = f[i] - z/2.0;
        ue[i] = big;
    }
}
if (kase == 3)
{
    for (i = 1; i <= n; i++)
    {
        el[i] = -big;
        ue[i] = f[i];
    }
}
if (kase == 4)
{
    cc1 = big;
    cc2 = -big;
    for (j = 1; j <= n; j++)
    {
        if (f[j] < cc1) cc1 = f[j];
        if (f[j] > cc2) cc2 = f[j];
    }
    for (i = 1; i <= n; i++)
    {
        el[i] = cc1;
        ue[i] = cc2;
    }
}

prn_algo_bnr ("Restch");

prn_example_delim();
PRN ("Example #%d: Size of matrix \"c\" %d by %d\n",
     Iexmpl, n, m);
prn_example_delim();
```

```
        if (kase == 1)
            PRN ("Ordinary Chebyshev Solution\n");
        else if (kase == 2)
            PRN ("Solution for arbitrary lower and upper "
                 "ranges\n");
        else if (kase == 3)
            PRN ("One-sided Chebyshev solution from above.\n");
        else if (kase == 4)
            PRN ("Solution for arbitrary lower and upper "
                         "ranges\n");
        prn_example_delim();
        PRN ("r.h.s. Vector \"f\"\n");
        prn_Vector_R (f, n);
        PRN ("Transpose of Coefficient Matrix, \"ct\"\n");
        prn_Matrix_R (ct, m, n);

        rc = LA_Restch (m, n, ct, f, el, ue, &irank, &iter, r, a,
                        &z);

        if (rc >= LaRcOk)
        {
            PRN ("\n");
            PRN ("Results of the Restricted Approximation\n");
            PRN ("Restricted solution vector, \"a\"\n");
            prn_Vector_R (a, m);
            PRN ("Restricted residual vector, \"r\"\n");
            prn_Vector_R (r, n);
            PRN ("Restricted Chebyshev norm, \"z\" = %8.4f\n",
                 z);
            PRN ("Rank of matrix \"c\" = %d, No. of "
                  "iterations = %d\n", irank, iter);
        }

        prn_la_rc (rc);
    }
}

free_Matrix_R (ct, MMc_COLS);
free_Vector_R (f);
free_Vector_R (r);
free_Vector_R (a);
free_Vector_R (el);
free_Vector_R (ue);
free_Matrix_R (c7, N7);
```

```
    uninit_Matrix_R (c1);
    uninit_Matrix_R (c2);
    uninit_Matrix_R (c3);
    uninit_Matrix_R (c4);
    uninit_Matrix_R (c5);
    uninit_Matrix_R (c6);
    uninit_Matrix_R (c8);
}
```

13.9 LA_Restch

```
/*--------------------------------------------------------------------
LA_Restch
--------------------------------------------------------------------
This program calculates the restricted Chebyshev solution of an
overdetermined system of linear equation. It uses a modified simplex
method to the linear programming formulation of the problem.

For purpose of numerical stability, this program uses a triangular
decomposition to the basis matrix.

The system of linear equations has the form

                        c*a = f

"c" is a given real n by m matrix of rank <= m < n.
"f" is a given real n vector.

The problem is to calculate the elements of the m vector "a" that
gives the minimum Chebyshev norm z,

                z = max|r[i]|,   i = 1, 2, ..., n

where r[i] is the ith residual and is given by

  r[i] = f[i] - (c[i][1]*a[1] + c[i][2]*a[2] +...+ c[i][m]*a[m]),

subject to the conditions

    el[i] <= c[i][1]*a[1] + ... + c[i][m]*a[m] <= ue[i]

where el[i] and ue[i] are given parameters for i = 1, 2, ..., n.
Also the following conditions should be satisfied.

         el[i] <= f[i] <= ue[i],   i = 1, 2, ..., n

Inputs
m       Number of columns of matrix "c" in the system c*a = f.
n       Number of rows of matrix "c" in the system c*a = f.
ct      A real (m+1) by n matrix. Its first m rows and its n
        columns contain the transpose of matrix "c" of the system
        c*a = f. Its (m+1)th row will be filled with ones by the
        program.
```

```
f       A real n vector containing the r.h.s. of the system c*a = f.
el      A real n vector containing the lower range vector "el".
ue      A real n vector containing the upper range vector "ue".

Local Variables
ginv    A real (m+1) square matrix that is the inverse of the basis
        matrix.
vb      A real (m+1) vector containing the basic solution in the
        linear programming problem.
ic      An integer m vector containing the column indices of matrix
        "ct" that form the columns of the basis matrix.
ir      An integer (m+1) vector containing the indices of the rows
        of matrix "ct".
ib      An integer n vector, each element of which has the value
        1, 2, 3 or 4.
p       An (((m+1)*((m+1)+3))/2) vector whose first
        (((irank+1)*(irank+4))/2)-1 elements contain the
        ((irank+1)*(irank+2))/2 elements of an upper triangular
        matrix + extra "irank" working locations.  See the comments
        in LA_pslvc().

Outputs
irank   The calculated rank of matrix "c".
iter    The number of iterations needed for the problem.
a       A real (m+1) vector whose first m elements are the solution
        vector "a".
r       A real n vector containing the residual vector r = f - c*a.
z       The restricted Chebyshev norm of the residual vector.

Returns one of
        LaRcSolutionUnique
        LaRcSolutionProbNotUnique
        LaRcSolutionDefNotUniqueRD
        LaRcNoFeasibleSolution
        LaRcInconsistentConstraints
        LaRcErrBounds
        LaRcErrNullPtr
        LaRcErrAlloc
-----------------------------------------------------------------------*/

#include "LA_Prototypes.h"

eLaRc LA_Restch (int m, int n, tMatrix_R ct, tVector_R f,
    tVector_R el, tVector_R ue, int *pIrank, int *pIter, tVector_R r,
    tVector_R a, tNumber_R *pZ)
```

```
{
    tMatrix_R   ginv    = alloc_Matrix_R (m + 1, m + 1);
    tVector_R   vb      = alloc_Vector_R (m + 1);
    tVector_R   zc      = alloc_Vector_R (n + 1);
    tVector_R   p       = alloc_Vector_R
                          (((m + 1) * ((m + 1) + 3)) / 2);
    tVector_R   g       = alloc_Vector_R (n + 1);
    tVector_R   u       = alloc_Vector_R (m + 1);
    tVector_R   v       = alloc_Vector_R (m + 1);
    tVector_R   w       = alloc_Vector_R (m + 1);
    tVector_I   ic      = alloc_Vector_I (m + 1);
    tVector_I   ir      = alloc_Vector_I (m + 1);
    tVector_I   ib      = alloc_Vector_I (n + 1);

    int         i = 0, k = 0, kd = 0, kk = 0, ko = 0, m1 = 0;
    int         ijk = 0, itest = 0, iout = 0, ivo = 0, jin = 0,
                nmm = 0, irank1 = 0;

    tNumber_R   ggg = 0.0, piv = 0.0;

    eLaRc       tempRc;

    /* Validation of the data before executing the algorithm */
    eLaRc       rc = LaRcSolutionUnique;
    VALIDATE_BOUNDS ((0 < m) && (m < n));
    VALIDATE_PTRS   (ct && f && el && ue && pIrank && pIter && r && a
                     && pZ);
    VALIDATE_ALLOC  (ginv && vb && zc && p && g && u && v && w && ic
                     && ir && ib);

    /* Initialization & checking some constraints */
    m1 = m + 1;
    nmm = (m1* (m1 + 3))/2;
    *pIrank = m;
    irank1 = *pIrank + 1;
    *pIter = 0;
    *pZ = 0.0;

    /* Initializing program data */
    tempRc = LA_restch_init (m, n, ct, f, el, ue, ir, ib, ic, g,
                             ginv, a);
    if ( tempRc < LaRcOk )
    {
        GOTO_CLEANUP_RC (tempRc);
    }
```

```
/* Part 1 of the algorithm */
iout = 0;

for (kk = 1; kk <= m1; kk++)
{
    iout = iout + 1;
    if (iout <= irank1)
    {
        /* Part 1 of the program */
        LA_restch_part_1 (&piv, iout, &jin, n, ct, ic, ginv);

        /* Detect the rank of matrix "c" */
        LA_restch_detect_rank (&piv, &iout, &jin, n, ct, ic, ir,
                              ib, g, v, ginv, pIrank, pIter);
    }
}

/* Part 2 of the algorithm.
Initial feasible basic solution and a positive norm z */
LA_restch_part_2 (n, ct, ic, ib, g, v, ginv, vb, pIrank);

/* Initializing the triangular matrix */
LA_restch_init_p (m, p, pIrank);

/* Calculating the marginal costs */
LA_restch_marg_costs (m, n, ct, f, el, ue, ic, ir, ib, g, u, v,
                      w, zc, p, ginv, pIrank, r, a, pZ);

/* Part 3 */
for (ijk = 1; ijk <= n*n; ijk++)
{
    irank1 = *pIrank + 1;
    ivo = 0;

    /* Determine the vector that enters the basis */
    LA_restch_vent (&ivo, &jin, &ggg, n, f, el, ue, ic, ib, zc,
                    pIrank, pZ);

    /* Calculate the results */
    if (ivo == 0)
    {
        tempRc = LA_restch_res (m, f, el, ue, ic, ib, v, p, vb,
                                pIrank, r, pZ);
        GOTO_CLEANUP_RC ( tempRc == LaRcOk ? rc : tempRc );
```

```
}

ko = ib[jin];
if (ko != ivo)
{
    kk = ko + ivo - 2;
    if (kk != 2 && kk != 4)
    {
        for (i = 1; i <= irank1; i++)
        {
            ct[i][jin] = -ct[i][jin];
        }
    }
    for (i = 1; i <= m1; i++)
    {
        k = ir[i];
        if (k == m1) ko = i;
    }
    if (ivo <= 2) ct[ko][jin] = 1.0;
    if (ivo > 2) ct[ko][jin] = 0.0;
}

/* Determine the vector that leaves the basis */
itest = 0;
ib[jin] = ivo;
g[jin] = ggg;

LA_pslvc (1, irank1, p, vb, v);

LA_restch_vleav (jin, &iout, &itest, ct, u, v, w, p, ginv,
                pIrank);

/* No feasible solution has been found */
if (itest != 1)
{
    GOTO_CLEANUP_RC (LaRcNoFeasibleSolution);
}

*pIter = *pIter + 1;

if (iout != irank1)
{
    /* Updating of the triangular factorization */
    LA_restch_update_p (iout, ic, p, pIrank);
}
```

```
        ic[irank1] = jin;
        k = irank1;
        kd = irank1;
        for (i = 1; i <= irank1; i++)
        {
            p[k] = u[i];
            k = k + kd;
            kd = kd - 1;
        }

        if (iout != irank1)
        {
            /* Update matrix ginv */
            LA_restch_update_ginv (iout, ct, ir, p, ginv, vb,
                                   pIrank);
        }

        /* Calculate the results; vector "a" and z */
        LA_restch_marg_costs (m, n, ct, f, el, ue, ic, ir, ib, g, u,
                              v, w, zc, p, ginv, pIrank, r, a, pZ);
    }

CLEANUP:

    free_Matrix_R (ginv, m + 1);
    free_Vector_R (vb);
    free_Vector_R (zc);
    free_Vector_R (p);
    free_Vector_R (g);
    free_Vector_R (u);
    free_Vector_R (v);
    free_Vector_R (w);
    free_Vector_I (ic);
    free_Vector_I (ir);
    free_Vector_I (ib);

    return rc;
}

/*-------------------------------------------------------------------
Initializing data for LA_Restch()
-------------------------------------------------------------------*/
eLaRc LA_restch_init (int m, int n, tMatrix_R ct, tVector_R f,
    tVector_R el, tVector_R ue, tVector_I ir, tVector_I ib,
    tVector_I ic, tVector_R g, tMatrix_R ginv, tVector_R a)
```

```
{
    int         i, j, m1;

    m1 = m + 1;
    for (j = 1; j <= n; j++)
    {
        if (el[j] > f[j] || ue[j] < f[j])
            return LaRcInconsistentConstraints;
        g[j] = f[j];
        ib[j] = 1;
        ct[m1][j] = 1.0;
    }

    for (j = 1; j <= m1; j++)
    {
        for (i = 1; i <= m1; i++)
        {
            ginv[i][j] = 0.0;
        }
        ginv[j][j] = 1.0;
        ic[j] = 0;
        ir[j] = j;
        a[j] = 0.0;
    }

    return LaRcOk;
}

/*-------------------------------------------------------------------
Part 1 of for LA_Restch()
-------------------------------------------------------------------*/
void LA_restch_part_1 (tNumber_R *pPiv, int iout, int *pJin, int n,
    tMatrix_R ct, tVector_I ic, tMatrix_R ginv)
{
    int         j, k, icb, ioutm1;
    tNumber_R   d, s;

    *pPiv = 0.0;
    if (iout == 1)
    {
        for (j = 1; j <= n; j++)
        {
            d = ct[iout][j];
            if (d < 0.0) d = -d;
            if (d > *pPiv)
```

```
            {
                *pJin = j;
                *pPiv = d;
            }
        }
    }
    else if (iout > 1)
    {
        ioutm1 = iout - 1;
        for (j = 1; j <= n; j++)
        {
            icb = 0;
            for (k = 1; k <= ioutm1; k++)
            {
                if (j == ic[k]) icb = 1;
            }
            if (icb == 0)
            {
                s = ct[iout][j];
                for (k = 1; k <= ioutm1; k++)
                {
                    s = s + ginv[iout][k] * (ct[k][j]);
                }
                d = s;
                if (d < 0.0) d = -d;
                if (d > *pPiv)
                {
                    *pJin = j;
                    *pPiv = d;
                }
            }
        }
    }
}

/*-----------------------------------------------------------------------
Detect the rank of matrix "c" in LA_Restch()
-----------------------------------------------------------------------*/
void LA_restch_detect_rank (tNumber_R *pPiv, int *pIout, int *pJin,
    int n, tMatrix_R ct, tVector_I ic, tVector_I ir, tVector_I ib,
    tVector_R g, tVector_R v, tMatrix_R ginv, int *pIrank,
    int *pIter)
{
    int         i, j, k, icb, ioutm1, irank1;
    tNumber_R   s, pivot;
```

```
/* Detection of rank deficiency */
irank1 = *pIrank + 1;
if ((*pPiv < EPS) && *pIout != irank1)
{
    for (j = 1; j <= n; j++)
    {
        ct[*pIout][j] = ct[*pIrank][j];
        ct[*pIrank][j] = 1.0;
        ct[irank1][j] = 0.0;
    }
    ioutm1 = *pIout - 1;
    for (j = 1; j <= ioutm1; j++)
    {
        ginv[*pIout][j] = ginv[*pIrank][j];
        ginv[*pIrank][j] = ginv[irank1][j];
        ginv[irank1][j] = 0.0;
    }
    ir[*pIout] = ir[*pIrank];
    ir[*pIrank] = ir[irank1];
    ir[irank1] = 0;
    *pIrank = *pIrank - 1;
    irank1 = *pIrank + 1;
    *pIout = *pIout - 1;
}

if ((*pPiv <= EPS) && *pIout == irank1)
{
    for (j = 1; j <= n; j++)
    {
        icb = 0;
        for (i = 1; i <= *pIrank; i++)
        {
            if (j == ic[i]) icb = 1;
        }
        if (icb == 0)
        {
            *pJin = j;
            *pPiv = 2.0;
            break;
        }
    }
    ib[*pJin] = 2;
    g[*pJin] = -g[*pJin];
    for (i = 1; i <= *pIrank; i++)
```

```
        {
            ct[i][*pJin] = -ct[i][*pJin];
        }
    }
    if (*pPiv > EPS)
    {
        for (i = 1; i <= irank1; i++)
        {
            s = ct[i][*pJin];
            if (*pIout != 1)
            {
                ioutm1 = *pIout -1;
                if (i < *pIout) s = 0.0;
                for (k = 1; k <= ioutm1; k++)
                {
                    s = s + ginv[i][k] * (ct[k][*pJin]);
                }
            }
            v[i] = s;
        }
        pivot = v[*pIout];
        for (j = 1; j <= *pIout; j++)
        {
            ginv[*pIout][j] = ginv[*pIout][j]/pivot;
        }

        for (i = 1; i <= irank1; i++)
        {
            if (i != *pIout)
            {
                for (j = 1; j <= *pIout; j++)
                {
                    ginv[i][j] = ginv[i][j] - v[i]
                                 * (ginv[*pIout][j]);
                }
            }
        }
        ic[*pIout] = *pJin;
        *pIter = *pIter + 1;
    }
}

/*-----------------------------------------------------------------
Part 2 of the algorithm for LA_Restch().
Obtaining a basic feasible solution and a positive norm z.
```

```
-------------------------------------------------------------------*/
void LA_restch_part_2 (int n, tMatrix_R ct, tVector_I ic,
    tVector_I ib, tVector_R g, tVector_R v, tMatrix_R ginv,
    tVector_R vb, int *pIrank)
{
    int         i, j, k, kk, l, ibv, jin, irank1;
    tNumber_R   d, s, z;

    irank1 = *pIrank + 1;
    for (i = 1; i <= irank1; i++)
    {
        v[i] = 0.0;
        vb[i] = ginv[i][irank1];
    }
    ibv = 0;
    d = 0.5;
    for (i = 1; i <= irank1; i++)
    {
        if (vb[i] <= -EPS)
        {
            vb[i] = -vb[i];
            ibv = ibv + 1;
            jin = ic[i];
            g[jin] = -g[jin];
            kk = 2;
            if (ib[jin] == 2) kk = 1;
            ib[jin] = kk;
            for (k = 1; k <= *pIrank; k++)
            {
                ct[k][jin] = -ct[k][jin];
            }
            d = d + vb[i];
            for (j = 1; j <= irank1; j++)
            {
                ginv[i][j] = -ginv[i][j];
                v[j] = v[j] + ginv[i][j];
            }
        }
    }
    if (ibv > 0)
    {
        for (j = 1; j <= irank1; j++)
        {
            v[j] = v[j]/d;
            for (i = 1; i <= irank1; i++)
```

```
            {
                ginv[i][j] = ginv[i][j] - v[j]*vb[i];
            }
        }
        for (j = 1; j <= irank1; j++)
        {
            vb[j] = ginv[j][irank1];
        }
    }
    s = 0.0;
    for (i = 1; i <= irank1; i++)
    {
        k = ic[i];
        s = s + g[k] * (vb[i]);
    }
    z = s;
    if (z <= -EPS)
    {
        for (j = 1; j <= n; j++)
        {
            g[j] = -g[j];
            kk = 2;
            if (ib[j] == 2) kk = 1;
            ib[j] = kk;
            for (i = 1; i <= *pIrank; i++)
            {
                ct[i][j] = -ct[i][j];
            }
            z = -z;
            for (l = 1; l <= *pIrank; l++)
            {
                for (i = 1; i <= irank1; i++)
                {
                    ginv[i][l] = -ginv[i][l];
                }
            }
        }
    }
}

/*-----------------------------------------------------------------
Initializing the triangular matrix for LA_Restch()
-----------------------------------------------------------------*/
void LA_restch_init_p (int m, tVector_R p, int *pIrank)
{
```

```
    int         i, k, kd, m1, nmm, irank1;

    m1 = m + 1;
    nmm = (m1 * (m1 + 3))/2;
    irank1 = *pIrank + 1;
    for (i = 1; i <= nmm; i++)
    {
        p[i] = 0.0;
    }
    k = 1;
    kd = *pIrank + 2;
    for (i = 1; i <= irank1; i++)
    {
        p[k] = 1.0;
        k = k + kd;
        kd = kd - 1;
    }
}

/*-------------------------------------------------------------------
Determine the vector that enters the basis in LA_Restch()
-------------------------------------------------------------------*/
void LA_restch_vent (int *pIvo, int *pJin, tNumber_R *pGgg, int n,
    tVector_R f, tVector_R el, tVector_R ue, tVector_I ic,
    tVector_I ib, tVector_R zc, int *pIrank, tNumber_R *pZ)
{
    int         i, j, kk;
    int         icb, irank1;

    tNumber_R   gg, ga, gb, gc, gd;

    irank1 = *pIrank + 1;

    gg = 1.0/ (EPS*EPS);
    for (j = 1; j <= n; j++)
    {
        ga = 0.0;
        gb = 0.0;
        gc = 0.0;
        gd = 0.0;
        icb = 0;
        kk = ib[j];
        for (i = 1; i <= irank1; i++)
        {
            if (j == ic[i]) icb = 1;
```

```
        }
        if (icb == 0)
        {
            if (kk == 1)
            {
                ga = zc[j];
                gb = *pZ + *pZ - ga;
                gc = ga - *pZ + f[j] - el[j];
                gd = *pZ - ga - f[j] + ue[j];
            }
            else if (kk == 2)
            {
                gb = zc[j];
                ga = *pZ + *pZ - gb;
                gc = *pZ - gb + f[j] - el[j];
                gd = gb - *pZ - f[j] + ue[j];
            }
            else if (kk == 3)
            {
                gc = zc[j];
                ga = *pZ + gc - f[j] + el[j];
                gb = *pZ - gc + f[j] - el[j];
                gd = ue[j] - el[j] - gc;
            }
            else if (kk == 4)
            {
                gd = zc[j];
                ga = *pZ - gd - f[j] + ue[j];
                gb = *pZ+ *pZ - ga;
                gc = ue[j] - el[j] - gd;
            }
            if ((ga <= -EPS) && ga < gg)
            {
                gg = ga;
                *pGgg = f[j];
                *pJin = j;
                *pIvo = 1;
            }
            else if ((gb <= -EPS) && gb < gg)
            {
                gg = gb;
                *pGgg = -f[j];
                *pJin = j;
                *pIvo = 2;
            }
```

```
            else if ((gc <= -EPS) && gc < gg)
            {
                gg = gc;
                *pGgg = el[j];
                *pJin = j;
                *pIvo = 3;
            }
            else if ((gd <= -EPS) && gd < gg)
            {
                gg = gd;
                *pGgg = -ue[j];
                *pJin = j;
                *pIvo = 4;
            }
        }

        else if (icb == 1)
        {
            if (kk == 1)
            {
                gc = -*pZ + f[j] - el[j];
                if ((gc <= -EPS) && gc < gg)
                {
                    gg = gc;
                    *pGgg = el[j];
                    *pJin = j;
                    *pIvo = 3;
                }
            }
            else if (kk == 2)
            {
                gd = -*pZ - f[j] + ue[j];
                if ((gd <= -EPS) && gd < gg)
                {
                    gg = gd;
                    *pGgg = -ue[j];
                    *pJin = j;
                    *pIvo = 4;
                }
            }
            else if (kk == 3)
            {
                ga = *pZ - f[j] + el[j];
                if ((ga <= -EPS) && ga < gg)
                {
```

```
                    gg = ga;
                    *pGgg = f[j];
                    *pJin = j;
                    *pIvo = 1;
                }
            }
            else if (kk == 4)
            {
                gb = *pZ + f[j] - ue[j];
                if ((gb <= -EPS) && gb < gg)
                {
                    gg = gb;
                    *pGgg = -f[j];
                    *pJin = j;
                    *pIvo = 2;
                }
            }
        }
    }
}

/*-----------------------------------------------------------------
Determine the vector that leaves the basis in LA_Restch()
-----------------------------------------------------------------*/
void LA_restch_vleav (int jin, int *pIout, int *pItest,
    tMatrix_R ct, tVector_R u, tVector_R v, tVector_R w,
    tVector_R p, tMatrix_R ginv, int *pIrank)
{
    int         i, k, irank1;
    tNumber_R   d, s, gg, thmax;

    thmax = 1.0/ (EPS*EPS);
    irank1 = *pIrank + 1;
    for (i = 1; i <= irank1; i++)
    {
        s = 0.0;
        for (k = 1; k <= irank1; k++)
        {
            s = s + ginv[i][k] * (ct[k][jin]);
        }
        u[i] = s;
    }

    LA_pslvc (1, irank1, p, u, w);
```

```
    for (i = 1; i <= irank1; i++)
    {
        d = w[i];
        if (d >= EPS)
        {
            gg = v[i]/d;
            if (gg <= thmax)
            {
                thmax = gg;
                *pIout = i;
                *pItest = 1;
            }
        }
    }
}

/*------------------------------------------------------------------
Update vector p in LA_Restch()
------------------------------------------------------------------*/
void LA_restch_update_p (int iout, tVector_I ic, tVector_R p,
    int *pIrank)
{
    int         i, j, k, kp1, kd, irank1;

    irank1 = *pIrank + 1;
    for (j = iout; j <= *pIrank; j++)
    {
        k = j;
        kp1 = k + 1;
        kd = irank1;

        /* Swap two elements of vector "ic" */
        swap_elems_Vector_I (ic, k, kp1);
        for (i = 1; i <= kp1; i++)
        {
            p[k] = p[k+1];
            k = k + kd;
            kd = kd - 1;
        }
    }
}

/*------------------------------------------------------------------
Update matrix ginv in LA_Restch()
------------------------------------------------------------------*/
```

```
void LA_restch_update_ginv (int iout, tMatrix_R ct, tVector_I ir,
    tVector_R p, tMatrix_R ginv, tVector_R vb, int *pIrank)
{
    int         i, j, k, l, kl, kd, kk, irank1;
    int         ii, ip1;

    tNumber_R   d, e;

    irank1 = *pIrank + 1;
    for (i = iout; i <= *pIrank; i++)
    {
        ii = i;
        ip1 = i + 1;
        k = 0;
        kd = *pIrank + 2;
        for (j = 1; j <= ii; j++)
        {
            k = k + kd;
            kd = kd - 1;
        }
        kk = k;
        kl = k - kd;
        l = kl;
        d = p[k];
        if (d < 0.0) d = -d;
        e = p[l];
        if (e < 0.0) e = -e;
        if (d > e)
        {
            for (j = ii; j <= irank1; j++)
            {
                /* Swap two elements of a real vector */
                swap_elems_Vector_R (p, l, k);
                k = k + 1;
                l = l + 1;
            }

            /* Swap two elements of an integer vector */
            swap_elems_Vector_I (ir, i, ip1);

            /* Swap two rows of matrix "ct" */
            swap_rows_Matrix_R (ct, i, ip1);

            /* Swap two rows of matrix "ginv" */
            swap_rows_Matrix_R (ginv, i, ip1);
```

```
            /* Swap two columns of matrix "ginv" */
            for (k = 1; k <= irank1; k++)
            {
                d = ginv[k][i];
                ginv[k][i] = ginv[k][ip1];
                ginv[k][ip1] = d;
            }

            /* Swap two elements of a real vector */
            swap_elems_Vector_R (vb, i, ip1);
        }
        e = p[kk]/p[kl];
        k = kk;
        l = kl;
        for (j = ii; j <= irank1; j++)
        {
            p[k] = p[k] - e * (p[l]);
            k = k + 1;
            l = l + 1;
        }
        for (j = 1; j <= irank1; j++)
        {
            ginv[ip1][j] = ginv[ip1][j] - e * (ginv[i][j]);
        }
        vb[ip1] = vb[ip1] - e * (vb[i]);
    }
}

/*-----------------------------------------------------------------
Calculate the results of LA_Restch()
-----------------------------------------------------------------*/
void LA_restch_marg_costs (int m, int n, tMatrix_R ct, tVector_R f,
    tVector_R el, tVector_R ue, tVector_I ic, tVector_I ir,
    tVector_I ib, tVector_R g, tVector_R u, tVector_R v,
    tVector_R w, tVector_R zc, tVector_R p, tMatrix_R ginv,
    int *pIrank, tVector_R r, tVector_R a, tNumber_R *pZ)
{
    int         i, j, k, kk, m1, icb, irank1;
    tNumber_R   d, s;

    irank1 = *pIrank + 1;
    m1 = m + 1;
    for (j = 1; j <= irank1; j++)
    {
```

```
        k = ic[j];
        u[j] = g[k];
    }
    LA_pslvc (2, irank1, p, u, w);

    for (j = 1; j <= irank1; j++)
    {
        s = 0.0;
        for (k = 1; k <= irank1; k++)
        {
            s = s + w[k] * (ginv[k][j]);
        }
        v[j] = s;
        k = ir[j];
        a[k] = s;
    }
    *pZ = a[m1];
    for (j = 1; j <= n; j++)
    {
        zc[j] = 0.0;
        icb = 0;
        for (i = 1; i <= irank1; i++)
        {
            if (j == ic[i]) icb = 1;
        }
        if (icb == 0)
        {
            kk = ib[j];
            s = f[j] + *pZ;
            if (kk == 2) s = f[j] - *pZ;
            if (kk == 3) s = f[j];
            d = -1.0;
            if (kk == 2 || kk == 4) d = 1.0;
            for (i = 1; i <= irank1; i++)
            {
                s = s + d * (v[i]) * (ct[i][j]);
            }
            r[j] = s;
            if (kk == 1) zc[j] = *pZ - r[j];
            if (kk == 2) zc[j] = r[j] + *pZ;
            if (kk == 3) zc[j] = f[j] - el[j] - r[j];
            if (kk == 4) zc[j] = r[j] + ue[j] - f[j];
        }
    }
}
```

```
/*------------------------------------------------------------------
Calculating the residuals in LA_Restch()
------------------------------------------------------------------*/
eLaRc LA_restch_res (int m, tVector_R f, tVector_R el, tVector_R ue,
    tVector_I ic, tVector_I ib, tVector_R v, tVector_R p,
    tVector_R vb, int *pIrank, tVector_R r, tNumber_R *pZ)
{
    int         i, j, kk, m1, irank1;

    eLaRc       rc = LaRcOk;

    irank1 = *pIrank + 1;
    m1 = m + 1;
    for (i = 1; i <= irank1; i++)
    {
        j = ic[i];
        kk = ib[j];
        if (kk == 1) r[j] = *pZ;
        if (kk == 2) r[j] = -*pZ;
        if (kk == 3) r[j] = f[j] - el[j];
        if (kk == 4) r[j] = f[j] - ue[j];
    }

    if (*pIrank < m)
    {
        rc = LaRcSolutionDefNotUniqueRD;
    }
    else if (*pIrank == m)
    {
        LA_pslvc (1, irank1, p, vb, v);
        for (i = 1; i <= m1; i++)
        {
            if (v[i] < EPS) rc = LaRcSolutionProbNotUnique;
        }
    }

    return rc;
}

/*------------------------------------------------------------------
LA_pslvc
------------------------------------------------------------------
This function solves the square non-singular system of linear
equations.
```

```
                     p*x = b

or the square non-singular system of linear equations

                  p(transpose)*x = b

where "p" is an upper triangular matrix, "b" is the right hand side
vector and "x" is the solution vector.

Inputs
id      An integer specifying the action to be performed.
        If id = 1 the equation "p*x = b" is solved.
        if id= any integer other than 1, the equation
        "p(transpose)*x = b" is solved.
k       The number of equations of the given system.
p       An (((m+1)*((m+1)+3))/2) vector.  Its first (k+1) elements
        contain the first k elements of row 1 of the upper triangular
        matrix + an extra location to the right.  Its next k elements
        contain the (k-1) elements of row 2 of the upper triangular
        matrix + an extra location to the right,..., etc.
b       An (m+1) vector.  Its first "k" elements contain the
        r.h.s. vector "b" of the given system.

Outputs
x       An (m+1) vector whose first k elements contain the solution
        to the given system.
-----------------------------------------------------------------*/
void LA_pslvc (int id, int k, tVector_R p, tVector_R b, tVector_R x)
{
    int         i, j, jj, l, ll, kd, kk;
    int         kkd, km1, kkm1;
    tNumber_R   s;

    /* Solution of the upper triangular system */
    if (id == 1)
    {
        l = (k-1) + (k* (k+1))/2;
        x[k] = b[k]/p[l];
        if (k > 1)
        {
            kd = 3;
            km1 = k - 1;
            for (i = 1; i <= km1; i++)
            {
                j = k - i;
```

```
                l = l - kd;
                kd = kd + 1;
                s = b[j];
                ll = l;
                jj = j;
                for (kk= 1; kk <= i; kk++)
                {
                    ll = ll + 1;
                    jj = jj + 1;
                    s = s - p[ll] * (x[jj]);
                }
                x[j] = s/p[l];
            }
        }
    }

    /* Solution of the lower triangular system */
    else if (id != 1)
    {
        x[1] = b[1]/p[1];
        if (k > 1)
        {
            l = 1;
            kd = k + 1;
            for (i = 2; i <= k; i++)
            {
                l = l + kd;
                kd = kd - 1;
                s = b[i];
                kk = i;
                kkm1 = i - 1;
                kkd = k;
                for (j = 1; j <= kkm1; j++)
                {
                    s = s - p[kk] * (x[j]);
                    kk = kk + kkd;
                    kkd = kkd - 1;
                }
                x[i] = s/p[l];
            }
        }
    }
}
```

Chapter 14

Strict Chebyshev Approximation

14.1 Introduction

Consider the overdetermined system of linear equations

$$\mathbf{Ca} = \mathbf{f}$$

$\mathbf{C} = (c_{ij})$ is a given real n by m matrix of rank k, $k \leq m < n$, $\mathbf{f} = (f_i)$ is a given real n-vector and $\mathbf{a} = (a_j)$ is the m-solution vector. The residual vector for this system is given by

$$\mathbf{r} = \mathbf{Ca} - \mathbf{f}$$

In the last four chapters, algorithms are presented for four kinds of linear Chebyshev approximations. In Chapter 10, the ordinary Chebyshev approximation of system **Ca** = **f** is presented, where the Chebyshev norm of the residual vector **r** is minimum [1]. In Chapter 11, the one-sided Chebyshev approximation of system **Ca** = **f** requires the additional constraints that all the elements of the residual vector **r** be either non-positive or non-negative. In Chapter 12, the bounded Chebyshev approximation is presented, where the additional constraints are that each element of the solution vector **a** is bounded between 1 and –1. In Chapter 13, different additional constraints are that the left hand side of system **Ca** = **f**, i.e., vector **Ca** be bounded between lower and upper rages. The solution vector **a** in any of the aforementioned Chebyshev algorithms, if it exists, may or may not be unique.

In this chapter, we study the uniqueness issue of the ordinary Chebyshev approximation. We consider what is known as the **Strict Chebyshev solution** of overdetermined systems of linear equations. A Strict Chebyshev solution is calculated only when the ordinary

Chebyshev solution is not unique. The Strict Chebyshev solution is always unique. Let us denote the Chebyshev solution of **Ca** = **f** by C.S., and the Strict Chebyshev solution by S.C.S.

Descloux [6] proved the important result that the L_p solution of system **Ca** = **f** converges to the S.C.S. as $p \to \infty$. Rice [10] introduced a certain C.S. defined on a finite set, that he called the S.C.S, which is always unique.

In this chapter, we use linear programming techniques to solve the Strict Chebyshev approximation problem. In Section 14.2, the problem is described as it was presented by Descloux. In Section 14.3, the linear programming analysis of the problem is given. This analysis provides a way to determine, for the majority of cases, all the equations that belong to the so-called **characteristic set**. It also gives an efficient method to calculate the inverse of the matrix needed to calculate the strict Chebyshev solution. In addition, it provides a way to recognize the elements of the solution vector of the ordinary Chebyshev solution that equal the corresponding elements of the strict Chebyshev solution. Necessary lemmas are presented. In Section 14.4, numerical results and comments on other algorithms to solve the same problem are given.

Let **E** denote the set of the n equations **Ca** = **f**. The C.S. to this system is the m-vector $\mathbf{a} = (a_i)$ that minimizes the Chebyshev norm of the residuals

$$z = \max|r_i|, \quad i \in \mathbf{E}$$

where r_i is the ith residual and is given by

$$(14.1.1) \qquad r_i = \sum_{j=1}^{m} c_{ij}a_j - f_i, \quad i \in \mathbf{E}$$

Let us denote the C.S. **a** by $(\mathbf{a})_{C.S.}$ It is known that if **C** satisfies the Haar condition, where every m by m sub-matrix of matrix **C** is nonsingular, the C.S. vector $(\mathbf{a})_{C.S.}$ is unique. Otherwise it may not be unique.

When the C.S. is not unique, there is a certain degree of freedom for some, but not all, of the residuals (14.1.1). For these residuals, the maximum absolute value is minimized over the C.S. and the resulting solution is the S.C.S. This is explained in the next section.

14.2 The problem as presented by Descloux

The following presentation of the problem is due to Descloux [6]. Assume that matrix **C** is of rank m and let the Chebyshev norm to the given system **Ca** = **f** be $z_1 = z$. The solution $(\mathbf{a})_{C.S.}$ and z_1 are obtained from (m + 1) equations of **E** known as the **reference equation set (RS)**. Equation (10.3.2) is a **RS**; let it be given by

$$\sum_{j=1}^{m} c_{ij}a_j + \delta_i z = f_i, \quad i \in \mathbf{RS} \tag{14.2.1}$$

where δ_i is either +1 or –1. It is better to write this equation as

$$\sum_{j=1}^{m} c_{ij}a_j = f_i - \delta_i z \quad, \quad i \in \mathbf{RS} \tag{14.2.1a}$$

Assume that $(\mathbf{a})_{C.S.}$ is not unique and let $\mathbf{W}_1$ be the set of all Chebyshev solutions $(\mathbf{a})_{C.S.}$ to the system **Ca** = **f**. Let $\mathbf{R}_1$ be the collection of all equations $i \in \mathbf{E}$ for which $|r_i| = z_1$. $\mathbf{R}_1$ is denoted by Descloux as the characteristic set of **f** relative to matrix **C**.

$$\sum_{j=1}^{m} c_{ij}a_j = f_i - \delta_i z, \quad i \in \mathbf{R}_1 \tag{14.2.2}$$

System (14.2.2) is of rank s_1, $s_1 \leq m$ and thus $\mathbf{W}_1$ is a subset of the solutions to (14.2.2).

It is required to obtain the C.S. of the system $(\mathbf{E} - \mathbf{R}_1)$ subject to the s_1 constraints (14.2.2). This may be done by eliminating certain s_1 elements of vector **a**, using (14.2.2). Then the C.S. of the obtained reduced system $(\mathbf{E} - \mathbf{R}_1)$ is calculated. To illustrate this point, see Example 14.2 below.

System **Ca** = **f** thus reduces to the system of $(\mathbf{E} - \mathbf{R}_1)$ equations in $(m - s_1)$ unknowns, where $\mathbf{C}^{(2)}$ is of rank $(m - s_1)$

$$\mathbf{C}^{(2)}\mathbf{a}^{(2)} = \mathbf{f}^{(2)} \tag{14.2.3}$$

The procedure is repeated for system (14.2.3). If the C.S. of (14.2.3) is not unique, the characteristic set $\mathbf{R}_2$ of (14.2.3) is obtained, and we eliminate s_2 elements of $\mathbf{a}^{(2)}$ from $(\mathbf{E} - \mathbf{R}_1)$ and obtain the C.S.

of the further reduced system ($\mathbf{E} - \mathbf{R}_1 - \mathbf{R}_2$). This process is repeated if necessary a finite number of times, until the C.S. of the most reduced system is unique.

At the end, an m equations in m unknowns is obtained. It consists of s_1 equations of $\mathbf{R}_1$ + s_2 equations of $\mathbf{R}_2$ + …, namely

$$\mathbf{Da} = \mathbf{d} \tag{14.2.4}$$

Duris and Temple [7] were the first to describe an algorithm for calculating the S.C.S. of $\mathbf{Ca} = \mathbf{f}$, following the presentation of Descloux.

Thiran and Thiry [12] did not follow the presentation of Descloux. They rather used an ascent exchange algorithm for computing the strict Chebyshev solution to overdetermined systems of equations. Branning [4, 5] proposed a direct method for computing the same problem. Comments on the works Thiran and Thiry [12] and of Branning [4, 5] are given in Section 14.4.

We present here an algorithm that is essentially that of Duris and Temple [7]. By using linear programming techniques, one obtains the solution in an efficient manner.

Our computational scheme [2] also differs in several significant respects with the result that normally, the computational effort is reduced considerably. Our method deals also with full rank as well as rank deficient coefficient matrix $\mathbf{C}$.

14.3 Linear programming analysis of the problem

The C.S. of system $\mathbf{Ca} = \mathbf{f}$ is obtained by using the linear programming algorithm described in Chapter 10. Let in the final tableau for the Chebyshev approximation matrix $\mathbf{B}$ be the basis matrix. Let also $\mathbf{b}_B$ be the optimal basic solution and $(z_i - f_i)$, $i = 1, 2, \ldots, 2n$, be the marginal costs for the optimum solution. By examining the final tableau of the linear programming problem, a simple procedure is used by which, for the majority of cases, we determine all the equations that belong to the characteristic set $\mathbf{R}_1$. If this procedure is not followed, many Gauss-Jordan iterations may be needed to obtain system (14.2.3) from system $\mathbf{Ca} = \mathbf{f}$.

From the equation in Lemma 10.2, matrix $\mathbf{B}$ is the transpose of the coefficient matrix on the left hand side of the reference set (14.2.1a) above [9]. Also, the residuals (14.1.1) in the **RS** are given by $r_i = \pm(z_i - f_i - z)$. See for example, the first equation in Lemma 12.5. So that if $(z_i - f_i) = 0$ or 2z, r_i is given by $|r_i| = z$.

From the final tableau of the programming problem for the Chebyshev approximation, we find out whether the C.S. of **Ca** = **f** is unique or not. Then, as indicated earlier, determine the characteristic set $\mathbf{R}_1$.

14.3.1 The characteristic set $\mathbf{R}_1$ and how to obtain it

The following lemma was proved in [2].

Lemma 14.1

If $\mathbf{b}_B$ has no zero components, i.e., $\mathbf{b}_B$ is non-degenerate, the C.S. of **Ca** = **f** is unique and the C.S. = S.C.S.

The proof also follows from the fact that no non-basic column can replace any one of the basic columns. The inverse is not always true. There are problems with degenerate basic solutions but the C.S. solution is unique. See Example 14.1 below.

Assume that C.S. of **Ca** = **f** is not unique. Assume that we have obtained all the optimal basic solutions $\mathbf{b}_B$ of the linear programming problem for system **Ca** = **f**. Let $\mathbf{b}_{B(1)}$, $\mathbf{b}_{B(2)}$, …, be such solutions. We expect that each of these solutions is degenerate, i.e., each has one or more zero elements. We deduce from the definition of $\mathbf{R}_1$ that $\mathbf{R}_1$ consists of the union of the equations in the reference sets (14.2.1) associated with the nonzero elements of the corresponding $\mathbf{b}_{B(i)}$, i = 1, 2, … .

How to obtain the characteristic set $\mathbf{R}_1$

To obtain the characteristic set $\mathbf{R}_1$, that is, to obtain all the optimal basic solutions, a simple procedure is followed, requiring the calculation of some of the optimal solutions without the need to change the simplex tableau.

To start with, the equations in **RS** of (14.2.1) corresponding to the nonzero elements of the $\mathbf{b}_B$ at hand belong to $\mathbf{R}_1$. The non-basic columns are then examined and those that have zero marginal costs or marginal costs of 2z are considered.

Remember that if the marginal cost $(z_i - f_i) = 0$ or 2z, r_i is given by $|r_i| = z$. Let i be one of such columns with zero marginal cost. Check in the usual manner, if column i may enter the basis with positive level. If so, calculate the new optimal solution $\mathbf{b}_B'$ without changing the simplex tableau. Then the equations of the new **RS** (14.2.1) that correspond to the nonzero elements of $\mathbf{b}_B'$, belong to $\mathbf{R}_1$. This of course includes column i itself, since it entered the basis with nonzero (positive) level.

This procedure is repeated for every non-basic column having zero marginal cost or a marginal cost of 2z. The reduced system (14.2.3) is then calculated as described. We call this the **simplified procedure**. An alternative procedure is given by Hadley ([8], pp. 166-168), which requires changing of the tableau.

Our simplified procedure does not, in general, calculate all the optimal solutions, but it is successful in the majority of cases in finding the set $\mathbf{R}_1$. In Table 1 in [2], this procedure worked for all examples, with the exception of one (Example 3a) in which two major iterations, instead of one, were needed to determine $\mathbf{R}_1$.

The following lemma has also been proved in [2].

Lemma 14.2

Assume that the optimal solutions $\mathbf{b}_{B(1)}, \ldots, \mathbf{b}_{B(r)}$ have q zero components in common. We mean that each $\mathbf{b}_{B(i)}$ has its jth component = 0. Then the equations corresponding to the nonzero elements of these solutions have rank $(m - q)$.

The proof follows directly from the fact that if the $\mathbf{b}_{B(i)}$ have q zeros in common, the columns in the simplex tableau associated with the nonzero components of these $\mathbf{b}_{B(i)}$, each has those q zero elements in common.

It follows that if $q = 0$, the C.S. is unique.

Example 14.1

Consider the C.S. of the equations

$$\begin{aligned} a_1 - 15a_2 &= -5 \\ -0.5a_1 + 7.5a_2 &= 17.5 \\ 2a_2 &= 12 \\ -4a_2 &= 6 \end{aligned}$$

Two of the optimal basic solutions, $\mathbf{b}_{B(1)}$ and $\mathbf{b}_{B(2)}$, obtained by the simplified procedure are $(0, 2/3, 1/3)^T$ and $(1/3, 0, 2/3)^T$ and there is no zero component in common between $\mathbf{b}_{B(1)}$ and $\mathbf{b}_{B(2)}$ and thus the

C.S. is unique; $(\mathbf{a})_{C.S.} = (0, 1)^T = (\mathbf{a})_{S.C.S.}$ and z = 10.

We note here that the two basic solutions $\mathbf{b}_{B(1)}$ and $\mathbf{b}_{B(2)}$ correspond respectively to equations (1, 3 and 4) and (1, 3 and 2), and that all 4 equations form $\mathbf{R}_1$ of (14.2.2) which is of rank 2.

14.3.2 Calculating matrix $(\mathbf{D}^T)^{-1}$

Instead of calculating matrix $\mathbf{D}$ of (14.2.4), we rather calculate the inverse of its transpose $(\mathbf{D}^T)^{-1}$. This is done by successively modifying matrix $\mathbf{B}^{-1}$, as we see shortly. In the product form, $\mathbf{B}^{-1}$ may be given by the matrices $\mathbf{E}_i$, i = 1, 2, …, m + 1.

(14.3.1) $$\mathbf{B}^{-1} = \mathbf{E}_{m+1}\mathbf{E}_m\ldots\mathbf{E}_1$$

The $\mathbf{E}_i$ are (m + 1)-square matrices and are given, for example, in ([8], p. 48). We need to calculate the matrices

$$(\mathbf{E}_{m+1}{}^{-1}\mathbf{B}^{-1}), (\mathbf{E}_m{}^{-1}\mathbf{E}_{m+1}{}^{-1}\mathbf{B}^{-1}), \ldots, (\mathbf{E}_{m+1-q}{}^{-1} \ldots \mathbf{E}_{m+1}{}^{-1}\mathbf{B}^{-1})$$

If necessary we exchange the rows of $\mathbf{B}^{-1}$ before calculating the matrix $E_{m+1}{}^{-1}$ and the columns of $(E_{m+1}{}^{-1}\mathbf{B}^{-1})$ before calculating $E_m{}^{-1}$, … . This is to achieve maximum numerical stability.

Let us assume that m = 5 and q = 2. Then the end result of this process is that the modified $\mathbf{B}^{-1}$ will have its right 3 columns become the right three columns of a 6-unit matrix.

It is not difficult to write (see the second equation in Lemma 11.2)

$$(\mathbf{a})_{C.S}{}^T = \mathbf{f}^T[\mathbf{E}_{m+1}{}^{-1}\mathbf{B}^{-1}]_{m\times m}$$

where the subscript m×m denotes the upper left sub-matrix.

If we assume that m = 5 and q = 2, and that $s_1 = 3$ and $s_2 = 2$, then the system $\mathbf{Da} = \mathbf{d}$ consists of s_1 (= 3) equations of $\mathbf{R}_1$ + s_2 (= 2) equations of $\mathbf{R}_2$. The elements of $\mathbf{a}^{(1)}$ are permuted such that $\mathbf{a}^{(2)}$ is obtained by eliminating the first s_1 elements of $\mathbf{a}^{(1)}$.

As in the case of $\mathbf{B}^{-1}$ given in the product form (14.3.1), matrix $(\mathbf{D}^T)^{-1}$ may also be given in the form

(14.3.2) $$(\mathbf{D}^T)^{-1} = (\mathbf{E}_m \ldots) \ldots (\mathbf{E}_{s1+s2} \ldots \mathbf{E}_{s1+1})(\mathbf{E}_{s1} \ldots \mathbf{E}_1)$$

where the $\mathbf{E}_i$ are suitable m by m matrices. We write (14.3.2) in the form

$$(\mathbf{D}^T)^{-1} = E_m \ldots E_{s1+1}[(\mathbf{E}_{s1} \ldots \mathbf{E}_1)]_{m\times m} = E_m \ldots E_{s1+1}[(\mathbf{G}_1)]_{m\times m}$$

from which, for $j \leq s_1$, we write

$$(a_j)_{S.C.S.} = \sum_{i=1}^{m} d_i(\mathbf{D}^T)_{ij}^{-1} = \sum_{i=1}^{s1} f_i[(\mathbf{G}_i)_{ij}]_{m \times m}$$

where f_i are the s_1 elements of (14.2.2). For $j \leq s_2$, a similar equation to the above is obtained, and so on. For more details, see [2].

The following lemma is proved in [2], and provides the important means to identify which elements of $(\mathbf{a})_{S.C.S.}$ equal their corresponding elements of $(\mathbf{a})_{C.S.}$. This is also illustrated by Example 14.2.

Lemma 14.3

Consider the first s_1 columns of matrix $\mathbf{G}_1$. Then if in column $j \leq s_1$, there exists q zero elements in the position of the q zero elements in $\mathbf{b}_B$, then

$$(a_j)_{S.C.S.} = (a_j)_{C.S.}$$

14.3.3 The case of a rank deficient coefficient matrix

For a system $\mathbf{Ca} = \mathbf{f}$, where matrix $\mathbf{C}$ is rank deficient, the columns of matrix $\mathbf{C}$ that are linearly dependent on the other columns are detected while obtaining the C.S. by the algorithm of Chapter 10, and are deleted. The parameters a_j associated with these columns are set equal to 0. In such cases, the calculated S.C.S. would be for the overdetermined system whose coefficient matrix $\mathbf{C}$ consists of the linearly independent columns of the given coefficient matrix.

14.4 Numerical results and comments

The algorithm described above may be illustrated by the following example from Duris and Temple ([7], p. 697).

Example 14.2

Obtain the strict Chebyshev solution of the following seven equations.

$$\begin{aligned}
a_1 \quad\quad + a_3 &= 1 \\
a_2 \quad\quad &= 1 \\
a_1 - a_2 + a_3 &= 1 \\
a_3 &= 3 \\
2a_3 &= 0 \\
a_1 - a_2 - a_3 &= -4 \\
2a_1 - a_2 \quad\quad &= 1
\end{aligned}$$

This example is solved by the linear programming technique of Chapter 10, which is incorporated in [2]. The inverse of the basis matrix and the basic solution in the final simplex tableau are shown for this problem.

$\mathbf{b}_B$	$\mathbf{B}^{-1}$			
0.33	−0.33	0.33	0.33	0.33
0	0	−1	−1	0
0	0	0	−1	0
0.67	0.33	0.67	1.67	0.67

Here, $\mathbf{b}_B$ is degenerate. Columns 5, 2, 6 and 4 in the final tableau form the basis. By examining the simplex tableau, we find that no non-basic column with zero marginal cost or a marginal cost of 2z, can enter the basis with a positive level. Thus $\mathbf{R}_1$ consists of equations 5 and 4, which correspond to the nonzero elements of $\mathbf{b}_B$. It has rank $s_1 = 1$ and the C.S. of $\mathbf{Ca} = \mathbf{f}$ is not unique with $(\mathbf{a})_{C.S.} = (2, 3, 1)^T$ and $z_1 = 2$.

It is observed that there exist two 0's in column 1 of $\mathbf{B}^{-1}$ which correspond to the two 0's of $\mathbf{b}_B$. Column 1 of $\mathbf{B}^{-1}$ correspond to a_3. Hence, according to Lemma 14.3, $(a_3)_{S.C.S.} = (a_3)_{C.S.} = 1$.

Since $s_1 = 1$ and one element of $(\mathbf{a})_{S.C.S.}$ is now known, system (14.2.3), namely $\mathbf{C}^{(2)}\mathbf{a}^{(2)} = \mathbf{f}^{(2)}$ is easily derived, as explained in Section 14.3.4. It is given by

$$\begin{aligned}
a_1 \quad\quad &= 1 - 1 = 0 \\
a_2 &= 1 \\
a_1 - a_2 &= 1 - 1 = 0 \\
a_1 - a_2 &= -4 + 1 = -3 \\
2a_1 - a_2 &= 1
\end{aligned}$$

Vector $\mathbf{b}_B$ of this system has one zero element with columns 7, 6 and 3 forming the basis and $z_2 = 1.5$. However, the non-basic column 2 has a zero marginal cost and can replace column 3 in the basis, with a positive level. The obtained optimal solution $\mathbf{b}_B'$ is not degenerate. Hence, the solution of this reduced system is unique. Matrix $(\mathbf{D}^T)^{-1}$ is then calculated as described at the end of Section 14.3.2, and the S.C.S. is obtained from (14.2.4).

The final result is $(\mathbf{a})_{S.C.S} = (1, 2.5, 1)^T$ and the Chebyshev solution has two systems with $z_1 = 2.0$ and $z_2 = 1.5$.

LA_Strict() implements this algorithm [3]. DR_Strict() tests 5 examples. Examples 2 and 5 each have 3 systems. Examples 1 and 4 each have 2 systems, and example 3 has one system; i.e., it has a unique C.S. Table 14.1 shows the results.

Table 14.1

Example	$\mathbf{C}(n \times m)$	Iterations	# systems	$z_1, z_2, z_3, \ldots$
1	25×10	24	2	0.0071, 0.0064
2	5 × 3	9	3	2.25, 1.5, 0.0
3	7 × 3	7	1	2.625
4	7 × 4	10	2	2.0, 1.5
5	25× 6	26	3	1.0, 0.5, 0.25

For each example, the number of iterations; i.e., the total number of times the simplex tableau is changed, the number of systems and their corresponding z values are given.

Example 1 is the same as example 2 in ([7], p. 698), solved by Duris and Temple. Example 4 is the same as Example 14.2. In Example 4, the 7 by 4 matrix $\mathbf{C}$ is of rank 3 and we get the same final vector $\mathbf{r}$, number of systems and corresponding z values, as those for $\mathbf{Ca} = \mathbf{f}$ when the linearly dependent column in $\mathbf{C}$ is discarded.

In [2] we experimented with 14 examples (Table 1 in [2]). The data and the number of points were taken from Watson ([13], Table 2), who used this data for a different purpose. The results in Table 1 in [2] were compared with the results of Duris and Temple [7]. The number of major iterations is the same as those of ours for all but two of the examples.

As we noted earlier, Thiran and Thiry [12] used an ascent exchange algorithm for computing the strict Chebyshev solution to overdetermined systems of equations. The numbers of major iterations agree with ours for all but two (examples 3a and 4a of Table 1 in [2]), where their numbers are slightly smaller. Their algorithm is adaptive to improvements in order to deal with very ill-conditioned systems ([12], p. 724). Branning [4, 5] proposed a direct method for computing the same problem. However, as reported by Thiran and Thiry [12], Branning's algorithm might lead to cycling, and convergence is not guaranteed.

References

1. Abdelmalek, N.N., Chebyshev solution of overdetermined systems of linear equations, *BIT*, 15(1975)117-129.
2. Abdelmalek, N.N., Computing the strict Chebyshev solution of overdetermined linear equations, *Mathematics of Computation*, 31(1977)974-983.
3. Abdelmalek, N.N., A computer program for the strict Chebyshev solution of overdetermined systems of linear equations, *International Journal for Numerical Methods in Engineering*, 13(1979)1715-1725.
4. Brannigan, M., Theory and computation of best strict constrained Chebyshev approximation of discrete data, *IMA Journal of Numerical Analysis*, 1(1980)169-184.
5. Brannigan, M., The strict Chebyshev solution of over-determined systems of linear equations with rank deficient matrix, *Numerische Mathematik*, 40(1982)307-318.
6. Descloux, J., Approximations in L^P and Chebyshev approximations, *Journal of Society of Industrial and Applied Mathematics*, 11(1963)1017-1026.
7. Duris, C.S. and Temple, M.G., A finite step algorithm for determining the "strict" Chebyshev solution to $\mathbf{Ax} = \mathbf{b}$, *SIAM Journal on Numerical Analysis*, 10(1973)690-699.
8. Hadley, G., *Linear Programming*, Addison-Wesley, Reading, MA, 1962.

9. Osborne, M.R. and Watson, G.A., On the best linear Chebyshev approximation, *Computer Journal*, 10(1967)172-177.
10. Rice, J.R., Tchebycheff approximation in a compact metric space, *Bulletin of American Mathematical Society*, 68(1962)405-410.
11. Rice, J.R., *The Approximation of Functions*, Vol. 2, Addison-Wesley, Reading, MA, 1969.
12. Thiran, J.P. and Thiry, S., Strict Chebyshev approximation for general systems of linear equations, *Numerische Mathematik*, 51(1987)701-725.
13. Watson, G.A., A multiple exchange algorithm for multivariate Chebyshev approximation, *SIAM Journal on Numerical Analysis*, 12(1975)46-52.

14.5 DR_Strict

```
/*-----------------------------------------------------------------
DR_Strict
-------------------------------------------------------------------
This program is a driver for the function LA_Strict(), which
calculates the "Strict" Chebyshev solution of an overdetermined
system of linear equations.  It uses a linear programming method.

The overdetermined system has the form

                        c*a = f

"c" is a given real n by m matrix of rank k, k <= m < n.
"f" is a given real n vector.
"a" is the solution m vector.

This driver contains the 5 examples whose results are given in the
text.  Example 3 is the one solved in detail in the text.
-----------------------------------------------------------------*/

#include "DR_Defs.h"
#include "LA_Prototypes.h"

#define N1s        25
#define M1s        10
#define N2s        5
#define M2s        3
#define N3s        7
#define M3s        3
#define N4s        7
#define M4s        4
#define N5s        25
#define M5s        6

void DR_Strict (void)
{
    /*---------------------------------------
      Constant matrices/vectors
    ---------------------------------------*/
    static tNumber_R c2init[N2s][M2s] =
    {
        { 1.0,  2.0, 0.0 },
        {-1.0, -1.0, 0.0 },
```

```
    { 1.0,  3.0, 0.0 },
    { 0.0,  1.0, 0.0 },
    { 0.0,  0.0, 1.0 }
};

static tNumber_R c3init[N3s][M3s] =
{
    { 1.0, 0.0,  1.0 },
    { 1.0, 2.0,  2.0 },
    { 1.0, 2.0,  0.0 },
    { 1.0, 1.0,  0.0 },
    { 1.0, 0.0, -1.0 },
    { 1.0, 0.0,  0.0 },
    { 1.0, 1.0,  1.0 }
};

static tNumber_R c4init[N4s][M4s] =
{
    { 1.0,  0.0,  1.0, 1.0 },
    { 0.0,  1.0,  0.0, 0.0 },
    { 1.0, -1.0,  1.0, 1.0 },
    { 0.0,  0.0,  1.0, 0.0 },
    { 0.0,  0.0,  2.0, 0.0 },
    { 1.0, -1.0, -1.0, 1.0 },
    { 2.0, -1.0,  0.0, 2.0 }
};

static tNumber_R c5init[N5s][M5s] =
{
    { 1.0, -2.0, -2.0, 4.0,  4.0, 4.0 },
    { 1.0, -1.0, -2.0, 1.0,  2.0, 4.0 },
    { 1.0,  0.0, -2.0, 0.0,  0.0, 4.0 },
    { 1.0,  1.0, -2.0, 1.0, -2.0, 4.0 },
    { 1.0,  2.0, -2.0, 4.0, -4.0, 4.0 },
    { 1.0, -2.0, -1.0, 4.0,  2.0, 1.0 },
    { 1.0, -1.0, -1.0, 1.0,  1.0, 1.0 },
    { 1.0,  0.0, -1.0, 0.0,  0.0, 1.0 },
    { 1.0,  1.0, -1.0, 1.0, -1.0, 1.0 },
    { 1.0,  2.0, -1.0, 4.0, -2.0, 1.0 },
    { 1.0, -2.0,  0.0, 4.0,  0.0, 0.0 },
    { 1.0, -1.0,  0.0, 1.0,  0.0, 0.0 },
    { 1.0,  0.0,  0.0, 0.0,  0.0, 0.0 },
    { 1.0,  1.0,  0.0, 1.0,  0.0, 0.0 },
    { 1.0,  2.0,  0.0, 4.0,  0.0, 0.0 },
    { 1.0, -2.0,  1.0, 4.0, -2.0, 1.0 },
```

```
    { 1.0, -1.0,  1.0, 1.0, -1.0, 1.0 },
    { 1.0,  0.0,  1.0, 0.0,  0.0, 1.0 },
    { 1.0,  1.0,  1.0, 1.0,  1.0, 1.0 },
    { 1.0,  2.0,  1.0, 4.0,  2.0, 1.0 },
    { 1.0, -2.0,  2.0, 4.0, -4.0, 4.0 },
    { 1.0, -1.0,  2.0, 1.0, -2.0, 4.0 },
    { 1.0,  0.0,  2.0, 0.0,  0.0, 4.0 },
    { 1.0,  1.0,  2.0, 1.0,  2.0, 4.0 },
    { 1.0,  2.0,  2.0, 4.0,  4.0, 4.0 }
};

static tNumber_R f1[N1s+1] =
{   NIL,
    0.0872673, 0.0872794, 0.0873029, 0.0873315, 0.0873576,
    0.3491184, 0.3498802, 0.3513824, 0.3532572, 0.3550109,
    0.6111334, 0.6150641, 0.6230824, 0.6336395, 0.6441493,
    0.8733883, 0.8841621, 0.9071868, 0.9400757, 0.9766021,
    1.135895,  1.157550,  1.206257,  1.283258,  1.384432
};

static tNumber_R f2[N2s+1] =
{   NIL,
    1.0, 2.0, 1.0, -3.0, 0.0
};

static tNumber_R f3[N3s+1] =
{   NIL,
    0.0, -2.0, 1.0, -1.0, 5.0, 7.0, 0.0
};

static tNumber_R f4[N4s+1] =
{   NIL,
    1.0, 1.0, 1.0, 3.0, 0.0, -4.0, 1.0
};

static tNumber_R f5[N5s+1] =
{   NIL,
    -1.0,    0.0,    1.0,    -1.0,    1.0,
    -1.5,   -0.875, -0.25,   -0.25,   0.375,
    -1.5,    0.0,    1.5,     2.25,   2.5,
     1.375,  3.0,    4.625,   6.0,    7.375,
     5.75,   8.0,   10.125,  12.125, 13.875
};

/*----------------------------------------
```

```
  Variable matrices/vectors
  -----------------------------------------*/
tMatrix_R   c       = alloc_Matrix_R (NN_ROWS, MMc_COLS);
tVector_R   f       = alloc_Vector_R (NN_ROWS);
tMatrix_R   cc      = alloc_Matrix_R (NN_ROWS, MMc_COLS);
tVector_R   fc      = alloc_Vector_R (NN_ROWS);
tVector_R   r       = alloc_Vector_R (NN_ROWS);
tVector_R   rch     = alloc_Vector_R (NN_ROWS);
tVector_R   z       = alloc_Vector_R (NN_ROWS);
tVector_R   a       = alloc_Vector_R (MMc_COLS);
tMatrix_R   c1      = alloc_Matrix_R (N1s, M1s);

tMatrix_R   c2      = init_Matrix_R (&(c2init[0][0]), N2s, M2s);
tMatrix_R   c3      = init_Matrix_R (&(c3init[0][0]), N3s, M3s);
tMatrix_R   c4      = init_Matrix_R (&(c4init[0][0]), N4s, M4s);
tMatrix_R   c5      = init_Matrix_R (&(c5init[0][0]), N5s, M5s);

int         iter, irank, ksys;
int         i, j, k, m, n, Iexmpl;
tNumber_R   d, dd, ddd, e, ee, eee;

eLaRc       rc = LaRcOk;

for (j = 1; j <= 5; j++)
{
    d = 0.15* (j-3);
    dd = d*d;
    ddd = d*dd;
    for (i = 1; i <= 5; i++)
    {
        e = 0.15* (i-3);
        ee = e*e;
        eee = e*ee;
        k = 5* (j-1) + i;
        c1[k][1] = 1.0;
        c1[k][2] = d;
        c1[k][3] = e;
        c1[k][4] = dd;
        c1[k][5] = ee;
        c1[k][6] = e*d;
        c1[k][7] = ddd;
        c1[k][8] = eee;
        c1[k][9] = dd*e;
        c1[k][10] = ee*d;
    }
```

```
}

prn_dr_bnr ("DR_Strict, Strict Chebyshev Solution "
            "of Overdetermined Systems");

for (Iexmpl = 1; Iexmpl <= 5; Iexmpl++)
{
    switch (Iexmpl)
    {
        case 1:
            n = N1s;
            m = M1s;
            for (i = 1; i <= n; i++)
            {
                f[i] = f1[i];
                fc[i] = f1[i];
                for (j = 1; j <= m; j++)
                {
                    c[i][j] = c1[i][j];
                    cc[i][j] = c1[i][j];
                }
            }
            break;

        case 2:
            n = N2s;
            m = M2s;
            for (i = 1; i <= n; i++)
            {
                f[i] = f2[i];
                fc[i] = f2[i];
                for (j = 1; j <= m; j++)
                {
                    c[i][j] = c2[i][j];
                    cc[i][j] = c2[i][j];
                }
            }
            break;

        case 3:
            n = N3s;
            m = M3s;
            for (i = 1; i <= n; i++)
            {
                f[i] = f3[i];
```

```
                fc[i] = f3[i];
                for (j = 1; j <= m; j++)
                {
                    c[i][j] = c3[i][j];
                    cc[i][j] = c3[i][j];
                }
            }
            break;

        case 4:
            n = N4s;
            m = M4s;
            for (i = 1; i <= n; i++)
            {
                f[i] = f4[i];
                fc[i] = f4[i];
                for (j = 1; j <= m; j++)
                {
                    c[i][j] = c4[i][j];
                    cc[i][j] = c4[i][j];
                }
            }
            break;

        case 5:
            n = N5s;
            m = M5s;
            for (i = 1; i <= n; i++)
            {
                f[i] = f5[i];
                fc[i] = f5[i];
                for (j = 1; j <= m; j++)
                {
                    c[i][j] = c5[i][j];
                    cc[i][j] = c5[i][j];
                }
            }
            break;
        default:
            break;
    }

    prn_algo_bnr ("Strict");
    prn_example_delim();
    PRN ("Example #%d: Size %d by %d\n", Iexmpl, n, m);
```

```
    prn_example_delim();
    PRN ("Strict Chebyshev Solution of an Overdetermined System "
         "of Linear Euations\n");
    prn_example_delim();
    PRN ("r.h.s. Vector \"f\"\n");
    prn_Vector_R (f, n);
    PRN ("Coefficient Matrix, \"c\"\n");
    prn_Matrix_R (c, n, m);

    rc = LA_Strict (m, n, c, f, &ksys, &irank, &iter, r, a, z);

    if (rc >= LaRcOk)
    {
        PRN ("\n");
        PRN ("Results of the Strict Chebyshev Solution\n");
        PRN ("Number of systems = %d, Iterations = %d\n",
             ksys, iter);
        PRN ("Solution vector, \"a\"\n");
        prn_Vector_R_nDec (a, m, 4);
        PRN ("Residual vector \"r\"\n");
        prn_Vector_R_nDec (r, n, 4);
        PRN ("System norms \"z\"\n");
        prn_Vector_R_nDec (z, ksys, 4);

        /* Here the residual vector is calculated from obtained
           vector "a" and from matrix "c" and vector "f".
           That is for checking the oalculated results.*/
        for (j = 1; j <= n; j++)
        {
            d = -fc[j];;
            for (i = 1; i <= m; i++)
            {
                d = d + a[i] * cc[j][i];
            }
            rch[j] = d;
        }
        PRN ("Residual vector 'rc'\n");
        prn_Vector_R_nDec (rch, n, 4);
    }

    prn_la_rc (rc);
}

free_Matrix_R (c, NN_ROWS);
free_Vector_R (f);
```

```
    free_Matrix_R (cc, NN_ROWS);
    free_Vector_R (fc);
    free_Vector_R (r);
    free_Vector_R (rch);
    free_Vector_R (z);
    free_Vector_R (a);
    free_Matrix_R (c1, N1s);

    uninit_Matrix_R (c2);
    uninit_Matrix_R (c3);
    uninit_Matrix_R (c4);
    uninit_Matrix_R (c5);
}
```

14.6 LA_Strict

```
/*-----------------------------------------------------------------
LA_Strict
-------------------------------------------------------------------
This program calculates the "Strict" Chebyshev solution of an
overdetermined system of linear equations.  It uses a modified
simplex method to the linear programming formulation of the problem.

The system of linear equations has the form

                        c*a = f

"c" is a given real n by m matrix of rank k, k <= m < n.
"f" is a given real n vector.

The strict Chebyshev solution of system c*a = f is the m
vector "a" that minimizes the Chebyshev norm

                   z = max|r[i]|,   i = 1, 2. ..., n

in the strict sense.

r[i] is the ith residual of system c*a = f and is given by

  r[i] = c[i][1]*a[1] + c[i][2]*a[2] + ... + c[i][m]*a[m] - f[i],
                         i = 1, 2, ..., n

Inputs
m       Number of columns of matrix "c" in the system c*a = f.
n       Number of rows of matrix "c" in the system c*a = f.
c       A real n by (m+1) matrix. Its n rows and first m columns
        contain matrix "c" of the system c*a = f.
f       A real n vector containing the r.h.s. of the system c*a = f.

Local Variables
ct      A real (m+1) by n matrix. Its first m rows and its n
        columns contain the transpose of matrix "c" of the system
        c*a = f. Its (m+1)th row will be filled with ones by the
        program.
binv    An (m + 1) square matrix containing the inverse of the basis
        matrix in the linear programming problem.
bv      An (m + 1) vector containing the basic solution in the
        linear programming problem.
```

```
icbas   An (m + 1) vector containing the indices of the columns of
        matrix "ct" that form the columns of the basis matrix.
irbas   An (m + 1) vector containing the row indices of "ct".

Outputs
irank   The calculated rank of matrix "c".
ksys    Number of involved systems.
        ksys = 1 ndicates that one system, namely c*a = f is solved.
        The obtained solution is the ordinary Chebyshev solution and
        it is unique.
        ksys = 2 indicates that 2 systems are involved, and so on.
iter    Total number of iterations, or total number of times the
        simplex tableau is changed by a Gauss-Jordan step.
a       A real (m+1) vector whose first m elements are the solution
        vector "a" of the problem.
r       A real n vector containing the residual vector r = c*a - f.
z       A real n vector whose first "ksys" elements are the
        Chebyshev "norms" of the "ksys" systems involved.

Returns one of
        LaRcSolutionUnique
        LaRcNoFeasibleSolution
        LaRcErrBounds
        LaRcErrNullPtr
        LaRcErrAlloc
-----------------------------------------------------------------*/

#include "LA_Prototypes.h"

eLaRc LA_Strict (int m, int n, tMatrix_R c, tVector_R f, int *pKsys,
   int *pIrank, int *pIter, tVector_R r, tVector_R a, tVector_R z)
{
   tMatrix_R   ct      = alloc_Matrix_R (m + 1, n);
   tMatrix_R   binv    = alloc_Matrix_R (m + 1, m + 1);
   tVector_R   bv      = alloc_Vector_R (m + 1);
   tVector_R   v       = alloc_Vector_R (m + 1);
   tVector_R   fs      = alloc_Vector_R (m + 1);
   tVector_I   ibound  = alloc_Vector_I (n);
   tVector_I   icbas   = alloc_Vector_I (m + 1);
   tVector_I   irbas   = alloc_Vector_I (m + 1);
   tVector_I   ib      = alloc_Vector_I (n);
   tVector_I   iv      = alloc_Vector_I (m + 1);
   tVector_I   ih      = alloc_Vector_I (m + 1);

   int         i = 0, j = 0;
```

```
int         ij = 0, ji = 0, kn = 0, kj = 0, kl = 0, km = 0,
            ld = 0, m1 = 0, ma = 0;
int         ivo = 0, jin = 0, ijk = 0, kln = 0, iter = 0,
            iout = 0, itest = 0;
int         ild1 = 0, klnm = 0;
tNumber_R   d = 0.0, zz = 0.0, piv = 0.0, bignum = 0.0;

/* Validation of data before executing the algorithm */
eLaRc       rc = LaRcSolutionUnique;
VALIDATE_BOUNDS ((0 < m) && (m < n));
VALIDATE_PTRS   (c && f && pKsys && pIrank && pIter && r && a &&
                 z);
VALIDATE_ALLOC  (ct && binv && bv && v && fs && ibound && icbas
                 && irbas && ib && iv && ih);

bignum = 1./ (EPS*EPS);
m1 = m + 1;
ma = m;
kn = n;
kl = 1;
zz = 0.0;
*pIrank = m;
*pIter = 0;
*pKsys = 0;

/* Initializing data (1) */
LA_strict_init_1 (m, n, binv, ib, iv, irbas, v, fs, a, r, z);

/* Loop for "ksys" */
for (ijk = 1; ijk <= n; ijk++)
{
    *pKsys = *pKsys + 1;

    /* Initializing data (2) */
    LA_strict_init_2 (kl, m, n, c, ct, ib, ibound, irbas);
    itest = 1;
    iout = kl - 1;

    LA_strict_part_1 (&kl, kn, &ma, m, n, c, ct, f, binv, bv, v,
                      fs, ib, ih, iv, ibound, icbas, irbas,
                      pIrank, pKsys, pIter, r, a, z, rc);

    /* Part 2 of the Chebyshev algorithm. Obtaining a basic
         feasible solution */
    LA_strict_part_2 (kl, m, n, ct, f, binv, bv, ib, ibound,
```

```
                 icbas, &iter);

/* Part 3 of the algorithm for the Chebyshev solution.
   Calculating the residuals and the Chebyshev norm zz */
LA_strict_part_3 (kl, m, n, c, ct, f, binv, bv, ib, ibound,
                 icbas, &zz);

/* Calculating the optimun Chebyshev solution of the system
   at hand */
for (ij = 1; ij <= n*n; ij++)
{
    ivo = 0;
    jin = 0;
    /* Determine the vector that enters the basis */
    LA_strict_vent (&ivo, &jin, kl, m, n, c, ib, icbas, zz);

    if (ivo == 0)
    {
        break;
    }
    if (ivo == -1)
    {
        for (i = kl; i <= m1; i++)
        {
            ct[i][jin] = bv[i] + bv[i] - ct[i][jin];
        }
        c[jin][m1] = zz + zz - c[jin][m1];
        f[jin] = - f[jin];
        ibound[jin] = - ibound[jin];
    }

    /* Determine the vector that leaves the basis */
    itest = 0;
    LA_strict_vleav (kl, &iout, jin, &itest, m, ct, bv);

    /* No feasible solution */
    if (itest != 1)
    {
        GOTO_CLEANUP_RC (LaRcNoFeasibleSolution);
    }

    /* A Gauss-Jordan elimination step to matrix "ct" */
    LA_strict_gauss_jordn (iout, jin, kl, m, n, ct, binv, bv,
                           ib, icbas);
    *pIter = *pIter + 1;
```

```
        d = c[jin][m1];
        /* Updating the residuals and zz of the system at hand */
        for (j = 1; j <= n; j++)
        {
            if (ib[j] != 0)
            {
                c[j][m1] = c[j][m1] - d * (ct[iout][j]);
            }
        }
        zz = zz - d * (bv[iout]);
    }

    kj = kl - 1;

    ld = 0;  /* Number of zero elements in the optimum basic
                solution vector "bv" */
    for (i = kl; i <= m1; i++)
    {
        ih[i] = 1;
        if (bv[i] < EPS)
        {
            ih[i] = 0;
            ld = ld + 1;
        }
    }

    /* Check for non-uniqueness of the Chebyshev solution */
    if (ld != 0)
    {
        LA_strict_uniquens (&ivo, &iout, kl, &kn, &ld, m, n, c,
                            ct, f, bv, v, ib, ih, ibound, icbas,
                            r, zz);
    }

    /* Eliminating zz */
    LA_strict_eliminat_zz (&iout, kl, m, c, binv, bv, ib, ih,
                           ibound, icbas, pKsys, r, z, zz);
    for (ji = 1; ji <= n; ji++)
    {
        ild1 = 0;

        /* A Gauss-Jordan step to matrix binv */
        LA_strict_gauss_jordn_binv (iout, kl, m, binv, iv);
        if (iout == m1)
```

```
        {
            if (ld != 0)
            {
                kln = m1 - ld;
                klnm = kln - 1;
                kn = kn + kl + ld - m1 - 1;
                km = m;
                LA_strict_permute_binv (kj, kl, &km, m, binv, bv,
                                        ih, icbas);
            }
            LA_strict_map (kl, m, f, fs, ih, icbas, r, zz);

            /* Calculate the solution vector "a" */
            LA_strict_calcul_a(m, fs, binv, iv, icbas, irbas, a);

            if ((ld == 0) && (*pKsys == 1) ||
                (rc == LaRcSolutionProbNotUnique))
            {
                LA_strict_calcul_r_1(m, n, c, ib, ibound, r, zz);
                GOTO_CLEANUP_RC (LaRcSolutionUnique);
            }
            if (ld == 0)
            {
                ild1 = 1;
                break;
            }
        }
        iout = iout - 1;
        if (iout < kln) break;
        piv = 0.0;
        for (j = kl; j <= iout; j++)
        {
            d = binv[iout][j];
            if (d < 0.0) d = -d;
            if (d <= piv) continue;
            piv = d;
            jin = j;
        }
        if (jin == iout) continue;
        for (i = kl; i <= m1; i++)
        {
            d = binv[i][jin];
            binv[i][jin] = binv[i][iout];
            binv[i][iout] = d;
        }
```

```
        /* Swap two elements of an integer vector */
        swap_elems_Vector_I (irbas, jin, iout);
        if (kj == 0) continue;
        for (j = 1; j <= kj; j++)
        {
            if (iv[j] == 0) continue;
            d = binv[jin][j];
            binv[jin][j] = binv[iout][j];
            binv[iout][j] = d;
        }
    }
    if (ild1 == 0)
    {
        for (j = 1; j <= n; j++)
        {
            if (ib[j] == 0) continue;
            if (ibound[j] == 1) continue;
            f[j] = -f[j];
        }

        /* calculating the reduced system */
        if (kl < kln)
        {
            /* Calculating the reduced system */
            LA_strict_reduce_sys (kl, kln, &ma, m, n, c, f, binv,
                                  ib, iv, irbas, a);
        }
    }
    if ((*pKsys != 1) && (kj != 0))
    {
        LA_strict_modify_binv (kl, kj, m, binv, v, iv);
    }
    if (ld == 0)
    {
        rc = LaRcSolutionProbNotUnique;
    }
    if (rc == LaRcSolutionProbNotUnique)
    {
        /* Calculating vector a */
        LA_strict_calcul_a (m, fs, binv, iv, icbas, irbas, a);
        break;
    }
    if (ma != ld)
    {
        if (kl != kln)
```

```
            {
                LA_strict_eliminate_ll (kl, kln, &ma, ld, m, n, c, f,
                                        ib, iv, ibound, icbas, irbas,
                                        r, zz);
            }
        }

        /* Calculating elements of residual vector r, part (c) */
        LA_strict_calcul_r_3 (kl, kln, m, c, ib, ibound, icbas, r,
                              zz);

        LA_strict_calcul_r_2 (&kn, kln, m, n, c, ib, ibound, irbas,
                              r, zz);

        if (kn != 0)
        {
            kl = kln;
        }
    }

    LA_strict_calcul_r_1 (m, n, c, ib, ibound, r, zz);
    rc = LaRcSolutionUnique;

CLEANUP:

    free_Matrix_R (ct, m + 1);
    free_Matrix_R (binv, m + 1);
    free_Vector_R (bv);
    free_Vector_R (v);
    free_Vector_R (fs);
    free_Vector_I (ibound);
    free_Vector_I (icbas);
    free_Vector_I (irbas);
    free_Vector_I (ib);
    free_Vector_I (iv);
    free_Vector_I (ih);

    return rc;
}

/*-------------------------------------------------------------------
Initializing the data in LA_Strict()
-------------------------------------------------------------------*/
void LA_strict_init_1 (int m, int n, tMatrix_R binv, tVector_I ib,
    tVector_I iv, tVector_I irbas, tVector_R v, tVector_R fs,
```

```
    tVector_R a, tVector_R r, tVector_R z)
{
    int         i, j, m1;

    m1 = m + 1;
    for (j = 1; j <= m1; j++)
    {
        for (i = 1; i <= m1; i++)
        {
            binv[i][j] = 0.0;
        }
        binv[j][j] = 1.0;
        iv[j] = 1;
        irbas[j] = j;
        v[j] = 0.0;
        a[j] = 0.0;
        fs[j] = 0.0;
    }
    for (j = 1; j <= n; j++)
    {
        r[j] = 0.0;
        ib[j] = 1;
        z[j] = -1.0;
    }
}

/*-----------------------------------------------------------------
Initializing data (2) in LA_Strict()
-----------------------------------------------------------------*/
void LA_strict_init_2 (int kl, int m, int n, tMatrix_R c,
    tMatrix_R ct, tVector_I ib, tVector_I ibound, tVector_I irbas)
{
    int         i, j, k, m1;

    m1 = m + 1;
    for (j = 1; j <= n; j++)
    {
        if (ib[j] == 0) continue;
        ibound[j] = 1;
        for (i = kl; i <= m; i++)
        {
            k = irbas[i];
            ct[i][j] = c[j][k];
        }
        ct[m1][j] = 1.0;
```

```
    }
}

/*-------------------------------------------------------------------
Part 1 in LA_Strict()
-------------------------------------------------------------------*/
void LA_strict_part_1 (int *pKl, int kn, int *pMa, int m, int n,
    tMatrix_R c, tMatrix_R ct, tVector_R f, tMatrix_R binv,
    tVector_R bv, tVector_R v, tVector_R fs, tVector_I ib,
    tVector_I ih, tVector_I iv, tVector_I ibound,
    tVector_I icbas, tVector_I irbas, int *pIrank,
    int *pKsys, int *pIter, tVector_R r, tVector_R a,
    tVector_R z, eLaRc rc)
{
    int         kj, ld, m1;
    int         iout, jin = 0;
    tNumber_R   zz, piv;

    m1 = m + 1;
    for (iout = *pKl; iout <= m1; iout++)
    {
        if ((kn == *pMa) && iout == m1)
        {
            zz = 0.0;
            z[*pKsys] = 0.0;
            kj = *pKl - 1;
            ld = 0;
            /* Mapping some data */
            LA_strict_map (*pKl, m, f, fs, ih, icbas, r, zz);

            /* Calculating vector a */
            LA_strict_calcul_a (m, fs, binv, iv, icbas, irbas, a);

            if ((ld == 0 && *pKsys == 1) ||
                (rc == LaRcSolutionProbNotUnique))
            {
                /* Calculating the residual vector r, part (c) */
                LA_strict_calcul_r_1 (m, n, c, ib, ibound, r, zz);
                break;
            }

            /* Modifying matrix binv */
            if (ld == 0)
            {
                /* Modifying matrix binv */
```

```
            LA_strict_modify_binv (*pKl, kj, m, binv, v, iv);
        }
    }

    if (iout <= m1)
    {
        piv = 0.0;
        /* Calculate pivot element */
        LA_strict_piv (iout, &jin, n, &piv, ct, ib);

        if (piv > EPS)
        {
            /* A Gauss-Jordan elimination step to matrix "ct" */
            LA_strict_gauss_jordn (iout, jin, *pKl, m, n, ct,
                                   binv, bv, ib, icbas);
            *pIter = *pIter + 1;
        }

        /* Detection of rank deficiency of matrix "c" */
        if (piv < EPS)
        {
            if (iout < m1)
            {
                /* Swapping process */
                LA_strict_swapping (*pKl, iout, m, n, ct, binv,
                                    ib, icbas, irbas);
                *pIrank = *pIrank - 1;
                *pMa = *pMa - 1;
                iv[*pKl] = 0;
                *pKl = *pKl + 1;
            }
            if (iout == m1)
            {
                /* Rank of Coefficient matrix "c" */
                LA_strict_detect_rank (*pKl, &jin, m, n, ct, f,
                                       ib, ibound, icbas);
                /* A Gauss-Jordan elimination step to matrix
                   "ct" */
                LA_strict_gauss_jordn (iout, jin, *pKl, m, n, ct,
                                       binv, bv, ib, icbas);
                *pIter = *pIter + 1;
            }
        }
    }
}
```

```
}

/*-----------------------------------------------------------------
Mapping some data in LA_Strict()
-----------------------------------------------------------------*/
void LA_strict_map (int kl, int m, tVector_R f, tVector_R fs,
   tVector_I ih, tVector_I icbas, tVector_R r, tNumber_R zz)
{
   int         i, k;

   for (i = kl; i <= m; i++)
   {
      k = icbas[i];
      r[k] = 0.0;
   }
   for (i = kl; i <= m; i++)
   {
      k = icbas[i];
      fs[i] = f[k] - zz;
      if (ih[i] == 1) f[k] = fs[i];
   }
}

/*-----------------------------------------------------------------
Calculating the pivot element in LA_Strict()
-----------------------------------------------------------------*/
void LA_strict_piv (int iout, int *pJin, int n, tNumber_R *pPiv,
   tMatrix_R ct, tVector_I ib)
{
   int         j;

   tNumber_R   d;

   for (j = 1; j <= n; j++)
   {
      if (ib[j] == 0) continue;
      d = ct[iout][j];
      if (d < 0.0) d = -d;
      if (d <= *pPiv) continue;
      *pJin = j;
      *pPiv = d;
   }
}

/*-----------------------------------------------------------------
```

```
Swapping processes in LA_Strict()
-----------------------------------------------------------------*/
void LA_strict_swapping (int kl, int iout, int m, int n,
    tMatrix_R ct, tMatrix_R binv, tVector_I ib, tVector_I icbas,
    tVector_I irbas)
{
    int         i, j, m1;

    tNumber_R   d;

    m1 = m + 1;
    irbas[iout] = irbas[kl];
    irbas[kl] = 0;
    icbas[iout] = icbas[kl];
    icbas[kl] = 0;

    for (j = 1; j <= n; j++)
    {
        if (ib[j] == 0) continue;
        d = ct[iout][j];
        ct[iout][j] = ct[kl][j];
        ct[kl][j] = d;
    }
    for (j = kl; j <= m; j++)
    {
        binv[iout][j] = binv[kl][j];
        binv[kl][j] = 0.0;
    }
    for (i = kl; i <= m1; i++)
    {
        binv[i][iout] = binv[i][kl];
        binv[i][kl] = 0.0;
    }
}

/*-----------------------------------------------------------------
Rank of matrix "c" in LA_Strict()
-----------------------------------------------------------------*/
void LA_strict_detect_rank (int kl, int *pJin, int m, int n,
    tMatrix_R ct, tVector_R f, tVector_I ib, tVector_I ibound,
    tVector_I icbas)
{
    int         i, j, m1, ibc;

    m1 = m + 1;
```

```
    for (j = 1; j <= n; j++)
    {
        if (ib[j] == 0) continue;
        ibc = 0;
        for (i = kl; i <= m; i++)
        {
            if (j == icbas[i]) ibc = 1;
        }
        if (ibc == 0)
        {
            *pJin = j;
            break;
        }
    }
    f[*pJin] = -f[*pJin];
    ibound[*pJin] = -ibound[*pJin];
    for (i = kl; i <= m; i++)
    {
        ct[i][*pJin] = -ct[i][*pJin];
    }
    ct[m1][*pJin] = 2.0 - ct[m1][*pJin];
}

/*------------------------------------------------------------------
Part 2 of the Chebyshev algorithm.
Obtaining a basic feasible solution in LA_Strict().
------------------------------------------------------------------*/
void LA_strict_part_2 (int kl, int m, int n, tMatrix_R ct,
    tVector_R f, tMatrix_R binv, tVector_R bv, tVector_I ib,
    tVector_I ibound, tVector_I icbas, int *pIter)
{
    int         i, k, m1, iout, jin;

    m1 = m + 1;
    for (i = kl; i <= m; i++)
    {
        if (bv[i] >= 0.0) continue;
        iout = i;
        jin = icbas[i];
        f[jin] = -f[jin];
        ibound[jin] = -ibound[jin];
        for (k = kl; k <= m1; k++)
        {
            ct[k][jin] = bv[k] + bv[k] - ct[k][jin];
        }
```

```
        /* A Gauss-Jordan elimination step to matrix "ct"
           for Part 2 of the Chebyshev algorithm */
        LA_strict_gauss_jordn (iout, jin, kl, m, n, ct, binv, bv,
            ib, icbas);
        *pIter = *pIter + 1;
    }
}

/*-----------------------------------------------------------------
Part 3 of the Chebyshev algorithm in LA_Strict()
-----------------------------------------------------------------*/
void LA_strict_part_3 (int kl, int m, int n, tMatrix_R c,
    tMatrix_R ct, tVector_R f, tMatrix_R binv, tVector_R bv,
    tVector_I ib, tVector_I ibound, tVector_I icbas, tNumber_R *pZz)
{
    int         i, j, k, m1, icb;
    tNumber_R   s;

    m1 = m + 1;
    for (j = 1; j <= n; j++)
    {
        if (ib[j] == 0) continue;
        icb = 0;
        for (i = kl; i <= m1; i++)
        {
            if (j == icbas[i]) icb = 1;
        }
        if (icb == 0)
        {
            s = -f[j];
            for (i = kl; i <= m1; i++)
            {
                k = icbas[i];
                s = s + ct[i][j] * (f[k]);
            }
            /* The residual of equation j is
            stored in c[j][m1] */
            c[j][m1] = s;
        }
        else if (icb == 1) c[j][m1] = 0.0;
    }
    s = 0.0;
    for (i = kl; i <= m1; i++)
    {
```

```
        k = icbas[i];
        s = s + bv[i] * (f[k]);
    }
    if (fabs (s) < EPS) s = 0.0;
    *pZz = s;
    if (*pZz < 0.0)
    {
        for (j = 1; j <= n; j++)
        {
            if (ib[j] == 0) continue;
            f[j] = -f[j];
            ibound[j] = -ibound[j];
            c[j][m1] = -c[j][m1];
        }
        *pZz = -*pZz;
        for (j = kl; j <= m; j++)
        {
            for (i = kl; i <= m1; i++)
            {
                binv[i][j] = -binv[i][j];
            }
        }
    }
}

/*------------------------------------------------------------------
Determine the vector that enters the basis in LA_Strict()
------------------------------------------------------------------*/
void LA_strict_vent (int *pIvo, int *pJin, int kl, int m, int n,
    tMatrix_R c, tVector_I ib, tVector_I icbas, tNumber_R zz)
{
    int         i, icb, j, m1;
    tNumber_R   d, e, g, tz;

    m1 = m + 1;
    g = 1.0/ (EPS*EPS);
    tz = zz + zz + EPS;
    for (j = 1; j <= n; j++)
    {
        if (ib[j] == 0) continue;
        icb = 0;
        for (i = kl; i <= m1; i++)
        {
            if (j == icbas[i]) icb = 1;
        }
```

```
        if (icb == 1) continue;
        d = c[j][m1];
        if (d < -EPS)
        {
            e = d;
            if (e < g)
            {
                g = e;
                *pJin = j;
                *pIvo = 1;
            }
        }
        if (d >= tz)
        {
            e = tz - d;
            if (e < g)
            {
                g = e;
                *pJin = j;
                *pIvo = -1;
            }
        }
    }
}

/*-----------------------------------------------------------------------
Determine the vector that leaves the basis in LA_Strict()
-----------------------------------------------------------------------*/
void LA_strict_vleav (int kl, int *pIout, int jin, int *pItest,
    int m, tMatrix_R ct, tVector_R bv)
{
    int         i, m1;
    tNumber_R   d, g, thmax;

    m1 = m + 1;
    thmax = 1.0/ (EPS*EPS);

    for (i = kl; i <= m1; i++)
    {
        d = ct[i][jin];
        if (d > EPS)
        {
            g = bv[i]/d;
            if (g <= thmax)
            {
```

```
                thmax = g;
                *pIout = i;
                *pItest = 1;
            }
        }
    }
}

/*-----------------------------------------------------------------------
A Gauss-Jordan elimination step to matrix "ct" in LA_Strict()
-----------------------------------------------------------------------*/
void LA_strict_gauss_jordn (int iout, int jin, int kl, int m,
    int n, tMatrix_R ct, tMatrix_R binv, tVector_R bv, tVector_I ib,
    tVector_I icbas)
{
    int         i, j, m1;
    tNumber_R   d, pivot;

    m1 = m + 1;
    pivot = ct[iout][jin];
    for (j = 1; j <= n; j++)
    {
        if (ib[j] == 0) continue;
        ct[iout][j] = ct[iout][j]/pivot;
    }
    for (j = kl; j <= m1; j++)
    {
        binv[iout][j] = binv[iout][j]/pivot;
    }
    for (i = kl; i <= m1; i++)
    {
        if (i == iout) continue;
        d = ct[i][jin];
        for (j = 1; j <= n; j++)
        {
            if (ib[j] == 0) continue;
            ct[i][j] = ct[i][j] - d * (ct[iout][j]);
        }
        for (j = kl; j <= m1; j++)
        {
            binv[i][j] = binv[i][j] - d * (binv[iout][j]);
        }
    }
    icbas[iout] = jin;
    for (i = kl; i <= m1; i++)
```

```
    {
        bv[i] = binv[i][m1];
    }
}

/*------------------------------------------------------------------
Check for non-uniqueness of the Chebyshev solution in LA_Strict()
------------------------------------------------------------------*/
void LA_strict_uniquens (int *pIvo, int *pIout, int kl, int *pKn,
    int *pLd, int m, int n, tMatrix_R c, tMatrix_R ct, tVector_R f,
    tVector_R bv, tVector_R v, tVector_I ib, tVector_I ih,
    tVector_I ibound, tVector_I icbas, tVector_R r, tNumber_R zz)
{
    int         i, j, k, icb, m1;
    tNumber_R   d, e, piv, bignum;

    m1 = m + 1;
    bignum = 1.0/ (EPS*EPS);
    for (j = 1; j <= n; j++)
    {
        if (ib[j] == 0) continue;
        icb = 0;
        for (i = kl; i <= m1; i++)
        {
            if (j == icbas[i]) icb = 1;
        }
        if (icb == 1) continue;
        d = c[j][m1];
        if (d >= EPS)
        {
            e = zz + zz - d;
            if (e > EPS) continue;
            for (i = kl; i <= m1; i++)
            {
                ct[i][j] = bv[i] + bv[i] - ct[i][j];
            }
            f[j] = -f[j];
            ibound[j] = -ibound[j];
            c[j][m1] = e;
        }
        piv = bignum;
        *pIvo = 0;
        for (i = kl; i <= m1; i++)
        {
            if ((bv[i] < EPS) || ct[i][j] < EPS) continue;
```

```
            d = bv[i]/ct[i][j];
            if (d > piv) continue;
            *pIvo = 1;
            *pIout = i;
            piv = d;
        }
        if (*pIvo == 0) continue;
        v[*pIout] = bv[*pIout]/ct[*pIout][j];
        for (i = kl; i <= ml; i++)
        {
            if (i == *pIout) continue;
            icb = 0;
            v[i] = bv[i] - v[*pIout] * (ct[i][j]);
            if (v[i] < -EPS)
            {
                icb = -1;
                break;
            }
        }
        if (icb == -1) continue;
        k = 0;
        for (i = kl; i <= ml; i++)
        {
            if (bv[i] > EPS) continue;
            if (v[i] < EPS) continue;
            ih[i] = 1;
            k = 1;
        }
        if (k == 0) continue;
        d = c[j][ml] - zz;
        if (ibound[j] == -1) d = -d;
        r[j] = d;
        ib[j] = 0;
        *pKn = *pKn - 1;
    }
    *pLd = 0;
    for (i = kl; i <= ml; i++)
    {
        if (ih[i] == 0) *pLd = *pLd + 1;
    }
}

/*-----------------------------------------------------------------
Eliminating zz in LA_Strict()
-----------------------------------------------------------------*/
```

```
void LA_strict_eliminat_zz (int *pIout, int kl, int m, tMatrix_R c,
    tMatrix_R binv, tVector_R bv, tVector_I ib, tVector_I ih,
    tVector_I ibound, tVector_I icbas, int *pKsys, tVector_R r,
    tVector_R z, tNumber_R zz)
{
    int         i, j, k, m1;
    tNumber_R   d, piv;

    m1 = m + 1;
    piv = 0.0;
    z[*pKsys] = zz;
    for (i = kl; i <= m1; i++)
    {
        d = bv[i];
        if (d > piv)
        {
            piv = d;
            *pIout = i;
        }
    }
    k = icbas[*pIout];
    d = c[k][m1] - zz;
    if (ibound[k] == -1) d = -d;
    r[k] = d;
    ib[k] = 0;
    if (*pIout != m1)
    {
        /* Swap two rows of matrix "binv" */
        for (j = kl; j <= m1; j++)
        {
            d = binv[*pIout][j];
            binv[*pIout][j] = binv[m1][j];
            binv[m1][j] = d;
        }
        bv[*pIout] = binv[*pIout][m1];
        bv[m1] = binv[m1][m1];

        /* Swap two elements of vector "icbas" */
        swap_elems_Vector_I (icbas, *pIout, m1);

        /* Swap two elements of vector "ih" */
        swap_elems_Vector_I (ih, *pIout, m1);
    }
    *pIout = m1;
}
```

```
/*------------------------------------------------------------------
A Gauss-Jordan step to matrix "binv" in LA_Strict()
------------------------------------------------------------------*/
void LA_strict_gauss_jordn_binv (int iout, int kl, int m,
    tMatrix_R binv, tVector_I iv)
{
    int         i, j, m1;
    tNumber_R   d, pivot;

    m1 = m + 1;
    pivot = binv[iout][iout];
    for (j = kl; j <= iout; j++)
    {
        if (iv[j] == 0) continue;
        binv[iout][j] = binv[iout][j]/pivot;
    }
    for (i = kl; i <= m1; i++)
    {
        if (i == iout) continue;
        d = binv[i][iout];
        for (j = kl; j <= iout; j++)
        {
            if (iv[j] == 0) continue;
            binv[i][j] = binv[i][j] - d * (binv[iout][j]);
        }
    }
}

/*------------------------------------------------------------------
Permuting matrix "binv" in LA_Strict()
------------------------------------------------------------------*/
void LA_strict_permute_binv (int kj, int kl, int *pKm, int m,
    tMatrix_R binv, tVector_R bv, tVector_I ih, tVector_I icbas)
{
    int         j, l, ij, ji, m1;
    tNumber_R   d;

    m1 = m + 1;

    l = kj;
    l = l + 1;
    for (ij = 1; ij <= m1; ij++)
    {
        if (l > *pKm) break;
```

```
        for (ji = l; ji <= *pKm - 1; ji++)
        {
            if (ih[ji] == 1) l = l + 1;
            if (ih[ji] != 1) break;
        }
        if ((ih[*pKm] != 0) && l != *pKm)
        {
            for (j = kl; j <= m; j++)
            {
                d = binv[l][j];
                binv[l][j] = binv[*pKm][j];
                binv[*pKm][j] = d;
            }
            swap_elems_Vector_R (bv, l, *pKm);

            swap_elems_Vector_I (icbas, l, *pKm);

            swap_elems_Vector_I (ih, l, *pKm);
        }
        *pKm = *pKm - 1;
    }
}

/*-----------------------------------------------------------------
Calculating vector "a" in LA_Strict()
-----------------------------------------------------------------*/
void LA_strict_calcul_a (int m, tVector_R fs,  tMatrix_R binv,
    tVector_I iv, tVector_I icbas, tVector_I irbas, tVector_R a)
{
    int         i, j, k;
    tNumber_R   s;

    for (j = 1; j <= m; j++)
    {
        if (iv[j] != 0)
        {
            s = 0.0;
            for (i = 1; i <= m; i++)
            {
                if (icbas[i] != 0)
                {
                    s = s + fs[i] * (binv[i][j]);
                }
            }
            k = irbas[j];
```

```
            a[k] = s;
        }
    }
}

/*--------------------------------------------------------------------
Calculating the reduced system in LA_Strict()
--------------------------------------------------------------------*/
void LA_strict_reduce_sys (int kl, int kln, int *pMa, int m, int n,
    tMatrix_R c, tVector_R f, tMatrix_R binv, tVector_I ib,
    tVector_I iv, tVector_I irbas, tVector_R a)
{
    int         i, j, k, l, ii;
    int         klnm;
    tNumber_R   d;

    klnm = kln - 1;
    if (kln > kl)
    {
        for (l = kl; l <= klnm; l++)
        {
            ii = 0;
            for (i = kln; i <= m; i++)
            {
                d = binv[i][l];
                if (d < 0.0) d = -d;
                if (d < EPS) continue;
                ii = 1;
                break;
            }
            if (ii == 1) continue;
            k = irbas[l];
            for (j = 1; j <= n; j++)
            {
                if (ib[j] == 0) continue;
                f[j] = f[j] - a[k] * (c[j][k]);
            }
            iv[l] = 0;
            irbas[l] = 0;
            *pMa = *pMa - 1;
        }
    }
}

/*--------------------------------------------------------------------
```

```
Modifying matrix "binv" in LA_Strict()
-----------------------------------------------------------------*/
void LA_strict_modify_binv (int kl, int kj, int m, tMatrix_R binv,
    tVector_R v, tVector_I iv)
{
    int         i, j, k;
    tNumber_R   s;

    for (j = 1; j <= kj; j++)
    {
        if (iv[j] == 0) continue;
        for (i = kl; i <= m; i++)
        {
            s = 0.0;
            for (k = kl; k <= m; k++)
            {
                s = s + binv[i][k] * (binv[k][j]);
            }
            v[i] = s;
        }
        for (i = kl; i <= m; i++)
        {
            binv[i][j] = v[i];
        }
    }
}

/*-----------------------------------------------------------------
Equation ll eliminates element a[k] in LA_Strict()
-----------------------------------------------------------------*/
void LA_strict_eliminate_ll (int kl, int kln, int *pMa, int ld,
    int m, int n, tMatrix_R c, tVector_R f, tVector_I ib,
    tVector_I iv, tVector_I ibound, tVector_I icbas, tVector_I irbas,
    tVector_R r, tNumber_R zz)
{
    int         i, j, k, l, ll = 0, kb, m1, klnm;
    tNumber_R   d, e, g, piv, pivot = 0;

    m1 = m + 1;
    klnm = kln - 1;
    for (i = kl; i <= klnm; i++)
    {
        if (iv[i] == 0) continue;
        k = irbas[i];
        piv = 0.0;
```

```
        for (j = kl; j <= klnm; j++)
        {
            l = icbas[j];
            if (ib[l] == 0) continue;
            g = c[l][k];
            d = g;
            if (d < 0.0) d = -d;
            if (d <= piv) continue;
            piv = d;
            pivot = g;
            ll = l;
        }
        /* Equation ll is used to eliminate element a[k] */
        if (piv < EPS) continue;
        for (j = kl; j <= m; j++)
        {
            l = irbas[j];
            if (l == 0) continue;
            c[ll][l] = c[ll][l]/pivot;
        }
        f[ll] = f[ll]/pivot;
        d = c[ll][m1] - zz;
        if (ibound[ll] == -1) d = -d;
        r[ll] = d;
        ib[ll] = 0;
        for (j = 1; j <= n; j++)
        {
            if (ib[j] == 0) continue;
            d = c[j][k];
            e = d;
            if (e < 0.0) e = -e;
            if (e < EPS) continue;
            for (l = 1; l <= m; l++)
            {
                kb = irbas[l];
                if (kb == 0) continue;
                c[j][kb] = c[j][kb] - d * (c[ll][kb]);
            }
            f[j] = f[j] - d * (f[ll]);
        }
        *pMa = *pMa - 1;
        if (*pMa == ld) break;
    }
}
```

```
/*-----------------------------------------------------------------
Calculating elements of residual vector r, part (c) in LA_Strict()
-----------------------------------------------------------------*/
void LA_strict_calcul_r_3 (int kl, int kln, int m, tMatrix_R c,
    tVector_I ib, tVector_I ibound, tVector_I icbas, tVector_R r,
    tNumber_R zz)
{
    int         i, k, m1, klnm;
    tNumber_R   d;

    m1 = m + 1;
    klnm = kln - 1;
    for (i = kl; i <= klnm; i++)
    {
        k = icbas[i];
        if (ib[k] == 0) continue;
        d = c[k][m1] - zz;
        if (ibound[k] == -1) d = -d;
        r[k] = d;
        ib[k] = 0;
    }
}

/*-----------------------------------------------------------------
Calculating elements of residual vector r, part (b) in LA_Strict()
-----------------------------------------------------------------*/
void LA_strict_calcul_r_2 (int *pKn, int kln, int m, int n,
    tMatrix_R c, tVector_I ib, tVector_I ibound, tVector_I irbas,
    tVector_R r, tNumber_R zz)
{
    int         i, j, k, ii, m1;
    tNumber_R   e, d, g;

    m1 = m + 1;
    for (j = 1; j <= n; j++)
    {
        ii = 0;
        if (ib[j] == 0) continue;
        d = c[j][m1];
        if (d >= EPS)
        {
            e = zz + zz - d;
            if (e >= EPS) continue;
        }
        for (i = kln; i <= m; i++)
```

```
        {
            k = irbas[i];
            g = c[j][k];
            if (g < 0.0) g = -g;
            if (g >= EPS)
            {
                ii = 1;
                break;
            }
        }
        if (ii == 1) continue;
        d = d - zz;
        if (ibound[j] == -1) d = -d;
        r[j] = d;
        ib[j] = 0;
        *pKn = *pKn - 1;
    }
}

/*-----------------------------------------------------------------
Calculating elements of residual vector r, part (a) in LA_Strict()
-----------------------------------------------------------------*/
void LA_strict_calcul_r_1 (int m, int n, tMatrix_R c, tVector_I ib,
    tVector_I ibound, tVector_R r, tNumber_R zz)
{
    int         j, m1;
    tNumber_R   d;

    m1 = m + 1;
    for (j = 1; j <= n; j++)
    {
        if (ib[j] == 0) continue;
        d = c[j][m1] - zz;
        if (ibound[j] == -1) d = -d;
        r[j] = d;
    }
}
```

Chapter 15

Piecewise Chebyshev Approximation

15.1 Introduction

In Chapter 9, two algorithms for the piecewise linear approximation of plane curves in the L_1 norm are described. In this chapter, we describe two corresponding algorithms for the piecewise linear approximation of plane curves in the Chebyshev norm.

The problem of piecewise approximation for plane curves in the Chebyshev norm, including polygonal approximation, received considerable attention for some time. Although the number of published works on the polygonal approximation of plane curve is considerably large (see the references in Chapter 8), yet there are relatively fewer references on piecewise approximation in the Chebyshev norm.

Lawson [8] derived characteristic properties of the segmented Chebyshev approximation problem and proposed an iterative algorithm for finding these approximation. Phillips [13] presented two simple algorithms for approximating a convex function (whose second derivative is positive), the first one is when the Chebyshev residual (error) norm in any segment is not to exceed a pre-assigned value, and the other one is when the number of segments is given and a balanced (equal) Chebyshev error norm solution is required. The Chebyshev residual was measured along the direction of the y-axis.

Pavlidis [9] reviewed various algorithms for piecewise linear Chebyshev segmentation and proposed a new one based on discrete optimization. His algorithm is for the near-balanced solution case. The Chebyshev residual norm is measured along the direction of the y-axis.

Pavlidis and Horowitz [11] then presented another algorithm for segmenting a digitized plane curve in the Chebyshev norm. They attempted to determine the minimum number of segments in the approximation. After an arbitrary initial segmentation of the given digitized curve, segments are split and merged to derive the Chebyshev residual norm below a specified value. The Chebyshev norm for each segment is measured perpendicular to the straight line approximating the segment.

Tomek [15] described two simple heuristic algorithms, using parallel lines technique for Chebyshev piecewise approximation by straight lines. One algorithm works well for smooth functions. Both algorithms are for the case where the Chebyshev residual norm for each segment is not to exceed a pre-assigned value. The Chebyshev error norm is measured along the y-axis direction.

Using functional iteration, Pavlidis and Maika [12] proposed a procedure for solving the near-balanced polynomial Chebyshev approximation problem. Their method compares favorably with Lawson's algorithm [8]. Again, the Chebyshev error norm is measured along the direction of the y-axis.

Kioustelidis [6] discussed the existence of segmented approximations with free knots for a given continuous function.

The majority of published methods consider only the special cases of piecewise approximation by constants (horizontal lines) or by straight lines. In this chapter, the approximating functions may be any kind of polynomials, not necessarily constants or straight lines. The Chebyshev error norm is measured along the direction of the y-axis. Two algorithms for the piecewise linear Chebyshev approximation of plane curves are presented [1]. They are for the 2 cases: (a) when the Chebyshev error norm in any segment is not to exceed a pre-assigned value and (b) when the number of segments is given and a near-balanced Chebyshev error norm solution is required.

The problem follows the same steps taken in Chapter 9 for the piecewise linear approximation of plane curves in the L_1 norm. The problem is solved by first digitizing the given curve into discrete points. Then each of the algorithms is applied to the discrete points. Both algorithms use the discrete linear Chebyshev approximation function LA_Linf() of Chapter 10.

In Section 15.2, the characteristic properties of the piecewise approximation are outlined. In Section 15.3, the Chebyshev approximation is described. In Section 15.4, the two Chebyshev piecewise approximations are presented. In Section 15.5, numerical results and comments are given.

15.1.1 Applications of piecewise approximation

We identify here some applications of piecewise approximation that were not presented in Chapters 8 and 9. In the field of numerical analysis, it is used in the reduction of non-linear problems to approximately equivalent linear problems. It is used also in the approximate solutions of linear differential equations with time varying parameters [14]. Applications are also used in the design of electronic analog and hybrid computers [7]. In the fields of image processing and pattern recognition it is used in feature extraction, noise filtering and data compression [4, 10].

15.2 Characteristic properties of piecewise approximation

Let $y = f(x)$ be a given plane curve and let it be defined on the interval [a, b]. Let this curve be digitized at the K points $(x_i, f(x_i))$, $i = 1, 2, \ldots, K$, where $x_1 = a$ and $x_K = b$. Let n be the number of pieces (segments) in the approximation and let (z_j), $j = 1, 2, \ldots, n$, be the Chebyshev error norms for the n pieces. For the first algorithm, the number of segments n is not known beforehand.

Assume f(x) is continuous and satisfies Lipschitz condition on [a, b]. Lawson [8] showed that the norm z_j, $1 \leq j \leq n$, has certain characteristics. Such characteristics are given in Section 9.2 for the piecewise linear approximation of plane curves in the L_1 norm. The same characteristics apply as well for the piecewise linear approximation in the Chebyshev norm and in the L_2 or the least squares norm (Chapter 18).

15.3 The discrete linear Chebyshev approximation problem

Consider any segment j, $1 \leq j \leq n$, of the given curve f(x). Let this segment consist of N digitized points with coordinates $(x_i, f(x_i))$, $i = 1, 2, \ldots, N$. Let these N points be approximated by the linear function

$$\mathbf{L}(\mathbf{a}, \mathbf{x}) = a_1\phi_1(x) + \ldots + a_M\phi_M(x)$$

which minimizes the Chebyshev norm z_j (or briefly z), of the residuals $r(x_i)$, where

$$z = \max|r(x_i)|$$

and

$$r(x_i) = \mathbf{L}(\mathbf{a}, x_i) - f(x_i), \quad i = 1, 2, \ldots, N \tag{15.3.1}$$

Here, $\{\phi_j(x)\}$, $j = 1, 2, \ldots, M$, $M < N$, is a given set of real linearly independent approximating functions and **a** is an M real vector to be calculated. For example, if the approximating function is a vertical parabola, $\mathbf{L}(\mathbf{a}, \mathbf{x}) = a_1 + a_2x + a_3x^2$, then a_1, a_2 and a_3 are the elements of vector **a**, which are to be calculated.

This problem reduces to the problem of obtaining the Chebyshev solution of the overdetermined system of linear equations

$$\mathbf{Ca} = \mathbf{f}$$

$\mathbf{C} = (c_{ij})$ is an N by M matrix given by $\mathbf{C} = (\phi_j(x_i))$, $i = 1, 2, \ldots, N$ and $j = 1, 2, \ldots, M$, and **f** is the N-vector $\mathbf{f} = f(x_i)$. The Chebyshev solution to this system is the real M-vector **a** that minimizes the Chebyshev norm

$$z = \max|r_i|, \quad i = 1, 2, \ldots, N$$

where r_i is the ith residual given by (15.3.1) or by

$$r_i = c_{i1}a_1 + \ldots + c_{im}a_m - f_i, \quad i = 1, 2, \ldots, N$$

15.4 Description of the algorithms

15.4.1 Piecewise linear Chebyshev approximation with pre-assigned tolerance

This algorithm uses the same steps as those of Section 9.4.1. However, each time a point is added to segment say j, we use the function LA_Linf() of Chapter 10 to re-calculate the new Chebyshev norm of the segment.

15.4.2 Piecewise linear Chebyshev approximation with near-balanced Chebyshev norms

This algorithm uses the same steps as those of Section 9.4.2. Again, each time we add a point or delete a point from segment j, we use the function LA_Linf() of Chapter 10 to re-calculate the new Chebyshev norm of the segment.

15.5 Numerical results and comments

Each of the functions LA_Linfpw1() and LA_Linfpw2() calculate the number of segments in the piecewise approximation (for LA_Linfpw2(), n is given), the starting points of the n segments, the coefficients of the approximating curves for the n segments, the residuals at each point of the digitized curve and finally the Chebyshev residual norms for the n segments.

LA_Linfpw1() computes for the case when the Chebyshev error norm in any segment is not to exceed a pre-assigned value. LA_Linfpw2() computes for the case when the number of segments is given and a near-balanced Chebyshev error norm solution is required. Each of these functions uses LA_Linf() [2, 3] of Chapter 10.

DR_Linfpw1() and DR_Linfpw2() were used to test the two algorithms in single-precision.

Recall that LA_Linfpw1() has the option of calculating connected or disconnected piecewise Chebyshev approximations, while LA_Linfpw2() can only calculate disconnected piecewise Chebyshev approximations. As explained at Section 9.2, in the connected piecewise approximation, the x-coordinate of the right end point of segment j is the x-coordinate of the starting (left) end point of segment (j + 1). In the disconnected piecewise approximation, the x-coordinate of the adjacent point to the right end point of segment j is the x-coordinate of the starting left point of segment (j + 1).

The numerical results of one example are presented here. They are for the same curve used in Chapter 9 for the piecewise linear approximation in the L_1 norm, and also for the L_2 norm of Chapter 18.

The given curve is digitized with equal x-intervals into 28 points and the data points are fitted with vertical parabolas, each is of the form $y = a_1 + a_2x + a_3x^2$.

The results of the first algorithm are shown in Figures 15-1 and 15-2. The results of the second algorithm are shown in Figure 15-3, where the number of segments was set to n = 4.

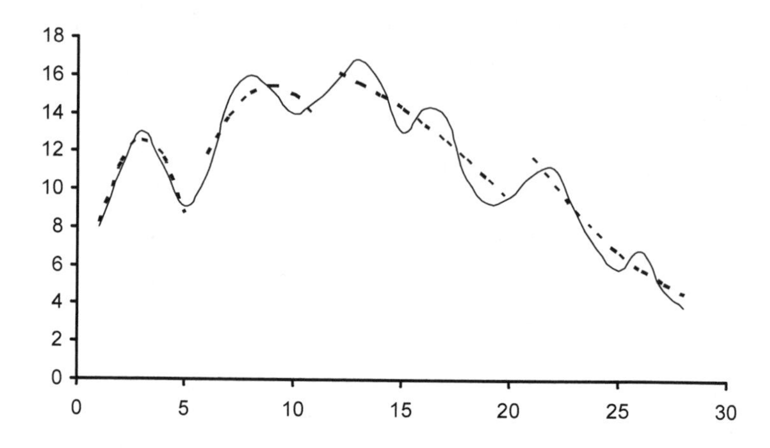

Figure 15-1: Disconnected linear Chebyshev piecewise approximation with vertical parabolas. Chebyshev residual norm in any segment ≤ 1.3

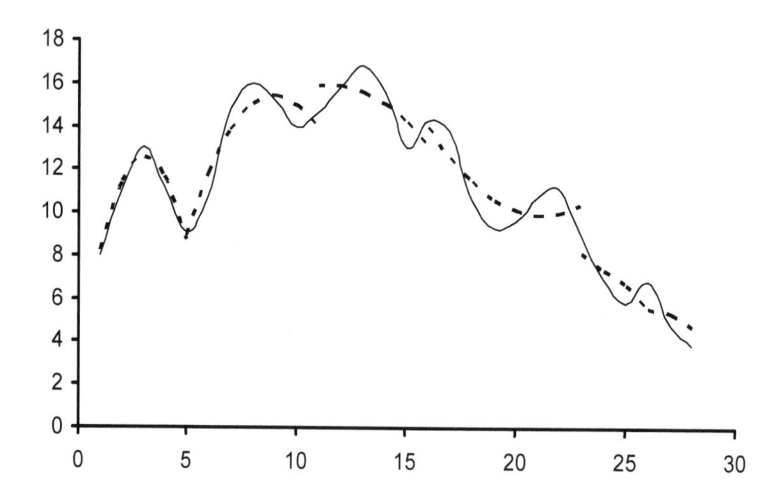

Figure 15-2: Connected linear Chebyshev piecewise approximation with vertical parabolas. Chebyshev residual norm in any segment ≤ 1.3

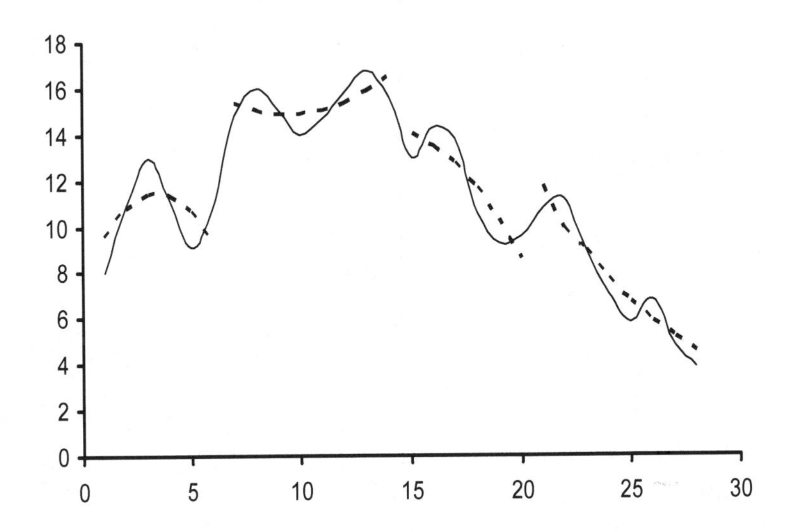

Figure 15-3: Near-balanced residual norm solution. Disconnected linear Chebyshev piecewise approximation with vertical parabolas. Number of segments = 4

The Chebyshev norms of Figures 15-1 and 15-2 are (0.438, 0.981, 1.238, 0.937) and n = 4, and (0.438, 0.981, 1.203, 1.285, 0.842) and n = 5, respectively. The Chebyshev norms of Figure 15-3 are (1.553, 0.920, 1.050, 0.937), for n = 4.

We observe that the second algorithm did not produce a balanced norm solution. However, by comparing Figures 15-1 and 15-3, we see that it gives an improved solution to that of the first algorithm.

In a previous version of our algorithms [1], we used parametric linear programming techniques [5] and made use of the equioscillation property of the Chebyshev approximation. This technique increases the code complexity but reduces the number of iterations.

References

1. Abdelmalek, N.N., Chebyshev solution of overdetermined systems of linear equations, *BIT*, 15(1975)117-129.

2. Abdelmalek, N.N., A computer program for the Chebyshev solution of overdetermined systems of linear equations, *International Journal for Numerical Methods in Engineering*, 10(1976)1197-1202.
3. Abdelmalek, N.N., Piecewise linear Chebyshev approximation of planar curves, *International Journal of Systems Science*, 14(1983)425-435.
4. Davison, L.D., Data compression using straight line interpolation, *IEEE Transactions on Information Theory*, IT-14(1968)390-394.
5. Hadley, G., *Linear Programming*, Addison-Wesley, Reading, MA, 1962.
6. Kioustelidis, J.B., Optical segmented approximations, *Computing*, 24(1980)1-8.
7. Korn, G.A. and Korn, T.M., *Electronic Analog and Hybrid Computers*, McGraw-Hill, New York, 1964.
8. Lawson, C.L., Characteristic properties of the segmented rational minimax approximation problem, *Numerische Mathematik*, 6(1964)293-301.
9. Pavlidis, T., Waveform segmentation through functional approximation, *IEEE Transactions on Computers*, 22(1973)689-697.
10. Pavlidis, T., The use of algorithms of piecewise approximations for picture processing applications, *ACM Transactions on Mathematical Software*, 2(1976)305-321.
11. Pavlidis, T. and Horowitz, S.L., Segmentation of plane curves, *IEEE Transactions on Computers*, 23(1974)860-870.
12. Pavlidis, T. and Maika, A.P., Uniform piecewise polynomial approximation with variable joints, *Journal of Approximation Theory*, 12(1974)61-69.
13. Phillips, G.M., Algorithms for piecewise straight line approximation, *Computer Journal*, 11(1968)211-212.
14. Solodov, A.V., *Linear Automatic Control Systems with Varying Parameters*, Blackie, London, 1966.
15. Tomek, I., Two algorithms for piecewise-linear continuous approximation of functions of one variable, *IEEE Transactions on Computers*, 23(1974)445-448.

15.6 DR_Linfpw1

```
/*-----------------------------------------------------------------
DR_Linfpw1
-------------------------------------------------------------------
This is a driver for the function LA_Linfpw1(), which calculates a
linear piecewise Chebyshev approximation of a given data point set
{x,y} that results from discretizing a given plane curve y = f(x).
The points of the set might not be equally spaced.

The approximation by LA_Linfpw1() is such that the Chebyshev norm for
any segment is not to exceed a pre-assigned tolerance donated by
"enorm".

LA_Linfpw1() calculates the connected or the disconnected piecewise
linear Chebyshev approximation according to the value of an integer
parameter denoted by "konect" set by the user.
See comments in LA_Linfpw1().

From the approximating curve we form the overdetermined system of
linear equations

                            c*a = f

"c" is a real n by m matrix of rank k, k <= m < n.
n is the number of digitized points of the given plane curve.

m is the number of terms in the approximating curves.  If for
example, the approximating curves are vertical parabolas of
the form
                        y = a1 + a2*x + a3*x*x
then m = 3.

"f" is a real n vector whose elements are the y coordinates of
the data set {x,y}.

"a" is the solution m vector.  There are different "a" solution
vectors for the different segments.

This driver contains 1 test example. A given curve is digitized
into 28 points at equal x intervals.  The points are piecewise
approximated by vertical parabolas of the form

                        y = a1 + a2*x + a3*x*x
```

```
The results for the disconnected and of the connected piecewise
Chebyshev approximation are given in the text.
-------------------------------------------------------------------*/

#include "DR_Defs.h"
#include "LA_Prototypes.h"

#define Nc          28

void DR_Linfpw1 (void)
{
    /*------------------------------------------
      Constant matrices/vectors
    ------------------------------------------*/
    static tNumber_R fc[Nc+1] =
    {   NIL,
         8.0, 11.0, 13.0, 11.2,  9.1, 10.8, 14.8, 16.0, 15.1, 14.0,
        14.7, 15.8, 16.8, 15.6, 13.0, 14.3, 13.8, 10.6,  9.3,  9.6,
        10.8, 11.2,  9.0,  7.0,  5.8,  6.8,  4.8,  3.9
    };

    /*------------------------------------------
      Variable matrices/vectors
    ------------------------------------------*/
    tMatrix_R   ct      = alloc_Matrix_R (MM_COLS, NN_ROWS);
    tMatrix_R   rp1     = alloc_Matrix_R (KK_PIECES, NN_ROWS);
    tMatrix_R   ap      = alloc_Matrix_R (KK_PIECES, MM_COLS);
    tVector_R   f       = alloc_Vector_R (NN_ROWS);
    tVector_R   zp      = alloc_Vector_R (KK_PIECES);
    tVector_I   irankp  = alloc_Vector_I (KK_PIECES);
    tVector_I   ixl     = alloc_Vector_I (KK_PIECES);

    int         j, k, m, n;
    int         konect, npiece;
    tNumber_R   enorm;
    eLaRc       prnRc;

    eLaRc       rc = LaRcOk;

    prn_dr_bnr ("DR_Linfpw1, Chebyshev Piecewise Approximation of "
                "a Plane with Pre-assigned Norm");

    for (k = 1; k <= 2; k++)
    {
```

```
switch (k)
{
case 1:
    enorm = 1.3;
    n = Nc;
    m = 3;
    for (j = 1; j <= n; j++)
    {
        f[j] = fc[j];
        ct[1][j] = 1.0;
        ct[2][j] = j;
        ct[3][j] = j*j;
    }
    break;
default:
    break;
}

prn_algo_bnr ("Linfpw1");
if (k == 1) konect = 0;
if (k == 2) konect = 1;
prn_example_delim();
PRN ("konect #%d: Size of matrix \"c\" %d by %d\n",
     konect, n, m);
PRN ("Chebyshev Piecewise Approximation with  "
     "Pre-assigned Norm\n");
PRN ("Pre-assigned Norm \"enorm\" = %8.4f\n", enorm);
prn_example_delim();
if (konect == 1)
    PRN ("Connected Piecewise Approximation\n");
else
    PRN ("Disconnected Piecewise Approximation\n");
prn_example_delim();
PRN ("r.h.s. Vector \"f\"\n");
prn_Vector_R (f, n);
PRN ("Transpose of Coefficient Matrix, \"ct\"\n");
prn_Matrix_R (ct, m, n);

rc = LA_Linfpw1 (m, n, enorm, konect, ct, f, ixl, irankp,
                 rp1, ap, zp, &npiece);

if (rc >= LaRcOk)
{
    PRN ("\n");
    PRN ("Results of the Chebyshev Piecewise "
```

```
                "Approximation\n");
            PRN ("Calculated number of segments (pieces) = %d\n",
                npiece);
            PRN ("Staring points of the \"npiece\" segments\n");
            prn_Vector_I (ixl, npiece);
            PRN ("Coefficients of the approximating curves\n");
            prn_Matrix_R (ap, npiece, m);
            PRN ("Residual vectors for the \"npiece\" segments\n");
            prnRc = LA_pw1_prn_rp1 (konect, npiece, n, ixl, rp1);
            PRN ("Chebyshev residuals for the \"npiece\" "
                "segments\n");
            prn_Vector_R (zp, npiece);
            if (prnRc < LaRcOk)
            {
                PRN ("Error printing PW1 results\n");
            }
        }

        prn_la_rc (rc);
    }

    free_Matrix_R (ct, MM_COLS);
    free_Matrix_R (rp1, KK_PIECES);
    free_Matrix_R (ap, KK_PIECES);
    free_Vector_R (f);
    free_Vector_R (zp);
    free_Vector_I (irankp);
    free_Vector_I (ixl);
}
```

15.7 LA_Linfpw1

```
/*-----------------------------------------------------------------
LA_Linfpw1
-------------------------------------------------------------------
This program calculates a linear piecewise Chebyshev (L-infinity)
approximation to a discrete point set {x,y}. The approximation is
such that the Chebyshev residual (error) norm for any segment is
not to exceed a given tolerance "enorm". The number of segments
(pieces) is not known before hand.

Given is a set of points {x,y}. From the approximating functions
(curves) one forms the overdetermined system of linear equations

                        c*a = f

This program uses program LA_Linf() for obtaining the Chebyshev
solution of overdetermined system of linear equations.

LA_Linfpw1() has the option of calculating connected or
disconnected piecewise Chebyshev approximation, according to
the value of an integer parameter "konect".

In the connected piecewise approximation the x-coordinate of the
end point of segment j say, is the x-coordinate of the starting
point of segment (j+1). In the disconnected piecewise
approximation, the x-coordinate of the adjacent point to the end
point of segment j is the x-coordinate of the starting point of
segment (j+1). See the comments on "ixl" below.

Inputs
m       Number of terms in the approximating functions.
n       Number of points to be piecewise approximated.
ct      A real (m+1) by n matrix. Its forst m rows and its n columns
        contain the transpose of matrix "c" of the system c*a = f.
        The (m + 1)th row of "ct" is a row of ones.
        Matrix "ct" is not destroyed in the computation.
f       An real n vector containing the r.h.s. of the system
        c*a = f. This vector contains the y-coordinates of the
        given point set. This vector is not destroyed in the
        computation.
enorm   A real pre-assigned parameter, such that the Chebyshev
        residual norm for any segment is <= enorm.
konect  An integer specifying the action to be performed.
```

```
        If konect = 1, the program calculates the connected
        Chebyshev piecewise approximation.
        If konect != 1, the program calculates the disconnected
        Chebyshev piecewise approximation.

Outputs
npiece  Obtained number of segments or pieces of the approximation.
ixl     An integer "npiece" vector containing the indices of the
        first elements of the "npiece" segments.
        For example, if ixl = (1,5,12,22,...), and if konect = 1,
        then the first segment contains points 1 to 5, the second
        segment contains points 5 to 12, the third segment contains
        points 12 to 22 and so on.
        Again if ixl = (1,5,12,22,...), and if konect !=1, then the
        first segment contains points 1 to 4, the second segment
        contains points 5 to 11, the third segment contains points 12
        to 21 ..., etc.
ap      A real "npiece" by m  matrix.  Its first row contains the
        the coefficients of the approximating curve for the first
        segment.  The second row contains the coefficients of the
        approximating curve for the second segment and so on.
        If any row j is all zeros, this indicates that vector "a" of
        segment j is not calculated as the number of points of
        segment j is <= m and there is a perfect fit by the
        approximating curve for segment j.
rpl     A real "npiece" by n matrix.  Its first row contains the
        residuals for the points of the first segment.  Its second
        row contains the residuals for the points of the second
        segment, and so on.
zp      A "npiece" real vector containing the "npiece" optimum values
        of the Chebyshev residual norms for the "npiece" segments.
        If zp[j] == 0.0, it indicates that there is a perfect fit for
        segment j.

Returns one of
        LaRcSolutionFound
        LaRcErrBounds
        LaRcErrNullPtr
        LaRcErrAlloc
-------------------------------------------------------------------*/

#include "LA_Prototypes.h"

eLaRc LA_Linfpwl (int m, int n, tNumber_R enorm, int konect,
    tMatrix_R ct, tVector_R f, tVector_I ixl, tVector_I irankp,
```

```
    tMatrix_R rp1, tMatrix_R ap, tVector_R zp, int *pNpiece)
{
    tMatrix_R   ctp     = alloc_Matrix_R (m + 1, n);
    tVector_R   fp      = alloc_Vector_R (n);
    tVector_R   r       = alloc_Vector_R (n);
    tVector_R   a       = alloc_Vector_R (m + 1);

    int         i = 0, j = 0, je = 0, ji = 0, jj = 0, is = 0, ie = 0,
                m1 = 0, nu = 0, irank = 0;
    int         ijk = 0, iter = 0;
    tNumber_R   z = 0.0;

    /* Validation of data before executing algorithm */
    eLaRc       rc = LaRcSolutionFound;
    VALIDATE_BOUNDS ((0 < m) && (m < n) && (0.0 < enorm));
    VALIDATE_PTRS   (ct && f && ixl && irankp && rp1 && ap && zp &&
                     pNpiece);
    VALIDATE_ALLOC  (ctp && fp && r && a);

    m1 = m + 1;
    *pNpiece = 1;

    /* "is" means i(start) for the segment at hand */
    is = 1;
    for (ijk = 1; ijk <= n; ijk++)
    {
        /* Initializing the data for "npiece".
           is means i (start)
           ie means i (end) for the segment at hand */
        ie = is + m1 - 1;
        LA_pwl_init (pNpiece, is, ie, m, rp1, ap, zp);

        if (ie > n)
        {
            GOTO_CLEANUP_RC (LaRcSolutionFound);
        }
        irank = m;
        z = 0;

        for (j = is; j <= ie; j++)
        {
            ji = j - is + 1;
            fp[ji] = f[j];
            for (i = 1; i <= m; i++)
            {
```

```
                ctp[i][ji] = ct[i][j];
            }
        }
        for (jj = 1; jj <= n; jj++)
        {
            nu = ie - is + 1;

            rc = LA_Linf (m, nu, ctp, fp, &irank, &iter, r, a, &z);
            if (rc < LaRcOk)
            {
                GOTO_CLEANUP_RC (rc);
            }

            if (z > enorm + EPS) break;
            else
            {
                LA_pw1_map (m, nu, r, a, z, rp1, ap, zp, pNpiece);
                ixl[*pNpiece] = is;
                je = ie + 1;
                if (je > n)
                {
                    GOTO_CLEANUP_RC (LaRcSolutionFound);
                }
                ie = je;
                nu = ie - is + 1;
                if (nu < m1)
                {
                    GOTO_CLEANUP_RC (LaRcSolutionFound);
                }
                for (j = is; j <= ie; j++)
                {
                    ji = j - is + 1;
                    fp[ji] = f[j];
                    for (i = 1; i <= m; i++)
                    {
                        ctp[i][ji] = ct[i][j];
                    }
                }
            }
        }
        is = ie;
        if (konect == 1) is = ie - 1;
        *pNpiece = *pNpiece + 1;
        ixl[*pNpiece] = is;
    }
```

```
CLEANUP:

    free_Matrix_R (ctp, m + 1);
    free_Vector_R (fp);
    free_Vector_R (r);
    free_Vector_R (a);

    return rc;
}
```

15.8 DR_Linfpw2

```
/*-------------------------------------------------------------------
DR_Linfpw2
---------------------------------------------------------------------
This is a driver for the function LA_Linfpw2() which calculates the
"near balanced" piecewise linear Chebyshev approximation  of a given
data point set {x,y} resulting from the discretization of a plane
curve y=f(x).

Given is an integer number "npiece" which is the number of segments
in the approximation.

The approximation by LA_Linfpw2() is such that the Chebyshev residual
norms for all segments are nearly equal, hence the name "near
balanced" piecewise approximation.

From the approximating curves we form the overdetermined system of
linear equations

                          c*a = f

"c" is a real n by m matrix of rank k, k <= m < n.
n is the number of digitized points of the given plane curve.
m is the number of terms in the approximating curves.  If for
example, the piecewise approximating curves are vertical parabolas
of the form
                    y = a1 + a2*x + a3*x*x
then m = 3.

"f" is a real n vector whose elements are the y coordinates of
the data set {x,y}.

"a" is the solution m vector.  There are different "a" solution
vectors for the different segments.

This driver contains 1 test example.
A given curve is digitized into 28 points at equal x intervals.  The
points are piecewise approximated by vertical parabolas of the form

                    y = a1 + a2*x + a3*x*x

The results for piecewise Chebyshev approximation are given in
the text.
```

```
----------------------------------------------------------------*/

#include "DR_Defs.h"
#include "LA_Prototypes.h"

#define Nc          28

void DR_Linfpw2 (void)
{
    /*----------------------------------------
      Constant matrices/vectors
    ----------------------------------------*/
    static tNumber_R fc[Nc+1] =
    {   NIL,
         8.0, 11.0, 13.0, 11.2,  9.1, 10.8, 14.8, 16.0, 15.1, 14.0,
        14.7, 15.8, 16.8, 15.6, 13.0, 14.3, 13.8, 10.6,  9.3,  9.6,
        10.8, 11.2,  9.0,  7.0,  5.8,  6.8,  4.8,  3.9
    };

    /*----------------------------------------
      Variable matrices/vectors
    ----------------------------------------*/
    tMatrix_R   ct      = alloc_Matrix_R (MM_COLS, NN_ROWS);
    tVector_R   f       = alloc_Vector_R (NN_ROWS);
    tVector_R   rp2     = alloc_Vector_R (NN_ROWS);
    tMatrix_R   ap      = alloc_Matrix_R (KK_PIECES, MM_COLS);
    tVector_R   zp      = alloc_Vector_R (KK_PIECES);
    tVector_I   ixl     = alloc_Vector_I (KK_PIECES);
    tVector_R   w       = alloc_Vector_R (NN_ROWS);

    int         j, m, n;
    int         Iexmpl, npiece;
    eLaRc       prnRc;

    eLaRc       rc = LaRcOk;

    prn_dr_bnr ("DR_Linfpw2, Chebyshev Piecewise Approximation of "
                "a Plane Curve with Near Equal Residual Norms");

    for (Iexmpl = 1; Iexmpl <= 1; Iexmpl++)
    {
        switch (Iexmpl)
        {
        case 1:
            npiece = 4;
```

```
            n = Nc;
            m = 3;
            for (j = 1; j <= n; j++)
            {
                f[j] = fc[j];
                ct[1][j] = 1.0;
                ct[2][j] = j;
                ct[3][j] = j*j;
            }
            break;
        default:
            break;
    }

    prn_algo_bnr ("Linfpw2");
    prn_example_delim();
    PRN ("Size of matrix \"c\" %d by %d\n", n, m);
    prn_example_delim();
    PRN ("Chebyshev Piecewise Approximation "
         "with Near Equal Norms\n");
    PRN ("Given number of segments (pieces) = %d\n", npiece);
    prn_example_delim();
    PRN ("r.h.s. Vector \"f\"\n");
    prn_Vector_R (f, n);
    PRN ("Transpose of Coefficient Matrix, \"ct\"\n");
    prn_Matrix_R (ct, m, n);

    rc = LA_Linfpw2 (m, n, npiece, ct, f, ap, rp2, zp, ixl);

    if (rc >= LaRcOk)
    {
        PRN ("\n");
        PRN ("Results of the Chebyshev Piecewise "
             "Approximation\n");
        PRN ("Starting points of the \"npiece\" segments\n");
        prn_Vector_I (ixl, npiece);
        PRN ("Chebyshev residual norms for the"
             " \"npiece\" segments\n");
        prn_Vector_R (zp, npiece);
        PRN ("Coefficients of the \"npiece\" approximating "
             " curves\n");
        prn_Matrix_R (ap, npiece, m);
        PRN ("Residuals at the given points\n");
        prnRc = LA_pw2_prn_rp2 (npiece, n, ixl, rp2);
        if (prnRc < LaRcOk)
```

```
            {
                PRN ("Error printing PW2 results:  ");
            }
        }

        prn_la_rc (rc);
    }

    free_Matrix_R (ct, MM_COLS);
    free_Vector_R (f);
    free_Vector_R (rp2);
    free_Matrix_R (ap, KK_PIECES);
    free_Vector_R (zp);
    free_Vector_I (ixl);
    free_Vector_R (w);
}
```

15.9 LA_Linfpw2

```
/*-----------------------------------------------------------------------
LA_Linfpw2
-------------------------------------------------------------------------
This program calculates the "near balanced" piecewise linear
Chebyshev approximation  of a given data point set {x,y} resulting
from the discretization of a plane curve y = f(x).

Given is an integer number "npiece" which is the number of segments
in the approximation.

The approximation by LA_Linfpw2() is such that the Chebyshev
residual norms for all segments are nearly equal, hence the name
"near balanced" piecewise approximation.

From the approximating functions (curves) one forms the
overdetermined system of linear equations

                    c*a = f

Inputs
npiece  Given umber of segments (pieces) of the approximation.
m       Number of terms in the approximating functions.
n       Number of points to be piecewise approximated
ct      A real (m+1) by n matrix. Its first m rows and its n columns
        contain the transpose of matrix "c" of the system c*a = f.
        The (m+1)th row of "ct" is a row of ones.
        Matrix "ct" is not destroyed in the computation.
f       A real n vector containing the r.h.s. of the system c*a = f.
        This vector contains the y-coordinates of the given point
        set.  This vector is not destroyed in the computation.

Outputs
ixl     An integer "npiece" vector containing the indices of the
        first elements of the "npiece" segments.
        For example, if ixl = (1,5,12,22,...), then the first
        segment contains points 1 to 4, the second segment  contains
        points 5 to 11, the third segment contains points 12
        to 21 ..., etc.
ap      A real "npiece" by m matrix.  Its first row contains the
        coefficients of the approximating curve for the first
        segment. The second row contains the coefficients of the
        approximating curve for the second segment and so on.
```

```
rp2     A real n vector containing the residual values of the n
        points of the given set {x,y}.
zp      A real npiece vector containing the optimum Chebyshev
        residual norms for the "npiece" segments.

Returns one of
        LaRcSolutionFound
        LaRcErrBounds
        LaRcErrNullPtr
        LaRcErrAlloc
-------------------------------------------------------------------*/

#include "LA_Prototypes.h"

eLaRc LA_Linfpw2 (int m, int n, int npiece, tMatrix_R ct,
    tVector_R f, tMatrix_R ap, tVector_R rp2, tVector_R zp,
    tVector_I ixl)
{
    tVector_R   al      = alloc_Vector_R (m + 1);
    tVector_R   rl      = alloc_Vector_R (n);
    tVector_R   bv      = alloc_Vector_R (m + 1);
    tMatrix_R   binv    = alloc_Matrix_R (m + 1, m + 1);
    tVector_I   icbas   = alloc_Vector_I (m + 1);
    tVector_I   irbas   = alloc_Vector_I (m + 1);
    tVector_I   ibound  = alloc_Vector_I (n);
    tVector_R   ar      = alloc_Vector_R (m + 1);
    tVector_R   rr      = alloc_Vector_R (n);
    tMatrix_R   ctp     = alloc_Matrix_R (m + 1, n);
    tVector_R   fp      = alloc_Vector_R (n);
    tVector_I   iflag   = alloc_Vector_I (npiece);

    int         i = 0, j = 0, k = 0, is = 0, ie = 0, ji = 0, nu = 0,
                kl = 0, klp1 = 0, ijk = 0;
    int         icb = 0, isl = 0, isr = 0, iel = 0, ier = 0,
                ieln = 0, isrn = 0;
    int         istart = 0, iend = 0, ipcp1 = 0, iterl = 0,
                iterr = 0;
    int         irankl = 0, irankr = 0;
    tNumber_R   con = 0.0, conn = 0.0, zl = 0.0, zln = 0.0,
                zrn = 0.0;

    /* Validation of the data before executing the algorithm */
    eLaRc       rc = LaRcSolutionFound;
    VALIDATE_BOUNDS ((0 < m) && (m < n) && (1 < npiece));
    VALIDATE_PTRS   (ct && f && ap && rp2 && zp && ixl);
```

```
VALIDATE_ALLOC  (al && rl && bv && binv && icbas && irbas &&
                 ibound && ar && rr && ctp && fp && iflag);

for (k = 1; k <= npiece; k++)
{
    iflag[k] = 1;

    /* Initializing LA_Linfpw2 */
    LA_pw2_init (k, npiece, m, n, &is, &ie, ct, f, ctp, fp, ixl);

    /* Calculating the Chebyshev solution of the "npiece"
       segment */
    nu = ie - is + 1;

    rc = LA_Linf (m, nu, ctp, fp, &irankl, &iterl, rl, al, &zl);
    if (rc < LaRcOk)
    {
        GOTO_CLEANUP_RC (rc);
    }

    zp[k] = zl;

    /* Mapping initial data for the "npiece" segments */
    for (i = 1; i <= m; i++)
    {
        ap[k][i] = al[i];
    }
    for (j = is; j <= ie; j++)
    {
        ji = j - is + 1;
        rp2[j] = rl[ji];
    }
}

/* Process of balancing the Chebyshev norms */
istart = 1;
iend = npiece - 1;
ipcp1 = npiece + 1;
ixl[ipcp1] = n + 1;

for (ijk = 1; ijk < n*n; ijk++)
{
    for (kl = istart; kl <= iend; kl=kl+2)
    {
        klp1 = kl + 1;
```

```
con = fabs (zp[klp1] - zp[kl]);
isl = ixl[kl];
isr = ixl[klp1];
iel = isr - 1;
if (kl != iend) ier = ixl[kl + 2] - 1;
if (kl == iend) ier = n;

/* The case where : -----z[i]<z[i+1] */
if (zp[kl] < zp[klp1])
{
    ieln = iel + 1;
    isrn = ieln + 1;
    nu = ieln - isl + 1;
    for (j = isl; j <= ieln; j++)
    {
        ji = j - isl + 1;
        fp[ji] = f[j];
        for (i = 1; i <= m; i++)
        {
            ctp[i][ji] = ct[i][j];
        }
    }

    rc = LA_Linf (m, nu, ctp, fp, &irankl, &iterl, rl,
                  al, &zln);
    if (rc < LaRcOk)
    {
        GOTO_CLEANUP_RC (rc);
    }

    nu = ier - isrn + 1;
    for (j = isrn; j <= ier; j++)
    {
        ji = j - isrn + 1;
        fp[ji] = f[j];
        for (i = 1; i <= m; i++)
        {
            ctp[i][ji] = ct[i][j];
        }
    }

    rc = LA_Linf (m, nu, ctp, fp, &irankr, &iterr, rr,
                  ar, &zrn);
    if (rc < LaRcOk)
    {
```

```
            GOTO_CLEANUP_RC (rc);
        }

        conn = fabs (zrn - zln);
        iflag[kl] = 1;
        iflag[klp1] = 1;
        if (conn >   con)
        {
            iflag[kl] = 0;
            iflag[klp1] = 0;
            continue;
        }
    }
    /* The case where : -----z[i]>z[i+1] */
    else if (zp[kl] > zp[klp1])
    {
        isrn = isr - 1;
        ieln = isrn - 1;

        nu = ieln - isl + 1;
        for (j = isl; j <= ieln; j++)
        {
            ji = j - isl + 1;
            fp[ji] = f[j];
            for (i = 1; i <= m; i++)
            {
                ctp[i][ji] = ct[i][j];
            }
        }

        rc = LA_Linf (m, nu, ctp, fp, &irankl, &iterl, rl,
                      al, &zln);
        if (rc < LaRcOk)
        {
            GOTO_CLEANUP_RC (rc);
        }

        nu = ier - isrn + 1;
        for (j = isrn; j <= ier; j++)
        {
            ji = j - isrn + 1;
            fp[ji] = f[j];
            for (i = 1; i <= m; i++)
            {
                ctp[i][ji] = ct[i][j];
```

```
                }
            }

            rc = LA_Linf (m, nu, ctp, fp, &irankr, &iterr, rr,
                          ar, &zrn);
            if (rc < LaRcOk)
            {
                GOTO_CLEANUP_RC (rc);
            }

            conn = fabs (zrn - zln);
            iflag[kl] = 1;
            iflag[klp1] = 1;
            if (conn > con)
            {
                iflag[kl] = 0;
                iflag[klp1] = 0;
                continue;
            }
        }
        for (j = isl; j <= ieln; j++)
        {
            ji = j - isl + 1;
            rp2[j] = rl[ji];
        }
        for (j = isrn; j <= ier; j++)
        {
            ji = j - isrn + 1;
            rp2[j] = rr[ji];
        }
        for (i = 1; i <= m; i++)
        {
            ap[kl][i] = al[i];
            ap[klp1][i] = ar[i];
        }
        zp[kl] = zln;
        zp[klp1] = zrn;
        ixl[klp1] = isrn;
        isr = isrn;
        iel = isr - 1;
    }

    is = 2;
    if (istart == 2) is = 1;
    istart = is;
```

```
        icb = 0;
        for (j = 1; j <= npiece; j++)
        {
            if (iflag[j] != 0) icb = 1;
        }
        if (icb == 0)
        {
            GOTO_CLEANUP_RC (LaRcSolutionFound);
        }
    }

CLEANUP:

    free_Vector_R (al);
    free_Vector_R (rl);
    free_Vector_R (bv);
    free_Matrix_R (binv, m + 1);

    free_Vector_I (icbas);
    free_Vector_I (irbas);
    free_Vector_I (ibound);

    free_Vector_R (ar);
    free_Vector_R (rr);

    free_Matrix_R (ctp, m + 1);
    free_Vector_R (fp);

    free_Vector_I (iflag);

    return rc;
}
```

Chapter 16

Solution of Linear Inequalities

16.1 Introduction

Until now, we have dealt with different solutions of overdetermined systems of linear equations, in the L_1 and in the Chebyshev norms. In this chapter, we deal with the solution of overdetermined systems of linear inequalities of the form

$$(16.1.1) \qquad \mathbf{Ca} > \mathbf{0}$$

C is a given real N by M matrix, and the **0** is a zero N-vector, N > M. It is required to calculate the M-vector **a** that satisfies this inequality.

Solution of linear systems of inequalities such as (16.1.1) has interesting applications for pattern classification problems, as described in detail in Section 16.2.

In short we will show in Section 16.4 that the solution of system (16.1.1) is none other than the one-sided solution of a system of linear equations of the form **Ca** = **f**, where **f** is a strictly positive vector.

As early as 1952, Hoffman [20] considered the solution of the consistent system of linear inequalities (using our notation)

$$(16.1.2) \qquad \mathbf{Ca} \le \mathbf{f}$$

By *consistent*, it was meant that a solution exists that satisfies the system of inequalities. He gave a quantitative formulation of the assertion that if vector **a** almost satisfies (16.1.2), then **a** is close to a solution.

Over forty years later, Guler et al. [15] considered the result of Hoffman and obtained a particularly simple proof of Hoffman's existence theorem. They also obtained a new representation for the corresponding Lipschitz bound in that theorem and provided geometric representation of these bounds.

In 1954, both Agmon [5] and Motzkin and Schoenberg [24] discussed an iterative method called the relaxation-projection method for solving the consistent system

$$\mathbf{Ca} \geq \mathbf{f} \tag{16.1.3}$$

Associating these inequalities to half-spaces, in which a point corresponding to a feasible solution lies, they proved that such a point can be reached from some arbitrary outside point. Forty-three years later, Labonte [21] examined the implementation of the relaxation projection method to solve sets of linear inequalities using artificial neural networks. He described the different versions of this method and reported on tests he ran with simulated realizations of neural networks.

Goffin [14] studied the rate of conversion of the relaxation method by Agmon [5] and Motzkin and Schoenberg [24] and the possible finiteness of the method.

Censor and Elfving [8] considered the iterative method of Cimmino for solving linear equations and generalized it to solve a general system of linear inequalities of the form $\mathbf{Ca} \leq \mathbf{f}$. They also showed how to modify a Richardson-type iterative least-squares algorithm for computing a solution for the linear inequalities. De Pierro and Iusem [11] proved the convergence of that method starting from any point, both for consistent and inconsistent systems. The convergence is to a feasible solution in the first case and to a weighted least squares type solution in the second case. He [18] presented a new method based on an iterative contraction technique to a convex minimization problem. A system of linear inequalities may be translated to an equivalent unconstrained smooth convex minimization problem to which the contraction method is applied.

None of the aforementioned authors presented an algorithm for solving the inequalities (16.1.2) or (16.1.3). Interesting enough, a solution to these equations is none other than a solution to the one-sided overdetermined system of equations of the form $\mathbf{Ca} = \mathbf{f}$ from above and from below respectively. We shall introduce that in Section 16.3.

Nagaraja and Krishna [25] developed an algorithm for solving the system of linear inequalities $\mathbf{Ca} > \mathbf{0}$ using the method of conjugate gradients for function minimization. They showed that the algorithm converges to a solution in a finite number of steps for both consistent and inconsistent cases. They stated that their algorithm converges faster than that of Ho and Kashyap [19] and the accelerated relaxation algorithms (which will be discussed shortly).

Nie and Xu [26] dealt with the inequality system $\mathbf{Ca} \geq \mathbf{f}$. If it is an inconsistent system, i.e., having an infeasible solution, they determine which inequality in the system causes the inconsistency, and correct its right hand side. They use the isometric plane method of linear programming.

Cohen and Megiddo [10] considered the linear systems of inequalities $\mathbf{Ca} \leq \mathbf{f}$, where each inequality involves, at-most, two out of the M elements (using our notation), i.e., each row of matrix **C** contains, at-most, two nonzero elements. They state that their algorithm is faster than previous algorithms.

Faigle et al. in their book ([13] Section 2.4) attempted to solve the linear system of inequalities $\mathbf{Ca} \leq \mathbf{f}$, by eliminating one variable after the other, until a solution is obtained, or decide that no solution is feasible. A solution, if it exists, is obtained by back substitution. However, in the elementary row operations, only multiples with strictly positive scalars are allowed. This is because multiplication of an inequality by a negative scalar reverses the inequality sign. They use the Fourier-Motzkin elimination method, which can be viewed as Gauss elimination with respect to the set of non-negative scalars. As a result, the Fourier-Motzkin elimination may considerably increase the number of inequalities in every elimination step.

Han [17] described an algorithm for solving the system of inequalities $\mathbf{Ca} \leq \mathbf{f}$ using a least squares solution technique. The algorithm employs a singular value decomposition to sub-matrices of matrix **C**. Later, Bramley and Winnicka [7] improved over Han's algorithm. They used the computationally efficient **QR** factorization instead, which allows updating and downdating of matrix **R**. As a result, their algorithm is much faster than that of Han [17]. For the **QR** factorization, see Chapter 17. For the updating and downdating techniques see Stewart [30] and for an algorithm using updating and downdating techniques, see Abdelmalek [4].

Ho and Kashyap [19] also used a least squares solution technique for solving the system $\mathbf{Ca} > \mathbf{0}$. They considered the following equivalent problem instead. It is required to calculate the M weight vector **a** and an N-vector **f** such that

(16.1.4) $$\mathbf{Ca} = \mathbf{f}, \quad \mathbf{f} > \mathbf{0}$$

The residual vector of (16.1.4) is

(16.1.5) $$\mathbf{r} = \mathbf{Ca} - \mathbf{f}$$

Starting from a guess vector $\mathbf{f} = \mathbf{f}^{(0)} > \mathbf{0}$, they used an iterative least squares minimization method, where the residual vector (16.1.5) is minimized in the Euclidean norm. In each iteration the r.h.s. vector **f** is changed to $\mathbf{f} + \delta\mathbf{f}$ and a new least squares solution vector is calculated. If there is a solution to the system $\mathbf{Ca} > \mathbf{0}$, the algorithm converges in a finite number of steps, due to the fact that in each iteration both vectors **a** and **f** change. Ho-Kashyap algorithm needs the calculation of the pseudo-inverse $\mathbf{C}^+$ of matrix **C**.

Abdelmalek [2] presented an algorithm analogous to that of Ho and Kashyap, except that the residual vector (16.1.5) is minimized in the Chebyshev norm rather than in the least squares sense. In the iterations of this method, parametric linear programming techniques [16] were used with the Chebyshev approximation algorithm. This resulted in a speed improvement to this method, and an algorithm that converges faster than that of Ho and Kashyap [19]. In this method, matrix **C** need not be a full rank matrix.

Pinar and Chen [27] presented an algorithm that calculated the L_1 solution for the system $\mathbf{Ca} = \mathbf{f}$ instead, where **f** is an n-vector each element of which is 1. In their method, the L_1 norm minimization problem is approximated by a piecewise quadratic smooth function.

Bahi and Sreedharan [6] presented an algorithm for the solution of the linear system of inequalities $\mathbf{Ca} \geq \mathbf{f}$, in the L_p norm, where, $1 < p < \infty$. They first characterized the solutions for the problem and then introduced a dual to this problem. They presented the results of their algorithm for 2 examples. Their numerical results for some cases show the number of iterations to be very large.

16.1.1 Linear programming techniques

Linear programming techniques have been applied for some time to the solution of the linear inequality $\mathbf{Ca} > \mathbf{0}$. Minnick [23] showed how linear programming could be used in the solution of the linear input logic problem, a linearly separable switching function, which reduces to a linear inequality problem. Mangasarian [22] suggested the use of linear programming to the solution of the problem, without actually solving it. Smith [29] presented a linear programming formulation of discriminant function design, which reduces to a pattern classifier design. Duda and Hart ([12] pp. 167-169) presented two different linear programming formulations to the problem $\mathbf{Ca} \geq \mathbf{f}$, without implementing them.

In our work, we solve problem (16.1.1) as a linear one-sided approximation problem in the L_p norm, for $p = \infty$ and for $p = 1$. We make use of the definition of the one-sided solution of overdetermined linear equations in the Chebyshev norm of Chapter 11 and in the L_1 norm of Chapter 6 respectively. We formulate the problem in such a way that we may apply either of these two algorithms to it. Both algorithms use linear programming techniques.

We also observe that there are one-sided problems whose solutions do not exist. Hence, there are linear systems of inequalities that do not have feasible solutions. See Example 16.2 in Section 16.6.

In Section 16.2, the pattern classification problem in formulated as a problem of linear inequalities. In Section 16.3, the solution of the system of linear inequalities $\mathbf{Ca} > \mathbf{0}$ is presented as a one-sided solution of the system of linear equations $\mathbf{Ca} = \mathbf{f}$, where $\mathbf{f}$ is a strictly positive vector. In Sections 16.4 and 16.5, the linear one-sided Chebyshev approximation algorithm and the linear one-sided L_1 approximation algorithm are outlined respectively. In Section 16.6, two numerical examples for solving the pattern classification problems are given with comments.

16.2 Pattern classification problem

Overdetermined linear inequalities form a basic problem in pattern classification. Given is a class $\mathbf{A}$ of s patterns and a class $\mathbf{B}$ of t patterns, where each pattern is a point in an m-dimensional Euclidean space. Let $n = (s + t)$ and usually $n >> m$.

The pattern classification problem is summarized as follows. It is required to find a surface in the m-dimensional space such that all points of class $\mathbf{A}$ be on one side of this surface and all points of class $\mathbf{B}$ be on the other side of the surface. Let the equation of this separating surface be

$$a_1\phi_1(x) + a_2\phi_2(x) + \dots + a_{m+1}\phi_{m+1}(x) = 0$$

Vector $\mathbf{a} = (a_1, a_2, \dots, a_{m+1})^T$ is to be calculated and $\{\phi_1(x), \phi_2(x), \dots, \phi_{m+1}(x)\}$ is a set of linearly independent functions to be specified according to the geometry of the problem. Without loss of generality, we may take $\phi_{m+1}(x) = 1$.

Following Tou and Gonzalez ([31], pp. 40-41, 48-49), a decision function d(x) of the form (taking $\phi_{m+1}(x) = 1$)

$$d(x) = a_1\phi_1(x) + a_2\phi_2(x) + \dots + a_m\phi_m(x) + a_{m+1}$$

is established. This function has the property that

(16.2.1) $$d(x_i) < 0, \quad x_i \in \mathbf{A}$$

and

(16.2.2) $$d(x_i) > 0, \quad x_i \in \mathbf{B}$$

The two classes of patterns **A** and **B** are linearly separable if and only if there exists an (m + 1)-vector **a** such that the above two inequalities are satisfied. If no such vector exists, then classes **A** and **B** are linearly inseparable. By multiplying the first set of inequalities by a –ve signs, we get

(16.2.3) $$-d(x_i) > 0, \quad x_i \in \mathbf{A}$$

The problem may now be posed as follows. Using (16.2.3) and (16.2.2), let **C** be an n by (m + 1) matrix whose i^{th} row $\mathbf{C}_i$ is

$$\mathbf{C}_i = (-\phi_1(x_i), -\phi_2(x_i), \dots, -\phi_m(x_i), -1), \quad 1 \le i \le s$$

and

$$\mathbf{C}_i = (\phi_1(x_i), \phi_2(x_i), \dots, \phi_m(x_i), 1), \quad (s + 1) \le i \le n$$

It is required to calculate the (m + 1)-vector **a** such that the system of inequalities

$$\mathbf{Ca} > \mathbf{0}$$

is satisfied. This is the same system of inequalities (16.1.1) above. In this case, N = n and M = (m + 1).

Clark and Gonzalez [9] presented an algorithm based on a search procedure where matrix **C** has to satisfy the Haar condition. That is, every (m + 1) by (m + 1) sub-matrix of **C** is of rank (m + 1). Sklansky and Michelotti [28] described an algorithm, called a locally trained piecewise linear classifier, for use with multi-dimensional data.

16.3 Solution of the system of linear inequalities Ca > 0

A common practice is to replace the inequalities $\mathbf{Ca} > \mathbf{0}$ by the inequalities

$$\mathbf{Ca} \geq \mathbf{f}$$

where **f** is an N-vector, each element of which is positive. It is easier to manipulate the system $\mathbf{Ca} \geq \mathbf{f}$ and that its solution is also a solution to the system $\mathbf{Ca} > \mathbf{0}$.

The solution of the system of linear inequalities $\mathbf{Ca} \geq \mathbf{f}$, reminds us of the one-sided solutions of overdetermined systems of linear equations in the Chebyshev norm (Chapter 11) or in the L_1 norm (Chapter 6).

Here, two algorithms for solving the overdetermined system of linear inequalities $\mathbf{Ca} > \mathbf{0}$ are presented [3]. The first algorithm calculates the one-sided Chebyshev solution from below of the system $\mathbf{Ca} = \mathbf{f}$, where **f** is a strictly positive vector. The second algorithm calculates the one-sided L_1 solution from below of the same system $\mathbf{Ca} = \mathbf{f}$, where **f** is a strictly positive vector. If a solution exists to either of the one-sided problems, it would be a solution to the given system of inequalities $\mathbf{Ca} > \mathbf{0}$.

16.4 Linear one-sided Chebyshev approximation algorithm

Consider the overdetermined system of linear equations derived from $\mathbf{Ca} \geq \mathbf{f}$, namely

$$\mathbf{Ca} = \mathbf{f} \tag{16.4.1}$$

where $\mathbf{C} = (c_{ij})$ is an N by M real matrix and $\mathbf{f} = (f_i)$ is an N real vector, N > M. The Chebyshev solution of (16.4.1) is vector **a** that minimizes the Chebyshev norm of the residuals

$$z = \max|r_i|, \quad i = 1, 2, \ldots, N$$

where r_i is the ith residual and is given by

$$r_i = \sum_{j=1}^{m} c_{ij}a_j - f_i, \quad i = 1, 2, \ldots, n$$

As in Chapter 11, when the solution vector **a** satisfies the additional conditions

$$r_i \geq 0, \quad i = 1, 2, \ldots, N$$

we have the one-sided Chebyshev solution from below [1] of system **Ca** = **f**.

Let $h = \max_i|r_i|$, then this problem is formulated in linear programming as follows

$$\text{minimize } h$$

subject to

$$\mathbf{0} \leq \mathbf{Ca} - \mathbf{f} \leq h\mathbf{e}$$

where **e** is an N-vector, each element of which is 1 and the **0** is an N-zero vector. From the left inequality we have

$$\mathbf{Ca} \geq \mathbf{f}$$

Since we assume that **f** is a strictly positive vector, the solution of the inequalities $\mathbf{Ca} \geq \mathbf{f}$, is a solution of the inequalities $\mathbf{Ca} > \mathbf{0}$. In Chapter 11, an algorithm for the one-sided Chebyshev solution of overdetermined systems of linear equations is presented.

16.5 Linear one-sided L_1 approximation algorithm

Consider the overdetermined system of linear equations **Ca** = **f** in (16.4.1). The L_1 solution of this equation is the solution vector **a** that minimizes the L_1 norm of the residuals

$$z = \sum_{i=1}^{N} |r_i|$$

where r_i is the ith residual, as defined in section 16.4.

As in Chapter 6, when the solution vector **a** satisfies the additional conditions

$$r_i \geq 0, \quad i = 1, 2, \ldots, N$$

we have the one-sided L_1 solution from below of system $\mathbf{Ca} = \mathbf{f}$; that is, for any equation i, i = 1, …, N, the observed value f_i is not greater than the calculated value of element i of **Ca**.

In other words, the inequalities $r_i \geq 0$, i = 1, 2, …, N, are translated to

$$\sum_{j=1}^{M} c_{ij}a_j \geq f_i, \quad i = 1, 2, \ldots, N$$

Or in vector-matrix form

$$\mathbf{Ca} \geq \mathbf{f}$$

Since we assume that **f** is a strictly positive vector, the solution vector **a** of $\mathbf{Ca} \geq \mathbf{f}$ is a solution to the inequalities $\mathbf{Ca} > \mathbf{0}$. In Chapter 6, an algorithm for the one-sided L_1 solution of overdetermined systems of linear equations was presented.

16.6 Numerical results and comments

DR_Chineq() calls the one-sided Chebyshev solution function LA_Linfside() of Chapter 11, and DR_L1ineq() calls the one-sided L_1 function LA_Loneside() of Chapter 6. Two pattern classification examples are solved here.

Example 16.1

Consider the two dimensional problem of 16 points that belong to the classes **A** and **B**. Their coordinates are as follows.

Class **A**:

{(1, 3), (1.5, 5), (2, 6), (2.5, 7), (–1, 2), (–1.5, 4), (–2, 6), (–2.5, 7)}

Class **B**:

{(2, 2), (4, 4), (5, 6), (6, 7), (–3, –1), (–4, 3), (–5, 4), (–6, 6)}

We assume that the separating curve of the two classes is the vertical parabola

$$a_1 + a_2\mathbf{x} + a_3\mathbf{x}^2 + a_4\mathbf{y} = 0 \quad (16.6.1)$$

We should remember that the decision function of (16.2.1) for class **A** is $d(x_i) < 0$, $x_i \in \mathbf{A}$. The decision function for this example is the l.h.s. of the equation of the vertical parabola in (16.6.1). In order to reverse the < sign for class **A**, we multiply $d(x_i) < 0$, $x_i \in \mathbf{A}$ by −1. Also, it is customary to take each element of vector **f** as 1. Hence, from **Ca** ≥ **f**, the following equation is formed

$$a_1\mathbf{1} + a_2\mathbf{x} + a_3\mathbf{x}^2 + a_4\mathbf{y} = \mathbf{f}$$

where **1** is a vector of 16 elements, each of which is 1. That is

$$\begin{bmatrix} -1 & -1 & -1 & -3 \\ -1 & -1.5 & -2.25 & -5 \\ -1 & -2 & -4 & -6 \\ -1 & -2.5 & -6.25 & -7 \\ -1 & 1 & -1 & -2 \\ -1 & 1.5 & -2.25 & -4 \\ -1 & 2 & -4 & -6 \\ -1 & 2.5 & -6.25 & -7 \\ 1 & 2 & 4 & 2 \\ 1 & 4 & 16 & 4 \\ 1 & 5 & 25 & 6 \\ 1 & 6 & 36 & 7 \\ 1 & -3 & 9 & -1 \\ 1 & -4 & 16 & 3 \\ 1 & -5 & 25 & 4 \\ 1 & -6 & 36 & 6 \end{bmatrix} \begin{bmatrix} a_1 \\ a_2 \\ a_3 \\ a_4 \end{bmatrix} = \begin{bmatrix} 1 \\ 1 \\ 1 \\ 1 \\ 1 \\ 1 \\ 1 \\ 1 \\ 1 \\ 1 \\ 1 \\ 1 \\ 1 \\ 1 \\ 1 \\ 1 \end{bmatrix}$$

The solution by the first algorithm, the one-sided Chebyshev approximation from below, is $\mathbf{a} = (1.42, 0.38, 0.287, -1.164)^T$, which is $y = 1.22 + 0.327x + 0.247x^2$. The solution by the second algorithm, the one-side L_1 approximation from below, is $\mathbf{a} = (0.733, 0.4, 0.267, -0.8)^T$, which is $y = 0.92 + 0.5x + 0.33x^2$. The separating curves for the two algorithms are shown in Figure 16-1.

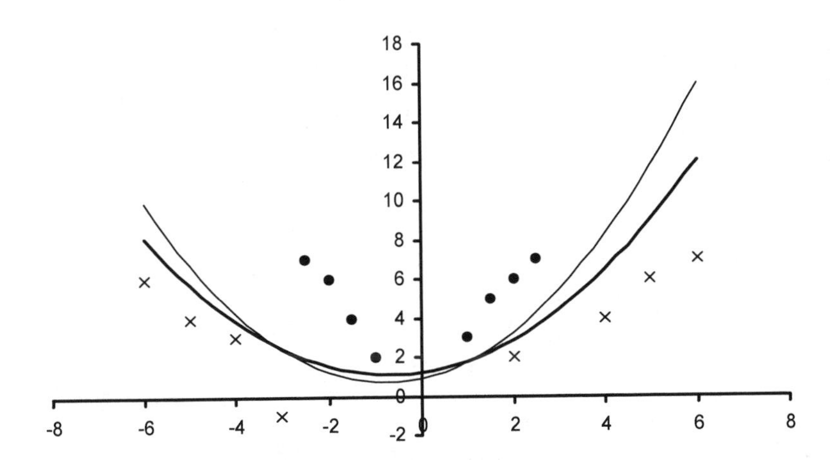

Figure 16-1: Two curves that each separate the patterns of class A from the patterns of class B

Algorithm 1 produces the thick parabola and algorithm 2 produces the thin parabola. Each parabola separates the patterns of class **A**, denoted by "**.**", from the patterns of class **B**, denoted by "×".

Example 16.2

Consider the two dimensional problem of 12 points that belong to the two classes **A** and **B**. Their coordinates are

Class **A**:

$$\{(2, 1), (2, -1), (4, 1), (4, -1), (-3, 1), (-3, -1)\}$$

Class **B**:

$$\{(3, -3), (5, -3), (4, -4), (-1, 3), (-3, 3), (-2, 5)\}$$

We assume that the separating curve of the two classes is the ellipse

$$a_1x^2 + a_2y^2 + a_3 = 0$$

Vector **f** is again a 12-vector, each element of which is 1. Hence, from the inequality $\mathbf{Ca} \geq \mathbf{f}$, the following equation is formed

$$a_1\mathbf{x}^2 + a_2\mathbf{y}^2 + a_3 = \mathbf{f}$$

$$\begin{bmatrix} -4 & -1 & -1 \\ -4 & -1 & -1 \\ -16 & -1 & -1 \\ -16 & -1 & -1 \\ -9 & -1 & -1 \\ -9 & -1 & -1 \\ 9 & 9 & 1 \\ 25 & 9 & 1 \\ 16 & 16 & 1 \\ 1 & 9 & 1 \\ 9 & 9 & 1 \\ 4 & 25 & 1 \end{bmatrix} \begin{bmatrix} a_1 \\ a_2 \\ a_3 \end{bmatrix} = \begin{bmatrix} 1 \\ 1 \\ 1 \\ 1 \\ 1 \\ 1 \\ 1 \\ 1 \\ 1 \\ 1 \\ 1 \\ 1 \end{bmatrix}$$

The solution vector by either of the two algorithms is $\mathbf{a} = (0.0, 0.25, -1.25)^T$, which gives the result $y^2 = 5$, or $y = \pm\text{sqrt}(5)$. Certainly, the two straight lines $y = \pm$ sqrt(5) is a solution to the inequality $\mathbf{Ca} > \mathbf{0}$.

For this example the second point in class **B** is then changed from (5, –3) to (5, 1) and the problem is re-solved. The obtained result by both algorithms is $\mathbf{a} = (0.222, 0.667, -5.222)^T$. This is an ellipse of semi-major and semi-minor axes 4.85 and 2.8 respectively.

Once more, for this example, the second point in class **B** is changed from (5, –3) to (4, 1). Both algorithms return, indicating that no feasible solution is obtained; that is, the two classes **A** and **B** are inseparable.

It should be noted that if we replace vector **f** in (16.4.1) by K**f**, where K is a positive scalar, the obtained solution vector would be K**a** and we would get the same separating surface.

Our two algorithms are comparable in speed. They are faster and simpler than other existing algorithms including a previous one by the author [2]. They also converge in a finite number of iterations. Being linear programs, the number of iterations is of the order of 2 to 3 times the smaller dimension of matrix **C** in $\mathbf{Ca} > \mathbf{0}$. Another important points is the following.

Finally, since we are using linear programming techniques, matrix **C** need not be of full rank.

References

1. Abdelmalek, N.N., The discrete linear one-sided Chebyshev approximation, *Journal of Institute of Mathematics and Applications*, 18(1976)361-370.
2. Abdelmalek, N.N., Chebyshev approximation algorithm for linear inequalities and its applications to pattern recognition, *International Journal of Systems Science*, 12(1981)963-975.
3. Abdelmalek, N.N., Linear one-sided approximation algorithms for the solution of overdetermined systems of linear inequalities, *International Journal of Systems Science*, 15(1984)1-8.
4. Abdelmalek, N.N., Piecewise linear least-squares approximation of planar curves, *International Journal of Systems Science*, 21(1990)1393-1403.
5. Agmon, S., The relaxation method for linear inequalities, *Canadian Journal of Mathematics*, 6(1954)382-392.
6. Bahi, S. and Sreedharan, V.P., An algorithm for a minimum norm solution of a system of linear inequalities, *International Journal of Computer Mathematics*, 80(2003)639-647.
7. Bramley, R. and Winnicka, B., Solving linear inequalities in a least squares sense, *SIAM Journal on Scientific Computation*, 17(1996)275-286.
8. Censor, Y. and Elfving, T., New methods for linear inequalities, *Linear Algebra and its Applications*, 42(1982)199-211.
9. Clark, D.C. and Gonzalez, R.C., Optimal solution of linear inequalities with applications to pattern recognition, *IEEE Transactions on Pattern Analysis and Machine Intelligence*, 6(1981)643-655.
10. Cohen, E. and Megiddo, N., Improved algorithms for linear inequalities with two variables per inequality, *SIAM Journal on Computing*, 23(1994)1313-1347.
11. De Pierro, A.R. and Iusem, A.N., A simultaneous projections method for linear inequalities, *Linear Algebra and its Applications*, 64(1985)243-253.
12. Duda, R.O. and Hart, P.E., *Pattern Classification and Scene Analysis*, John Wiley & Sons, New York, 1973.

13. Faigle, U., Kern, W. and Still, G., *Algorithmic Principles of Mathematical Programming*, Kluwer Academic Publishers, London, 2002.
14. Goffin, J.L., The relaxation method for solving systems of linear inequalities, *Mathematics of Operations Research*, 5(1980)388-414.
15. Guler, O., Hoffman, A.J. and Rothblum, U.G., Approximations to solutions to systems of linear inequalities, *SIAM Journal on Matrix Analysis and Applications*, 16(1995)688-696.
16. Hadley, G., *Linear Programming*, Addison-Wesley, Reading, MA, 1962.
17. Han, S-P., Least squares solution of linear inequalities, Technical Report TR-2141, Mathematics Research Center, University of Wisconsin-Madison, 1980.
18. He, B., New contraction methods for linear inequalities, *Linear Algebra and its Applications*, 207(1994)115-133.
19. Ho, Y-C. and Kashyap, R.L., An algorithm for linear inequalities and its applications, *IEEE Transactions on Electronic Computers*, 14(1965)683-688.
20. Hoffman, A.J., On approximate solutions of systems of linear inequalities, *Journal of Research of the National Bureau of Standards*, 49(1952)263-265.
21. Labonte, G., On solving systems of linear inequalities with artificial neural networks, *IEEE Transactions on Neural Networks*, 8(1997)590-600.
22. Mangasarian, O.L., Iterative solution of linear programs, *SIAM Journal on Numerical Analysis*, 18(1981)606-614.
23. Minnick, R.C., Linear-input logic, *IRE Transactions on Electronic Computers*, 10(1961)6-16.
24. Motzkin, T.S. and Schoenberg, I.J., The relaxation method for linear inequalities, *Canadian Journal of Mathematics*, 6(1954)393-404.
25. Nagaraja, G. and Krishna, G., An algorithm for the solution of linear inequalities, *IEEE Transactions on Computers*, 23(1974)421-427.

26. Nie, Y.Y. and Xu, S.R., Determination and correction of an inconsistent system of linear inequalities, *Journal of Computational Mathematics*, 13(1995)211-217.
27. Pinar, M.C. and Chen, B., l_1 solution of linear inequalities, *IMA Journal of Numerical Analysis*, 19(1999)19-37.
28. Sklansky, J. and Michelotti, L., Locally trained piecewise linear classifier, *IEEE Transactions on Pattern Analysis and Machine Intelligence*, 2(1980)101-110.
29. Smith, F.W., Pattern classifier design by linear programming, *IEEE Transactions on Computers*, 17(1968)367-372.
30. Stewart, G.W., The effects of rounding residual on an algorithm for downdating a Cholesky factorization, *Journal of Institute of Mathematics and Applications*, 23(1979)203-213.
31. Tou, J.L. and Gonzalez, R.C., *Pattern Recognition Principles*, Addison-Wesley, Reading, MA, 1974.

16.7 DR_Chineq

```
/*-----------------------------------------------------------------
DR_Chineq
-------------------------------------------------------------------
Given is an overdetermined system of linear inequalities of the form

(1)                     c*a > 0

"c" is a given real n by m matrix of rank k, k <= m < n.  Usually,
n>>m.  It is required to obtain the m solution vector "a" for this
system.

System (1) is first replaced by the overdetermined system of linear
equations

(2)                     c*a = f

where "f" is a strictly real positive n vector.

This program is a driver for the function LA_Chineq(), which
calculates the one-sided "Chebyshev" solution from below of system
(2).  This solution itself would be a solution of the inequality (1).

This driver contains examples for problems of pattern classification
in which there are two classes of patterns.  It is required to
calculate the parameters of a plane curve that separates the two
classes. The results of these examples appear in the text.

Example 1:
    class a: { (1,3), (1.5,5), (2,6), (2.5,7), (-1,2), (-1.5,4),
               (-2,6), (-2.5,7) }
    class b: { (2,2), (4,4), (5,6), (6,7), (-3,-1), (-4,3), (-5,4),
               (-6,6) }
    separating curve is the parabola

               a1 + a2*x + a3*x*x + a4*y = 0

Example 2:
    class A: { (2,1), (2,-1), (4,1), (4,-1), (-3,1), (-3,-1) }
    class B: { (3,-3), (5,-3), (4,-4), (-1,3), (-3,3), (-2,5) }
    Separating curve is the ellipse

               a1*x*x + a2*y*y + a3 = 0
```

```
Example 3:
    class A: { (2,1), (2,-1), (4,1), (4,-1), (-3,1), (-3,-1) }
    class B: { (3,-3), (5,1), (4,-4), (-1,3), (-3,3), (-2,5) }
    Separating curve is the ellipse

                a1*x*x + a2*y*y + a3 = 0

    This example is itself example 2, but with the second point
    in B (5,-3) replaced by (5,1).

Example 4:
    class A: { (2,1), (2,-1), (4,1), (4,-1), (-3,1), (-3,-1) }
    class B: { (3,-3), (4,1), (4,-4), (-1,3), (-3,3), (-2,5) }
    Separating curve is the ellipse

                a1*x*x + a2*y*y + a3 = 0

    This example is itself example 2, but with the second point in
    B (5,-3) replaced by (4,1).
--------------------------------------------------------------------*/

#include "DR_Defs.h"
#include "LA_Prototypes.h"

#define N1q   16
#define M1q    4
#define N2q   12
#define M2q    3

void DR_Chineq (void)
{
    /*-----------------------------------------
      Constant matrices/vectors
    -----------------------------------------*/
    static tNumber_R Cainit[N1q][M1q] =
    {
        { -1.0,  -1.0,   -1.0,  -3.0 },
        { -1.0,  -1.5,   -2.25, -5.0 },
        { -1.0,  -2.0,   -4.0,  -6.0 },
        { -1.0,  -2.5,   -6.25, -7.0 },
        { -1.0,   1.0,   -1.0,  -2.0 },
        { -1.0,   1.5,   -2.25, -4.0 },
        { -1.0,   2.0,   -4.0,  -6.0 },
        { -1.0,   2.5,   -6.25, -7.0 },
```

```
    { 1.0,   2.0,    4.0,   2.0 },
    { 1.0,   4.0,   16.0,   4.0 },
    { 1.0,   5.0,   25.0,   6.0 },
    { 1.0,   6.0,   36.0,   7.0 },
    { 1.0,  -3.0,    9.0,  -1.0 },
    { 1.0,  -4.0,   16.0,   3.0 },
    { 1.0,  -5.0,   25.0,   4.0 },
    { 1.0,  -6.0,   36.0,   6.0 }
};


static tNumber_R Cbinit[N2q][M2q] =
{
    {  -4.0, -1.0, -1.0 },
    {  -4.0, -1.0, -1.0 },
    { -16.0, -1.0, -1.0 },
    { -16.0, -1.0, -1.0 },
    {  -9.0, -1.0, -1.0 },
    {  -9.0, -1.0, -1.0 },
    {   9.0,  9.0,  1.0 },
    {  25.0,  9.0,  1.0 },
    {  16.0, 16.0,  1.0 },
    {   1.0,  9.0,  1.0 },
    {   9.0,  9.0,  1.0 },
    {   4.0, 25.0,  1.0 }
};

static tNumber_R fa[N1q+1] =
{   NIL,
    1.0, 1.0, 1.0, 1.0, 1.0, 1.0, 1.0, 1.0, 1.0, 1.0, 1.0, 1.0,
    1.0, 1.0, 1.0, 1.0
};

static tNumber_R fb[N2q+1] =
{   NIL,
    1.0, 1.0, 1.0, 1.0, 1.0, 1.0, 1.0, 1.0, 1.0, 1.0, 1.0, 1.0
};

/*----------------------------------------
  Variable matrices/vectors
----------------------------------------*/
tMatrix_R   ct  = alloc_Matrix_R (MMc_COLS, NN_ROWS);
tVector_R   f   = alloc_Vector_R (NN_ROWS);
tVector_R   r   = alloc_Vector_R (NN_ROWS);
tVector_R   a   = alloc_Vector_R (MMc_COLS);
```

```
tMatrix_R   Ca  = init_Matrix_R (&(Cainit[0][0]), N1q, M1q);
tMatrix_R   Cb  = init_Matrix_R (&(Cbinit[0][0]), N2q, M2q);

int         irank, iter, iside;
int         i, j, m, n, Iexmpl;
tNumber_R   z = 0.0;

eLaRc       rc = LaRcOk;

prn_dr_bnr ("DR_Chineq, Solving an Overdetermined System of"
            " Linear Inequalities");

for (Iexmpl = 1; Iexmpl <= 4; Iexmpl++)
{
    switch (Iexmpl)
    {
        case 1:
            n = N1q;
            m = M1q;
            for (i = 1; i <= n; i++)
            {
                f[i] = fa[i];
                for (j = 1; j <= m; j++)
                {
                    ct[j][i] = Ca[i][j];
                }
            }
            break;

        case 2:
            n = N2q;
            m = M2q;
            for (i = 1; i <= n; i++)
            {
                f[i] = fb[i];
                for (j = 1; j <= m; j++)
                {
                    ct[j][i] = Cb[i][j];
                }
            }
            break;

        case 3:
            n = N2q;
```

```
            m = M2q;
            for (i = 1; i <= n; i++)
            {
                f[i] = fb[i];
                for (j = 1; j <= m; j++)
                {
                    ct[j][i] = Cb[i][j];
                }
                ct[2][8] = 1.0;
            }
            break;

        case 4:
            n = N2q;
            m = M2q;
            for (i = 1; i <= n; i++)
            {
                f[i] = fb[i];
                for (j = 1; j <= m; j++)
                {
                    ct[j][i] = Cb[i][j];
                }
                ct[1][8] = 16.0;
                ct[2][8] = 1.0;
            }
            break;

        default:
            break;
    }
    prn_algo_bnr("Chineq");
    prn_example_delim();
    PRN ("Example #%d: Size of matrix \"c\" %d by %d\n",
         Iexmpl, n, m);
    prn_example_delim();
    PRN ("\"Linfside\" for Solving an Overdetermined System of"
         " Linear Inequalities\n");
    prn_example_delim();
    PRN ("r.h.s. Vector \"f\"\n");
    prn_Vector_R (f, n);
    PRN ("Transpose of Coefficient Matrix, \"ct\"\n");
    prn_Matrix_R (ct, m, n);
    iside = 0;

    rc = LA_Linfside (iside, m, n, ct, f, &irank, &iter, r, a,
```

```
                    &z);

        if (rc >= LaRcOk)
        {
            PRN ("\n");
            PRN ("Results of the Solution of Linear Inequalities\n");
            PRN ("One-sided Chebyshev solution vector \"a\"\n");
            prn_Vector_R (a, m);
            PRN ("One-sided Chebyshev residual vector \"r\"\n");
            prn_Vector_R (r, n);
            PRN ("One-sided Chebyshev norm \"z\" = %8.4f\n", z);
            PRN ("Rank of of matrix \"c\" = %d, No. of "
                 " Iterations = %d\n", irank, iter);
        }

        prn_la_rc (rc);
    }

    free_Matrix_R (ct, MMc_COLS);
    free_Vector_R (f);
    free_Vector_R (r);
    free_Vector_R (a);

    uninit_Matrix_R (Ca);
    uninit_Matrix_R (Cb);
}
```

16.8 DR_L1ineq

```
/*------------------------------------------------------------------
DR_L1ineq
--------------------------------------------------------------------
Given is an overdetermined system of linear inequalities of the form

(1)                     c*a > 0

"c" is a given real n by m matrix of rank k, k <= m < n.  Usually,
n>>m.  It is required to obtain the m solution vector "a" for this
system.

System (1) is first replaced by the overdetermined system of linear
equations

(2)                     c*a = f

where "f" is a strictly real positive n vector.

This program is a driver for the function LA_L1ineq(), which
calculates the one-sided L-One solution from below of system (2).
This solution itself would be a solution of the inequality (1).

This driver contains examples for problems of pattern classification
in which there are two classes of patterns.  It is required to
calculate the parameters of a plane curve that separates the two
classes. The results of these examples appear in the text.

Example 1:
   class a: { (1,3), (1.5,5), (2,6), (2.5,7), (-1,2), (-1.5,4),
              (-2,6), (-2.5,7) }
   class b: { (2,2), (4,4), (5,6), (6,7), (-3,-1), (-4,3), (-5,4),
              (-6,6) }
   separating curve is the parabola

              a1 + a2*x + a3*x*x + a4*y = 0

Example 2:
   class A: { (2,1), (2,-1), (4,1), (4,-1), (-3,1), (-3,-1) }
   class B: { (3,-3), (5,-3), (4,-4), (-1,3), (-3,3), (-2,5) }
   Separating curve is the ellipse

              a1*x*x + a2*y*y + a3 = 0
```

```
Example 3:
    class A: { (2,1), (2,-1), (4,1), (4,-1), (-3,1), (-3,-1) }
    class B: { (3,-3), (5,1), (4,-4), (-1,3), (-3,3), (-2,5) }
    Separating curve is the ellipse

                a1*x*x + a2*y*y + a3 = 0

    This example is itself example 2, but with the second point
    in B (5,-3) replaced by (5,1).

Example 4:
    class A: { (2,1), (2,-1), (4,1), (4,-1), (-3,1), (-3,-1) }
    class B: { (3,-3), (4,1), (4,-4), (-1,3), (-3,3), (-2,5) }
    Separating curve is the ellipse

                a1*x*x + a2*y*y + a3 = 0.

    This example is itself example 2, but with the second point in
    B (5,-3) replaced by (4,1).
----------------------------------------------------------------------*/

#include "DR_Defs.h"
#include "LA_Prototypes.h"

#define Na 16
#define Ma 4
#define Nb 12
#define Mb 3

void DR_L1ineq (void)
{
    /*----------------------------------------
      Constant matrices/vectors
    ----------------------------------------*/
    static tNumber_R Cainit[Na][Ma] =
    {
        { -1.0,  -1.0,   -1.0,  -3.0 },
        { -1.0,  -1.5,   -2.25, -5.0 },
        { -1.0,  -2.0,   -4.0,  -6.0 },
        { -1.0,  -2.5,   -6.25, -7.0 },
        { -1.0,   1.0,   -1.0,  -2.0 },
        { -1.0,   1.5,   -2.25, -4.0 },
        { -1.0,   2.0,   -4.0,  -6.0 },
        { -1.0,   2.5,   -6.25, -7.0 },
```

```
    { 1.0,   2.0,    4.0,   2.0 },
    { 1.0,   4.0,   16.0,   4.0 },
    { 1.0,   5.0,   25.0,   6.0 },
    { 1.0,   6.0,   36.0,   7.0 },
    { 1.0,  -3.0,    9.0,  -1.0 },
    { 1.0,  -4.0,   16.0,   3.0 },
    { 1.0,  -5.0,   25.0,   4.0 },
    { 1.0,  -6.0,   36.0,   6.0 }
};

static tNumber_R Cbinit[Nb][Mb] =
{
    {  -4.0, -1.0, -1.0 },
    {  -4.0, -1.0, -1.0 },
    { -16.0, -1.0, -1.0 },
    { -16.0, -1.0, -1.0 },
    {  -9.0, -1.0, -1.0 },
    {  -9.0, -1.0, -1.0 },
    {   9.0,  9.0,  1.0 },
    {  25.0,  9.0,  1.0 },
    {  16.0, 16.0,  1.0 },
    {   1.0,  9.0,  1.0 },
    {   9.0,  9.0,  1.0 },
    {   4.0, 25.0,  1.0 }
};

static tNumber_R fa[Na+1] =
{   NIL,
    1.0, 1.0, 1.0, 1.0, 1.0, 1.0, 1.0, 1.0, 1.0,
    1.0, 1.0, 1.0, 1.0, 1.0, 1.0, 1.0
};

static tNumber_R fb[Nb+1] =
{   NIL,
    1.0, 1.0, 1.0, 1.0, 1.0, 1.0, 1.0, 1.0, 1.0, 1.0, 1.0, 1.0
};

/*----------------------------------------
  Variable matrices/vectors
----------------------------------------*/
tMatrix_R   ct   = alloc_Matrix_R (MM_COLS, NN_ROWS);
tVector_R   f    = alloc_Vector_R (NN_ROWS);
tVector_R   r    = alloc_Vector_R (NN_ROWS);
tVector_R   a    = alloc_Vector_R (MM_COLS);
```

```
tMatrix_R   Ca  = init_Matrix_R (&(Cainit[0][0]), Na, Ma);
tMatrix_R   Cb  = init_Matrix_R (&(Cbinit[0][0]), Nb, Mb);

int         irank, iter, iside;
int         i, j, m, n, Iexmpl;
tNumber_R   z;

eLaRc       rc = LaRcOk;

prn_dr_bnr ("DR_L1ineq, Solving an Overdetermined System of"
            " Linear Inequalities");

for (Iexmpl = 1; Iexmpl <= 4; Iexmpl++)
{
    switch (Iexmpl)
    {
        case 1:
            n = Na;
            m = Ma;
            for (i = 1; i <= n; i++)
            {
                f[i] = fa[i];
                for (j = 1; j <= m; j++)
                {
                    ct[j][i] = Ca[i][j];
                }
            }
            break;

        case 2:
            n = Nb;
            m = Mb;
            for (i = 1; i <= n; i++)
            {
                f[i] = fb[i];
                for (j = 1; j <= m; j++)
                {
                    ct[j][i] = Cb[i][j];
                }
            }
            break;

        case 3:
            n = Nb;
            m = Mb;
```

```
            for (i = 1; i <= n; i++)
            {
                f[i] = fb[i];
                for (j = 1; j <= m; j++)
                {
                    ct[j][i] = Cb[i][j];
                }
                ct[2][8] = 1.0;
            }
            break;

        case 4:
            n = Nb;
            m = Mb;
            for (i = 1; i <= n; i++)
            {
                f[i] = fb[i];
                for (j = 1; j <= m; j++)
                {
                    ct[j][i] = Cb[i][j];
                }
                ct[1][8] = 16.0;
                ct[2][8] = 1.0;
            }
            break;

        default:
            break;
    }

    prn_algo_bnr ("L1ineq");
    prn_example_delim();
    PRN ("Example #%d: Size of matrix \"c\": %d by %d\n",
         Iexmpl, n, m);
    prn_example_delim();
    PRN ("\"Loneside\" for Solving an Overdetermined System of"
         " Linear Inequalities\n");
    prn_example_delim();
    PRN ("r.h.s. Vector \"f\"\n");
    prn_Vector_R (f, n);
    PRN ("Transpose of Coefficient Matrix, \"ct\"\n");
    prn_Matrix_R (ct, m, n);
    iside = 0;

    rc = LA_Loneside (iside, m, n, ct, f, &irank, &iter, r, a,
```

```
                    &z);

        if (rc >= LaRcOk)
        {
            PRN ("\n");
            PRN ("Results of the Solution of Linear Inequalities\n");
            PRN ("One-sided L-One solution vector \"a\"\n");
            prn_Vector_R (a, m);
            PRN ("One-sided L-One residual vector \"r\"\n");
            prn_Vector_R (r, n);
            PRN ("One-sided L-One norm \"z\" = %8.4f\n", z);
            PRN ("Rank of of matrix \"c\" = %d, No. of "
                 " Iterations =  %d\n", irank, iter);
        }

        prn_la_rc (rc);
    }

    free_Matrix_R (ct, MM_COLS);
    free_Vector_R (f);
    free_Vector_R (r);
    free_Vector_R (a);

    uninit_Matrix_R (Ca);
    uninit_Matrix_R (Cb);
}
```

PART 4

The Least Squares Approximation

Chapter 17

Least Squares and Pseudo-Inverses of Matrices

17.1 Introduction

It is expected that the reader of this chapter is familiar with the concept of the least squares solution of systems of linear equations. This chapter is the extension of Chapter 4, and is also a tutorial one. Consider the systems of linear equations

$$\mathbf{Ax} = \mathbf{b}$$

$\mathbf{A}$ is a real n by m matrix, $\mathbf{b}$ is a real n-vector and $\mathbf{x}$ is the solution m-vector. The residual vector for the system $\mathbf{Ax} = \mathbf{b}$, is given here by

$$\mathbf{r} = \mathbf{b} - \mathbf{Ax}$$

It is known that $\mathbf{A}$ has an inverse $\mathbf{A}^{-1}$ if and only if $\mathbf{A}$ is a square nonsingular matrix. The solution of the system $\mathbf{Ax} = \mathbf{b}$, is given by $\mathbf{x} = \mathbf{A}^{-1}\mathbf{b}$. However, if $\mathbf{A}$ is a rectangular matrix, or a square singular matrix, $\mathbf{Ax} = \mathbf{b}$ may have an approximate solution. A least squares solution $\mathbf{x}$ minimizes the L_2 norm $\|\mathbf{r}\|_2$ of the residual vector. The least squares solution is given by $\mathbf{x} = \mathbf{A}^{+}\mathbf{b}$, where $\mathbf{A}^{+}$ is known as the pseudo-inverse of $\mathbf{A}$.

We state here some theorems concerning the least squares solution of a system of linear equations whose coefficient matrix is a real general one. Such theorems make it easy to understand the term **unique least squares solution** or **minimal length least squares solution** of a system of linear equations.

The pseudo-inverse of $\mathbf{A}$ is calculated by factorizing matrix $\mathbf{A}$ into the product of 2 or 3 matrices, each is of full rank. The pseudo-inverse of $\mathbf{A}$ is calculated in terms of the pseudo-inverses of the factorized matrices.

Two efficient matrix factorization methods are considered. These are the Gauss **LU** factorization with complete pivoting and the Householder's **QR** factorization method with pivoting.

In Section 17.2, theorems are given concerning necessary and sufficient conditions for the square of the residual vector to have a minimum value. This leads to the definition and calculation of the pseudo-inverse of matrix **A** and the definition of the minimal length least squares solution.

In Section 17.3, the two mentioned factorizations of matrix **A** are presented. A review of other factorization methods, such as Givens' plane rotation and the classical and modified Gram-Schmidt methods, is also presented.

In Section 17.4, explicit expressions for the pseudo-inverse in terms of the factorization matrices are presented. The singular value decomposition (SVD) of matrix **A** is given in Section 17.5, together with the spectral condition number of **A** and some properties of the pseudo-inverses of **A**. In Section 17.6, practical considerations in computing the linear least squares solution of $\mathbf{Ax} = \mathbf{b}$ are given. An introduction of linear spaces and the pseudo-inverses is given in Section 17.7.

The interesting subject of multicollinearity, collinearity of the columns of matrix **A**, or the ill-conditioning of matrix **A**, is presented in Section 17.8. This leads to the subject of dealing with multicollinearity via the **Principal components analysis** (PCA), the **Partial least squares method** (PLS) and the **Ridge regression** or **Ridge equation** technique. These are given in Sections 17.9-17.11.

In Section 17.12, numerical results using the Gauss **LU** factorization method with complete pivoting and the Householder's **QR** factorization method with pivoting are presented with comments.

17.2 Least squares solution of linear equations

Let us assume that we have real matrices and vectors. Let $\mathbf{A}^T$ and $\mathbf{x}^T$ refer to the transpose of matrix **A** and of vector **x** respectively. The vector norm $||.||$ denotes the L_2 norm.

Theorem 17.1

A necessary and sufficient condition for the square of the residual vector, $(\mathbf{r}, \mathbf{r}) = \|\mathbf{r}\|^2$, to be minimum is that [18]

(17.2.1) $$\mathbf{A}^T(\mathbf{b} - \mathbf{A}\mathbf{x}) = \mathbf{0}$$

Proof:

Sufficiency: From the definition of **r**, we have

$$(\mathbf{r}, \mathbf{r}) = ((\mathbf{b} - \mathbf{A}\mathbf{x}), \ (\mathbf{b} - \mathbf{A}\mathbf{x})) = (\mathbf{b}, \mathbf{b}) - (\mathbf{b}, \mathbf{A}\mathbf{x}) - (\mathbf{A}\mathbf{x}, \mathbf{b}) + (\mathbf{A}\mathbf{x}, \mathbf{A}\mathbf{x})$$

Let us differentiate $(\mathbf{r}, \mathbf{r})$ w.r.t. $\mathbf{x}$ and equate to 0. This gives

$$\mathbf{A}^T\mathbf{A}\mathbf{x} = \mathbf{A}^T\mathbf{b} \text{ or } \mathbf{A}^T(\mathbf{b} - \mathbf{A}\mathbf{x}) = \mathbf{0}$$

Necessity: Suppose $\mathbf{x}$ gives the minimum $(\mathbf{r}, \mathbf{r})$ but

$$\mathbf{A}^T(\mathbf{b} - \mathbf{A}\mathbf{x}) = \mathbf{w} \neq \mathbf{0}$$

The square of the residual corresponding to $\mathbf{x} + \varepsilon\mathbf{w}$, where ε is a small positive parameter is easily found to be

$$\|\mathbf{b} - \mathbf{A}\mathbf{x} - \varepsilon\mathrm{A}\mathbf{w}\|^2 = \|\mathbf{b} - \mathbf{A}\mathbf{x}\|^2 - 2\varepsilon\|\mathbf{w}\|^2 + \varepsilon^2\|\mathrm{A}\mathbf{w}\|^2$$

For sufficiently small ε, the r.h.s. is $< \|\mathbf{b} - \mathbf{A}\mathbf{x}\|^2$. The hypothesis is thus contradicted and we should have $\mathbf{w} = \mathbf{0}$ and the theorem is proved.

Theorem 17.2

Let **A** be an n by m matrix (n > m) of rank m. Then the linear least squares solution to the system $\mathbf{A}\mathbf{x} = \mathbf{b}$ is given by

(17.2.2) $$\mathbf{x} = (\mathbf{A}^T\mathbf{A})^{-1}\mathbf{A}^T\mathbf{b}$$

Proof:

From (17.2.1), we write

(17.2.3) $$\mathbf{A}^T\mathbf{A}\mathbf{x} = \mathbf{A}^T\mathbf{b}$$

Since **A** is of rank m, the m by m matrix $(\mathbf{A}^T\mathbf{A})$ is nonsingular and equation (17.2.3) has the solution $\mathbf{x} = (\mathbf{A}^T\mathbf{A})^{-1}\mathbf{A}^T\mathbf{b}$ and the theorem is proved.

Equation (17.2.3) is known as the **normal equation** to the linear least squares problem of $\mathbf{A}\mathbf{x} = \mathbf{b}$.

In (17.2.2), we may write $\mathbf{x} = \mathbf{A}^+\mathbf{b}$, where we define

(17.2.4) $$\mathbf{A}^{+} = (\mathbf{A}^{T}\mathbf{A})^{-1}\mathbf{A}^{T}$$

as the pseudo-inverse of **A**. The motivation behind this is that if **A** is square nonsingular, $\mathbf{A}^{+}$ in (17.2.4) reduces to $\mathbf{A}^{-1}$ and $\mathbf{x} = \mathbf{A}^{-1}\mathbf{b}$.

17.2.1 Minimal-length least squares solution

In (17.2.2) the solution $\mathbf{x} = \mathbf{A}^{+}\mathbf{b}$, where $\mathbf{A}^{+}$ is given by (17.2.4) is unique, since **A** is an n by m matrix, n > m, of rank m and thus $\mathbf{A}^{T}\mathbf{A}$ is nonsingular. However, in the general case, when $n < m$ or when $\text{rank}(\mathbf{A}) = k < \min(n, m)$, the least squares solution of $\mathbf{Ax} = \mathbf{b}$ is not unique.

It is desired that the solution given by $\mathbf{x} = \mathbf{A}^{+}\mathbf{b}$, where $\mathbf{A}^{+}$ has a suitable definition, for a general n by m matrix, be unique. On the other hand, it was found that the solution of $\mathbf{Ax} = \mathbf{b}$ that gives the minimum residual norm $\|\mathbf{r}\|_2$ and also the minimum norm $\|\mathbf{x}\|_2$, is unique. It is known as the **minimal-length** (or minimum norm) least squares solution.

For this reason, the general derivation of $\mathbf{A}^{+}$ is based on the fact that the least squares solution **x** of $\mathbf{Ax} = \mathbf{b}$ is of minimum length. This point is the basis of the next two theorems.

Theorem 17.3

If **A** is an n by m matrix ($n < m$) of rank n, then the minimal length least squares solution **x** of $\mathbf{Ax} = \mathbf{b}$ is given by

(17.2.5) $$\mathbf{x} = \mathbf{A}^{T}(\mathbf{A}\mathbf{A}^{T})^{-1}\mathbf{b}$$

Proof:

Since $n < m$, the system of equations $\mathbf{Ax} = \mathbf{b}$ is an underdetermined system and thus has an infinite number of solutions (Chapter 4). Hence, in order to obtain the unique minimal length least squares solution, the problem is formulated as follows.

Find **x** that minimizes the inner product $(\mathbf{x}, \mathbf{x})$ subject to the condition $\mathbf{b} - \mathbf{Ax} = \mathbf{0}$. This problem is solved by the method of **Lagrange multipliers**. Let $\mathbf{c} = (c_1, c_2, \ldots, c_n)^{T}$ be the Lagrange multiplier vector. Let the function F be

$$F = (\mathbf{x}, \mathbf{x}) + (\mathbf{c}, (\mathbf{b} - \mathbf{Ax}))$$

For minimum F, we differentiate F w.r.t. x_i, $i = 1, \ldots, m$ and also w.r.t.

c_i, i = 1, 2, …, n, and equate each of the m + n equations to 0. This gives the two vector equations

$$\mathbf{A}^T\mathbf{c} - 2\mathbf{x} = \mathbf{0} \text{ and } \mathbf{b} - \mathbf{A}\mathbf{x} = \mathbf{0}$$

Thus $(\mathbf{A}\mathbf{A}^T)\mathbf{c} = 2\mathbf{b}$ or $\mathbf{c} = 2(\mathbf{A}\mathbf{A}^T)^{-1}\mathbf{b}$ (since $(\mathbf{A}\mathbf{A}^T)$ is non-singular). Substituting this value of **c** into the first equation gives $\mathbf{x} = \mathbf{A}^T(\mathbf{A}\mathbf{A}^T)^{-1}\mathbf{b}$ and the theorem is proved.

We write the solution (17.2.5) in the form

$$\mathbf{x} = \mathbf{A}^+\mathbf{b}, \text{ where } \mathbf{A}^+ = \mathbf{A}^T(\mathbf{A}\mathbf{A}^T)^{-1}$$

Also, $\mathbf{A}^+$ reduces to $\mathbf{A}^{-1}$ if **A** is square nonsingular. As indicated in Section 17.1, in order to be able to calculate the pseudo-inverse $\mathbf{A}^+$ efficiently, we first factorize matrix **A** into the product of 2 or 3 matrices, each of full rank. Then we get the pseudo-inverse of **A** in terms of the pseudo-inverses of the factorized matrices.

17.3 Factorization of matrix A

Assume that **A** is a real n by m matrix of rank k, $k \le \min(n, m)$. Then **A** may be factorized into two matrices, each of rank k. We shall consider the factorization by the Gauss **LU** and by Householder's **QR** methods.

17.3.1 Gauss LU factorization

Let $(\mathbf{A}|\mathbf{b})$ denote the n by (m + 1) matrix whose first m columns are matrix **A** and its $(m + 1)^{th}$ column is vector **b**. For maximum numerical stability and also in order to help determine the rank of $(\mathbf{A}|\mathbf{b})$, we obtain the **LU** factorization of $(\mathbf{A}|\mathbf{b})$ with complete pivoting. This is done by permuting the rows of $(\mathbf{A}|\mathbf{b})$ and/or the columns of **A**. The **LU** factorization will not be for matrix **A**, but rather for matrix $\underline{\mathbf{A}}$, where, if **S** and **P** are permutation matrices, $\underline{\mathbf{A}} = \mathbf{SAP}$. See Chapter 4. The formal factorization proceeds exactly as in the nonsingular case in Chapter 4. However, in the current case, if matrix **A** is of rank k, after the kth step we shall have the factorization

$$\underline{\mathbf{A}} = \mathbf{L}_1\mathbf{U}_1 \tag{17.3.1}$$

$\mathbf{L}_1$ is an n by n-unit lower triangular matrix, while $\mathbf{U}_1$ is an n by m matrix. The upper k rows of $\mathbf{U}_1$ form an upper trapezoidal matrix and its remaining rows consist of zero elements. See the structure of $\mathbf{L}_1$ and of $\mathbf{U}_1$ as illustrated by the following, where n = 6, m = 4 and k = 2.

It is clear that the last 4 columns of $\mathbf{L}_1$ and the last 4 rows of $\mathbf{U}_1$, in this case, play no part in the factorization. They might be discarded and instead we get the two trapezoidal matrices **L** and **U**.

Obviously, the elements a_{ij} of $\mathbf{U}_1$ or of **U** are not the elements of matrix **A**; that is, we get the factorization

(17.3.2) $$\underline{\mathbf{A}} = \mathbf{LU}$$

where **L** is an n by k-unit lower trapezoidal (its diagonal elements are 1's) and **U** is a k by m upper trapezoidal matrix

$$\mathbf{L}_1\mathbf{U}_1 = \overset{\mathbf{L}_1}{\begin{bmatrix} 1 & & & & & \\ m_{21} & 1 & & & & \\ m_{31} & m_{32} & 1 & & & \\ m_{41} & m_{42} & 0 & 1 & & \\ m_{51} & m_{52} & 0 & 0 & 1 & \\ m_{61} & m_{62} & 0 & 0 & 0 & 1 \end{bmatrix}} \overset{\mathbf{U}_1}{\begin{bmatrix} a_{11} & a_{12} & a_{13} & a_{14} \\ & a_{22} & a_{23} & a_{24} \\ & & & \\ & & & \\ & & & \\ & & & \end{bmatrix}}$$

$$\mathbf{LU} = \overset{\mathbf{L}}{\begin{bmatrix} 1 & \\ m_{21} & 1 \\ m_{31} & m_{32} \\ m_{41} & m_{42} \\ m_{51} & m_{52} \\ m_{61} & m_{62} \end{bmatrix}} \overset{\mathbf{U}}{\begin{bmatrix} a_{11} & a_{12} & a_{13} & a_{14} \\ & a_{22} & a_{23} & a_{23} \end{bmatrix}}$$

Note 17.1

For practical considerations whenever matrix **U** is a trapezoidal (not a triangular) matrix, **U** may be factorized into

$$\mathbf{U} = \mathbf{D}\mathit{U}$$

and thus

(17.3.3) $$\underline{\mathbf{A}} = \mathbf{LD}\boldsymbol{U}$$

where $\mathbf{D}$ is a k by k diagonal matrix whose elements are the diagonal elements of $\mathbf{U}$ and $\boldsymbol{U}$ is a unit trapezoidal matrix (with unit diagonal elements).

17.3.2 Householder's factorization

This factorization proceeds in the same manner as in the nonsingular case of Chapter 4. Once more, the factorization is not for matrix $\mathbf{A}$, but for matrix $\underline{\mathbf{A}} = \mathbf{AP}$, and $\mathbf{P}$ is a permutation matrix. If matrix $\mathbf{A}$ is of rank k, $k < \min(n, m)$, then after the kth step, we shall get

(17.3.4) $$\underline{\mathbf{A}} = \mathbf{Q}_1\mathbf{T}_1$$

Here, $\mathbf{Q}_1$ is an n by n orthonormal matrix ($\mathbf{Q}_1{}^T\mathbf{Q}_1 = \mathbf{I}_n$), while $\mathbf{T}_1$ is an n by m matrix whose first k rows form an upper trapezoidal and the remaining $(n-k)$ rows consist of zero elements. The last $(n-k)$ columns of $\mathbf{Q}_1$ and the last $(m-k)$ rows of $\mathbf{T}_1$ play no part in the factorization of $\underline{\mathbf{A}}$ and should be discarded. We get

(17.3.5) $$\underline{\mathbf{A}} = \mathbf{QT}$$

where $\mathbf{Q}$ is an n by k orthonormal matrix ($\mathbf{Q}^T\mathbf{Q} = \mathbf{I}_k$) and $\mathbf{T}$ is a k by m upper trapezoidal. In (17.3.5), we might also factorize $\mathbf{T}$ into

$$\mathbf{T} = \mathbf{RW}^T$$

where $\mathbf{R}$ is a k by k upper triangular and $\mathbf{W}$ is an m by k orthonormal matrix. This factorization might be achieved by a modified version of the Householder's method [10]. In this case, (17.3.5) becomes

(17.3.6) $$\underline{\mathbf{A}} = \mathbf{QRW}^T$$

where $\mathbf{Q}^T\mathbf{Q} = \mathbf{I}_k$ and $\mathbf{W}^T\mathbf{W} = \mathbf{I}_k$.

To simplify the notation, we shall assume from now on that the permuted $\underline{\mathbf{A}}$ is denoted by $\mathbf{A}$, $\underline{\mathbf{A}}^+$ is denoted by $\mathbf{A}^+$, where $\underline{\mathbf{A}}^+$ is the pseudo-inverse of $\underline{\mathbf{A}}$ and the permuted $\underline{\mathbf{x}}$ is denoted by $\mathbf{x}$.

There are other orthogonal factorization methods that give equally-accurate results as the Gauss **LU** factorization and Householder's factorization methods, and also have their own merits.

These are the Givens' transformation (known as plane rotations) and the Gram-Schmidt methods.

For the next two sections, we shall assume that the n by m matrix **A**, n > m, is of rank m. These methods are described here briefly and they give the **QR** factorization of the un-permuted matrix **A** as the Householder's factorization method, apart possibly from numerical signs [1].

17.3.3 Givens' transformation (plane rotations)

Givens' transformation factorizes **A** into, **A** = **QR**. To achieve this, matrix **A** is pre-multiplied by the so-called rotation matrices, producing a triangular matrix **R**. Givens' method makes 0's of desired elements of matrix **A** while preserving existing 0 elements. Pre-multiplying **A** by rotation matrices produces 0's below the diagonal of **A**. This is done column by column or row by row. As in Householder's factorization, the orthonormal matrix **Q** is not calculated explicitly; rather, vector $\mathbf{Q}^T\mathbf{b}$ is calculated.

There are two advantages for this method. It is best suited for sparse matrices, where the existing 0's in matrix **A** are preserved and need not be treated [9]. Another advantage is that Givens' transformation is best suited for updating the least squares solution of the problem when equations are added to or deleted from the matrix equation **Ax** = **b**. Stewart calls such a technique, the **updating and downdating** of a system of linear equations [3, 23, 24].

17.3.4 Classical and modified Gram-Schmidt methods

Let vectors $\mathbf{a}_1, \mathbf{a}_2, \ldots, \mathbf{a}_m$ be the columns of matrix **A**, assumed linearly independent. It is required to construct a set of orthonormal columns $\mathbf{q}_1, \mathbf{q}_2, \ldots, \mathbf{q}_m$ of a matrix **Q** such that for each $k \leq m$, $\mathbf{q}_k$ is a linear combination of $\mathbf{a}_1, \mathbf{a}_2, \ldots, \mathbf{a}_k$. Again, by orthonormal, we mean, $(\mathbf{q}_s, \mathbf{q}_s) = 1$ and $(\mathbf{q}_s, \mathbf{q}_k) = 0$, $s \neq k$. This is done by the Gram-Schmidt algorithm. As a result, **A** is factorized as **A** = **QR**, where $\mathbf{Q}^T\mathbf{Q} = \mathbf{I}_m$, and **R** is an upper triangular m by m matrix.

For the classical Gram-Schmidt algorithm, the elements of **R** are calculated one column at a time. For the modified Gram-Schmidt, the elements of **R** are instead calculated one row at a time [1, 5]. For the classical Gram-Schmidt with re-orthogonalization, intermediate vectors are formed and the re-orthogonalization is then carried out. Such extra computation is a drawback for this last method. The classical Gram-Schmidt algorithm is derived in Section 17.7.3.

In [1], following the presentation of Bjorck [5], a round-off error analysis for the classical Gram-Schmidt algorithm with re-orthogonalization was presented. A numerical example for an ill-conditioned case was solved by the Householder's factorization (without pivoting), the classical Gram-Schmidt, the modified Gram-Schmidt and the classical Gram-Schmidt with re-orthogonalization. The last two methods compared favorably with Householder's method, while the classical Gram-Schmidt lacks accuracy for the solution vector $\mathbf{x}$, matrix $\mathbf{R}$ and for the orthogonality of the columns of matrix $\mathbf{Q}$, meaning $\mathbf{Q}^T\mathbf{Q} \approx \mathbf{I}_m$, instead of $\mathbf{Q}^T\mathbf{Q} = \mathbf{I}_m$ (neglecting the rounding error).

Eventually, Longley [15] modified the presentation of Bjorck for both the classical and the modified Gram-Schmidt methods.

A merit for the Gram-Schmidt algorithms is that in calculating the least squares solution, the number of basis functions can be added without recalculating the problem. Barrodale and Stuart [4] gave a FORTRAN program for the modified Gram-Schmidt method that allows the number of basis functions to be increased without recalculating the problem form scratch.

17.4 Explicit expression for the pseudo-inverse

Let $\mathbf{A}$ be a real n by m matrix of rank k, $k \le \min(n, m)$. Then in general, $\mathbf{A}$ may be factorized into the form

$$\mathbf{A} = \mathbf{BC}$$

where $\mathbf{B}$ is an n by k matrix and $\mathbf{C}$ is a k by m matrix, each of rank k.

Theorem 17.4

The minimal-length least squares solution of $\mathbf{Ax} = \mathbf{b}$, is given by $\mathbf{x} = \mathbf{A}^+\mathbf{b}$, where

$$\mathbf{A}^+ = \mathbf{C}^+\mathbf{B}^+ = \mathbf{C}^T(\mathbf{CC}^T)^{-1}(\mathbf{B}^T\mathbf{B})^{-1}\mathbf{B}^T \tag{17.4.1}$$

Proof:

From Theorem 17.1, the linear least squares solution $\mathbf{x}$ satisfies $\mathbf{A}^T(\mathbf{b} - \mathbf{Ax}) = \mathbf{0}$, or since $\mathbf{A}^T = \mathbf{C}^T\mathbf{B}^T$

$$\mathbf{C}^T\mathbf{B}^T\mathbf{BCx} = \mathbf{C}^T\mathbf{B}^T\mathbf{b}$$

From this we have $\mathbf{CC}^T\mathbf{B}^T\mathbf{BCx} = \mathbf{CC}^T\mathbf{B}^T\mathbf{b}$ and since each of $(\mathbf{CC}^T)$ and $(\mathbf{B}^T\mathbf{B})$ is a k by k square nonsingular matrix, by pre-multiplying both sides by $(\mathbf{B}^T\mathbf{B})^{-1}(\mathbf{CC}^T)^{-1}$, we get

$$\mathbf{Cx} = (\mathbf{B}^T\mathbf{B})^{-1}\mathbf{B}^T\mathbf{b}$$

Finally, by applying Theorem 17.3, (17.4.1) is obtained and the theorem is proved.

17.4.1 $\mathbf{A}^+$ in terms of Gauss factorization

In view of (17.4.1), for $\mathbf{A} = \mathbf{LU}$ by Gauss factorization, in terms of $\mathbf{L}$ and $\mathbf{U}$, we get

$$\mathbf{A}^+ = \mathbf{U}^T(\mathbf{UU}^T)^{-1}(\mathbf{L}^T\mathbf{L})^{-1}\mathbf{L}^T \tag{17.4.2}$$

and thus the least squares solution $\mathbf{x}$ is given by

$$\mathbf{x} = \mathbf{U}^T(\mathbf{UU}^T)^{-1}(\mathbf{L}^T\mathbf{L})^{-1}\mathbf{L}^T\mathbf{b} \tag{17.4.3}$$

If $\mathbf{A}$ is factorized into $\mathbf{A} = \mathbf{LD}\boldsymbol{U}$, as in (17.3.3), we get respectively

$$\mathbf{A}^+ = \boldsymbol{U}^T(\boldsymbol{UU}^T)^{-1}\mathbf{D}^{-1}(\mathbf{L}^T\mathbf{L})^{-1}\mathbf{L}^T$$

and

$$\mathbf{x} = \boldsymbol{U}^T(\boldsymbol{UU}^T)^{-1}\mathbf{D}^{-1}(\mathbf{L}^T\mathbf{L})^{-1}\mathbf{L}^T\mathbf{b}$$

For the case $k = m < n$, in the factorization $\mathbf{A} = \mathbf{LU}$, $\mathbf{U}$ is an upper triangular matrix and instead of (17.4.2) we get

$$\mathbf{A}^+ = \mathbf{U}^{-1}\mathbf{L}^+ = \mathbf{U}^{-1}(\mathbf{L}^T\mathbf{L})^{-1}\mathbf{L}^T$$

For the case $k = n < m$, $\mathbf{L}$ is a lower triangular matrix. We have

$$\mathbf{A}^+ = \mathbf{U}^+\mathbf{L}^{-1} = \mathbf{U}^T(\mathbf{UU}^T)^{-1}\mathbf{L}^{-1}$$

17.4.2 $\mathbf{A}^+$ in terms of Householder's factorization

For the factorization $\mathbf{A} = \mathbf{QT}$ of (17.3.5), in view of (17.4.1) and the fact that $\mathbf{Q}$ is orthonormal, $\mathbf{Q}^+ = \mathbf{Q}^T$

(17.4.4) $$\mathbf{A}^+ = \mathbf{T}^T(\mathbf{T}\mathbf{T}^T)^{-1}\mathbf{Q}^T$$

and

$$\mathbf{x} = \mathbf{T}^T(\mathbf{T}\mathbf{T}^T)^{-1}\mathbf{Q}^T\mathbf{b}$$

For the case k = m < n, **T** is an upper triangular matrix and instead of (17.4.4), we get

$$\mathbf{A}^+ = \mathbf{T}^{-1}\mathbf{Q}^+ = \mathbf{T}^{-1}\mathbf{Q}^T$$

Also, from (17.3.6)

$$\mathbf{A}^+ = \mathbf{W}^{T+}\mathbf{R}^{-1}\mathbf{Q}^+ = \mathbf{W}\mathbf{R}^{-1}\mathbf{Q}^T$$

Hence

(17.4.5) $$\mathbf{x} = \mathbf{W}\mathbf{R}^{-1}\mathbf{Q}^T\mathbf{b}$$

17.5 The singular value decomposition (SVD)

One of the most important factorizations for an n by m matrix **A** of rank k, k ≤ min(n, m), is the singular value decomposition, given by

(17.5.1) $$\mathbf{A} = \mathbf{V}\mathbf{S}\mathbf{W}^T$$

V is a real n by n orthonormal matrix, $\mathbf{V}^T\mathbf{V} = \mathbf{I}_n$, **W** is a real m by m orthonormal matrix, $\mathbf{W}^T\mathbf{W} = \mathbf{I}_m$ and **S** is a n by m diagonal matrix, $\mathbf{S} = \text{diag}(s_i)$. [Matrix **W** in (17.5.1) is different from matrix **W** in (17.3.6)]. The positive real numbers s_i, i = 1, 2, ..., k are known as the singular values of **A**. They are often ordered so that $s_1 \geq s_2 \geq \ldots \geq s_k > 0$.

The singular value decomposition of **A** is closely related to the eigenvalues and eigenvectors of matrix $(\mathbf{A}^T\mathbf{A})$, where from (17.5.1), $(\mathbf{A}^T\mathbf{A})$ is decomposed as

(17.5.2) $$(\mathbf{A}^T\mathbf{A}) = \mathbf{W}\mathbf{S}^2\mathbf{W}^T$$

Theorem 17.5

The orthonormal matrix **W** diagonalizes $(\mathbf{A}^T\mathbf{A})$ and hence the diagonal elements of $\mathbf{S}^2$ are the eigenvalues of $(\mathbf{A}^T\mathbf{A})$ and are the squares of the singular values of **A**.

Proof:

Matrix **W** is orthonormal. From (17.5.2)

$$(\mathbf{A}^T\mathbf{A})\mathbf{W} = \mathbf{W}\mathbf{S}^2$$

If we now write $\mathbf{W} = [\mathbf{w}_1, \mathbf{w}_2, \ldots, \mathbf{w}_m]$, where the $(\mathbf{w}_i)$ are the m columns of **W**, the above equation becomes

$$(\mathbf{A}^T\mathbf{A})[\mathbf{w}_1, \mathbf{w}_2, \ldots, \mathbf{w}_m] = [s_1^2\mathbf{w}_1, s_2^2\mathbf{w}_2, \ldots, s_m^2\mathbf{w}_m]$$

and the theorem is proved.

Though, the SVD of matrix **A** provides information about the eigensystem of $(\mathbf{A}^T\mathbf{A})$. More-importantly, it provides information that relates directly to matrix **A** itself; that is, concerning the notion of the (spectral) condition number of **A**, which is given in terms of the largest and smallest singular values of **A**.

For the n by m matrix **A** of rank k, $k \le \min(n, m)$, since **V** and **W** are each orthonormal, and hence are of full rank, rank(**A**) = rank(**S**) and the SVD in (17.5.1) may be partitioned as

$$\mathbf{A} = \mathbf{V}\mathbf{S}\mathbf{W}^T = \mathbf{V}\begin{bmatrix}\mathbf{S}_{11} & \mathbf{0}\\ \mathbf{0} & \mathbf{0}\end{bmatrix}\mathbf{W}^T$$

where $\mathbf{S}_{11}$ is a k by k diagonal non-singular matrix. The zeros denoted by "**0**" are zero matrices. In this case, both **V** and **W** may be partitioned as follows

$$\mathbf{V} = [\mathbf{V}_1\ \mathbf{V}_2] \text{ and } \mathbf{W} = [\mathbf{W}_1\ \mathbf{W}_2]$$

where $\mathbf{V}_1$ is an n by k orthonormal matrix and $\mathbf{V}_2$ is an n by $(n-k)$ zero matrix. $\mathbf{W}_1$ is an m by k orthonormal matrix and $\mathbf{W}_2$ is an m by $(m-k)$ zero matrix. By post multiplying **A** by **W**, we deduce that

$$\mathbf{A}\mathbf{W}_1 = \mathbf{V}_1\mathbf{S}_{11} \text{ and } \mathbf{A}\mathbf{W}_2 = \mathbf{0}$$

from which the n by k matrix $\mathbf{V}_1$ provides an orthonormal basis for the range of **A** (Section 17.7.4).

If we take the transpose of **A**, we get similar relations for $\mathbf{A}^T$, namely

$$\mathbf{A}^T\mathbf{V}_1 = \mathbf{W}_1\mathbf{S}_{11} \text{ and } \mathbf{A}^T\mathbf{V}_2 = \mathbf{0}$$

from which the m by k matrix $\mathbf{W}_1$ provides an orthonormal basis for the range of $\mathbf{A}^T$.

From now on, the SVD given by (17.5.1) is used, where, **S**, **V** and **W** denote respectively $\mathbf{S}_{11}$, $\mathbf{V}_1$ and $\mathbf{W}_1$.

17.5.1 Spectral condition number of matrix A

Because each of **V** and **W** is an orthonormal matrix and **S** is diagonal, from (17.5.1), the pseudo-inverse

(17.5.3) $$\mathbf{A}^{+} = \mathbf{W}\mathbf{S}^{-1}\mathbf{V}^{T}$$

Since **V** and **W** are orthonormal, the spectral norms of **A** and $\mathbf{A}^{+}$ are given respectively by (see Theorem 4.3)

$$\|\mathbf{A}\|_2 = \|\mathbf{S}\|_2 = s_1 \text{ and } \|\mathbf{A}^{+}\|_2 = \|\mathbf{S}^{-1}\|_2 = s_k^{-1}$$

The **spectral condition number** for **A**, denoted by K is defined as

(17.5.4) $$K = \|\mathbf{A}\|_2 \, \|\mathbf{A}^{+}\|_2 = [s_1]/[s_k]$$

The condition number of a matrix defines the condition of the matrix with respect to a computing problem. When the condition number is very large, in most cases, s_k is very small. As a result, the round-off error, or a small variation of vector **b** and/or of matrix **A**, will affect the accuracy of the obtained solution **x** of $\mathbf{Ax} = \mathbf{b}$. The resulting error in the solution **x** is proportional to the condition number of the matrix. See Section 4.2.8 and also [2].

The least squares solution of the system $\mathbf{Ax} = \mathbf{b}$, by the singular value decomposition, where $\text{rank}(\mathbf{A}) = k \le \min(m, n)$ is given by $\mathbf{x} = \mathbf{A}^{+}\mathbf{b}$, or

$$\mathbf{x} = \mathbf{W}\mathbf{S}^{-1}\mathbf{V}^{T}\mathbf{b}$$

17.5.2 Main properties of the pseudo-inverse $\mathbf{A}^{+}$

Theorem 17.6

$\mathbf{A}^{+}$ satisfies the following relations

(i) $\mathbf{A}^{+}\mathbf{A}\mathbf{A}^{+} = \mathbf{A}^{+}$
(ii) $\mathbf{A}\mathbf{A}^{+}\mathbf{A} = \mathbf{A}$
(iii) $(\mathbf{A}\mathbf{A}^{+})^{T} = (\mathbf{A}\mathbf{A}^{+})$
(iv) $(\mathbf{A}^{+}\mathbf{A})^{T} = (\mathbf{A}^{+}\mathbf{A})$

The proof of each of these four equalities follow easily from the definition of **A** and $\mathbf{A}^{+}$ in (17.5.1) and (17.5.3) respectively.

If the n by m matrix **A** is of full rank, i.e., $\text{rank}(\mathbf{A}) = k = \min(n, m)$, then:

(i) If $k = m < n$, $\mathbf{A}^+ = (\mathbf{A}^T\mathbf{A})^{-1}\mathbf{A}^T$, $\mathbf{A}^+\mathbf{A} = \mathbf{I}_k$
(ii) If $k = n < m$, $\mathbf{A}^+ = \mathbf{A}^T(\mathbf{A}\mathbf{A}^T)^{-1}$, $\mathbf{A}\mathbf{A}^+ = \mathbf{I}_k$

where $\mathbf{I}_k$ is a k-unit matrix. Finally, if **A** is a square nonsingular matrix $\mathbf{A}^+ = \mathbf{A}^{-1}$ the ordinary inverse of **A**.

17.6 Practical considerations in computing

When writing program code, a sequence of intermediate calculations are performed in computing a final solution. The following sections describe some practical considerations in this regard. Assume that **A** and **b** are real.

17.6.1 Cholesky's decomposition

Given a real k by k symmetric positive definite matrix **B**, Cholesky's decomposition method decomposes **B** into

$$\mathbf{B} = \mathbf{L}\mathbf{L}^T$$

where **L** is a k by k lower triangular matrix. As it is assumed that rank(**B**) = k, the Cholesky's decomposition will not break down in the process of the decomposition of **B**. See for example, Lau ([14], p. 101).

17.6.2 Solution of the normal equation

To calculate the least squares solution of $\mathbf{Ax} = \mathbf{b}$ from the normal equation $\mathbf{A}^T\mathbf{Ax} = \mathbf{A}^T\mathbf{b}$ (equation (17.2.3)), the following practical steps are taken. We assume that the n by m matrix **A** is of rank m. Then the m by m matrix $(\mathbf{A}^T\mathbf{A})$ is symmetric positive definite and may be decomposed by Cholesky's decomposition into $\mathbf{A}^T\mathbf{A} = \mathbf{G}\mathbf{G}^T$, where **G** is m by m lower triangular. The normal equation becomes

$$\mathbf{G}\mathbf{G}^T\mathbf{x} = \mathbf{A}^T\mathbf{b}$$

We write $\mathbf{u} = \mathbf{A}^T\mathbf{b}$, $\mathbf{y} = \mathbf{G}^T\mathbf{x}$ and we have $\mathbf{Gy} = \mathbf{u}$. Then **y** is obtained by forward substitution and from $\mathbf{G}^T\mathbf{x} = \mathbf{y}$, **x** is obtained by backward substitution.

Spath ([22], pp. 22-24) presented a FORTRAN routine for this method and stated that it is very fast. However, if the columns of matrix $\mathbf{A}$ are nearly linearly dependent, the obtained solution $\mathbf{x}$ would not be accurate.

17.6.3 Solution via Gauss LU factorization method

To calculate the least squares solution $\mathbf{x}$ of $\mathbf{Ax} = \mathbf{b}$, given by

$$\mathbf{x} = \mathbf{U}^T(\mathbf{UU}^T)^{-1}(\mathbf{L}^T\mathbf{L})^{-1}\mathbf{L}^T\mathbf{b}$$

as in (17.4.3), we calculate the intermediate vectors $\mathbf{u}$, $\mathbf{y}$ and $\mathbf{z}$ first, as follows (assuming that matrix $\mathbf{A}$ is of rank k)

$$(17.6.1) \quad \mathbf{u} = \mathbf{L}^T\mathbf{b}, \ \ (\mathbf{L}^T\mathbf{L})\mathbf{y} = \mathbf{u}, \ \ (\mathbf{UU}^T)\mathbf{z} = \mathbf{y} \ \text{ and } \ \mathbf{x} = \mathbf{U}^T\mathbf{z}$$

Each of $\mathbf{L}^T\mathbf{L}$ and $\mathbf{UU}^T$ is a positive definite k by k symmetric matrix. They may be decomposed by Cholesky's decomposition into $(\mathbf{L}^T\mathbf{L}) = \mathbf{YY}^T$ and $(\mathbf{UU}^T) = \mathbf{ZZ}^T$, where each of $\mathbf{Y}$ and $\mathbf{Z}$ is a k by k lower triangular matrix.

To calculate $\mathbf{y}$ and $\mathbf{z}$ we calculate the intermediate vectors $\mathbf{x}_1$ and $\mathbf{x}_2$ first as follows. The solution of $(\mathbf{L}^T\mathbf{L})\mathbf{y} = \mathbf{u}$ is done in two steps, by solving the two triangular systems $\mathbf{Yx}_1 = \mathbf{u}$ and $\mathbf{Y}^T\mathbf{y} = \mathbf{x}_1$. The solution of $(\mathbf{UU}^T)\mathbf{z} = \mathbf{y}$ is given by solving the two triangular systems $\mathbf{Zx}_2 = \mathbf{y}$ and $\mathbf{Z}^T\mathbf{z} = \mathbf{x}_2$. Triangular systems are solved by backward or by forward substitutions, depending on the triangular matrix at hand.

17.6.4 Solution via Householder's method

The solution $\mathbf{x}$ of $\mathbf{Ax} = \mathbf{b}$ by the Householder's method

$$\mathbf{x} = \mathbf{WR}^{-1}\mathbf{Q}^T\mathbf{b}$$

as given by (17.4.5) is obtained by calculating the intermediate vectors $\mathbf{u}$ and $\mathbf{v}$, where

$$(17.6.2) \qquad \mathbf{u} = \mathbf{Q}^T\mathbf{b}, \ \ \mathbf{Rv} = \mathbf{u} \ \text{ and } \ \mathbf{x} = \mathbf{Wv}$$

We note that $\mathbf{Rv} = \mathbf{u}$ is a triangular system and the elements of $\mathbf{v}$ are calculated by back substitution.

17.6.5 Calculation of $\mathbf{A}^+$

To calculate the m by n pseudo-matrix $\mathbf{A}+$, we calculate its columns $\mathbf{x}_1, \mathbf{x}_2, \ldots, \mathbf{x}_n$. That by taking in (17.6.1) (or in (17.6.2)),

$\mathbf{b} = \mathbf{e}_1, \mathbf{e}_2, \ldots, \mathbf{e}_n$, in succession, where $\mathbf{e}_i$ is the ith column of the unit matrix $\mathbf{I}_n$. The proof follows the proof given in Section 4.5.4, for calculating the matrix inverse $\mathbf{A}^{-1}$.

17.6.6 Influence of the round-off error

In the presence of round-off error, the computed parameters will differ slightly from the actual (correct) parameters. Equations (17.3.1) and (17.3.4) may be given by (**A** is assumed of rank k)

(17.6.3) $$\underline{\mathbf{A}} + \mathbf{F}_1 = \underline{\mathbf{L}}_1\underline{\mathbf{U}}_1 \text{ and } \underline{\mathbf{A}} + \mathbf{G}_1 = \underline{\mathbf{Q}}_1\underline{\mathbf{T}}_1$$

where $\mathbf{F}_1$ and $\mathbf{G}_1$ are two error matrices. Also, the matrices on the r.h.s. of each equation are not the matrices that correspond to the exact factorizations of $\underline{\mathbf{A}}$. Expected zero elements in the last (m – k) rows of $\underline{\mathbf{U}}_1$ and of $\underline{\mathbf{T}}_1$ are small numbers.

Let **H** denote either $\mathbf{F}_1$ or $\mathbf{G}_1$. The singular values s_i of matrices **A** and (**A** + **H**), i = 1, 2, ..., are related by the following inequalities ([27], p. 102 - see also Theorem 4.8)

$$s_i(\mathbf{A}) - ||\mathbf{H}||_2 \leq s_i(\mathbf{A} + \mathbf{H}) \leq s_i(\mathbf{A}) + ||\mathbf{H}||_2$$

In particular, for i = k, $|s_k(\mathbf{A} + \mathbf{H}) - s_k(\mathbf{A})| \leq ||\mathbf{H}||_2$ and for i = k + 1, since $s_{k+1}(\mathbf{A}) = 0$, $s_{k+1}(\mathbf{A} + \mathbf{H}) \leq ||\mathbf{H}||_2$.

When **H** is a perturbation (slight error) matrix, $s_k(\mathbf{A}) >> ||\mathbf{H}||_2$, $s_k(\mathbf{A} + \mathbf{H}) \approx s_k(\mathbf{A})$ and $s_{k+1}(\mathbf{A} + \mathbf{H}) << s_k(\mathbf{A})$, we have a well division line in each of the factorizations (17.6.3) between the very small numbers and the other numbers in $\underline{\mathbf{U}}_1$ and $\underline{\mathbf{T}}_1$.

When these small numbers are detected and discarded, instead of (17.6.3), we shall have

(17.6.3a) $$\underline{\mathbf{A}} + \mathbf{F} = \underline{\mathbf{L}}_1\underline{\mathbf{U}}_1 \text{ and } \underline{\mathbf{A}} + \mathbf{G} = \underline{\mathbf{Q}}_1\underline{\mathbf{T}}_1$$

where $||\mathbf{F}|| < ||\mathbf{F}_1||$ and $||\mathbf{G}|| < ||\mathbf{G}_1||$.

Let d**A**, d**b** and d**x** be perturbation terms. Then the computed solution of **Ax** = **b** will satisfy

$$(\mathbf{A} + d\mathbf{A})(\mathbf{x} + d\mathbf{x}) = (\mathbf{b} + d\mathbf{b})$$

For error analysis for the Gauss **LU** factorization method and for the Householder's method, the reader is referred to a detailed analysis in [2].

17.7 Linear spaces and the pseudo-inverses

The simple idea of vector spaces, known as linear spaces, is presented. The null and range spaces of a general n by m matrix **A** are introduced. Finally, the orthogonal projection operators onto the range and null spaces of **A** and of the pseudo-inverse $\mathbf{A}^+$ are derived. The following definitions are needed to introduce this subject

17.7.1 Definitions, notations and related theorems

A **linear combination of vectors**: A vector **x** is said to be a linear combination of vectors $\mathbf{x}_1, \mathbf{x}_2, \ldots, \mathbf{x}_k$, if it can be written in the form

$$\mathbf{x} = c_1\mathbf{x}_1 + c_2\mathbf{x}_2 + \ldots + c_k\mathbf{x}_k$$

where the c_i are scalars. Obviously, vectors **x** and $(\mathbf{x}_i)$ are of the same dimensions.

A **vector space** (or a **linear space**): denoted by **V**, is the collection of all vectors that are closed under the operations of addition and multiplication by a scalar. That is, if vectors $\mathbf{x}_1$ and $\mathbf{x}_2$ belong to **V**, so do vectors $(\mathbf{x}_1 + \mathbf{x}_2)$, $c\mathbf{x}_1$ and $d\mathbf{x}_2$, where c and d are scalars. The vectors $-\mathbf{x}_1$, $-\mathbf{x}_2$ and the zero vector **0** also belong to **V**. Symbolically we write

$$\mathbf{x} \in \mathbf{V}$$

meaning **x** is a vector in **V** or **x** belongs to **V**. Vectors $\mathbf{x}_1, \mathbf{x}_2, \ldots,$ in **V** are also known as points in the space **V**. Let $\mathbf{x}_1, \mathbf{x}_2, \ldots, \mathbf{x}_k$, be m-dimensional vectors. Then all possible combinations of this set of vectors form a vector space **V**.

Spanning of a vector space: If a vector space consists of all linear combinations of a set of vectors $\mathbf{x}_1, \mathbf{x}_2, \ldots, \mathbf{x}_k$, this set of vectors is said to span the linear space.

Linear dependence and linear independence: Let $\mathbf{x}_1, \mathbf{x}_2, \ldots, \mathbf{x}_j$, be a set of vectors. Then these vectors are linearly dependent if there exist scalars $c_1, c_2, \ldots, c_j$, not all 0's such that

$$c_1\mathbf{x}_1 + c_2\mathbf{x}_2 + \ldots + c_j\mathbf{x}_j = \mathbf{0}$$

If this equality is satisfied only when all the c_i are 0's, the given set of vectors are linearly independent.

If $\mathbf{x}_1, \mathbf{x}_2, \ldots, \mathbf{x}_k$, span a vector space and one or more of the $\mathbf{x}_i$ is linearly dependent on the others, then the vector space is spanned by the given set of vectors after omitting the dependent vectors.

Basis: Given a set of vectors $\mathbf{x}_1, \mathbf{x}_2, \ldots, \mathbf{x}_k$, they form a basis for the vector space if they are linearly independent and they span the vector space.

Theorem 17.7

Every vector in a vector space may be expressed uniquely as a linear combination of a given basis.

Dimension of a linear space: The dimension of a linear space equals the number of vectors in its basis. If a space has a basis consisting of a finite number of vectors, the space is known to be of finite dimensions.

Theorem 17.8

The maximum number of linearly independent m-dimensional vectors in the linear space $\mathbf{V}$ is m. In this case, we denote $\mathbf{V}$ by $\mathbf{V}_m$.

17.7.2 Subspaces and their dimensions

Theorem 17.9

Let $\mathbf{V}_m$ be a vector space of dimension m and let $\mathbf{x}_1, \mathbf{x}_2, \ldots, \mathbf{x}_r$, $r < m$, be linearly independent vectors in $\mathbf{V}_m$. Then there exist vectors $\mathbf{x}_{r+1}, \mathbf{x}_{r+2}, \ldots, \mathbf{x}_m$ such that $\mathbf{x}_1, \mathbf{x}_2, \ldots, \mathbf{x}_r, \mathbf{x}_{r+1}, \ldots, \mathbf{x}_m$, form a basis for $\mathbf{V}_m$. Let $\mathbf{x}_1, \mathbf{x}_2, \ldots, \mathbf{x}_r$, span the space $\mathbf{U}$ and $\mathbf{x}_{r+1}, \mathbf{x}_{r+2}, \ldots, \mathbf{x}_m$ span the space $\mathbf{W}$.

Subspaces: Each of $\mathbf{U}$ and $\mathbf{W}$ is called a subspace of the linear space $\mathbf{V}_m$. The dimension of $\mathbf{U}$ is r and the dimension of $\mathbf{W}$ is $(m - r)$. Every element of $\mathbf{U}$ is an element of $\mathbf{V}_m$, but the converse is not necessarily true. It is said that $\mathbf{U}$ is contained in $\mathbf{V}_m$ or $\mathbf{V}_m$ contains $\mathbf{U}$. This is symbolized by $\mathbf{U} \subseteq \mathbf{V}_m$ or $\mathbf{V}_m \supseteq \mathbf{U}$.

A proper subspace: If $\mathbf{U} \neq \mathbf{V}_m$, we write $\mathbf{U} \subset \mathbf{V}_m$ or $\mathbf{V}_m \supset \mathbf{U}$ and $\mathbf{U}$ is called a proper subspace of $\mathbf{V}_m$. The same is said about the subspace $\mathbf{W}$.

Direct sum of subspaces: In Theorem 17.9, we say that $\mathbf{V}_m$ is the direct sum of **U** and **W**. This is symbolized by

$$\mathbf{V}_m = \mathbf{U} \oplus \mathbf{W}$$

In this case, the following conditions are satisfied:

(i) For every vector $\mathbf{x} \in \mathbf{V}_m$, there exists $\mathbf{x}_1 \in \mathbf{U}$ and $\mathbf{x}_2 \in \mathbf{W}$ such that $\mathbf{x} = \mathbf{x}_1 + \mathbf{x}_2$, and

(ii) If $\mathbf{x} \in \mathbf{U}$ and $\mathbf{x} \in \mathbf{W}$, then $\mathbf{x} = \mathbf{0}$. This condition indicates that the only vector common to **U** and **W** is the zero vector.

Complements of subspaces: If $\mathbf{V}_m = \mathbf{U} \oplus \mathbf{W}$, it is said that **U** and **W** are complements of each other.

Inner product spaces: If in a vector space **V**, the inner product of two of its vectors satisfies:

(i) $(\mathbf{x}, c\mathbf{y}+d\mathbf{z}) = c(\mathbf{x}, \mathbf{y}) + d(\mathbf{x}, \mathbf{z})$,

(ii) $(\mathbf{x}, \mathbf{y}) = (\mathbf{y}, \mathbf{x})$, and

(iii) $(\mathbf{x}, \mathbf{x}) > 0$, if $\mathbf{x} \neq \mathbf{0}$, where c and d are scalars, **V** is known as an inner product space.

Orthogonality: If $\mathbf{q}_1$ and $\mathbf{q}_2$ are two vectors in an inner product vector space **V**, then $\mathbf{q}_1$ and $\mathbf{q}_2$ are **orthogonal** if the inner product $(\mathbf{q}_1, \mathbf{q}_2) = 0$.

Orthogonal set of vectors: If $\mathbf{q}_1, \mathbf{q}_2, \ldots, \mathbf{q}_k$ is a set of vectors in **V**, then they form an orthogonal set if $(\mathbf{q}_i, \mathbf{q}_j) = 0$, $i \neq j$. This set of vectors are also known as orthonormal set if $||\mathbf{q}_i||_2 = 1$, $i = 1, 2, \ldots, k$.

Theorem 17.10

Let **V** be an inner product space and $\mathbf{q}_1, \mathbf{q}_2, \ldots, \mathbf{q}_k$ form an orthonormal set in **V**. Then assuming this set is real:

(i) $\mathbf{q}_1, \mathbf{q}_2, \ldots, \mathbf{q}_k$ are linearly independent,

(ii) $k \leq$ the dimension of **V**, and

(iii) if a vector **x** is a linear combination of this set, then

$$\mathbf{x} = \sum_{i=1}^{k} (\mathbf{q}_i, \mathbf{x})\mathbf{q}_i = \sum_{i=1}^{k} \mathbf{q}_i \mathbf{q}_i^T \mathbf{x}$$

Proof:

The proof of (i) and (ii) is immediate since any $\mathbf{q}_i$ cannot be a linear combination of the rest. To prove (iii), we write **x** in the form

$$\mathbf{x} = c_1\mathbf{q}_1 + c_2\mathbf{q}_2 + \ldots + c_k\mathbf{q}_k$$

Then by taking the inner products with $\mathbf{q}_i$, $i = 1, 2, \ldots, k$, $c_i = (\mathbf{q}_i, \mathbf{x}) = \mathbf{q}_i^T\mathbf{x}$ and the theorem is proved.

17.7.3 Gram-Schmidt orthogonalization

Theorem 17.11

(i) Given a set of linearly independent vectors $\mathbf{a}_1, \mathbf{a}_2, \ldots, \mathbf{a}_k$, in an inner product space $\mathbf{V}_m$, $k \leq m$, the Gram-Schmidt orthogonalization algorithm constructs a set of orthonormal vectors $\mathbf{q}_1, \mathbf{q}_2, \ldots, \mathbf{q}_k$ such that for each $i \leq k$, $\mathbf{q}_i$ is a linear combination of $\mathbf{a}_1, \mathbf{a}_2, \ldots, \mathbf{a}_i$.

(ii) Every m-dimensional inner product space $\mathbf{V}_m$ has an orthonormal basis.

Proof:

We start by choosing $\mathbf{q}_1$ in the direction of $\mathbf{a}_1$ and write

$$t_{11}\mathbf{q}_1 = \mathbf{a}_1$$

t_{11} is a normalization factor, chosen such that $||\mathbf{q}_1|| = 1$, thus $t_{11} = ||\mathbf{a}_1||$. That is

$$\mathbf{q}_1 = \mathbf{a}_1/||\mathbf{a}_1||$$

Next, we write $\mathbf{q}_2$ as a linear combination of $\mathbf{a}_1$ and $\mathbf{a}_2$, or in other words, in terms of $\mathbf{q}_1$ and $\mathbf{a}_2$. We write

$$t_{22}\mathbf{q}_2 = \mathbf{a}_2 - t_{12}\mathbf{q}_1$$

We calculate t_{12} such that $\mathbf{q}_2$ is orthogonal to $\mathbf{q}_1$. Also, t_{22} is a normalization factor. By taking the inner product of this equation with $\mathbf{q}_1$, we get

$$(\mathbf{q}_1, t_{22}\mathbf{q}_2) = 0 = (\mathbf{q}_1, \mathbf{a}_2) - t_{12}(\mathbf{q}_1, \mathbf{q}_1)$$

Hence, since $(\mathbf{q}_1, \mathbf{q}_1) = 1$, $t_{12} = (\mathbf{q}_1, \mathbf{a}_2)$ and $t_{22} = ||\mathbf{a}_2 - t_{12}\mathbf{q}_1||$.

In general, for $i = 2, 3, \ldots, k$, the vectors $\mathbf{q}_i$ are given by

$$t_{ii}\mathbf{q}_i = \mathbf{a}_i - \sum_{j=1}^{i-1} t_{ji}\mathbf{q}_j, \quad 2 \leq i \leq m$$

where $t_{ji} = (\mathbf{q}_j, \mathbf{a}_i)$, $j = 1, 2, \ldots, i-1$ and

$$t_{ii} = \left\| \mathbf{a}_i - \sum_{j=1}^{i-1} t_{ji}\mathbf{q}_j \right\|$$

The proof of (ii) follows immediately.

The algorithm described above is in fact a factorization method. Let the vectors $\mathbf{a}_1, \mathbf{a}_2, \ldots, \mathbf{a}_k$ be the columns of the m by k matrix **A** and the vectors $\mathbf{q}_1, \mathbf{q}_2, \ldots, \mathbf{q}_k$ be the columns of the m by k matrix **Q**. Then we have $\mathbf{A} = \mathbf{QT}$, where $\mathbf{Q}^T\mathbf{Q} = \mathbf{I}_k$ and **T** an upper triangular k by k matrix whose elements are the t_{ji} parameters.

As noted earlier, the Gram-Schmidt **QT** factorization of **A** gives the **QR** factorization of the un-permuted matrix **A**, as does the Householder's factorization method, apart possibly from numerical signs. The difference also is in the accuracies of the calculated parameters, as indicated at the end of Section 17.3.4.

Orthonormal basis: If further $\mathbf{q}_1, \mathbf{q}_2, \ldots, \mathbf{q}_k$ form the basis of a vector space **V**, then the vector space is spanned by an orthonormal basis.

The Euclidean space: An m-dimensional Euclidean space denoted by $\mathbf{E}_m$ is an inner product space associated with any two vectors **x** and **y** in $\mathbf{E}_m$. A non-negative number called the distance between **x** and **y** is given by

$$\|\mathbf{x} - \mathbf{y}\| = [(\mathbf{x} - \mathbf{y}), (\mathbf{x} - \mathbf{y})]^{1/2}$$

The familiar 2 and 3 dimensional geometric spaces are Euclidean spaces. Let the 3 dimensional vectors $\mathbf{x}_1, \mathbf{x}_2$ and $\mathbf{x}_3$, be linearly independent. Then they span $\mathbf{E}_3$. The two vectors $\mathbf{x}_1$ and $\mathbf{x}_2$ will span a plane passing through the origin. This plane is a subspace of $\mathbf{E}_3$. Likewise the vector $\mathbf{x}_3$ spans a straight line through the origin. This line is also a subspace of $\mathbf{E}_3$. The vectors $\mathbf{e}_1$, $\mathbf{e}_2$ and $\mathbf{e}_3$, where $\mathbf{e}_i$ is the ith column of the unit matrix $\mathbf{I}_3$ may be taken as an orthonormal basis for $\mathbf{E}_3$.

Orthogonal complements: If, in Section 17.7.2, for every vector $\mathbf{u} \in \mathbf{U}$ and for every vector $\mathbf{w} \in \mathbf{W}$, the inner product $(\mathbf{u}, \mathbf{w}) = 0$, **U** and **W** are known as orthogonal complements of each other. In this case, we write

$$\mathbf{W} = \mathbf{U}^{\perp} \text{ or } \mathbf{U} = \mathbf{W}^{\perp}$$

and also

$$\mathbf{V}_m = \mathbf{U} \oplus \mathbf{U}^{\perp} \text{ and } \mathbf{V}_m = \mathbf{W} \oplus \mathbf{W}^{\perp}$$

Null space of a matrix A: Consider the solution of the system of linear equations $\mathbf{Ax} = \mathbf{0}$, where $\mathbf{A}$ is an m by m matrix and $\mathbf{x}$ is an m-vector. It is known that if rank($\mathbf{A}$) < m, $\mathbf{Ax} = \mathbf{0}$ has an infinite number of solutions (Theorem 4.14).

Theorem 17.12

The set of all solutions $\mathbf{x}$ of $\mathbf{Ax} = \mathbf{0}$ is a vector space.

Proof:

If the nonzero vectors $\mathbf{x}_1$ and $\mathbf{x}_2$ are two solutions to $\mathbf{Ax} = \mathbf{0}$, i.e., $\mathbf{Ax}_1 = \mathbf{0}$ and $\mathbf{Ax}_2 = \mathbf{0}$, then $\mathbf{A}(c\mathbf{x}_1 + d\mathbf{x}_2) = \mathbf{0}$. Thus $(c\mathbf{x}_1 + d\mathbf{x}_2)$ is also a solution to $\mathbf{Ax} = \mathbf{0}$. All such solutions with different values of the scalars c and d constitute a vector space.

The vector space containing all the solutions $\mathbf{x}$ of $\mathbf{Ax} = \mathbf{0}$, is known as the null space of $\mathbf{A}$ and is denoted by $\mathbf{N}(\mathbf{A})$. Symbolically we write

$$\mathbf{N}(\mathbf{A}) = \{\mathbf{x} \in \mathbf{E}_m \mid \mathbf{Ax} = \mathbf{0}\}$$

17.7.4 Range spaces of A and A^T and their orthogonal complements

The following definitions apply to range spaces and their orthogonal complements:

(1) **Range space of a matrix A**: Let $\mathbf{A}$ be a general n by m matrix of rank k. Then the vector space spanned by the columns of $\mathbf{A}$ is known as the column space or the range space of $\mathbf{A}$ and is denoted by $\mathbf{R}(\mathbf{A})$. Obviously, $\mathbf{R}(\mathbf{A})$ is a subspace of the Euclidean space $\mathbf{E}_n$ and is of dimension k. Symbolically we write

$$\mathbf{R}(\mathbf{A}) = \{\mathbf{y} \in \mathbf{E}_n \mid \mathbf{y} = \mathbf{Ax}, \mathbf{x} \in \mathbf{E}_m\}$$

The space of all vectors that are orthogonal to the columns of $\mathbf{A}$ is the orthogonal complement of $\mathbf{R}(\mathbf{A})$ and is denoted by $\mathbf{R}(\mathbf{A})^{\perp}$.

Let us recall from Section 17.3 that a general n by m matrix of rank k, $k \leq \min(n, m)$ may be factorized in different forms. In the orthogonal factorization of equation (17.3.4), we get (**A** denotes the permuted **A**)

$$\mathbf{A} = \mathbf{Q}_1\mathbf{T}_1 \tag{17.7.1}$$

where $\mathbf{Q}_1$ is an n by n orthonormal and $\mathbf{T}_1$ is an n by m matrix whose first k rows form an upper trapezoidal and its last (m – k) rows consist of zero elements. When the last (n – k) columns of $\mathbf{Q}_1$ and the last (m – k) rows of $\mathbf{T}_1$ are discarded, we get

$$\mathbf{A} = \mathbf{QT} \tag{17.7.2}$$

where **Q** is an n by k orthonormal matrix and **T** is a k by m upper trapezoidal matrix. Factorization (17.7.2) indicates that every column $\mathbf{a}_i$ of matrix **A** may be written as a linear combination of the k columns $\mathbf{q}_i$ of **Q**.
That is, $\mathbf{a}_i = \Sigma_j\, t_{ji}\mathbf{q}_j$, where t_{ji} are the elements of matrix **T**. Hence, the k columns of **Q** span **R**(**A**) and may be taken as its basis.
In the singular value decomposition of Section 17.5 for matrix **A** of rank k, we have

$$\mathbf{A} = \mathbf{VSW}^T \tag{17.7.3}$$

The columns of **V** span **R**(**A**) and may be taken as another basis for **R**(**A**).

(2) **The orthogonal complement of R(A)**: In (17.7.1) the first k columns of $\mathbf{Q}_1$ form matrix **Q** of (17.7.2) and span **R**(**A**). Then since $\mathbf{Q}_1$ is orthonormal, the last (n – k) columns of $\mathbf{Q}_1$ that form say $\mathbf{Q}_2$, span $\mathbf{R}(\mathbf{A})^{\perp}$; the orthogonal complement of **R**(**A**).

(3) **Range space of matrix $\mathbf{A}^T$**: We note that by taking the transpose of (17.7.3) for instance, the columns of **W** span $\mathbf{R}(\mathbf{A}^T)$.

Let us now revisit a variation of Theorem 4.13. Let **A** be an n by m matrix and **b** be an n-vector. Then [17]:

(i) The system of linear equations $\mathbf{Ax} = \mathbf{b}$ has a solution if and only if $\text{rank}(\mathbf{A}|\mathbf{b}) = \text{rank}(\mathbf{A})$.

(ii) If rank($\mathbf{A}|\mathbf{b}$) = rank($\mathbf{A}$) = m, the solution is unique.
(iii) If rank($\mathbf{A}|\mathbf{b}$) = rank($\mathbf{A}$) < m, the solution is not unique.

We may now restate (i), (ii) and (iii) of this theorem as follows:
(i) The system $\mathbf{Ax} = \mathbf{b}$ has a solution if and only if $\mathbf{b} \in \mathbf{R(A)}$,
(ii) If $\mathbf{b} \in \mathbf{R(A)}$, and the columns of $\mathbf{A}$ form a basis for $\mathbf{R(A)}$, i.e., they are linearly independent, the solution is unique, and
(iii) If $\mathbf{b} \in \mathbf{R(A)}$, and the columns of $\mathbf{A}$ are linearly dependent, the solution is not unique.

We may also state the following theorem.

Theorem 17.13

Assuming that we are dealing with real matrices. Then:
(i) $\mathbf{N(A^T)} = \mathbf{R(A)}^\perp$
(ii) $\mathbf{N(A)} = \mathbf{R(A^T)}^\perp$
As a consequence, we have:
(iii) $\mathbf{E}_n = \mathbf{R(A)} + \mathbf{R(A)}^\perp$
(iv) $\mathbf{E}_m = \mathbf{R(A^T)} + \mathbf{R(A^T)}^\perp$

Proof:

Let $\mathbf{Q}_1$ in (17.7.1) be written as

$$\mathbf{Q}_1 = [\mathbf{Q}|\mathbf{Q}_2]$$

where the columns of $\mathbf{Q}$ span $\mathbf{R(A)}$, the columns of $\mathbf{Q}_2$ span $\mathbf{R(A)}^\perp$ and $\mathbf{Q}^T\mathbf{Q}_2 = \mathbf{0}$.

To prove (i), let $\mathbf{x} \in \mathbf{R(A)}^\perp$; then there exists a vector $\mathbf{y}$ such that $\mathbf{x} = \mathbf{Q}_2\mathbf{y}$. Hence from (17.7.2), $\mathbf{A}^T\mathbf{x} = \mathbf{T}^T\mathbf{Q}^T\mathbf{Q}_2\mathbf{y} = \mathbf{0}$ and thus $\mathbf{x} \in \mathbf{N(A^T)}$. Therefore, $\mathbf{N(A^T)} \subseteq \mathbf{R(A)}^\perp$.

On the other hand, if $\mathbf{x} \in \mathbf{N(A^T)}$, $\mathbf{T}^T\mathbf{Q}^T\mathbf{x} = \mathbf{0}$ and $\mathbf{x}$ is orthogonal to the columns of $\mathbf{Q}$. That is, $\mathbf{x}$ may be written as $\mathbf{x} = \mathbf{Q}_2\mathbf{y}$. Hence, $\mathbf{x} \in \mathbf{R(A)}^\perp$ and therefore, $\mathbf{R(A)}^\perp \subseteq \mathbf{N(A^T)}$. We conclude that $\mathbf{R(A)}^\perp = \mathbf{N(A^T)}$.

In the same way we prove (ii). Then (iii) and (iv) follow.

17.7.5 Representation of vectors in $\mathbf{V}_m$

It is shown in Theorem 17.10 that if $\mathbf{V}_m$ is an inner product space of dimension m and the m-vectors $\mathbf{q}_1, \mathbf{q}_2, \ldots, \mathbf{q}_m$ form an orthonormal basis for $\mathbf{V}_m$, then any vector $\mathbf{x} \in \mathbf{V}_m$ may be expressed as a linear combination of this basis, in the form

$$\mathbf{x} = \sum_{i=1}^{m} (\mathbf{q}_i, \mathbf{x})\mathbf{q}_i = \sum_{i=1}^{m} \mathbf{q}_i\mathbf{q}_i^T\mathbf{x}$$

Hence, if $\mathbf{Q}$ is an m by m matrix that has the $\mathbf{q}_i$ as its columns, we have the alternative expression to this summation as

$$\mathbf{x} = \mathbf{Q}\mathbf{Q}^T\mathbf{x}$$

17.7.6 Orthogonal projection onto range and null spaces

Let $\mathbf{V}_m$ be expressed as the sum of two spaces $\mathbf{U}$ and $\mathbf{W}$; that is, $\mathbf{V}_m = \mathbf{U} \oplus \mathbf{W}$. Let $\mathbf{q}_1, \mathbf{q}_2, \ldots, \mathbf{q}_k$ span the subspace $\mathbf{U} \subseteq \mathbf{V}_m$ and $\mathbf{q}_{k+1}, \ldots, \mathbf{q}_m$, span the subspace $\mathbf{W} \subseteq \mathbf{V}_m$. Hence, if $\mathbf{x} \in \mathbf{V}_m$, then we may write

$$\mathbf{x} = \sum_{i=1}^{k} \alpha_i\mathbf{q}_i + \sum_{i=k+1}^{m} \alpha_i\mathbf{q}_i = \mathbf{x}_1 + \mathbf{x}_2$$

Then $\mathbf{x}_1 \in \mathbf{U}$ and $\mathbf{x}_2 \in \mathbf{W}$.

Theorem 17.14

Let $\mathbf{Q}_1$ be the m by k matrix whose columns are $\mathbf{q}_1, \mathbf{q}_2, \ldots, \mathbf{q}_k$ and $\mathbf{Q}_2$ be the m by (m – k) matrix whose columns are $\mathbf{q}_{k+1}, \ldots, \mathbf{q}_m$. Let also $\mathbf{I}_m$ be an m-unit matrix. Then:

(i) $\mathbf{Q}_1\mathbf{Q}_1^T = (\mathbf{I}_m - \mathbf{Q}_2\mathbf{Q}_2^T)$

(ii) $\mathbf{Q}_2\mathbf{Q}_2^T = (\mathbf{I}_m - \mathbf{Q}_1\mathbf{Q}_1^T)$

(iii) $\mathbf{x}_1 = \mathbf{Q}_1\mathbf{Q}_1^T\mathbf{x} = (\mathbf{I}_m - \mathbf{Q}_2\mathbf{Q}_2^T)\mathbf{x}$

(iv) $\mathbf{x}_2 = \mathbf{Q}_2\mathbf{Q}_2^T\mathbf{x} = (\mathbf{I}_m - \mathbf{Q}_1\mathbf{Q}_1^T)\mathbf{x}$

Proof:

Since $\mathbf{Q}$ is an m by m orthonormal, $\mathbf{Q}^T\mathbf{Q} = \mathbf{Q}\mathbf{Q}^T = \mathbf{I}_m$ and we have $\mathbf{I}_m = \mathbf{Q}_1\mathbf{Q}_1^T + \mathbf{Q}_2\mathbf{Q}_2^T$ and the proof follows.

Definition 17.1

We define $\mathbf{Q}_1\mathbf{Q}_1^T$ as the orthogonal projection operator that projects any vector $\mathbf{x} \in \mathbf{V}_m$ onto the space $\mathbf{U}$. Likewise let $\mathbf{Q}_2\mathbf{Q}_2^T$ be the orthogonal projection operator that projects any vector $\mathbf{x} \in \mathbf{V}_m$

onto the space **W**. This notion is continued as follows.

Orthogonal projection operators onto R(A) and R(A^T)

From (17.7.3), $\mathbf{A} = \mathbf{VSW}^T$ and thus $\mathbf{A}^+ = \mathbf{WS}^{-1}\mathbf{V}^T$ from which

$$\mathbf{AA}^+ = \mathbf{VV}^T \text{ and } \mathbf{A}^+\mathbf{A} = \mathbf{WW}^T$$

Theorem 17.15

$$\mathbf{AA}^+, (\mathbf{I}_n - \mathbf{AA}^+), \mathbf{A}^+\mathbf{A} \text{ and } (\mathbf{I}_m - \mathbf{A}^+\mathbf{A})$$

or respectively

$$\mathbf{VV}^T, (\mathbf{I}_n - \mathbf{VV}^T), \mathbf{WW}^T \text{ and } (\mathbf{I}_m - \mathbf{WW}^T)$$

are

(i) Orthogonal projector operators,
(ii) They project any vector **x** onto $\mathbf{R(A)}$, $\mathbf{R(A)}^\perp$, $\mathbf{R(A}^T)$ and $\mathbf{R(A}^T)^\perp$ respectively.

Consider

$$\mathbf{AA}^+\mathbf{A} = (\mathbf{AA}^+)\mathbf{A} = (\mathbf{VV}^T)\mathbf{A} = \mathbf{A}$$

This is interpreted as the columns of **A** belong to the range space of **A**, which is a trivial result. Similarly

$$\mathbf{A}^+\mathbf{AA}^+ = (\mathbf{A}^+\mathbf{A})\mathbf{A}^+ = (\mathbf{WW}^T)\mathbf{A}^+ = \mathbf{A}^+$$

is interpreted as the columns of $\mathbf{A}^+$ belong to the range space of $\mathbf{A}^+$. The other two operators are interpreted in the same way.

Theorem 17.16

Let $\mathbf{Ax} = \mathbf{b}$, and **b** be resolved into two vectors $\mathbf{b} = \mathbf{b}_1 + \mathbf{b}_2$, where $\mathbf{b}_1 \in \mathbf{R(A)}$ and $\mathbf{b}_2 \in \mathbf{R(A)}^\perp$. Then the least squares solution **x** is the exact solution of $\mathbf{Ax} = \mathbf{b}_1$, and the residual $\mathbf{r} = \mathbf{b}_2$.

Proof:

The least squares solution vector **x** of $\mathbf{Ax} = \mathbf{b}$ is given by $\mathbf{x} = \mathbf{A}^+\mathbf{b}$ and by pre-multiplying by **A**, we get

$$\mathbf{Ax} = \mathbf{AA}^+\mathbf{b}$$

and from the previous theorem, the right hand side is the projection of **b** onto **R(A)**, i.e., $= \mathbf{b}_1$. Then $\mathbf{b}_2 = \mathbf{b} - \mathbf{b}_1 = (\mathbf{I}_n - \mathbf{AA}^+)\mathbf{b} = \mathbf{b} - \mathbf{Ax} = \mathbf{r}$ and the theorem is proved, emphasizing that $\mathbf{r} \in \mathbf{R(A)}^\perp$.

17.7.7 Singular values of the orthogonal projection matrices

Theorem 17.17

Assuming that **A** is of rank k, the singular values of $\mathbf{AA}^+$ are k 1's and (n – k) 0's. The singular values of $(\mathbf{I}_n - \mathbf{AA}^+)$ are (n – k) 1's and k 0's. Similar results are for the two other operators $\mathbf{A}^+\mathbf{A}$, and $(\mathbf{I}_m - \mathbf{A}^+\mathbf{A})$.

Proof:

From Theorem 4.10 (Wilkinson ([27], p. 54), given **C** as an m by k matrix and **D** as a k by m matrix, $m \geq k$, **CD** and **DC** have k identical eigenvalues and the remaining (m – k) eigenvalues of **CD** are 0's. The proof of the theorem follows from the fact that the eigenvalues of $\mathbf{AA}^+$ are themselves their singular values and by observing that $\mathbf{AA}^+ = \mathbf{VV}^T$ and that $\mathbf{V}^T\mathbf{V} = \mathbf{I}_k$.

Corollary

$$||\mathbf{AA}^+||_2 = 1, \quad ||\mathbf{I}_n - \mathbf{AA}^+||_2 = 1, \text{ etc.}$$

Let us now consider the interesting subject of the ill-conditioning of the coefficient matrix **A** in the matrix equation $\mathbf{Ax} = \mathbf{b}$.

17.8 Multicollinearity, collinearity or the ill-conditioning of matrix A

In analyzing their data, statisticians build a linear model of the form

$$\mathbf{Ax} = \mathbf{b}$$

A is a given n by m matrix containing the data of the problem, where in most cases, n > m, and **b** is a given observation n-vector. The least squares solution of this equation is mostly obtained by forming the normal equation

$$\mathbf{A}^T\mathbf{Ax} = \mathbf{A}^T\mathbf{b}$$

In case matrix **A** is of rank m, $\mathbf{A}^T\mathbf{A}$ is non-singular and the least squares solution is given by

$$\mathbf{x} = (\mathbf{A}^T\mathbf{A})^{-1}\mathbf{A}^T\mathbf{b}$$

When matrix **A** is well conditioned, its columns are almost orthogonal. However, this is not the case in most applications. In many problems the lack of orthogonality is not serious enough to affect the analysis of the result.

The state of strong non-orthogonality is referred to as the problem of **collinear data**, also known as **multicollinearity**, and is a condition of deficient matrix **A**. The calculated elements of **x** are very sensitive to perturbations (slight errors) in the data. Hence, it is important to know when multicollinearity occurs and how to remedy it. Let us consider the following example by Chatterjee and Price ([6], pp. 144-151).

Example 17.1

The congress of the United Sates ordered a survey concerning the lack of availability of equal educational opportunities for students. Data were collected from 70 schools across the country. A linear model of the form **Ax** = **b** is constructed with the solution vector **x** to be calculated.

Vector **b** is the measure of the level of student achievement. Matrix **A** consists of 4 columns; $\mathbf{a}_1$, $\mathbf{a}_2$, $\mathbf{a}_3$ and $\mathbf{a}_4$. Column $\mathbf{a}_1$ is a constant column of 1's, $\mathbf{a}_2$ is a measure of the school facilities, $\mathbf{a}_3$ consists of measures of home environments and column $\mathbf{a}_4$ is a measure of the school credentials. Equation **Ax** = **b**, may written as

$$x_1\mathbf{1} + x_2\mathbf{a}_2 + x_3\mathbf{a}_3 + x_4\mathbf{a}_4 = \mathbf{b}$$

The results of this equation were disappointing. Some of the calculated coefficients had negative values, which was not expected. That is besides the failed statistical tests carried out by the analysts.

These results indicate the existence of extreme multicollinearity between the last three columns of **A**. There exist simple relationships between every pair of the three columns $\mathbf{a}_2$, $\mathbf{a}_3$ and $\mathbf{a}_4$. It is concluded that these three columns may be replaced by only one of them.

In Section 17.7.1, we defined linear dependence and linear independence of a given set of vectors. We define that here again for the columns of a given matrix **A**.

Definition 17.2

Let $\mathbf{a}_1, \mathbf{a}_2, \ldots, \mathbf{a}_m$ denote the m columns of matrix **A**. Then these columns are linearly dependent if there exist a vector **c** of constants whose elements are $c_1, c_2, \ldots, c_m$, not all 0's, such that

$$\sum_{i=1}^{m} c_i \mathbf{a}_i = \mathbf{0}$$

Definition 17.3

Collinearity exists among the columns of matrix **A** if for some specified small constant $\rho > 0$

$$\sum_{i=1}^{m} c_i \mathbf{a}_i = \mathbf{d}, \quad \|\mathbf{d}\| \leq \rho \|\mathbf{c}\|$$

where **c** is defined above.

If definition 17.2 holds for a subset of the columns of **A**, then rank(**A**) < m and $(\mathbf{A}^T\mathbf{A})$ is singular and its inverse does not exist. However, if definition 17.3 holds for a subset of the columns of **A**, we have near linear dependency among these columns and collinearity exists as matrix $(\mathbf{A}^T\mathbf{A})$ is ill-conditioned.

17.8.1 Sources of multicollinearity

Montgomery and Peck ([16], p. 306) cited two primary sources of multicollinearity. The first is an inappropriate choice of columns of matrix **A**, thus causing ill-conditioning of $(\mathbf{A}^T\mathbf{A})$. They also suggested how to remedy this situation. For example, if columns $\mathbf{a}_1$, $\mathbf{a}_2$ and $\mathbf{a}_3$ of **A** are nearly linearly dependent, one might replace the three columns by one column whose i^{th} element is the product of the ith elements of the three columns. This preserves the information given by the three original columns. This method is a kind of re-specifying the given problem. Another technique of re-specifying the problem is to eliminate one of the columns $\mathbf{a}_1$, $\mathbf{a}_2$ and $\mathbf{a}_3$. Such techniques might not be totally satisfactory.

The second source of multicollinearity, is the improbably-defined model, which means that there are not enough data points. In this case, the usual approach is to increase the data points (increase n) or eliminate some of the columns of matrix **A** (decrease m).

17.8.2 Detection of multicollinearity

Multicollinearity may be detected when there occurs instability in the elements of the solution vector **x**. Large changes occur in the elements of **x** when a variable is added or is deleted in search of a better matrix equation, i.e., when a column is added to or a column is deleted from matrix **A**.

A second indication of multicollinearity is large changes in the elements of **x** when a data point is added or is dropped; i.e., when n is increased or decreased by one.

A third indication is the numerical signs and/or the magnitude of the elements of **x** that are not consistent with the physical problem. A calculated number of students in a school, for example, is expected to be a positive number.

A statistical detection of the presence of multicollinearity is the size of correlation coefficients (measure of dependence) that exist between the columns of matrix **A**. A large correlation between two of the columns indicates a strong linear relationship between those two columns.

More reliable means are needed to detect the existence of multicollinearity. One of the powerful means is by calculating the singular values of matrix $(\mathbf{A}^T\mathbf{A})$, where we get the factorization $(\mathbf{A}^T\mathbf{A}) = \mathbf{W}\mathbf{S}^2\mathbf{W}^T$(equations (17.5.2)). $\mathbf{S}^2$ is a diagonal matrix, whose elements $s_1^2 \geq s_2^2 \geq \ldots \geq s_m^2$ are the singular values of $(\mathbf{A}^T\mathbf{A})$. If one or more of the singular values are very small, that implies the near linear dependency among a subset of the columns of matrix **A**. However, some authors argue by saying, we are not informed what "small" is and there is a natural tendency to compare "small" to 0.

The condition number of matrix $(\mathbf{A}^T\mathbf{A})$; $\kappa = s_1^2/s_m^2$, is a better indication of the presence of multicollinearity. Montgomery and Peck ([16], p. 319) argue saying, if this condition number is < 100, there is no serious problem with multicollinearity. Yet, if this condition number is between 100 and 1000, moderate to strong multicollinearity exists and if it exceeds 1000, severe multicollinearity exists. They also defined the **condition indices** of $(\mathbf{A}^T\mathbf{A})$ as

$$\kappa_i = s_1^2/s_i^2, \quad i = 1, 2, \ldots, m$$

Example 17.2

Montgomery and Peck ([16], pp. 319) recorded the condition indices of $(\mathbf{A}^T\mathbf{A})$ for an example of m = 9 as $\kappa_1 = 1, \ldots, \kappa_6 = 84.96$, $\kappa_7 = 309.18$, $\kappa_8 = 824.47$ and $\kappa_9 = 42{,}048$. Since one of the condition indices exceed 1000, it is concluded that there exists strong near dependency in the data of this example.

Existence of multicollinearity may also be detected by standard statistical means. In the following sections, we describe three non-statistical methods to combat multicollinearity; **principal components analysis** (PCA), **partial least squares** (PLS) and the **Ridge equation**.

17.9 Principal components analysis (PCA)

Principal components analysis is a method that rewrites matrix **A** in the equation $\mathbf{Ax} = \mathbf{b}$ in terms of a set of orthogonal columns. These new columns are obtained as linear combinations of the columns of **A**. They are known as the **principal components** of the columns of **A**.

17.9.1 Derivation of the principal components

Consider the equations $\mathbf{Ax} = \mathbf{b}$. The principal components are realized by the help of the singular value decomposition of the n by m matrix **A** of rank m ([6], pp. 172-174). From (17.5.1)

$$\mathbf{A} = \mathbf{VSW}^T$$

From this we write

$$\mathbf{VS} = \mathbf{AW} = (\mathbf{a}_1, \mathbf{a}_2, \ldots, \mathbf{a}_m)\mathbf{W} \tag{17.9.1}$$

Let us write $\mathbf{VS} = \boldsymbol{V}$ whose columns are $(\boldsymbol{v}_1, \boldsymbol{v}_2, \ldots, \boldsymbol{v}_m)$ and the diagonal elements of **S** be $(s_1, s_2, \ldots, s_m)$. The columns of vector $\boldsymbol{V}$ are orthogonal, $(\boldsymbol{v}_i, \boldsymbol{v}_i) = s_i^2$ and $(\boldsymbol{v}_i, \boldsymbol{v}_j) = 0$, $i \neq j$.

The columns $\boldsymbol{v}_i$ are referred to as the “principal components”. Equation $\mathbf{Ax} = \mathbf{b}$ may be rewritten in terms of the principal components as follows. Since $\mathbf{WW}^T = \mathbf{I}_m$, $\mathbf{Ax} = \mathbf{b}$ is

$$\mathbf{AWW}^T\mathbf{x} = \mathbf{b}$$

Let

(17.9.2) $$\mathbf{y} = \mathbf{W}^T\mathbf{x}$$

then from (17.9.1), we get

(17.9.3) $$\mathbf{VSy} = \mathbf{b}$$

Thus

(17.9.4) $$\mathbf{y} = \mathbf{S}^{-1}\mathbf{V}^T\mathbf{b} \text{ or } \mathbf{y} = \sum_{i=1}^{m} s_i^{-1}\mathbf{v}_i^T b_i$$

Equation (17.9.3) is a re-presentation of equation $\mathbf{Ax} = \mathbf{b}$ in terms of the orthogonal columns of $\mathcal{V} = \mathbf{VS}$. Also, from (17.9.1), the ith principal component $\boldsymbol{v}_i = \mathbf{v}_i s_i$, i = 1, 2, …, m, is a linear function of the columns $\mathbf{a}_j$, namely

$$\mathbf{v}_i s_i = \sum_{j=1}^{m} w_{ij}\mathbf{a}_j$$

It follows that when $s_i = 0$, the l.h.s. of the above equation is 0. From definition 17.2 linear dependences exists among the columns of **A**. If s_i is very small, from definition 17.3, there is an approximate linear dependence among the given columns of **A**.

Suppose that the first k singular values of **A** are significant and the last (m – k) are either 0's or are considered 0's. The last (m – k) principal components are then considered 0's. The solution of (17.9.4) will instead be

$$\mathbf{y} = \sum_{i=1}^{k} s_i^{-1}\mathbf{v}_i^T b_i$$

and from (17.9.2) $\mathbf{x} = \mathbf{Wy}$, or

$$\mathbf{x} = \sum_{i=1}^{k} s_i^{-1}\mathbf{w}_i\mathbf{v}_i^T b_i$$

where $(\mathbf{w}_i)$ are the columns of matrix **W**.

The principal components analysis may be used as a linear dimensionality reduction technique. It determines a set of orthogonal vectors, which are the principal components, ordered by the largest singular values which describe most of the state of the problem.

The portion of the PCA space corresponding to the smaller singular values describe the random noise. By properly determining the number of principal components, to maintain the PCA model, the system can be decoupled from the random noise. The principal components, corresponding to the large singular values of $(\mathbf{A}^T\mathbf{A})$ are retained and the smallest singular values are set equal to 0's. To illustrate this point, consider the following example.

Example 17.3

Data was collected by Cheng et al. ([7], p. 107) for pattern analysis in industry and was used by Russell et al. ([20], pp. 36, 37). For this data, the n by m matrix **A** in the equation $\mathbf{Ax} = \mathbf{b}$ is a 50 by 4 matrix. The 4 by 4 matrix $(\mathbf{A}^T\mathbf{A})$ is decomposed by the singular value decomposition into $(\mathbf{A}^T\mathbf{A}) = \mathbf{W}\mathbf{S}^2\mathbf{W}^T$, where the elements of $\mathbf{S}^2$ are the singular values of $(\mathbf{A}^T\mathbf{A})$.

The singular values of $(\mathbf{A}^T\mathbf{A})$ are (1.92, 0.96, 0.88, 0.24) and their sum = 4. If the first two columns of the principal components matrix V are retained, the ratio of the sum of their singular values to the sum of the total singular values is (1.92 + 0.96)/4 = 72%, contributed to this problem. That means dimension reduction to the problem with advantages explained by Russell et al. ([20], p. 37).

An alternative to the principal components analysis is the **partial least squares** (PLS) technique, presented next.

17.10 Partial least squares method (PLS)

Given is the linear model or the overdetermined matrix equation $\mathbf{Ax} = \mathbf{b}$. Like the PCA method, the partial least squares (PLS) method is a dimensionality reduction technique. It is particularly appealing because the calculation is performed on the given data **A** (not on $\mathbf{A}^T\mathbf{A}$). This makes it suitable for large problems.

There is a number of variations of the algorithm for this method. The most popular one is known as the **non-iterative partial least squares** (NIPALS). It is described by Geladi and Kowalski [8] and by Hoskuldsson [13]. See also Wold et al. [28] and Russell et al. [20]. The NIPALS does not calculate all the principal components at once.

In the matrix equation $\mathbf{Ax} = \mathbf{b}$, there might be more than one right hand side vector **b**; $\mathbf{b}_1, \mathbf{b}_2, \ldots, \mathbf{b}_r$, and in this case, there will be more than one solution vector **x**; $\mathbf{x}_1, \mathbf{x}_2, \ldots, \mathbf{x}_r$. Then we may write $\mathbf{Ax} = \mathbf{b}$ in the form

$$\mathbf{AX} = \mathbf{B} \tag{17.10.1}$$

where $\mathbf{B} = [\mathbf{b}_1, \mathbf{b}_2, \ldots, \mathbf{b}_r]$ is an n by r matrix and $\mathbf{X} = [\mathbf{x}_1, \mathbf{x}_2, \ldots, \mathbf{x}_r]$ is an m by r matrix. It is important that **A** and **B** be mean-centered and scaled [8].

Matrix **A** may be given in terms of the sum of m matrices, each is of rank 1.

$$\mathbf{A} = \mathbf{t}_1\mathbf{p}_1^T + \mathbf{t}_2\mathbf{p}_2^T + \ldots + \mathbf{t}_m\mathbf{p}_m^T$$

Each of the $\mathbf{t}_s\mathbf{p}_s^T$ is an outer product of $\mathbf{t}_s$ (known as score vector) and $\mathbf{p}_s$ (known as loading vector). We may write **A** as

$$\mathbf{A} = \mathbf{TP}^T + \mathbf{E} = \sum_{j=1}^{k} \mathbf{t}_j\mathbf{p}_j^T + \mathbf{E}$$

where $\mathbf{T} = [\mathbf{t}_1, \mathbf{t}_2, \ldots, \mathbf{t}_k]$ and $\mathbf{P} = [\mathbf{p}_1, \mathbf{p}_2, \ldots, \mathbf{p}_k]$. **E** is a residual matrix.

Similarly, **B** in (17.10.1) may be decomposed into a score matrix **U** and a loading matrix **Q** added to a residual matrix **F***

$$\mathbf{B} = \mathbf{UQ}^T + \mathbf{F}^* = \sum_{j=1}^{k} \mathbf{u}_j\mathbf{p}_j^T + \mathbf{F}^*$$

Let k be the rank of matrix **A**. If k is set equal to min(m, n), then **E** and **F*** are 0's and the PLS reduces to the ordinary least squares solution. Setting k < min(m, n) reduces noise and colliniarity. The aim of using the PLS is to describe matrix **A** in terms of a smaller number of the $\mathbf{t}_j\mathbf{p}_j^T$ components. The aim of the following analysis is to get a useful relation between matrices **A** and **B**. The simplest relation is via taking

$$\mathbf{u}_h = b_h\mathbf{t}_h$$

where $b_h = \mathbf{u}_h^T\mathbf{t}_h/\|\mathbf{t}_h\|$.

17.10.1 Model building for the PLS method

The PLS model may be considered as consisting of outer products of matrices in **A** and in **B** and inner relations linking such matrices. A simplified model is as follows: For matrix **A**, calculating the outer product matrices is described in 5 steps, below.

To start with, one calculates $\mathbf{t}_1$ and $\mathbf{p}_1^T$ from matrix **A**. Then $\mathbf{t}_1\mathbf{p}_1^T$ is subtracted from **A** and the residual $\mathbf{E}_1$ is calculated. Then $\mathbf{E}_1$ is used to calculate $\mathbf{t}_2$ and $\mathbf{p}_2^T$ and so on. Thus

$$\mathbf{E}_1 = \mathbf{A} - \mathbf{t}_1\mathbf{p}_1^T,\ \mathbf{E}_2 = \mathbf{E}_1 - \mathbf{t}_2\mathbf{p}_2^T,\ \ldots,\ \mathbf{E}_h = \mathbf{E}_{h-1} - \mathbf{t}_h\mathbf{p}_h^T,\ \ldots$$

For matrix **A**, the algorithm is:

(1) $\mathbf{t}_{start}$ = a column $\mathbf{a}_j$ of **A**
(2) $\mathbf{p}_{old} = \mathbf{A}^T\mathbf{t}/\|\mathbf{t}\|\ (=\mathbf{A}^T\mathbf{u}/\|\mathbf{u}\|)$
(3) Scale **p** to be of length one; $\mathbf{p}_{new} = \mathbf{p}_{old}/\|\mathbf{p}_{old}\|$, $\mathbf{p} = \mathbf{p}_{new}$
(4) $\mathbf{t} = \mathbf{A}\mathbf{p}/\|\mathbf{p}\|$
(5) Compare **t** used in step (2) with **t** obtained in step (4). If the difference is of the order of rounding error, the iteration has converged. If not go to step (2).

For matrix **B**, the procedure is similar, namely

(1) $\mathbf{u}_{start}$ = a column $\mathbf{b}_j$ of **B**
(2) $\mathbf{q}_{old} = \mathbf{B}^T\mathbf{u}/\|\mathbf{u}\|\ (= \mathbf{B}^T\mathbf{t}/\|\mathbf{t}\|)$
(3) Scale **q** to be of length one; $\mathbf{q}_{new} = \mathbf{q}_{old}/\|\mathbf{q}_{old}\|$, $\mathbf{q} = \mathbf{q}_{new}$
(4) $\mathbf{u} = \mathbf{B}\mathbf{q}/\|\mathbf{q}\|$
(5) Compare **u** used in step (2) with **u** obtained in step (4). If the difference is of the order of rounding error, the iteration has converged. If not go to step (2).

The above two separate algorithms are applied to **A** and **B** respectively. However, in order that each can get information about the other, let **t** and **u** change places in step (2) and the two algorithms would be written in sequence. The following algorithm is due to Geladi and Kowalski [8] and also to Hoskuldsson [13].

For each major iteration

(1) $\mathbf{u}_{start}$ = a column $\mathbf{b}_j$ of **B**

For matrix **A**

(2) $\mathbf{w} = \mathbf{A}^T\mathbf{u}/\|\mathbf{u}\|$
(3) Normalize **w**, $\mathbf{w}_{new} = \mathbf{w}_{old}/\|\mathbf{w}_{old}\|$

(4) $\mathbf{t} = \mathbf{Aw}/||\mathbf{w}||$

For matrix **B**

(5) $\mathbf{q} = \mathbf{B}^T\mathbf{t}/||\mathbf{t}||$

(6) Normalize **q**; $\mathbf{q}_{new} = \mathbf{q}_{old}/||\mathbf{q}_{old}||$

(7) $\mathbf{u} = \mathbf{Bq}/||\mathbf{q}||$

(8) Compare **t** in step (4) with the one in the previous iteration. If the difference is of the order of rounding error, the iteration has converged and go to step (9). If not, go to step (2).

If **B** consists of one column only, steps (5)-(8) can be omitted and no more iterations are required.

(9) $\mathbf{p} = \mathbf{A}^T\mathbf{t}/||\mathbf{t}||$

(10) Normalize **p**; $\mathbf{p}_{new} = \mathbf{p}_{old}/||\mathbf{p}_{old}||$

(11) $\mathbf{t}_{new} = \mathbf{t}_{old}||\mathbf{p}_{old}||$

(12) $\mathbf{w}_{new} = \mathbf{w}_{old}||\mathbf{p}_{old}||$

Obtain the regression coefficient b for the inner relation

(13) $b = \underline{\mathbf{u}}^T\mathbf{t}/||\mathbf{t}||$

The residual for the **A** matrix (block) for component h is

$$\mathbf{E}_h = \mathbf{E}_{h-1} - \mathbf{t}_{h-1}\mathbf{p}_{h-1}{}^T, \quad \mathbf{E}_0 = \mathbf{A}$$

and for the **B** matrix (block), for component h is

$$\mathbf{F}_h = \mathbf{F}_{h-1} - b_{h-1}\mathbf{t}_{h-1}\mathbf{q}_{h-1}{}^T, \quad \mathbf{F}_0 = \mathbf{B}$$

For the next major iteration, meaning for the next component, go to step (1). However, after the first component, **A** is replaced by $\mathbf{E}_h$ and **B** is replaced $\mathbf{F}_h$. The iterations continue until a stopping criteria is used or $\mathbf{E}_h$ becomes the zero matrix.

The following analysis is due to Hoskuldsson [13]. Using steps (7), (5), (4) and (2) above in succession, we get

$$\begin{aligned}
\mathbf{u}_h &= \mathbf{Bq}_h/||\mathbf{q}_h|| \\
&= \mathbf{BB}^T\mathbf{t}_h/[||\mathbf{q}_h||\ ||\mathbf{t}_h||] \\
&= \mathbf{BB}^T\mathbf{Aw}_h/[||\mathbf{q}_h||\ ||\mathbf{t}_h||\ ||\mathbf{w}||] \\
&= \mathbf{BB}^T\mathbf{AA}^T\mathbf{u}_{h-1}/[||\mathbf{q}_h||\ ||\mathbf{t}_h||\ ||\mathbf{w}_h||\ ||\mathbf{u}_{h-1}||]
\end{aligned}$$

These equations show that the algorithm described above performs like calculating the largest eigenvalue of a matrix by the power point ([12], pp. 330-332). In most practical cases, convergence is obtained in less than ten iterations, unless there are equal eigenvalues that are the largest. Similar equations are derived for $\mathbf{q}_h$, $\mathbf{t}_h$ and $\mathbf{w}_h$.

At convergence we may write

$$\mathbf{BB}^T\mathbf{AA}^T\mathbf{u} = \mathrm{a}\mathbf{u}$$

$$\mathbf{B}^T\mathbf{AA}^T\mathbf{B}\,\mathbf{q} = \mathrm{a}\mathbf{q}$$

$$\mathbf{AA}^T\mathbf{BB}^T\mathbf{t} = \mathrm{a}\mathbf{t}$$

$$\mathbf{A}^T\mathbf{BB}^T\mathbf{Aw} = \mathrm{a}\mathbf{w}$$

where a is the largest eigenvalue and vectors **u**, **q**, **t** and **w** are eigenvectors of the appropriate matrices corresponding to the largest eigenvalue.

The next equation relates a residual matrix to its previous residual matrix

$$\begin{aligned}\mathbf{E}_h &= \mathbf{E}_{h-1} - \mathbf{t}_{h-1}\mathbf{p}_{h-1}^T \\ &= \mathbf{E}_{h-1} - \mathbf{t}_{h-1}\mathbf{t}_{h-1}^T\mathbf{E}_{h-1}/\|\mathbf{t}_{h-1}\| \\ &= [\mathbf{I} - \mathbf{t}_{h-1}\mathbf{t}_{h-1}^T/\|\mathbf{t}_{h-1}\|]\,\mathbf{E}_{h-1}\end{aligned}$$

The basic properties of vectors **w**, **t** and **p** are derived from the residual matrices [13]

(i) The vectors $\mathbf{w}_i$ are mutually orthogonal, $(\mathbf{w}_i, \mathbf{w}_j) = 0$, $i \neq j$.

(ii) The vectors $\mathbf{t}_i$ are mutually orthogonal, $(\mathbf{t}_i, \mathbf{t}_j) = 0$, $i \neq j$.

(iii) The vectors $\mathbf{w}_i$ are orthogonal to the vectors $\mathbf{p}_j$, $(\mathbf{w}_i, \mathbf{p}_j)$, $i < j$.

What is important is that these properties do not depend on the way a new **t** vector is constructed.

There are other more interesting properties for the PLS algorithm. For this see the tutorial paper by Hoskuldsson [13]. The subject of PLS is a large one and the analysis given above is a brief description for the non-iterative partial least squares (NIPALS) version.

Example 17.4

The data of Example 17.3, which illustrated the use of the principal components analysis (PCA), was solved again by Russell et al. ([20], p. 57) to illustrate the use of the PLS method. For the NIPALS algorithm, they took $\mathbf{E}_0 = \mathbf{A}$, $\mathbf{F}_0{}^* = \mathbf{B}$. After 12 iterations for the first component $\mathbf{t}_1$, convergence was obtained with an error

$< 10^{-10}$. One more major iteration was needed to obtain a satisfactory result for the problem.

In the next section, we describe a third technique to overcome the problem of multicollinearity or the ill-conditioning of matrix **A**; the **Ridge equation**, which is a stabilized normal equation.

17.11 Ridge equation

As indicated in Section 17.5, the condition number of a matrix defines the condition of the matrix with respect to a computing problem. The spectral condition number K of matrix **A** of rank k is given in terms of its largest and smallest singular values s_1 and s_k, respectively

$$K = ||\mathbf{A}||_2 \, ||\mathbf{A}^+||_2 = [s_1]/[s_k]$$

When the condition number is very large, that implies in most cases that s_k is very small. On the other hand, a singular value s_k that is very small is an indicator of multicollinearity.

Given the system of linear equations $\mathbf{Ax} = \mathbf{b}$, the original idea behind the Ridge equation is to reduce the condition number of matrix $\mathbf{A}^T\mathbf{A}$ in the normal equation $\mathbf{A}^T\mathbf{Ax} = \mathbf{A}^T\mathbf{b}$. Normally, $0 \leq \varepsilon \leq 1$ and the Ridge equation is given by

(17.11.1) $$(\mathbf{A}^T\mathbf{A} + \varepsilon \mathrm{I})\mathbf{x} = \mathbf{A}^T\mathbf{b}$$

I is an m-unit matrix. The solution of this equation is

$$\mathbf{x} = (\mathbf{A}^T\mathbf{A} + \varepsilon \mathrm{I})^{-1}\mathbf{A}^T\mathbf{b}$$

It can be shown ([22], pp. 207-208, [25], p. 102) that (17.11.1) gives

$$\mathbf{x} = \min_{\mathbf{x}}(||\mathbf{Ax} - \mathbf{b}||^2 + \varepsilon||\mathbf{x}||^2)$$

The singular values of the m by m matrix $\mathbf{A}^T\mathbf{A}$ are non-negative and if they are denoted by $\sigma_1, \sigma_2, \ldots, \sigma_m$, the singular values of $(\mathbf{A}^T\mathbf{A} + \varepsilon \mathrm{I})$ are $(\sigma_1 + \varepsilon), (\sigma_2 + \varepsilon), \ldots, (\sigma_m + \varepsilon)$. The condition number of $(\mathbf{A}^T\mathbf{A} + \varepsilon \, \mathrm{I})$ is

$$[(\sigma_1 + \varepsilon)/(\sigma_m + \varepsilon)] < (\sigma_1/\sigma_m)$$

As a result, the solution vector **x** of the Ridge equation is stable with respect to slight perturbations in the given data **A**.

A modified version of (17.11.1) was used by Varah [25, 26] in the solution of the Fredholm integral equation of the first kind (Chapter 19), namely

$$(\mathbf{A}^T\mathbf{A} + \varepsilon\mathbf{L}^T\mathbf{L})\mathbf{x} = \mathbf{A}^T\mathbf{b}$$

L is some discrete approximation to the first or the second derivative operator. However, we shall only consider equation (17.11.1).

Example 17. 5

The data for this example is given in Chatterjee and Price ([6], p. 152). This example is also presented in Section 1.2.6. This example is now solved by the Ridge equation. The linear model is $\mathbf{Ax} = \mathbf{b}$, or

(17.11.2) $$x_0\mathbf{1} + x_1\mathbf{a}_1 + x_2\mathbf{a}_2 + x_3\mathbf{a}_3 = \mathbf{b}$$

The **1** denotes an n-column of 1's, $\mathbf{a}_1$ refers to domestic production, $\mathbf{a}_2$ refers to stock formation and $\mathbf{a}_3$ refers to domestic consumption, while **b** refers to the imports, all measured in millions of French francs.

The Ridge equation (17.11.1) for this model was solved many times for values of ε from $\varepsilon = 0$ to $\varepsilon = 1$ in increments of 0.001. The results of x_1, x_2 and x_3 were plotted and also were given as a function of ε ([6], pp. 184, 185). The value of x_1 changes –0.339 for $\varepsilon = 0$ to a stable value of 0.42 for $\varepsilon = 0.14$, x_3 changes from 1.302 for $\varepsilon = 0$ to 0.525 for $\varepsilon = 0.14$, which are big changes for x_1 and x_3. The value of x_2 was not affected and remains stable at about 0.21.

17.11.1 Estimating the Ridge parameter

One of the early ways to estimate ε that is still in-use today is to construct a ridge trace (graph). In this graph, the value of each element of the solution vector **x** is plotted against ε, $0 \le \varepsilon \le 1$, at small increments, as was done in the above example. The value ε at which the elements of **x** seem to stabilize is selected. This is of course a subjective way of selection. There are other techniques that use statistical factors. For these techniques, see for example, Montgomery and Peck ([16], pp. 339-343) and the references in Ryan ([21], pp. 400, 401).

17.11.2 Ridge equation and variable selection

In some linear models **Ax** = **b**, a subset of the columns of **A** cause multicollinearity. It is thus desirable to find these columns and to delete some of them. This process is known as **variable selection.**

This is mainly done by examining the ridge trace mentioned above. It is done by removing the columns of matrix **A** that correspond to unstable **x** element, which changes sign as ε increase from 0 towards 1, and also delete the columns of **A** that correspond to the **x** elements that decrease in value towards 0, again as ε increase from 0 towards 1. This is illustrated by the following example, due to Montgomery and Peck ([16], pp. 343-345). This example is also discussed in Section 1.3.2.

Example 17.6

The matrix equation for this example is given by

$$\mathbf{P} = x_0\mathbf{1} + x_1\mathbf{T} + x_2\mathbf{H} + x_3\mathbf{C} + x_4\mathbf{TH} + x_5\mathbf{TC} + x_6\mathbf{HC} + x_7\mathbf{T}^2 + x_8\mathbf{H}^2 + x_9\mathbf{C}^2$$

Here, $\mathbf{b} = \mathbf{P}$, $\mathbf{a}_1 = \mathbf{1}$, an n-column of 1's, $\mathbf{a}_1 = \mathbf{T}$, $\mathbf{a}_2 = \mathbf{H}$, …, etc. **TH** means each element of **T** is multiplied by the corresponding element of **H**. Similarly, in $\mathbf{T}^2$ each element of **T** is squared, …, etc. (see the notations in Section 1.3.2).

The ridge trace is plotted as a function of the ridge variable ε, from ε = 0 to ε = 1, at small increments. As ε increases the coefficients x_5 and x_9 decease rapidly toward 0. Coefficients x_6 changed sign at ε = 0.32 and x_8 decreased towards 0 but not so rapidly.

The four columns of matrix **A** that correspond to the above 4 coefficients are deleted and the ridge model is calculated again for the remaining 5 parameters, x_1, x_2, x_3, x_4 and x_7. The new ridge trace seems more stable than when all 9 terms were considered. Also, an increase of the ridge parameter ε from ε = 0 to ε = 1, did not change the values of the remaining 5 variables much.

17.12 Numerical results and comments

The purpose of this final section is to show that the Gauss **LU** factorization and the Householder's **QR** factorization methods are among the most efficient methods for calculating the least squares solution of system $\mathbf{Ax} = \mathbf{b}$ and the pseudo-inverse matrix $\mathbf{A}^+$. The two methods give results with comparable accuracies. See also [1].

LA_Eluls() computes the least squares solution and pseudo inverse of matrices using the Gauss **LU** factorization method with complete pivoting. LA_Hhls() implements the Householder's **QR** factorization method with pivoting. We note that both Lau [15] and Press et al. [19] presented C programs for the **QR** factorization method, but for a matrix with full rank only. Both DR_Eluls() and DR_Hhls() test the same 3 examples.

We also noted earlier that the factorizations by the Gauss **LU** method and by Householder's **QR** method are not for matrix **A**, but for the permuted matrix $\underline{\mathbf{A}}$ defined in these methods. Yet, the final results computed by LA_Eluls() and LA_Hhls() are for the given non-permuted equation $\mathbf{Ax} = \mathbf{b}$.

Note 17.2

Both LA_Eluls() and LA_Hhls() are designed to solve linear systems with several r.h.s. vectors **b**(s).

Example 17.7

This example was solved by Golub, and Reinsch ([11], p. 412) by the singular value decomposition method. The system of linear equations is

$$\begin{bmatrix} 22 & 10 & 2 & 3 & 7 \\ 14 & 7 & 10 & 0 & 8 \\ -1 & 13 & -1 & -11 & 3 \\ -3 & -2 & 13 & -2 & 4 \\ 9 & 8 & 1 & -2 & 4 \\ 9 & 1 & -7 & 5 & -1 \\ 2 & -6 & 6 & 5 & 1 \\ 4 & 5 & 0 & -2 & 2 \end{bmatrix} \begin{bmatrix} x_{11} & x_{12} & x_{13} \\ x_{21} & x_{22} & x_{23} \\ x_{31} & x_{32} & x_{33} \\ x_{41} & x_{42} & x_{43} \\ x_{51} & x_{52} & x_{53} \end{bmatrix} = \begin{bmatrix} -1 & 1 & 0 \\ 2 & -1 & 1 \\ 1 & 10 & 11 \\ 4 & 0 & 4 \\ 0 & -6 & -6 \\ -3 & 6 & 3 \\ 1 & 11 & 12 \\ 0 & -5 & -5 \end{bmatrix}$$

Matrix **A** is a well conditioned 8 by 5 matrix of rank 3. There are 3 r.h.s. vectors **b**. The 3 r.h.s. vectors **b**(s) are chosen so that the minimum norm least squares solutions $\mathbf{x}_1$, $\mathbf{x}_2$ and $\mathbf{x}_3$ are respectively, $\mathbf{x}_1 = (-1/12, 0, 1/4, -1/12, 1/12)^T$, $\mathbf{x}_2 = (0, 0, 0, 0, 0)^T$ and $\mathbf{x}_3 = \mathbf{x}_1$.

As expected, matrix **A** is found by the two routines to be of rank 3. Results obtained in single-precision by the two methods are as follows.

Solution by Gauss LU factorization method

$$\mathbf{x}_1 = (-0.083333, 1.1\text{E}{-}06, 0.25, -0.083333, 0.083333)^T$$
$$\mathbf{x}_2 = (9.4\text{E}{-}07, 1.8\text{E}{-}06, -3.5\text{E}{-}07, 2.1\text{E}{-}06, 4.9\text{E}{-}07)^T$$
$$\mathbf{x}_3 = (-0.083333, 1.1176\text{E}{-}06, 0.25, -0.083333, 0.083333)^T$$

A measure of the accuracy of calculation of the pseudo-inverse $\mathbf{A}^+$ is the expression $||\mathbf{AA}^+\mathbf{A} - \mathbf{A}||_E/||\mathbf{A}||_E$, where $||.||_E$ refers to the Euclidean matrix norm.

The calculated $||\mathbf{AA}^+\mathbf{A} - \mathbf{A}||_E/||\mathbf{A}||_E = 1.3\text{E}{-}05$.

Solution by Householder's QR factorization

$$\mathbf{x}_1 = (-0.083333, 9.8\text{E}{-}07, 0.25, -0.083333, 0.083333)^T$$
$$\mathbf{x}_2 = (-1.54168\text{E}{-}06, 4.8\text{E}{-}06, -8.9\text{E}{-}07, -4.6\text{E}{-}06, 7.2\text{E}{-}07)^T$$
$$\mathbf{x}_3 = (-0.083333, 5.68441\text{E}{-}06, 0.25, -0.083333, 0.083333)^T$$

The calculated $||\mathbf{AA}^+\mathbf{A} - \mathbf{A}||_E/||\mathbf{A}||_E = 9.7\text{E}{-}06$.

Example 17.8

This example was solved by Golub ([10], p. 211) using an iterative scheme. **A** is a 6 by 5 badly conditioned matrix. It consists of the first 5 columns of the inverse of the 6×6 Hilbert matrix. There are 2 r.h.s.columns **b**(s).

Each column **b** is chosen such that the exact minimum norm least squares solution for each of them is $x = (1, 1/2, 1/3, 1/4, 1/5)^T$.

$$\mathbf{A} = \begin{bmatrix} 0.360\text{D}02 & -0.630\text{D}03 & 0.336\text{D}04 & -0.756\text{D}04 & 0.756\text{D}04 \\ -0.630\text{D}03 & 0.147\text{D}05 & -0.882\text{D}05 & 0.21168\text{D}06 & -0.2205\text{D}06 \\ 0.336\text{D}04 & -0.882\text{D}05 & 0.56448\text{D}06 & -0.14112\text{D}07 & 0.1512\text{D}07 \\ -0.756\text{D}04 & 0.21168\text{D}06 & -0.14112\text{D}07 & 0.36288\text{D}07 & -0.3969\text{D}07 \\ 0.756\text{D}04 & -0.2205\text{D}06 & 0.1512\text{D}07 & -0.3969\text{D}07 & 0.441\text{D}07 \\ -0.2772\text{D}04 & 0.8316\text{D}05 & -0.58212\text{D}06 & 0.155232\text{D}07 & -0.174636\text{D}07 \end{bmatrix}$$

$$\mathbf{b}_1 = (0.463\text{D}03, -0.1386\text{D}05, 0.9702\text{D}05, -0.25872\text{D}06, 0.29106\text{D}06, -0.116424\text{D}06)^T$$
$$\mathbf{b}_2 = (-0.4157\text{D}04, -0.1782\text{D}05, 0.93555\text{D}05, -0.2618\text{D}06, 0.288288\text{D}06, -0.118944\text{D}06)^T$$

The computation is performed in double-precision. Rank(A) is found by the two routines to be 5.

Solution by Gauss LU factorization method

Calculated Solutions $\mathbf{x}_1 = \mathbf{x}_2 = (1, 0.5, 0.3333333333, 0.25, 0.2)^T$ for each of $\mathbf{b}_1$ and $\mathbf{b}_2$.
Measure of accuracy for the calculated pseudo-inverse $\mathbf{A}^+$ is $||\mathbf{AA}^+\mathbf{A} - \mathbf{A}||_E/||\mathbf{A}||_E = 0.1633277774\text{E}{-}11$.

Solution by Householder's QR factorization

$\mathbf{x} = (1, 0.5, 0.3333333333, 0.25, 0.2)^T$ for each of **b**1 and **b**2.
Measure of accuracy for the calculated pseudo-inverse $\mathbf{A}^+$ is
$||\mathbf{AA}^+\mathbf{A} - \mathbf{A}||_E/||\mathbf{A}||_E = 0.2715803665\text{E}{-}11$.

Note 17.3

As noted earlier, the calculated triangular matrix **R** in the **QR** factorization is exactly as displayed by Golub ([10], p. 211) but different from matrix **R** in ([1], p. 364). The reason is that in [1], **QR** factorization was done without pivoting, whereas in [10] and in our implementation for this chapter, pivoting was used.

Example 17.9

In this example, matrix **A** is a badly conditioned 4 by 5 matrix. There is no r.h.s. vector **b**. The pseudo-inverse $\mathbf{A}^+$ was computed in both single- and double-precision.

$$\mathbf{A} = \begin{bmatrix} 0.4087 & 0.1593 & 0.6594 & 0.4302 & 0.3516 \\ 0.6246 & 0.3383 & 0.6591 & 0.9342 & 0.9038 \\ 0.0661 & 0.9112 & 0.6898 & 0.1931 & 0.1498 \\ 0.2112 & 0.815 & 0.7983 & 0.3406 & 0.2803 \end{bmatrix}$$

Solution by Gauss LU factorization method

In single-precision, the calculated rank(**A**) = 3 and
$||\mathbf{AA}^+\mathbf{A} - \mathbf{A}||_E/||\mathbf{A}||_E = 0.1141155\text{E}{-}04$.
In double-precision, the calculated rank(**A**) = 4 and
$||\mathbf{AA}^+\mathbf{A} - \mathbf{A}||_E/||\mathbf{A}||_E = 0.8727612201\text{E}{-}11$.

Solution by Householder's QR factorization

In single-precision, the calculated rank(**A**) = 3 and
$\|\mathbf{AA}^{+}\mathbf{A} - \mathbf{A}\|_E/\|\mathbf{A}\|_E = 0.8185643\text{E}{-05}$.
In double-precision, the calculated rank(**A**) = 4 and
$\|\mathbf{AA}^{+}\mathbf{A} - \mathbf{A}\|_E/\|\mathbf{A}\|_E = 0.1190011279\text{E}{-11}$.

Recall that LA_Eluls() and LA_Hhls() can also solve non-singular square systems. The following paragraphs also confirm the superiority of the Householder's **QR** factorization (and the Gauss **LU** factorization) methods.

Spath [22] collected 42 test examples, which he used to evaluate different existing algorithms, after converting them to FORTRAN 77. For the linear least squares solution of system **Ax** = **b**, he compares between the solution by the normal equation (NGL) using Cholesky's decomposition of $\mathbf{A}^T\mathbf{A}$, the modified Gram-Schmidt (MGS), Givens' (GIVR), Householder's transformation (HFTI) and the Singular Value Decomposition (SVD) methods. For computer storage requirements, for n > m (using our notation), all subroutines need more or less the same amount of storage except the SVD, which needs almost 3 times the storage.

Spath ([22], pp. 43-46) compares the behavior of the different subroutines with respect to the accuracy of the results. As expected, the results by the NGL method are not very accurate for ill-conditioned cases (only few decimal places agree with the results of other subroutines). In one example, only the MGS and the HFTI gave reasonable results, while the others failed. In another example, the MGS, GIVR, and the HFTI gave the most accurate results. As for the CPU times for the 42 examples, and for matrices randomly generated, the fastest was the NGL, followed by MGS, GIVR, HFTI and the SVD.

References

1. Abdelmalek, N.N., Round-off error analysis for Gram-Schmidt method and solution of linear least squares problems, *BIT*, 11(1971)345-367.

2. Abdelmalek, N.N., On the solution of linear least squares problems and pseudo-inverses, *Computing*, 13(1974)215-228.
3. Abdelmalek, N.N., Piecewise linear least-squares approximation of planar curves, *International Journal of Systems Science*, 21(1990)1393-1403.
4. Barrodale, I. and Stuart, G.F., A Fortran program for linear least squares problems of variable degree, *Proceedings of the Fourth Manitoba Conference on Numerical Mathematics*, Hartnell, B.L. and Williams, H.C. (eds.), Winnipeg, Manitoba, Canada, pp. 191-204, 1975.
5. Bjorck, A., Solving linear least squares problems by Gram-Schmidt orthogonalization, *BIT*, 7(1967)1-21.
6. Chatterjee, S. and Price, B., *Regression Analysis by Example*, John Wiley & Sons, New York, 1977.
7. Cheng, Y-Q., Zhuang, Y-M. and Yang, J-Y., Optimal Fisher discriminant analysis using the rank decomposition, *Pattern Recognition*, 25(1992)101-111.
8. Geladi, P. and Kowalski, B.R., Partial least-squares regression: A tutorial, *Analytica Chimica Acta*, 185(1986)1-17.
9. Gentleman, W.M., Algorithm AS 75: Basic procedures for large, sparse or weighted linear least squares problems, *Applied Statistics*, 22(1974)448-454.
10. Golub, G.H., Numerical methods for solving linear least squares problems, *Numerische Mathematik*, 7(1965)206-216.
11. Golub, G.H. and Reinsch, C., Singular value decomposition and the least squares solutions, *Numerische Mathematik*, 14(1970)403-420.
12. Golub, G.H. and Van Loan, C.F., *Matrix Computation*, The Johns Hopkins University Press, Third Edition, London, 1996.
13. Hoskuldsson, A., PLS regression methods, *Journal of Chemometrics*, 2(1988)211-228.
14. Lau, H.T., *A Numerical Library in C for Scientists and Engineers*, CRC Press, Ann Arbor, 1995.
15. Longley, J.M., *Least Squares Computations Using Orthogonalization Methods*, Marcel Dekker, New York, 1984.
16. Montgomery, D.C. and Peck, E.A., *Introduction to Linear Regression Analysis*, John Wiley & Sons, New York, 1992.

17. Noble, B., *Applied Linear Algebra*, Prentice-Hall, Englewood Cliffs, NJ, 1969.
18. Peters, G. and Wilkinson, J.H., The least squares problem and pseudo-inverses, *Computer Journal*, 13(1970)309-316.
19. Press, W.H., Flannery, B.P., Teukolsky, S.A. and Vetterling, W.T., *Numerical Recipes in C, The Art of Scientific Computing*, Cambridge University Press, Second Edition, Cambridge, 1992.
20. Russell, E., Chiang, L.H. and Braatz, R.D., *Data-Driven Methods for Fault Detection and Diagnosis in Chemical Processes*, Springer-Verlag, London, 2000.
21. Ryan, T.P., *Modern Regression Methods*, John Wiley & Sons, New York, 1997.
22. Spath, H., *Mathematical Algorithms for Linear Regression*, Academic Press, English Edition, London, 1991.
23. Stewart, G.W., *Introduction to Matrix Computations*, Academic Press, New York, 1973.
24. Stewart, G.W., The effects of rounding residual on an algorithm for downdating a Cholesky factorization, *Journal of Institute of Mathematics and Applications*, 23(1979)203-213.
25. Varah, J.M., A practical examination of some numerical methods for linear discrete ill-posed problems, *SIAM Review*, 21(1979)100-111.
26. Varah, J.M., Pitfalls in the numerical solution of linear ill-posed problems, *SIAM Journal on Scientific Statistical Computation*, 4(1983)164-176.
27. Wilkinson, J.H., *The Algebraic Eigenvalue Problem*, Clarendon Press, Oxford, 1965.
28. Wold, S., Ruhe, A., Wold, H. and Dunn III, W.J., The collinearity problem in linear regression. The partial least squares (PLS) approach to generalized inverses, *SIAM Journal on Scientific and Statistical Computing*, 5(1984)735-743.

17.13 DR_Eluls

```
/*-----------------------------------------------------------------
DR_Eluls
-------------------------------------------------------------------
This program is a driver for the function LA_Eluls(), which
calculates the minimal-length least squares solution of a system of
linear equations and/or calculates the pseudo-inverse of the
coefficient matrix.  It uses Gauss "LU" decomposition method with
complete pivoting.

The system of linear equations is of the form

                         a*xs = bs

"a"  is a given real n by m matrix  of rank k, k <= [min(n,m)].
"bs" is (are) given real r.h.s. n vector(s).
"xs" is (are) the m solution vector(s).

The system of linear equations might be overdetermined,
determined or underdetermined.

The required results are obtained according to a parameter "ientry"
specified by the user:
    0   LA_Eluls() calculates matrices "l" and "u" in the "lu"
        decomposition of (the permuted) matrix "a(permuted)"
        as a(permuted) = l*u.
        or as a(permuted) = l*diag*u; "diag" is a diagonal matrix
        and "u" has unit diagonal elements.
    1   LA_Eluls() calculates matrices "l" and "u" in the "lu"
        decomposition + the least squares solution
        vector(s) "xs" (if "bs" != 0).
    2   LA_Eluls() calculates matrices "l" and "u" in the "lu"
        decomposition + the pseudo-inverse of "a", "apsudo".
    3   LA_Eluls() calculates matrices "l" and "u" in the "lu"
        decomposition + the pseudo-inverse of matrix "a" +
        the least squares solution(s) (if "bs" != 0).

This program contains 3 examples whose results appear in the text.

Example 1:
    matrix "a" is 8 by 5 well conditioned of rank 3.  There
    are 3 r.h.s. vectors "bs". It is solved in double precision.
```

```
Example 2:
    matrix "a" is 6 by 5 badly conditioned of rank 5.  There
    are 2 r.h.s. vectors "bs". It is solved in double precision.

Example 3:
    matrix "a" is 4 by 5 badly conditioned.  There are no
    r.h.s. vectors "bs".  In single precision, calculated rank (a)
    = 3.  In double precision, calculated rank (a) = 4.
-------------------------------------------------------------------*/

#include "DR_Defs.h"
#include "LA_Prototypes.h"

#define Na2         8
#define Ma2         5
#define Irhsa       3
#define Nb2         6
#define Mb2         5
#define Irhsb       2
#define Nc2         4
#define Mc2         5
#define Irhsc       0

void DR_Eluls (void)
{
    /*----------------------------------------
      Constant matrices/vectors
    ----------------------------------------*/
    static tNumber_R a1init[Na2][Ma2] =
    {
        { 22.0, 10.0,   2.0,   3.0,  7.0 },
        { 14.0,  7.0,  10.0,   0.0,  8.0 },
        { -1.0, 13.0,  -1.0, -11.0,  3.0 },
        { -3.0, -2.0,  13.0,  -2.0,  4.0 },
        {  9.0,  8.0,   1.0,  -2.0,  4.0 },
        {  9.0,  1.0,  -7.0,   5.0, -1.0 },
        {  2.0, -6.0,   6.0,   5.0,  1.0 },
        {  4.0,  5.0,   0.0,  -2.0,  2.0 }
    };

    static tNumber_R bs1init[Na2][Irhsa] =
    {
        { -1.0,  1.0,   0.0 },
        {  2.0, -1.0,   1.0 },
        {  1.0, 10.0,  11.0 },
```

```
    {  4.0,  0.0,   4.0 },
    {  0.0, -6.0,  -6.0 },
    { -3.0,  6.0,   3.0 },
    {  1.0, 11.0,  12.0 },
    {  0.0, -5.0,  -5.0 }
};

static tNumber_R a2init[Nb2][Mb2] =
{
    {    36.0,    -630.0,     3360.0,    -7560.0,      7560.0 },
    {  -630.0,   14700.0,   -88200.0,   211680.0,   -220500.0 },
    {  3360.0,  -88200.0,   564480.0, -1411200.0,   1512000.0 },
    { -7560.0,  211680.0, -1411200.0,  3628800.0,  -3969000.0 },
    {  7560.0, -220500.0,  1512000.0, -3969000.0,   4410000.0 },
    { -2772.0,   83160.0,  -582120.0,  1552320.0,  -1746360.0 }
};

static tNumber_R bs2init[Nb2][Irhsb] =
{
    {     463.0,   -4157.0 },
    {  -13860.0,  -17820.0 },
    {   97020.0,   93555.0 },
    { -258720.0, -261800.0 },
    {  291060.0,  288288.0 },
    { -116424.0, -118944.0 }
};

static tNumber_R a3init[Nc2][Mc2] =
{
    { 0.4087, 0.1593, 0.6594, 0.4302, 0.3516 },
    { 0.6246, 0.3383, 0.6591, 0.9342, 0.9038 },
    { 0.0661, 0.9112, 0.6898, 0.1931, 0.1498 },
    { 0.2112, 0.8150, 0.7983, 0.3406, 0.2803 }
};

/*----------------------------------------
  Variable matrices/vectors
----------------------------------------*/
tMatrix_R  aa     = alloc_Matrix_R (NN2_ROWS, MM2_COLS);
tMatrix_R  bs     = alloc_Matrix_R (NN2_ROWS, KK2_COLS);
tMatrix_R  xs     = alloc_Matrix_R (MM2_COLS, KK2_COLS);
tMatrix_R  l      = alloc_Matrix_R (NN2_ROWS, NN2_ROWS);
tMatrix_R  u      = alloc_Matrix_R (NN2_ROWS, MM2_COLS);
tMatrix_R  apsudo = alloc_Matrix_R (MM2_COLS, NN2_ROWS);
tMatrix_R  rres   = alloc_Matrix_R (NN2_ROWS, KK2_COLS);
```

```
    tMatrix_R  aux    = alloc_Matrix_R (NN2_ROWS, NN2_ROWS);
    tMatrix_R  auy    = alloc_Matrix_R (NN2_ROWS, MM2_COLS);
    tVector_R  diag   = alloc_Vector_R (MM2_COLS);
    tVector_R  bb     = alloc_Vector_R (NN2_ROWS);

    tMatrix_R  a1     = init_Matrix_R (&(a1init[0][0]), Na2, Ma2);
    tMatrix_R  bs1    = init_Matrix_R (&(bs1init[0][0]), Na2, Irhsa);
    tMatrix_R  a2     = init_Matrix_R (&(a2init[0][0]), Nb2, Mb2);
    tMatrix_R  bs2    = init_Matrix_R (&(bs2init[0][0]), Nb2, Irhsb);
    tMatrix_R  a3     = init_Matrix_R (&(a3init[0][0]), Nc2, Mc2);

    int        irank, irhs, ientry;
    int        i, j, k, m, n, Iexmpl;

    tNumber_R  s, sa, sum1, sum2, aerr;

    eLaRc      rc = LaRcOk;

    prn_dr_bnr ("DR_Eluls, Minimal Length L2 Solution of a System of"
                " Linear Equations by Gauss \"LU\" Factorization");

    for (Iexmpl = 1; Iexmpl <= 3; Iexmpl++)
    {
        switch (Iexmpl)
        {
            case 1:
                ientry = 3;
                n = Na2;
                m = Ma2;
                irhs = Irhsa;
                for (i = 1; i <= n; i++)
                {
                    for (j = 1; j <= m; j++)
                    {
                        aa[i][j] = a1[i][j];
                    }
                }
                for (i = 1; i <= n; i++)
                {
                    for (j = 1; j <= irhs; j++)
                    {
                        bs[i][j] = bs1[i][j];
                    }
                }
                break;
```

```
    case 2:
        ientry = 3;
        n = Nb2;
        m = Mb2;
        irhs = Irhsb;
        for (i = 1; i <= n; i++)
        {
            for (j = 1; j <= m; j++)
            {
                aa[i][j] = a2[i][j];
            }
        }
        for (i = 1; i <= n; i++)
        {
            for (j = 1; j <= irhs; j++)
            {
                bs[i][j] = bs2[i][j];
            }
        }
        break;
    case 3:
        ientry = 2;
        n = Nc2;
        m = Mc2;
        irhs = 0;
        for (i = 1; i <= n; i++)
        {
            for (j = 1; j <= m; j++)
            {
                aa[i][j] = a3[i][j];
            }
        }
        break;

    default:
        break;
}

prn_algo_bnr ("LA_Eluls");

prn_example_delim();
PRN ("Example #%d: Size of coefficient matrix "
    "\"a\" %d by %d\n", Iexmpl, n, m);
prn_example_delim();
PRN ("Minimal Least Squares Solution(s) of a System of"
```

```
     " Linear Equations Using  Gauss\"LU\" Decomposition\n");
prn_example_delim();
PRN ("Coefficient Matrix, \"a\"\n");
prn_Matrix_R (aa, n, m);
if (irhs != 0)
{
    PRN ("\n");
    PRN ("Right Hand Vector(s), \"bs\"\n");
    prn_Matrix_R (bs, n, irhs);
}

rc = LA_Eluls (ientry, n, m, irhs, aa, bs, l, u, diag, bb,
               &irank, apsudo, xs, rres);

if (rc >= LaRcOk)
{
    PRN ("\n");
    PRN ("Results of the Least Squares Problem\n");
    PRN ("Rank of the coefficient matrix = %d\n\n", irank);

    PRN ("Lower Triangular (Trapezoidal) Matrix \"l\"\n");
    prn_Matrix_R (l, n, irank);
    if (irank != m)
    {
        PRN ("Elements of Diagonal Vector \"diag\"\n");
        prn_Vector_R (diag, irank);
    }
    PRN ("Upper Triangular (Trapezoidal) Matrix \"u\"\n");
    prn_Matrix_R (u, irank, m);

    if (ientry >= 2)
    {
        PRN ("Pseudo-inverse Matrix \"apsudo\"\n");
        prn_Matrix_R (apsudo, m, n);

        s = 0.0;
        for (j = 1; j <= m; j++)
        {
            for (i = 1; i <= n; i++)
            {
                s = s + aa[i][j] * (aa[i][j]);
            }
        }
        sum1 = sqrt (s);
        for (j = 1; j <= n; j++)
```

```
        {
            for (i = 1; i <= n; i++)
            {
                s = 0.0;
                for (k = 1; k <= m; k++)
                {
                    s = s + aa[i][k] * (apsudo[k][j]);
                }
                aux[i][j] = s;
            }
        }
        for (j = 1; j <= m; j++)
        {
            for (i = 1; i <= n; i++)
            {
                s = 0.0;
                for (k = 1; k <= n; k++)
                {
                    s = s + aux[i][k] * (aa[k][j]);
                }
                auy[i][j] = s - aa[i][j];
            }
        }
        sa = 0.0;
        for (j = 1; j <= m; j++)
        {
            for (i = 1; i <= n; i++)
            {
                sa = sa + auy[i][j] * (auy[i][j]);
            }
        }
        sum2 = sqrt (sa);
        aerr = sum2/sum1;
        PRN ("||a*apsudo*a - a||/||a|| = %22.15f\n", aerr);
    }

    if ((ientry == 1) || (ientry == 3))
    {
        if (irhs != 0)
        {
            PRN ("\n");
            PRN ("The Least Squares Solution(s)\n");
            for (i = 1; i <= irhs; i++)
            {
                for (j = 1; j <= n; j++)
```

```
                        {
                            bb[j] = bs[j][i];
                        }
                        PRN ("Right Hand Vector \"b\"\n");
                        prn_Vector_R (bb, n);
                        for (j = 1; j <= m; j++)
                        {
                            bb[j] = xs[j][i];
                        }
                        PRN ("Solution vector \"x\"\n");
                        prn_Vector_R (bb, m);
                        for (j = 1; j <= n; j++)
                        {
                            bb[j] = rres[j][i];
                        }
                        s = 0.0;
                        for (j = 1; j <= n; j++)
                        {
                            s = s + bb[j] * (bb[j]);
                        }
                        sum1 = sqrt (s);
                        PRN ("Residual vector \"rres\"\n");
                        prn_Vector_R (bb, n);
                    }
                }
            }
        }

        prn_la_rc (rc);
    }

    free_Matrix_R (aa, NN2_ROWS);
    free_Matrix_R (bs, NN2_ROWS);
    free_Matrix_R (xs, MM2_COLS);
    free_Matrix_R (l, NN2_ROWS);
    free_Matrix_R (u, NN2_ROWS);
    free_Matrix_R (apsudo, MM2_COLS);
    free_Matrix_R (rres, NN2_ROWS);
    free_Matrix_R (aux, NN2_ROWS);
    free_Matrix_R (auy, NN2_ROWS);
    free_Vector_R (diag);
    free_Vector_R (bb);

    uninit_Matrix_R (a1);
    uninit_Matrix_R (bs1);
```

```
    uninit_Matrix_R (a2);
    uninit_Matrix_R (bs2);
    uninit_Matrix_R (a3);
}
```

17.14 LA_Eluls

```
/*-----------------------------------------------------------------
LA_Eluls
-------------------------------------------------------------------
This program calculates the minimal-length least squares solution(s)
of a system of linear equations using the Gauss "LU" factorization
method with complete pivoting.

The system of linear equations is of the form

                          a*xs = bs

"a"  is a given real n by m matrix  of rank k,  k <= [min (n,m)].
"bs" is (are) the given r.h.s. n vector(s).
"xs" is (are) the m solution vector(s).

The problem is to calculate the elements of the "irhs" vector(s)
xs[i][j], i = 1, 2, ..., m, j = 1, 2, ..., "irhs" of the system
a*xs = bs.

Inputs
n       Number of rows of matrix "a" in the system a*xs = bs.
m       Number of columns of matrix "a" in the system a*xs = bs.
irhs    Number of columns in matrix "bs" in the system a*xs = bs.
a       A real n by m matrix of the system a*xs = bs.
bs      An n rows by "irhs" columns matrix in the system a*xs = bs.
ientry  An integer specifying the action by the user:
        0   LA_Eluls() calculates matrices "l" and "u" in the "lu"
            decomposition of (the permuted) matrix "a(permuted)"
            into a(permuted) = l*u
            or a(permuted) =l*diag*u; "diag" is a diagonal matrix
            and "u" has unit diagonal elements.
        1   LA_Eluls() calculates matrices "l" and "u" in the "lu"
            decomposition + the least squares solution
            vector(s) "xs" (if "bs" != 0).
        2   LA_Eluls() calculates matrices "l" and "u" in the "lu"
            decomposition + the pseudo-inverse of "a", "apsudo".
        3   LA_Eluls() calculates matrices "l" and "u" in the "lu"
            decomposition + the pseudo-inverse of matrix "a" +
            the least squares solution(s) (if "bs" != 0).

Local Data
t, v, tempp
```

```
        Real working matrices.
b, x, w, y temp
        Real working vectors.
ir      An n vector containing the indices of the rows of the
        permuted matrix "a".
ic      An m vector containing the indices of the columns of the
        permuted matrix "a".

Outputs
irank   The calculated rank of matrix "a".
xs      An m by "irhs" matrix containing the "irhs" columns that
        are the "irhs" solution m vector(s) xs.
rres    An n by "irhs" matrix.  (rres[1][j], rres[2][j], ...,
        rres[n][j]) is the least squares residual vector for the
        jth solution, j = 1, ..., "irhs".
        rres[i][j] = a[i][1]*xs[1][j] + ... + a[i][m]*xs[m][j]
                   - bs[i][j],        for j = 1, 2,..., "irhs".
l       An n by n real lower triangular matrix whose n rows and
        first "irank" columns contain the lower triangular
        (trapezoidal) matrix "l" in the decomposition of
        a(permuted) = l*u  or a(permuted) = l*diag*u.
diag    An m real vector whose first "irank" elements contain the
        diagonal elements of matrix "u".
u       An m by m real matrix whose first "irank" rows and m columns
        contain the upper triangular (trapezoidal) matrix "u" in the
        decomposition of a(permuted) = l*u or = l*diag*u.
apsudo  An m by n real matrix containing the pseudo inverse of
        matrix "a".

NOTE: The calculated results "xs" and "apsudo" are for the given
      equation a*xs = bs, (not for the permuted one).

Returns one of
        LaRcSolutionFound
        LaRcErrBounds
        LaRcErrNullPtr
        LaRcErrAlloc
-----------------------------------------------------------------*/

#include "LA_Prototypes.h"

eLaRc LA_Eluls (int ientry, int n, int m, int irhs, tMatrix_R aa,
    tMatrix_R bs, tMatrix_R l, tMatrix_R u, tVector_R diag,
    tVector_R b, int *pIrank, tMatrix_R apsudo, tMatrix_R xs,
    tMatrix_R rres)
```

```
{
    tMatrix_R   t       = alloc_Matrix_R (m, m);
    tMatrix_R   v       = alloc_Matrix_R (m, m);
    tMatrix_R   tempp   = alloc_Matrix_R (m, m);
    tVector_R   x       = alloc_Vector_R (m);
    tVector_R   y       = alloc_Vector_R (m);
    tVector_R   w       = alloc_Vector_R (m);
    tVector_R   temp    = alloc_Vector_R (m);
    tVector_I   ir      = alloc_Vector_I (n);
    tVector_I   ic      = alloc_Vector_I (m);

    int         iend = 0, iout = 0, iwish = 0, ifirst = 0,
                isecnd = 0;
    int         i = 0, j = 0, ij = 0, irank = 0;

    /* Validation of the data before executing the algorithm */
    eLaRc       rc = LaRcSolutionFound;
    VALIDATE_BOUNDS ((0 < n) && (0 < m) && !((n == 1) && (m == 1))
                     && (0 <= irhs) && IN_BOUNDS (ientry, 0, 3));
    VALIDATE_PTRS   (aa && bs && l && u && diag && b && pIrank &&
                     apsudo && xs && rres);
    VALIDATE_ALLOC  (t && v && tempp && x && y && w && temp && ir &&
                     ic );
    iwish = ientry;
    iout = 1;
    irank = n;

    if (m <= n) irank = m;

    /* lu decomposition */
    LA_eluls_lu_decomp (&iout, n, m, aa, l, u, diag, ir, ic);

    *pIrank = iout - 1;
    if (*pIrank != n)
    {
        /* l factorizing */
        LA_eluls_l_decomp (n, m, tempp, l, u, pIrank);

        /* Cholesky decomposition of matrix tempp */
        LA_chols (irank, tempp, t);
    }
    if (*pIrank != m)
    {
        /* u factorizing */
        LA_eluls_u_decomp (m, tempp, u, diag, pIrank);
```

```
        /* Cholesky decomposition of matrix tempp */
        LA_chols (irank, tempp, v);
    }
    /* End of l-u factorizations */

    if (iwish == 0)
    {
        GOTO_CLEANUP_RC (LaRcSolutionFound);
    }

    iend = n;
    if (iwish != 2)
    {
      iend = irhs;
    }

    /* Calculating the solution vector(s) and the psudo-inverse
         of the coefficient matrix "a" */
    for (ij = 1; ij <= 2; ij++)
    {
        for (ifirst = 1; ifirst <= iend; ifirst++)
        {
            if (iwish == 2)
            {
                for (isecnd = 1; isecnd <= n; isecnd++)
                {
                    b[isecnd] = 0.0;
                }
                b[ifirst] = 1.0;
            }
            else if (iwish != 2)
            {
                for (i = 1; i <= n; i++)
                {
                    j = ir[i];
                    b[i] = bs[j][ifirst];
                }
            }

            /* Calculating vector y */
            LA_calcul_y (pIrank, n, t, y, w, temp, l, b);

            /* Calculating vector x */
            LA_calcul_x (pIrank, m, v, y, w, temp, u, diag, x);
```

```
            /* Permuting ellements of vector x and calculating
                 pseudo-inverse matrix "apsudo" */
            LA_permute_x (iwish, ifirst, isecnd, n, m, aa, bs, w, x,
                          ir, ic, apsudo, xs, rres);
        }
        if ((iwish == 1) || (iwish == 2))
        {
            GOTO_CLEANUP_RC (LaRcSolutionFound);
        }
        iwish = 2;
        iend = n;
    }

CLEANUP:

    free_Matrix_R (t, m);
    free_Matrix_R (v, m);
    free_Matrix_R (tempp, m);
    free_Vector_R (x);
    free_Vector_R (y);
    free_Vector_R (w);
    free_Vector_R (temp);
    free_Vector_I (ir);
    free_Vector_I (ic);

    return rc;
}

/*-------------------------------------------------------------------
Part 1 of LA_Eluls()
-------------------------------------------------------------------*/
void LA_eluls_lu_decomp (int *pIout, int n, int m, tMatrix_R aa,
    tMatrix_R l, tMatrix_R u, tVector_R diag, tVector_I ir,
    tVector_I ic)
{
    int         i, j, ij, iout1, kp = 0, kp1, kdp = 0, nless;
    tNumber_R   e, gh, piv, pivot;

    for (j = 1; j <= m; j++)
    {
        ic[j] = j;
        for (i = 1; i <= n; i++)
        {
            u[i][j] = aa[i][j];
```

```
        }
    }
    for (i = 1; i <= n; i++)
    {
        for (j = 1; j <= n; j++)
        {
            l[i][j] = 0.0;
        }
        ir[i] = i;
        l[i][i] = 1.0;
    }
    *pIout = 1;
    nless = n;
    if (m < n) nless = m;
    for (ij = 1; ij <= nless; ij++)
    {
        piv = 0.0;

        /* Complete pivoting for elements of matrix u */
        for (j = *pIout; j <= m; j++)
        {
            for (i = *pIout; i <= n; i++)
            {
                e = u[i][j];
                if (e < 0.0) e = -e;
                if (e > piv)
                {
                    kp = i;
                    kdp = j;
                    piv = e;
                }
            }
        }

        if (piv < EPS)
            return;
        else
        {
            if (*pIout != kdp)
            {
                /* Swap two elements of integer vector "ic" */
                swap_elems_Vector_I (ic, kdp, *pIout);

                /* Swap two columns of real matrix "u" */
                for (i = 1; i <= n; i++)
```

```
        {
            e = u[i][*pIout];
            u[i][*pIout] = u[i][kdp];
            u[i][kdp] = e;
        }
    }
    if (*pIout != kp)
    {
        /* Swap two elements of integer vector "ir" */
        swap_elems_Vector_I (ir, kp, *pIout);

        /* Swap two partial rows of real matrix "u" */
        for (j = *pIout; j <= m; j++)
        {
            e = u[*pIout][j];
            u[*pIout][j] = u[kp][j];
            u[kp][j] = e;
        }
        if (*pIout != 1)
        {
            iout1 = *pIout - 1;
            for (j = 1; j <= iout1; j++)
            {
                e = l[*pIout][j];
                l[*pIout][j] = l[kp][j];
                l[kp][j] = e;
            }
        }
    }

    pivot = u[*pIout][*pIout];
    diag[*pIout] = pivot;
    if (*pIout != n)
    {
        kp1 = *pIout + 1;
        for (i = kp1; i <= n; i++)
        {
            e = u[i][*pIout];
            gh = e/pivot;
            u[i][*pIout] = 0.0;
            l[i][*pIout] = gh;
            for (j = kp1; j <= m; j++)
            {
                u[i][j] = u[i][j] - gh * (u[*pIout][j]);
            }
```

```
                }
            }
            *pIout = *pIout + 1;
        }
    }
}

/*------------------------------------------------------------------
Calculating vector y for LA_Eluls()
------------------------------------------------------------------*/
void LA_calcul_y (int *pIrank, int n, tMatrix_R t, tVector_R y,
    tVector_R w, tVector_R temp, tMatrix_R l, tVector_R b)
{
    int         i, j, k, ki, im1, ip1;
    tNumber_R   sm;

    if (*pIrank == n)
    {
        y[1] = b[1];
        if (*pIrank == 1) return;
        for (i = 2; i <= *pIrank; i++)
        {
            im1 = i - 1;
            sm = -b[i];
            for (k = 1; k <= im1; k++)
            {
                sm = sm + l[i][k] * (y[k]);
            }
            y[i] = -sm;
        }
        return;
    }
    else if (*pIrank != n)
    {
        for (i = 1; i <= *pIrank; i++)
        {
            sm = 0.0;
            for (k = 1; k <= n; k++)
            {
                sm = sm + l[k][i] * (b[k]);
            }
            temp[i] = sm;
        }
        w[1] = temp[1]/t[1][1];
        if (*pIrank > 1)
```

```
        {
            for (i = 2; i <= *pIrank; i++)
            {
                im1 = i - 1;
                sm = -temp[i];
                for (k = 1; k <= im1; k++)
                {
                    sm = sm + t[i][k] * (w[k]);
                }
                w[i] = -sm/t[i][i];
            }
        }
        y[*pIrank] = w[*pIrank]/t[*pIrank][*pIrank];
        if (*pIrank == 1) return;
        for (j = 2; j <= *pIrank; j++)
        {
            i = *pIrank - j + 1;
            ip1 = i + 1;
            sm = -w[i];
            ki = j - 1 + ip1 - 1;
            for (k = ip1; k <= ki; k++)
            {
                sm = sm + t[k][i] * (y[k]);
            }
            y[i] = -sm/t[i][i];
        }
    }
}

/*-----------------------------------------------------------------------
Calculating vector x for LA_Eluls()
-----------------------------------------------------------------------*/
void LA_calcul_x (int *pIrank, int m, tMatrix_R v, tVector_R y,
    tVector_R w, tVector_R temp, tMatrix_R u, tVector_R diag,
    tVector_R x)
{
    int         i, j, im1, ip1, k, ki;

    tNumber_R   sm;

    if (*pIrank == m)
    {
        x[*pIrank] = y[*pIrank]/u[*pIrank][*pIrank];
        if (*pIrank == 1) return;
        for (j = 2; j <= *pIrank; j++)
```

```
        {
            i = *pIrank - j + 1;
            ip1 = i + 1;
            sm = -y[i];
            ki = j - 1 + ip1 - 1;
            for (k = ip1; k <= ki; k++)
            {
                sm = sm + u[i][k] * (x[k]);
            }
            x[i] = -sm/u[i][i];
        }
        return;
    }
    else if (*pIrank != m)
    {
        for (i = 1; i <= *pIrank; i++)
        {
            y[i] = y[i]/diag[i];
        }
        w[1] = y[1]/v[1][1];
        if (*pIrank > 1)
        {
            for (i = 2; i <= *pIrank; i++)
            {
                im1 = i - 1;
                sm = -y[i];
                for (k = 1; k <= im1; k++)
                {
                    sm = sm + v[i][k] * (w[k]);
                }
                w[i] = -sm/v[i][i];
            }
        }
        temp[*pIrank] = w[*pIrank]/v[*pIrank][*pIrank];
        if (*pIrank > 1)
        {
            for (j = 2; j <= *pIrank; j++)
            {
                i = *pIrank - j + 1;
                ip1 = i + 1;
                sm = -w[i];
                ki = j - 1 + ip1 - 1;
                for (k = ip1; k <= ki; k++)
                {
                    sm = sm + v[k][i] * (temp[k]);
```

```
                }
                temp[i] = -sm/v[i][i];
            }
        }
        for (i = 1; i <=  m; i++)
        {
            sm = 0.0;
            for (k = 1; k <= *pIrank; k++)
            {
                sm = sm + u[k][i] * (temp[k]);
            }
            x[i] = sm;
        }
    }
}

/*------------------------------------------------------------------
Permute elements of vector x and calculate apsudo for LA_Eluls()
------------------------------------------------------------------*/
void LA_permute_x (int iwish, int ifirst, int isecnd, int n, int m,
    tMatrix_R aa, tMatrix_R bs, tVector_R w, tVector_R x,
    tVector_I ir, tVector_I ic, tMatrix_R apsudo, tMatrix_R xs,
    tMatrix_R rres)
{
    int         i, j, k;

    tNumber_R   sm;

    for (i = 1; i <= m; i++)
    {
        k = ic[i];
        w[k] = x[i];
    }
    if (iwish != 2)
    {
        for (i = 1; i <= m; i++)
        {
            xs[i][ifirst] = w[i];
        }
        for (j = 1; j <= n; j++)
        {
            sm = -bs[j][ifirst];
            for (k = 1; k <= m; k++)
            {
                sm = sm + aa[j][k] * (w[k]);
```

```
            }
            rres[j][ifirst] = -sm;
        }
    }
    else if (iwish == 2)
    {
        for (isecnd= 1; isecnd <= m; isecnd++)
        {
            k = ir[ifirst];
            apsudo[isecnd][k] = w[isecnd];
        }
    }
}

/*-----------------------------------------------------------------
lu factorization for LA_Eluls
-----------------------------------------------------------------*/
void LA_eluls_l_decomp (int n, int m, tMatrix_R tempp, tMatrix_R l,
    tMatrix_R u, int *pIrank)
{
    int         i, j, k, irank1;
    tNumber_R   sm;

    if (*pIrank < n)
    {
        irank1 = *pIrank + 1;
        for (i = irank1; i <= n; i++)
        {
            for (j = *pIrank; j <= m; j++)
            {
                u[i][j] = 0.0;
            }
        }
    }
    for (i = 1;  i <= *pIrank; i++)
    {
        for (j = 1; j <= *pIrank; j ++)
        {
            sm = 0.0;
            for (k = 1; k <= n; k++)
            {
                sm = sm + l[k][i] * (l[k][j]);
            }
            tempp[i][j] = sm;
        }
```

```
    }
}

/*-----------------------------------------------------------------
Factorizing matrix u for LA_Eluls()
-----------------------------------------------------------------*/
void LA_eluls_u_decomp (int m, tMatrix_R tempp, tMatrix_R u,
    tVector_R diag, int *pIrank)
{
    int         i, j, k;
    tNumber_R   con, sm;

    for (i = 1; i <= *pIrank; i++)
    {
        u[i][i] = 1.0;
        k = i + 1;
        con = diag[i];
        for (j = k; j <= m; j++)
        {
            u[i][j] = u[i][j]/con;
        }
    }
    for (i = 1; i <= *pIrank; i++)
    {
        for (j = 1; j <= *pIrank; j++)
        {
            sm = 0.0;
            for (k = 1; k <= m; k++)
            {
                sm = sm + u[i][k] * (u[j][k]);
            }
            tempp[i][j] = sm;
        }
    }
}

/*-----------------------------------------------------------------
LA_chols
-----------------------------------------------------------------
This program calculates the Cholesky decomposition of a positive
definite symmetric matrix q.

Matrix "q" is decomposed into q = el*el(transpose) where "el" is a
lower triangular matrix.
```

```
Inputs
irank   The dimension of the square matrix q.
q       An m by m matrix whose first "irank" rows and "irank"
        columns contain the positive definite symmetric matrix "q".

Outputs
el      An "irank" by "irank" matrix whose "irank" rows and "irank"
        columns contain the lower triangular matrix "el" of the
        decomposition
                 q = el*el (transpose)
-------------------------------------------------------------------*/
void LA_chols (int irank, tMatrix_R q, tMatrix_R el)
{
    int         i, j, k, im1, jp1;
    tNumber_R   c, sm;

    j = 1;
    c = q[1][1];
    el[1][1] = sqrt (c);
    if (irank == 1) return;
    for (j = 1; j <= irank-1; j++)
    {
        jp1 = j + 1;
        el[jp1][1] = q[1][jp1]/el[1][1];
        if (j > 1)
        {
            for (i = 2; i <= j; i++)
            {
                im1 = i - 1;
                c = -q[i][jp1];
                sm = c;
                for (k = 1; k <= im1; k++)
                {
                    sm = sm + el[i][k] * (el[jp1][k]);
                }
                el[jp1][i] = -sm/el[i][i];
            }
        }
        for (i = 1; i <= j; i++)
        {
            el[i][jp1] = 0.0;
        }
        c = -q[jp1][jp1];
        sm = c;
        for (k = 1; k <= j; k++)
```

```
        {
            sm = sm + el[jp1][k] * (el[jp1][k]);
        }
        el[jp1][jp1] = sqrt (-sm);
    }
}
```

17.15 DR_Hhls

```
/*-----------------------------------------------------------------------
DR_Hhls
-------------------------------------------------------------------------
This program is a driver for the function LA_Hhls(), which
calculates the minimal-length least squares solution of a system of
linear equations and/or calculates the pseudo-inverse of the
coefficient matrix. It uses Householder's "qrp" decomposition method
with pivoting.

The system of linear equations is of the form

                          a*xs = bs

"a"  is a given real n by m matrix  of rank k, k <= [min (n,m)]
"bs" is (are) given real r.h.s. n vector(s).
"xs" is (are) the m solution vector(s).

The system of linear equations might be overdetermined, determined or
underdetermined.

The required results are obtained according to a parameter "ientry"
specified by the user:
    0   LA_Hhls() calculates the factorization matrices q, r
        and p of "a(permuted)" = q * r * p(transpose).
    1   LA_Hhls() calculates the matrices "q", "r" and "p" +
        the least squares solution vector(s) "xs" (if "bs" != 0).
    2   LA_Hhls() calculates the matrices q", "r" and "p" + the
        pseudo-inverse of "a", "apsudo".
    3   LA_Hhls() calculates the matrices q", "r" and "p" +
        the pseudo-inverse of matrix "a" + the least squares
        solution(s) (if "bs" != 0).

This program contains 3 examples whose results appear in the text.

Example 1:
    matrix "a" is 8 by 5 well conditioned matrix of rank 3.  There
    are 3 r.h.s. vectors "bs".

Example 2:
    matrix "a" is 6 by 5 badly conditioned matrix of rank 5.  There
    are 2 r.h.s. vectors "bs".
```

```
Example 3:
    matrix "a" is 4 by 5 badly conditioned matrix.  There are no
    r.h.s. vectors "bs".
    In single precision, the calculated rank of coefficient matrix
    "a" = 3.
    In double precision, the calculated rank of coefficient matrix
    "a" = 4.
-------------------------------------------------------------------*/

#include "DR_Defs.h"
#include "LA_Prototypes.h"

#define Na2         8
#define Ma2         5
#define irhsa       3
#define Nb2         6
#define Mb2         5
#define irhsb       2
#define Nc2         4
#define Mc2         5
#define irhsc       0

void DR_Hhls (void)
{
    /*-------------------------------------
      Constant matrices/vectors
    -------------------------------------*/
    static tNumber_R a1init[Na2][Ma2] =
    {
        { 22.0, 10.0,   2.0,   3.0,  7.0 },
        { 14.0,  7.0,  10.0,   0.0,  8.0 },
        { -1.0, 13.0,  -1.0, -11.0,  3.0 },
        { -3.0, -2.0,  13.0,  -2.0,  4.0 },
        {  9.0,  8.0,   1.0,  -2.0,  4.0 },
        {  9.0,  1.0,  -7.0,   5.0, -1.0 },
        {  2.0, -6.0,   6.0,   5.0,  1.0 },
        {  4.0,  5.0,   0.0,  -2.0,  2.0 }
    };

    static tNumber_R bs1init[Na2][irhsa] =
    {
        { -1.0,  1.0,   0.0 },
        {  2.0, -1.0,   1.0 },
        {  1.0, 10.0,  11.0 },
        {  4.0,  0.0,   4.0 },
```

```
    {  0.0, -6.0,  -6.0 },
    { -3.0,  6.0,   3.0 },
    {  1.0, 11.0,  12.0 },
    {  0.0, -5.0,  -5.0 }
};

static tNumber_R a2init[Nb2][Mb2] =
{
    {    36.0,    -630.0,     3360.0,    -7560.0,      7560.0 },
    {  -630.0,   14700.0,   -88200.0,   211680.0,   -220500.0 },
    {  3360.0,  -88200.0,   564480.0, -1411200.0,   1512000.0 },
    { -7560.0,  211680.0, -1411200.0,  3628800.0,  -3969000.0 },
    {  7560.0, -220500.0,  1512000.0, -3969000.0,   4410000.0 },
    { -2772.0,   83160.0,  -582120.0,  1552320.0,  -1746360.0 }
};

static tNumber_R bs2init[Nb2][irhsb] =
{
    {     463.0,   -4157.0 },
    {  -13860.0,  -17820.0 },
    {   97020.0,   93555.0 },
    { -258720.0, -261800.0 },
    {  291060.0,  288288.0 },
    { -116424.0, -118944.0 }
};

static tNumber_R a3init[Nc2][Mc2] =
{
    { 0.4087, 0.1593,  0.6594, 0.4302, 0.3516 },
    { 0.6246, 0.3383,  0.6591, 0.9342, 0.9038 },
    { 0.0661, 0.9112,  0.6898, 0.1931, 0.1498 },
    { 0.2112, 0.8150,  0.7983, 0.3406, 0.2803 }
};

/*----------------------------------------
  Variable matrices/vectors
----------------------------------------*/
tMatrix_R   aa     = alloc_Matrix_R (NN2_ROWS, MM2_COLS);
tMatrix_R   bs     = alloc_Matrix_R (NN2_ROWS, KK2_COLS);
tMatrix_R   xs     = alloc_Matrix_R (MM2_COLS, KK2_COLS);
tMatrix_R   q      = alloc_Matrix_R (NN2_ROWS, NN2_ROWS);
tMatrix_R   r      = alloc_Matrix_R (MM2_COLS, MM2_COLS);
tMatrix_R   p      = alloc_Matrix_R (MM2_COLS, MM2_COLS);
tMatrix_R   apsudo = alloc_Matrix_R (MM2_COLS, NN2_ROWS);
tMatrix_R   res    = alloc_Matrix_R (NN2_ROWS, KK2_COLS);
```

```
tMatrix_R   aux     = alloc_Matrix_R (NN2_ROWS, NN2_ROWS);
tMatrix_R   auy     = alloc_Matrix_R (NN2_ROWS, MM2_COLS);
tVector_R   bb      = alloc_Vector_R (NN2_ROWS);

tMatrix_R   a1      = init_Matrix_R (&(a1init[0][0]), Na2, Ma2);
tMatrix_R   bs1     = init_Matrix_R (&(bs1init[0][0]), Na2,irhsa);
tMatrix_R   a2      = init_Matrix_R (&(a2init[0][0]), Nb2, Mb2);
tMatrix_R   bs2     = init_Matrix_R (&(bs2init[0][0]), Nb2,irhsb);
tMatrix_R   a3      = init_Matrix_R (&(a3init[0][0]), Nc2, Mc2);

int         irank, irhs, ientry;
int         i, j, k, m, n, Iexmpl;

tNumber_R   s, sa, sum1, sum2, aerr;

eLaRc       rc = LaRcOk;

prn_dr_bnr ("DR_Hhls, Minimal Length L2 Solution of a System of"
            " Linear Equations by Householder's \"QR\""
            " Decomposition");

for (Iexmpl = 1; Iexmpl <= 3; Iexmpl++)
{
    switch (Iexmpl)
    {
        case 1:
            ientry = 3;
            n = Na2;
            m = Ma2;
            irhs = irhsa;
            for (i = 1; i <= n; i++)
            {
                for (j = 1; j <= m; j++)
                {
                    aa[i][j] = a1[i][j];
                }
            }
            for (i = 1; i <= n; i++)
            {
                for (j = 1; j <= irhs; j++)
                {
                    bs[i][j] = bs1[i][j];
                }
            }
            break;
```

```
    case 2:
        ientry = 3;
        n = Nb2;
        m = Mb2;
        irhs = irhsb;
        for (i = 1; i <= n; i++)
        {
            for (j = 1; j <= m; j++)
            {
                aa[i][j] = a2[i][j];
            }
        }
        for (i = 1; i <= n; i++)
        {
            for (j = 1; j <= irhs; j++)
            {
                bs[i][j] = bs2[i][j];
            }
        }
        break;

    case 3:
        ientry = 2;
        n = Nc2;
        m = Mc2;
        irhs = 0;
        for (i = 1; i <= n; i++)
        {
            for (j = 1; j <= m; j++)
            {
                aa[i][j] = a3[i][j];
            }
        }
        break;

    default:
        break;
}

prn_algo_bnr ("Hhls");

prn_example_delim();
PRN ("Example #%d: Size of matrix \"a\" %d by %d\n",
    Iexmpl, n, m);
```

```
prn_example_delim();
PRN ("Minimal Least Squares Solution(s) of a System of"
     " Linear Equations Using Householder's \"QR\""
     " Decomposition\n");
prn_example_delim();

PRN ("Coefficient Matrix \"a\"\n");
prn_Matrix_R (aa, n, m);

if (irhs != 0)
{
    PRN ("\n");
    PRN ("Right Hand Vector(s) \"bs\"\n");
    prn_Matrix_R (bs, n, irhs);
}

rc = LA_Hhls (ientry, n, m, irhs, aa, bs, q, r, p, &irank,
              apsudo, xs, res);

if (rc >= LaRcOk)
{
    PRN ("\n");
    PRN ("Results of the Least Squares Problem\n");
    PRN ("Rank of the coefficient matrix = %d\n\n", irank);
    PRN ("Orthogonal Matrix \"q\"\n");
    prn_Matrix_R (q, n, irank);

    PRN ("Upper Triangular Matrix \"r\"\n");
    prn_Matrix_R (r, irank, irank);

    PRN ("Orthogonal Matrix \"p\"\n");
    prn_Matrix_R (p, m, irank);

    if (ientry >= 2)
    {
        PRN ("Pseudo-inverse Matrix \"apsudo\"\n");
        prn_Matrix_R (apsudo, m, n);

        s = 0.0;
        for (j = 1; j <= m; j++)
        {
            for (i = 1; i <= n; i++)
            {
                s = s + aa[i][j] * (aa[i][j]);
```

```
        }
    }
    sum1 = sqrt (s);

    for (j = 1; j <= n; j++)
    {
        for (i = 1; i <= n; i++)
        {
            s = 0.0;
            for (k = 1; k <= m; k++)
            {
                s = s + aa[i][k] * (apsudo[k][j]);
            }
            aux[i][j] = s;
        }
    }

    for (j = 1; j <= m; j++)
    {
        for (i = 1; i <= n; i++)
        {
            s = 0.0;
            for (k = 1; k <= n; k++)
            {
                s = s + aux[i][k] * (aa[k][j]);
            }
            auy[i][j] = s - aa[i][j];
        }
    }

    sa = 0.0;
    for (j = 1; j <= m; j++)
    {
        for (i = 1; i <= n; i++)
        {
            sa = sa + auy[i][j] * (auy[i][j]);
        }
    }
    sum2 = sqrt (sa);
    aerr = sum2/sum1;
    PRN ("||a*apsudo*a - a||/||a|| = %22.15f\n", aerr);
}

if ((ientry == 1) || (ientry == 3))
{
```

```
            if (irhs != 0)
            {
                PRN ("\n");
                PRN ("The Least Squares Solution(s)\n");
                for (i = 1; i <= irhs; i++)
                {
                    for (j = 1; j <= n; j++)
                    {
                        bb[j] = bs[j][i];
                    }

                    PRN ("Right Hand Vector \"b\"\n");
                    prn_Vector_R (bb, n);
                    for (j = 1; j <= m; j++)
                    {
                        bb[j] = xs[j][i];
                    }

                    PRN ("Solution vector \"x\"\n");
                    prn_Vector_R (bb, m);
                    for (j = 1; j <= n; j++)
                    {
                        bb[j] = res[j][i];
                    }
                    s = 0.0;
                    for (j = 1; j <= n; j++)
                    {
                        s = s + bb[j] * (bb[j]);
                    }
                    sum1 = sqrt (s);

                    PRN ("Residual vector \"res\"\n");
                    prn_Vector_R (bb, n);
                }
            }
        }
    }

    prn_la_rc (rc);
}

free_Matrix_R (aa, NN2_ROWS);
free_Matrix_R (bs, NN2_ROWS);
free_Matrix_R (xs, MM2_COLS);
free_Matrix_R (q, NN2_ROWS);
```

```
    free_Matrix_R (r, MM2_COLS);
    free_Matrix_R (p, MM2_COLS);
    free_Matrix_R (apsudo, MM2_COLS);
    free_Matrix_R (res, NN2_ROWS);
    free_Matrix_R (aux, NN2_ROWS);
    free_Matrix_R (auy, NN2_ROWS);
    free_Vector_R (bb);

    uninit_Matrix_R (a1);
    uninit_Matrix_R (bs1);
    uninit_Matrix_R (a2);
    uninit_Matrix_R (bs2);
    uninit_Matrix_R (a3);
}
```

17.16 LA_Hhls

```
/*-----------------------------------------------------------------
LA_Hhls
-------------------------------------------------------------------
This program calculates the minimal-length least squares solution(s)
of a system of linear equations and/or the pseudo-inverse of the
coefficient matrix.  It uses the Householder's "QR" decomposition
method with pivoting.

The system of linear equations is of the form

                    a*xs = bs

"a"  is a given real n by m matrix  of rank k,  k <= [min (n,m)].
"bs" is (are) the given r.h.s. n vector(s).
"xs" is (are) the m solution vector(s).

In this method the permuted matrix "a(permuted)" is factorized into

(1)     "a(permuted)" = q*r*p(transpose)

See description of "q", "r" and "p" below".

Inputs
n       Number of rows of matrix "a" in the system a*xs = bs.
m       Number of columns of matrix "a" in the system a*xs = bs.
irhs    Number of columns of the r.h.s matrix "bs".
a       A real n by m matrix of the given equation a*xs = bs.
        This matrix is not destroyed in the computation.
bs      A real n by "irhs" matrix of the r.h.s. of a*xs = bs.
ientry  An integer specifying the action by the user:
        0   LA_Hhls() calculates the factorization matrices q, r
            and p of "a(permuted)" = q * r * p(transpose).
        1   LA_Hhls() calculates the matrices "q", "r" and "p" +
            the least squares solution vector(s) "xs" (if "bs" != 0).
        2   LA_Hhls() calculates the matrices q", "r" and "p" + the
            pseudo-inverse of "a", "apsudo".
        3   LA_Hhls() calculates the matrices q", "r" and "p" +
            the pseudo-inverse of matrix "a" + the least squares
            solution(s) (if "bs" != 0).

Local Data
ic      An integer n vector containing the column permutation of
```

```
        matrix "a".
t       A real working matrix.
b, x, w, dm
        Real working vectors.

Outputs
irank   The calculated rank of matrix "a".
q       An n by n matrix whose first n rows and "irank" columns
        contain the orthogonal matrix "q" in the factorization (1)
        above.
r       A real m by m matrix whose first "irank" rows and "irank"
        columns contain the triangular matrix "r" in the
        factorization (1) above.
p       An m by m matrix whose first "irank" rows and m columns
        contain matrix "p(transpose)" of the factorization (1) above.
xs      A real m by "irhs" matrix containing the minimal length
        least squares solution matrix "xs" of system a*xs = bs.
res     A real n by "irhs" matrix containing the residual vectors
        given by (bs - a*xs) for the "irhs" solutions of a*xs = bs.
apsudo  A real m by n matrix containing the pseudo-inverse of matrix
        "a".

NOTE: The calculated results "xs" and "apsudo" are for the given
      equation a*xs = bs, (not for the permuted one).

Returns one of
        LaRcSolutionFound
        LaRcErrBounds
        LaRcErrNullPtr
        LaRcErrAlloc
-------------------------------------------------------------------*/

#include "LA_Prototypes.h"

eLaRc LA_Hhls (int ientry, int n, int m, int irhs, tMatrix_R aa,
    tMatrix_R bs, tMatrix_R q, tMatrix_R r, tMatrix_R p, int *pIrank,
    tMatrix_R apsudo, tMatrix_R xs, tMatrix_R res)
{
    tMatrix_R   t   = alloc_Matrix_R (n, m);
    tVector_R   x   = alloc_Vector_R (m);
    tVector_R   w   = alloc_Vector_R (n + 1);
    tVector_R   dm  = alloc_Vector_R (n + 1);
    tVector_I   ic  = alloc_Vector_I (m);
    tVector_R   b   = alloc_Vector_R (n);
```

```
int         iout = 0, iwish = 0, in = 0, ifirst = 0, isecnd = 0,
            irank1 = 0;
int         i = 0, ij = 0;
tNumber_R   eps2 = 0.0;

/* Validation of the data before executing the algorithm */
eLaRc       rc = LaRcSolutionFound;
VALIDATE_BOUNDS ((0 < n) && (0 < m) && !((n == 1) && (m == 1))
                && (0 <= irhs) && IN_BOUNDS (ientry, 0, 3));
VALIDATE_PTRS   (aa && bs && q && r && p && pIrank && apsudo &&
                xs && res);
VALIDATE_ALLOC  (t && x && w && dm && ic && b);

/* Initialization */
eps2 = EPS*EPS;
iwish = ientry;

LA_hhls_init (n, m, aa, q, t, ic);
iout = 1;

/* Calculation of matrix q */
LA_hhls_calcul_q (&iout, n, m, q, t, ic, w, dm);

*pIrank = iout - 1;

irank1 = *pIrank + 1;

/* Intialize matrix p */
LA_hhls_init_p (n, m, pIrank, r, t, p);

/* Calculation of matrix r */
LA_hhls_calcul_r_p (pIrank, m, r, p, w, dm);

if (iwish == 0)
{
    GOTO_CLEANUP_RC (LaRcSolutionFound);
}

in = n;
if (iwish != 2) in = irhs;

for (ij = 1; ij <= 2; ij++)
{
    for (ifirst = 1; ifirst <= in; ifirst++)
    {
```

```
            if (iwish != 2)
            {
                for (i = 1; i <= n; i++)
                {
                    b[i] = bs[i][ifirst];
                }
            }
            else if (iwish == 2)
            {
                for (isecnd = 1; isecnd <= n; isecnd++)
                {
                    b[isecnd] = 0.0;
                }
                b[ifirst] = 1.0;
            }
            /* Calculate the results */
            LA_hhls_calcul_res (iwish, ifirst, isecnd, pIrank, n, m,
                                aa, bs, xs, apsudo, q, r, p, ic, b,
                                x, w, dm, res);
        }

        if ((iwish == 1) || (iwish == 2))
        {
            GOTO_CLEANUP_RC (LaRcSolutionFound);
        }
        iwish = 2;
        in = n;
    }

CLEANUP:

    free_Matrix_R (t, n);
    free_Vector_R (x);
    free_Vector_R (w);
    free_Vector_R (dm);
    free_Vector_I (ic);
    free_Vector_R (b);

    return rc;
}

/*--------------------------------------------------------------------
Calculation of matrix q in LA_Hhls()
--------------------------------------------------------------------*/
void LA_hhls_calcul_q (int *pIout, int n, int m, tMatrix_R q,
```

```
    tMatrix_R t, tVector_I ic, tVector_R w, tVector_R dm)
{
    int          nless, kdp, iout1;
    int          i, ij, j, k;

    tNumber_R    beta, e, eps2, piv, s, sqsg;

    eps2 = EPS*EPS;
    nless = n;
    if (m < n) nless = m;

    /* Calculation of matrix q */
    for (ij = 1; ij <= nless; ij++)
    {
        /* Permuting the columns of matrix t */
        piv = 0.0;
        kdp = *pIout;
        for (i = *pIout; i <= m; i++)
        {
            s = 0.0;
            for (k = *pIout; k <= n; k++)
            {
                s = s + t[k][i] * t[k][i];
            }
            if (s >= piv)
            {
                piv = s;
                kdp = i;
            }
        }

        if (piv >= eps2)
        {
            if (*pIout != kdp)
            {
                /* Swap two elements of integer vector "ic" */
                swap_elems_Vector_I (ic, kdp, *pIout);

                /* Swap two columns of real matrix "t" */
                for (i = 1; i <= n; i++)
                {
                    e = t[i][*pIout];
                    t[i][*pIout] = t[i][kdp];
                    t[i][kdp] = e;
                }
```

```
    }

    if (*pIout < n)
    {
        e = t[*pIout][*pIout];
        if (e < 0.0) e = -e;
        sqsg = sqrt (piv);
        beta = 1.0/ (piv + sqsg * e);
        for (i = 1;  i <= n; i++)
        {
            w[i] = 0.0;
            if (i > *pIout)
            {
                w[i] = t[i][*pIout];
            }
            else if (i == *pIout)
            {
                e = sqsg;
                if (t[*pIout][*pIout] < 0.0) e = -e;
                w[i] = t[*pIout][*pIout] + e;
            }
        }

        for (j = *pIout; j <= m; j++)
        {
            s = 0.0;
            for (k = *pIout; k <= n; k++)
            {
                s = s + w[k] * (t[k][j]);
            }
            dm[j] = s * beta;
        }

        for (j = *pIout; j <= m; j++)
        {
            for (i = *pIout; i <= n; i++)
            {
                t[i][j] = t[i][j] - w[i] * (dm[j]);
            }
        }
        iout1 = *pIout + 1;

        for (i = iout1; i <= n; i++)
        {
            t[i][*pIout] = 0.0;
```

```
                }

                for (i = 1; i <= n; i++)
                {
                    s = 0.0;
                    for (k = *pIout; k <= n; k++)
                    {
                        s = s + w[k] * (q[i][k]);
                    }
                    dm[i] = s * beta;
                }

                for (j = *pIout; j <= n; j++)
                {
                    for (i = 1; i <= n; i++)
                    {
                        q[i][j] = q[i][j] - dm[i] * (w[j]);
                    }
                }
            }
            *pIout = *pIout + 1;
        }
    }
}

/*-------------------------------------------------------------------
Intialize matrix p in LA_Hhls()
-------------------------------------------------------------------*/
void LA_hhls_init_p (int n, int m, int *pIrank, tMatrix_R r,
    tMatrix_R t, tMatrix_R p)
{
    int         i, j, irank1;

    irank1 = *pIrank + 1;

    if (*pIrank != n)
    {
        for (j = *pIrank; j <= m; j++)
        {
            for (i = irank1; i <= n; i++)
            {
                t[i][j] = 0.0;
            }
        }
    }
```

```
    for (j = 1; j <= m; j++)
    {
        for (i = 1; i <= *pIrank; i++)
        {
            r[i][j] = t[i][j];
        }
    }

    for (j = 1; j <= m; j++)
    {
        for (i = 1; i <= m; i ++)
        {
            p[i][j] = 0.0;
        }
        p[j][j] = 1.0;
    }
}

/*-----------------------------------------------------------------------
Calculating matrix r in LA_Hhls()
-----------------------------------------------------------------------*/
void LA_hhls_calcul_r_p (int *pIrank, int m, tMatrix_R r,
    tMatrix_R p, tVector_R w, tVector_R dm)
{
    int          iout, irank1;
    int          i, ij, j, k;

    tNumber_R    e, s, abass, beta, sigma, sqsg;

    irank1 = *pIrank + 1;
    if (*pIrank != m)
    {
        iout = *pIrank;
        for (ij = 1; ij <= *pIrank; ij++)
        {
            e = r[iout][iout];
            if (e < 0.0) e = -e;
            abass = e;
            s = abass*abass;

            for (k = irank1; k <= m; k++)
            {
                s = s + r[iout][k] * (r[iout][k]);
            }
```

```
sigma = s;
sqsg = sqrt (sigma);
beta = 1.0/ (sigma + abass * sqsg);

for (j = 1; j <= m; j++)
{
    w[j] = 0.0;
    if (j > *pIrank)
    {
        w[j] = r[iout][j];
    }
    else if (j == *pIrank)
    {
        e = sqsg;
        if (r[iout][iout] < 0.0) e = -e;
        w[iout] = r[iout][iout] + e;
    }
}

for (i = 1; i <= m; i++)
{
    s = 0.0;
    for (k = 1; k <= m; k++)
    {
        s = s + p[i][k] * (w[k]);
    }
    dm[i] = s * beta;
}

for (j = 1; j <= m; j++)
{
    for (i = 1; i <= m; i++)
    {
        p[i][j] = p[i][j] - dm[i] * (w[j]);
    }
}

for (i = 1; i <= *pIrank; i++)
{
    s = 0.0;
    for (k = 1; k <= m; k++)
    {
        s = s + r[i][k] * (w[k]);
    }
```

```
                dm[i] = s * beta;
            }

            for (j = 1; j <= m; j++)
            {
                for (i = 1; i <= *pIrank; i++)
                {
                    r[i][j] = r[i][j] - dm[i] * (w[j]);
                }
            }

            iout = iout - 1;
            if (iout < 1)
            {
                return;
            }
        }
    }
}

/*--------------------------------------------------------------------
Calculate the results of LA_Hhls()
--------------------------------------------------------------------*/
void LA_hhls_calcul_res (int iwish, int ifirst, int isecnd,
    int *pIrank, int n, int m, tMatrix_R aa, tMatrix_R bs,
    tMatrix_R xs, tMatrix_R apsudo, tMatrix_R q, tMatrix_R r,
    tMatrix_R p, tVector_I ic, tVector_R b, tVector_R x, tVector_R w,
    tVector_R dm, tMatrix_R res)
{
    int         ip1;
    int         i, ii, j, k;

    tNumber_R   s;

    for (i = 1; i <= *pIrank; i++)
    {
        s = 0.0;
        for (k = 1; k <= n; k++)
        {
            s = s + b[k] * (q[k][i]);
        }
        dm[i] = s;
    }

    w[*pIrank] = dm[*pIrank]/r[*pIrank][*pIrank];
```

```
if (*pIrank != 1)
{
    for (ii = 2; ii <= *pIrank; ii++)
    {
        i = *pIrank - ii + 1;
        ip1 = i + 1;
        s = -dm[i];
        for (k = ip1; k <= *pIrank; k++)
        {
            s = s + r[i][k] * (w[k]);
        }
        w[i] = -s/r[i][i];
    }
}

for (i = 1; i <= m; i++)
{
    s = 0.0;
    for (k = 1; k <= *pIrank; k++)
    {
        s = s + w[k] * (p[i][k]);
    }
    dm[i] = s;
    k = ic[i];
    x[k] = dm[i];
}

if (iwish == 2)
{
    for (isecnd = 1; isecnd <= m; isecnd++)
    {
        apsudo[isecnd][ifirst] = x[isecnd];
    }
}
else if (iwish != 2)
{
    for (i = 1; i <= m; i++)
    {
        xs[i][ifirst] = x[i];
    }
    for (j = 1; j <= n; j++)
    {
        s = -bs[j][ifirst];
        for (k = 1; k <= m; k++)
```

```
            {
                s = s + aa[j][k] * (x[k]);
            }
            res[j][ifirst] = -s;
        }
    }
}

/*--------------------------------------------------------------------
Initialization of LA_Hhls()
--------------------------------------------------------------------*/
void LA_hhls_init (int n, int m, tMatrix_R aa, tMatrix_R q,
    tMatrix_R t, tVector_I ic)
{
    int         i, j;

    for (j = 1; j <= m; j++)
    {
        ic[j] = j;
        for (i = 1; i <= n; i++)
        {
            t[i][j] = aa[i][j];
        }
    }

    for (j = 1; j <= n; j++)
    {
        for (i = 1; i <= n; i++)
        {
            q[i][j] = 0.0;
        }
        q[j][j] = 1.0;
    }
}


/*--------------------------------------------------------------------
LA_Hhlsro
--------------------------------------------------------------------
Given is the real n by m matrix "c" of rank m <= n and a real n
vector "f" for the equation

    (a)     c*a = f
```

```
This program applies householder"s transformation to obtain the upper
triangular matrix, which is the r.h.s. in the orthogonal
factorization
                                    [ r | h ]
    (b)      q(transpose)*[c|f] = [---|---]
                                    [ 0 |rho]

where
"q" is a real n by n orthogonal matrix
                     q(transpose)*q = I(unit matrix)
    "q" is not calculated nor it is stored.
    The matrix on the r.h.s. of (b) is an upper (m+1)
    by (m+1) matrix.
"r" is a real m by m upper triangular matrix.
"h" is a real m vector given by q(transpose)*b=h.
"rho" is a real scalar that equals plus or minus of the L-Two norm
of the residual vector res = (b - a*x), i.e.
    |rho| = ||c*a-f|| where ||.|| denotes the L-Two norm.

This program then calculates the least squares solution of the
overdetermined system (a) above, given by

                        a = (r**-1)*h

Returns one of
        LaRcSolutionFound
        LaRcInconsistentSystem
        LaRcErrBounds
        LaRcErrNullPtr
-----------------------------------------------------------------*/
eLaRc LA_Hhlsro (int n, int m, tMatrix_R c, tVector_R f, tVector_R a,
    tMatrix_R r, tVector_R res, tMatrix_R t, tVector_R dm,
    tVector_R w, tNumber_R *pRho)
{
    int          iend, mp1;
    int          i, j;
    tNumber_R    e, eps2;
    eLaRc        tempRc;

    eLaRc        rc = LaRcSolutionFound;
    VALIDATE_BOUNDS ((0 < n) && (0 < m) && !((n == 1) && (m == 1)));

    mp1 = m + 1;

    eps2 = EPS * EPS;
```

```
/* Initialization, mapping matrix "c" on matrix t */
for (j = 1; j <= m; j++)
{
    for (i = 1; i <= n; i++)
    {
        t[i][j] = c[i][j];
    }
}

for (i = 1; i <= n; i++)
{
    t[i][mp1] = f[i];
}
iend = mp1;

if (n == mp1)
{
    iend = m;
}
if (n == m)
{
    iend = m - 1;
}

/* Householder"s transformation of matrix "t"
   This will never return a failure code */
tempRc = LA_hhlsro_hh_t (iend, n, m, t, dm, w);
if (tempRc < LaRcOk)
{
    return tempRc;
}

for (j = 1; j <= mp1; j++)
{
    for (i = 1; i <= mp1; i++)
    {
        r[i][j] = t[i][j];
    }
}

e = r[m][m];
if (e < 0.0)
{
    e = -e;
```

```
    }

    if (e < EPS)
    {
        GOTO_CLEANUP_RC (LaRcInconsistentSystem);
    }

    /* Calculating the least squares solution vector "a" */
    LA_hhlsro_x_res (n, m, c, f, a, r, res, pRho);

CLEANUP:

    return rc;
}

/*-----------------------------------------------------------------
Householder"s transformation of matrix "t" in LA_Hhlsro()
-----------------------------------------------------------------*/
eLaRc LA_hhlsro_hh_t (int iend, int n, int m, tMatrix_R t,
    tVector_R dm, tVector_R w)
{
    int         i, j, k, iout, kdp, mp1;
    tNumber_R   e, s, beta, piv, eps2, sqsg;

    eps2 = EPS * EPS;
    mp1 = m + 1;
    for (iout = 1; iout <= iend; iout++)
    {
        piv = 0.0;
        kdp = iout;
        s = 0.0;
        for (k = iout; k <= n; k++)
        {
            s = s + t[k][iout] * (t[k][iout]);
        }
        piv = s;

        /* Pemature exit due to rank deficiency of coefficient
           matrix */
        if ((iout <= m) && (piv < eps2))
            return LaRcNoFeasibleSolution;

        /* Householder"s transformation of matrix "t" */
        e = t[iout][iout];
        if (e < 0.0) e = - e;
```

```
        sqsg = sqrt (piv);
        beta = 1.0 / (piv + sqsg * e);
        for (i = 1; i <= n; i++)
        {
            w[i] = 0.0;
            if (i >= iout)
            {
                if (i > iout) w[i] = t[i][iout];
                if (i == iout)
                {
                    e = sqsg;
                    if (t[iout][iout] < 0.0) e = - e;
                    w[i] = t[iout][iout] + e;
                }
            }
        }

        for (j = iout; j <= mp1; j++)
        {
            s = 0.0;
            for (k = iout; k <= n; k++)
            {
                s = s + w[k] * (t[k][j]);
            }
            dm[j] = s * beta;
        }

        for (j = iout; j <= mp1; j++)
        {
            for (i = iout; i <= n; i++)
            {
                t[i][j] = t[i][j] - w[i] * (dm[j]);
            }
        }
    }

    return LaRcOk;
}

/*------------------------------------------------------------------
Calculating the least squares solution vector in LA_Hhlsro()
------------------------------------------------------------------*/
void LA_hhlsro_x_res (int n, int m, tMatrix_R c, tVector_R f,
    tVector_R a, tMatrix_R r, tVector_R res, tNumber_R *pRho)
{
```

```
    int          i, j, k, ii, ip1, mp1;
    tNumber_R    s;

    mp1 = m + 1;
    a[m] = r[m][mp1]/r[m][m];

    if (m != 1)
    {
        for (ii = 2; ii <= m; ii++)
        {
            i = m - ii + 1;
            ip1 = i + 1;
            s = -r[i][mp1];
            for (k = ip1; k <= m; k++)
            {
                s = s + r[i][k] * a[k];
            }
            a[i] = -s/r[i][i];
        }
    }

    for (j = 1; j <= n; j++)
    {
        s = -f[j];
        for (k = 1; k <= m; k++)
        {
            s = s + c[j][k] * (a[k]);
        }
        res[j] = -s;
    }

    if (n == m) r[mp1][mp1] = 0.0;

    /* L2 norm of residual vector |rho| */
    *pRho = r[mp1][mp1];
    if (*pRho < 0.0) *pRho = -*pRho;
}
```

Chapter 18

Piecewise Linear Least Squares Approximation

18.1 Introduction

In Chapter 9, two algorithms for the piecewise linear approximation of plane curves in the L_1 norm are described [2]. In Chapter 15, two corresponding algorithms in the Chebyshev norm are given [1]. In this chapter, we describe two corresponding algorithms for the piecewise linear approximation of plane curves in the L_2 or the least squares norm [4].

The problem of piecewise linear approximation of plane curves in the least squares sense has been dealt with by a number of authors. As early as 1961, Stone [18] presented an algorithm for finding best approximation to a given convex function f(x) on a finite interval by K straight line segments using the least squares norm. He formulated the problem giving a closed form solution when the approximated function is quadratic. Bellman [5] showed that when the approximated function is quadratic, dynamic programming also provides a solution. For both Stone and Bellman, the L_2 errors between an approximated segment and the approximating straight line are measured in the y-axis direction.

Cantoni [6] described a method for finding the break points for segment approximation using weighted least squares norm. Pavlidis [11] derived necessary and sufficient conditions and suggested simple functional iteration algorithms for locating the break points of the L_2 piecewise approximation of continuous differentiable functions of one and two variables.

Pavlidis [13] also used Newton's method to optimally locate the breaking points for a continuous differentiable function for L_2 piecewise linear approximation. In [11], Pavlidis measured the errors between the approximated segments and the approximating straight lines in the y-axis direction, while in [13] Pavlidis measured the errors in the Euclidean distance, i.e., perpendicular to the approximating line for each segment.

Salotti [15] presented a new algorithm for an optimal polygonal approximation for digitized curves using the least squares norm criterion. By optimal is meant using the minimum number of segments.

Sarkar et al. [16] presented a genetic algorithm-based approach for detection of significant vertices for polygonal approximation in the L_2 norm of digital curves. The error is the Euclidean distances between the given points and the line segments.

For the applications and the merits of piecewise linear approximation of plane curves, see Section 9.1.1 and also Pavlidis [10, 12].

In our work, the following two algorithms for the piecewise linear approximation in the least squares sense are presented:

(a) when the L_2 residual norm in any segment is not to exceed a pre-assigned value, and

(b) when the number of segments is given and a near-balanced L_2 residual norm solution is required. That is, all the segments would have nearly the same L_2 residual norm.

The errors are measured along the y-axis direction.

As usual, the problem is solved by first digitizing the given curve into discrete points. Then each algorithm is applied to the discrete points. No conditions are imposed on the approximating functions. The approximating functions are any polynomials. They need not be constants or straight lines as is the case for the majority of the published works.

In Section 18.2, the characteristics of the piecewise approximation are outlines. In Section 18.3, the discrete linear L_2 approximation problem is presented. In Section 18.4, the two algorithms listed above are outlined. In Section 18.5, numerical results and comments are given.

Section 18.6 offers a brief tutorial describing the technique of updating and downdating systems of linear equations, used for calculating the piecewise linear approximation of plane curves in the least squares sense.

18.1.1 Preliminaries and notation

For convenience sake, the following notation is used again. Let us be given the plane curve $y = f(x)$ defined on the interval [a, b]. Let this curve be digitized at the K points $(x_i, f(x_i))$, $i = 1, 2, \ldots, K$, where $x_1 = a$ and $x_K = b$. Let n be the number of segments (pieces) in the piecewise approximation. For the first algorithm, n is not known beforehand. Finally, let (z_j), $j = 1, 2, \ldots, n$, be the L_2 residual norms of the n segments.

18.2 Characteristics of the approximation

As indicated in Chapters 9 and 15, assume that f(x) is continuous and satisfies Lipschitz condition on [a, b]. Following the analysis of Lawson [9] for the Chebyshev norm, one shows that for segment k, $1 \leq k \leq n$, the L_2 norm z_k, has the following properties:

(a) z_k is a continuous function of the end points of the segment,
(b) z_k is non-increasing in the segment left end point, and
(c) z_k is non-decreasing in the segment right end point.

For the meanings of these characteristics, see Section 9.2. One also shows that if a piecewise L_2 approximation is calculated for the curve f(x) (not for the discrete points of the curve) and if $z_1 = z_2 = \ldots = z_n$, then the solution is optimal. Such a solution always exists and is known as a balanced residual norm solution. However, in the case of the approximation of the digitized curve, a balanced residual norm solution may not exist. One may in this case attempt to minimize the variance of the residual norms (z_k) and get a near-balanced L_2 solution.

18.3 The discrete linear least squares approximation problem

Consider any segment k, $1 \leq k \leq n$, of the given curve f(x). Let this segment consist of N digitized points with coordinates $(x_i, f(x_i))$, $i = 1, 2, \ldots, N$. Let these N points be approximated by the function

$$\mathbf{L}(\mathbf{a}, \mathbf{x}) = \Sigma_j a_j \phi_j(x)$$

where, $\{\phi_j(x)\}$, $j = 1, 2, \ldots, M$, is a given set of real linearly independent approximating functions. The number of points N is usually far more than the number of the M parameters (a_j); N > M. The linear least squares (approximate) solution to the data set by this curve is vector **a** that minimizes the L_2 norm

$$z = \text{sqrt}\left[\sum_{i=1}^{N} [r(x_i)]^2\right]$$

where $r(x_i)$ is i^{th} residual and is given by

$$r(x_i) = f(x_i) - \mathbf{L}(\mathbf{a}, x_i), \quad i = 1, 2, \ldots, N \qquad (18.3.1)$$

Note that the residual $r(x_i)$ in (18.3.1) is the –ve of that defined in equation (9.3.1) and also in equation (15.3.1). Such definitions are arbitrary.

As in Section 2.2, this problem reduces to the problem of obtaining the least squares solution of the overdetermined system of linear equations

$$\mathbf{Ca} = \mathbf{f}$$

C is an N by M matrix given by $\mathbf{C} = (\phi_j(x_i))$, i = 1, 2, …, N and j = 1, 2, …, M, and **f** is the N-vector $(f(x_1), \ldots, f(x_N))^T$.

The least squares solution to system **Ca** = **f** is the real M-vector **a** that minimizes the L_2 norm

$$z = \text{sqrt}\left[\sum_{i=1}^{N} [r_i]^2\right]$$

where r_i is residual i, given by (18.3.1), which is

$$r_i = f_i - \sum_{i=1}^{M} c_{ij} a_j, \quad i = 1, 2, \ldots, N$$

and c_{ij} are the elements of matrix **C**.

18.4 Description of the algorithms

The C implementations of the algorithms are similar to those of Chapters 9 and 15, as follows.

18.4.1 Piecewise linear L_2 approximation with pre-assigned tolerance

This algorithm uses the same steps as those of Section 9.4.1. However, each time a point is added to segment say j, we use the function LA_Hhls() of Chapter 17 to re-calculate the new L_2 norm of the segment.

18.4.2 Piecewise linear L_2 approximation with near-balanced L_2 norms

This algorithm uses the same steps as those of Section 9.4.2. Again, each time we add a point or delete a point from segment say j, we use the function LA_Hhls() of Chapter 17 to re-calculate the new Chebyshev norm of the segment.

18.5 Numerical results and comments

Each of the functions LA_L2pw1() and LA_L2pw2() calculate the number of segments in the piecewise approximation (for LA_L2pw2(), n is given), the starting points of the n segments, the coefficients of the approximating curves for the n segments, the residuals at each point of the digitized curve and finally the L_2 residual norms for the n segments.

LA_L2pw1() computes for the case when the L_2 error norm in any segment is not to exceed a pre-assigned value. LA_L2pw2() computes for the case when the number of segments is given and a near-balanced L_2 error norm solution is required.

DR_L2pw1() and DR_L2pw2() were used to test the algorithms on a number of problems in single-precision.

Again, as in Chapters 9 and 15, LA_L2pw1() has the option of calculating connected or disconnected piecewise linear L_2 approximations, while LA_L2pw2() can only calculate disconnected piecewise linear L_2 approximations. This is explained at the end of Section 9.2.

In order to compare the results of these algorithms with those of the algorithms of Chapters 9 and 15 for the L_1 and the Chebyshev norms respectively, we chose the same curve used in those chapters.

The curve is digitized with equal x-intervals into 28 points and the data points are fitted with vertical parabolas. Each is of the form $y = a_1 + a_2x + a_3x^2$. The results of LA_L2pw1() are shown in figures 18-1 and 18-2. The results of LA_L2pw2() are shown in figure 18-3, where the number of segments was set to n = 4.

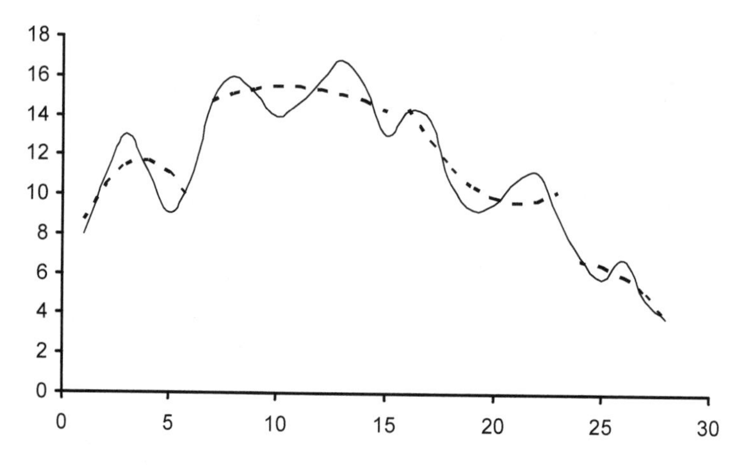

Figure 18-1: Disconnected linear L_2 piecewise approximation with vertical parabolas. The L_2 residual norm in any segment ≤ 3

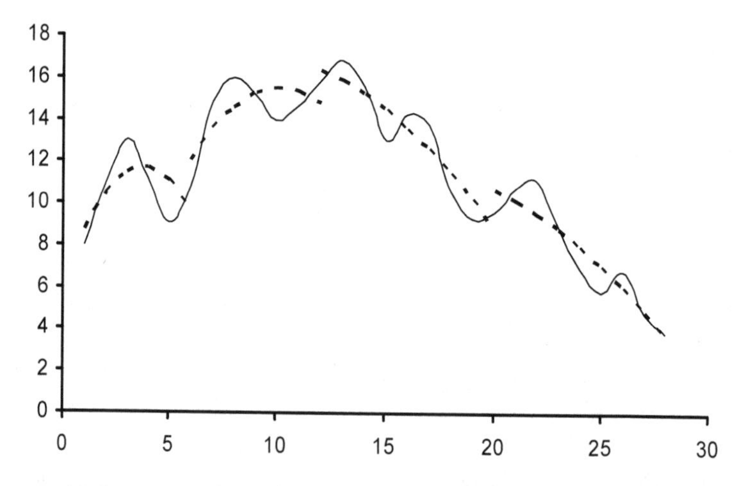

Figure 18-2: Connected linear L_2 piecewise approximation with vertical parabolas. The L_2 residual norm in any segment ≤ 3

The L_2 norms of Figures 18-1 and 18-2 are (2.905, 3.000, 2.867, 1.165) and $n = 4$, and (2.905, 2.984, 2.762, 2.756) and $n = 4$, respectively. The L_2 norms of Figure 18-3 are (2.905, 1.990, 2.390, 2.028), for $n = 4$. We observe that LA_L2pw2() did not produce a balanced norm solution. But it did give an improved approximation to that of figure 18-1.

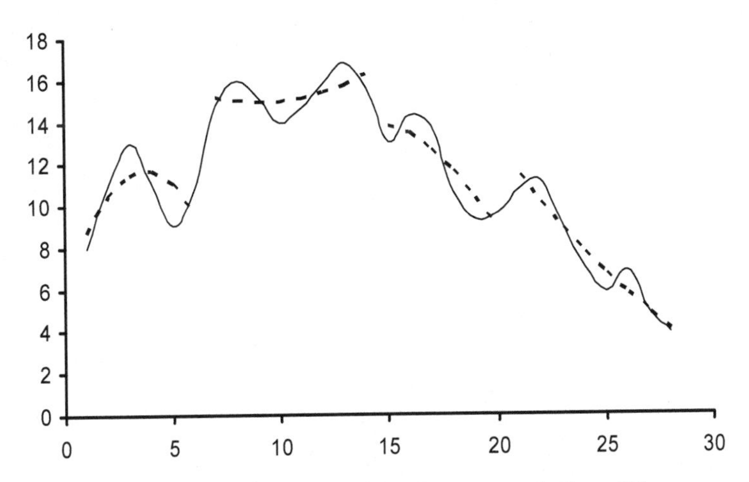

Figure 18-3: Near-balanced residual norm solution. Disconnected linear L_2 piecewise approximation with vertical parabolas. Number of segments = 4

In a previous version of these algorithms [4], we utilized the techniques of updating and downdating the systems of linear equations, which solve the piecewise approximation in the least squares sense. These techniques use information from a current iteration and carry them into the calculation of the next iteration without recomputing the least squares solution of a modified system from scratch. This increases the code complexity but reduces the number of iterations.

In the next section we present a brief review of the updating and downdating technique.

18.6 The updating and downdating techniques

For any segment at hand in the piecewise approximation, let the system of linear equations be $\mathbf{Ca} = \mathbf{f}$, where as usual, $\mathbf{C}$ is an N by M matrix, N > M, and **f** is an N-vector. The augmented matrix [**C**|**f**] of this system is factorized by the Householder's **QR** factorization method [14, 19] (Chapter 17). The least squares solutions are then calculated in terms of the **R** matrices [8]. The **Q** matrices are not explicitly calculated. The algorithms work in an iterative manner by updating the least squares solutions via updating the **R** matrices of the segments.

Assume that the least squares solution of system **Ca** = **f** has been obtained. Let one new data point be added to or an old point be deleted from the existing point set and a new least squares solution be re-calculated for the new set. Stewart [17] denotes these two problems as the problems of updating and of downdating of the least squares problem respectively.

These two problems are equivalent to updating the least squares solution vector **a** when an extra equation is augmented to or when an existing equation is deleted from the system **Ca** = **f** respectively.

For the updating problem, we make use of the scheme of Gill et al. [7] and for the downdating problem, we use the technique of Stewart [17]. This is explained here as follows.

As indicated earlier, the least squares solution of system **Ca** = **f** may be calculated without explicitly storing matrix **Q**. We pre-multiply the matrix [**C**|**f**] by Householder's elementary orthogonal matrices (Section 5.4.1). We get the upper triangular matrix

$$\underline{\mathbf{R}} = \begin{bmatrix} \mathbf{R} & \mathbf{h} \\ \mathbf{0} & \rho \end{bmatrix} \tag{18.6.1a}$$

where

$$\mathbf{h} = \mathbf{Q}^T\mathbf{f} \text{ and } |\rho| = \|\mathbf{r}\| \tag{18.6.1b}$$

R and **Q** are the factorization of matrix **C** by Householder's decomposition into **C** = **QR**.

The least squares solution to system **Ca** = **f** is given by

$$\mathbf{a} = \mathbf{R}^{-1}\mathbf{h} \tag{18.6.1c}$$

18.6.1 The updating algorithm

Assume that we have added an extra equation to the system **Ca** = **f** and that we want to re-calculate the least squares solution of the new system. Assume without loss of generality that the added equation is augmented to the bottom of the given system $\mathbf{Ca} = \mathbf{f}$. That is, the new system is given by

$$\begin{bmatrix} \mathbf{C} \\ \mathbf{c} \end{bmatrix} \mathbf{a} = \begin{bmatrix} \mathbf{f} \\ \eta \end{bmatrix} \tag{18.6.2}$$

$\mathbf{c}$ is a row vector and η is the corresponding $\mathbf{f}$ element.

Calculating the least squares solution of system (18.6.2) amounts to calculating the upper triangular matrix $\underline{\mathbf{R}}'$ in the $\underline{\mathbf{Q}}'\underline{\mathbf{R}}'$ factorization of the augmented system. This is done by updating matrix $\underline{\mathbf{R}}$ of (18.6.1a) in the $\underline{\mathbf{QR}}$ factorization of matrix $[\mathbf{C}|\mathbf{f}]$ as follows.

From (18.6.2), it is easy to verify that

$$\begin{bmatrix} \mathbf{Q} & \mathbf{0} \\ \mathbf{0} & 1 \end{bmatrix} \begin{bmatrix} \mathbf{C} & \mathbf{f} \\ \mathbf{c} & \eta \end{bmatrix} = \begin{bmatrix} \mathbf{R} & \mathbf{h} \\ \mathbf{0} & \rho \\ \mathbf{c} & \eta \end{bmatrix} \tag{18.6.3}$$

By considering (18.6.3), we first augment $[\mathbf{c}|\eta]$ to the bottom of matrix $\underline{\mathbf{R}}$ in (18.6.1a). Then we pre-multiply the augmented matrix by the product of either Householder or Givens matrices to reduce the r.h.s. matrix in (18.6.3) to an upper triangular matrix. The resulting matrix is

$$\underline{\mathbf{R}}' = \begin{bmatrix} \mathbf{R}' & \mathbf{h}' \\ \mathbf{0} & \rho' \end{bmatrix} \tag{18.6.4}$$

The least squares solution for the new system is $\mathbf{a}' = \mathbf{R}'^{-1}\mathbf{h}'$, where $|\rho'| = ||\mathbf{r}'||$; $\mathbf{r}'$ is the new residual vector.

18.6.2 The downdating algorithm

Downdating matrix $\underline{\mathbf{R}}$ after deleting an existing data point is done as follows. Let the row in matrix $[\mathbf{C}|\mathbf{f}]$ that corresponds to this deleted point be $[\mathbf{c}|\eta]$. Stewart [17] shows that the algorithm is stable in the presence of rounding errors. However, the new matrix $\mathbf{R}'$ can be a very ill-conditioned function of $\mathbf{R}$ and $\mathbf{c}$. The algorithm by Stewart reduces the obtained system to (18.6.1a) or (18.6.4) and the result is given by (18.6.1b, c). See also [4].

We applied the updating and downdating techniques described in this section [4] to the problems that we solved by the algorithms of Section 18.4, and obtained the same results.

18.6.3 Updating and downdating in the L_1 norm

We note that the updating and downdating techniques for linear equations in the L_1 norm could also be achieved. Such techniques use parametric linear programming together with making use of the characteristic of solution of the L_1 approximation problem. For details, see [3].

References

1. Abdelmalek, N.N., Piecewise linear Chebyshev approximation of planar curves, *International Journal of Systems Science*, 14(1983)425-435.
2. Abdelmalek, N.N., Piecewise linear L_1 approximation of planar curves, *International Journal of Systems Science*, 16(1985)447-455.
3. Abdelmalek, N.N., A recursive algorithm for discrete L_1 linear estimation using the dual simplex method, *IEEE Transactions on Systems, Man and Cybernetics*, SMC-15(1985)737-742.
4. Abdelmalek, N.N., Piecewise linear least-squares approximation of planar curves, *International Journal of Systems Science*, 21(1990)1393-1403.
5. Bellman, R., On the approximation of curves by line segments using dynamic programming, *Communications of the ACM*, 4(1961)284.
6. Cantoni, A., Optimal curve fitting with piecewise linear functions, *IEEE Transactions on Computers*, 20(1971)59-67.
7. Gill, P.E., Golub, G.H., Murray, W. and Saunders, M.A., Methods for modifying matrix factorizations, *Mathematics of Computation*, 28(1974)505-535.

8. Golub, G., Numerical methods for solving linear least squares problems, *Numerische Mathematik*, 7(1965)206-216.
9. Lawson, C.L., Characteristic properties of segmented rational minimax approximation problem, *Numerische Mathematik*, 6(1964)293-301.
10. Pavlidis, T., Waveform segmentation through functional approximation, *IEEE Transactions on Computers*, 22(1973)689-697.
11. Pavlidis, T., Optimal piecewise polygonal L_2 approximation of functions of one and two variables, *IEEE Transactions on Computers*, 24(1975)98-102.
12. Pavlidis, T., The use of algorithms of piecewise approximations for picture processing, *ACM Transactions on Mathematical Software*, 2(1976)305-321.
13. Pavlidis, T., Polygonal approximation by Newton's method, *IEEE Transactions on Computers*, 26(1977)800-807.
14. Peters, G. and Wilkinson, J.H., The least squares problem and pseudo-inverses, *Computer Journal*, 13(1970)309-316.
15. Salotti, M., Optimal polygonal approximation of digitized curves using the sum of squares deviation criterion, *Pattern Recognition*, 35(2002)435-443.
16. Sarkar, B., Singh, L.K. and Sarkar, D., A genetic algorithm-based approach for detection of significant vertices for polygonal approximation of digital curves, *International Journal of Image and Graphics*, 4(2004)223-239.
17. Stewart, G.W., The effects of rounding residual on an algorithm for downdating a Cholesky factorization, *Journal of Institute of Mathematics and Applications*, 23(1979)203-213.
18. Stone, H., Approximation of curves by line segments, *Mathematics of Computation*, 15(1961)40-47.
19. Wilkinson, J.H., *The Algebraic Eigenvalue Problem*, Clarendon Press, Oxford, 1965.

18.7 DR_L2pw1

```
/*--------------------------------------------------------------------
DR_L2pw1
--------------------------------------------------------------------
This is a driver for the function LA_L2pw1(), which calculates a
linear piecewise L2 (L-Two) approximation of given data point set
{x,y} that results from the discretization of a given plane curve
y = f(x). The points of the set might not be equally spaced.

The approximation by LA_L2pw1() is such that the L2 norm for any
segment is not to exceed a pre-assigned tolerance denoted by
"enorm".

LA_L2pw1() calculates the connected or the disconnected linear
piecewise L2 approximation, according to the value of an integer
parameter "konect" set by the user.
See comments in program LA_L2pw1().

From the approximating curve we form the overdetermined system of
linear equations

                            c*a = f

"c" is a real n by m matrix of rank k, k <= m < n.
n is the number of digitized points of the given plane curve.

m is the number of terms in the approximating curves.  If for
example, the approximating curves are vertical parabolas of
the form
                      y = a1 + a2*x + a3*x*x
then m = 3.

"f" is a real n vector whose elements are the y coordinates of
the data set {x,y}.

"a" is the solution m vector.  There are different "a" solution
vectors for the different segments.

This driver contains 1 test example. A given curve is digitized
into 28 points at equal x intervals.  The points are piecewise
approximated by vertical parabolas of the form

                      y = a1 + a2*x + a3*x*x
```

```
The results for the disconnected and of the connected piecewise
L2 approximation are given in the text.
-------------------------------------------------------------------*/

#include "DR_Defs.h"
#include "LA_Prototypes.h"

#define Nc          28

void DR_L2pw1 (void)
{
    /*-----------------------------------------
      Constant matrices/vectors
    -----------------------------------------*/
    static tNumber_R fc[Nc+1] =
    {   NIL,
         8.0, 11.0, 13.0, 11.2,  9.1, 10.8, 14.8, 16.0, 15.1, 14.0,
        14.7, 15.8, 16.8, 15.6, 13.0, 14.3, 13.8, 10.6,  9.3,  9.6,
        10.8, 11.2,  9.0,  7.0,  5.8,  6.8,  4.8,  3.9
    };

    /*-----------------------------------------
      Variable matrices/vectors
    -----------------------------------------*/
    tMatrix_R   c      = alloc_Matrix_R (NN_ROWS, MM_COLS);
    tMatrix_R   ap     = alloc_Matrix_R (KK_PIECES, MM_COLS);
    tMatrix_R   rp1    = alloc_Matrix_R (KK_PIECES, NN_ROWS);
    tVector_R   f      = alloc_Vector_R (NN_ROWS);
    tVector_R   zp     = alloc_Vector_R (KK_PIECES);
    tVector_I   ixl    = alloc_Vector_I (KK_PIECES);

    int         j, k, n, m, konect, npiece;
    tNumber_R   enorm;
    eLaRc       prnRc;

    eLaRc       rc = LaRcOk;

    prn_dr_bnr ("DR_L2pw1, L2 Piecewise Approximation of a Plane "
                "with Pre-assigned Norm");

    for (k = 1; k <= 2; k++)
    {
        switch (k)
        {
```

```
        case 1:
            enorm = 3.0;
            n = Nc;
            m = 3;
            for (j = 1; j <= n; j++)
            {
                f[j] = fc[j];
                c[j][1] = 1.0;
                c[j][2] = j;
                c[j][3] = j*j;
            }
            break;
        default:
            break;
    }

    prn_algo_bnr ("L2pw1");
    if (k == 1) konect = 0;
    if (k == 2) konect = 1;
    prn_example_delim();
    PRN ("konect #%d: Size of matrix \"c\" %d by %d\n",
         konect, n, m);
    PRN ("L2 Piecewise Approximation with Pre-assigned Norm\n");
    PRN ("Pre-assigned Norm \"enorm\" = %8.4f\n", enorm);
    prn_example_delim();
    if (konect == 1)
        PRN ("Connected Piecewise Approximation\n");
    else
        PRN ("Disconnected Piecewise Approximation\n");
    prn_example_delim();
    PRN ("r.h.s. Vector \"f\"\n");
    prn_Vector_R (f, n);
    PRN ("Coefficient Matrix, \"c\"\n");
    prn_Matrix_R (c, n, m);

    rc = LA_L2pw1 (n, m, enorm, konect, c, f, ixl, rp1, ap, zp,
                   &npiece);

    if (rc >= LaRcOk)
    {
        PRN ("\n");
        PRN ("Results of the L2 Piecewise Approximation\n");
        PRN ("Calculated number of segments (pieces) = %d\n",
             npiece);
        PRN ("Staring points of the \"npiece\" segments\n");
```

```
            prn_Vector_I (ixl, npiece);
            PRN ("Coefficients of the approximating curves\n");
            prn_Matrix_R (ap, npiece, m);
            PRN ("Residual vectors for the \"npiece\" segments\n");
            prnRc = LA_pw1_prn_rp1 (konect, npiece, n, ixl, rp1);
            PRN ("L2 residual norms for the \"npiece\" segments\n");
            prn_Vector_R (zp, npiece);
            if (prnRc < LaRcOk)
            {
                PRN ("Error printing PW1 results\n");
            }
        }

        prn_la_rc (rc);
    }

    free_Matrix_R (c, NN_ROWS);
    free_Matrix_R (ap, KK_PIECES);
    free_Matrix_R (rp1, KK_PIECES);
    free_Vector_R (f);
    free_Vector_R (zp);
    free_Vector_I (ixl);
}
```

18.8 LA_L2pw1

```
/*-------------------------------------------------------------------
LA_L2pw1
---------------------------------------------------------------------
This program calculates a linear piecewise L2 (L-Two) approximation
to a discrete point set {x,y}.  The approximation is such that the
L2 residual (error) norm for any segment is not to exceed a given
tolerance "enorm".  The number of segments (pieces) is not known
before hand.

Given is a set of points {x,y}.  From the approximating functions
of the piecewise approximation, one forms the overdetermined system
of linear equations

                          c*a = f

This program uses LA_Hhls() for obtaining the L2 solution of
overdetermined system of linear equations.

LA_L2pw1() has the option of calculating connected or disconnected
piecewise L2 approximation, according to the value of an integer
parameter "konect".

In the connected piecewise approximation the x-coordinate of the
end point of segment j say, is the x-coordinate of the starting
point of segment (j+1).  In the disconnected piecewise
approximation, the x-coordinate of the adjacent point to the end
point of segment j is the x-coordinate of the starting point of
segment (j+1).  See the comments on "ixl" below.

Inputs
m       Number of terms in the approximating functions.
n       Number of points to be piecewise approximated.
c       A real n by m matrix of containing matric "c" of the
        system c*a = f.
        Matrix "c" is not destroyed in the computation.
f       A real n vector containing the r.h.s. of the system
        c*a = f. This vector contains the y-coordinates of the
        given point set.  This vector is not destroyed in the
        computation.
enorm   enorm   A real pre-assigned parameter, such that the L2
        residual norm for any segment is <= enorm.
konect  An integer specifying the action to be performed.
```

```
        If konect = 1, the program calculates the connected L2
        piecewise approximation.
        If konect != 1, the program calculates the disconnected
        L2 piecewise approximation.

Outputs
npiece  Obtained number of segments or pieces of the approximation.
ixl     An integer "npiece" vector containing the indices of the
        first elements of the "npiece" segments.
        For example, if ixl = (1,5,12,22,...), and if konect = 1,
        then the first segment contains points 1 to 5, the second
        segment contains points 5 to 12, the third segment contains
        points 12 to 22 and so on.
        Again if ixl = (1,5,12,22,...), and if konect !=1, then the
        first segment contains points 1 to 4, the second segment
        contains points 5 to 11, the third segment contains points 12
        to 21 ..., etc.
irankp  An integer "npiece" vector containing the rank values of the
        approximating curves for the "npiece" segments.
ap      A real "npiece" by m  matrix.  Its first row contains the
        the coefficients of the approximating curve for the first
        segment.  The second row contains the coefficients of the
        approximating curve for the second segment and so on.
        If any row j is all zeros, this indicates that vector "a" of
        segment j is not calculated as the number of points of
        segment j is <= m and there is a perfect fit by the
        approximating curve for segment j.
rpl     A real "npiece" by n matrix.  Its first row contains the
        residuals for the points of the first segment.  Its second
        row contains the residuals for the points of the second
        segment, and so on.
zp      A real "npiece" vector containing the "npiece" optimum
        values of the L2 residual norms for the "npiece" segments.
        If zp[j] == 0.0, it indicates that there is a perfect fit for
        segment j.

Returns one of
        LaRcSolutionFound
        LaRcErrBounds
        LaRcErrNullPtr
        LaRcErrAlloc
-------------------------------------------------------------------*/

#include "LA_Prototypes.h"
```

```
eLaRc LA_L2pw1 (int n, int m, tNumber_R enorm, int konect,
    tMatrix_R c, tVector_R f, tVector_I ixl, tMatrix_R rp1,
    tMatrix_R ap, tVector_R zp, int *pNpiece)
{
    tMatrix_R   cp      = alloc_Matrix_R (n, m);
    tMatrix_R   t       = alloc_Matrix_R (n, m + 1);
    tMatrix_R   r       = alloc_Matrix_R (m + 1, m + 1);
    tVector_R   fp      = alloc_Vector_R (n);
    tVector_R   res     = alloc_Vector_R (n);
    tVector_R   dm      = alloc_Vector_R (n);
    tVector_R   w       = alloc_Vector_R (n);
    tVector_R   a       = alloc_Vector_R (m);

    int         i = 0, j = 0, je = 0, ii = 0, jj = 0, is = 0, ie = 0,
                nu = 0, irank = 0;
    int         ijk = 0;
    tNumber_R   z = 0.0;

    /* Validation of data before executing algorithm */
    eLaRc       rc = LaRcSolutionFound;
    VALIDATE_BOUNDS ((0 < m) && (m <= n) && !((n == 1) && (m == 1))
                    && (0.0 < enorm));
    VALIDATE_PTRS   (c && f && ixl && rp1 && ap && zp && pNpiece);
    VALIDATE_ALLOC  (cp && t && r && fp && res && dm && w && a);

    *pNpiece = 1;
    /* "is" means i(start) for the segment at hand */
    is = 1;
    for (ijk = 1; ijk <= n; ijk++)
    {
        /* Initializing the data for "npiece" */
        ie = is + m - 1;
        LA_pw1_init (pNpiece, is, ie, m, rp1, ap, zp);

        irank = m;
        z = 0;
        for (i = is; i <= ie; i++)
        {
            ii = i - is + 1;
            fp[ii] = f[i];
            for (j = 1; j <= m; j++)
            {
                cp[ii][j] = c[i][j];
            }
        }
```

```
for (jj = 1; jj <= n; jj++)
{
    nu = ie - is + 1;
    if (nu < m)
    {
        GOTO_CLEANUP_RC (LaRcSolutionFound);
    }

    rc = LA_Hhlsro (nu, m, cp, fp, a, r, res, t, dm, w, &z);
    if (rc < LaRcOk)
    {
        GOTO_CLEANUP_RC (rc);
    }

    if (z > enorm + EPS)
    {
        break;
    }
    else
    {
        LA_pwl_map (m, nu, res, a, z, rpl, ap, zp, pNpiece);

        ixl[*pNpiece] = is;
        je = ie + 1;
        if (je > n)
        {
            GOTO_CLEANUP_RC (LaRcSolutionFound);
        }

        ie = je;
        nu = ie - is + 1;
        for (i = is; i <= ie; i++)
        {
            ii = i - is + 1;
            fp[ii] = f[i];
            for (j = 1; j <= m; j++)
            {
                cp[ii][j] = c[i][j];
            }
        }
    }
}
is = ie;
if (konect == 1) is = ie - 1;
```

```
        *pNpiece = *pNpiece + 1;
        ixl[*pNpiece] = is;
    }

CLEANUP:

    free_Matrix_R (cp, n);
    free_Matrix_R (t, n);
    free_Matrix_R (r, m + 1);
    free_Vector_R (fp);
    free_Vector_R (res);
    free_Vector_R (dm);
    free_Vector_R (w);
    free_Vector_R (a);

    return rc;
}
```

18.9 DR_L2pw2

```
/*-----------------------------------------------------------------
DR_L2pw2
-------------------------------------------------------------------
This is a driver for the function LA_L2pw2() which calculates the
"near balanced" piecewise linear L2 approximation problem of a given
data point set {x,y} resulting from the discretization of a plane
curve y=f(x).

Given is an integer number "npiece" which is the number of segments
in the approximation.

The approximation by LA_L2pw2() is such that the L2 residual norms
for all segments are nearly equal, hence the name "near balanced"
piecewise approximation.

From the approximating curves we form the overdetermined system of
linear equations

                          c*a = f

"c" is a real n by m matrix of rank k, k <= m < n.
n is the number of digitized points of the given plane curve.
m is the number of terms in the approximating curves.  If for
example, the piecewise approximating curves are vertical parabolas
of the form
                     y = a1 + a2*x + a3*x*x
then m = 3.

"f" is a real n vector whose elements are the y coordinates of
the data set {x,y}.

"a" is the solution m vector.  There are different "a" solution
vectors for the different segments.

This driver contains 1 test example.
A given curve is digitized into 28 points at equal x intervals.  The
points are piecewise approximated by vertical parabolas of the form

                     y = a1 + a2*x + a3*x*x

The results for piecewise L2 approximation are given in the text.
-----------------------------------------------------------------*/
```

```
#include "DR_Defs.h"
#include "LA_Prototypes.h"

#define Nc          28

void DR_L2pw2 (void)
{
    /*-----------------------------------------
      Constant matrices/vectors
    -----------------------------------------*/
    static tNumber_R fc[Nc+1] =
    {   NIL,
         8.0, 11.0, 13.0, 11.2,  9.1, 10.8, 14.8, 16.0, 15.1, 14.0,
        14.7, 15.8, 16.8, 15.6, 13.0, 14.3, 13.8, 10.6,  9.3,  9.6,
        10.8, 11.2,  9.0,  7.0,  5.8,  6.8,  4.8,  3.9
    };

    /*-----------------------------------------
      Variable matrices/vectors
    -----------------------------------------*/
    tMatrix_R   c       = alloc_Matrix_R (NN_ROWS, MM_COLS);
    tVector_R   f       = alloc_Vector_R (NN_ROWS);
    tMatrix_R   ap      = alloc_Matrix_R (KK_PIECES, MM_COLS);
    tVector_R   rp2     = alloc_Vector_R (NN_ROWS);
    tVector_R   zp      = alloc_Vector_R (KK_PIECES);
    tVector_I   ixl     = alloc_Vector_I (KK_PIECES);

    int         j, m, n;
    int         Iexmpl, npiece;
    eLaRc       prnRc;

    eLaRc       rc = LaRcOk;

    prn_dr_bnr ("DR_L2pw2, L2 Piecewise Approximation of a Plane "
                "Curve with Near Equal Residual Norms");

    for (Iexmpl = 1; Iexmpl <= 1; Iexmpl++)
    {
        switch (Iexmpl)
        {
            case 1:
                npiece = 4;
                n = Nc;
                m = 3;
```

```
            for (j = 1; j <= n; j++)
            {
                f[j] = fc[j];
                c[j][1] = 1.0;
                c[j][2] = j;
                c[j][3] = j*j;
            }
            break;
        default:
            break;
    }

    prn_algo_bnr ("L2pw2");
    prn_example_delim();
    PRN ("Size of matrix \"c\" %d by %d\n", n, m);
    prn_example_delim();
    PRN ("L2 Piecewise Approximation with Near Equal Norms\n");
    PRN ("Given number of segments (pieces) = %d\n", npiece);
    prn_example_delim();
    PRN ("r.h.s. Vector \"f\"\n");
    prn_Vector_R (f, n);
    PRN ("Coefficient Matrix, \"c\"\n");
    prn_Matrix_R (c, n, m);

    rc = LA_L2pw2 (m, n, npiece, c, f, ap, rp2, zp, ixl);

    if (rc >= LaRcOk)
    {
        PRN ("\n");
        PRN ("Results of the L2 Piecewise Approximation\n");
        PRN ("Starting points of the \"npiece\" segments\n");
        prn_Vector_I (ixl, npiece);
        PRN ("L2 residual norms for the \"npiece\" segments\n");
        prn_Vector_R (zp, npiece);
        PRN ("Coefficients of the \"npiece\" approximating "
             "curves\n");
        prn_Matrix_R (ap, npiece, m);

        PRN ("Residuals at the given points\n");
        prnRc = LA_pw2_prn_rp2 (npiece, n, ixl, rp2);
        if (prnRc < LaRcOk)
        {
            PRN ("Error printing PW2 results:  ");
        }
    }
```

```
        prn_la_rc (rc);
    }

    free_Matrix_R (c, NN_ROWS);
    free_Vector_R (f);
    free_Matrix_R (ap, KK_PIECES);
    free_Vector_R (rp2);
    free_Vector_R (zp);
    free_Vector_I (ixl);
}
```

18.10 LA_L2pw2

```
/*-----------------------------------------------------------------------
LA_L2pw2
-------------------------------------------------------------------------
This program calculates the "near balanced" piecewise linear
L2 (L-Two) approximation  of a given data point set {x,y} resulting
from the discretization of a plane curve y = f(x).

Given is an integer number "npiece" which is the number of segments
in the approximation.

The approximation by LA_L2pw2() is such that the L2 residual norms
for all segments are nearly equal, hence the name "near balanced"
piecewise approximation.

From the approximating functions (curves) one forms the
overdetermined system of linear equations

                        c*a = f

Inputs
npiece  Given umber of segments (pieces) of the approximation.
m       Number of terms in the approximating functions.
n       Number of points to be piecewise approximated.
c       A real n by m matrix of the system c*a = f.  This matrix is
        not destroyed in the computation.
f       A real n vector containing the r.h.s. of the system c*a = f.
        This vector contains the y-coordinates of the given point
        set.

Outputs
ixl     An integer "npiece' vector containing the indices of the
        first elements of the "npiece" segments.
        For example, if ixl = (1,5,12,22,...), then the first
        segment contains points 1 to 4, the second segment  contains
        points 5 to 11, the third segment contains points 12
        to 21 ..., etc.
ap      A real "npiece" by m matrix.  Its first row contains the
        coefficients of the approximating curve for the first
        segment. The second row contains the coefficients of the
        approximating curve for the second segment and so on.
rp2     A real n vector containing the residual values of the n
        points of the given set {x,y}.
```

```
zp      A real npiece vector containing the npiece optimum values of
        the L-Two residual norms for the "npiece" segments.

Returns one of
        LaRcSolutionFound
        LaRcErrBounds
        LaRcErrNullPtr
        LaRcErrAlloc
------------------------------------------------------------------*/

#include "LA_Prototypes.h"

eLaRc LA_L2pw2 (int m, int n, int npiece, tMatrix_R c, tVector_R f,
    tMatrix_R ap, tVector_R rp2, tVector_R zp, tVector_I ixl)
{
    tMatrix_R   cp      = alloc_Matrix_R (n, m);
    tMatrix_R   t       = alloc_Matrix_R (n + 1, m + 1);
    tVector_R   fp      = alloc_Vector_R (n);
    tVector_R   a       = alloc_Vector_R (m);
    tVector_R   al      = alloc_Vector_R (m);
    tVector_R   ar      = alloc_Vector_R (m);
    tVector_R   dm      = alloc_Vector_R (n);
    tVector_R   w       = alloc_Vector_R (n);
    tVector_I   iflag   = alloc_Vector_I (npiece);
    tVector_R   resl    = alloc_Vector_R (n);
    tVector_R   resr    = alloc_Vector_R (n);
    tMatrix_R   r       = alloc_Matrix_R (m + 1, m + 1);
    tVector_I   ir      = alloc_Vector_I (n);

    int         i = 0, j = 0, k = 0, ii = 0, ji = 0, is = 0, ie = 0,
                isl = 0, iel = 0, isr = 0, ier = 0, nu = 0;
    int         istart = 0, iend = 0, kl = 0, ijk = 0, klp1 = 0,
                ipcp1 = 0, ibc = 0;
    int         ieln = 0, isrn = 0;
    tNumber_R   con = 0.0, conn = 0.0, zl = 0.0, zln = 0.0,
                zrn = 0.0;

    /* Validation of the data before executing the algorithm */
    eLaRc       rc = LaRcSolutionFound;
    VALIDATE_BOUNDS ((0 < m) && (m <= n) && !((n == 1) && (m == 1))
                     && (1 < npiece));
    VALIDATE_PTRS   (c && f && ap && rp2 && zp && ixl);
    VALIDATE_ALLOC  (cp && t && fp && a && al && ar && dm && w &&
                     iflag && resl && resr && r && ir);
```

```
/* Calculating the residuals for initial segments */
for (k = 1; k <= npiece; k++)
{
    iflag[k] = 1;

    LA_l2pw2_init (k, npiece, n, m, &is, &ie, c, f, cp, fp, ixl);

    /* Triangularizing the coefficient matrix using
       Householder's transformation */
    nu = ie - is + 1;

    rc = LA_Hhlsro (nu, m, cp, fp, al, r, resl, t, dm, w, &zl);
    if (rc < LaRcOk)
    {
        GOTO_CLEANUP_RC (rc);
    }

    zp[k] = fabs (zl);
    /* Map initial data for the "npiece" segments */
    for (j = 1; j <= m; j++)
    {
        ap[k][j] = al[j];
    }
    for (j = is; j <= ie; j++)
    {
        ji = j - is + 1;
        rp2[j] = resl[ji];
    }
}

/* Process of balancing the L2 norms */
istart = 1;
iend = npiece - 1;
ipcp1 = npiece + 1;
ixl[ipcp1] = n + 1;
for (ijk = 1; ijk <= n; ijk++)
{
    for (kl = istart; kl <= iend; kl=kl+2)
    {
        klp1 = kl + 1;
        con = fabs (zp[klp1] - zp[kl]);
        isl = ixl[kl];
        isr = ixl[klp1];
        iel = isr - 1;
        if (kl != iend) ier = ixl[kl+2] - 1;
```

```
if (kl == iend) ier = n;

/* The case where : -----z[i]<z[i+1] */
if (zp[kl] < zp[klp1])
{
    ieln = iel + 1;
    isrn = isr + 1;
    for (i = isl; i <= ieln; i++)
    {
        ii = i - isl + 1;
        fp[ii] = f[i];
        for (j = 1; j <= m; j++)
        {
            cp[ii][j] = c[i][j];
        }
    }

    /* Updating the L2 approximation for left segment */
    nu = ieln - isl + 1;
    rc = LA_Hhlsro (nu, m, cp, fp, al, r, resl, t, dm, w,
                    &zln);
    if (rc < LaRcOk)
    {
        GOTO_CLEANUP_RC (rc);
    }

    for (i = isrn; i <= ier; i++)
    {
        ii = i - isrn + 1;
        fp[ii] = f[i];
        for (j = 1; j <= m; j++)
        {
            cp[ii][j] = c[i][j];
        }
    }

    /* Updating the L2 approximation for right segment */
    nu = ier - isrn + 1;
    rc = LA_Hhlsro (nu, m, cp, fp, ar, r, resr, t, dm, w,
                    &zrn);
    if (rc < LaRcOk)
    {
        GOTO_CLEANUP_RC (rc);
    }
```

```
            zln = fabs (zln);
            zrn = fabs (zrn);
            conn = fabs (zrn - zln);
            iflag[kl] = 1;
            iflag[klp1] = 1;
            if (conn >= con)
            {
                iflag[kl] = 0;
                iflag[klp1] = 0;
                continue;
            }
        }
        /* The case where : ---------- z[i]>z[i+1] */
        else if (zp[kl] > zp[klp1])
        {
            isrn = isr - 1;
            ieln = isrn - 1;
            for (i = isl; i <= ieln; i++)
            {
                ii = i - isl + 1;
                fp[ii] = f[i];
                for (j = 1; j <= m; j++)
                {
                    cp[ii][j] = c[i][j];
                }
            }
            /* Updating the l2 approximation for left segment */
            nu = ieln - isl + 1;
            rc = LA_Hhlsro (nu, m, cp, fp, al, r, resl, t, dm, w,
                            &zln);
            if (rc < LaRcOk)
            {
                GOTO_CLEANUP_RC (rc);
            }

            for (i = isrn; i <= ier; i++)
            {
                ii = i - isrn + 1;
                fp[ii] = f[i];
                for (j = 1; j <= m; j++)
                {
                    cp[ii][j] = c[i][j];
                }
            }
            /* Updating the L2 approximation for right segment */
```

```
            nu = ier - isrn + 1;
            rc = LA_Hhlsro (nu, m, cp, fp, ar, r, resr, t, dm, w,
                            &zrn);
            if (rc < LaRcOk)
            {
                GOTO_CLEANUP_RC (rc);
            }

            zln = fabs (zln);
            zrn = fabs (zrn);
            conn = fabs (zrn - zln);
            iflag[kl] = 1;
            iflag[klp1] = 1;
            if (conn >= con)
            {
                iflag[kl] = 0;
                iflag[klp1] = 0;
                continue;
            }
        }

        for (j = isl; j <= ieln; j++)
        {
            ji = j - isl + 1;
            rp2[j] = resl[ji];
        }
        for (j = isrn; j <= ier; j++)
        {
            ji = j - isrn + 1;
            rp2[j] = resr[ji];
        }
        for (i = 1; i <= m; i++)
        {
            ap[kl][i] = al[i];
            ap[klp1][i] = ar[i];
        }
        zp[kl] = zln;
        zp[klp1] = zrn;
        ixl[klp1] = isrn;
        isr = isrn;
        iel = isr - 1;
    }
    is = 2;
    if (istart == 2) is = 1;
    istart = is;
```

```
        ibc = 0;
        for (j = 1; j <= npiece; j++)
        {
            if (iflag[j] != 0) ibc = 1;
        }
        if (ibc == 0)
        {
            GOTO_CLEANUP_RC (LaRcSolutionFound);
        }
    }

CLEANUP:

    free_Matrix_R (cp, n);
    free_Matrix_R (t, n + 1);
    free_Vector_R (fp);
    free_Vector_R (a);
    free_Vector_R (al);
    free_Vector_R (ar);
    free_Vector_R (dm);
    free_Vector_R (w);
    free_Vector_I (iflag);

    free_Vector_R (resl);
    free_Vector_R (resr);
    free_Matrix_R (r, m + 1);

    free_Vector_I (ir);

    return rc;
}

/*-----------------------------------------------------------------------
Initializing LA_L2pw2()
-----------------------------------------------------------------------*/
void LA_l2pw2_init (int k, int npiece, int n, int m, int *pIs,
    int *pIe, tMatrix_R c, tVector_R f, tMatrix_R cp, tVector_R fp,
    tVector_I ixl)
{
    int         i, j, ii, jj, kp1;

    jj = n/npiece;
    ixl[k] = (k-1) * jj + 1;
    *pIs = ixl[k];
    if (k == npiece) *pIe = n;
```

```
    if (k != npiece)
    {
        kp1 = k + 1;
        ixl[kp1] = k * jj + 1;
        *pIe = ixl[kp1] - 1;
    }
    for (i = *pIs; i <= *pIe; i++)
    {
        ii = i - *pIs + 1;
        fp[ii] = f[i];
        for (j = 1; j <= m; j++)
        {
            cp[ii][j] = c[i][j];
        }
    }
}
```

Chapter 19

Solution of Ill-Posed Linear Systems

19.1 Introduction

This chapter presents the solution of ill-posed linear systems such as those arising from the discretization of Fredholm integral equation of the first kind.

Fredholm integral equation of the first kind arises in the mathematical analysis of many physics, chemistry and biology problems. It also arises in the restoration of blurred images [4, 5, 6] and in several classical mathematical problems, such as the numerical inversion of Laplace transform and many other problems [28, 30].

Consider the Fredholm integral equation of the first kind

$$\int_{\alpha}^{\beta} A(t, t')x(t')dt' = b(t)$$

A is a continuous **kernel** and ill-posed in the sense that the solution vector **x** does not depend continuously on the data **b**. In other words, the problem is not stable with respect to small perturbation on vector **b**. In a real situation, the elements of **b** are the output of some measurements contaminated with errors.

Using numerical integration, the discretization of the above integral equation results in the system of linear equations

$$\mathbf{Ax} = \mathbf{b}$$

A is a real n by m matrix of rank k, $k \le \min(n, m)$ and **b** is a real n-vector. Assume that $n \ge m$.

Again, system $\mathbf{Ax} = \mathbf{b}$ is ill-posed and ill-posed systems are also ill-conditioned. An ill-conditioned system such as $\mathbf{Ax} = \mathbf{b}$ has the smallest singular values of **A** very small. The elements of the solution vector **x**, would be of large and alternate positive and negative values. Computational difficulties arise when one attempts to solve this system.

We shall only consider the case that kernel **A** is smooth. As a result, one expects a reasonable approximate solutions to this problem. Varah [30] examined the existence and uniqueness questions of such reasonable approximate solutions. He defined a reasonable solution as follows. Vector **x** is a reasonable solution to $\mathbf{Ax} = \mathbf{b}$ with respect to noise level ε, if the norm of the residual, $||\mathbf{Ax} - \mathbf{b}|| = O(\varepsilon)$, provided that both **A** and **b** are properly scaled. O(ε) denotes *of the order of* ε.

In Section 19.2, different methods of solution of ill-posed systems of linear equations are discussed. For each method of solution, a free parameter is to be estimated. In Section 19.3, the estimation of the free parameter is outlined. In Section 19.4, our method of solution of ill-posed systems is described. This method utilizes linear programming techniques and the free parameter to be estimated is the rank of the system of equations. The optimum value of the rank is estimated in Section 19.5, and the linear programming techniques are described in Section 19.6. In Section 19.7, numerical results and comments are given.

19.2 Solution of ill-posed linear systems

There exist several methods for solving ill-posed linear systems. These include the methods of steepest descents and conjugate gradients by Squire [24], without using any **regularization** technique (explained below). As reported by Squire, when proper termination criteria are used, the conjugate gradients methods require less computational time than the steepest descents. He also noted that a noticeable increase in the residual during the calculation of the conjugate gradients indicates that the limiting accuracy of the method has been reached. Kammerer and Nashed [18] discussed the convergence of the conjugate gradient method. They then studied iterative solutions for integral equations of the first and second kinds [19].

Hanson and Phillips [17] presented a method where the approximate solution of $\mathbf{Ax} = \mathbf{b}$ is expressed as a continuous piecewise linear spline function. The procedure may be used iteratively to improve the accuracy of the approximate solution.

However, the most widely used approaches for solving the ill-posed system $\mathbf{Ax} = \mathbf{b}$ are the following:

(1) The first approach is to use the regularization methods, such as the original one of Tikhonov [27] and Phillips [23] and the modified regularization method [30].

(2) The second approach is to replace matrix $\mathbf{A}$ in system $\mathbf{Ax} = \mathbf{b}$ by an approximate matrix of smaller rank k, $k < m$. This approach is illustrated by Hanson [16] and by Varah [28], using a truncated singular value decomposition (SVD) expansion of matrix $\mathbf{A}$. This approach also includes the truncated $\mathbf{QR}$ method [29], where $\mathbf{Q}$ is an orthogonal matrix and $\mathbf{R}$ an upper triangular. The method used in this chapter is of this nature but quite a different one.

The methods listed in (1) and (2) both require the estimation of a free parameter. We shall give a brief description of these methods:

(1a) The original regularization methods [27, 23] are called by Varah [29, 30] damped least squares. Assuming that the n by m matrix $\mathbf{A}$ is of rank k, $k < m \leq n$, these methods calculate a solution $\mathbf{x}_\alpha$ such that (the $||.||$ refers to the Euclidean vector norm)

$$\mathbf{x}_\alpha = \min_\mathbf{x} (||\mathbf{Ax} - \mathbf{b}||^2 + \alpha^2||\mathbf{x}||^2)$$

The free parameter in this method is α, which is to be estimated. This method is equivalent to the solution of the modified normal equation in the form of the Ridge equation [29] (discussed in Chapter 17)

$$(\mathbf{A}^T\mathbf{A} + \alpha^2\mathbf{I}_m)\mathbf{x}_\alpha = \mathbf{A}^T\mathbf{b}$$

where $\mathbf{I}_m$ is an m-unit matrix. Matrix $(\mathbf{A}^T\mathbf{A} + \alpha^2\mathbf{I}_m)$ is matrix $\mathbf{A}^T\mathbf{A}$ with α^2 added to each of its singular values, and thus $(\mathbf{A}^T\mathbf{A} + \alpha^2\mathbf{I}_m)$ is square nonsingular. The solution of this equation is obtained efficiently via the singular value decomposition of $\mathbf{A}$, as $\mathbf{A} = \mathbf{VSW}^T$, $\mathbf{V}$ is a real n by m orthonormal matrix, $\mathbf{V}^T\mathbf{V} = \mathbf{I}_m$, $\mathbf{W}$ is a real m by m orthonormal matrix, $\mathbf{W}^T\mathbf{W} = \mathbf{I}_m$ and $\mathbf{S}$ is an m by m diagonal matrix, $\mathbf{S} = \text{diag}(s_1, s_2, \ldots, s_m)$, the s_i are the singular values of $\mathbf{A}$, ordered as $s_1 \geq s_2 \geq \ldots s_m \geq 0$. With little manipulation

$$\mathbf{x}_\alpha = \sum_{i=1}^{m} s_i \beta_i \mathbf{w}_i / (s_i^2 + \alpha^2) \tag{19.2.1}$$

where $\mathbf{V}^T\mathbf{b} = \beta$ and $(\mathbf{w}_i)$ are the columns of $\mathbf{W}$.

For a proper choice of the parameter α, most of the terms in the above summation are damped and the summation would be from 1 to say k, not from 1 to m, where $k < m$. The solution would be very close to the truncated SVD, given by (19.2.2) in approach (2a) below.

To see this, for $i > k$, $s_i \to 0$, $|\beta_i| \to \varepsilon$, so if we chose $\alpha \geq \varepsilon$, in (19.2.1), $s_i\beta_i/(s_i^2 + \alpha^2) \approx 0$, since $s_i \to 0$. For $i \leq k$, if we choose $\alpha \approx \varepsilon$ in (19.2.1), and since $s_i > \varepsilon$, $s_i\beta_i/(s_i^2 + \alpha^2) \approx \beta_i/s_i$. Thus (19.2.1) becomes

$$\mathbf{x}_\alpha = \sum_{i=1}^{k} (\beta_i \mathbf{w}_i)/s_i$$

Compare this result with (19.2.2) of the truncated SVD in approach (2a).

(1b) The modified regularization or the modified least squares method produces the modified normal equation [29, 30]

$$(\mathbf{A}^T\mathbf{A} + \alpha^2\mathbf{L}^T\mathbf{L})\mathbf{x}_\alpha = \mathbf{A}^T\mathbf{b}$$

$\mathbf{L}$ is usually some discrete approximation to the first or the second derivative operator. The solution $\mathbf{x}_\alpha$ has the same form of $\mathbf{x}_\alpha$ of the previous method except that the solution is given in terms of some orthogonal vectors other than the $(\mathbf{w}_i)$. te Riele [26] presented a program with numerical results for the solution of Fredholm integral equation of the first kind, using this method.

(2a) The truncated SVD method [16] is described as follows. Again using the SVD of $\mathbf{A} = \mathbf{VSW}^T$, $\mathbf{Ax} = \mathbf{b}$, reduces to

$$\mathbf{VSW}^T\mathbf{x} = \mathbf{b}$$

In this method, the smallest (m – k) singular values of matrix **A** are replaced by 0's. The free parameter in this method is k, and the solution is

$$\mathbf{x}_k = \sum_{i=1}^{k} (\beta_i \mathbf{w}_i)/s_i \tag{19.2.2}$$

This approach requires the estimation of the rank k of the matrix that approximates matrix **A**. The truncated SVD method is an appropriate method, but it is computationally expensive.

(2b) The truncated **QR** method [29] is an alternative to the truncated SVD method. As we observed in the above 3 approaches, the solution $\mathbf{x}_\alpha$ or $\mathbf{x}_k$ are given in the form of an expansion in terms of orthogonal vectors, as in (19.2.2) for example. Now, assume some expansion of the form

$$\mathbf{x} = \sum_{i=1}^{m} c_i \mathbf{y}_i$$

where the $\{\mathbf{y}_i\}$ is a set of m orthogonal vectors, the columns of some orthogonal matrix **Y**. The factors (c_i) are elements of an m-vector **c**, are to be calculated by solving a least squares problem. Let us write this equation as

$$\mathbf{x} = \mathbf{Y}\mathbf{c}$$

Varah [29] suggested the factorization **AY**= **QR**, where **Q** is an orthogonal matrix and **R** an upper triangular [22]. Then by solving the first k equations of

$$\mathbf{R}\mathbf{c} = \mathbf{Q}^T\mathbf{b}$$

we get the k-vector **c**. However, Varah pointed out that the singular values of **AY** are not always reflected in the size of the diagonal elements of **R**. That is, the truncated **QR** system may not be a good approximation to the system to **AY** whose rank is estimated by $k < m$.

Hansen [13] compared the truncated SVD method with the regularization method and defined necessary conditions for which the two methods give similar results.

19.3 Estimation of the free parameter

Each of the above methods involves a free parameter α for the regularization techniques, and the rank k for the truncated SVD and the **QR** methods. It is extremely difficult to calculate the value of this free parameter that results in a "good" approximate solution to the "true" one. It is a trade-off between accuracy and smoothness of the solution. Among the methods for calculating this free parameter are the cross-validation, the discrepancy principle and the L-curve methods. See also Section 19.5.

For the cross-validation method, for choosing α, see Wahba [32] and for the truncated SVD method when data are noisy, see Vogel [31]. The last method is based on statistical assumptions about the data error.

For the discrepancy principle method, see Morozov [20]. It is based on the coupling of the regularization parameter α and the noise level in the given data vector **b**. The L-curve method [14, 15] seems to strike a balance between the size of the solution $||\mathbf{x}_\alpha||$ and the corresponding residual norm $||\mathbf{A}\mathbf{x}_\alpha - \mathbf{b}||$.

In our work [2, 3], we use an alternative to the truncated SVD method. The ill-posed system $\mathbf{Ax} = \mathbf{b}$ is first replaced by an equivalent consistent system of linear equations. The algorithm calculates the minimal length least squares solution of the consistent system. Starting from rank $k = 1$ of the consistent system, the rank k is increased by one in succession and a new solution is calculated. This is repeated until a simple criterion is satisfied. Then the obtained minimal length least squares solution would be the approximate (smooth) solution of the system $\mathbf{Ax} = \mathbf{b}$.

Our method may also be viewed as the analog of the so-called stepwise regression, described by Albert ([8] Section 4.4). Linear programming techniques are used for which the successive solutions of the consistent system are themselves the basic solutions in the successive simplex tableaux. The algorithm is numerically stable. Results show that this method gives comparable accuracy to the truncated SVD method. Yet it is 2 to 5 times faster [2].

19.4 Description of the new algorithm

Let the system $\mathbf{Ax} = \mathbf{b}$ be pre-multiplied by $\mathbf{A}^T$. One gets the consistent system of linear equations

(19.4.1) $$\mathbf{A}^T\mathbf{Ax} = \mathbf{A}^T\mathbf{b}$$

Let (19.4.1) be written as

(19.4.1a) $$\mathbf{C}_{(m)}\mathbf{x}^{(m)} = \mathbf{c}_{(m)}$$

where $\mathbf{C}_{(m)} = \mathbf{A}^T\mathbf{A}$ and $\mathbf{c}_{(m)} = \mathbf{A}^T\mathbf{b}$. If rank($\mathbf{A}$) = k, $k \le m$, (19.4.1a) would have k linearly independent equations and (m – k) linearly dependent ones.

Assume that the equations in (19.4.1a) are properly permuted and that the first k equations, $k \le m$ are linearly independent. Let $\mathbf{C}_{(k)}$ denotes the first k rows of matrix $\mathbf{C}_{(m)}$ and $\mathbf{c}_{(k)}$ denotes the first k elements of vector $\mathbf{c}_{(m)}$. Then the first k equations in (19.4.1a) are

(19.4.2) $$\mathbf{C}_{(k)}\mathbf{x}^{(k)} = \mathbf{c}_{(k)}$$

System (19.4.2) is an underdetermined system of rank k. The pseudo-inverse of $\mathbf{C}_{(k)}$ is given by (Chapter 17)

$$[\mathbf{C}_{(k)}]^+ = [\mathbf{C}_{(k)}]^T([\mathbf{C}_{(k)}][\mathbf{C}_{(k)}]^T)^{-1}$$

and its minimal length least squares solution is

(19.4.3) $$\mathbf{x}^{(k)} = [\mathbf{C}_{(k)}]^+\mathbf{c}_{(k)}$$

19.4.1 Steps of the algorithm

(1) First we show that the least squares solution of $\mathbf{Ax} = \mathbf{b}$ is itself the least squares solution of $\mathbf{A}^T\mathbf{Ax} = \mathbf{A}^T\mathbf{b}$ of (19.4.1) (Lemma 19.1).

(2) We then show that matrix $\mathbf{C}_{(k)}$ of (19.4.2) is a good approximation to matrix $(\mathbf{A}^T\mathbf{A})$, which is assumed of rank k, $k \le m$ (Lemma 19.2). As a result, $\mathbf{x}^{(k)}$ of (19.4.3) is the least squares solution of $\mathbf{A}^T\mathbf{Ax} = \mathbf{A}^T\mathbf{b}$, which is the least squares solution of $\mathbf{Ax} = \mathbf{b}$ (Lemma 19.3).

(3) We measure the error between matrix $(\mathbf{A}^T\mathbf{A})^2$ and the matrix that approximates it (Lemma 19.4). As a result, we illustrate that $|\text{pivot}(i)|^{1/4}$ is of the order of magnitude of the singular

value $s_i(\mathbf{A})$, where pivot(i) is the pivot element in the ith simplex step (Lemma 19.5).

(4) Meanwhile, for estimating the rank value k, we need to show that the norm of residual $(\mathbf{b} - \mathbf{Ax})$ decreases monotonically as k increases (Lemma 19.6).

Lemma 19.1

The least squares solution of $\mathbf{Ax} = \mathbf{b}$ is itself the least squares solution of $\mathbf{A}^T\mathbf{Ax} = \mathbf{A}^T\mathbf{b}$.

Proof:

The proof follows directly from the SVD of matrix $\mathbf{A}$, namely $\mathbf{A} = \mathbf{VSW}^T$ and its pseudo-inverse, $\mathbf{A}^+ = \mathbf{WS}^{-1}\mathbf{V}^T$. The solution of $\mathbf{Ax} = \mathbf{b}$ is $\mathbf{x}_{(1)} = \mathbf{A}^+\mathbf{b}$ and the solution of $\mathbf{A}^T\mathbf{Ax} = \mathbf{A}^T\mathbf{b}$ is $\mathbf{x}_{(2)} = (\mathbf{A}^T\mathbf{A})^+(\mathbf{A}^T\mathbf{b})$. By substituting the expressions of $\mathbf{A}$ and $\mathbf{A}^+$ in the two equations, $\mathbf{x}_{(1)} = \mathbf{x}_{(2)}$ and the lemma is proved.

Lemma 19.2

The following relations between the singular values of $(\mathbf{A}^T\mathbf{A})^2$ and $(\mathbf{C}_k[\mathbf{C}_k]^T)$ exist

$$s_k(\mathbf{A}^T\mathbf{A})^2 \le s_k(\mathbf{C}_k[\mathbf{C}_k]^T) \le s_{k-1}(\mathbf{A}^T\mathbf{A})^2$$

Proof:

Each of $(\mathbf{A}^T\mathbf{A})^2$ and $(\mathbf{C}_{(k)}[\mathbf{C}_{(k)}]^T)$ is Hermitian (symmetric), positive semi-definite (Theorem 4.1). Their eigenvalues are themselves their singular values. The lemma is proved from the fact that $(\mathbf{C}_{(k)}[\mathbf{C}_{(k)}]^T)$ is the k by k leading principal sub-matrix of matrix $(\mathbf{A}^T\mathbf{A})^2$ ([25], pp. 317). Note that in [25], the eigenvalues of the Hermitian matrices are ordered in an increasing manner while in our case the singular values are ordered in a decreasing manner.

This lemma means that if $s_{k+1}(\mathbf{A}^T\mathbf{A})^2$ and the smaller singular values of $(\mathbf{A}^T\mathbf{A})^2$ are very small and considered 0's, $(\mathbf{C}_k[\mathbf{C}_k]^T)$ would be good approximation of $(\mathbf{A}^T\mathbf{A})^2$, meaning $\mathbf{C}_k$ is a good approximation of $(\mathbf{A}^T\mathbf{A})$.

Lemma 19.3

The minimal length least squares solution $\mathbf{x}^{(k)}$ of the system of k equations $\mathbf{C}_{(k)}\mathbf{x}^{(k)} = \mathbf{c}_{(k)}$ of (19.4.2), is itself the least squares solution of $\mathbf{C}_{(m)}\mathbf{x}^{(m)} = \mathbf{c}_{(m)}$, assuming that rank($\mathbf{A}$) = k.

Proof:

Since $\mathbf{C}_{(m)} = (\mathbf{A}^T\mathbf{A})$ is assumed of rank k, it has k linearly independent rows and its last (m – k) rows are linearly dependent on the first k rows. The same is said for vector $\mathbf{c}_{(m)}$. We partition each of $\mathbf{C}_{(m)}$ and $\mathbf{c}_{(m)}$ and we may write $\mathbf{Cx}_{(m)} = \mathbf{c}_{(m)}$ as

$$(\mathbf{I}_k/\mathbf{P})\mathbf{C}_{(k)}\mathbf{x} = (\mathbf{I}_k/\mathbf{P})\mathbf{c}_{(k)}$$

where $\mathbf{I}_k$ is a k-unit matrix and $\mathbf{P}$ is an (m – k) by k matrix. See for example, Noble ([21], pp. 144, 145). The minimal length least squares solution of this equation is

$$\mathbf{x} = [\mathbf{C}_{(k)}]^+(\mathbf{I}_k/\mathbf{P})^+(\mathbf{I}_k/\mathbf{P})\mathbf{c}_{(k)} = [\mathbf{C}_{(k)}]\mathbf{c}_{(k)} = \mathbf{x}^{(k)}$$

since $(\mathbf{I}_k/\mathbf{P})$ is of full rank and $(\mathbf{I}_k/\mathbf{P})^+(\mathbf{I}_k/\mathbf{P}) = \mathbf{I}_k$. This proves the lemma.

This lemma shows that system $\mathbf{C}_{(k)}\mathbf{x}^{(k)}=\mathbf{c}_{(k)}$ is equivalent to $\mathbf{A}^T\mathbf{Ax} = \mathbf{A}^T\mathbf{b}$ in the sense that the two systems have the same minimal length least squares solution, assuming rank($\mathbf{A}$) = k.

Our algorithm begins as follows. We calculate $\mathbf{x}^{(k)}$ for k = 1 of system (19.4.2). That is, system $\mathbf{C}_{(k)}\mathbf{x}^{(k)} = \mathbf{c}_{(k)}$ consists of only one equation. We then increase the rank k by 1 at a time so that $\mathbf{C}_{(k)}\mathbf{x}^{(k)} = \mathbf{c}_{(k)}$ consists of 2, 3, …, equations in succession. In each case the solution $\mathbf{x}^{(k)}$ is calculated. This is repeated until a certain simple criterion is satisfied. Linear programming techniques are used for which the successive solution vectors $\mathbf{x}^{(k)}$, for k = 1, 2, …, appear as the basic solutions in the successive simplex tableaux. The criterion for choosing the rank value k is described in Section 19.5.

Before every simplex iteration, the program pivots on the diagonal elements of matrix $-([\mathbf{C}_{(m)}][\mathbf{C}_{(m)}]^T) = -(\mathbf{A}^T\mathbf{A})^2$ and its updates in the simplex tableaux (see Tableau (a) in Section 19.6). That is, complete pivoting is used before each simplex iteration. Since this matrix is symmetric semi-definite, the algorithm is numerically stable.

It is easy to illustrate that the pivot elements in the Gauss-Jordan steps are themselves the diagonal element of matrix $\mathbf{D}$ in the Cholesky's factorization $-(\mathbf{A}^T\mathbf{A})^2 = -\mathbf{LDL}^T$, where $\mathbf{L}$ is a lower triangular matrix with unit diagonals and $\mathbf{D}$ is diagonal, assuming that rank($\mathbf{A}$) = m ([10], pp. 146, 147). Let us illustrate this point by following example. Let a matrix $\mathbf{C}$ be decomposed by Cholesky's factorization into $\mathbf{C} = \mathbf{LDL}^T$ as follows

$$\mathbf{C} = \begin{bmatrix} 1 & 2 & 1 & 1 \\ 2 & 8 & 4 & 4 \\ 1 & 4 & 11 & 5 \\ 1 & 4 & 5 & 19 \end{bmatrix} = \begin{bmatrix} 1 & & & \\ 2 & 1 & & \\ 1 & 0.5 & 1 & \\ 1 & 0.5 & 0.33 & 1 \end{bmatrix} \begin{bmatrix} 1 & & & \\ & 4 & & \\ & & 9 & \\ & & & 16 \end{bmatrix} \begin{bmatrix} 1 & 2 & 1 & 1 \\ & 1 & 0.5 & 0.5 \\ & & 1 & 0.33 \\ & & & 1 \end{bmatrix}$$

This matrix is a modification of that of ([11], p. 26). Let us now perform 4 Gauss-Jordan steps to matrix $\mathbf{C}$ (without pivoting). It is easy to show that the pivot elements are going to be (1, 4, 9, 16), which are the diagonal elements of $\mathbf{D}$.

This idea would allow us to obtain an upper bound to $s_{k+1}[\mathbf{C}_{(k+1)}][\mathbf{C}_{(k+1)}]^T$, had we stopped the simplex method after step (k + 1) instead of step k.

Consider now matrix $[\mathbf{C}_{(m)}][\mathbf{C}_{(m)}]^T = (\mathbf{A}^T\mathbf{A})^2$ and assume that we decompose it by Cholesky's factorization with complete pivoting. We get

$$(\mathbf{A}^T\mathbf{A})^2 = \mathbf{LDL}^T + \mathbf{B} \tag{19.4.4}$$

$\mathbf{L}$ is a lower m by k trapezoidal matrix with unit diagonal elements and $\mathbf{D}$ is k-diagonal matrix of diagonal elements $d_1 \geq d_2 \geq \ldots \geq d_k > 0$. The sum of the singular values of $\mathbf{LDL}^T$ is the trace (sum of diagonal elements) of $\mathbf{LDL}^T$ or of $\mathbf{D}^{1/2}\mathbf{L}^T\mathbf{L}\mathbf{D}^{1/2}$ (Section 4.2.6). The factorization is stopped after step k, when there is no significant addition to the trace ([10], p. 146); in other words, when d_{k+1} is not significant. The next lemma gives a measure of error matrix $\mathbf{B}$.

Lemma 19.4

$$||\mathbf{B}||_2 \leq (m - k)\, d_{k+1} \tag{19.4.5}$$

Proof:

In (19.4.4), each of $(\mathbf{A}^T\mathbf{A})^2$ and $\mathbf{LDL}^T$ is symmetric positive semi-definite of the same dimensions and their eigenvalues are their singular values. We have ([25], p. 315)

$$s_k(\mathbf{A}^T\mathbf{A})^2 \leq s_k(\mathbf{LDL}^T) + ||\mathbf{B}||_2$$

Yet

$$||\mathbf{B}||_2 \le ||\mathbf{B}||_E \le (m - k)d_{k+1}$$

where $||.||_E$ denotes the Euclidean matrix norm, and the lemma is proved.

In obtaining the upper bound of $||\mathbf{B}||_E$, we assumed that the absolute value of each element in the right lower $(m - k)$ square submatrix $\mathbf{B}$ is d_{k+1}. That is unrealistic; a more realistic factor is $(m - k)^{1/2}$, and (19.4.5) would be replaced by

$$||\mathbf{B}||_2 \le (m - k)^{1/2}d_{k+1} \quad (19.4.6)$$

This lemma shows that the diagonal elements (d_i) of $\mathbf{D}$, $i = 1, 2, \ldots$, in the factorization $(\mathbf{A}^T\mathbf{A})^2 \approx \mathbf{LDL}^T$, are good measures of the size of $s_i(\mathbf{A}^T\mathbf{A})^2$. Or that the pivot elements in the Gauss-Jordan elimination steps are good measures of the size of $s_i(\mathbf{A}^T\mathbf{A})^2$. Thus we shall replace the parameter d_{k+1} in the above inequality by $|\text{pivot}(k + 1)|$, where pivot(i) is the ith pivot element in the simplex iteration. We get

$$||\mathbf{B}||_E \le (m - k)^{1/2}|\text{pivot}(k + 1)|$$

Lemma 19.5

The value $|\text{pivot}(i)|^{1/4}$ is of the order of magnitude of $s_i(\mathbf{A})$.

Proof:

From Lemma 19.2, $s_{k+1}(\mathbf{C}_{k+1}[\mathbf{C}_{k+1}]^T) \le s_k(\mathbf{A}^T\mathbf{A})^2$ and we have $s_k(\mathbf{A}) = (s_k(\mathbf{A}^T\mathbf{A})^2)^{1/4}$.

By taking the fourth power of $(m - k)^{1/2}|\text{pivot}(k + 1)|$ in (19.4.6), we get $(m - k)^{1/8}|\text{pivot}(k)|^{1/4}$, $k = 1, 2, \ldots$, a measure of $s_k(\mathbf{A})$.

Consider now the factor $(m - k)^{1/8}$. For all practical purposes, we have $(m - k)^{1/8} < 2$. For example, for large m, say $m = 200$ and $k = 30$, $(m - k)^{1/8} \approx 1.9$ and for smaller m, say $m = 15$ and $k = 3$, we get $(m - k)^{1/8} \approx 1.36$. The lemma is thus proved.

Lemma 19.6

Let $\mathbf{x}^{(k)}$ and $\mathbf{x}^{(k+1)}$ be the respective minimal length least squares solutions of the systems $\mathbf{C}_{(k)}\mathbf{x}^{(k)} = \mathbf{c}_{(k)}$ and $\mathbf{C}_{(k+1)}\mathbf{x}^{(k+1)} = \mathbf{c}_{(k+1)}$ of (19.4.2). Then $||\mathbf{r}^{(k+1)}|| \le ||\mathbf{r}^{(k)}||$, where $\mathbf{r}^{(k+1)}$ and $\mathbf{r}^{(k)}$ are the residuals

$(\mathbf{b} - \mathbf{A}\mathbf{x}^{(k+1)})$ and $(\mathbf{b} - \mathbf{A}\mathbf{x}^{(k)})$ respectively. That is, the residual of $\mathbf{A}\mathbf{x}^{(k)} = \mathbf{b}$ decreases monotonically as k increases.

19.5 Optimum value of the rank

For the estimation of the rank k of matrix **A**, which gives best or near-best solution to system $\mathbf{A}\mathbf{x} = \mathbf{b}$, we adopt a simple criterion similar to that used by Hanson [16] and also by Squire [24]. It is based on Lemma 19.6 that the Euclidean norm of the residual vector of equation $\mathbf{A}\mathbf{x} = \mathbf{b}$; $\|\mathbf{b} - \mathbf{A}\mathbf{x}^{(k)}\|$ decreases monotonically as k increases.

For each solution vector $\mathbf{x}^{(k)}$, $k = 1, 2, \ldots$, the program calculates the residual norm $\|\mathbf{b} - \mathbf{A}\mathbf{x}^{(k)}\|$. The program exits when this residual norm is less than a specified tolerance TOLER, or if the magnitude of the pivot element for the next linear programming tableau is less than a machine-dependent tolerance EPS.

19.5.1 The parameters TOLER and EPS

The parameter TOLER is estimated as follows. If the right hand side vector **b** in system $\mathbf{A}\mathbf{x} = \mathbf{b}$ is expected to be contaminated by an error vector $\delta\mathbf{b}$, then an appropriate value of the parameter TOLER is of the order of the Euclidean norm $\|\delta\mathbf{b}\|$. If $\delta\mathbf{b} = 0$, a suitable value is TOLER = 1.0E–04 or 1.0E–03, the justification for which is supported only by numerical experimentation.

As described in Chapter 2, EPS, on the other hand, is a machine-dependent tolerance such that a calculated parameter z is considered 0 if $|z| < \text{EPS}$. EPS is set to 10^{-4} for single-precision and 10^{-11} for double-precision (see Section 4.7.1).

We might be able to improve the accuracy of the solution vector **x** by making a better estimate of the parameter EPS. This depends on the order of magnitude of the pivot elements in the simplex tableaux. In such cases we might let EPS take smaller values. We examine the absolute values of the pivot elements in the simplex tableaux that are say < 1.0E–03. We take EPS as the mean of the absolute values of two consecutive pivot elements where there is a large separation between them. See numerical experimentation for a similar problem ([6], p. 1418) and also ([7], p. 374).

In example 2 in [2], the absolute values of the pivot elements of the simplex tableaux, accurate to 3 decimal places, are 0.304E+04, 0.242E+01, 0.148E–03, 0.272E–06, 0.107E–08, … . There is a large separation between the third and the fourth values. We take EPS as the mean of these 2 values; 0.741E–04. This gives the calculated $\text{rank}(\mathbf{A}^T\mathbf{A})^2 = 3$ and $\|\mathbf{x}(\text{exact}) - \mathbf{x}(\text{calculated})\| = 0.299$ which is a slight improvement to the result given in [2], which is 0.377 for $\text{rank}(A) = 5$. This example is solved as Example 19.1, using a different discretization method to the Fredholm integral equation of the first kind.

19.6 Use of linear programming techniques

The method described here uses linear programming techniques. It achieves the result that the calculated successive minimal length least squares solutions are themselves the basic solutions in the successive simplex tableaux. It is a hybrid of a technique by the author for a different problem [1]. See also Chapter 23.

Assume that we have the set of the 2m underdetermined linear equations, where **v**, **x** and **y** are m unknown vectors.

$$[\mathbf{C}_{(m)}]^T\mathbf{v} + \mathbf{I}_m\mathbf{x} + \mathbf{0} = \mathbf{0} \tag{19.6.1a}$$

$$-\mathbf{C}_{(m)}[\mathbf{C}_{(m)}]^T\mathbf{v} + \mathbf{0} + \mathbf{I}_m\mathbf{y} = \mathbf{c}_{(m)} \tag{19.6.1b}$$

$\mathbf{C}_{(m)}$ and $\mathbf{c}_{(m)}$ are those of (19.4.1a) and $\mathbf{I}_m$ is an m-unit matrix. A solution to system (19.6.1) is $\mathbf{v} = \mathbf{x} = \mathbf{0}$ and $\mathbf{y} = \mathbf{c}_{(m)}$. Let us write down this system in a simplex tableau format [1, 12].

Tableau (a)

B	$\mathbf{b}_B$	$\mathbf{v}^T$	$\mathbf{x}^T$	$\mathbf{y}^T$
x	$\mathbf{0}$	$[\mathbf{C}_{(m)}]^T$	$\mathbf{I}_m$	$\mathbf{0}$
y	$\mathbf{c}_{(m)}$	$-\mathbf{C}_{(m)}[\mathbf{C}_{(m)}]^T$	$\mathbf{0}$	$\mathbf{I}_m$

In Tableau (a), **B** and $\mathbf{b}_B$ denote respectively the initial basis matrix and the initial basic solution. In linear programming terminology, the elements of vectors **x** and **y** are basic variables.

Assume that $\text{rank}(\mathbf{C}_{(m)}) = m$. Then $\mathbf{C}_{(m)}[\mathbf{C}_{(m)}]^T$ is nonsingular.

Let now the elements of vector $\mathbf{v}$ replace the corresponding elements of vector $\mathbf{y}$ as basic variables. This gives Tableau (b).

Tableau (b)

B	$\mathbf{b}_B$	$\mathbf{v}^T$	$\mathbf{x}^T$	$\mathbf{y}^T$
$\mathbf{x}$	$[\mathbf{C}_{(m)}]^+\mathbf{c}_{(m)}$	$\mathbf{0}$	$\mathbf{I}_m$	$[\mathbf{C}_{(m)}]^+$
$\mathbf{v}$	$-(\mathbf{C}_{(m)}[\mathbf{C}_{(m)}]^T)^{-1}\mathbf{c}_{(m)}$	$\mathbf{I}_m$	$\mathbf{0}$	$-(\mathbf{C}_{(m)}[\mathbf{C}_{(m)}]^T)^{-1}$

In effect, Tableau (b) is obtained by premultiplying Tableau (a) by matrix $\mathbf{E}$, where

$$\mathbf{E} = \begin{bmatrix} \mathbf{I}_m & [\mathbf{C}_{(m)}]^T \\ 0 & -\mathbf{C}_{(m)}[\mathbf{C}_{(m)}]^T \end{bmatrix}^{-1} = \begin{bmatrix} \mathbf{I}_m & [\mathbf{C}_{(m)}]^+ \\ 0 & -(\mathbf{C}_{(m)}[\mathbf{C}_{(m)}]^T)^{-1} \end{bmatrix}$$

However, in practice, Tableau (b) is obtained by applying m Gauss-Jordan elimination steps to Tableau (a) and its updates. Since under $\mathbf{v}^T$ in Tableau (a), $-\mathbf{C}_{(m)}[\mathbf{C}_{(m)}]^T$ is symmetric semi-definite, before each Gauss-Jordan elimination step, we pivot over the diagonal elements of $-\mathbf{C}_{(m)}[\mathbf{C}_{(m)}]^T$ in Tableau (a) and its updates, in the successive tableaux between Tableaux (a) and (b). Under the basic solution $\mathbf{b}_B$ in Tableau (b), we find that $\mathbf{x} = [\mathbf{C}_{(m)}]^+\mathbf{c}_{(m)}$, which is the required minimal length least squares solution (19.4.3) of the system $\mathbf{Ax} = \mathbf{b}$, for $k = m$.

However, if $\text{rank}(\mathbf{C}_{(m)}) = k < m$, the absolute value of the pivot element in the $(k + 1)^{th}$ elimination step is assumed very small < EPS and is replaced by 0. The columns of $[\mathbf{C}_{(m)}]^T$ under $\mathbf{v}^T$ in Tableau (a), used in the calculation so far, are matrix $[\mathbf{C}_{(k)}]^T$. The obtained basic solution in this case is $\mathbf{x}^{(k)} = [\mathbf{C}_{(k)}]^+\mathbf{c}_{(k)}$, which is the approximate least squares solution of system $\mathbf{Ax} = \mathbf{b}$.

Let us denote the successive tableaux between Tableau (a) and Tableau (b), as Tableau (1), Tableau (2), …, Tableau (k). Tableau (0) is Tableau (a) and Tableau (k) is Tableau (b). Then the columns under $\mathbf{y}^T$ in Tableau (p), p = 1, 2, …, is $[\mathbf{C}_{(p)}]^+$ and the first m elements of $\mathbf{b}_B$ is $\mathbf{x}^{(p)} = [\mathbf{C}_{(p)}]^+\mathbf{c}_{(p)}$. In other words, in Tableaux (1), (2), …, (k), the calculated basic solutions are the repeated solutions

$\mathbf{x}^{(1)}, \mathbf{x}^{(2)}, \ldots, \mathbf{x}^{(k)}$ respectively. The calculation may also be done in a smaller (condensed) tableaux, as we did in [1].

19.7 Numerical results and comments

In this section, we show that our algorithm obtains a smooth solution to the Fredholm integral equation of the first kind and also may be applied in solving other ill-conditioned linear systems.

LA_Mls() implements the Minimum norm Least Squares algorithm. DR_Mls() tests two examples, the results of which were computed in double-precision, as follows.

Example 19.1

Solve the following Fredholm integral equation of the first kind

$$\int_0^1 (x^2 + y^2)^{0.5} f(y)dy = [(1 + x^2)^{1.5} - x^3]/3$$

This example was solved by Baker ([9], pp. 664-667) and also by Baker et al. [10] and in [2, 3]. The exact solution is f(x) = x.

Table 19.1

f(x)	x(exact)	x(calculated)	error
0.0000	0.0000	–0.0066	0.0066
0.0714	0.0714	0.1249	–0.0535
0.1429	0.1429	0.1135	0.0294
0.2143	0.2143	0.1628	0.0515
0.2857	0.2857	0.2590	0.0267
0.3571	0.3571	0.3690	–0.0119
0.4286	0.4286	0.4720	–0.0434
0.5000	0.5000	0.5585	–0.0585
0.5714	0.5714	0.6259	–0.0544
0.6429	0.6429	0.6749	–0.0321
0.7143	0.7143	0.7080	0.0063
0.7857	0.7857	0.7279	0.0578
0.8571	0.8571	0.7372	0.1200
0.9286	0.9286	0.7382	0.1904
1.0000	1.0000	0.7329	0.2671

In discretizing this equation, we use here the rectangular rule rather than the Chebyshev rule, as in [2, 3]. We used the parameters n = 15, m = 15, EPS = 10^{-11}, TOLER = 10^{-4}. The calculated rank(**A**) = 5.

The results are shown in Table 19.1.

Residual vector (**b** – **Ax**) =

(–0.00001515, 0.00009018, –0.00008601, –0.00007357, 0.00000561, 0.00005994, 0.00006877, 0.00004385, 0.00000466, –0.00003197, –0.00005458, –0.00005681, –0.00003611, 0.00000753, 0.00007266)^T

$$||\mathbf{b} - \mathbf{Ax}|| = 0.21293071\text{E}{-03}$$

$$||\mathbf{x}(\text{exact}) - \mathbf{x}(\text{calculated})|| = 0.37677$$

The pivots in the simplex tableaux to 8 decimal places are

(–3039.73005711, –2.41835358, –0.00014840, –0.00000027, 0.00000000)

$|\text{pivot}|^{1/4}$ =

(7.42521023, 1.24703875, 0.11037181, 0.02283408, 0.00571954)

The pivots and $|\text{pivot}|^{1/4}$ shown here differ slightly from those in ([2], p. 108) because here we use the rectangular rule in the discretization of the given equation while in [2], we used the Chebyshev rule. It is well-known (Baker [9]) that different quadrature rules for discretizing a given integral equation give results with different accuracies. In ([2], p. 107), where the discretization of this example is done by the Chebyshev rule, slightly better results are obtained and the estimated rank(**A**) = 3.

In ([2], p. 108) we displayed the results of this problem together with the results by the truncated SVD method. The calculated values of the elements of **x** were almost the same for both methods. However, our method was 4 times faster than the truncated SVD method.

Example 19.2

Obtain the least squares solution of the ill-conditioned system **Ax** = **b**, where matrix **A** (a scaled version of Hilbert matrix) and vector **b** are given such that the exact solution is **x** = (1, 1, 1, 1, 1, 1)^T. We have n = 7 and m = 6. The results are shown in Table 19.2.

Table 19.2

x(exact)	x(calculated)	error
1.0000	0.9999	0.0001
1.0000	1.0016	–0.0016
1.0000	0.9950	0.0050
1.0000	1.0032	–0.0032
1.0000	1.0044	–0.0044
1.0000	0.9959	0.0041

Residual vector ($\mathbf{b} - \mathbf{A}\mathbf{x}$) =
(0.00000001, –0.00000016, 0.00000046, –0.00000019, –0.00000036, 0.00000001, 0.00000025)^T

$$||\mathbf{b} - \mathbf{A}\mathbf{x}|| = 0.6798635653\text{E}{-06}$$

$$||\mathbf{x}(\text{exact}) - \mathbf{x}(\text{calculated})|| = 0.008639$$

Matrix **A** and vector **b** are given in ([3], p. 292) and they are not multiplied by 1.0E05. Their elements should be of reasonable sizes.

The pivots in the simplex tableaux to 8 decimal places are

(–673.99536685, –0.28826497, –0.00001107, 0.00000000)

$|\text{pivot}|^{1/4}$ = (5.09523510, 0.73273674, 0.05767962, 0.00366241)

For these results, we used EPS = 10^{-11} and TOLER = 10^{-4}. This example was also solved in ([22], p. 315).

We now illustrate that $|\text{pivot(i)}|^{1/4}$ in the simplex tableau are of the order of magnitude of the elements of $s_i(\mathbf{A})$. In example 2 in ([2], p. 108), the values ($s_i(\mathbf{A})$) and ($|\text{pivot(i)}|^{1/4}$) were compared, where rank(**A**) was found to be 3

$$(s_i(\mathbf{A})) = (12.16, 1.435, 0.100)$$

which is compared with

$$(|\text{pivot(i)}|^{1/4}) = (7.275, 1.132, 0.088)$$

Compare also $|\text{pivot(i)}|^{1/4}$ and $s_i(\mathbf{A})$ in example 1 in ([2], p. 107) and the numerical example given in Baker et al. ([10], p. 147).

This makes our method a good alternative to the truncated singular value decomposition method. It is also concluded that matrix $\mathbf{C}_{(k)}$ in $\mathbf{C}_{(k)}\mathbf{x} = \mathbf{c}_{(k)}$ of (19.4.2) is a (good) approximation to matrix

$\mathbf{C}_{(m)} = (\mathbf{A}^T\mathbf{A})$, where it is assumed that rank($\mathbf{A}$) = k.

For this example, the results here agree with those of Peters and Wilkinson ([22], p. 315), where TOLER was set to 10^{-4} and rank($\mathbf{A}$) = 4. We solved this example also for EPS = 10^{-11} and TOLER = 10^{-3}. The results were less accurate, giving the estimated rank($\mathbf{A}$) = 3.

Finally, one observes that the matrix from which we chose the pivot elements in the linear programming problem is matrix $-\mathbf{C}_{(m)}[\mathbf{C}_{(m)}]^T$ and its updates in the simplex tableaux. Thus if **A** has large elements, the tableaux will contain very large numbers. Also, if **A** has many elements of very small sizes, this may result in some very small pivot elements of the order of EPS. A proper scaling of system **Ax** = **b** before calling LA_Mls() would solve this problem.

In conclusion, the singular values of matrix **A** give an accurate representation of the condition of matrix **A**. In our method, however, the absolute values of the pivot elements in the successive simplex tableaux, raised to the power (1/4), reflect the condition of matrix **A** in the system **Ax** = **b**, (Lemma 19.5).

In the simplex tableaux, the number of arithmetic operations for each iteration is of the order of m^2 multiplications/divisions. The computation time is approximately proportional to $m^2(n + k/2)$, where k is the estimated rank of matrix **A**. The actual code of our subroutine (in FORTRAN IV, not including comment lines) consists of less than 100 lines.

References

1. Abdelmalek, N.N., Minimum energy problem for discrete linear admissible control systems, *International Journal of Systems Science*, 10(1979)77-88.
2. Abdelmalek, N.N., An algorithm for the solution of ill-posed linear systems arising from the discretization of Fredholm integral equation of the first kind, *Journal of Mathematical Analysis and Applications*, 97(1983)95-111.
3. Abdelmalek, N.N., A program for the solution of ill-posed linear systems arising from the discretization of Fredholm

integral equation of the first kind, *Computer Physics Communications*, 58(1990)285-292.
4. Abdelmalek, N.N., Kasvand, T. and Croteau, J.P., Image restoration for space invariant pointspread functions, *Applied Optics*, 19(1980)1184-1189.
5. Abdelmalek, N.N., Kasvand, T., Olmstead, J. and Tremblay, M.M., Direct algorithm for digital image restoration, *Applied Optics*, 20(1981)4227-4233.
6. Abdelmalek, N.N. and Otsu, N., Restoration of images with missing high-frequency components by minimizing the L_1 norm of the solution vector, *Applied Optics*, 24(1985)1415-1420.
7. Abdelmalek, N.N. and Otsu, N., Speed comparison among methods for restoring signals with missing high-frequency components using two different low-pass-filter matrix dimensions, *Optics Letters*, 10(1985)372-374.
8. Albert, A., *Regression and More-Penrose Pseudoinverse*, Academic Press, New York, 1972.
9. Baker, C.T.H., *The Numerical Treatment of Integral Equations*, Oxford University Press, Clarendon, Oxford, 1977.
10. Baker, C.T.H., Fox, L., Mayers, D.F. and Wright, K., Numerical solution of Fredholm integral equations of the first kind, *Computer Journal*, 7(1964)141-148.
11. Fisher, M.E., *Introductory Numerical Methods with the NAG Software Library*, The University of Western Australia, Crawley, 1988.
12. Hadley, G., *Linear Programming*, Addison-Wesley, Reading, MA, 1962.
13. Hansen, P.C., The truncated SVD as a method for regularization, *BIT*, 27(1987)534-553.
14. Hansen, P.C., Analysis of discrete ill-posed problems by means of the L-curve, *SIAM Review*, 34(1992)561-580.
15. Hansen, P.C. and O'Leary, D., The use of the L-curve in the regularization of discrete ill-posed problems, *SIAM Journal on Scientific Computation*, 14(1993)1487-1502.
16. Hanson, R.J., A numerical method for solving Fredholm integral equations of the first kind using singular values, *SIAM Journal on Numerical Analysis*, 8(1971)616-622.

17. Hanson, R.J. and Phillips, J.L., An adaptive numerical method for solving linear Fredholm integral equations of the first kind, *Numerische Mathematik*, 24(1975)291-307.
18. Kammerer, W.J. and Nashed, M.Z., On the convergence of the conjugate gradient method for singular linear equations, *SIAM Journal on Numerical Analysis*, 8(1971)65-101.
19. Kammerer, W.J. and Nashed, M.Z., Iterative methods for best approximate solution of linear integral equations of the first and second kinds, *Journal of Mathematical Analysis and Applications*, 40(1972)547-573.
20. Morozov, V.A., *Methods for Solving Incorrectly Posed Problems*, Springer-Verlag, New York, 1984.
21. Noble, B., *Applied Linear Algebra*, Prentice-Hall, Englewood Cliffs, NJ, 1969.
22. Peters, G. and Wilkinson, J.H., The least squares problem and pseudo-inverses, *Computer Journal*, 13(1970)309-316.
23. Phillips, D.L., A technique for the numerical solution of certain integral equations of the first kind, *Journal of ACM*, 9(1962)84-97.
24. Squire, W., The solution of ill-conditioned linear systems arising from Fredholm equations of the first kind by steepest descents and conjugate gradients, *International Journal for Numerical Methods in Engineering*, 10(1976)607-617.
25. Stewart, G.W., *Introduction to Matrix Computations*, Academic Press, New York, 1973.
26. te Riele, H.J.J., A program for solving first kind Fredholm integral equations by means of regularization, *Computer Physics Communications*, 36(1985)423-432.
27. Tikhonov, A.N., Solution of incorrectly formulated problems and method of regularization, *Soviet Mathematics*, 4(1963)1035-1038.
28. Varah, J.M., On the numerical solution of ill-conditioned linear systems with applications to ill-posed problems, *SIAM Journal on Numerical Analysis*, 10(1973)257-267.
29. Varah, J.M., A practical examination of some numerical methods for linear discrete ill-posed problems, *SIAM Review*, 21(1979)100-111.

30. Varah, J.M., Pitfalls in the numerical solution of linear ill-posed problems, *SIAM Journal on Scientific Statistical Computation*, 4(1983)164-176.
31. Vogel, C.R., Optimal choice of a truncation level for the truncated SVD solution of linear first kind integral equations when data are noisy, *SIAM Journal on Numerical Analysis*, 23(1986)109-134.
32. Wahba, G., Practical approximate solutions to linear operator equations when the data are noisy, *SIAM Journal on Numerical Analysis*, 14(1977)651-667.

19.8 DR_Mls

```
/*--------------------------------------------------------------------
DR_Mls
----------------------------------------------------------------------
This program is a driver for the function LA_Mls(), which calculates
a smooth solution vector "x" for an ill-posed / ill-conditioned
system of linear equations

                        a*x = b

"a" is a real n by m matrix of rank k, k <= m <= n.
"b" is a real n vector.

Remark 1:
    It is recommended to solve this program in double precision only.
    That is due to the ill-conditioning of the derived system of
    equations a*x = b.

This  program  contains  2 examples.

Example 1:
    This example is for getting the least squares solution of
    Fredholm integral equation of the first kind.
    For this example the real solution is y = x.
    The integration is done by the rectangular rule.

Example 2:
    This example is for getting the least squares solution of an
    ill-conditioned system of linear equations.

Each example is solved twice, once for the parameter
"toler" = 1.0e-04 and once for "toler" = 1.0e-03
--------------------------------------------------------------------*/

#include "DR_Defs.h"
#include "LA_Prototypes.h"

#define TOLER1      1.0E-04
#define TOLER2      1.0E-03

#define NN1_ROWS    60
#define MM1_COLS    40
```

```
#define N1m        15
#define M1m        15
#define N2m         7
#define M2m         6

void DR_Mls (void)
{
    /*-----------------------------------------
      Constant matrices/vectors
    -----------------------------------------*/
    static tNumber_R a2init[N2m][M2m] =
    {
        { 3.6036,  1.8018,  1.2012,  0.9009,  0.72072, 0.6006  },
        { 1.8018,  1.2012,  0.9009,  0.72072, 0.6006,  0.5148  },
        { 1.2012,  0.9009,  0.72072, 0.6006,  0.5148,  0.45045 },
        { 0.9009,  0.72072, 0.6006,  0.5148,  0.45045, 0.4004  },
        { 0.72072, 0.6006,  0.5148,  0.45045, 0.4004,  0.36036 },
        { 0.6006,  0.5148,  0.45045, 0.4004,  0.36036, 0.3276  },
        { 0.5148,  0.45045, 0.4004,  0.36036, 0.3276,  0.3003  }
    };

    static tNumber_R b2[N2m+1] =
    {   NIL,
        8.82882, 5.74002, 4.38867, 3.58787, 3.04733, 2.65421, 2.35391
    };

    /*-----------------------------------------
      Variable matrices/vectors
    -----------------------------------------*/
    tMatrix_R  a       = alloc_Matrix_R (NN1_ROWS, MM1_COLS);
    tVector_R  b       = alloc_Vector_R (NN1_ROWS);
    tVector_R  x       = alloc_Vector_R (MM1_COLS);
    tVector_R  xp      = alloc_Vector_R (M1m + 1);
    tVector_R  yp      = alloc_Vector_R (M1m + 1);
    tVector_R  xexact  = alloc_Vector_R (M1m + 1);
    tVector_R  xerror  = alloc_Vector_R (M1m + 1);
    tVector_R  r       = alloc_Vector_R (N1m + 1);

    tMatrix_R  a2      = init_Matrix_R (&(a2init[0][0]), N2m, M2m);

    int        krun, irank;
    int        i, j, k, m, n, Iexmpl;
    tNumber_R  pi, dx, dy, e, e1, s, sum, toler;
```

```
    eLaRc        rc = LaRcOk;

    pi = 4.0*atan (1.0);
    dx = 2.*pi/99;

    pi = 4.0*atan (1.0);

    prn_dr_bnr ("DR_Mls, "
                "Solution of Ill-posed Systems of Equations");

    for (krun = 1; krun <= 2; krun++)
    {
        if (krun == 1) toler = TOLER1;
        if (krun == 2) toler = TOLER2;

        PRN ("Test run output # = %d, toler = %8.4f\n", krun, toler);

        for (Iexmpl = 1; Iexmpl <= 2; Iexmpl++)
        {
            switch (Iexmpl)
            {
            case 1:
                n = N1m;
                m = M1m;
                dx = 1.0 / (tNumber_R)(n - 1);
                for (i = 1; i <= n; i++)
                {
                    xp[i] = dx * ((tNumber_R)(i - 1));
                }
                dy = 1.0 / (tNumber_R)(m - 1);
                for (j = 1;  j <= m; j++)
                {
                    yp[j] = dy * (tNumber_R)(j - 1);
                }
                for (i = 1; i <= n; i++)
                {
                    e = xp[i];
                    xexact[i] = e;
                    e1 = (pow ((1.0 + e*e), 1.5) - e*e*e) / 3.0;
                    b[i] = e1/dx;
                    for (j = 1; j <= m; j++)
                    {
                        a[i][j] = pow ((xp[i]*xp[i] + yp[j]*yp[j]),
                                       0.50);
                    }
```

```
        }
        break;

    case 2:
        n = N2m;
        m = M2m;
        for (i = 1; i <= n; i++)
        {
            b[i] = b2[i];
            xexact[i] = 1.0;
            for (j = 1; j <= m; j++)
            {
                a[i][j] = a2[i][j];
            }
        }
        break;
    default:
        break;
    }

    prn_algo_bnr ("Mls");

    prn_example_delim();
    PRN ("Example #%d: Size of coefficient matrix "
        "%d by %d\n", Iexmpl, n, m);
    prn_example_delim();
    if (Iexmpl == 1)
    {
        PRN ("Least Squares Solution of a Fredholm "
             "Integral Equation of the First Kind\n");
        prn_example_delim();
        PRN ("Exact solution is f(x) = x\n");
    }
    if (Iexmpl == 2)
    {
        PRN ("Least Squares Solution of an Ill-conditioned "
             "System of Linear Equations\n");
        prn_example_delim();
        PRN ("r.h.s. Vector \"b\"\n");
        prn_Vector_R (b, n);
        PRN ("Coefficient Matrix, \"a\"\n");
        prn_Matrix_R (a, n, m);
    }
    PRN ("Parameters \"toler\" = %11.4f, \"EPS\" "
        "= %13.11f\n", toler, EPS);
```

```
rc = LA_Mls (m, n, a, b, toler, x, &irank);

if (rc >= LaRcOk)
{
    PRN ("Calculated rank of the system of equations "
         " = %d\n", irank);
    PRN ("Calculated solution vector, \"x\"\n");
    prn_Vector_R (x, m);
    PRN ("Exact solution vector, \"xexact\"\n");
    prn_Vector_R (xexact, m);
    sum = 0.0;
    for (i = 1; i <= m; i++)
    {
        xerror[i] = xexact[i] - x[i];
        sum = sum + xerror[i] * (xerror[i]);
    }
    sum = pow (sum, 0.5);

    if (Iexmpl == 1)
    {
        PRN (" Coordinate         x(exact)       "
             "x(calculated)      xerror\n");
        for (i = 1; i <= m; i++)
        {
            PRN (" %8.4f,         %8.4f,"
                 "        %8.4f,       %8.4f\n",
                 xp[i], xexact[i], x[i], xerror[i]);
        }
    }
    if (Iexmpl == 2)
    {
        PRN ("  x(exact)       x(calculated)        "
             "xerror\n");
        for (i = 1; i <= m; i++)
        {
            PRN (" %8.4f,         %8.4f,         %8.4f\n",
                 xexact[i], x[i], xerror[i]);
        }
    }

    sum = 0.0;
    for (i = 1; i <= m; i++)
    {
        xerror[i] = xexact[i] - x[i];
```

```
                    sum = sum + xerror[i] * (xerror[i]);
                }
                sum = pow (sum, 0.5);
                PRN ("L2 Norm ||x(exact) - x(calculated)|| "
                    "= %17.10f\n", sum);
                sum = 0.0;
                for (i = 1; i <= n; i++)
                {
                    s = b[i];
                    for (k = 1; k <= m; k++)
                    {
                        s = s - a[i][k] * (x[k]);
                    }
                    r[i] = s;
                    sum = sum + s*(s);
                }
                sum = pow (sum, 0.5);
                PRN ("Residual vector, (b-a*x)\n");
                prn_Vector_R_exp (r, m);
                PRN ("L2 Norm ||b-a*x|| = %17.10f\n", sum);
            }

            prn_la_rc (rc);
        }
    }

    free_Matrix_R (a, NN1_ROWS);
    free_Vector_R (b);
    free_Vector_R (x);
    free_Vector_R (xp);
    free_Vector_R (yp);
    free_Vector_R (xexact);
    free_Vector_R (xerror);
    free_Vector_R (r);

    uninit_Matrix_R (a2);
}
```

19.9 LA_Mls

```
/*-----------------------------------------------------------------
LA_Mls
-------------------------------------------------------------------
Given is the system of linear equations

                         a*x = b

"a" is a given real n by m matrix of rank k, k <= m <= n.
"b" is a given real n vector.

The system a*x = b may result from the discretization of "Fredholm
integral equation of the first kind".  In this case, it is known that
a*x = b is ill-posed; a small perturbation in the r.h.s. vector "b"
results in a large change to the solution vector "x". An ill-posed
system is also ill-conditioned.

The problem is to calculate a smooth solution m vector "x" to the
(ill-posed / ill-conditioned) system a*x = b.

The program pre-multiplies the equation a*x = b by a(transpose).
One gets the system of linear equations

                         c*x = f

The m by m matrix c = a(transpose)*a and the m vector f =
a(transpose)*b.

The program then, using linear programming techniques calculates the
minimum length least squares solution x(k) of k equations of the
system c*x=f, where k = 1, 2, ... .

Meanwhile after each simplex iteration (solution) the Euclidean norm
||b-a*x|| is calculated.  Computation stops when this norm is less
than a given tolerance "toler" or when the absolute value of the
pivot element for the next simplex iteration is less than "EPS", a
computer-dependent parameter.

The obtained solution vector x[k] is itself the smooth solution "x"
of the system a*x = b.

Inputs
m       Number of columns of matrix "a" of the system a*x = b.
```

```
n       Number of rows of matrix "a" of a*x = b.
a       An n by m real matrix "a" of the system a*x = b.
b       An n real vector of the system a*x = b.  Both matrix "a"
        and vector "b" are not destroyed in the computation.
toler   A real specified tolerance; Euclidian ||b - a*x|| < toler.

Local Variables
c       An (m + m) by n real matrix.
f       An m real vector.

Outputs
irank   The estimated rank of matrix "c" of c*x = f that results
        in a best solution "x" for the system a*x = b.  Also,
        "irank" is the number of simplex iterations in the linear
        programming problem.
x       Am m real vector containing the minimum length least squares
        solution to the "irank" equations of the system c*x = f.
        "x" is the smooth solution to the system a*x = b.
diag    A real m vector containing the pivot elements in the "irank"
        simplex iterations.
ic      An m integer vector whose first "irank" elements are the
        indices of the "irank" linearly independent equations of
        the system c*x = f.

Returns one of
        LaRcSolutionFound
        LaRcErrBounds
        LaRcErrNullPtr
        LaRcErrAlloc
-------------------------------------------------------------------*/

#include "LA_Prototypes.h"

eLaRc LA_Mls (int m, int n, tMatrix_R a, tVector_R b,
    tNumber_R toler, tVector_R x, int *pIrank)
{
    tVector_R   f       = alloc_Vector_R (m);
    tMatrix_R   c       = alloc_Matrix_R (m + m, m);
    tVector_I   ic      = alloc_Vector_I (m);
    tVector_R   diag    = alloc_Vector_R (m);

    int         i = 0, j = 0, k = 0, kd = 0, kdp = 0, mpi = 0,
                mpj = 0, mpm = 0;
    int         iout = 0, mout = 0;
    tNumber_R   d = 0.0, piv = 0.0, pivot = 0.0, s = 0.0, sum = 0.0;
```

```
/* Validation of the data before executing the algorithm */
eLaRc        rc = LaRcSolutionFound;
VALIDATE_BOUNDS ((0 < m) && (0 < n) && !((n == 1) && (m == 1)));
VALIDATE_PTRS   (a && b && toler && x && pIrank);
VALIDATE_ALLOC  (f && c && ic && diag);

/* Initialization */
mpm = m + m;
*pIrank = m;
for (j = 1; j <= m; j++)
{
    ic[j] = j;
    diag[j] = 0.0;
    x[j] = 0.0;
    s = 0.0;
    for (k = 1; k <= n; k++)
    {
        s = s + b[k] * (a[k][j]);
    }
    f[j] = s;
}
for (j = 1; j <= m; j++)
{
    for (i = j; i <= m; i++)
    {
        sum = 0.0;
        for (k = 1; k <= n; k++)
        {
            sum = sum + a[k][i] * (a[k][j]);
        }
        c[j][i] = sum;
        c[i][j] = sum;
    }
}

for (j = 1; j <= m; j++)
{
    mpj = m + j;
    for (i = j; i <= m; i++)
    {
        mpi = m + i;
        s = 0.0;
        for (k = 1; k <= m; k++)
        {
```

```
                s = s + c[k][i] * (c[k][j]);
            }
            c[mpj][i] = - s;
            c[mpi][j] = - s;
        }
    }

    for (iout = 1; iout <= m; iout++)
    {
        mout = m + iout;
        piv = 0.0;
        /* Pivoting along the diagonal elements of matrix "c" */
        for (j = iout; j <= m; j++)
        {
            mpj = m + j;
            d = fabs (c[mpj][j]);
            if (d > piv)
            {
                kd = j;
                kdp = mpj;
                piv = d;
            }
        }

        if (piv < EPS)
        {
            *pIrank = iout - 1;
            GOTO_CLEANUP_RC (LaRcSolutionFound);
        }

        if (kdp != mout)
        {
            /* Swap two rows of matrix "c" */
            swap_rows_Matrix_R (c, kdp, mout);

            /* Swap of two elements of vector "f" */
            swap_elems_Vector_R (f, kd, iout);

            /* Swap two columns of matrix "c" */
            for (i = 1; i <= mpm; i++)
            {
                d = c[i][iout];
                c[i][iout] = c[i][kd];
                c[i][kd] = d;
            }
```

```
        swap_elems_Vector_I (ic, kd, iout);
    }

    /* A Gauss-Jorden elimination step */
    pivot = c[mout][iout];
    diag[iout] = pivot;
    for (j = 1; j <= m; j++)
    {
        c[mout][j] = c[mout][j]/pivot;
    }
    f[iout] = f[iout]/pivot;
    for (i = 1; i <= mpm; i++)
    {
        if (i != mout)
        {
            d = c[i][iout];
            for (j = 1; j <= m; j++)
            {
                if (j != iout)
                {
                    c[i][j] = c[i][j] - d * (c[mout][j]);
                }
            }
            c[i][iout] = - c[i][iout]/pivot;
            if (i > m)
            {
                k = i - m;
                f[k] = f[k] - d * (f[iout]);
            }
            else if (i <= m)
            {
                x[i] = x[i] - d * (f[iout]);
            }
        }
    }
    c[mout][iout] = 1.0/pivot;
    sum = 0.0;
    for (i = 1; i <= n; i++)
    {
        s = b[i];
        for (k = 1;  k <= m; k++)
        {
            s = s - a[i][k] * (x[k]);
        }
        sum = sum + s*s;
```

```
        }
        sum = sqrt (sum);
        if (sum < toler)
        {
            *pIrank = iout;
            break;
        }
    }

CLEANUP:

    free_Vector_R (f);
    free_Matrix_R (c, m + m);
    free_Vector_I (ic);
    free_Vector_R (diag);

    return rc;
}
```

PART 5

Solution of Underdetermined Systems Of Linear Equations

Chapter 20

L_1 Solution of Underdetermined Linear Equations

20.1 Introduction

This is the first of four chapters on the solution of underdetermined consistent systems of linear equations subject to certain types of constraints. For consistent underdetermined systems of linear equations, the residual vector is zero. Hence, the constraints are on the solution vector of the system, not on the residual vector. In this chapter, the constraint is that the L_1 norm of the solution vector be as small as possible. Consider the underdetermined system of linear equations

$$\mathbf{Ca} = \mathbf{f}$$

$\mathbf{C} = (c_{ij})$ is a given real n by m matrix of rank k, $k \leq n < m$, and $\mathbf{f} = (f_i)$ is a given real n-vector. It is required to calculate a solution m-vector $\mathbf{a} = (a_j)$ for this system.

It is known that system $\mathbf{Ca} = \mathbf{f}$ has a solution if and only if $\text{rank}(\mathbf{C}|\mathbf{f}) = \text{rank}(\mathbf{C})$. If $\text{rank}(\mathbf{C}|\mathbf{f}) > \text{rank}(\mathbf{C})$, the system is inconsistent and it has no solution (Theorem 4.13). We shall assume throughout our work that the system $\mathbf{Ca} = \mathbf{f}$ is consistent.

Also, because the number of equations is less than the number of unknowns, system $\mathbf{Ca} = \mathbf{f}$ has an infinite number of solutions. In this problem, among these infinite solutions, we seek a solution vector $\mathbf{a}$ whose L_1 norm

$$z = \|\mathbf{a}\|_1 = \sum_{i=1}^{m} |a_i| \qquad (20.1.1)$$

is as small as possible.

Using basic theorems of functional analysis, Cadzow [4] developed algorithmic procedures for solving both the problem of the minimum L_1 solution and the problem of the minimum L_∞ solution (Chapter 22), of the underdetermined system $\mathbf{Ca} = \mathbf{f}$.

Kolev [7] presented an iterative algorithm to solve both these problems based on the steepest descent method for constrained optimization. His algorithm results in a procedure of theoretically infinite number of iterations. Kolev [8] then developed his algorithm such that it required a finite number of iterations. In both of Kolev's methods, the minimum energy solution (Chapter 23) is used to start the iterations. A third algorithm was presented by Kolev [9] based on the simplex method of linear programming [6], but unfortunately, this method was not properly developed. See the comments in [2].

In this context, Dax [5] presented a method for calculating the L_p norm of the solution vector of the consistent system $\mathbf{Ca} = \mathbf{f}$, where $1 < p < \infty$. He presented a primal Newton method for $p > 2$ and a dual Newton method for $1 < p < 2$.

In our work, this problem is formulated as a linear programming problem [2] for which a modified simplex algorithm is described. In this algorithm minimum computer storage is required, where an initial basic feasible solution is available with no artificial variables needed. This method applies to full rank as well as rank deficient cases i.e., when $\text{rank}(\mathbf{C}|\mathbf{f}) = \text{rank}(\mathbf{C}) < n$. Our method is initiated by the method that we described for the Chebyshev solution of overdetermined systems of linear equations [1].

In Section 20.2, the linear programming formulation of the problem is given. In Section 20.3, the algorithm is described and in Section 20.4, numerical results and comments are given.

20.1.1 Applications of the algorithm

In control theory this problem is known as the **Minimum fuel problem for discrete linear control systems** [11, 9, 3]. The parameter a_i represents the rate of fuel consumption in time interval i and thus z in (20.1.1) represents the total fuel consumption per unit time.

20.2 Linear programming formulation of the problem

This problem is reduced to a linear programming problem as follows

$$\text{minimize } z = \sum_{j=1}^{m} |a_j|$$

subject to

$$\mathbf{Ca} = \mathbf{f}$$

$$a_j,\ j = 1, 2, \ldots, m, \text{ unrestricted in sign}$$

Since the variables a_j are unrestricted in sign, we may write

$$a_j = v_j - w_j$$

Hence

$$|a_j| = v_j + w_j$$

$$v_j \geq 0 \text{ and } w_j \geq 0, \quad j = 1, 2, \ldots, m$$

Let the m-vectors $\mathbf{v} = (v_j)$ and $\mathbf{w} = (w_j)$. Hence, the problem now reduces to

(20.2.1a) $$\text{minimize } z = \sum_{j=1}^{m} v_j + \sum_{j=1}^{m} w_j$$

(20.2.1b) $$\begin{bmatrix} \mathbf{C} & -\mathbf{C} \end{bmatrix} \begin{bmatrix} \mathbf{v} \\ \mathbf{w} \end{bmatrix} = \mathbf{f}$$

(20.2.1c) $$v_j \geq 0 \text{ and } w_j \geq 0, \quad j = 1, 2, \ldots, m$$

Let us denote the matrix on the l.h.s. of (20.2.1b) by $\mathbf{D}$ and let the vectors $\mathbf{v}$ and $\mathbf{w}$ be augmented to form the 2m-vector $\mathbf{b}$. Problem (20.2.1) may be rewritten as

(20.2.2a) $$\text{minimize } z = \mathbf{e}_{2m}{}^T\mathbf{b}$$

subject to

(20.2.2b) $$\mathbf{Db} = \mathbf{f}$$

(20.2.2c) $b_j \geq 0, \ \ j = 1, 2, \ldots, 2m$

where $\mathbf{e}_{2m}$ is a 2m-vector each element of which is 1.

Problem (20.2.2) is a linear programming problem of n constraints in 2m variables. Assume without loss of generality that rank($\mathbf{C}$) = n. Let the basis matrix for this problem be denoted by the n-square matrix $\mathbf{B}$. Let us, as usual, construct the simplex tableau for problem (20.2.2) by calculating the vectors ($\mathbf{y}_j$)

(20.2.3) $\mathbf{y}_j = \mathbf{B}^{-1}\mathbf{D}_j, \ \ j = 1, 2, \ldots, 2m$

where $\mathbf{D}_j$ is the jth column of matrix $\mathbf{D}$.

From (20.2.2a), the marginal costs are given by $(z_j - 1)$

(20.2.4) $z_j - 1 = \mathbf{e}_n^T\mathbf{y}_j - 1, \ \ j = 1, 2, \ldots, 2m$

where $\mathbf{e}_n$ is an n-vector each element of which is 1. Let $\mathbf{b}_B$ denote the basic solution to problem (20.2.2)

$$\mathbf{b}_B = \mathbf{B}^{-1}\mathbf{f}$$

The objective function denoted by z is given by

$$z = \mathbf{e}_n^T\mathbf{b}_B$$

20.2.1 Properties of the matrix of constraints

The analysis of this problem is initiated by the asymmetry matrix $\mathbf{D}$ of (20.2.2b) has. For any column j, $1 \leq j \leq m$, we have

(20.2.5) $\mathbf{D}_j = \mathbf{C}_j$ and $\mathbf{D}_{j+m} = -\mathbf{C}_j, \ \ j = 1, 2, \ldots, m$

where again $\mathbf{D}_j$ and $\mathbf{C}_j$ are the jth columns of matrix $\mathbf{D}$ and matrix $\mathbf{C}$ respectively. This asymmetry enables us to use a condensed simplex tableau of n constraints in only m variables.

Definition

Because of this asymmetry in matrix $\mathbf{D}$, we define any column j, $1 \leq j \leq m$, of $\mathbf{D}$ and the column (j + m) as two corresponding columns. Consider the following lemmas [2].

Lemma 20.1

Any two corresponding columns could not appear together in any basis. Otherwise, the basis matrix would be singular.

Lemma 20.2

Let i and j be any two corresponding columns in the simplex tableau, i.e., $|i - j| = m$, $1 \le i, j \le 2m$. Then from (20.2.3-5)

$$\mathbf{y}_i + \mathbf{y}_j = \mathbf{0}$$

$$(z_i - 1) + (z_j - 1) = -2$$

Lemma 20.3

Assume that we have obtained an infeasible basic solution $\mathbf{b}_B$, to the linear programming problem (20.2.2); that is, one or more elements of $\mathbf{b}_B$ is < 0. Let $\mathbf{b}_{Bs}$, $1 \le s \le n$, be one of such elements. Let i be the column associated with $\mathbf{b}_{Bs}$. Let j be the corresponding column to column i. Then if column j replaces column i in the basis, the new basic solution has the new $\mathbf{b}_{Bs} = -\mathbf{b}_{Bs}$ and all the other elements of $\mathbf{b}_B$ are unchanged.

The proof of this lemma is similar to the proof of Lemma 10.6.

It is easily seen, in this case, that the simplex tableau is changed as follows. Row s of the tableau and the element $\mathbf{b}_{Bs}$ are multiplied by –1, while the rest of the tableau as well as the other elements of $\mathbf{b}_B$ are left unchanged.

Lemma 20.4

The elements of the solution vector **a** to the system $\mathbf{Ca} = \mathbf{f}$ are given in terms of the elements of the optimal basic solution $\mathbf{b}_B$ to problem (20.2.2) as follows. For any element $\mathbf{b}_{Bs}$, s = 1, 2, …, n, of the optimal basic solution, let j be the column in the simplex tableau associated with $\mathbf{b}_{Bs}$. Then

$$a_j = \mathbf{b}_{Bs}, \quad \text{if} \quad 1 \le j \le m$$

$$a_{j-m} = -\mathbf{b}_{Bs}, \quad \text{if instead } m < j \le 2m$$

The remaining (m – n) elements of **a** are zero elements and thus the L_1 norm z of (20.1.1) is the sum of the elements of $\mathbf{b}_B$, being all non-negative.

Proof:

The proof follows directly from $a_j = v_j - w_j$ and $\mathbf{b} = [\mathbf{v}^T\ \mathbf{w}^T]^T$. Also, since the non-basic elements of vector **b** are 0's, the remaining elements of vector **a** are 0's and the objective function z = the sum of the elements of the basic solution $\mathbf{b}_B$.

Note 1

If rank($\mathbf{C}|\mathbf{f}$) = rank($\mathbf{C}$) = k < n, then (m – k) elements of **a** are zero elements instead.

Lemma 20.5

If the system of equation $\mathbf{Ca} = \mathbf{f}$ is not consistent, the linear programming solution would have one or more zero rows in the simplex tableau but the corresponding $\mathbf{b}_B$ elements would be nonzero. The solution of the programming problem is not feasible.

20.3 Description of the algorithm

From Lemma 20.2, the corresponding columns $\mathbf{y}_i$ and $\mathbf{y}_j$ are the negative of one another, $|i-j| = m$. Also, the marginal costs of the corresponding columns are related to one another. Hence, we need to construct simplex tableaux for problem (20.2.2), for n constraints in only m variables, and we call this the **condensed tableaux**. We start with the first half of matrix **D** of (20.2.2b).

In part 1 of this algorithm, we obtain a basic solution, feasible or not, without needing any artificial variables. This is simply done by performing a finite number of Gauss-Jordan elimination steps to the initial data, with partial pivoting. We choose the pivot as the largest element in absolute value in row i in tableau (i – 1), i = 1, 2, …, n.

If rank($\mathbf{C}|\mathbf{f}$) = rank($\mathbf{C}$) = k < n, this is indicated by the presence of 0 rows and corresponding 0 $\mathbf{b}_B$ elements in the simplex tableaux. These rows are deleted from the following tableaux.

If one or more elements of the basic solution is < 0, the basic solution is not feasible. For each of these elements $\mathbf{b}_{Bs}$, we apply Lemma 20.3, which implies replacing column i associated with $\mathbf{b}_{Bs}$ by its corresponding column.

If $\mathbf{Ca} = \mathbf{f}$ is inconsistent, this is detected, as explained by Lemma 20.5. In this case the calculation is terminated as the solution is not feasible, with an indication that the system is inconsistent.

We end part 1 by calculating the marginal costs in the condensed tableau using (20.2.4).

Part 2 of the algorithm is the ordinary simplex method. The choice of the non-basic column that enters the basis would be the column with the largest positive marginal cost of the non-basic columns and their corresponding columns. That is, by using Lemma 20.2 for calculating the marginal costs of the corresponding columns. The solution vector **a** of system **Ca** = **f** and the optimum norm z of (20.1.1) are obtained from Lemma 20.4.

The steps described here are explained by the following detailed numerical example. This example is a simplified version of the example solved by Cadzow ([3], pp. 489, 490) and by Kolev ([7], p. 783).

Example 20.1

Obtain the minimum L_1 norm of the solution vector **a** of the following underdetermined system

$$\begin{aligned} 2a_1 - a_2 - 4a_3 - 3a_4 + a_5 &= 2 \\ 2a_1 - a_2 - 4a_3 - 3a_4 + a_5 &= 2 \\ 5a_1 + a_2 + 3a_3 - 2a_4 + 0 &= 1 \end{aligned} \tag{20.3.1}$$

Here, matrix **C** is a 3 by 5 matrix of rank 2 and system (20.3.1) is consistent. The second equation in (20.3.1) is the same as the first one. The Initial Data and the condensed simplex tableaux are shown. The pivot in each tableau is bracketed.

Initial Data

B	b_B	D_1	D_2	D_3	D_4	D_5
	2	2	–1	(–4)	–3	1
	2	2	–1	–4	–3	1
	1	5	1	3	–2	0

Tableau 20.3.1 (part 1)

B	b_B	D_1	D_2	D_3	D_4	D_5
D_3	–0.5	–0.5	0.25	1	0.75	–0.25
	0	0	0	0	0	0
	2.5	(6.5)	0.25	0	–4.25	0.75

Tableau 20.3.2

B	b_B	D_1	D_2	D_3	D_4	D_5
D_3	–0.308	0	0.270	1	0.423	–0.192
D_1	0.385	1	0.039	0	–0.654	0.115

Tableau 20.3.2*

B	b_B	D_1	D_2	D_8	D_4	D_5
D_8	0.308	0	–0.270	1	–0.423	0.192
D_1	0.385	1	0.039	0	–0.654	0.115
		0	–1.231	0	–2.077	–0.693

Tableau 20.3.3 (part 2)

B	b_B	D_1	D_2	D_8	D_9	D_5
D_8	0.308	0	–0.270	1	0.423	0.192
D_{10}	0.385	1	0.039	0	(0.654)	0.115
		0	–1.231	0	0.077	–0.693

Tableau 20.3.4

B	b_B	D_1	D_2	D_8	D_9	D_5
D_8	0.059	–0.647	–0.295	1	0	–0.266
D_{10}	0.588	1.529	0.060	0	1	0.176
	z = 0.647	–0.118	–1.235	0	0	–1.090

Note that the prices are the coefficients of vector **b** in (20.2.2a), which are the elements of $\mathbf{e}_{2m}$, which are all 1's, so we need not write these prices on the top of each tableau.

In Tableau 20.3.1, the whole of the second row consists of zero element, indicating rank deficiency. This row is deleted from the following tableaux. Tableau 20.3.2 gives a basic solution that is not feasible, since $b_{B1} < 0$.

In Tableau 20.3.2*, $\mathbf{y}_8$ replaces its corresponding column $\mathbf{y}_3$ (which is associated with b_{B1}) in the basis. In effect, the simplex tableau is left unchanged except for the first row and the first element in $\mathbf{b}_B$. They are multiplied by –1. The element y_{18} is corrected to +1. This gives Tableau 20.3.2* in which the marginal costs $(z_j - 1)$ are also calculated from (20.2.4).

In Tableau 20.3.2*, $\mathbf{y}_9$, the corresponding column to $\mathbf{y}_4$, has the largest positive marginal cost. We exchange $\mathbf{y}_4$ by $\mathbf{y}_9$ (Tableau 20.3.3) and perform a Gauss-Jordan elimination step. This gives Tableau 20.3.4, which is optimal.

From Lemma 20.4, the solutions **a** and z of (20.1.1) are obtained, $\mathbf{a} = (0, 0, -0.059, -0.588, 0)^T$ and $z = (0.059 + 0.588) = 0.647$.

Since rank(**C**) = 2, only 2 elements of **a** are nonzeros and the other 3 are zeros (Lemma 20.4). Finally, since $\mathbf{b}_B$ is not degenerate and none of the marginal costs for the non-basic columns is zero, **a** is unique ([6], p. 166).

20.4 Numerical results and comments

LA_Fuel() implements the minimum fuel algorithm, and DR_Fuel() demonstrates 7 test cases.

Table 20.4.1 shows the results, calculated in single-precision.

Table 20.4.1

Example	C(n×m)	Iterations	z
1	4×5	2	2.12
2	4×8	7	2.055
3	2×3	2	4.975
4	4×6	4	1.359
5	4×100	13	10.930
6	5×100	19	14.778
7	7×100	22	20.464

For each example, the size of matrix **C**, the number of iterations and the optimum L_1 norm z are given.

This method can be characterized as follows. All the calculations are done in the condensed simplex tableaux. The inverse of the basis matrix, i.e., matrix $\mathbf{B}^{-1}$, is never calculated. The elements of the solution vector **a** and the optimum norm z are obtained directly from the optimal solution $\mathbf{b}_B$.

References

1. Abdelmalek, N.N., Chebyshev solution of overdetermined systems of linear equations, *BIT*, 15(1975)117-129.
2. Abdelmalek, N.N., A simplex algorithm for minimum fuel problems of linear discrete control systems, *International Journal of Control*, 26(1977)635-642.
3. Cadzow, J.A., Functional analysis and the optimal control of linear discrete systems, *International Journal of Control*, 17(1973)481-495.
4. Canon, M.D., Cullum Jr., C.D. and Polak, E., *Theory of Optimal Control and Mathematical Programming*, Mcgraw-Hill, New York, 1970.
5. Dax, A., Methods for calculating *l*p-minimum norm solutions of consistent linear systems, *Journal of Optimization Theory and Applications*, 83(1994)333-354.
6. Hadley, G., *Linear Programming*, Addison-Wesley, Reading, MA, 1962.
7. Kolev, L.V., Iterative algorithm for the minimum fuel and minimum amplitude problems, *International Journal of Control*, 21(1975)779-784.
8. Kolev, L.V., Algorithm of finite number of iterations for the minimum fuel and minimum amplitude control problems, *International Journal of Control*, 22(1975)97-102.
9. Kolev, L.V., Minimum-fuel control of linear discrete systems, *International Journal of Control*, 23(1976)207-216.
10. Luenberger, D.G., *Optimization by Vector Space Methods*, John Wiley, New York, 1969.
11. Porter, W.A., *Modern Foundations of System Engineering*, Macmillan, New York, 1966.

20.5 DR_Fuel

```
/*-----------------------------------------------------------------------
DR_Fuel
-------------------------------------------------------------------------
This program is a driver for the function LA_Fuel(), which
calculates the L-One solution of an underdetermined system of
consistent linear equations.

Given is the underdetermined consistent system

                         c*a = f

"c" is a given real n by m matrix of rank k <= n <= m.
"f" is a given real n vector.

It is required to calculate the m vector "a" for this system such
that the L-One norm z of "a"

               z = |a[1]|+ |a[2]|+ ... + |a[m]|

is as small as possible.

In control theory, this problem is known as the "Minimum Fuel"
problem for linear discrete control systems.

This program has 7 examples whose results appear in the text.
-----------------------------------------------------------------------*/

#include "DR_Defs.h"
#include "LA_Prototypes.h"

#define N1e         4
#define M1e         5
#define N2e         4
#define M2e         8
#define N3e         2
#define M3e         3
#define N4e         4
#define M4e         6
#define N5e         4
#define M5e         100
#define N6e         5
#define M6e         100
```

```
#define N7e          7
#define M7e          100

void DR_Fuel (void)
{
    /*---------------------------------------
      Constant matrices/vectors
    ---------------------------------------*/
    static tNumber_R b1init[N1e][M1e] =
    {
        { 2.0,  1.0, -3.0,  5.0, 3.0  },
        { 2.0,  1.0, -3.0,  5.0, 3.0  },
        { 2.0,  1.0, -3.0,  5.0, 3.0  },
        {-8.0, -4.0,  2.0,  5.0, 6.0  }
    };

    static tNumber_R f1[N1e+1] =
    {   NIL,
        10.0, 10.0, 10.0, 8.0
    };

    static tNumber_R b2init[N2e][M2e] =
    {
        {-6.0,  4.0,  2.0, -8.0,  5.0, -1.0,  6.0,  3.0  },
        { 7.0, -3.0,  9.0,  5.0, -9.0,  8.0,  7.0, -4.0  },
        { 9.0,  7.0, -5.0,  2.0,  7.0,  0.0,  1.0,  8.0  },
        { 9.0, -3.0,  2.0,  4.0,  0.0,  3.0,  0.0,  1.0  }
    };

    static tNumber_R f2[N2e+1] =
    {   NIL,
        5.0, 7.0, -9.0, 1.0
    };

    static tNumber_R b3init[N3e][M3e] =
    {
        {-0.7182, -3.6706, -11.6961 },
        {-1.7182, -4.6706, -12.6961 }
    };

    static tNumber_R f3[N3e+1] =
    {   NIL,
        24.8218, 24.4218
    };
```

```
static tNumber_R b4init[N4e][M4e] =
{
    {2.0, -1.0, -4.0,  0.0, -3.0, 1.0 },
    {2.0, -1.0, -4.0,  0.0, -3.0, 1.0 },
    {5.0,  1.0,  3.0,  1.0, -2.0, 0.0 },
    {1.0, -2.0, -1.0, -5.0,  1.0, 4.0 }
};

static tNumber_R f4[N4e+1] =
{   NIL,
    2.0, 2.0, 1.0, -4.0
};

/*----------------------------------------
  Variable matrices/vectors
----------------------------------------*/
tMatrix_R   c       = alloc_Matrix_R (Ne_ROWS, Me_COLS);
tVector_R   f       = alloc_Vector_R (Ne_ROWS);
tVector_R   a       = alloc_Vector_R (Me_COLS);
tMatrix_R   fay     = alloc_Matrix_R (N7e, M7e);
tVector_R   fy      = alloc_Vector_R (N7e + 1);

tMatrix_R   b1      = init_Matrix_R (&(b1init[0][0]), N1e, M1e);
tMatrix_R   b2      = init_Matrix_R (&(b2init[0][0]), N2e, M2e);
tMatrix_R   b3      = init_Matrix_R (&(b3init[0][0]), N3e, M3e);
tMatrix_R   b4      = init_Matrix_R (&(b4init[0][0]), N4e, M4e);

int         irank, iter;
int         i, j, m, n, Iexmpl;
tNumber_R   pi, dx, x1, x2, x3, z;

eLaRc       rc = LaRcOk;

pi = 4.0*atan (1.0);
dx = 2.*pi/99;
for (j = 1; j <= 100; j++)
{
   x1 = (j - 1)* dx;
   x2 = x1 + x1;
   x3 = x2 + x1;
   fay[1][j] = 1.0;
   fay[2][j] = sin (x1);
   fay[3][j] = cos (x1);
   fay[4][j] = sin (x2);
   fay[5][j] = cos (x2);
```

```
      fay[6][j] = sin (x3);
      fay[7][j] = cos (x3);
   }
   for (i = 1; i <= 7; i++)
   {
      fy[i] = (10.0+15.0*fay[2][i]-7.0*fay[3][i]+9.0*fay[4][i]) *
              (1.0 + 0.01*fay[4][i]);
   }

   prn_dr_bnr ("DR_Fuel, L1 Solution of an Underdetermined System");

   for (Iexmpl = 1; Iexmpl <= 7; Iexmpl++)
   {
      switch (Iexmpl)
      {
         case 1:
            n = N1e;
            m = M1e;
            for (i = 1; i <= n; i++)
            {
               f[i] = f1[i];
               for (j = 1; j <= m; j++)
                  c[i][j] = b1[i][j];
            }
            break;

         case 2:
            n = N2e;
            m = M2e;
            for (i = 1; i <= n; i++)
            {
               f[i] = f2[i];
               for (j = 1; j <= m; j++)
                  c[i][j] = b2[i][j];
            }
            break;

         case 3:
            n = N3e;
            m = M3e;
            for (i = 1; i <= n; i++)
            {
               f[i] = f3[i];
               for (j = 1; j <= m; j++)
                  c[i][j] = b3[i][j];
```

```
            }
            break;
        case 4:
            n = N4e;
            m = M4e;
            for (i = 1; i <= n; i++)
            {
                f[i] = f4[i];
                for (j = 1; j <= m; j++)
                    c[i][j] = b4[i][j];
            }
            break;
        case 5:
            n = N5e;
            m = M5e;
            for (i = 1; i <= n; i++)
            {
                f[i] = fy[i];
                for (j = 1; j <= m; j++)
                    c[i][j] = fay[i][j];
            }
            break;
        case 6:
            n = N6e;
            m = M6e;
            for (i = 1; i <= n; i++)
            {
                f[i] = fy[i];
                for (j = 1; j <= m; j++)
                    c[i][j] = fay[i][j];
            }
            break;
        case 7:
            n = N7e;
            m = M7e;
            for (i = 1; i <= n; i++)
            {
                f[i] = fy[i];
                for (j = 1; j <= m; j++)
                    c[i][j] = fay[i][j];
            }
            break;
        default:
            break;
    }
}
```

```
        prn_algo_bnr ("Fuel");
        prn_example_delim();
        PRN ("Example #%d: Size of matrix \"c\" %d by %d\n",
             Iexmpl, n, m);
        prn_example_delim();
        PRN ("L1 Solution of an Underdetermined System\n");
        prn_example_delim();
        PRN ("r.h.s. Vector \"f\"\n");
        prn_Vector_R (f, n);
        PRN ("Coefficient Matrix, \"c\"\n");
        prn_Matrix_R (c, n, m);

        rc = LA_Fuel (m, n, c, f, &irank, &iter, a, &z);

        if (rc >= LaRcOk)
        {
            PRN ("\n");
            PRN ("Results of the Minimum Fuel Solution\n");
            PRN ("L1 solution vector \"a\"\n");
            prn_Vector_R (a, m);
            PRN ("L1 Norm of vector \"a\", ||a|| = %8.4f\n", z);
            PRN ("Rank of matrix \"c\" = %d, No. of Iterations "
                 "= %d\n", irank, iter);
        }

        LA_check_rank_def (n, irank);
        prn_la_rc (rc);
    }

    free_Matrix_R (c, Ne_ROWS);
    free_Vector_R (f);
    free_Vector_R (a);
    free_Matrix_R (fay, N7e);
    free_Vector_R (fy);

    uninit_Matrix_R (b1);
    uninit_Matrix_R (b2);
    uninit_Matrix_R (b3);
    uninit_Matrix_R (b4);
}
```

20.6 LA_Fuel

```
/*--------------------------------------------------------------------
LA_Fuel
----------------------------------------------------------------------
This program solves an underdetermined system of consistent linear
equations whose solution vector has a minimum L1 norm.
The underdetermined system is of the form

                            c*a = f

"c" is a given real n by m matrix of rank k, k <= n <= m.
"f" is a given real n vector.

It is required to calculate the m vector "a" for this system such
that the L-One norm "z" of vector "a"

              z = |a[1]| + |a[2]| + ... + |a[m]|

is as small as possible.

In control theory, this problem is known as the "Minimum Fuel"
problem for linear discrete control systems.

Inputs
m       Number of columns of matrix "c" in the system c*a = f.
n       Number of rows of matrix "c" in the system c*a = f.
c       A real n by m matrix containing matrix "c" of system c*a = f.
f       A real n vector containing the r.h.s. of the system c*a = f.

Outputs
irank   The calculated rank of matrix "c".
iter    The number of iterations or the number of times the simplex
        tableau is changed by a Gauss-Jordan step.
a       A real m vector whose elements are the solution of system
        c*a = f.
z       The minimum L-One norm of the solution vector "a".

Local Data
icbas   An m vector containing the indices of the columns of matrix
        "c" that form the basis matrix.

Returns one of
        LaRcSolutionUnique
```

```
        LaRcSolutionProbNotUnique
        LaRcNoFeasibleSolution
        LaRcInconsistentSystem
        LaRcErrBounds
        LaRcErrNullPtr
        LaRcErrAlloc
-------------------------------------------------------------------*/

#include "LA_Prototypes.h"

eLaRc LA_Fuel (int m, int n, tMatrix_R c, tVector_R f, int *pIrank,
    int *pIter, tVector_R a, tNumber_R *pZ)
{
    tVector_I   icbas   = alloc_Vector_I (n);
    tVector_I   irbas   = alloc_Vector_I (n);
    tVector_I   ibound  = alloc_Vector_I (m);
    tVector_R   zc      = alloc_Vector_R (m);

    int         itest = 0, i = 0, ij = 0, iout = 0, ivo = 0, j = 0,
                jin = 0, kl = 0;
    tNumber_R   d = 0.0;
    eLaRc       tempRc;

    /* Validation of data data before executing the algorithm */
    eLaRc       rc = LaRcSolutionUnique;
    VALIDATE_BOUNDS ((0 < n) && (n <= m) && !((n == 1) && (m == 1)));
    VALIDATE_PTRS   (c && f && pIrank && pIter && a && pZ);
    VALIDATE_ALLOC  (icbas && irbas && ibound && zc);

    /* Initialization */
    kl = 1;
    *pZ = 0.0;
    *pIter = 0;
    *pIrank = n;

    LA_fuel_init (m, n, icbas, irbas, ibound, a);

    iout= 0;

    /* Part 1 of the algorithm */
    for (ij = 1; ij <= n; ij++)
    {
        iout = iout + 1;
        tempRc = LA_fuel_part_1 (iout, &jin, &kl, m, n, c, f, icbas,
                                 irbas, pIrank, pIter);
```

```
        if (tempRc < LaRcOk)
        {
            GOTO_CLEANUP_RC (tempRc);
        }
    }

    for (i = kl; i <= n; i++)
    {
        if (f[i] <= -EPS)
        {
            jin = icbas[i];
            ibound[jin] = -ibound[jin];
            f[i] = -f[i];
            for (j = 1; j <= m; j++)
            {
                c[i][j] = -c[i][j];
            }
            c[i][jin] = 1.0;
        }
    }

    /* Part 2 of the algorithm.
    Calculating the marginal costs */
    LA_fuel_part_2 (kl, m, n, c, icbas, zc);

    /* Part 3 of the algorithm, */
    for (ij = 1; ij <= m*m; ij++)
    {
        ivo = 0;
        jin = 0;
        /* Determine the vector that enters the basis */
        LA_fuel_vent (&ivo, &jin, kl, m, n, icbas, zc);

        /* Calculate the results */
        if (ivo == 0)
        {
            rc = LA_fuel_res (kl, n, f, icbas, ibound, a, pZ);
            GOTO_CLEANUP_RC (rc);
        }

        if (ivo != 1)
        {
            for (i = kl; i <= n; i++)
            {
                c[i][jin] = -c[i][jin];
```

```
            }
            zc[jin] = -2.0 - zc[jin];
            ibound[jin] = -ibound[jin];
        }
        itest= 0;

        /* Determine the vector that leaves the basis */
        LA_fuel_leav (&itest, jin, &iout, kl, n, c, f);

        /* No feasible solution is possible */
        if (itest != 1)
        {
            GOTO_CLEANUP_RC (LaRcNoFeasibleSolution);
        }

        LA_fuel_gauss_jordn (iout, jin, kl, m, n, c, f, icbas);
        *pIter = *pIter + 1;

        d = zc[jin];
        for (j = 1; j <= m; j++)
        {
            zc[j] = zc[j] - d*c[iout][j];
        }
    }

CLEANUP:

    free_Vector_I (icbas);
    free_Vector_I (irbas);
    free_Vector_I (ibound);
    free_Vector_R (zc);

    return rc;
}

/*-----------------------------------------------------------------
Initialization of LA_fuel()
-----------------------------------------------------------------*/
void LA_fuel_init (int m, int n, tVector_I icbas, tVector_I irbas,
    tVector_I ibound, tVector_R a)
{
    int         i, j;

    for (j = 1; j <= m; j++)
    {
```

```
        a[j] = 0.0;
        ibound[j] = 1;
    }
    for (i = 1; i <= n; i++)
    {
        icbas[i] = 0;
        irbas[i] = i;
    }
}

/*-----------------------------------------------------------------
Part 1 of LA_Fuel()
-----------------------------------------------------------------*/
eLaRc LA_fuel_part_1 (int iout, int *pJin, int *pKl, int m, int n,
    tMatrix_R c, tVector_R f, tVector_I icbas, tVector_I irbas,
    int *pIrank, int *pIter)
{
    int         j;
    tNumber_R   d, piv;

    piv = 0.0;
    for (j = 1; j <= m; j++)
    {
        d = fabs (c[iout][j]);
        if (d > piv)
        {
            *pJin = j;
            piv = d;
        }
    }

    /* Detection of rank deficiency of matrix "c" */
    if (piv < EPS)
    {
        /* Inconsistent system */
        if (fabs (f[iout]) > EPS)
            return LaRcInconsistentSystem;

        /* Swap two rows of matrix "c" */
        swap_rows_Matrix_R (c, *pKl, iout);

        /* Swap two elements of vector "f" */
        swap_elems_Vector_R (f, *pKl, iout);
        irbas[iout] = irbas[*pKl];
        irbas[*pKl] = 0;
```

```
        icbas[iout] = icbas[*pKl];
        icbas[*pKl] = 0;

        *pIrank = *pIrank - 1;
        *pKl = *pKl + 1;
    }
    else
    {
        icbas[iout] = *pJin;
        /* A Gauss-Jordon elimination step */
        LA_fuel_gauss_jordn (iout, *pJin, *pKl, m, n, c, f, icbas);
        *pIter = *pIter + 1;
    }

    return LaRcOk;
}

/*-------------------------------------------------------------------
A Gauss-Jordan elimination step in LA_Fuel()
-------------------------------------------------------------------*/
void LA_fuel_gauss_jordn (int iout, int jin, int kl, int m, int n,
    tMatrix_R c, tVector_R f, tVector_I icbas)
{
    tNumber_R   pivot, d;
    int         i, j;

    pivot = c[iout][jin];
    for (j = 1; j <= m; j++)
        c[iout][j] = c[iout][j]/pivot;

    f[iout] = f[iout]/pivot;
    for (i = kl; i <= n; i++)
    {
        if (i != iout)
        {
            d = c[i][jin];
            for (j = 1; j <= m; j++)
            {
                c[i][j] = c[i][j] - d * (c[iout][j]);
            }
            f[i] = f[i] - d * (f[iout]);
        }
    }
    icbas [iout] = jin;
```

```
}

/*------------------------------------------------------------------
Part 2 of LA_Fuel()
------------------------------------------------------------------*/
void LA_fuel_part_2 (int kl, int m, int n, tMatrix_R c, tVector_I
    icbas, tVector_R zc)
{
    int         i, j, icb;
    tNumber_R   s;

    for (j = 1; j <= m; j++)
    {
        zc[j] = 0.0;
        icb = 0;
        for (i = kl; i <= n; i++) if (j == icbas[i]) icb = 1;
        if (icb == 0)
        {
            s = -1.0;
            for (i = kl; i <= n; i++)
            {
                s = s + c[i][j];
            }
            zc[j] = s;
        }
    }
}

/*------------------------------------------------------------------
Determine the vector that enters the basis in LA_Fuel()
------------------------------------------------------------------*/
void LA_fuel_vent (int *pIvo, int *pJin, int kl, int m, int n,
    tVector_I icbas, tVector_R zc)
{
    int         i, j, icb;
    tNumber_R   d, e, g;

    g = - 1.0/ (EPS*EPS);
    for (j = 1; j <= m; j++)
    {
        icb = 0;
        for (i = kl; i <= n; i++)
        {
            if (j == icbas[i]) icb = 1;
        }
```

```
        if (icb == 0)
        {
            d = zc[j];
            if (d > EPS)
            {
                e = d;
                if (e > g)
                {
                    *pIvo = 1;
                    g = e;
                    *pJin = j;
                }
            }
            else
            {
                e = -2.0 - d;
                if (e > EPS && e > g)
                {
                    *pIvo = -1;
                    g = e;
                    *pJin = j;
                }
            }
        }
    }
}

/*-------------------------------------------------------------------
Determine the vector that leaves the basis in LA_Fuel()
-------------------------------------------------------------------*/
void LA_fuel_leav (int *pItest, int jin, int *pIout, int kl, int n,
    tMatrix_R c, tVector_R f)
{
    int         i;
    tNumber_R   d, g, thmax;

    thmax = 1.0/ (EPS*EPS);
    for (i = kl; i <= n; i++)
    {
        d = c[i][jin];
        if (d > EPS)
        {
            g = f[i]/d;
            if (g <= thmax)
            {
```

```
                thmax = g;
                *pIout = i;
                *pItest = 1;
            }
        }
    }
}

/*--------------------------------------------------------------------
Calculate the results of LA_fuel()
--------------------------------------------------------------------*/
eLaRc LA_fuel_res (int kl, int n, tVector_R f, tVector_I icbas,
  tVector_I ibound, tVector_R a, tNumber_R *pZ)
{
    int         i, k;
    tNumber_R   e;

    *pZ = 0.0;
    for (i = kl; i <= n; i++)
    {
        k = icbas[i];
        e = f[i];
        *pZ = *pZ + e;

        if (ibound[k] == -1) e = -e;
        a[k] = e;
        if (fabs (a[k]) < EPS)
            return LaRcSolutionProbNotUnique;
    }

    return LaRcSolutionUnique;
}
```

Chapter 21

Bounded and L_1 Bounded Solutions of Underdetermined Linear Equations

21.1 Introduction

In the previous chapter, the L_1 solution of an underdetermined system of consistent linear equations was obtained. The underdetermined system was solved subject to the constraint that the L_1 norm of the solution vector be as small as possible.

In this chapter, the constraints are (a) each element of the solution vector would be bounded between –1 and +1 or (b) each element of the solution vector would be bounded between –1 and +1 and that the L_1 norm of the solution vector be as small as possible.

Consider the underdetermined consistent system

$$\mathbf{Ca} = \mathbf{f}$$

$\mathbf{C} = (c_{ij})$ is a given real n by m matrix of rank k, $k \leq n < m$, $\mathbf{f} = (f_i)$ is a given real n-vector and $\mathbf{a} = (a_i)$ is the required m-solution vector.

The first problem may be formulated as follows. Among the infinite number of solutions that the underdetermined system $\mathbf{Ca} = \mathbf{f}$ has, a solution satisfying the conditions

$$-1 \leq a_i \leq 1, \quad i = 1, 2, \ldots, m \tag{21.1.1}$$

may or may not exist. We call this problem (A).

The second problem may be formulated as follows. Given the system $\mathbf{Ca} = \mathbf{f}$, find vector $\mathbf{a}$ that satisfies (21.1.1) and such that the L_1 norm of the solution vector $\mathbf{a}$

$$z = \|\mathbf{a}\|_1 = \sum_{i=1}^{m} |a_i|$$

be as small as possible. We call this problem (B).

If instead of the constraints (21.1.1), we require the elements of vector **a** to satisfy the constraints

$$b_i \leq a_i \leq c_i, \quad i = 1, 2, \ldots, m$$

where vectors $\mathbf{b} = (b_i)$ and $\mathbf{c} = (c_i)$ are given m-vectors, by substituting variables, the above constraints reduce to the constraints (21.1.1) in the new variables. See Section 7.1.

Lin [8] attempted to construct boundary hyper-planes of a reachable set for problem (A). Outrata [9] stated necessary conditions for the solution of problem (A) and developed a simple algorithm to get an approximate solution to it. Weischedel [11] used the minimum energy problem (Chapter 23), as a subsidiary problem, with an iterative technique to solve problem (A). Torng [10] used the simplex method of linear programming and solved both problems (A) and (B). Bashein [5] also used an efficient simplex method to solve problem (A). We shall comment on both Torng's and Bashein's methods in Section 21.4.

Here, we formulate both problems (A) and (B) as one linear programming problem but with two different objective functions [1]. This linear programming problem is very similar to the dual form of the linear programming problem for the discrete linear L_1 approximation problem [1] given in Chapter 5. The algorithm for solving the L_1 approximation problem is modified and used here to obtain the solution of either problem (A) or problem (B).

Unlike the problem of the previous chapter, problems (A) and (B) may have no solution. If problem (A) has no solution, then problem (B) also has no solution. Condition (21.1.1), which requires the elements of the solution vector to be bounded, may not be achieved for some given problems.

If problem (A) and problem (B) have no solution, this is determined by the algorithm. In this case, the corresponding linear programming problem has an unbounded solution.

In Section 21.2, the linear programming formulation of the problem is presented. In Section 21.3, the algorithm is described and the problem of degeneracy in this algorithm is discussed in detail. Also, the uniqueness of the solution for problems (A) and (B) is outlined. In Section 21.4, numerical results and comments are given.

21.1.1 Applications of the algorithms

Our algorithm for problem (B) has been applied to problems of restoration of images with missing high frequency components [3, 4].

However, the two algorithms have applications in control theory, where problems (A) and (B) are known respectively as the **minimum time problem** and the **minimum fuel problem** for discrete linear admissible control systems [6].

21.2 Linear programming formulation of the two problems

Problem (A) is formulated as follows. Obtain a basic feasible solution to system $\mathbf{Ca} = \mathbf{f}$ subject to the conditions $|a_j| \leq 1$, $j = 1, 2, \ldots, m$. By introducing n artificial variables $a_{Si} \geq 0$, $i = 1, 2, \ldots, n$, problem (A) is stated as

$$\text{minimize } Z = \sum_{i=1}^{n} a_{Si}$$

subject to the conditions $\mathbf{Ca} = \mathbf{f}$ and $|a_j| \leq 1$, $j = 1, \ldots, m$.

Problem (B) is stated as

$$\text{minimize } z = \|\mathbf{a}\|_1 = \sum_{j=1}^{m} |a_j| \tag{21.2.1a}$$

subject to $\mathbf{Ca} = \mathbf{f}$ and $|a_j| \leq 1$, $j = 1, \ldots, m$.

The elements of vector $\mathbf{a}$ in $\mathbf{Ca} = \mathbf{f}$ are unrestricted in sign. We may write

$$a_i = v_i - w_i \tag{21.2.1b}$$

Hence

$$|a_i| = v_i + w_i \tag{21.2.1c}$$

and

$$0 \leq v_i \leq 1 \text{ and } 0 \leq w_i \leq 1, \quad i = 1, 2, \ldots, m \tag{21.2.1d}$$

We shall now reformulate problem (B) first. Using (21.2.1), problem (B) may be reformulated as

$$\text{minimize } z = \sum_{j=1}^{m} v_j + \sum_{j=1}^{m} w_j$$

subject to

$$\begin{bmatrix} \mathbf{C} & -\mathbf{C} \end{bmatrix} \begin{bmatrix} \mathbf{v} \\ \mathbf{w} \end{bmatrix} = \mathbf{f}$$

$$0 \le v_i \le 1 \text{ and } 0 \le w_i \le 1, \quad i = 1, 2, \ldots, m$$

Let us rewrite this problem as

(21.2.2a)
$$\text{maximize } z = -\sum_{i=1}^{2m} b_i$$

subject to

(21.2.2b)
$$\mathbf{Db} = \mathbf{f}$$

(21.2.2c)
$$0 \le b_i \le 1, \quad i = 1, 2, \ldots, 2m$$

where $\mathbf{D} = [\mathbf{C} \ \ -\mathbf{C}]$ and the 2m augmented vector $\mathbf{b} = [\mathbf{v}^T \ \ \mathbf{w}^T]^T$.

A negative sign is introduced in (21.2.2a) so that the problem is posed as a maximization problem. Reformulation of problem (A) is

(21.2.2a')
$$\text{maximize } Z = -\sum_{i=1}^{n} a_{Si}$$

subject to

$$\mathbf{Db} = \mathbf{f}$$

and

$$0 \le b_i \le 1, \quad i = 1, 2, \ldots, 2m$$

Problem (21.2.2) is very similar to the problem described by equations (5.2.3). Both problems are linear programming problems with bounded variables. The following theorem states the necessary and sufficient conditions for the existence of a solution to problem (21.2.2).

Theorem 21.1

A necessary and sufficient condition for a nonzero program for problem (A) or for problem (B) to be optimal, is that m elements of the 2m-vector **b** each has the value 0 (lower bound), (m – n) elements of **b**, each has the value 0 (lower bound) or 1 (upper bound) and the other n elements of **b** are basic variables ([7], pp. 387-394). See also Lemma 21.3 below.

Without loss of generality, assume that matrix **C** is of rank n. Hence, matrix **D** of (21.2.2b) is also of rank n. A simplex tableau of n constraints in 2m variables for problem (A) or problem (B) is constructed as if the elements of **b** were not bounded from above. Let **B** denote a basis matrix. Then vectors ($\mathbf{y}_j$) in the simplex tableau are given by

(21.2.3) $$\mathbf{y}_j = \mathbf{B}^{-1}\mathbf{D}_j, \quad j = 1, 2, \ldots, 2m$$

where $\mathbf{D}_j$ is the jth column of matrix **D**.

Denote the marginal costs in this problem by (zc_j), j = 1, 2, …, 2m. When the artificial variables a_{Si} of (21.2.2a') are driven out, for problem (A), the marginal costs $zc_j = 0$, for all j. For problem (B), in part 2 of the algorithm, since from (21.2.2a) the prices are all 1's

(21.2.4) $$zc_j = -\sum_{i=1}^{n} y_{ij} + 1, \quad j = 1, 2, \ldots, 2m$$

where y_{ij} is the ith element of $\mathbf{y}_j$.

21.2.1 Properties of the matrix of constraints

Again, as in the previous chapter, the analysis for this problem is initiated by the asymmetry matrix **D** has. From (21.2.2b) we have

$$\mathbf{D}_j = \mathbf{C}_j \text{ and } \mathbf{D}_{j+m} = -\mathbf{C}_j, \quad j = 1, 2, \ldots, m$$

That is

(21.2.5) $$\mathbf{D}_j + \mathbf{D}_{j+m} = 0, \quad j = 1, 2, \ldots, m$$

where again $\mathbf{D}_j$ and $\mathbf{C}_j$ are the jth columns of **D** and **C** respectively. This asymmetry enables us to use a condensed simplex tableau of n constraints in only m variables.

Definition

Let us define any column j, $1 \le j \le m$, and the column $(j + m)$ of matrix **D** as two corresponding columns.

Lemma 21.1

Any two corresponding columns could not appear together in any basis.

The proof is obvious. If two corresponding columns appear together in the basis matrix **B**, the basis matrix would be singular, since one column is the negative of the other.

Lemma 21.2

$$\mathbf{y}_j + \mathbf{y}_{j+m} = 0, \quad j = 1, 2, \ldots, m$$

and for their marginal costs we have the following.

For problem (A), as indicated earlier, in part 2 of the algorithm

$$zc_j = zc_{j+m} = 0, \quad j = 1, 2, \ldots, m$$

and for problem (B), in part 2 of the algorithm, we have

$$zc_j + zc_{j+m} = 2, \quad j = 1, 2, \ldots, m$$

The lemma is proved from (21.2.3-5).

Lemma 21.3

Let i and j be two corresponding columns, meaning $|i - j| = m$, $1 \le i, j \le 2m$. Then if $\mathbf{D}_i$ is in the basis, $\mathbf{D}_j$ is at zero level, i.e., $b_j = 0$.

Lemma 21.4

The elements of the solution vector **a** of the system $\mathbf{Ca} = \mathbf{f}$ are obtained as follows. As in Lemma 20.4, for any element $\mathbf{b}_{Bs}$, $s = 1, 2, \ldots, n$, of the optimal basic solution, let j be the column in the simplex tableau associated with $\mathbf{b}_{Bs}$. Then

$$a_j = \mathbf{b}_{Bs}, \text{ if } 1 \le j \le m$$

$$a_{j-m} = -\mathbf{b}_{Bs}, \text{ if instead } m < j \le 2m$$

However, the remaining $(m - n)$ elements of **a** are either at their lower bound (zero) or at their upper bound (one). The upper bound means $a_j = 1$ for $1 \le j \le m$ and $a_{j-m} = -1$ for $m < j \le 2m$. The L_1 norm z is the sum of the absolute values of all of these elements.

21.3 Description of the algorithms

Since from Lemma 21.2, if any vector $\mathbf{y}_j$, j = 1, 2, …, m, and its marginal cost in the simplex tableau are known, the corresponding vector $\mathbf{y}_{j+m}$ and its marginal cost are easily derived. Accordingly, we shall use a simplex tableau in n constraints in only m variables. We call this the **condensed tableau**. We start this tableau with columns 1, 2, …, m, of matrix **D**.

The computation for either problem (A) or (B) is divided into two parts. In part 1, a numerically stable initial basic solution, feasible or not, is obtained. This is done simply by performing to the initial data a finite number of Gauss-Jordan elimination steps with partial pivoting.

As in the previous chapter, the program will detect the cases when $\text{rank}(\mathbf{C}|\mathbf{f}) = \text{rank}(\mathbf{C}) = k < n$, i.e., the consistent rank deficient case. It will also detect an inconsistent system, i.e., when $\text{rank}(\mathbf{C}|\mathbf{f}) > \text{rank}(\mathbf{C})$.

For problem (A), part 1 is equivalent to driving the artificial variables a_{Si} of (21.2.2a') out, but having an initial basic solution that might not be feasible. At the end of part 1 for problem (A), $zc_j = 0$, for all j, and the objective function Z = 0. The (zc_j) and Z remain 0's in part 2. For problem (B), the marginal costs (zc_j), j = 1, 2, …, m, are calculated from (21.2.4).

Part 2 of the algorithm is almost identical to the algorithm for the discrete linear L_1 approximation problem of Chapter 5 [1]. In this algorithm certain intermediate iterations are skipped. Since in our algorithm, only one half of the simplex tableau is stored, we account for the second un-stored half from Lemma 21.2.

21.3.1 Occurrence of degeneracy

In part 2, for problem (A), the marginal costs are all 0's. Also, for problem (B), some non-basic marginal costs may acquire zero values. As in Chapter 5, the degeneracy is resolved as follows. Every zero non-basic marginal cost zc_j is replaced by either δ or $-\delta$ according to whether b_j is 0 or 1 respectively. The parameter δ is the round-off error of the computer; δ is set to 10^{-6} and 10^{-16} for single- and double-precision computation respectively.

However, for problem (A), another difficulty exists. This is when two non-basic corresponding columns, one at its zero level and the other at its upper level (= 1), are both eligible to replace a column in the basis. It should be decided which one enters the basis so that Lemma 21.3 is not violated. Let us consider $\mathbf{D}_i$ and $\mathbf{D}_j$ as two non-basic corresponding columns, $|i - j| = m$ such that $b_i = 0$ and $b_j = 1$.

According to the previous paragraph we replace respectively zc_i by δ and replace zc_j by $-\delta$. Yet, from Lemma 21.2, $\mathbf{y}_i = -\mathbf{y}_j$ and according to the selection rules for the column to enter the basis, we find that either $\mathbf{D}_i$ or $\mathbf{D}_j$ may enter the basis. However, if $\mathbf{D}_i$ (with $b_i = 0$) enters the basis, Lemma 21.3 would be violated and the algorithm breaks down since its corresponding column $\mathbf{D}_j$ (with $b_j = 1$) is at its upper bound. This lemma states that if a column is in (or enters) the basis, its corresponding column is (should be) at zero level, i.e., $b_j = 0$. Hence, we always chose the column which is at its upper bound to enter the basis.

21.3.2 Uniqueness of the solution

For problem (A), since in the final condensed tableau, the non-basic marginal costs are all 0's, it is almost certain that the solution vector **a** is not unique.

Problem (B) has a unique solution **a** if in the final condensed tableau none of the elements of the basic solution $\mathbf{b}_B$ is 0 (lower bound) or 1 (upper bound) and none of the non-basic marginal costs is 0 or 1. Otherwise the solution **a** might not be unique.

21.4 Numerical results and comments

LA_Tmfuel() implements the Minimum Time Minimum Fuel algorithm. DR_Tmfuel() demonstrates 7 test cases. These are the same test cases shown in Table 20.4.1, for the L_1 solution of underdetermined linear systems. For each of the test cases, DR_Tmfuel() solves the Minimum Time case, problem (A), then the Minimum Fuel case, problem (B).

Table 21.4.1 shows the results of the 7 test examples calculated in single-precision.

Table 21.4.1

		Minimum Time		Minimum Fuel	
Example	**C**(n×m)	Iterations	z	Iterations	z
1	4×5	3	2.750	4	2.667
2	4×8	6	2.851	7	2.055
3	2×3	no solution		no solution	
4	4×6	4	1.546	5	1.359
5	4×100	9	14.492	14	11.094
6	5×100	9	18.864	17	15.286
7	7×100	22	27.647	23	23.130

For each example, the size of matrix **C**, the number of iterations and the L_1 norm z are given. Also, "no solution" refers to the case when the given problem has no feasible solution.

Again, the obtained optimum z and the number of iterations for each example are comparable to the results of Table 20.4.1.

The results of the first example in Table 21.4.1 are given here in more detail in Example 21.1.

Example 21.1

Find the bounded and L_1 bounded solutions of the system

$$\begin{aligned} 2a_1 + a_2 - 3a_3 + 5a_4 + 3a_5 &= 10 \\ 2a_1 + a_2 - 3a_3 + 5a_4 + 3a_5 &= 10 \\ 2a_1 + a_2 - 3a_3 + 5a_4 + 3a_5 &= 10 \\ -8a_1 - 4a_2 + 2a_3 + 5a_4 + 6a_5 &= 8 \end{aligned}$$

This system of 4 equations is consistent but of rank 2. Each of the second and third equations is the same as the first one.

The solution for problem (A) is

$$\mathbf{a} = (0.25, 0, -0.5, 1, 1)^T \text{ and } z = 2.75$$

The solution for problem (B) is

$$\mathbf{a} = (0, 0, -0.875, 1, 0.7917)^T \text{ and } z = 2.6667$$

We observe that the L_1 norm z for problem (B) is smaller than that of problem (A). This is besides the fact that each element of **a** is bounded between -1 and $+1$.

Finally, we note that the nearest method to our algorithm is that of Torng [10] and of Bashein [5]. Torng's method uses m slack variables and no intermediate simplex iterations are skipped. Bashein's method, which applies to a problem similar to problem (A), preserves the minimum computer storage, but no intermediate iterations could be skipped.

References

1. Abdelmalek, N.N., Efficient solution for the discrete linear L_1 approximation problem, *Mathematics of Computation*, 29(1975)844-850.
2. Abdelmalek, N.N., Solutions of minimum time problem and minimum fuel problem for discrete linear admissible control systems, *International Journal of Systems Science*, 9(1978)849-855.
3. Abdelmalek, N.N. and Otsu, N., Restoration of Images with missing high-frequency components by minimizing the L_1 norm of the solution vector, *Applied Optics*, 24(1985)1415-1420.
4. Abdelmalek, N.N. and Otsu, N., Speed comparison among methods for restoring signals with missing high-frequency components using two different low-pass-filter matrix dimensions, *Optics Letters*, 10(1985)372-374.
5. Bashein, G., A simplex algorithm for on-line computation of time optimal controls, *IEEE Transactions on Automatic Control*, 16(1971)479-482.
6. Canon, M.D., Cullum Jr., C.D. and Polak, E., *Theory of Optimal Control and Mathematical Programming*, McGraw-Hill, New York, 1970.
7. Hadley, G., *Linear Programming*, Addison-Wesley, Reading, MA, 1962.
8. Lin, J.N., Determination of reachable set for a linear discrete system, *IEEE Transactions on Automatic Control*, 15(1970)339-342.

9. Outrata, J.V., On the minimum time problem in linear discrete systems with the discrete set of admissible controls, *Kybernetika*, 11(1975)368-374.
10. Torng, H.C., Optimization of discrete control systems through linear programming, *Journal of Franklin Institute*, 278(1964)28-44.
11. Weischedel, H.R., A solution of the discrete minimum-time control problem, *Journal of Optimization Theory and Applications*, 5(1970)81-96.

21.5 DR_Tmfuel

```
/*------------------------------------------------------------------
DR_Tmfuel
--------------------------------------------------------------------
This program is a driver for the function LA_Tmfuel(), which
calculates the "bounded" or the "L-One" bounded solutions of an
underdetermined consistent system of linear equations.

Given is the underdetermined consistent system

                        c*a = f

"c" is a given real n by m matrix of rank k, k <=n <= m.
"f" is a given real n vector.

It is required to calculate the m vector "a" for either of the
following two problems:

Problem (a):
   The m elements a[i] of "a"  satisfy the  conditions
                |a[i]| < 1,   i = 1, 2, ..., m

or

Problem (b):
   The m elements of "a" satisfy (1) and the L1 norm z of vector "a"
                z = |a[1]| + |a[2]| + ... + |a[m]|
   is as small as possible.

In control theory, problem (a) is known as the "minimum time" problem
and problem (b) is known as the "minimum fuel problem" for linear
discrete admissible control systems.

This program has 7 examples whose results are given in the text.
The program solves the 7 examples twice, once for the minimum time
case and once for the minimum fuel case.
------------------------------------------------------------------*/

#include "DR_Defs.h"
#include "LA_Prototypes.h"

#define N1e         4
#define M1e         5
```

```
#define N2e        4
#define M2e        8
#define N3e        2
#define M3e        3
#define N4e        4
#define M4e        6
#define N5e        4
#define M5e        100
#define N6e        5
#define M6e        100
#define N7e        7
#define M7e        100

void DR_Tmfuel (void)
{
    /*--------------------------------------
      Constant matrices/vectors
    --------------------------------------*/
    static tNumber_R b1init[N1e][M1e] =
    {
        { 2.0,  1.0, -3.0,  5.0, 3.0  },
        { 2.0,  1.0, -3.0,  5.0, 3.0  },
        { 2.0,  1.0, -3.0,  5.0, 3.0  },
        {-8.0, -4.0,  2.0,  5.0, 6.0  }
    };

    static tNumber_R f1[N1e+1] =
    {   NIL,
    10.0, 10.0, 10.0, 8.0
    };

    static tNumber_R b2init[N2e][M2e] =
    {
        {-6.0,  4.0,  2.0, -8.0,  5.0, -1.0,  6.0,  3.0  },
        { 7.0, -3.0,  9.0,  5.0, -9.0,  8.0,  7.0, -4.0  },
        { 9.0,  7.0, -5.0,  2.0,  7.0,  0.0,  1.0,  8.0  },
        { 9.0, -3.0,  2.0,  4.0,  0.0,  3.0,  0.0,  1.0  }
    };

    static tNumber_R f2[N2e+1] =
    {   NIL,
        5.0, 7.0, -9.0, 1.0
    };

    static tNumber_R b3init[N3e][M3e] =
```

```
{
    {-0.7182, -3.6706, -11.6961 },
    {-1.7182, -4.6706, -12.6961 }
};

static tNumber_R f3[N3e+1] =
    {NIL,
    24.8218, 24.4218
};

static tNumber_R b4init[N4e][M4e] =
{
    {2.0, -1.0, -4.0,  0.0, -3.0, 1.0 },
    {2.0, -1.0, -4.0,  0.0, -3.0, 1.0 },
    {5.0,  1.0,  3.0,  1.0, -2.0, 0.0 },
    {1.0, -2.0, -1.0, -5.0,  1.0, 4.0 }
};

static tNumber_R f4[N4e+1] =
{   NIL,
    2.0, 2.0, 1.0, -4.0
};

/*---------------------------------------
  Variable matrices/vectors
---------------------------------------*/
tMatrix_R   c       = alloc_Matrix_R (N_ROWS, M_COLS);
tVector_R   f       = alloc_Vector_R (N_ROWS);
tVector_R   a       = alloc_Vector_R (M_COLS);
tMatrix_R   fay     = alloc_Matrix_R (N7e, M7e);
tVector_R   fy      = alloc_Vector_R (N7e + 1);

tMatrix_R   b1      = init_Matrix_R (&(b1init[0][0]), N1e, M1e);
tMatrix_R   b2      = init_Matrix_R (&(b2init[0][0]), N2e, M2e);
tMatrix_R   b3      = init_Matrix_R (&(b3init[0][0]), N3e, M3e);
tMatrix_R   b4      = init_Matrix_R (&(b4init[0][0]), N4e, M4e);

int         irank, iter, itf;
int         i, j, kase, m, n, Iexmpl;
tNumber_R   pi, dx, x1, x2, x3, znorm;

eLaRc       rc = LaRcOk;

pi = 4.0*atan (1.0);
dx = 2.*pi/99;
```

```
for (j = 1; j <= 100; j++)
{
    x1 = (j - 1)* dx;
    x2 = x1 + x1;
    x3 = x2 + x1;
    fay[1][j] = 1.0;
    fay[2][j] = sin (x1);
    fay[3][j] = cos (x1);
    fay[4][j] = sin (x2);
    fay[5][j] = cos (x2);
    fay[6][j] = sin (x3);
    fay[7][j] = cos (x3);
}
for (i = 1; i <= 7; i++)
{
    fy[i] = (10.0+15.0*fay[2][i]-7.0*fay[3][i]+9.0*fay[4][i])
            * (1.0+0.01*fay[4][i]);
}

prn_dr_bnr ("DR_Tmfuel, Bounded & L1 Bounded Solutions of "
            "an Underdetermined System of Linear Equations");

for (kase = 1; kase <= 2; kase++)
{
    if (kase == 1) itf = 0;
    if (kase == 2) itf = 1;
    for (Iexmpl = 1; Iexmpl <= 7; Iexmpl++)
    {
        switch (Iexmpl)
        {
            case 1:
                n = N1e;
                m = M1e;
                for (i = 1; i <= n; i++)
                {
                    f[i] = f1[i];
                    for (j = 1; j <= m; j++) c[i][j] = b1[i][j];
                }
                break;
            case 2:
                n = N2e;
                m = M2e;
                for (i = 1; i <= n; i++)
                {
                    f[i] = f2[i];
```

```
            for (j = 1; j <= m; j++) c[i][j] = b2[i][j];
        }
        break;
    case 3:
        n = N3e;
        m = M3e;
        for (i = 1; i <= n; i++)
        {
            f[i] = f3[i];
            for (j = 1; j <= m; j++) c[i][j] = b3[i][j];
        }
        break;
    case 4:
        n = N4e;
        m = M4e;
        for (i = 1; i <= n; i++)
        {
            f[i] = f4[i];
            for (j = 1; j <= m; j++) c[i][j] = b4[i][j];
        }
        break;
    case 5:
        n = N5e;
        m = M5e;
        for (i = 1; i <= n; i++)
        {
            f[i] = fy[i];
            for (j = 1; j <= m; j++)
                c[i][j] = fay[i][j];
        }
        break;
    case 6:
        n = N6e;
        m = M6e;
        for (i = 1; i <= n; i++)
        {
            f[i] = fy[i];
            for (j = 1; j <= m; j++)
                c[i][j] = fay[i][j];
        }
        break;
    case 7:
        n = N7e;
        m = M7e;
        for (i = 1; i <= n; i++)
```

```
                {
                    f[i] = fy[i];
                    for (j = 1; j <= m; j++)
                        c[i][j] = fay[i][j];
                }
                break;
            default:
                break;
        }
        prn_algo_bnr ("Tmfuel");
        prn_example_delim();
        PRN ("Example #%d: Size of matrix \"c\" %d by %d\n",
             Iexmpl, n, m);
        prn_example_delim();
        if (itf != 1)
            PRN ("Minimum Time Problem (Bounded Solution) "
                 " for an Underdetermined System\n");
        else
            PRN ("Minimum Fuel Problem (L1 Bounded Solution) "
                 " for an Underdetermined System\n");
        prn_example_delim();
        PRN ("r.h.s. Vector \"f\"\n");
        prn_Vector_R (f, n);
        PRN ("Coefficient Matrix, \"c\"\n");
        prn_Matrix_R (c, n, m);

        rc = LA_Tmfuel (itf, m, n, c, f, &irank, &iter, a,
                        &znorm);

        if (rc >= LaRcOk)
        {
            PRN ("\n");
            PRN ("Results of the Problem\n");
            PRN ("L1 Solution vector, \"a\"\n");
            prn_Vector_R (a, m);
            PRN ("L1 Norm of vector \"a\", ||a|| = %8.4f\n",
                 znorm);
            PRN ("Rank of matrix \"c\" = %d, No. of"
                 "iterations = %d\n", irank, iter);
        }

        LA_check_rank_def (n, irank);
        prn_la_rc (rc);
    }
}
```

```
    free_Matrix_R (c, N_ROWS);
    free_Vector_R (f);
    free_Vector_R (a);
    free_Matrix_R (fay, N7e);
    free_Vector_R (fy);

    uninit_Matrix_R (b1);
    uninit_Matrix_R (b2);
    uninit_Matrix_R (b3);
    uninit_Matrix_R (b4);
}
```

21.6 LA_Tmfuel

```
/*-----------------------------------------------------------------
LA_Tmfuel
-------------------------------------------------------------------
Given is the underdetermined consistent system of linear equations

                          c*a = f

"c" is a given real n by m matrix  of rank k, k <= n <= m.
"f" is a given real n vector.

This program calculates the m vector "a" for either of the following
two problems:

Problem (a):
    The m elements a[i] of "a"  satisfy the  conditions
                 |a[i]| < 1,   i = 1, 2, ..., m

or

Problem (b):
    The m elements of "a" satisfy (1) and the L1 norm z of vector "a"
                 z = |a[1]| + |a[2]| + ... + |a[m]|
    is as small as possible.

In control theory, problem (a) is known as the "minimum time" problem
and problem (b) is known as the "minimum fuel problem" for linear
discrete admissible control systems.

This program uses a dual simplex algorithm to the linear programming
formulation of the given problem.  In this algorithm certain simplex
intermediate iterations are skipped.

Inputs
itf     An integer set by the user.
        If itf = 1, the program calculates a solution to the minimum
        fuel problem; problem (b).
        If itf != 1, the program calculates a solution to the minimum
        time problem; problem (a).
m       Number of columns of matrix "c" in the system c*a = f.
n       Number of rows of matrix "c" in the system c*a = f.
c       A real n by m matrix containing matrix "c" of system c*a = f.
f       A real n vector containing the r.h.s. of the system c*a = f.
```

```
Local Variables
icbs    An integer n vector whose elements are the indices of the
        columns of matrix "c" that form the columns of the basis
        matrix.
irbs    An integer n vector whose elements are the row indices of
        matrix "c".

Outputs
irank   The calculated rank of matrix "c".
iter    The number of iterations, or the number of times the simplex
        tableau is changed by a Gauss-Jordan step.
a       A real m vector whose elements are the solution of c*a = f.
znorm   The L1 norm of the solution vector "a".

Returns one of
        LaRcSolutionUnique
        LaRcSolutionProbNotUnique
        LaRcNoFeasibleSolution
        LaRcInconsistentSystem
        LaRcErrBounds
        LaRcErrNullPtr
        LaRcErrAlloc
-----------------------------------------------------------------------*/

#include "LA_Prototypes.h"

eLaRc LA_Tmfuel (int itf, int m, int n, tMatrix_R c, tVector_R f,
    int *pIrank, int *pIter, tVector_R a, tNumber_R *pZnorm)
{
    tVector_I   icbs    = alloc_Vector_I (n);
    tVector_I   irbs    = alloc_Vector_I (n);
    tVector_I   ibnd    = alloc_Vector_I (m + m);
    tVector_I   kbnd    = alloc_Vector_I (m + m);
    tVector_I   ib      = alloc_Vector_I (m);
    tVector_R   th      = alloc_Vector_R (m);
    tVector_R   tu      = alloc_Vector_R (m);
    tVector_R   zc      = alloc_Vector_R (m);

    int         ii = 0, ij = 0, ijk = 0, iout = 0, itest = 0,
                ivo = 0;
    int         j = 0, jin = 0, jout = 0, ktest = 0;
    tNumber_R   d = 0, oneps = 0, pivot = 0, pivoto = 0, xb = 0;
    eLaRc       tempRc;
```

```
/* Validation of the data before executing the algorithm */
eLaRc       rc = LaRcSolutionUnique;
VALIDATE_BOUNDS ((0 < n) && (n <= m) && !((n == 1) && (m == 1)));
VALIDATE_PTRS   (c && f && pIrank && pIter && a && pZnorm);
VALIDATE_ALLOC  (icbs && irbs && ibnd && kbnd && ib && th && tu
                 && zc);

*pIter = 0;
*pIrank = n;
oneps = 1.0 + EPS;
*pZnorm = 0.0;
iout = 0;

LA_tmfuel_init (m, n, icbs, irbs, ibnd, kbnd, ib, zc, a);

/* Part 1 of the algorithm */
for (ij = 1; ij <= n; ij++)
{
    iout = iout + 1;
    if (iout <= *pIrank)
    {
        jin = 0;
        tempRc = LA_tmfuel_part_1 (iout, &jin, m, c, f, icbs,
                                   irbs, pIrank, pIter);
        if (tempRc < LaRcOk)
        {
            GOTO_CLEANUP_RC (tempRc);
        }
    }
}

/* Obtain a basic feasible solution */
if (itf == 1)
{
    /* Calculating the marginal costs */
    LA_tmfuel_marg_costs (m, c, f, icbs, ibnd, kbnd, zc, pIrank);
}

/* Part 2 of the algorithm */
for (ijk = 1; ijk <= m; ijk++)
{
    /* Determine the vector that leaves the basis */
    ivo = 0;
    xb = 0.0;
    LA_tmfuel_vleav (pIrank, &iout, &ivo, f, &xb);
```

```
/* Calculate the results */
if (ivo == 0)
{
    rc = LA_tmfuel_res (itf, m, f, icbs, ibnd, kbnd, ib, zc,
                        pIrank, a, pZnorm);
    GOTO_CLEANUP_RC (rc);
}

/* Calculating of the possible ratios th[j] and tu[j] */
LA_tmfuel_th_tu (itf, m, ivo, iout, &jout, c, icbs, ibnd,
                 kbnd, th, tu, zc);

pivoto = 1.0;
ktest = 1;
for (ii = 1; ii <=m; ii++)
{
    if (ktest == 0)
    {
        break;
    }

    itest = 0;
    /* Determine the vector that enters the basis */
    LA_tmfuel_vent (m, &jin, ivo, &itest, th, tu);

    /* No feasible solution has been found */
    if (itest == 0)
    {
        GOTO_CLEANUP_RC (LaRcNoFeasibleSolution);
    }

    if (kbnd[jin] != 1)
    {
        if (jin > m)
        {
            jin = jin - m;
        }

        if (ibnd[jin] == 1)
        {
            LA_tmfuel_swap (itf, pIrank, &jin, c, ibnd, kbnd,
                            ib, th, tu, zc);
        }
    }
```

```
            /* Cascading vectors, each enters then leaves the basis*/
            LA_tmfuel_cascade (iout, jin, &jout, xb, &pivot, pivoto,
                               c, f, ibnd, pIrank);
            LA_tmfuel_test (iout, jin, &itest, &xb, pivot, f, icbs,
                            th);

            if (itest == 0)
            {
                pivoto = pivot;
                jout = jin;
                ktest = 1;
            }

            if (itest == 1)
            {
                ktest = 0;
                /* A Gauss-Jordan elimination step */
                LA_tmfuel_gauss_jordn (iout, jin, pIrank, m, c, f);
                *pIter = *pIter + 1;

                if (itf == 1)
                {
                    d = zc[jin];
                    for (j = 1; j <= m; j++)
                    {
                        zc[j] = zc[j] - d * (c[iout][j]);
                    }
                }
            }
        }
    }

CLEANUP:

    free_Vector_I (icbs);
    free_Vector_I (irbs);
    free_Vector_I (ibnd);
    free_Vector_I (kbnd);
    free_Vector_I (ib);
    free_Vector_R (th);
    free_Vector_R (tu);
    free_Vector_R (zc);

    return rc;
```

```
}

/*-----------------------------------------------------------------
LA_Tmfuel() initialization
-----------------------------------------------------------------*/
void LA_tmfuel_init (int m, int n, tVector_I icbs, tVector_I irbs,
    tVector_I ibnd, tVector_I kbnd, tVector_I ib, tVector_R zc,
    tVector_R a)
{
    int         i, j;

    for (j = 1; j <= m; j++)
    {
        ib[j] = 1;
        a[j] = 0.0;
        zc[j] = 0.0;
        ibnd[j] = 1;
        kbnd[j] = 1;
    }

    for (i = 1; i <= n; i++)
    {
        icbs[i] = 0;
        irbs[i] = i;
    }
}

/*-----------------------------------------------------------------
Part 1 of LA_Tmfuel()
-----------------------------------------------------------------*/
eLaRc LA_tmfuel_part_1 (int iout, int *pJin, int m, tMatrix_R c,
    tVector_R f, tVector_I icbs, tVector_I irbs, int *pIrank,
    int *pIter)
{
    int         i, j, li = 0;
    tNumber_R   e, d, piv, pivot;

    piv = 0.0;
    for (j = 1; j <= m; j++)
    {
        for (i = iout; i <= *pIrank; i++)
        {
            d = fabs (c[i][j]);
            if (d > piv)
            {
```

```
                li = i;
                *pJin = j;
                piv = d;
            }
        }
    }

    if (piv < EPS)
    {
        /* Detection of rank deficiency of matrix "c" */
        for (i = iout; i <= *pIrank; i++)
        {
            e = f[i];
            if (fabs (e) > EPS)
                return LaRcInconsistentSystem;
        }
        *pIrank = iout - 1;
    }
    else
    {
        pivot = c[li][*pJin];
        icbs[iout] = *pJin;

        if (li != iout)
        {
            /* Swap two rows of matrix "c" */
            swap_rows_Matrix_R (c, li, iout);
            /* Swap two elements of vectors "f" and "irbs" */
            swap_elems_Vector_R (f, li, iout);
            swap_elems_Vector_I (irbs, li, iout);
        }
        /* A Gauss-Jordan elimination step */
        LA_tmfuel_gauss_jordn (iout, *pJin, pIrank, m, c, f);
        *pIter = *pIter + 1;
    }

    return LaRcOk;
}

/*-----------------------------------------------------------------
Calculating the marginal costs in LA_Tmfuel()
-----------------------------------------------------------------*/
void LA_tmfuel_marg_costs (int m, tMatrix_R c, tVector_R f,
    tVector_I icbs, tVector_I ibnd, tVector_I kbnd, tVector_R zc,
    int *pIrank)
```

```
{
    int          i, j, icb;
    tNumber_R    d, g, s;

    for (j = 1; j <= m; j++)
    {
        icb = 0;
        for (i = 1; i <= *pIrank; i++)
        {
            if (j == icbs[i]) icb = 1;
        }
        if (icb == 0)
        {
            s = -1.0;
            for (i = 1; i <= *pIrank; i++)
            {
                s = s + c[i][j];
            }
            zc[j] = -s;
            d = -s;
            if (d < 0.0)
            {
                for (i = 1; i <= *pIrank; i++)
                {
                    f[i] = f[i] - c[i][j];
                }
                ibnd[j] = -1;
            }
            else
            {
                g = 2.0 - d;
                if (g < 0.0)
                {
                    for (i = 1; i <= *pIrank; i++)
                    {
                        f[i] = f[i] + c[i][j];
                    }
                    kbnd[j] = -1;
                }
            }
        }
    }
}

/*-----------------------------------------------------------------
```

```
Determine the vector that leaves the basis in LA_Tmfuel()
-----------------------------------------------------------------*/
void LA_tmfuel_vleav (int *pIrank, int *pIout,int *pIvo, tVector_R f,
    tNumber_R *pXb)
{
    int         i;
    tNumber_R   e, d, g, oneps;

    oneps = 1.0 + EPS;
    *pIvo = 0;
    g = 1.0;
    for (i = 1; i <= *pIrank; i++)
    {
        e = f[i];
        if (e >= oneps)
        {
            d = 1.0 - e;
            if (d < g)
            {
                g = d;
                *pIvo = 1;
                *pIout = i;
                *pXb = e;
            }
        }

        if (e < -EPS)
        {
            d = e;
            if (d < g)
            {
                g = d;
                *pIvo = -1;
                *pIout = i;
                *pXb = e;
            }
        }
    }
}

/*-----------------------------------------------------------------
Calculating possible ratios th[j] and tu[j] in LA_Tmfuel()
-----------------------------------------------------------------*/
void LA_tmfuel_th_tu (int itf, int m, int ivo, int iout,
    int *pJout, tMatrix_R c, tVector_I icbs, tVector_I ibnd,
```

```
    tVector_I kbnd, tVector_R th, tVector_R tu, tVector_R zc)
{
    tNumber_R   d, e, g, gg, thmax;
    int         j;

    *pJout = icbs[iout];
    thmax = 0.0;
    for (j = 1; j <= m; j++)
    {
        tu[j] = 0.0;
        th[j] = 0.0;
        e = c[iout][j];
        if (fabs (e) > EPS)
        {
            d = zc[j];
            g = 2.0 - d;
            if (itf != 1)
            {
                g = 0.0;
            }

            if (fabs (g) <= EPS)
            {
                g = PREC * (kbnd[j]);
                if (kbnd[j] == 1) g = g + g;
            }

            tu[j] = -g/e;
            g = ivo * (tu[j]);
            if (g < 0.0) tu[j] = 0.0;
            if (j != *pJout)
            {
                gg = tu[j];
                if (gg < 0.0) gg = -gg;
                if (gg > thmax)
                {
                    thmax = gg;
                }

                if (fabs (d) < EPS)
                {
                    d = PREC * (ibnd[j]);
                    if (ibnd[j] == 1) d = d + d;
                }
```

```
                th[j] = d/e;
                d = ivo * (th[j]);
                if (d < 0.0) th[j] = 0.0;
                gg = th[j];
                if (gg < 0.0) gg = -gg;
                if (gg > thmax) thmax = gg;
            }
        }
    }
    if (itf == 1 && ivo == -1) tu[*pJout] = -2.0;
    if (itf != 1 && ivo == -1) tu[*pJout] = -PREC;
}

/*-------------------------------------------------------------------
Determine the vector that enters the basis in LA_Tmfuel()
-------------------------------------------------------------------*/
void LA_tmfuel_vent (int m, int *pJin, int ivo, int *pItest,
    tVector_R th, tVector_R tu)
{
    tNumber_R   e, thmax, thmin;
    int         j;
    thmax = 1.0/ (EPS * EPS);
    thmin = -thmax;
    for (j = 1; j <= m; j++)
    {
        e = th[j];
        if (e != 0.0)
        {
            if (ivo == 1)
            {
                if (e < thmax)
                {
                    thmax = e;
                    *pJin = j;
                    *pItest = 1;
                }
            }

            if (ivo != 1)
            {
                if (e > thmin)
                {
                    thmin = e;
                    *pJin = j;
                    *pItest = 1;
```

```
                }
            }
        }
    }
    for (j = 1; j <= m; j++)
    {
        e = tu[j];
        if (e != 0.0)
        {
            if (ivo == 1)
            {
                if (e < thmax)
                {
                    thmax = e;
                    *pJin = j + m;
                    *pItest = 1;
                }
            }

            if (ivo != 1)
            {
                if (e > thmin)
                {
                    thmin = e;
                    *pJin = j + m;
                    *pItest = 1;
                }
            }
        }
    }
}

/*-----------------------------------------------------------------------
Swap elements and vectors in LA_Tmfuel()
-----------------------------------------------------------------------*/
void LA_tmfuel_swap (int itf, int *pIrank, int *pJin, tMatrix_R c,
    tVector_I ibnd, tVector_I kbnd, tVector_I ib, tVector_R th,
    tVector_R tu, tVector_R zc)
{
    tNumber_R   d;
    int         i, k;

    k = ibnd[*pJin];
    ibnd[*pJin] = kbnd[*pJin];
    kbnd[*pJin] = k;
```

```
    d = th[*pJin];
    th[*pJin] = tu[*pJin];
    tu[*pJin] = d;
    ib[*pJin] = -ib[*pJin];
    for (i = 1; i <= *pIrank; i++)
    {
        c[i][*pJin] = -c[i][*pJin];
    }
    zc[*pJin] = 2.0 - zc[*pJin];
    if (itf != 1) zc[*pJin] = 0.0;
}

/*-----------------------------------------------------------------
Cascades vectors, each enters and then leaves the basis in
LA_Tmfuel()
-----------------------------------------------------------------*/
void LA_tmfuel_cascade (int iout, int jin, int *pJout, tNumber_R xb,
    tNumber_R *pPivot, tNumber_R pivoto, tMatrix_R c, tVector_R f,
    tVector_I ibnd, int *pIrank)
{
    tNumber_R   oneps, pivotn;
    int         i;

    oneps = 1.0 + EPS;
    *pPivot = c[iout][jin];
    pivotn = *pPivot/pivoto;
    if (xb > oneps)
    {
        for (i = 1; i <= *pIrank; i++)
        {
            f[i] = f[i] - c[i][*pJout];
        }

        ibnd[*pJout] = -1;
        if (pivotn <= 0.0)
        {
            for (i = 1; i <= *pIrank; i++)
            {
                f[i] = f[i] + c[i][jin];
            }
            ibnd[jin] = 1;
        }
     }
     else
     {
```

```
        if (pivotn > 0.0)
        {
            for (i = 1; i <= *pIrank; i++)
            {
                f[i] = f[i] + c[i][jin];
            }
            ibnd[jin] = 1;
        }
    }
}

/*-----------------------------------------------------------------
Test if xb is not bounded in LA_Tmfuel()
-----------------------------------------------------------------*/
void LA_tmfuel_test (int iout, int jin, int *pItest, tNumber_R *pXb,
    tNumber_R pivot, tVector_R f, tVector_I icbs, tVector_R th)
{
    tNumber_R   oneps;

    oneps = 1.0 + EPS;
    *pXb = f[iout]/ (pivot);
    if (*pXb <= -EPS || *pXb >= oneps)
    {
        *pItest = 0;
    }

    th[jin] = 0.0;
    icbs[iout] = jin;
}

/*-----------------------------------------------------------------
A Gauss-Jordan elimination step in LA_Tmfuel()
-----------------------------------------------------------------*/
void LA_tmfuel_gauss_jordn (int iout, int jin, int *pIrank, int m,
    tMatrix_R c, tVector_R f)
{
    int         i, j;
    tNumber_R   d, pivot;

    pivot = c[iout][jin];
    for (j = 1; j <= m; j++)
    {
        c[iout][j] = c[iout][j]/pivot;
    }
    f[iout] = f[iout]/pivot;
```

```
    for (i = 1; i <= *pIrank; i++)
    {
        if (i != iout)
        {
            d = c[i][jin];
            for (j = 1; j <= m; j++)
            {
                c[i][j] = c[i][j] - d * (c[iout][j]);
            }
            f[i] = f[i] - d * (f[iout]);
        }
    }
}

/*--------------------------------------------------------------------
Calculate the results of LA_Tmfuel()
------------------------------------------------------------------*/
eLaRc LA_tmfuel_res (int itf, int m, tVector_R f, tVector_I icbs,
    tVector_I ibnd, tVector_I kbnd, tVector_I ib, tVector_R zc,
    int *pIrank, tVector_R a, tNumber_R *pZnorm)
{
    int         i, j, i0, icb;
    tNumber_R   e, d;
    eLaRc       rc = LaRcSolutionUnique;

    e = 1.0 - EPS;
    for (j = 1; j <= m; j++)
    {
        a[j] = 0.0;
        i0 = 0;
        for (i = 1; i <= *pIrank; i++)
        {
            if (j == icbs[i]) i0 = i;
        }

        if (i0 == 0)
        {
            d = 0.0;
            if (ibnd[j] == -1) d = 1.0;
            if (kbnd[j] == -1) d = -1.0;
        }
        else
        {
            d = f[i0];
            if (d < EPS || d > e)
```

```
                {
                    rc = LaRcSolutionProbNotUnique;
                }
            }

            if (ib[j] == -1) d = -d;
            a[j] = d;

            if (d < 0.0) d = -d;
            *pZnorm = *pZnorm + d;
        }

        if (itf != 1)
        {
            rc = LaRcSolutionProbNotUnique;
        }

        for (j = 1; j <= m; j++)
        {
            icb = 0;
            for (i = 1; i <= *pIrank; i++)
            {
                if (j == icbs[i])
                {
                    icb = 1;
                }
            }
            if (icb == 0)
            {
                e = zc[j];
                d = e;
                if (e < 0.0) d = -d;
                if (d < EPS)
                {
                    rc = LaRcSolutionProbNotUnique;
                }
            }
        }

        return rc;
}
```

Chapter 22

Chebyshev Solution of Underdetermined Linear Equations

22.1 Introduction

In Chapter 20, an algorithm was presented for the solution of underdetermined systems of consistent linear equations, where the L_1 norm of the solution vector is minimum. In Chapter 21, algorithms for the bounded and for the L_1 bounded solution of underdetermined systems of consistent linear equations were presented as problems (A) and (B) respectively. In problem (A), each element of the solution vector is bounded between 1 and –1, and in problem (B), each element of the solution vector is bounded between 1 and –1 and the L_1 norm of the solution vector is as small as possible.

This chapter is the third chapter for the solution of underdetermined consistent systems where the constraint is to have the Chebyshev or the L_∞ norm of the solution vector be as small as possible.

Consider the underdetermined system of linear equations

$$\mathbf{Ca} = \mathbf{f}$$

$\mathbf{C} = (c_{ij})$ is a given real n by m matrix of rank k, $k \leq n < m$, and $\mathbf{f} = (f_i)$ is a given real n-vector. We are seeking the m-solution vector $\mathbf{a} = (a_i)$ whose L_∞ or Chebyshev norm

$$\|\mathbf{a}\|_\infty = \max(|a_1|, |a_2|, \ldots, |a_m|) \tag{22.1.1}$$

is minimized.

Based on the steepest descent method, Kolev [11] presented an iterative algorithm for solving both the L_1 (Chapter 20) and L_∞ (this chapter) problems for consistent underdetermined linear equations.

His algorithm results in a procedure of theoretically infinite number of iterations. Kolev [12] then developed his algorithm such that it requires a finite number of iterations. In both of Kolev's methods, the minimum energy (Chapter 23) solution is used to start the iterations.

Based on the duality principle from functional analysis, Cadzow [3, 4] developed a number of important properties related to the solution of the consistent underdetermined systems **Ca** = **f**. As noted in Chapter 20, algorithmic procedures for both the minimum L_1 solution and the minimum L_∞ solution were then developed. Cadzow [5] further refined his algorithm using a column exchange method. This necessitates that matrix **C** satisfies the Haar condition. Next, Cadzow [6] described another algorithm that handles the non-Haar cases but for the full rank case. He also outlined a linear programming scheme for the problem of this chapter. See also Cadzow [7].

As noted in Chapter 20, Dax [8] presented a method for calculating the L_p norm of the solution vector of the consistent system **Ca** = **f**, where $1 < p < \infty$. He presented a primal Newton method for $p > 2$ and a dual Newton method for $1 < p < 2$.

Lucchetti and Mignanego [13] studied the variational perturbations of the minimum effort problem in control theory. They gave a sufficiency characterization for the convergence of the sequence of solutions of the perturbed problem to the original problem.

Gravagne and Walker [9] explored the details of a related subject to the minimum effort problem. In their exploration, they noted that they introduced for the first time a closed-form expression for the minimum effort solution of underdetermined consistent linear systems. They also gave an extended discussion of the minimum effort solutions from a geometric point of view.

In our work, this is reduced to a linear programming problem. No conditions are imposed on matrix **C**, such as the Haar condition or the full rank condition.

Our algorithm [2] has two parts. In part 1, an initial basic feasible solution for the linear programming problem is obtained. The objective function z is calculated next. If $z < 0$, it is easily made positive. The marginal costs are then calculated. Part 2 consists of a slightly modified simplex method The algorithm needs minimum computer storage. The elements of the solution vector **a** are calculated from the objective function and the marginal costs in the final tableau of the programming problem.

In Section 22.2, the linear programming formulation of the problem is presented. In Section 22.3, the algorithm is described, and in Section 22.4, numerical results and comments are given.

22.1.1 Applications of the algorithm

In control theory, this problem is known as the **minimum effort** or **minimum amplitude** problem for discrete linear control systems. See the references in [5, 6, 9].

22.2 The linear programming problem

Let in (22.1.1), $||\mathbf{a}||_\infty = h > 0$. Then this problem may be reduced to a linear programming problem as follows

$$\text{minimize } h$$

subject to

$$-h \le a_i \le h, \quad i = 1, 2, \ldots, m$$

and

$$\mathbf{Ca} = \mathbf{f}$$

The last set of constraints may be replaced by

$$\mathbf{f} \le \mathbf{Ca} \le \mathbf{f}$$

After rearranging the constraints, this problem is conventionally formulated as follows

$$\text{(22.2.1a)} \qquad \text{minimize } Z = \mathbf{e}_{m+1}{}^T \begin{bmatrix} \mathbf{a} \\ h \end{bmatrix}$$

subject to

$$\text{(22.2.1b)} \qquad \begin{aligned} \mathbf{Ca} \quad\quad &\ge \mathbf{f} \\ \mathbf{a} + h\mathbf{e} &\ge \mathbf{0} \\ -\mathbf{Ca} \quad\quad &\ge -\mathbf{f} \\ -\mathbf{a} + h\mathbf{e} &\ge \mathbf{0} \end{aligned}$$

and

(22.2.1c) $$h \geq \mathbf{0}$$

In (22.2.1a), $\mathbf{e}_{m+1}$ is an (m + 1)-vector each element of which is 0 except the $(m + 1)^{th}$ element, which is 1. Again, $\mathbf{e}_{m+1}$ is the last column in an (m + 1)-unit matrix. In (22.2.1b), $\mathbf{e}$ is an m-vector each element of which is 1.

It is more efficient to use the dual of problem (22.2.1). For a related case see ([14], p. 174). The dual formulation of (22.2.1) is

(22.2.2a) $$\text{maximize } z = [\mathbf{f}^T \ \mathbf{0}^T \ -\mathbf{f}^T \ \mathbf{0}^T]\mathbf{b} = \mathbf{g}^T\mathbf{b}$$

subject to

$$\begin{bmatrix} \mathbf{C}^T & \mathbf{I} & -\mathbf{C}^T & -\mathbf{I} \\ \mathbf{0}^T & \mathbf{e}^T & \mathbf{0}^T & \mathbf{e}^T \end{bmatrix} \mathbf{b} = \mathbf{e}_{m+1}$$

For convenience, we write the above equation in the form

(22.2.2b) $$\mathbf{Db} = \mathbf{e}_{m+1}$$

where $\mathbf{D}$ is the coefficient matrix on the l.h.s. of the previous equation.

A simplex tableau for (m + 1) constraints in 2(n + m) variables is to be constructed for this problem. However, we show later that we need to store only n columns of the constraint matrix $\mathbf{D}$.

Let the basis matrix be denoted by the (m + 1) square matrix $\mathbf{B}$. Then the vectors $\mathbf{y}_j$ in the simplex tableau are given by

(22.2.3) $$\mathbf{y}_j = \mathbf{B}^{-1}\mathbf{D}_j, \quad j = 1, 2, \ldots, 2(n + m)$$

$\mathbf{D}_j$ is the jth column of matrix $\mathbf{D}$ of (22.2.2b).

Let the elements of vector $\mathbf{g}$ of (22.2.2a), associated with the basic variables be the (m + 1) vector $\mathbf{g}_B$. Then the marginal costs are

(22.2.4) $$z_j - g_j = \mathbf{g}_B{}^T\mathbf{y}_j - g_j, \quad j = 1, 2, \ldots, 2(n + m)$$

The basic solution $\mathbf{b}_B$ and the objective function z are given by

(22.2.5) $$\mathbf{b}_B = \mathbf{B}^{-1}\mathbf{e}_{m+1}$$

(22.2.6) $$z = \mathbf{g}_B{}^T\mathbf{b}_B$$

22.2.1 Properties of the matrix of constraints

Again, as in the previous two chapters, as well as in other chapters in this book, the analysis in this problem is initiated by the kind of asymmetry matrix **D** has.

The first n columns of the left half of matrix **D**, are the negative of the first n columns of the right half. Again the unit matrix **I** exists in the last m columns of the left half of **D**, and –**I** exists in the second n columns of the right half. That kind of asymmetry in **D** and the existence of unit matrices in matrix **D** will enable us to use a simplex tableau for this problem of only $(m + 1)$ constraints in only n variables, instead of $(m + 1)$ constraints in $2(n + m)$ variables. We call this the **reduced tableau**. This is explained in Section 22.3. Consider first the following analysis.

Definition

Because of the kind of asymmetry of matrix **D** described above, we define any column i, $1 \leq i \leq (n + m)$, and the column $j = (i + (n + m))$ in this matrix as two corresponding columns.

We can show that any two corresponding columns should not appear together in any basis.

Let z and **B** be respectively the optimum objective function and the basis matrix associated with the optimum solution. Then the optimum Chebyshev norm $\|\mathbf{a}\|_\infty = z = Z$, where Z is the optimum solution of problem (22.2.1).

Again, if in the dual problem (22.2.2), a column is in the basis for the optimal solution, its corresponding inequality in the primal (22.2.1b) is an equality ([10], p. 239). As a result, $(\mathbf{a}^T\ z)$ is the solution of the system

(22.2.7) $$\mathbf{B}^T \begin{bmatrix} \mathbf{a} \\ z \end{bmatrix} = \mathbf{g}_B$$

where $\mathbf{g}_B$ is associated with the basic variables for the optimum solution of (22.2.2). See also Lemma 10.2. For further use, we write (22.2.7) in the form

(22.2.7a) $$(\mathbf{a}^T\ z) = \mathbf{g}_B{}^T(\mathbf{B})^{-1}$$

Consider an example of obtaining the minimum L_∞ solution of the underdetermined system $\mathbf{Ca} = \mathbf{f}$, for 2 equations of rank 2 in 4 unknowns. In this case (22.2.7) consists of 5 equations in 5 unknowns, the 4 elements of $\mathbf{a}$ and z. The equations would be the first 2 and a suitable 3 of the remaining 4 of the following system, where $\rho_i = +1$ or -1. If column i, i = 1, 2, …, n+m of the matrix of constraints $\mathbf{D}$ in (22.2.2b), is in the basis, $\rho_i = +1$, and $\rho_i = -1$ if instead its corresponding column is in the basis

(22.2.7b)
$$\begin{bmatrix} \rho_1 c_{11} & \rho_1 c_{12} & \rho_1 c_{13} & \rho_1 c_{14} & 0 \\ \rho_2 c_{21} & \rho_2 c_{22} & \rho_2 c_{23} & \rho_2 c_{24} & 0 \\ \rho_3 & 0 & 0 & 0 & 1 \\ 0 & \rho_4 & 0 & 0 & 1 \\ 0 & 0 & \rho_5 & 0 & 1 \\ 0 & 0 & 0 & \rho_6 & 1 \end{bmatrix} \begin{bmatrix} a_1 \\ a_2 \\ a_3 \\ a_4 \\ z \end{bmatrix} = \begin{bmatrix} \rho_1 f_1 \\ \rho_2 f_2 \\ 0 \\ 0 \\ 0 \\ 0 \end{bmatrix}$$

In this example, the first 2 equations are in system (22.2.7a) and thus system $\mathbf{Ca} = \mathbf{f}$ is satisfied. Also, since the 3 last equations but one in (22.2.7b) are in (22.2.7a), 3 out of the 4 elements of $\mathbf{a}$, each equals +z or –z.

It is clear from this example that in general, (m + 1 – n) elements of $\mathbf{a}$, each equals +z or –z and therefore the remaining (n – 1) elements of $\mathbf{a}$ in absolute value, each ≤ z. See Lemmas 22.5 and 22.6 below.

Lemma 22.1

In any stage of the computation, we have the following relations between the corresponding columns i and j, $1 \le i \le (n+m)$

(22.2.8a)
$$\mathbf{y}_i + \mathbf{y}_j = 0, \quad i = 1, 2, \ldots, n$$

and

(22.2.8b)
$$\mathbf{y}_i + \mathbf{y}_j = 2\mathbf{b}_B, \quad i = n+1, n+2, \ldots, n+m$$

Also, for the marginal costs

(22.2.9a)
$$(z_i - f_i) + (z_j - f_j) = 0, \quad i = 1, 2, \ldots, n$$

and

(22.2.9b)
$$(z_i - f_i) + (z_j - f_j) = 2z, \quad i = n+1, n+2, \ldots, n+m$$

Lemma 22.2

(a) For each basic solution, feasible or not, there correspond two bases $\mathbf{B}_{(1)}$ and $\mathbf{B}_{(2)}$, each of which determines the same basic solution. Every column in one of the bases has its corresponding column in the other basis, arranged in the same order.

(b) The two values of z are equal in magnitude but opposite in sign. This lemma corresponds to Lemma 10.7.

Lemma 22.3

Consider the two bases $\mathbf{B}_{(1)}$ and $\mathbf{B}_{(2)}$, defined in the previous Lemma, and let us use (22.2.3-5) to construct two simplex tableaux $T_{(1)}$ and $T_{(2)}$ that correspond respectively to $\mathbf{B}_{(1)}$ and $\mathbf{B}_{(2)}$. Let i be the corresponding column to column j, where $1 \leq j \leq 2(n+m)$. Then we have $\mathbf{y}_i$ in $T_{(1)} = \mathbf{y}_j$ in $T_{(2)}$.

This lemma corresponds to Lemma 10.8.

Lemma 22.4

At any stage of the computation, the $(m+1)^{th}$ column of matrix $\mathbf{B}^{-1}$ equals the basic solution $\mathbf{b}_B$. This is obvious from (22.2.5) since $\mathbf{e}_{m+1}$ is the $(m+1)^{th}$ column in an $(m+1)$-unit matrix.

Assume that we have obtained an optimum basic feasible solution to the linear programming problem (22.2.2). Then it is concluded earlier that $(m+1-n)$ elements of $\mathbf{a}$, each equals +z or –z and the remaining $(n-1)$ elements of $\mathbf{a}$ in absolute value, each $\leq$ z.

Note 1

If $\text{rank}(\mathbf{C}|\mathbf{f}) = \text{rank}(\mathbf{C}) = k < n$, then $(m+1-k)$ elements of $\mathbf{a}$, each equals +z or –z and the remaining $(k-1)$ elements of $\mathbf{a}$ in absolute value, each $\leq$ z.

Lemma 22.5

If column j, $(n+1) \leq j \leq (n+m)$, of the matrix of constraints of (22.2.2b) is in the final basis, then $a_{j-n} = -z$. But if instead, the corresponding column of j is in the final basis, $a_{j-n} = z$.

Proof:

The proof of the lemma is established from the structure of equation (22.2.7b).

Lemma 22.6

The (n – 1) elements of **a** whose absolute value $\leq z$, each is calculated from the marginal cost of a non-basic column in the final (condensed) simplex tableau. These non-basic columns are not corresponding columns of any column in the basis. Let column j, $(n + 1) \leq j \leq (n + m)$ of the matrix of constraints be such non-basic column. Then

(22.2.10a) $a_{j-n} = (z_j - g_j) - z, \quad (n + 1) \leq j \leq (n + m)$

(22.2.10b) $a_{j-2n-m} = z - (z_j - g_j), \quad (2n + m + 1) \leq j \leq 2(n + m)$

Proof:

Consider the case $(n + 1) \leq j \leq (n + m)$. From (22.2.2b) and (22.2.3-4)

$$(z_j - g_j) = z_j = \mathbf{g}_B{}^T \mathbf{B}^{-1} \begin{bmatrix} u_{j-n} \\ 1 \end{bmatrix}$$

where u_j is the jth column of an m-unit matrix. Hence

$$(z_j - g_j) = \mathbf{g}_B{}^T[(\mathbf{B})_{j-n}{}^{-1} + (\mathbf{B})_{m+1}{}^{-1}]$$

where $(\mathbf{B})_{j-n}{}^{-1}$ and $(\mathbf{B})_{m+1}{}^{-1}$ are respectively the (j – n)th and the $(m + 1)^{th}$ column of $(\mathbf{B})^{-1}$. Finally, from (22.2.7) and (22.2.6) and Lemma 22.5, (22.2.10a) is proved. In the same way, (22.2.10b) is proved.

Lemma 22.7

If the set of equations $\mathbf{Ca} = \mathbf{f}$ is inconsistent, i.e., $\text{rank}(\mathbf{C}|\mathbf{f}) > \text{rank}(\mathbf{C})$, then the solution of the linear programming problem (22.2.2) would be unbounded.

22.3 Description of the algorithm

Lemma 22.1 relates the columns $\mathbf{y}_i$, $1 \leq i \leq (n + m)$ in the simplex tableau and their corresponding column $\mathbf{y}_j$, $j = i + (n + m)$. The same is true for their marginal costs. If one column and its marginal cost is known, the corresponding column and its marginal cost is easily derived. We start by constructing a simplex tableau for problem

(22.2.2), for (m + 1) constraints in only (n + m) variables. Obviously, let these be the first (n + m) elements of the 2(n + m)-vector **b**. This tableau is the condensed tableaux.

In part 1 of the algorithm, an initial basic feasible solution to the programming problem (22.2.2) is obtained. We also calculate the initial objective function z (≥ 0) and the initial marginal costs ($z_i - g_i$).

We take advantage of the existence of the m-unit sub-matrices **I** in matrix **D** in (22.2.2b) in obtaining an initial basic feasible solution, and we need no artificial variables.

The m columns (n + 1), (n + 2), …, (n + m), in matrix **D**, each is a column in an m-unit matrix augmented by a 1 as the $(m + 1)^{th}$ element. We chose these m columns, or their corresponding columns, to form the first m columns in the initial basis matrix **B**. This is simply done by performing m Gauss-Jordan eliminations for each of these columns, which consists of one step only. This is the step needed to eliminate the 1 in the $(m + 1)^{th}$ position of each column. The choice between column (n + i), i = 1, 2 …, m, or its corresponding column to form the first m columns in the initial basis matrix **B** is given next.

Consider any one of the first n columns in matrix **D**. Denote this column by **X**. Consider element i, i = 1, 2, …, m, in succession of column **X**. If in **X**, element i is ≤ 0, we chose column (n + i) in matrix **D** to form the ith column of **B**. If element i in **X** is > 0, we chose instead the corresponding column to column (n + i) to form the ith column of **B**. When all these m columns enter the basis, the first m elements of **X**, each would keep its value, with a negative sign and the $(m + 1)^{th}$ element of **X** would be > 0. In fact, this $(m + 1)^{th}$ element would equal the sum of the absolute values of the first m elements in **X** = 1. Column **X** will now be chosen to be the $(m + 1)^{th}$ column of **B**. The process described so far, guarantees that when column **X** enters the basis as the $(m + 1)^{th}$ column of **B**, the initial basic solution would be feasible. That is, with all the element of $\mathbf{b}_B \geq 0$.

The objective function z is then calculated from (22.2.6). If z < 0, we use Lemmas 22.2 and 22.3 and replace the basis matrix by its corresponding one. We also replace the columns of the condensed simplex tableau by their corresponding columns. In effect, we keep the simplex tableau unchanged, except for the **f** values and z. Such parameters have their signs reversed. See [1] and also Chapter 10. The marginal costs ($z_j - g_j$) are then calculated from (22.2.4). We now have an initial basic feasible solution and $z \geq 0$.This ends part 1 of the algorithm.

Part 2 of the algorithm is the ordinary simplex algorithm. However, the column to enter the basis is that which has the most negative marginal cost among the non-basic columns in the current tableau and their corresponding columns. Lemma 22.1 is used for calculating the marginal costs of the corresponding columns. Finally, the elements of the solution vector **a** to the given problem **Ca** = **f** are calculated from Lemmas 22.5 and 22.6.

The steps described here are explained by the following detailed numerical example.

Example 22.1

Obtain the minimum L_∞ solution of the underdetermined system of linear equations.

$$\begin{aligned} 7a_1 - 4a_2 + 5a_3 + 3a_4 &= -30 \\ -2a_1 + a_2 + 5a_3 + 4a_4 &= 15 \end{aligned}$$

Here, matrix C is a 2 by 4 matrix of rank 2 and the system is consistent. This example is a hybrid of example 1 in ([2], p. 65). The Initial Data and the condensed simplex tableaux are shown. Again D_j, $1 \leq j \leq 2(n + m)$ is the jth column of the matrix of constraints D in (22.2.2b). The pivot elements are shown between brackets, as first seen in Tableau 22.3.1**.

Initial Data

g	–30	15	0	0	0	0
$\mathbf{b}_B$	$\mathbf{D}_1$	$\mathbf{D}_2$	$\mathbf{D}_3$	$\mathbf{D}_4$	$\mathbf{D}_5$	$\mathbf{D}_6$
0	7	–2	1	0	0	0
0	–4	1	0	1	0	0
0	5	5	0	0	1	0
0	3	4	0	0	0	1
1	0	0	1	1	1	1

Let column **X** be the first column in the initial data, column $\mathbf{D}_1$. We see that elements i = 1, 3, and 4 of $\mathbf{D}_1$ are > 0. Hence, we chose columns [(i + n) + (n + m)], i.e, columns 9, 11, and 12 to be columns

1, 3 and 4 of the basis matrix **B**. Again since the second element of $\mathbf{D}_1$ is < 0, we chose column (i + n), i.e., column 4 to form column 2 of the basis matrix **B**.

The basis (m + 1)-matrix **B** is now formed from columns 9, 4, 11, 12 and 1 of the matrix of constraints **D**. As indicated above, we perform 4 Gauss-Jordan steps to eliminate the 1 in the 5th position of each of $\mathbf{D}_3$, $\mathbf{D}_4$, $\mathbf{D}_5$ and $\mathbf{D}_6$. We then perform another Gauss-Jordan step to make column $\mathbf{D}_1$ the fifth column in the basis matrix **B**. This gives Tableau 22.3.1, where we also calculate the objective function z from (22.2.5-6).

Tableau 22.3.1 (part 1)

g	–30	15	0	0	0	0
$\mathbf{b}_B$	$\mathbf{D}_1$	$\mathbf{D}_2$	$\mathbf{D}_9$	$\mathbf{D}_4$	$\mathbf{D}_{11}$	$\mathbf{D}_{12}$
0.368	0	4.21	1	0	0	0
0.211	0	2.26	0	1	0	0
0.263	0	–3.42	0	0	1	0
0.158	0	–3.05	0	0	0	1
0.053	1	0.316	0	0	0	0

z = –1.59

Tableau 22.3.1*

g	30	–15	0	0	0	0
$\mathbf{b}_B$	$\mathbf{D}_7$	$\mathbf{D}_8$	$\mathbf{D}_3$	$\mathbf{D}_{10}$	$\mathbf{D}_5$	$\mathbf{D}_6$
0.368	0	4.21	1	0	0	0
0.211	0	2.26	0	1	0	0
0.263	0	–3.42	0	0	1	0
0.158	0	–3.05	0	0	0	1
0.053	1	0.316	0	0	0	0
z = 1.59	0	24.47	0	0	0	0

In Tableau 22.3.1, however, the initial basic solution is feasible but z < 0. Hence, we make use of Lemmas 22.2 and 22.3 and replace Tableau 22.3.1 by Tableau 22.3.1*. This tableau is the same as

Tableau 22.3.1, except for the **f** values and z; their signs are reversed. Also, the columns in Tableau 22.3.1* are the corresponding columns of Tableau 22.3.1. This ends part 1 of the algorithm.

In Tableau 22.3.1*, $\mathbf{D}_2$ (the corresponding column to $\mathbf{D}_8$) has the most negative marginal cost. Hence, we replace $\mathbf{y}_8$ by $\mathbf{y}_2$. From (22.2.8a) $\mathbf{y}_2 = -\mathbf{y}_8$, and from (22.2.9a), $(z_2 - g_2) = -(z_8 - g_8)$. This gives Tableau 22.3.1**. In Tableau 22.3.1**, $\mathbf{y}_2$ replaces $\mathbf{y}_6$ in the basis. This gives Tableau 22.3.2.

Tableau 22.3.1** (part 2)

g	30	15	0	0	0	0
$\mathbf{b}_B$	$\mathbf{D}_7$	$\mathbf{D}_2$	$\mathbf{D}_3$	$\mathbf{D}_{10}$	$\mathbf{D}_5$	$\mathbf{D}_6$
0.368	0	–4.21	1	0	0	0
0.211	0	–2.26	0	1	0	0
0.263	0	3.42	0	0	1	0
0.158	0	(3.05)	0	0	0	1
0.053	1	–0.316	0	0	0	0
z = 1.59	0	–24.47	0	0	0	0

Tableau 22.3.2

g	30	15	0	0	0	0
$\mathbf{b}_B$	$\mathbf{D}_7$	$\mathbf{D}_2$	$\mathbf{D}_3$	$\mathbf{D}_{10}$	$\mathbf{D}_5$	$\mathbf{D}_6$
0.586	0	0	1	0	0	1.379
0.328	0	0	0	1	0	0.741
0.086	0	0	0	0	1	–1.121
0.052	0	1	0	0	0	0.328
0.069	1	0	0	0	0	0.103
z = 2.845	0	0	0	0	0	8.017

In Tableau 22.3.2, $\mathbf{y}_{12}$ has the most negative marginal cost and it replaces its corresponding column $\mathbf{y}_6$. This gives Tableau 22.3.2*. In Tableau 22.3.2*, $\mathbf{y}_{12}$ has the most negative marginal cost. It replaces $\mathbf{y}_5$ in the basis. This gives Tableau 22.3.3, which gives the optimal objective function z = 3. From Lemmas 22.5 and 22.6, the optimum solution z = 3 and $\mathbf{a} = (-3, 3, -1.2, 3)^T$.

Tableau 22.3.2*

g	30	15	0	0	0	0
$\mathbf{b}_B$	$\mathbf{D}_1$	$\mathbf{D}_2$	$\mathbf{D}_3$	$\mathbf{D}_{10}$	$\mathbf{D}_5$	$\mathbf{D}_{12}$
0.586	0	0	1	0	0	–0.207
0.328	0	0	0	1	0	–0.086
0.086	0	0	0	0	1	(1.293)
0.052	0	1	0	0	0	–0.224
0.069	1	0	0	0	0	0.035
z = 2.845	0	0	0	0	0	–2.328

Tableau 22.3.3

g	30	15	0	0	0	0
$\mathbf{b}_B$	$\mathbf{D}_7$	$\mathbf{D}_2$	$\mathbf{D}_3$	$\mathbf{D}_{10}$	$\mathbf{D}_5$	$\mathbf{D}_{12}$
0.6	0	0	1	0	0.16	0
0.333	0	0	0	1	0.067	0
0.067	0	0	0	0	0.773	1
0.067	0	1	0	0	0.173	0
0.067	1	0	0	0	–0.027	0
z = 3	0	0	0	0	1.8	0

22.3.1 The reduced tableaux

A careful look to the simplex Tableaux in this solved example, we see that 5 out of 6, i.e., (m + 1) out of (n + m) columns in the condensed tableaux are actually 5 columns in a 6-unit matrix. Such columns need not be accounted for nor they need to be stored. The condensed simplex tableaux may be condensed more. We denote such tableaux as the reduced tableaux. In the reduced tableaux we calculate only $\mathbf{b}_B$ and (n – 1) columns and their marginal costs. These (n – 1) columns are the non-basic columns in the condensed tableaux. An index indicator is used for the columns that we actually use in the tableaux.

22.4 Numerical results and comments

LA_Effort() implements this Minimum Effort algorithm, and DR_Effort() demonstrates 7 test cases.

Table 22.4.1 shows the results calculated in single-precision. These are the same 7 examples that were solved in Chapters 20 and 21.

Table 22.4.1

Example	C(n×m)	Iterations	z
1	4×5	3	0.906
2	4×8	8	0.537
3	3×5	4	3.0
4	4×6	7	0.372
5	4×100	49	0.244
6	5×100	51	0.303
7	7×100	60	0.508

For each example, the size of matrix **C**, the number of iterations and the optimum L_∞ norm z are given. This method can be characterized by the following features. All the calculations are done in the reduced (not even in the condensed) simplex tableaux. An initial basic feasible solution for the linear programming problem as well as the initial objective function z (≥ 0) are obtained with minimum effort. We do not need to calculate the marginal costs until the end of part 1 of the algorithm. The inverse of the basis matrix, $\mathbf{B}^{-1}$, is never calculated. The elements of the solution vector **a** of the given system **Ca** = **f** are calculated from z and the marginal costs of the final reduced tableau. Rank(**C**) is known at the end of the solution. See example 1 in ([2], p. 65).

The closest technique to our algorithm is that of Cadzow [6]. Using our notation, we compared the number of arithmetic operations of Cadzow's method with that of ours. The number of multiplications/divisions (m/d) per iteration in [6] is > (3nm + m). In our method, the number of (m/d) required per iteration, i.e., to change a simplex tableau, is n(m + 3). This is about 1/3 of the iterations in [6]. Again, the numerical example solved by Cadzow ([5], p. 616) was solved by our program and the execution time was half the time given by Cadzow, who used a faster computer.

References

1. Abdelmalek, N.N., Chebyshev solution of overdetermined systems of linear equations, *BIT*, 15(1975)117-129.
2. Abdelmalek, N.N., Minimum L_∞ solution of underdetermined systems of linear equations, *Journal of Approximation Theory*, 20(1977)57-69.
3. Cadzow, J.A., Algorithm for the minimum-effort problem, *IEEE Transactions on Automatic Control*, 16(1971)60-63.
4. Cadzow, J.A., Functional analysis and the optimal control of linear discrete systems, *International Journal of Control*, 17(1973)481-495.
5. Cadzow, J.A., A finite algorithm for the minimum l_∞ solution to a system of consistent linear equations, *SIAM Journal on Numerical Analysis*, 10(1973)607-617.
6. Cadzow, J.A., An efficient algorithmic procedure for obtaining a minimum l_∞-norm solution to a system of consistent linear equations, *SIAM Journal on Numerical Analysis*, 11(1974)1151-1165.
7. Cadzow, J.A., Minimum-amplitude control of linear discrete systems, *International Journal of Control*, 19(1974)765-780.
8. Dax, A., Methods for calculating l_p-minimum norm solutions of consistent linear systems, *Journal of Optimization Theory and Applications*, 83(1994)333-354.
9. Gravagne, I.A. and Walker, I.D., On the structure of minimum effort solutions with application to kinematic redundancy resolution, *IEEE Transactions on Robotics and Automation*, 16(2000)855-863.
10. Hadley, G., *Linear Programming*, Addison-Wesley, Reading, MA, 1962.
11. Kolev, L.V., Iterative algorithm for the minimum fuel and minimum amplitude problems, *International Journal of Control*, 21(1975)779-784.

12. Kolev, L.V., Algorithm of finite number of iterations for the minimum fuel and minimum amplitude control problems, *International Journal of Control*, 22(1975)97-102.
13. Lucchetti, R. and Mignanego, F., Variational perturbation of the minimum effort problem, *Journal of Optimization Theory and Applications*, 30(1980)485-499.
14. Osborne, M.R. and Watson, G.A., On the best linear Chebyshev approximation. *Computer Journal*, 10(1967)172-177.

22.5 DR_Effort

```
/*-------------------------------------------------------------------
DR_Effort
---------------------------------------------------------------------
This program is a driver for the function LA_Effort(), which
calculates the Chebyshev solution of an underdetermined system
of consistent linear equations.

Given is the underdetermined consistent system

                        c*a = f

"c" is a given real n by m matrix of rank k, k <= n <= m.
"f" is a given real n vector.

It is required to calculate the m vector "a" for this system, such
that the Chebyshev norm z of "a"

             z = max|a[j]|,   j = 1, 2, ..., m

is as small as possible.

In control theory, this problem is known as the "Minimum Effort" or
"Minimum Amplitude" problem for discrete linear control systems.

This program contains 7 examples whose results appear in the text.
-------------------------------------------------------------------*/

#include "DR_Defs.h"
#include "LA_Prototypes.h"

#define Nf_ROWS     25
#define Mf_COLS     151
#define N1f         4
#define M1f         5
#define N2f         4
#define M2f         8
#define N3f         3
#define M3f         5
#define N4f         4
#define M4f         6
#define N5f         4
#define M5f         100
```

```
#define N6f         5
#define M6f         100
#define N7f         7
#define M7f         100

void DR_Effort (void)
{
    /*-----------------------------------------
      Constant matrices/vectors
    -----------------------------------------*/
    static tNumber_R b1init[N1f][M1f] =
    {
        { 2.0,  1.0, -3.0,  5.0, 3.0  },
        { 2.0,  1.0, -3.0,  5.0, 3.0  },
        { 2.0,  1.0, -3.0,  5.0, 3.0  },
        {-8.0, -4.0,  2.0,  5.0, 6.0  }
    };

    static tNumber_R f1[N1f+1] =
    {   NIL,
        10.0, 10.0, 10.0, 8.0
    };

    static tNumber_R b2init[N2f][M2f] =
    {
        {-6.0,  4.0,  2.0, -8.0,  5.0, -1.0,  6.0,  3.0  },
        { 7.0, -3.0,  9.0,  5.0, -9.0,  8.0,  7.0, -4.0  },
        { 9.0,  7.0, -5.0,  2.0,  7.0,  0.0,  1.0,  8.0  },
        { 9.0, -3.0,  2.0,  4.0,  0.0,  3.0,  0.0,  1.0  }
    };

    static tNumber_R f2[N2f+1] =
    {   NIL,
        5.0, 7.0, -9.0, 1.0
    };

    static tNumber_R b3init[N3f][M3f] =
    {
        { 7.0, -4.0,  5.0, 3.0, 1.0},
        {-2.0,  1.0,  5.0, 4.0, 1.0},
        { 5.0, -3.0, 10.0, 7.0, 2.0}
    };

    static tNumber_R f3[N3f+1] =
    {   NIL,
```

```
        -30.0, 15.0, -15
    };

    static tNumber_R b4init[N4f][M4f] =
    {
        {2.0, -1.0, -4.0,  0.0, -3.0, 1.0 },
        {2.0, -1.0, -4.0,  0.0, -3.0, 1.0 },
        {5.0,  1.0,  3.0,  1.0, -2.0, 0.0 },
        {1.0, -2.0, -1.0, -5.0,  1.0, 4.0 }
    };

    static tNumber_R f4[N4f+1] =
    {   NIL,
        2.0, 2.0, 1.0, -4.0
    };

    static tNumber_R b5init[N5f][M5f] =
    {
        {2.0, -1.0, -4.0,  0.0, -3.0, 1.0 },
        {2.0, -1.0, -4.0,  0.0, -3.0, 1.0 },
        {5.0,  1.0,  3.0,  1.0, -2.0, 0.0 },
        {1.0, -2.0, -1.0, -5.0,  1.0, 4.0 }
    };

    static tNumber_R f5[N5f+1] =
    {   NIL,
        2.0, 2.0, 1.0, -4.0
    };

    /*---------------------------------------
      Variable matrices/vectors
    ---------------------------------------*/
    tMatrix_R   ct  = alloc_Matrix_R (Mf_COLS, Nf_ROWS);
    tVector_R   f   = alloc_Vector_R (Nf_ROWS);
    tVector_R   a   = alloc_Vector_R (Mf_COLS);
    tMatrix_R   fay = alloc_Matrix_R (N7f, M7f);
    tVector_R   fy  = alloc_Vector_R (N7f + 1);

    tMatrix_R   b1  = init_Matrix_R (&(b1init[0][0]), N1f, M1f);
    tMatrix_R   b2  = init_Matrix_R (&(b2init[0][0]), N2f, M2f);
    tMatrix_R   b3  = init_Matrix_R (&(b3init[0][0]), N3f, M3f);
    tMatrix_R   b4  = init_Matrix_R (&(b4init[0][0]), N4f, M4f);

    tNumber_R   pi, dx, x1, x2, x3, z;
    int         irank, iter;
```

```
int         i, j, m, n, Iexmpl;

eLaRc       rc = LaRcOk;

pi = 4.0*atan (1.0);
dx = 2.*pi/99;

for (j = 1; j <= 100; j++)
{
    x1 = (j - 1)* dx;
    x2 = x1 + x1;
    x3 = x2 + x1;
    fay[1][j] = 1.0;
    fay[2][j] = sin (x1);
    fay[3][j] = cos (x1);
    fay[4][j] = sin (x2);
    fay[5][j] = cos (x2);
    fay[6][j] = sin (x3);
    fay[7][j] = cos (x3);
}

for (i = 1; i <= 7; i++)
{
    fy[i] = (10.0+15.0*fay[2][i]-7.0*fay[3][i]+
            9.0*fay[4][i]) * (1.0 + 0.01*fay[4][i]);
}

prn_dr_bnr ("DR_Effort, "
            "Chebyshev Solution of an Underdetermined System");

for (Iexmpl = 1; Iexmpl <= 7; Iexmpl++)
{
    switch (Iexmpl)
    {
        case 1:
            n = N1f;
            m = M1f;
            for (i = 1; i <= n; i++)
            {
                f[i] = f1[i];
                for (j = 1; j <= m; j++)
                    ct[j][i] = b1[i][j];
            }
            break;
```

```
case 2:
    n = N2f;
    m = M2f;
    for (i = 1; i <= n; i++)
    {
        f[i] = f2[i];
        for (j = 1; j <= m; j++)
            ct[j][i] = b2[i][j];
    }
    break;

case 3:
    n = N3f;
    m = M3f;
    for (i = 1; i <= n; i++)
    {
        f[i] = f3[i];
        for (j = 1; j <= m; j++)
            ct[j][i] = b3[i][j];
    }
    break;
case 4:
    n = N4f;
    m = M4f;
    for (i = 1; i <= n; i++)
    {
        f[i] = f4[i];
        for (j = 1; j <= m; j++)
            ct[j][i] = b4[i][j];
    }
    break;
case 5:
    n = N5f;
    m = M5f;
    for (i = 1; i <= n; i++)
    {
        f[i] = fy[i];
        for (j = 1; j <= m; j++)
            ct[j][i] = fay[i][j];
    }
    break;
case 6:
    n = N6f;
    m = M6f;
    for (i = 1; i <= n; i++)
```

```
            {
                f[i] = fy[i];
                for (j = 1; j <= m; j++)
                    ct[j][i] = fay[i][j];
            }
            break;
        case 7:
            n = N7f;
            m = M7f;
            for (i = 1; i <= n; i++)
            {
                f[i] = fy[i];
                for (j = 1; j <= m; j++)
                    ct[j][i] = fay[i][j];
            }
            break;
        default:
            break;
    }

    prn_algo_bnr ("Effort");

    prn_example_delim();
    PRN ("Example #%d: Size of matrix \"c\" %d by %d\n",
         Iexmpl, n, m);
    prn_example_delim();
    PRN ("Chebyshev Solution of an Underdetermined System\n");
    prn_example_delim();
    PRN ("r.h.s. Vector \"f\"\n");
    prn_Vector_R (f, n);
    PRN ("Transpose of Coefficient Matrix, \"ct\"\n");
    prn_Matrix_R (ct, m, n);

    rc = LA_Effort (m, n, ct, f, &irank, &iter, a, &z);

    if (rc >= LaRcOk)
    {
        PRN ("\n");
        PRN ("Results of the Minimum Effort Solution\n");
        PRN ("Minimum Effort solution vector \"a\"\n");
        prn_Vector_R (a, m);
        PRN ("Chebyshev norm of vector \"a\", ||a|| "
            "= %8.4f\n", z);
        PRN ("Rank of matrix \"c\" = %d, No. of Iterations "
            "= %d\n", irank, iter);
```

```
        }

        LA_check_rank_def (n, irank);
        prn_la_rc (rc);
    }

    free_Matrix_R (ct, Mf_COLS);
    free_Vector_R (f);
    free_Vector_R (a);
    free_Matrix_R (fay, N7f);
    free_Vector_R (fy);

    uninit_Matrix_R (b1);
    uninit_Matrix_R (b2);
    uninit_Matrix_R (b3);
    uninit_Matrix_R (b4);
}
```

22.6 LA_Effort

```
/*-----------------------------------------------------------------------
LA_Effort
-------------------------------------------------------------------------
This program solves an underdetermined system of consistent linear
equations whose solution vector has a minimum Chebyshev norm.  The
underdetermined system is of the form

                          c*a = f

"c" is a given real n by m matrix of rank k, k <= n <= m.
"f" is a given real n vector.

It is required to calculate the m vector "a" for this system such
that the Chebyshev norm "z" of vector "a"

               z = max|a[j]|,   j = 1, 2, ..., m

is as small as possible.

In control theory, this problem is known as the "Minimum Effort"
or "Minimum Amplitude" problem for discrete linear control systems.

Inputs
m       Number of columns of matrix "c" in the system c*a = f.
n       Number of rows of matrix "c" in the system c*a = f.
ct      A real (m + 1) by n matrix, Its first m row and its n columns
        contain the transpose of matrix "c" of the system c*a = f.
f       A real n vector containing the r.h.s. of the system c*a = f.

Outputs
irank   The calculated rank of matrix "c".
iter    The number of iterations or the number of times the simplex
        tableau is changed by a Gauss-Jordan step.
a       A real m vector whose elements are the solution of system
        c*a = f.
z       The minimum Chebyshev norm of the solution vector "a".

Returns one of
        LaRcSolutionUnique
        LaRcSolutionProbNotUnique
        LaRcNoFeasibleSolution
        LaRcErrBounds
```

```
        LaRcErrNullPtr
        LaRcErrAlloc
-------------------------------------------------------------------*/

#include "LA_Prototypes.h"

eLaRc LA_Effort (int m, int n, tMatrix_R ct, tVector_R f,
    int *pIrank, int *pIter, tVector_R a, tNumber_R *pZ)
{
    tVector_I   ic  = alloc_Vector_I (m + 1);
    tVector_I   ip  = alloc_Vector_I (n);
    tVector_I   ib  = alloc_Vector_I (m + n);
    tVector_I   nb  = alloc_Vector_I (n);
    tVector_R   zc  = alloc_Vector_R (n);

    int         i = 0, ii = 0, ijk = 0, iout = 0, itest = 0, ivo = 0;
    int         j = 0, l, jc = 0, jin = 0, m1 = 0, nm = 0;
    tNumber_R   bignum = 0.0, d = 0.0, pivot = 0.0;

    /* Validation of the data before executing the algorithm */
    eLaRc       rc = LaRcSolutionUnique;
    VALIDATE_BOUNDS ((0 < n) && (n < m));
    VALIDATE_PTRS   (ct && f && pIrank && pIter && a && pZ);
    VALIDATE_ALLOC  (ic && ip && ib && nb && zc);

    /* Part 1 of the algorithm, */
    bignum = 1.0 / (EPS*EPS);
    *pIter = 0;
    *pIrank = n;
    m1 = m + 1;
    nm = n + m;
    *pZ = 0.0;

    /* Program initialization */
    LA_effort_init (m, n, ct, ib, ic, ip, nb);
    iout = m1;
    jin = 1;
    ic[iout] = 0;
    jc = nb[jin];
    ii = 1;

    /* A Gauss-Jordan elimination step */
    pivot = ct[iout][jin];
    LA_effort_gauss_jordn (iout, jin, m, n, pivot, ct, ic, nb);
    *pIter = *pIter + 1;
```

```
    /* Part 2 of the algorithm, */
    LA_effort_marg_costs (m, n, ct, f, ib, zc);

    /* Calculate the results */
    if (n == 1)
    {
        rc = LA_effort_res (m, n, ct, ib, ic, ip, nb, zc, pIrank, a,
                            pZ);
        GOTO_CLEANUP_RC (rc);
    }

    /* Part 3 of the algorithm */
    for (ijk = 1; ijk <= m*m; ijk++)
    {
        /* Determine the vector that enters the basis */
        ivo = 0;
        LA_effort_vent (&ivo, &jin, n, nb, zc, pZ);

        /* Calculate the results */
        if (ivo == 0)
        {
            rc = LA_effort_res (m, n, ct, ib, ic, ip, nb, zc, pIrank,
                                a, pZ);
            GOTO_CLEANUP_RC (rc);
        }
        jc = nb[jin];
        if (ivo != 1)
        {
            ib[jc] = -ib[jc];
            if (jc > n)
            {
                for (i = 1;  i <= m1; i++)
                {
                    ct[i][jin] = ct[i][1] + ct[i][1] - ct[i][jin];
                }
                zc[jin] = zc[1] + zc[1] - zc[jin];
            }
            else
            {
                for (i = 1; i <= m1; i++)
                {
                    ct[i][jin] = -ct[i][jin];
                }
                zc[jin] = -zc[jin];
```

```
                f[jc] = -f[jc];
            }
        }

        itest = 0;
        /* Determine the vector that leaves the basis */
        LA_effort_vleav (&itest, jin, &iout, m, ct);

        /* No feasible solution */
        if (itest != 1)
        {
            GOTO_CLEANUP_RC (LaRcNoFeasibleSolution);
        }

        /* A Gauss-Jordan elimination step */
        pivot = ct[iout][jin];

        LA_effort_gauss_jordn (iout, jin, m, n, pivot, ct, ic, nb);
        *pIter = *pIter + 1;
        d = zc[jin];
        for (j = 1; j <= n; j++)
        {
            if (j != jin)
            {
                zc[j] = zc[j] - d * (ct[iout][j]);
            }
        }
        zc[jin] = -zc[jin]/pivot;
        l = ic[m1];
        if (l != ii)
        {
            swap_elems_Vector_I (ip, l, ii);
            ii = l;
        }
    }

CLEANUP:

    free_Vector_I (ic);
    free_Vector_I (ip);
    free_Vector_I (ib);
    free_Vector_I (nb);
    free_Vector_R (zc);

    return rc;
```

```
}

/*-------------------------------------------------------------------
A Gauss-Jordan elimination step in LA_Effort()
-------------------------------------------------------------------*/
void LA_effort_gauss_jordn (int iout, int jin, int m, int n,
    tNumber_R pivot, tMatrix_R ct, tVector_I ic, tVector_I nb)
{
    int         i, j, k, m1;

    m1 = m + 1;
    k = nb[jin];
    nb[jin] = ic[iout];
    ic[iout] = k;
    for (j = 1; j <= n; j++)
        ct[iout][j] = ct[iout][j]/pivot;
    for (i = 1; i <= m1; i++)
    {
        if (i != iout)
        {
            for (j = 1; j <= n; j++)
            {
                if (j != jin)
                {
                    ct[i][j] = ct[i][j] - ct[i][jin] * (ct[iout][j]);
                }
            }
            ct[i][jin] = -ct[i][jin]/pivot;
        }
        ct[iout][jin] = 1.0/pivot;
    }
}

/*-------------------------------------------------------------------
Initialization of LA_Effort()
-------------------------------------------------------------------*/
void LA_effort_init (int m, int n, tMatrix_R ct, tVector_I ib,
    tVector_I ic, tVector_I ip, tVector_I nb)
{
    int     i, j, k, m1, nm;

    m1 = m + 1;
    nm = n + m;
    for (j = 1; j <= n; j++)
    {
```

```
        nb[j] = j;
        ip[j] = j;
        ct[m1][j] = 0.0;
    }
    for (j = 1; j <= nm; j++)
    {
        ib[j] = 1;
    }
    for (i = 1; i <= m; i++)
    {
        k = i + n;
        ic[i] = k;
        if (ct[i][1] >= EPS)
        {
            for (j = 1; j <= n; j++)
            {
                ct[i][j] = -ct[i][j];
                ib[k] = -1;
            }
        }
        for (j = 1; j <= n; j++)
        {
            ct[m1][j] = ct[m1][j] - ct[i][j];
        }
    }
}

/*-----------------------------------------------------------------
Calculating the initial marginal costs in LA_Effort()
-----------------------------------------------------------------*/
void LA_effort_marg_costs (int m, int n, tMatrix_R ct, tVector_R f,
    tVector_I ib, tVector_R zc)
{
    int         i, m1, nm;
    tNumber_R   d;
    m1 = m + 1;
    nm = n + m;
    d = f[1];
    zc[1] = d * (ct[m1][1]);
    if (zc[1] <= -EPS)
    {
        for (i = 1; i <= n; i++)
        {
            f[i] = -f[i];
        }
```

```
        for (i = 1; i <= nm; i++)
        {
            ib[i] = -ib[i];
        }
        zc[1] = -zc[1];
    }
    if (n > 1)
    {
        for (i = 2; i <= n; i++)
        {
            zc[i] = -f[i] + f[1] * (ct[m1][i]);
        }
    }
}

/*-------------------------------------------------------------------
Determine the vector that enters the basis in LA_Effort()
-------------------------------------------------------------------*/
void LA_effort_vent (int *pIvo, int *pJin, int n, tVector_I nb,
    tVector_R zc, tNumber_R *pZ)
{
    int         j, jc;
    tNumber_R   bignum, d, e, g, gg, tz, tze;

    bignum = 1.0/ (EPS*EPS);
    *pIvo = 0;
    g = bignum;
    *pZ = zc[1];
    tz = *pZ + *pZ;
    tze = tz + EPS;
    for (j = 2; j <= n; j++)
    {
        d = zc[j];
        jc = nb[j];
        if (jc <= n)
        {
            gg = d;
            if (gg < 0.0) gg = -gg;
            if (gg > EPS)
            {
                if (d >= 0.0)
                {
                    e = -d;
                    if (e < g)
```

```
                    {
                        *pIvo = -1;
                        g = e;
                        *pJin = j;
                    }
                }
                else if (d < 0.0)
                {
                    e = d;
                    if (e < g)
                    {
                        *pIvo = 1;
                        g = e;
                        *pJin = j;
                    }
                }
            }
        }
        else if (jc > n)
        {
            if (d < -EPS)
            {
                e = d;
                if (e < g)
                {
                    *pIvo = 1;
                    g = e;
                    *pJin = j;
                }
            }
            else if (d >= tze)
            {
                e = tz - d;
                if (e < g)
                {
                    *pIvo = -1;
                    g = e;
                    *pJin = j;
                }
            }
        }
    }
}

/*------------------------------------------------------------------
```

```
Determine the vector that leaves the basis in LA_Effort()
-----------------------------------------------------------------*/
void LA_effort_vleav (int *pItest, int jin, int *pIout, int m,
    tMatrix_R ct)
{
    int         i, m1;
    tNumber_R   d, g, thmax;

    m1 = m + 1;
    thmax = 1.0/ (EPS*EPS);
    for (i = 1;  i <= m1; i++)
    {
        d = ct[i][jin];
        if (d > EPS)
        {
            g = ct[i][1]/d;
            if (g <= thmax)
            {
                thmax = g;
                *pIout = i;
                *pItest = 1;
            }
        }
    }
}

/*-----------------------------------------------------------------
Calculate the results of LA_Effort()
-----------------------------------------------------------------*/
eLaRc LA_effort_res (int m, int n, tMatrix_R ct, tVector_I ib,
    tVector_I ic, tVector_I ip, tVector_I nb, tVector_R zc, int
    *pIrank, tVector_R a, tNumber_R *pZ)
{
    int         i, k, kj, l, m1;
    tNumber_R   gg;

    m1 = m + 1;
    for (i = 1; i <= m; i++)
    {
        k = ic[i];
        if (k <= n)
        {
            l = ip[k];
            k = nb[l];
            kj = k - n;
```

```
            a[kj] = zc[l] - *pZ;
            if (ib[k] == -1) a[kj] = -a[kj];
        }
        else if (k > n)
        {
            kj = k - n;
            a[kj] = -*pZ;
            if (ib[k] == -1) a[kj] = *pZ;
        }
    }
    for (i = 1; i <= n; i++)
    {
        gg = fabs (zc[i]);
        if (gg <= EPS)
        {
            *pIrank = *pIrank - 1;
        }
    }
    for (i = 1; i <= m1; i++)
    {
        if (ct[i][1] < EPS)
            return LaRcSolutionProbNotUnique;
    }

    return LaRcSolutionUnique;
}
```

Chapter 23

Bounded Least Squares Solution of Underdetermined Linear Equations

23.1 Introduction

This is the last of four chapters on the solution of underdetermined systems of consistent linear equations subject to constraints on the solution vectors. This chapter differs from the previous 3 chapters, where we used linear programming techniques, in that we use quadratic programming techniques. Here, it is required to calculate the least squares solution of an underdetermined consistent linear system subject to the constraints that each element of the solution vector be bounded between –1 and 1.

Consider the underdetermined system of linear equations

(23.1.1) $$\mathbf{Ca} = \mathbf{f}$$

$\mathbf{C} = (c_{ij})$ is a given n by m real matrix of rank k, $k \leq n < m$, $\mathbf{f} = (f_i)$ is a given real n-vector and $\mathbf{a} = (a_j)$ is the solution m-vector.

Assume throughout that rank($\mathbf{C}|\mathbf{f}$) = rank($\mathbf{C}$) meaning that system $\mathbf{Ca} = \mathbf{f}$ in (23.1.1) is consistent. Since $n < m$, system $\mathbf{Ca} = \mathbf{f}$ by itself has an infinite number of solutions. In Chapter 17, among these infinite number of solutions, we obtained a solution vector **a** that minimizes the least square or the L_2 norm $||\mathbf{a}||_2$ of vector **a**. We call this the **minimal length least squares solution**.

In this chapter, the problem is stated as follows. It is required to minimize half of the square of the L_2 norm of vector **a**

(23.1.2) $$(1/2)||\mathbf{a}||^2 = (1/2)(\mathbf{a}^T\mathbf{a})$$

subject to the constraints

(23.1.3) $$-1 \leq a_j \leq 1, \quad j = 1, 2, \ldots, m$$

The problem is dealt with in two steps. The first step is to calculate the minimal length least squares solution of the undetermined system **Ca** = **f**, without applying the constraints (23.1.3). Let this be called problem (E0).

We then examine the elements of the solution vector. If they satisfy the given constraints (23.1.3), then the solution of problem (E0) is itself the solution of the given problem. If not, we proceed to the second step, where the elements of vector **a** are to be bounded between –1 and 1. We call this problem (E).

Problem (E0): A solution for problem (E0) always exists and is given by

(23.1.4) $$\mathbf{a} = \mathbf{C}^{+}\mathbf{f}$$

where $\mathbf{C}^{+}$ is the pseudo-inverse of matrix **C**. If rank(**C**) = n < m, the solution is unique [9] and

(23.1.5) $$\mathbf{C}^{+} = \mathbf{C}^{T}(\mathbf{C}\mathbf{C}^{T})^{-1}$$

Problem (E) is to find a solution m-vector **a** that minimizes (23.1.2), or in effect, minimizes the L_2 norm $\|\mathbf{a}\|$ and whose elements are bounded between 1 and –1. As indicated in Chapter 21, if instead of the constraints (23.1.3), we require vector **a** to satisfy the constraints

$$b_i \le a_i \le c_i, \quad i = 1, 2, \ldots, m$$

where vectors $\mathbf{b} = (b_i)$ and $\mathbf{c} = (c_i)$ are given m-vectors, by substituting variables, the above constraints reduce to the constraints (23.1.3) in the new variables. See Section 7.1.

23.1.1 Applications of the algorithm

This algorithm has been used in digital image restoration [2, 3]. However, this problem is known in control theory as the **minimum energy problem for discrete linear admissible control systems** [4, 5].

A dynamical control system may be described by the vector difference equation

(23.1.6) $$\mathbf{x}[i + 1] = \mathbf{A}\mathbf{x}[i] + \mathbf{B}\mathbf{a}[i], \quad i = 0, 1, 2, \ldots$$

A and **B** are real constant matrices of dimensions n by n and n by r respectively, $\mathbf{x}[i]$ is the state n-vector and $\mathbf{a}[i]$ is the control vector at time interval i. The minimum energy problem for this control system is to find $\mathbf{a}[0], \mathbf{a}[1], \ldots, \mathbf{a}[N-1]$, $N \geq n$, that brings the system from an initial state $\mathbf{x}[1] = \mathbf{b}$ to a desired state $\mathbf{x}[N] = \mathbf{c}$ such that

$$E = (1/2) \sum_{i=0}^{N-1} \mathbf{a}[i]^T \mathbf{a}[i]$$

is as small as possible. The recurrence solution of equation (23.1.6) with the conditions $\mathbf{x}[1] = \mathbf{b}$ and $\mathbf{x}[N] = \mathbf{c}$, gives

$$\mathbf{c} = \mathbf{A}^N \mathbf{b} + \sum_{i=0}^{N-1} \mathbf{A}^{N-1-i} \mathbf{B} \mathbf{a}[i]$$

This equation could finally be written in the form $\mathbf{Ca} = \mathbf{f}$, which is equation (23.1.1). For **admissible control systems**, the elements of vector $\mathbf{a}$ satisfy the inequalities (23.1.3), which yields problem (E). A solution of system $\mathbf{Ca} = \mathbf{f}$ that satisfies the above constraints may or may not exist.

Canon and Eaton [4] introduced canonical representation of the control system and solved problem (E) as a quadratic programming problem. Again Polak and Deparis [10] used optimal control and convex programming ideas for solving this problem.

Stark and Parker [11] introduced a FORTRAN subroutine named BVLS (bounded-variable least-squares) that is modeled on the non-negative least squares (NNLS) algorithm of Lawson and Hanson [8]. BVLS solves the least squares problem of the system $\mathbf{Ca} = \mathbf{f}$ subject to the constraints $\mathbf{l} \leq \mathbf{a} \leq \mathbf{u}$, where $\mathbf{l}$ and $\mathbf{u}$ are given vectors. Using our notation, the relative size of the n by m matrix $\mathbf{C}$ of (23.1.1), is typically $n \ll m$.

In this chapter, problem (E) is solved as a quadratic programming problem for bounded variables. It is based on the simplex method for quadratic programming introduced by Dantzig [6] and implemented by van de Panne and Whinston [12, 13].

We take advantage of the special structure of the simplex tableaux for this problem. Hence, minimum computer storage is needed and the computation time is considerably reduced. As a result, this method can deal with fairly large size problems without exhausting the capacity of the computer.

Problem (E0) is solved first as a quadratic programming problem. Its solution requires k simplex iterations, where k = rank(**C**). It needs a small number of arithmetic operations. If the obtained solution satisfies the inequality constraints (23.1.3), then this solution is also a solution of problem (E). If not, the simplex tableau is enlarged and the problem at hand is solved as a quadratic programming problem for bounded variables.

If $\mathbf{Ca} = \mathbf{f}$ is consistent but rank deficient; that is, one or more equations in $\mathbf{Ca} = \mathbf{f}$ are redundant, the redundant equation(s) are deleted. The algorithm also detects the case when system $\mathbf{Ca} = \mathbf{f}$ is inconsistent, i.e., rank($\mathbf{C}|\mathbf{f}$) > rank(**C**) and a solution for problems (E0) and (E) is not feasible. In this case, the calculation terminates.

In section 23.2, the quadratic programming formulation of the problem is given. In section 23.3, the solution of problem (E0) as a quadratic programming problem is obtained. Also in section 23.3, the case of rank deficient consistent system $\mathbf{Ca} = \mathbf{f}$ and the case of inconsistent system $\mathbf{Ca} = \mathbf{f}$ are detected. In section 23.4, the solution of problem (E) as a quadratic programming problem with bounded variables is obtained. In section 23.5, numerical results and comments are given.

23.2 Quadratic programming formulation of the problems

Problem (E0) is formulated as a quadratic programming problem

(23.2.1a) $$\text{maximize } z = -(1/2)\mathbf{a}^T\mathbf{a}$$

subject to

(23.2.1b) $$\mathbf{Ca} = \mathbf{f}$$

(23.2.1c) a_j, $j = 1, 2, \ldots, m$, unrestricted in sign

The problem is posed as a maximization problem by introducing a –ve sign in (23.2.1a).

Problem (E) is formulated by (23.2a, b) and

(23.2.1d) $$-1 \le a_j \le 1, \quad j = 1, 2, \ldots, m$$

23.3 Solution of problem (E0)

To solve problem (E0) as a quadratic programming problem, we first write down the **Lagrange function**

$$L = -(1/2)\mathbf{a}^T\mathbf{a} - \mathbf{v}^T(\mathbf{Ca} - \mathbf{f})$$

v is an n-vector of **Lagrange multipliers** for the constraints (23.2.1b), namely $\mathbf{Ca} = \mathbf{f}$. For optimum L, we differentiate L w.r.t. a_i, $i = 1, 2, \ldots, m$ and also w.r.t. v_j, $j = 1, 2, \ldots, n$, and equate each of the (m + n) equations to 0. However, we shall not equate the (m + n) equations to 0's right away. We do that in two steps, as follows:

(1) The differentiation produces [12, 13]

(23.3.1a) $$\mathbf{I}_m\mathbf{u}^0 - \mathbf{C}^T\mathbf{v} - \mathbf{I}_m\mathbf{a} = \mathbf{0}$$

(23.3.1b) $$\mathbf{Ca} + \mathbf{I}_n\mathbf{y} = \mathbf{f}$$

(23.3.1c) a_j, $j = 1, 2, \ldots, m$, and v_i, $i = 1, 2, \ldots, n$, unrestricted in sign

In (23.3.1a), $\mathbf{u}^0$ is the m-vector whose elements are the negative of the partial derivatives of the Lagrange function with respect to the **a**-variables. In (23.3.1b) we added the n-vector **y**, which is assumed to be an n-vector of artificial variables used to form the initial basic solution of the problem. Also, $\mathbf{I}_m$ and $\mathbf{I}_n$ are unit matrices of order m and n respectively.

(2) Solve for **a** and **v** by making both **u** and **y** as non-basic. This is done via putting equations (23.3.1a, b) in a simplex tableau format (Tableau 23.3.1), which is a setup tableau for problem (E0).

Tableau 23.3.1 (Setup tableau for problem (E0))

B	$\mathbf{b}_B$	$\mathbf{u}^{0T}$	$\mathbf{v}^T$	$\mathbf{a}^T$	$\mathbf{y}^T$
$\mathbf{u}^0$	$\mathbf{0}$	$\mathbf{I}_m$	$-\mathbf{C}^T$	$-\mathbf{I}_m$	$\mathbf{0}$
$\mathbf{y}$	$\mathbf{f}$	$\mathbf{0}$	$\mathbf{0}$	$\mathbf{C}$	$\mathbf{I}_n$

In this tableau, **B** denotes the (m + n) basis matrix and $\mathbf{b}_B$ denotes the basic solution. The elements of $\mathbf{u}^0$ and of **y** form the (initial) basic variables.

To solve problem (E0), we change this tableau, by applying

Gauss-Jordan elimination steps so that the elements of vector **a** replace the corresponding elements of vector $\mathbf{u}^0$ as basic variables. In effect, Tableau 23.3.2 is obtained by pre-multiplying the main body of Tableau 23.3.1 by the inverse of matrix $\mathbf{B}_1^{-1}$, which is its own inverse

$$\begin{bmatrix} -\mathbf{I}_m & \mathbf{0} \\ \mathbf{C} & \mathbf{I}_n \end{bmatrix}^{-1} = \begin{bmatrix} -\mathbf{I}_m & \mathbf{0} \\ \mathbf{C} & \mathbf{I}_n \end{bmatrix}$$

Then the elements of vector **v** replace the corresponding elements of vector **y** as basic variables making **y** a non-basic, again by applying Gauss-Jordan elimination steps. This gives Tableau 23.3.3. Again, in effect, Tableau 23.3.3 is obtained by pre-multiplying the main body of Tableau 23.3.2 by matrix $\mathbf{B}_2^{-1}$, where

$$\mathbf{B}_2^{-1} = \begin{bmatrix} \mathbf{I}_m & \mathbf{C}^T \\ \mathbf{0} & -\mathbf{C}\mathbf{C}^T \end{bmatrix}^{-1} = \begin{bmatrix} \mathbf{I}_m & \mathbf{C}^+ \\ \mathbf{0} & -(\mathbf{C}\mathbf{C}^T)^{-1} \end{bmatrix}$$

where $\mathbf{C}^+$ is given by (23.1.5) for rank($\mathbf{C}$) = n < m.

Tableau 23.3.2 (For problem (E0))

B	$\mathbf{b}_B$	$\mathbf{u}^{0T}$	$\mathbf{v}^T$	$\mathbf{a}^T$	$\mathbf{y}^T$
a	**0**	$-\mathbf{I}_m$	$\mathbf{C}^T$	$\mathbf{I}_m$	**0**
y	**f**	**C**	$-\mathbf{C}\mathbf{C}^T$	**0**	$\mathbf{I}_n$

Tableau 23.3.3 (Optimal solution for problem (E0))

B	$\mathbf{b}_B$	$\mathbf{u}^{0T}$	$\mathbf{v}^T$	$\mathbf{a}^T$	$\mathbf{y}^T$
a	$\mathbf{C}^+\mathbf{f}$	$(-\mathbf{I}_m + \mathbf{C}^+\mathbf{C})$	**0**	$\mathbf{I}_m$	$\mathbf{C}^+$
v	$-(\mathbf{C}\mathbf{C}^T)^{-1}\mathbf{f}$	$-\mathbf{C}^{+T}$	$\mathbf{I}_n$	**0**	$-(\mathbf{C}\mathbf{C}^T)^{-1}$

In (23.3.1c), the elements of vectors **a** and **v** are unrestricted in sign. Therefore, Tableau 23.3.3 would be the tableau for the optimal solution. In Tableau 23.3.3, $\mathbf{C}^+$ is the pseudo-inverse of matrix **C** as given by (23.1.5). Hence, by comparing with (23.1.4), the basic solution in Tableau 23.3.3, $\mathbf{a} = \mathbf{C}^{+}\mathbf{f}$, is the optimal solution for problem (E0).

To ensure the numerical stability of the solution, before changing the simplex tableau, pivoting is performed as follows. To obtain Tableau 23.3.2 from Tableau 23.3.1, Gauss-Jordan elimination steps are used to reduce matrix **C** in the column of $\mathbf{a}^T$ to 0.

Again, to obtain Tableau 23.3.3 from Tableau 23.3.2, Gauss-Jordan elimination steps with pivoting are used to reduce the columns under $\mathbf{v}^T$ to a matrix **0** augmented vertically by matrix $\mathbf{I}_n$. Since matrix $\mathbf{CC}^T$ is positive semi-definite, pivoting is done along the diagonal elements of $-\mathbf{CC}^T$.

Because of using partial pivoting in obtaining Tableau 23.3.3, the calculated $\mathbf{C}^+$ in Tableau 23.3.3 would be the pseudo-inverse of matrix $\boldsymbol{C}$ (not of matrix **C**), where $\boldsymbol{C}$ is matrix **C** whose rows are permuted. The permutation of the rows of **C** are recorded in an n-vector of indices.

If rank(**C**) = k < n, this is determined in the process of applying the Gauss-Jordan elimination with pivoting. In this case one or more of the intermediate tableaux between Tableaux 23.3.2 and 23.3.3 has a row of 0's. If the corresponding element of the basic solution $\mathbf{b}_{Bi}$ to this row is also 0, then system (23.1.2) is consistent but rank deficient. The row and its zero $\mathbf{b}_{Bi}$ element are deleted and the rows of the current tableau are then rearranged. Again, a row of 0's indicates that a corresponding column of $\mathbf{C}^T$ is linearly dependent on one or more of the previous columns. The calculated $\mathbf{C}^+$ would be of dimensions m by k and it is the pseudo-inverse of matrix $\underline{\boldsymbol{C}}$ of dimension k by m, where $\underline{\boldsymbol{C}}$ contains the linearly independent rows of matrix **C**, properly permuted.

If the corresponding element of $\mathbf{b}_B$ to a row of 0's is nonzero, the system is inconsistent and the calculation is terminated, as system $\mathbf{Ca} = \mathbf{f}$ would have no feasible solution.

Again, if the elements of the obtained **a**-solution of problem (E0) are bounded between −1 and +1, this solution would also be the solution of problem (E). If the solution vector **a** of problem (E0) does not satisfy (23.1.3), we proceed to solve problem (E), as a quadratic programming problem with bounded variables.

23.4 Solution of problem (E)

Slack variables $x_j^1 \geq 0$ and the surplus variables $x_j^2 \geq 0$, $j = 1, 2, \ldots, m$, are needed to convert respectively the right and left inequalities in (23.2.1d) into equalities.

(23.4.1a) $\quad a_j + x_j^1 = 1, \quad j = 1, 2, \ldots, m$

(23.4.1b) $\quad -a_j + x_j^2 = 1, \quad j = 1, 2, \ldots, m$

The Lagrange function for problem (E) is then given by

$$L = -(1/2)\mathbf{a}^T\mathbf{a} - \mathbf{v}^T(\mathbf{Ca} - \mathbf{f}) - \mathbf{u}^{1T}(\mathbf{a} + \mathbf{x}^1 - \mathbf{e}) - \mathbf{u}^{2T}(\mathbf{a} + \mathbf{x}^2 - \mathbf{e})$$

The vectors $\mathbf{u}^0$, $\mathbf{v}$ and $\mathbf{y}$ are defined as in (23.3.1a, b), $\mathbf{e}$ is an m-vector each element of which is 1, and $\mathbf{u}^1$ and $\mathbf{u}^2$ are m-vectors of Lagrange multipliers for (23.4.1a, b) respectively. The **Kuhn-Tucker** conditions for the solution of problem (E) are the following [14].

(a) $\mathbf{I}_m\mathbf{u}^0 - \mathbf{I}_m\mathbf{u}^1 + \mathbf{I}_m\mathbf{u}^2 - \mathbf{C}^T\mathbf{v} - \mathbf{I}_m\mathbf{a} = \mathbf{0}$

(b) $\mathbf{I}_m\mathbf{a} + \mathbf{I}_m\mathbf{x}^1 = \mathbf{e}$

(c) $-\mathbf{I}_m\mathbf{a} + \mathbf{I}_m\mathbf{x}^2 = \mathbf{e}$

(d) $\mathbf{Ca} + \mathbf{I}_n\mathbf{y} = \mathbf{f}$

(e) $\mathbf{u}^{1T}\mathbf{x}^1 + \mathbf{u}^{2T}\mathbf{x}^2 = \mathbf{0}$

(f) $x_j^1, x_j^2, u_j^1, u_j^2 \geq 0, \quad j = 1, 2, \ldots, m$

(g) $a_j, j = 1, 2, \ldots, m$ and $v_i, \quad i = 1, 2, \ldots, n$, unrestricted in sign

From these conditions, the setup tableau for problem (E) is constructed (Tableau 23.4.1), where the elements of vectors $\mathbf{u}^0$, $\mathbf{x}^1$, $\mathbf{x}^2$ and $\mathbf{y}$ form the initial basic variables.

Tableau 23.4.1 (Setup Tableau for problem (E))

B	$\mathbf{b}_B$	$\mathbf{u}^{0T}$	$\mathbf{u}^{1T}$	$\mathbf{u}^{2T}$	$\mathbf{v}^T$	$\mathbf{a}^T$	$\mathbf{x}^{1T}$	$\mathbf{x}^{2T}$	$\mathbf{y}^T$
$\mathbf{u}^0$	$\mathbf{0}$	$\mathbf{I}_m$	$-\mathbf{I}_m$	$\mathbf{I}_m$	$-\mathbf{C}^T$	$-\mathbf{I}_m$	$\mathbf{0}$	$\mathbf{0}$	$\mathbf{0}$
$\mathbf{x}^1$	$\mathbf{e}$	$\mathbf{0}$	$\mathbf{0}$	$\mathbf{0}$	$\mathbf{0}$	$\mathbf{I}_m$	$\mathbf{I}_m$	$\mathbf{0}$	$\mathbf{0}$
$\mathbf{x}^2$	$\mathbf{e}$	$\mathbf{0}$	$\mathbf{0}$	$\mathbf{0}$	$\mathbf{0}$	$-\mathbf{I}_m$	$\mathbf{0}$	$\mathbf{I}_m$	$\mathbf{0}$
$\mathbf{y}$	$\mathbf{f}$	$\mathbf{0}$	$\mathbf{0}$	$\mathbf{0}$	$\mathbf{0}$	$\mathbf{C}$	$\mathbf{0}$	$\mathbf{0}$	$\mathbf{I}_n$

The setup tableau is now changed as in problem (E0), where the elements of vector $\mathbf{a}$ replace the corresponding elements of vector $\mathbf{u}^0$ as basic variables. The elements of vector $\mathbf{v}$ then replace the corresponding elements of vector **y** as basic variables. That gives Tableau 23.4.2 (not shown), then Tableau 23.4.3.

Tableau 23.4.3 (For problem (E))

$\mathbf{B}$	$\mathbf{b}_B$	$\mathbf{u}^{0T}$	$\mathbf{u}^{1T}$	$\mathbf{u}^{2T}$	$\mathbf{v}^T$	$\mathbf{a}^T$	$\mathbf{x}^{1T}$	$\mathbf{x}^{2T}$	$\mathbf{y}^T$
$\mathbf{a}$	$\mathbf{C}^+\mathbf{f}$	$\mathbf{G}$	$-\mathbf{G}$	$\mathbf{G}$	$\mathbf{0}$	$\mathbf{I}_m$	$\mathbf{0}$	$\mathbf{0}$	$\mathbf{C}^+$
$\mathbf{x}^1$	$\mathbf{e}-\mathbf{C}^+\mathbf{f}$	$-\mathbf{G}$	$\mathbf{G}$	$-\mathbf{G}$	$\mathbf{0}$	$\mathbf{0}$	$\mathbf{I}_m$	$\mathbf{0}$	$-\mathbf{C}^+$
$\mathbf{x}^2$	$\mathbf{e}+\mathbf{C}^+\mathbf{f}$	$\mathbf{G}$	$-\mathbf{G}$	$\mathbf{G}$	$\mathbf{0}$	$\mathbf{0}$	$\mathbf{0}$	$\mathbf{I}_m$	$\mathbf{C}^+$
$\mathbf{v}$	$-(\mathbf{CC}^T)^{-1}\mathbf{f}$	$-\mathbf{C}^{+T}$	$\mathbf{C}^{+T}$	$-\mathbf{C}^{+T}$	$\mathbf{I}_n$	$\mathbf{0}$	$\mathbf{0}$	$\mathbf{0}$	$-(\mathbf{CC}^T)^{-1}$

In Tableau 23.4.3, $\mathbf{G} = (-\mathbf{I}_m + \mathbf{C}^+\mathbf{C})$. This tableau is then changed in such a way as to satisfy the Kuhn-Tucker conditions (e) and (f).

23.4.1 Asymmetries in the simplex tableau

By examining Tableau 23.4.3, the following asymmetries exist.

(i) The columns under $\mathbf{u}^{0T}$ and under $\mathbf{u}^{1T}$ are the –ve of one another.

(ii) The columns under $\mathbf{u}^{1T}$ and under $\mathbf{u}^{2T}$ are the –ve of one another.

Lemma 23.1

Let $\mathbf{w}_j^{\,1}$ and $\mathbf{w}_j^{\,2}$ be respectively the columns under $u_j^{\,1}$and $u_j^{\,2}$ in any tableau after Tableau 23.4.2. Then

$$\mathbf{w}_j^{\,1} + \mathbf{w}_j^{\,2} = \mathbf{0}, \quad j = 1, 2, \ldots, m$$

and as a result, we have the following lemma.

Lemma 23.2

In any tableau after Tableau 23.4.3, for any j, $1 \le j \le m$, $u_j^{\,1}$ and $u_j^{\,2}$, cannot together be basic variables.

Proof:

If $u_j^{\,1}$ and $u_j^{\,2}$ appear together in any basis, the basis matrix would be singular.

Lemma 23.3

Assume in a tableau after Tableau 23.4.3, x_j^1 and x_j^2, $1 \leq j \leq m$, are both basic variables and (g_{ij}) and (h_{ij}) are the elements of the tableau opposite x_j^1 and x_j^2 respectively in the non-basic columns. Then

$$g_{ij} = -h_{ij}$$

For the proof of a similar lemma, see ([14], pp. 176, 177)

Standard and non-standard tableau

If in any of the tableaux following Tableau 23.4.3, condition (e) of Kuhn-Tucker is satisfied, the tableau is defined as a standard tableau.

For condition (e) to be satisfied, if u_j^1, $j = 1, 2, \ldots, m$, is a basic variable, x_j^1 is non-basic, and vice versa. The same is true for u_j^2 and x_j^2, $j = 1, 2, \ldots, m$. If condition (e) is not satisfied, the tableau is nonstandard.

For any standard tableau and from Lemma 23.2, one of the following pairs of variables is basic, (x_j^1 and u_j^2), (u_j^1 and x_j^2) or (x_j^1 and x_j^2), $j = 1, 2, \ldots, m$.

Lemma 23.4

In any standard tableau the following are satisfied:

(a) If x_j^1 and u_j^2 are basic variables, $x_j^2 = 0$, $x_j^1 = 2$ and $a_j = -1$.
(b) If u_j^1 and x_j^2 are basic variables, $x_j^1 = 0$, $x_j^2 = 2$ and $a_j = 1$.
(c) If x_j^1 and x_j^2 are both basic variables, $x_j^1 + x_j^2 = 2$.

Lemma 23.5

In any tableau, standard or not, where only x_j^1 or x_j^2 is a basic variable, $1 \leq j \leq m$, we have the following. If x_j^1 say is non-basic, the row opposite x_j^2 has zero elements in the non-basic columns, except under x_j^1, where the element is 1 and $x_j^2 = 2$. The same is true when the roles of x_j^1 and x_j^2 are exchanged.

23.4.2 The condensed tableau for problem (E)

Because of the Kuhn-Tucker condition (g), the elements of vectors **a** and **v** are unrestricted in sign, the elements of their respective corresponding vectors $\mathbf{u}^0$ and **y** should be 0's. They are so in Tableau

23.4.3, and they should stay non-basic in all the following tableaux. In other words, the elements of vectors **a** and **v** should stay basic variables in all the following tableaux.

Tableau 23.4.4 (Condensed tableau for problem (E))

B	b_B	$\mathbf{u}^{0T}$	$\mathbf{v}^T$	$\mathbf{a}^T$	$\mathbf{y}^T$
$\mathbf{x}^1$	$\mathbf{e} - \mathbf{C}^+\mathbf{f}$	$\mathbf{G}$	$-\mathbf{G}$	$\mathbf{I}_m$	$\mathbf{0}$
$\mathbf{x}^2$	$\mathbf{e} + \mathbf{C}^+\mathbf{f}$	$-\mathbf{G}$	$\mathbf{G}$	$\mathbf{0}$	$\mathbf{I}_m$

where **G** was defined earlier and is $\mathbf{G} = (-\mathbf{I}_m + \mathbf{C}^+\mathbf{C})$.

In effect, we may disregard in Tableau 23.4.3 and the following tableaux, the m and n rows that correspond to the elements of vectors **a** and **v** respectively. We also disregard the columns that correspond to the elements of vectors **a**, $\mathbf{u}^{0T}$, $\mathbf{v}^T$ and **y**. The calculation in the following simplex steps is confined to the remaining part of Tableau 23.4.3, which we denote by Tableau 23.4.4. We call this a **condensed tableau.** Finally, for optimal solution, the elements x_j^1, x_j^2, u_j^1, u_j^2, $j = 1, 2, \ldots, m$, are to be non-negative in a standard condensed tableau.

23.4.3 The dual method of solution

Let from now on, **u** denote either of the vectors $\mathbf{u}^1$ or $\mathbf{u}^2$ and similarly let **x** denote either of the vectors $\mathbf{x}^1$ or $\mathbf{x}^2$. The x-variables are known as the primal variables and the corresponding u-variables are the dual variables.

A new obtained tableau may or may not be a standard one. In a non-standard tableau, if an x-variable and its corresponding u-variable are both basic, they are known as a basic pair. If they are both non-basic, they are known as a non-basic pair. It has been shown [12, 13], that the rules of changing the simplex tableau may be such that there is only one basic pair and one non-basic pair in a nonstandard tableau.

Tableau 23.4.4 is a standard tableau, and where $\mathbf{u} \geq \mathbf{0}$, the dual variables have feasible solution. Thus Tableau 23.4.4 is a suitable initial tableau for the solution of problem (E) using the dual method for quadratic programming. The rules of changing the simplex tableau are found in ([12], p. 296).

23.4.4 The reduced tableau for problem (E)

We show here that it is sufficient to store 1/8 of the condensed tableau. That is, 1/2 the non-basic columns and 1/2 the rows opposite the basic variables. The columns of the basic variables are also not stored, as each is a column in a 2m-unit matrix. The remaining part of the tableau can easily be obtained from the stored part. We call this m by m tableau, the **reduced tableau**.

Let x_j^1 and u_j^2, $1 \leq j \leq m$, both be basic variables. In this case the column under x_j^2 and the row opposite u_j^2 are stored. From Lemma 23.1, in the condensed tableau, the column under u_j^1 is $\mathbf{e}_j$, where $\mathbf{e}_j$ is the jth column in an 2m-unit matrix, and from Lemma 23.5, the row opposite x_j^1 consists of zero elements, except under x_j^2, where the element is 1.

Let x_j^1 and x_j^2 both be basic variables, $1 \leq j \leq m$. If $0 \leq x_j^1 \leq 2$, then from the Kuhn Tucker conditions (b) and (c), $0 \leq x_j^2 \leq 2$. In this case the row opposite either x_j^1 or x_j^2 and the column under the corresponding non-basic u-variable are stored.

From Lemmas 23.3 and 23.1, the un-stored row and column respectively are known. If, on the other hand, we let $x_j^1 < 0$, then $x_j^2 > 2$. In this case the row opposite x_j^1 and the column under its corresponding non-basic u-variable are stored.

In a nonstandard tableau, in which u_i^1 say, did replace u_j^1 in the basis, we store the column under x_j^1 and the row opposite u_j^1. Again from Lemma 23.1, the column under u_i^2 is $-\mathbf{e}_i$, and from Lemma 23.5, the row opposite x_j^2 has zero elements, except under x_j^1, where it is 1. Hence, in all cases, 1/8 of the condensed tableau is stored. This includes the data needed for the following iterations.

23.4.5 The method of solution of problem (E)

All calculation is done in the reduced tableaux [1], using the dual method for quadratic programming, in a manner very similar to that used in [13]. Index indicator vector accounts for the corresponding part of the tableau.

If a solution for problem (E) does not exist, in the simplex step, no non-basic variable at a positive level is found to replace a basic variable at a negative level. However, if a solution exists, the convergence of the method is guaranteed [12].

23.5 Numerical results and comments

LA_Energy() implements this algorithm, and DR_Energy() tests 3 examples. The examples were solved in both single- and double-precision, the results of which are given here.

Also, the 7 test cases of Chapter 20 were demonstrated for this algorithm. The third example, whose matrix **C** is a 2 by 3 matrix, has no feasible solution. The one whose matrix **C** is a 4 by 5 matrix is given here as Example 23.1. For the other 5 test cases of Chapter 20, the elements of the solution vector **a** satisfy the boundedness constraints (23.1.3); that is, the solution of problem (E0) is itself the solution of the given problem (E).

Example 23.1

Matrix **C** is a well conditioned 4 by 5 matrix of rank 2. The solution is obtained after 3 (= rank + 1) iterations; 2 for problem (E0) and 1 for problem (E). The result is the same, regardless of single- or double-precision computation.

$$\begin{aligned} 2a_1 + a_2 - 3a_3 + 5a_4 + 3a_5 &= 10 \\ 2a_1 + a_2 - 3a_3 + 5a_4 + 3a_5 &= 10 \\ 2a_1 + a_2 - 3a_3 + 5a_4 + 3a_5 &= 10 \\ -8a_1 - 4a_2 + 2a_3 + 5a_4 + 6a_5 &= 8 \end{aligned}$$

$$a = (0.139, 0.070, -0.614, 1.0, 0.937)^T$$

$$z = 0.5\|\mathbf{a}\|^2 = 1.139$$

Example 23.2

Matrix **C** is a 4 by 19 badly conditioned matrix of rank 4. Using single-precision computation, the problem has no feasible solution. The following solution, computed in double-precision, is obtained after 20 iterations.

$$a = (-1, -1, -1, -1, -1, -0.534, 1, 1, 1, 1, 1, 1, 0.393, -1, -1, -1, -1, 0.139, 1)^T$$

$$z = 0.5\|\mathbf{a}\|^2 = 8.230$$

Example 23.3

Matrix **C** is a 4 by 20 matrix of rank 4, the extension of matrix **C** of the previous example. It is not badly conditioned. The obtained results are the same, regardless of single- or double-precision computation. It converges in 8 iterations. The results are

$$\mathbf{a} = (-1, -1, -1, -0.799, -0.410, -0.058, 0.247, 0.491, 0.661, 0.744, 0.732, 0.618, 0.403, 0.102, -0.251, 0.596, -0.832, -0.798, -0.254, 1)^T$$

$$z = 0.5\|\mathbf{a}\|^2 = 4.502$$

We note here that the calculation for problem (E0) is numerically stable. This is because of using pivoting in the simplex steps. As for problem (E), the numerical results indicate the following.

If the problem has a solution, the parameter θ

$$\theta = \min(b_{Bi}/g_i \geq 0), \quad i \in \boldsymbol{I}$$

where $\boldsymbol{I}$ is the set of indices associated with the basic u-variables and x_l, are less than or of the order of 1, the calculation is also numerically stable. If the problem does not have a solution, the parameters θ becomes larger and larger in the successive iterations, and as a result, the build up of the round-off error increases. The algorithm terminates when the pivot in the simplex step is of wrong sign or is a very small number of the order of the round-off error of the computer.

References

1. Abdelmalek, N.N, Minimum energy problem for discrete linear admissible control systems, *International Journal of Systems Science*, 10(1979)77-88.
2. Abdelmalek, N.N., Restoration of images with missing high-frequency components using quadratic programming, *Applied Optics*, 22(1983)2182-2188.
3. Abdelmalek, N.N. and Kasvand, T., Digital image restoration using quadratic programming, *Applied Optics*, 19(1980)3407-3415.

4. Canon, M.D. and Eaton, J.H., A new algorithm for a class of quadratic programming problems with application to control, *SIAM Journal on Control*, 4(1966)34-45.
5. Canon, M.D., Cullum Jr., C.D. and Polak, E., *Theory of Optimal Control and Mathematical Programming*, McGraw-Hill, New York, 1970.
6. Dantzig, G.B., *Linear Programming and Extensions*, Princeton University Press, Princeton, NJ, 1963.
7. Hadley, G., *Linear Programming*, Addison-Wesley, Reading, MA, 1962.
8. Lawson, C.W. and Hanson, R.J., *Solving Least Squares Problems*, Prentice Hall, Englewood Cliffs, NJ, 1974.
9. Peters, G. and Wilkinson, J.H., The least squares problem and pseudo-inverses, *Computer Journal*, 13(1970)309-316.
10. Polak, E. and Deparis, M., An algorithm for minimum energy control, *IEEE Transactions on Automatic Control*, 14(1969)367-377.
11. Stark, P.B. and Parker, R.L., Bounded-variable least-squares: An algorithm and applications, *Computational Statistics*, 10(1995)129-141.
12. van de Panne, C. and Whinston, A., Simplicial methods for quadratic programming, *Naval Research Logistic Quarterly*, 11(1964)273-302.
13. van de Panne, C. and Whinston, A., Simplex and the dual method for quadratic programming, *Operations Research Quarterly*, 15(1964)355-388.
14. Whinston, A., The bounded variable problem – An application of the dual method for quadratic programming, *Naval Research Logistics Quarterly*, 12(1965)173-179.

23.6 DR_Energy

```
/*--------------------------------------------------------------------
DR_Energy
----------------------------------------------------------------------
This program is a driver for the function LA_Energy(), which
calculates the minimum bounded least squares solution of an
underdetermined consistent system of linear equations.

Given is the underdetermined consistent system

                         c*a = f

"c" is a given real n by m matrix of rank k, k <= n <= m.
"f" is a given real n vector.

It is required to calculate the m vector "a" for this system such
that the elements of "a" satisfy

              -1 <= a[j] <= 1,   j = 1, 2, ..., m

and that half of the square of the L2 (L-Two) norm of vector "a"

         z = 0.5*[sum[a[j]*a[j]], sum over j from 1 to m

is as small as possible.

In control theory, this problem is known as the "Minimum Energy"
problem for discrete linear admissible control systems.

This program carries 3 examples.whose results appear in the text.
--------------------------------------------------------------------*/

#include "DR_Defs.h"
#include "LA_Prototypes.h"

#define MNe_ROWS    (Ne_ROWS + Me_COLS) /* From DR_Defs.h */
#define MNe_COLS    (Ne_ROWS + Me_COLS)
#define N1e         4
#define M1e         5
#define N2e         4
#define M2e         19
#define M3e         20
```

```
void DR_Energy (void)
{
    /*------------------------------------------
      Constant matrices/vectors
    ------------------------------------------*/
    static tNumber_R b1init[N1e][M1e] =
    {
        { 2.0,  1.0, -3.0,  5.0, 3.0  },
        { 2.0,  1.0, -3.0,  5.0, 3.0  },
        { 2.0,  1.0, -3.0,  5.0, 3.0  },
        {-8.0, -4.0,  2.0,  5.0, 6.0  }
    };

    static tNumber_R f1[N1e+1] =
    {   NIL,
        10.0, 10.0, 10.0, 8.0
    };

    /*------------------------------------------
      Variable matrices/vectors
    ------------------------------------------*/
    tMatrix_R   ct      = alloc_Matrix_R (MNe_ROWS, MNe_COLS);
    tVector_R   f       = alloc_Vector_R (Ne_ROWS);
    tVector_R   a       = alloc_Vector_R (Me_COLS);
    tMatrix_R   fay     = alloc_Matrix_R (N2e, M3e);
    tVector_R   fy      = alloc_Vector_R (N2e + 1);

    tMatrix_R   b1      = init_Matrix_R (&(b1init[0][0]), N1e, M1e);

    int         irank, iter;
    int         i, j, m, n, i2, Iexmpl;
    tNumber_R   z;

    eLaRc       rc = LaRcOk;

    for (i = 1; i <= 2; i++)
    {
        i2 = i + 2;
        fy[i] = -2.0;
        fy[i2] = -4.0;
    }
    for (j = 1; j <= 20; j++)
    {
        i = j;
        fay[1][j] = 1.0;
```

```
    fay[2][j] = pow (1.1, i);
    fay[3][j] = pow (1.2, i);
    fay[4][j] = pow (1.3, i);
}

prn_dr_bnr ("DR_Energy, "
            "Bounded L2 Solution of an Underdetermined System");

for (Iexmpl = 1; Iexmpl <= 3; Iexmpl++)
{
    switch (Iexmpl)
    {
        case 1:
            n = N1e;
            m = M1e;
            for (i = 1; i <= n; i++)
            {
                f[i] = f1[i];
                for (j = 1; j <= m; j++)
                    ct[j][i] = b1[i][j];
            }
            break;

        case 2:
            n = N2e;
            m = M2e;
            for (i = 1; i <= n; i++)
            {
                f[i] = fy[i];
                for (j = 1; j <= m; j++)
                    ct[j][i] = fay[i][j];
            }
            break;

        case 3:
            n = N2e;
            m = M3e;
            for (i = 1; i <= n; i++)
            {
                f[i] = fy[i];
                for (j = 1; j <= m; j++)
                    ct[j][i] = fay[i][j];
            }
            break;
        default:
```

```
            break;
        }
        prn_algo_bnr ("Energy");
        prn_example_delim();
        PRN ("Example #%d: Size of matrix \"c\" %d by %d\n",
             Iexmpl, n, m);
        prn_example_delim();
        PRN ("Bounded Least Squares Solution of an Underdetermined "
             "System\n");
        prn_example_delim();
        PRN ("r.h.s. Vector \"f\"\n");
        prn_Vector_R (f, n);
        PRN ("Transpose of Coefficient Matrix, \"ct\"\n");
        prn_Matrix_R (ct, m, n);

        rc = LA_Energy (m, n, ct, f, &irank, &iter, a, &z);

        if (rc >= LaRcOk)
        {
            PRN ("\n");
            PRN ("Results of Minimum Energy Solution\n");
            PRN ("Minimum Energy solution vector \"a\"\n");
            prn_Vector_R (a, m);
            PRN ("Norm z = 0.5||a||*||a||= %8.4f\n", z);
            PRN ("Rank of matrix \"c\" = %d, No. of Iterations "
                "= %d\n", irank, iter);
        }

        LA_check_rank_def (n, irank);
        prn_la_rc (rc);
    }

    free_Matrix_R (ct, MNe_ROWS);
    free_Vector_R (f);
    free_Vector_R (a);
    free_Matrix_R (fay, N2e);
    free_Vector_R (fy);

    uninit_Matrix_R (b1);
}
```

23.7 LA_Energy

```
/*-----------------------------------------------------------------
LA_Energy
-------------------------------------------------------------------
Given is the consistent underdetermined system of linear equations

                       c*a = f

"c" is a given real n by m matrix of rank k, k <= n <= m.
"f" is a given real n vector.

It is required to calculate the m vector "a" for this system that
satisfies the conditions

            -1 <= a[j] <= 1,   j = 1, 2, ..., m

and that half of square of the L2 norm of vector "a"

          z = 0.5*sum([a[j]*a[j]), sum over j from 1 to m

is as small as possible.

In control theory, this is known as the "Minimum Energy" problem for
discrete linear admissible control systems.

Inputs
m       Number of columns of matrix "c" in the system c*a = f.
n       Number of rows of matrix "c" in the system c*a = f.
ct      An (m + n) by (m + n) matrix whose first m rows and first n
        columns contain the transpose of matrix "c" of the system
        c*a = f.
f       An n vector that contains the r.h.s. of the system c*a = f.

Outputs
irank   The calculated rank of matrix "c".
iter    The number of iterations or the number of times the simplex
        tableau is changed by a Gauss-Jordan step.
a       A real m vector that is the minimum energy solution of the
        system c*a = f.
z       Half the square of the minimum L2 norm of the solution
        vector "a".

Returns one of
```

```
        LaRcSolutionFound
        LaRcNoFeasibleSolution
        LaRcInconsistentSystem
        LaRcErrBounds
        LaRcErrNullPtr
        LaRcErrAlloc
-----------------------------------------------------------------*/

#include "LA_Prototypes.h"

eLaRc LA_Energy (int m, int n, tMatrix_R ct, tVector_R f,
    int *pIrank, int *pIter, tVector_R a, tNumber_R *pZ)
{
    tVector_I   ic      = alloc_Vector_I (n);
    tVector_I   ir      = alloc_Vector_I (m);
    tVector_I   ik      = alloc_Vector_I (m);

    int         i = 0, jin = 0, ibc = 0, iijj = 0, ijk = 0, iout = 0,
                irjin = 0;
    int         ikj0 = 0, ikj1 = 0, istd = 0, ivo = 0;
    int         itest = 0, ikj2 = 0, ikk = 0, iriout = 0, j = 0,
                k = 0, j0 = 0, j1 = 0, j2 = 0;
    int         mn = 0, np1 = 0, nj0 = 0;
    tNumber_R   e = 0.0, opeps = 0.0, pivot = 0.0;
    eLaRc       tempRc;

    /* Validation of the data before executing algorithm */
    eLaRc       rc = LaRcSolutionFound;
    VALIDATE_BOUNDS ((0 < n) && (n <= m) && !((n == 1) && (m == 1)));
    VALIDATE_PTRS   (ct && f && pIrank && pIter && a && pZ);
    VALIDATE_ALLOC  (ic && ir && ik);

    /* Initialization */
    *pIrank = n;
    *pIter = 0;
    *pZ = 0.0;
    np1 = n + 1;
    mn = m + n;
    opeps = 1.0 + EPS;

    /* Initializing the data */
    LA_energy_init (m, n, ct, ic, ir, ik, a);

    /* Phase 1 of the algorithm */
    tempRc = LA_energy_phase_1 (m, n, ct, f, ic, a, pIrank, pIter);
```

```
    if (tempRc < LaRcOk)
    {
        GOTO_CLEANUP_RC (tempRc);
    }

    ibc = 0;
    for (i = 1; i <= m; i++)
    {
        if (fabs (a[i]) > opeps) ibc = 1;
    }

    if (ibc == 0)
    {
        /* The norm z */
        LA_energy_norm (m, a, pZ);
        GOTO_CLEANUP_RC (LaRcSolutionFound);
    }

    /* Phase 2 of the program */
    LA_energy_phase_2 (m, n, ct, ir, ik, pIrank, a);

    istd = 1;
    j1 = 0;
    for (ijk = 1; ijk <= m*m; ijk++)
    {
        if (istd != 1)
        {
            j0 = j1;
            nj0 = n + j1;
            iijj = 1;
        }

        if (istd == 1)
        {
            /* Determine the vector that enters the basis */
            ivo = 0;
            LA_energy_vent (&ivo, &jin, m, n, ir, a);

            /* Calculate the results */
            if (ivo == 0)
            {
                LA_energy_res (m, n, ct, ir, a);
                /* The specified norm */
                LA_energy_norm (m, a, pZ);
```

```
        GOTO_CLEANUP_RC (LaRcSolutionFound);
    }

    if (ivo == 1)
    {
        for (j = np1; j <= mn; j++)
        {
            ct[jin][j] = -ct[jin][j];
        }
        a[jin] = 2.0 - a[jin];
        ir[jin] = -ir[jin];
    }
    irjin = ir[jin];
    ikj1 = irjin + mn;

    if (irjin < 0) ikj1 = irjin - mn;
    iijj = 0;

    for (j = 1; j <= m; j++)
    {
        j0 = j;
        nj0 = n + j0;
        ikj0 = ik[j0];
        if (ikj0 == ikj1)
        {
            iijj = 1;
            break;
        }

        if (ikj0 == -ikj1)
        {
            for (i = 1; i <= m; i++)
            {
                ct[i][nj0] = -ct[i][nj0];
            }
            ik[j0] = -ik[j0];
            ikj0 = -ikj0;
            iijj = 1;
            break;
        }
    }

    if (iijj == 0)
    {
        GOTO_CLEANUP_RC (LaRcNoFeasibleSolution);
```

```
        }
    }

    itest = 0;
    iout = 0;
    /* Determine the vector that leaves the basis */
    LA_energy_vleav (&itest, jin, m, n,  &iout, j0, ct, ir, a);

    /* The problem has no feasible solution */
    if (itest == 0)
    {
            GOTO_CLEANUP_RC (LaRcNoFeasibleSolution);
    }
    if ((istd != 1) || (iout != jin))
    {
        istd = 0;
        iriout = ir[iout];
        ikj1 = iriout - mn;
        if (iriout < 0) ikj1 = iriout + mn;
        iijj = 0;
        for (j = 1; j <= m; j++)
        {
            j2 = j;
            ikj2 = ik[j2];
            if (ikj2 == ikj1)
            {
                for (k = 1; k <= m; k++)
                {
                    ikk = ik[k];
                    if (ikk == ikj2)
                    {
                        j1 = k;
                        iijj = 1;
                        break;
                    }
                }
                if (iijj == 1) break;
            }
        }
        if (iijj == 0)
        {
            /* The problem has no feasible solution */
            GOTO_CLEANUP_RC (LaRcNoFeasibleSolution);
        }
    }
```

```
        istd = 1;
        pivot = ct[iout][nj0];
        e = fabs (pivot);
        LA_energy_gauss_jordn_e (m, n, iout, j0, ct, ir, ik, a);
        *pIter = *pIter + 1;
    }

CLEANUP:

    free_Vector_I (ic);
    free_Vector_I (ir);
    free_Vector_I (ik);

    return rc;
}

/*--------------------------------------------------------------------
LA_Energy() initialization
--------------------------------------------------------------------*/
void LA_energy_init (int m, int n, tMatrix_R ct, tVector_I ic,
    tVector_I ir, tVector_I ik, tVector_R a)
{
    int         i, j, k, np1, mi, mj, nj, mn;
    tNumber_R   s;

    np1 = n + 1;
    mn = m + n;

    for (j = np1; j <= mn; j++)
    {
        for (i = 1; i <= m; i++)
        {
            ct[i][j] = 0.0;
        }
    }

    for (j = 1; j <= m; j++)
    {
        ir[j] = 0;
        ik[j] = 0;
        a[j] = 0.0;
        nj = j + n;
        for (i = 1; i <= n; i++)
        {
            mi = m + i;
```

```
            ct[mi][nj] = ct[j][i];
        }
    }

    /* Calculating matrix  [-(c*c(transpose))] */
    for (j = 1; j <= n; j++)
    {
        ic[j] = j;
        mj = m + j;
        for (i = j; i <= n; i++)
        {
            mi = m + i;
            s = 0.0;
            for (k = 1;  k <= m; k++)
            {
                s = s + ct[k][i] * (ct[k][j]);
            }
            ct[mj][i] = -s;
            ct[mi][j] = -s;
        }
    }
}

/*-----------------------------------------------------------------
Phase 1 of LA_Energy()
-----------------------------------------------------------------*/
eLaRc LA_energy_phase_1 (int m, int n, tMatrix_R ct, tVector_R f,
    tVector_I ic, tVector_R a, int *pIrank, int *pIter)
{
    int         i, j, ij, iout, kd = 0, kdp = 0, mj, mn, mout;
    tNumber_R   d, piv;

    mn = m + n;
    iout = 0;

    for (ij = 1; ij <= n; ij++)
    {
        iout = iout + 1;
        if (iout <= *pIrank)
        {
            /* Pivoting along the diagonal elements of
               ct*ct(transpose) */
            mout = m + iout;
            piv = 0.0;
            for (j = iout; j <= n; j++)
```

```
        {
            mj = m + j;
            d = fabs (ct[mj][j]);
            if (d > piv)
            {
                kd = j;
                kdp = mj;
                piv = d;
            }
        }

        if (piv > EPS)
        {
            if (kdp != mout)
            {
                swap_rows_Matrix_R (ct, kdp, mout);
            }
            if (kd != iout)
            {
                swap_elems_Vector_R (f, kd, iout);
                swap_elems_Vector_I (ic, kd, iout);
                /* Swap two columns of matrix "ct" */
                for (i = 1; i <= mn; i++)
                {
                    d = ct[i][iout];
                    ct[i][iout] = ct[i][kd];
                    ct[i][kd] = d;
                }
            }

            /* A Gauss-Jordan eliminaion step */
            LA_energy_gauss_jordn_e0 (iout, m, n, ct, f, a);
            *pIter = *pIter + 1;
        }
        else if (piv < EPS)
        {
            for (i = iout; i <= n; i ++)
            {
                if (fabs (f[i]) > EPS)
                    return LaRcInconsistentSystem;
            }
            *pIrank = iout - 1;
        }
    }
}
```

```
    return LaRcOk;
}

/*-----------------------------------------------------------------------
Phase 2 of LA_Energy()
-----------------------------------------------------------------------*/
void LA_energy_phase_2 (int m, int n, tMatrix_R ct, tVector_I ir,
    tVector_I ik, int *pIrank, tVector_R a)
{
    int         i, j, k, mk, mn, ni, nj;
    tNumber_R   s;

    /* Calculating (c * pseuedo-inverse * c) */
    mn = m + n;
    for (j = 1; j <= m; j++)
    {
        nj = n + j;
        for (i = j; i <= m; i++)
        {
            ni = n + i;
            s = 0.0;
            for (k = 1; k <= *pIrank; k++)
            {
                mk = m + k;
                s = s + ct[i][k] * (ct[mk][nj]);
            }
            ct[i][nj] = s;
            ct[j][ni] = s;
        }
    }

    /* Calculating (c * pseuedo-inverse * c - m-Unit matrix) */
    for (i = 1; i <= m; i++)
    {
        ir[i] = i;
        ik[i] = mn + i;
        a[i] = 1.0 - a[i];
        ni = n + i;
        ct[i][ni] = ct[i][ni] - 1.0;
    }
}

/*-----------------------------------------------------------------------
Norm z in LA_Energy()
```

```
------------------------------------------------------------------*/
void LA_energy_norm (int m, tVector_R a, tNumber_R *pZ)
{
    int         i;
    tNumber_R   s;

    s = 0.0;
    for (i = 1; i <= m; i++)
    {
        s = s + a[i] * (a[i]);
    }
    *pZ = 0.5 * s;
}

/*------------------------------------------------------------------
A Gauss-Jordan elimination step for problem E0 in LA_Energy()
------------------------------------------------------------------*/
void LA_energy_gauss_jordn_e0 (int iout, int m, int n, tMatrix_R ct,
    tVector_R f, tVector_R a)
{
    int         i, j, jin, mn, mout, k;
    tNumber_R   d, pivot;

    mn = m + n;
    jin = iout;
    mout = m + iout;
    pivot = ct[mout][jin];
    for (j = 1; j <= n; j++)
    {
        ct[mout][j] = ct[mout][j]/pivot;
    }
    f[iout] = f[iout]/pivot;
    for (i = 1; i <= mn; i++)
    {
        if (i != mout)
        {
            d = ct[i][jin];
            for (j = 1; j <= n; j++)
            {
                if (j != jin)
                {
                    ct[i][j] = ct[i][j] - d * (ct[mout][j]);
                }
            }
```

```
                ct[i][jin] = -ct[i][jin]/pivot;
                if (i > m)
                {
                    k = i - m;
                    f[k] = f[k] - d * (f[iout]);
                }
                else if (i <= m)
                {
                    a[i] = a[i] - d * (f[iout]);
                }
            }
        }
        ct[mout][jin] = 1.0/pivot;
}

/*-----------------------------------------------------------------
A Gauss-Jordan elimination step for problem E in LA_Energy()
-----------------------------------------------------------------*/
void LA_energy_gauss_jordn_e (int m, int n, int iout, int j0,
    tMatrix_R ct, tVector_I ir, tVector_I ik, tVector_R a)
{
    int         i, j, k, mn, np1, nj0;
    tNumber_R   d, pivot;

    mn = m + n;
    np1 = n + 1;
    nj0 = n + j0;
    pivot = ct[iout][nj0];

    for (j = np1; j <= mn; j++)
    {
        ct[iout][j] = ct[iout][j]/pivot;
    }

    a[iout] = a[iout]/pivot;

    for (i = 1; i <= m; i++)
    {
        if (i != iout)
        {
            d = ct[i][nj0];
            ct[i][nj0] = -ct[i][nj0]/pivot;
            a[i] = a[i] - d * (a[iout]);
            for (j = np1; j <= mn; j++)
            {
```

```
                if (j != nj0)
                {
                    ct[i][j] = ct[i][j] - d * (ct[iout][j]);
                }
            }
        }
    }

    ct[iout][nj0] = 1.0/pivot;
    k = ik[j0];
    ik[j0] = ir[iout];
    ir[iout] = k;
}

/*-----------------------------------------------------------------
Determine the vector that enters the basis in LA_Energy()
-----------------------------------------------------------------*/
void LA_energy_vent (int *pIvo, int *pJin, int m, int n, tVector_I ir,
    tVector_R a)
{
    int         i, k, mn;
    tNumber_R   d, e, g, tpeps;

    mn = m + n;
    tpeps = 2.0 + EPS;
    g = 1.0;
    for (i = 1;  i <= m; i++)
    {
        k = ir[i];
        if (k < 0) k = -k;
        if (k <= mn)
        {
            e = a[i];
            if (e < -EPS)
            {
                d = e;
                if (d < g)
                {
                    *pIvo = -1;
                    g = d;
                    *pJin = i;
                }
            }

            if (e >= tpeps)
```

```
                {
                    d = 2.0 - e;
                    if (d < g)
                    {
                        *pIvo = 1;
                        g = d;
                        *pJin = i;
                    }
                }
            }
        }
    }

/*-----------------------------------------------------------------
Determine the vector that leaves the basis in LA_Energy()
-----------------------------------------------------------------*/
void LA_energy_vleav (int *pItest, int jin, int m, int n, int *pIout,
    int j0, tMatrix_R ct, tVector_I ir, tVector_R a)
{
    int          i, iri, mn, nj0;
    tNumber_R    d, e, g, thmax;

    mn = m + n;
    nj0 = n + j0;
    thmax = 1.0/ (EPS*EPS);
    for (i = 1; i <= m; i++)
    {
        if (i != jin)
        {
            iri = ir[i];
            if (iri < 0) iri = -iri;
            if (iri < mn) continue;
        }
        d = ct[i][nj0];
        e = d;

        if (e < 0.0)
        {
            e = -e;
        }

        if (e >= EPS)
        {
            g = a[i]/d;
            if (g >= 0.0 && g < thmax)
```

```
            {
                thmax = g;
                *pItest = 1;
                *pIout = i;
            }
        }
    }
}

/*-----------------------------------------------------------------------
Calculate the results of LA_Energy()
-----------------------------------------------------------------------*/
void LA_energy_res (int m, int n, tMatrix_R ct, tVector_I ir,
    tVector_R a)
{
    int         i, iri, mn;
    tNumber_R   g;

    mn = m + n;

    for (i = 1; i <= m; i++)
    {
        ct[mn][i] = a[i];
    }

    for (i = 1; i <= m; i++)
    {
        g = ct[mn][i];
        iri = ir[i];
        if (iri > mn)
        {
            g = 1.0;
            iri = iri - mn;
            a[iri] = g;
            continue;
        }

        if (iri < -mn)
        {
            g = -1.0;
            iri = -mn - iri;
            a[iri] = g;
            continue;
        }
```

```
        if (iri > 0)
        {
            g = 1.0 - g;
            a[iri] = g;
            continue;
        }

        if (iri <= 0)
        {
            g = g - 1.0;
            iri = -iri;
            a[iri] = g;
            continue;
        }
    }
}
```

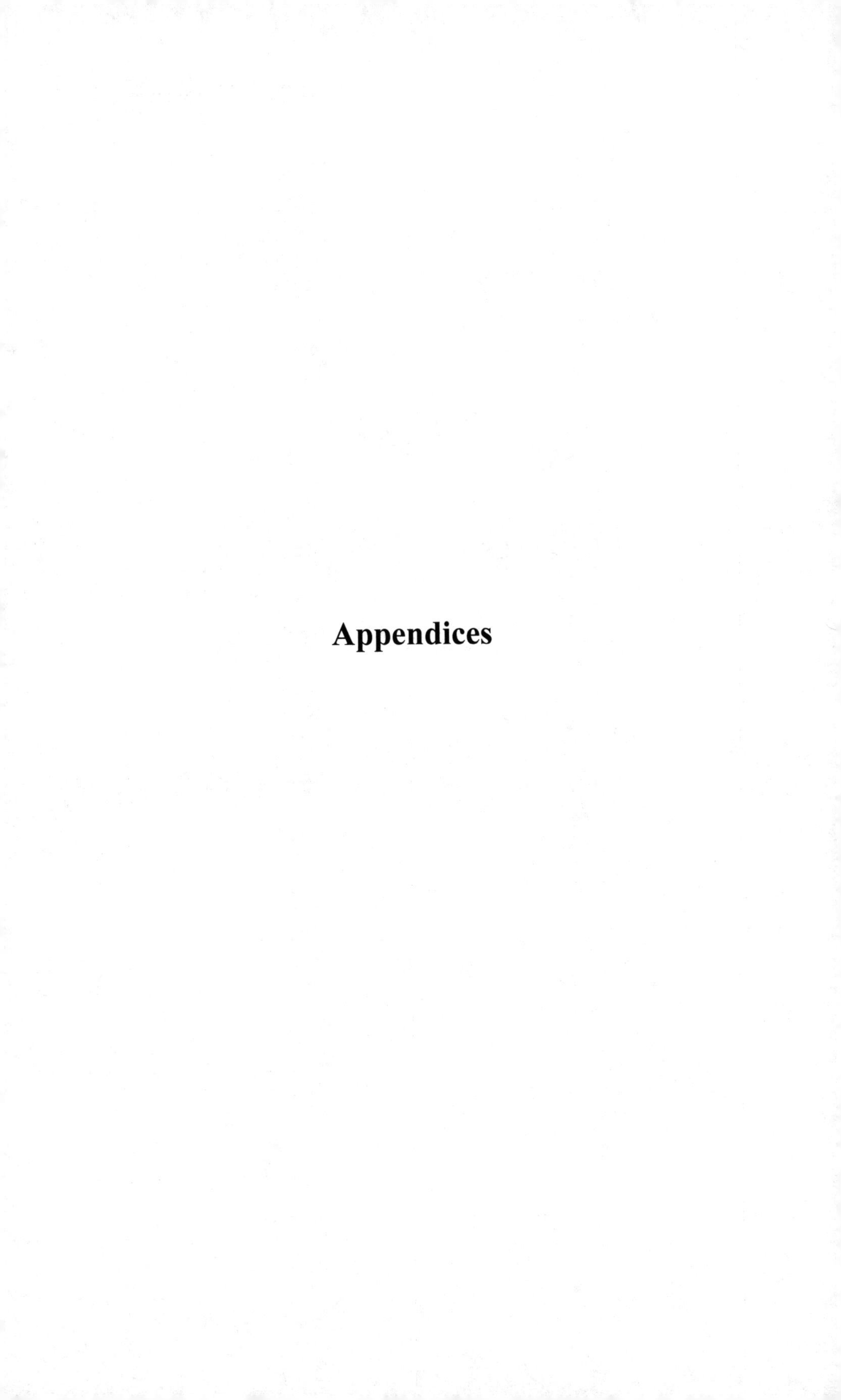

Appendices

Appendix A

References

1. Abdelmalek, N.N., Linear L_1 approximation for a discrete point set and L_1 solutions of overdetermined linear equations, *Journal of ACM*, 18(1971)41-47.
2. Abdelmalek, N.N., Round-off error analysis for Gram-Schmidt method and solution of linear least squares problems, *BIT*, 11(1971)345-367.
3. Abdelmalek, N.N., On the discrete L_1 approximation and L_1 solutions of overdetermined linear equations, *Journal of Approximation Theory*, 11(1974)38-53.
4. Abdelmalek, N.N., On the solution of linear least squares problems and pseudo-inverses, *Computing*, 13(1974)215-228.
5. Abdelmalek, N.N., Chebyshev solution of overdetermined systems of linear equations, *BIT*, 15(1975)117-129.
6. Abdelmalek, N.N., An efficient method for the discrete linear L_1 approximation problem, *Mathematics of Computation*, 29(1975)844-850.
7. Abdelmalek, N.N. A computer program for the Chebyshev solution of overdetermined systems of linear equations, *International Journal for Numerical Methods in Engineering*, 10(1976)1197-1202.
8. Abdelmalek, N.N., The discrete linear one-sided Chebyshev approximation, *Journal of Institute of Mathematics and Applications*, 18(1976)361-370.
9. Abdelmalek, N.N., Minimum L_∞ solution of underdetermined systems of linear equations, *Journal of Approximation Theory*, 20(1977)57-69.
10. Abdelmalek, N.N., A simplex algorithm for minimum fuel problems of linear discrete control systems, *International Journal of Control*, 26(1977)635-642.
11. Abdelmalek, N.N., The discrete linear restricted Chebyshev approximation, *BIT*, 17(1977)249-261.

12. Abdelmalek, N.N., Computing the strict Chebyshev solution of overdetermined linear equations, *Mathematics of Computation*, 31(1977)974-983.
13. Abdelmalek, N.N., Solutions of minimum time problem and minimum fuel problem for discrete linear admissible control systems, *International Journal of Systems Science*, 9(1978)849-855.
14. Abdelmalek, N.N., Minimum energy problem for discrete linear admissible control systems, *International Journal of Systems Science*, 10(1979)77-88.
15. Abdelmalek, N.N., A computer program for the strict Chebyshev solution of overdetermined systems of linear equations, *International Journal for Numerical Methods in Engineering*, 13(1979)1715-1725.
16. Abdelmalek, N.N., L_1 solution of overdetermined systems of linear equations, *ACM Transactions on Mathematical Software*, 6(1980)220-227.
17. Abdelmalek, N.N., Algorithm 551: A FORTRAN subroutine for the L_1 solution of overdetermined systems of linear equations [F4], *ACM Transactions on Mathematical Software*, 6(1980)228-230.
18. Abdelmalek, N.N., Computer program for the discrete linear restricted Chebyshev approximation, *Journal of Computational and Applied Mathematics*, 7(1981)141-150.
19. Abdelmalek, N.N., Chebyshev approximation algorithm for linear inequalities and its applications to pattern recognition, *International Journal of Systems Science*, 12(1981)963-975.
20. Abdelmalek, N.N., Piecewise linear Chebyshev approximation of planar curves, *International Journal of Systems Science*, 14(1983)425-435.
21. Abdelmalek, N.N., An algorithm for the solution of ill-posed linear systems arising from the discretization of Fredholm integral equation of the first kind, *Journal of Mathematical Analysis and Applications*, 97(1983)95-111.
22. Abdelmalek, N.N., Restoration of images with missing high-frequency components using quadratic programming, *Applied Optics*, 22(1983)2182-2188.

23. Abdelmalek, N.N., Linear one-sided approximation algorithms for the solution of overdetermined systems of linear inequalities, *International Journal of Systems Science*, 15(1984)1-8.
24. Abdelmalek, N.N., Piecewise linear L_1 approximation of planar curves, *International Journal of Systems Science*, 16(1985)447-455.
25. Abdelmalek, N.N., A recursive algorithm for discrete L_1 linear estimation using the dual simplex method, *IEEE Transactions on Systems, Man and Cybernetics,* SMC-15 (1985)737-742.
26. Abdelmalek, N.N., Chebyshev and L_1 solutions of overdetermined systems of linear equations with bounded variables, *Numerical Functional Analysis and Optimization*, 8(1985-86)399-418.
27. Abdelmalek, N.N., Noise filtering in digital images and approximation theory, *Pattern Recognition*, 19(1986)417-424.
28. Abdelmalek, N.N., Polygonal approximation of planar curves in the L_1 norm, *International Journal of Systems Science*, 17(1986)1601-1608.
29. Abdelmalek, N.N., Heuristic procedure for segmentation of 3-D range images, *International Journal of Systems Science*, 21(1990)225-239.
30. Abdelmalek, N.N., Piecewise linear least-squares approximation of planar curves, *International Journal of Systems Science*, 21(1990)1393-1403.
31. Abdelmalek, N.N., A program for the solution of ill-posed linear systems arising from the discretization of Fredholm integral equation of the first kind, *Computer Physics Communications*, 58(1990)285-292.
32. Abdelmalek, N.N. and Kasvand, T., Image restoration by Gauss LU decomposition, *Applied Optics*, 18(1979)1684-1686.
33. Abdelmalek, N.N., Kasvand, T. and Croteau, J.P., Image restoration for space invariant pointspread functions, *Applied Optics*, 19(1980)1184-1189.
34. Abdelmalek, N.N. and Kasvand, T., Digital image restoration using quadratic programming, *Applied Optics*, 19(1980)3407-3415.

35. Abdelmalek, N.N., Kasvand, T., Olmstead, J. and Tremblay, M.M., Direct algorithm for digital image restoration, *Applied Optics*, 20(1981)4227-4233.
36. Abdelmalek, N.N. and Otsu, N., Restoration of images with missing high-frequency components by minimizing the L_1 norm of the solution vector, *Applied Optics*, 24(1985)1415-1420.
37. Abdelmalek, N.N. and Otsu, N., Speed comparison among methods for restoring signals with missing high-frequency components using two different low-pass-filter matrix dimensions, *Optics Letters*, 10(1985)372-374.
38. Agmon, S., The relaxation method for linear inequalities, *Canadian Journal of Mathematics*, 6(1954)382-392.
39. Albert, A., *Regression and More-Penrose Pseudoinverse*, Academic Press, New York, 1972.
40. Andrews, H.C. and Hunt, B.R., *Digital Image Restoration*, Prentice-Hall, Englewood Cliffs, NJ, 1977.
41. Ansari, N. and Delp, E.J., On detecting dominant points, *Pattern Recognition*, 24(1991)441-451.
42. Appa, G. and Smith, C., On L_1 and Chebyshev estimation, *Mathematical Programming*, 5(1973)73-87.
43. Arkin, E.M., Chew, L.P., Huttenlocher, D.P., Kedem, K. and Mitchell, J.S.B., An efficient computable metric for comparing polygonal shapes, *IEEE Transactions on Pattern Analysis and Machine Intelligence*, 13(1991)209-215.
44. Armitage, D.H., Gardiner, S.J., Haussmann, W. and Rogge, L., Best one-sided L^1 – approximation by harmonic functions, *Manuscripta Mathematica*, 96(1998)181-194.
45. Armstrong, R.D., Frome, E.L. and Kung, D.S., Algorithm 79-01: A revised simplex algorithm for the absolute deviation curve fitting problem, *Communications on Statistics-Simulation and Computation*, B8(1979)175-190.
46. Armstrong, R.D. and Godfrey, J., Two linear programming algorithms for the linear discrete L_1 norm problem, *Mathematics of Computation*, 33(1979)289-300.
47. Armstrong, R.D. and Hultz, J.W., An algorithm for a restricted discrete approximation problem in the L_1 norm, *SIAM Journal on Numerical Analysis*, 14(1977)555-565.

48. Armstrong, R.D. and Kung, D.S., Algorithm AS 135: Min-Max estimates for a linear multiple regression problem, *Applied Statistics*, 28(1979)93-100.
49. Armstrong, R.D. and Sklar, M.G., A linear programming algorithm for curve fitting in the L_∞ norm, *Numerical Functional Analysis and Optimization*, 2(1980)187-218.
50. Babenko, V.F. and Glushko, V.N., On the uniqueness of elements of the best approximation and the best one-sided approximation in the space L_1, *Ukrainian Mathematical Journal*, 46(1994)503-513.
51. Badi'i, F. and Peikari, B., Functional approximation of planar curves via adaptive segmentation, *International Journal of Systems Science*, 13(1982)667-674.
52. Bahi, S. and Sreedharan, V.P., An algorithm for a minimum norm solution of a system of linear inequalities, *International Journal of Computer Mathematics*, 80(2003)639-647.
53. Baker, C.T.H., *The Numerical Treatment of Integral Equations*, Clarendon Press, Oxford, 1977.
54. Baker, C.T.H., Fox, L., Mayers, D.F. and Wright, K., Numerical solution of Fredholm integral equations of the first kind, *Computer Journal*, 7(1964)141-148.
55. Barrodale, I., L_1 approximation and analysis of data, *Applied Statistics*, 17(1968)51-57.
56. Barrodale, I. and Phillips, C., An improved algorithm for discrete Chebyshev linear approximation, *Proceedings of the Fourth Manitoba conference on Numerical Mathematics*, Hartnell, B.L. and Williams, H.C. (eds.), Winnipeg, Manitoba, Canada, pp. 177-190, 1975.
57. Barrodale, I. and Phillips, C., Algorithm 495: Solution of an overdetermined system of linear equations in the Chebyshev norm, *ACM Transactions on Mathematical Software*, 1(1975)264-270.
58. Barrodale, I. and Roberts, F.D.K., An improved algorithm for discrete l_1 approximation, *SIAM Journal on Numerical Analysis*, 10(1973)839-848.
59. Barrodale, I. and Roberts, F.D.K., Algorithm 478, Solution of an overdetermined system of equations in the l_1 norm [F4], *Communications of ACM*, 17(1974)319-320.

60. Barrodale, I. and Roberts, F.D.K., Algorithms for restricted least absolute value estimation, *Communications on Statistics-Simulation and Computation*, B6(1977)353-363.
61. Barrodale, I. and Roberts, F.D.K., An efficient algorithm for discrete l_1 linear approximation with linear constraints, *SIAM Journal on Numerical Analysis*, 15(1978)603-611.
62. Barrodale, I. and Roberts, F.D.K., Algorithm 552: Solution of the constrained L_1 linear approximation problem, *ACM Transactions on Mathematical Software*, 6(1980)231-235.
63. Barrodale, I. and Stuart, G.F., A Fortran program for linear least squares problems of variable degree, *Proceedings of the Fourth Manitoba Conference on Numerical Mathematics*, Hartnell, B.L. and Williams, H.C. (eds.), Winnipeg, Manitoba, Canada, pp. 191-204, 1975.
64. Bartels, R.H. and Conn, A.R., Linearly constrained discrete l_1 problems, *ACM Transactions on Mathematical Software*, 6(1980)594-608.
65. Bartels, R.H. and Conn, A.R., Algorithm 563: A program for linearly constrained discrete l_1 problems, *ACM Transactions on Mathematical Software*, 6(1980)609-614.
66. Bartels, R.H, Conn, A.R. and Charalambous, C., Minimization techniques for piecewise differentiable functions: The l_∞ solution to overdetermined linear system, The Johns Hopkins University, Baltimore, MD, Technical report no. 247, May 1976.
67. Bartels, R.H., Conn, A.R. and Sinclair, J.W., Minimization techniques for piecewise differentiable functions: The l_1 solution of an overdetermined linear system, *SIAM Journal on Numerical Analysis*, 15(1978)224-241.
68. Bartels, R.H. and Golub, G.H., Stable numerical methods for obtaining the Chebyshev solution of an overdetermined system of equations, *Communications of ACM*, 11(1968)401-406.
69. Bartels, R.H. and Golub, G.H., Algorithm 328: Chebyshev solution to an overdetermined linear system, *Communications of ACM*, 11(1968)428-430.

70. Bartels, R.H. and Golub, G.H., The simplex method of linear programming using LU decomposition, *Communications of ACM*, 12(1969)266-268.
71. Bartels, R.H., Golub, G.H. and Saunders, M.A., Numerical techniques in mathematical programming, *Nonlinear Programming*, Rosen, J.B., Mangasarian, O.L. and Ritter, K. (eds.), Academic Press, New York, 1970.
72. Bartels, R.H, Stoer, J. and Zenger, Ch., A realization of the simplex method based on triangular decomposition, *Handbook for Automatic Computation, Vol. II: Linear Algebra*, Wilkinson, J.H. and Reinsch, C. (eds.), Springer-Verlag, New York, pp. 152-190, 1971.
73. Bashein, G., A simplex algorithm for on-line computation of time optimal controls, *IEEE Transactions on Automatic Control*, 16(1971)479-482
74. Bellman, R., On the approximation of curves by line segments using dynamic programming, *Communications of ACM*, 4(1961)284.
75. Bellman, R., *Introduction to Matrix Analysis*, McGraw-Hill, New York, 1970.
76. Belsley, D.A., Kuh, E. and Welch, R.E., *Regression Diagnostics Identifying Influential Data and Sources of Collinearity*, John Wiley & Sons, New York, 1980.
77. Bjorck, A., Solving linear least squares problems by Gram-Schmidt orthogonalization, *BIT*, 7(1967)1-21.
78. Bloomfield, P. and Steiger, W.L., Least absolute deviations curve-fitting, *SIAM Journal on Scientific and Statistical Computing*, 1(1980)290-301.
79. Bloomfield, P. and Steiger, W.L., *Least Absolute Deviations, Theory, Applications, and Algorithms*, Birkhauser, Boston, 1983.
80. Bramley, R. and Winnicka, B., Solving linear inequalities in a least squares sense, *SIAM Journal on Scientific Computation*, 17(1996)275-286.
81. Brannigan, M., Theory and computation of best strict constrained Chebyshev approximation of discrete data, *IMA Journal of Numerical Analysis*, 1(1980)169-184.

82. Brannigan, M., The strict Chebyshev solution of overdetermined systems of linear equations with rank deficient matrix, *Numerische Mathematik*, 40(1982)307-318.
83. Businger, P. and Golub, G., Linear least squares solution by Householder transformation, *Numerische Mathematik*, 7(1965)269-276.
84. Cadzow, J.A., Algorithm for the minimum-effort problem, *IEEE Transactions on Automatic Control*, 16(1971)60-63.
85. Cadzow, J.A., Functional analysis and the optimal control of linear discrete systems, *International Journal of Control*, 17(1973)481-495
86. Cadzow, J.A., A finite algorithm for the minimum l_∞ solution to a system of consistent linear equations, *SIAM Journal on Numerical Analysis*, 10(1973)607-617.
87. Cadzow, J.A., An efficient algorithmic procedure for obtaining a minimum l_∞-norm solution to a system of consistent linear equations, *SIAM Journal on Numerical Analysis*, 11(1974)1151-1165.
88. Cadzow, J.A., Minimum-amplitude control of linear discrete systems, *International Journal of Control*, 19(1974)765-780.
89. Canon, M.D., Cullum Jr., C.D. and Polak, E., *Theory of Optimal Control and Mathematical Programming*, McGraw-Hill, New York, 1970.
90. Canon, M.D. and Eaton, J.H., A new algorithm for a class of quadratic programming problems with application to control, *SIAM Journal on Control*, 4(1966)34-45.
91. Cantoni, A., Optimal curve fitting with piecewise linear functions, *IEEE Transactions on Computers*, 20(1971)59-67.
92. Censor, Y. and Elfving, T., New methods for linear inequalities, *Linear Algebra and its Applications*, 42(1982)199-211.
93. Chatterjee, S. and Price, B., *Regression Analysis by Example*, John Wiley & Sons, New York, 1977.
94. Cheney, E.W., *Introduction to Approximation Theory*, McGraw-Hill, New York, 1966.
95. Cheng, Y-Q., Zhuang, Y-M. and Yang, J-Y., Optimal Fisher discriminant analysis using the rank decomposition, *Pattern Recognition*, 25(1992)101-111.

96. Clark, D.C. and Gonzalez, R.C., Optimal solution of linear inequalities with applications to pattern recognition, *IEEE Transactions on Pattern Analysis and Machine Intelligence*, 6(1981)643-655.

97. Clasen, R.J., Techniques for automatic tolerance control in linear programming, *Communications of ACM*, 9(1966)802-803.

98. Cline, A.K., A descent method for the uniform solution to overdetermined systems of linear equations, *SIAM Journal on Numerical Analysis*, 13(1976)293-309.

99. Cohen, E. and Megiddo, N., Improved algorithms for linear inequalities with two variables per inequality, *SIAM Journal on Computing*, 23(1994)1313-1347.

100. Coleman, T.F. and Li, Y., A global and quadratically convergent method for linear l_∞ problems, *SIAM Journal on Numerical Analysis*, 29(1992)1166-1186.

101. Cook, R.D. and Weisberg, S., *Residuals and Influence in Regression*, Chapman-Hall, London, 1982.

102. Dantzig, G.B., *Linear Programming and Extensions*, Princeton University Press, Princeton, NJ, 1963.

103. Davis, L.S., Shape matching using relaxation techniques, *IEEE Transactions on Pattern Analysis and Machine Intelligence*, 1(1979)60-72.

104. Davison, L.D., Data compression using straight line interpolation, *IEEE Transactions on Information Theory*, IT-14(1968)390-394.

105. Dax, A., The l_1 solution of linear equations subject to linear constraints, *SIAM Journal on Scientific and Statistical Computation*, 10(1989)328-340.

106. Dax, A., The minimax solution of linear equations subject to linear constraints, *IMA Journal of Numerical Analysis*, 9(1989)95-109.

107. Dax, A., Methods for calculating lp-minimum norm solutions of consistent linear systems, *Journal of Optimization Theory and Applications*, 83(1994)333-354.

108. Deng, J., Feng, Y. and Chen, F., Best one-sided approximation of polynomials under L_1 norm, *Journal of Computational and Applied Mathematics*, 144(2002)161-174.

109. De Pierro, A.R. and Iusem, A.N., A simultaneous projections method for linear inequalities, *Linear Algebra and its Applications*, 64(1985)243-253.
110. Descloux, J., Approximations in L^P and Chebyshev approximations, *Journal of Society of Industrial and Applied Mathematics*, 11(1963)1017-1026.
111. Duda, R.O. and Hart, P.E., *Pattern Classification and Scene Analysis*, John Wiley & Sons, New York, 1973.
112. Dunham, J.G., Optimum uniform piecewise linear approximation of planar curves, *IEEE Transactions on Pattern Analysis and Machine Intelligence*, 8(1986)67-75.
113. Duris, C.S., An exchange method for solving Haar and non-Haar overdetermined linear equations in the sense of Chebyshev, *Proceedings of Summer ACM Computer Conference*, (1968)61-65.
114. Duris, C.S. and Sreedharan, V.P., Chebyshev and l_1-solutions of linear equations using least squares solutions, *SIAM Journal on Numerical Analysis*, 5(1968)491-505.
115. Duris, C.S. and Temple, M.G., A finite step algorithm for determining the "strict" Chebyshev solution to Ax = b, *SIAM Journal on Numerical Analysis*, 10(1973)690-699.
116. Easton, M.C., A fixed point method for Tchebycheff solution of inconsistent linear equations, *Journal of Institute of Mathematics and Applications*, 12(1973)137-159.
117. Faigle, U., Kern, W. and Still, G., *Algorithmic Principles of Mathematical Programming*, Kluwer Academic Publishers, London, 2002.
118. Fisher, M.E., *Introductory Numerical Methods with the NAG Software Library*, The University of Western Australia, Crawley, 1988.
119. Forsythe, G.E. and Moler, C.B., *Computer Solution of Linear Algebraic Systems*, Prentice-Hall, Englewood Cliffs, NJ, 1967.
120. Geladi, P. and Kowalski, B.R., Partial least-squares regression: A tutorial, *Analytica Chimica Acta*, 185(1986)1-17.
121. Gentleman, W.M., Algorithm AS 75: Basic procedures for large, sparse or weighted linear least squares problems, *Applied Statistics*, 22(1974)448-454.

122. Gill, P.E., Golub, G.H., Murray, W. and Saunders, M.A., Methods for modifying matrix factorizations, *Mathematics of Computation*, 28(1974)505-535.
123. Gimlin, D.R., Cavin, III, R.K. and Budge, Jr., M.C., A multiple exchange algorithm for calculation of best restricted approximations, *SIAM Journal on Numerical Analysis*, 11(1974)219-231.
124. Givens, J.W., Numerical computation of the characteristic values of a real matrix, *Oak Ridge National Laboratory*, ORNL-1574, 1954.
125. Givens, J.W., Computation of plane unitary rotations transforming a general matrix to triangular form, *Journal of SIAM*, 6(1958)26-50.
126. Glashoff, K. and Gustafson, S.A., Numerical treatment of a parabolic boundary-value control problem, *Journal of Optimization Theory and Applications*, 19(1976)645-663.
127. Goffin, J.L., The relaxation method for solving systems of linear inequalities, *Mathematics of Operations Research*, 5(1980)388-414.
128. Golden, D.E., *Genetic Algorithms in Search, Optimization, and Machine Learning*, Addison-Wesley Publishing Company, Reading, MA, 1989.
129. Goldstein, A.A., Levine, N. and Hereshoff, J.B., On the best and least q^{th} approximation of an overdetermined system of linear equations, *Journal of ACM*, 4(1957)341-347.
130. Golub, G., Numerical methods for solving linear least squares problems, *Numerische Mathematik*, 7(1965)206-216.
131. Golub, G.H. and Reinsch, C., Singular value decomposition and the least squares solutions, *Numerische Mathematik*, 14(1970)403-420.
132. Golub, G.H. and Van Loan, C.F., *Matrix Computation*, Third Edition, The Johns Hopkins University Press, Baltimore, 1996.
133. Graham, R.E., Snow removal – A noise-stripping process for picture signals, *IEEE Transactions on Information Theory*, IT-8(1962)129-144.

134. Grant, P.M. and Hopkins, T.R., A remark on algorithm AS 135: Min-Max estimates for linear multiple regression problems, *Applied Statistics*, 32(1983)345-347.
135. Gravagne, I.A. and Walker, I.D., On the structure of minimum effort solutions with application to kinematic redundancy resolution, *IEEE Transactions on Robotics and Automation*, 16(2000)855-863.
136. Guler, O., Hoffman, A.J. and Rothblum, U.G., Approximations to solutions to systems of linear inequalities, *SIAM Journal on Matrix Analysis and Applications*, 16(1995)688-696.
137. Gunst, R.F. and Mason, R.L., *Regression Analysis and its Application: A Data Oriented Approach*, Marcel Dekker, Inc., New York, 1980.
138. Gustafson, S.A. and Kortanek, K.O., Numerical treatment of a class of semi-infinite programming problems, *Naval Research Logistics Quarterly*, 20(1973)477-504.
139. Hadley, G., *Linear Programming*, Addison-Wesley, Reading, MA, 1962.
140. Hamann, B. and Chen, J-L., Data point selection for piecewise linear curve approximation, *Computer Aided Geometric Design*, 11(1994)289-301.
141. Han, S-P., Least squares solution of linear inequalities, Technical Report TR-2141, *Mathematics Research Center, University of Wisconsin-Madison*, 1980.
142. Hansen, P.C., The truncated SVD as a method for regularization, *BIT*, 27(1987)534-553
143. Hansen, P.C., Analysis of discrete ill-posed problems by means of the L-curve, *SIAM Review*, 34(1992)561-580.
144. Hansen, P.C. and O'Leary, D., The use of the L-curve in the regularization of discrete ill-posed problems, *SIAM Journal on Scientific Computation,* 14(1993)1487-1502.
145. Hanson, R.J., A numerical method for solving Fredholm integral equations of the first kind using singular values, *SIAM Journal on Numerical Analysis*, 8(1971)616-622.
146. Hanson, R.J. and Phillips, J.L., An adaptive numerical method for solving linear Fredholm integral equations of the first kind, *Numerische Mathematik*, 24(1975)291-307.

147. Haralick, M.R. and Watson, L., A facet model for image data, *Computer Graphics and Image Processing*, 15(1981)113-129.
148. He, B., New contraction methods for linear inequalities, *Linear Algebra and its Applications*, 207(1994)115-133.
149. Hettich, R., A Newton method for nonlinear Chebyshev approximation, *Lecture Notes in Mathematics No. 556*, Dold, A. and Eckmann, B. (eds.) Springer-Verlag, Berlin, pp. 222-236 (1976).
150. Ho, S-Y. and Chen, Y-C., An efficient evolutionary algorithm for accurate polygonal approximation, *Pattern Recognition*, 34(2001)2305-2317.
151. Ho, Y-C. and Kashyap, R.L., An algorithm for linear inequalities and its applications, *IEEE Transactions on Electronic Computers*, 14(1965)683-688.
152. Hoffman, A.J., On approximate solutions of systems of linear inequalities, *Journal of Research of the National Bureau of Standards*, 49(1952)263-265.
153. Hoskuldsson, A., PLS regression methods, *Journal of Chemometrics*, 2(1988)211-228.
154. Householder, A.S., Unitary triangularization of a non-symmetric matrix, *Journal of ACM*, 5(1958)339-342.
155. Huang, S-C. and Sun, Y-N., Polygonal approximation using genetic algorithms, *Pattern Recognition*, 32(1999)1409-1420.
156. Ignizio, J.P. and Cavalier, T.M., *Linear Programming*, Prentice Hall, Englewood Cliffs, NJ, 1993.
157. Joe, B. and Bartels, R., An exact penalty method for constrained, discrete, linear l_∞ data fitting, *SIAM Journal on Scientific and Statistical Computation*, 4(1983)76-84.
158. Johnson, H.H. and Vogt, A., A geometric method for approximating convex arcs, *SIAM Journal on Applied Mathematics*, 38(1980)317-325.
159. Jones, R.C. and Karlovitz, L.A., Iterative construction of constrained Chebyshev approximation of continuous functions, *SIAM Journal on Numerical Analysis*, 5(1968)574-585.
160. Kammerer, W.J. and Nashed, M.Z., On the convergence of the conjugate gradient method for singular linear equations, *SIAM Journal on Numerical Analysis*, 8(1971)65-101.

161. Kammerer, W.J. and Nashed, M.Z., Iterative methods for best approximate solution of linear integral equations of the first and second kinds, *Journal of Mathematical Analysis and Applications*, 40(1972)547-573.
162. Kioustelidis, J.B., Optical segmented approximations, *Computing*, 24(1980)1-8.
163. Kleinbaum, D.G., Kupper, L.L. and Muller, K.E., *Applied Regression Analysis and Other Multivariate Methods*, Second Edition, PWS-Kent Publishing Company, Boston, 1988.
164. Kolesnikov, A. and Franti, P., Polygonal Approximation of Closed Contours, *Lecture Notes in Computer Science*, 2749(2003)778-785.
165. Kolev, L.V., Iterative algorithm for the minimum fuel and minimum amplitude problems, *International Journal of Control*, 21(1975)779-784.
166. Kolev, L.V., Algorithm of finite number of iterations for the minimum fuel and minimum amplitude control problems, *International Journal of Control*, 22(1975)97-102.
167. Kolev, L.V., Minimum-fuel control of linear discrete systems, *International Journal of Control*, 23(1976)207-216.
168. Korn, G.A. and Korn, T.M., *Electronic Analog and Hybrid Computers*, McGraw-Hill, New York, 1964.
169. Kurita, T. and Abdelmalek, N.N., An edge based approach for the segmentation of 3-D range images of small industrial-like objects, *International Journal of Systems Science*, 23(1992)1449-1461.
170. Kurozumi, Y. and Davis, W.A., Polygonal approximation by the minimax method, *Computer Graphics and Image Processing*, 19(1982)248-264.
171. Labonte, G., On solving systems of linear inequalities with artificial neural networks, *IEEE Transactions on Neural Networks*, 8(1997)590-600.
172. Lancaster, P., *Theory of Matrices*, Academic Press, New York, 1969.
173. Lau, H.T., *A Numerical Library in C for Scientists and Engineers*, CRC Press, Ann Arbor, 1995.

174. Lawson, C.L., Characteristic properties of the segmented rational minimax approximation problem, *Numerische Mathematik*, 6(1964)293-301.
175. Lawson, C.W. and Hanson, R.J., *Solving Least Squares Problems*, Prentice Hall, Englewood Cliffs, NJ, 1974.
176. Lenze, B., Uniqueness in best one-sided L_1 – approximation by algebraic polynomials on unbounded intervals, *Journal of Approximation Theory*, 57(1989)169-177.
177. Lewis, J.T., Computation of best one-sided L_1 approximation, *Mathematics of Computation*, 24(1970)529-536.
178. Lewis, J.T., Restricted range approximation and its application to digital filter design, *Mathematics of Computation*, 29 (1975)522-539.
179. Lin, J.N., Determination of reachable set for a linear discrete system, *IEEE Transactions on Automatic Control*, 15(1970)339-342.
180. Longley, J.M., *Least Squares Computations Using Orthogonalization Methods*, Marcel Dekker, New York, 1984.
181. Lopes, H., Oliveira, J.B. and de Figueiredo, L.H., Robust adaptive polygonal approximation of implicit curves, *Computers and Graphics*, 26(2002)841-852.
182. Lucchetti, R. and Mignanego, F., Variational perturbation of the minimum effort problem, *Journal of Optimization Theory and Applications*, 30(1980)485-499.
183. Luenberger, D.G., *Optimization by Vector Space Methods*, John Wiley, New York, 1969.
184. Madsen, K., Nielsen, H.B. and Pinar, M.C., New characterizations of l_1 solutions to overdetermined systems of linear equations, *Operations Research Letters*, 16(1994)159-166.
185. Madsen, K. and Powell, M.J.D., A FORTRAN subroutine that calculates the minimax solution of linear equations subject to bounds on the variables, *United Kingdom Atomic Energy Research Establishment*, AERE-R7954, February 1975.
186. Mangasarian, O.L., Iterative solution of linear programs, *SIAM Journal on Numerical Analysis*, 18(1981)606-614.
187. Mason, R.L. and Gunst, R.F., Outlier-induced collinearities, *Technometrics*, 27(1985)401-407.

188. Michalewicz, Z., *Genetic Algorithms + Data Structures = Evolution Programs*, Second Extended Edition, Springer-Verlag, New York, 1992.
189. Minnick, R.C., Linear-input logic, *IRE Transactions on Electronic Computers*, 10(1961)6-16.
190. Montgomery, D.C. and Peck, E.A., *Introduction to Linear Regression Analysis*, John Wiley & Sons, New York, 1992.
191. Morozov, V.A., *Methods for Solving Incorrectly Posed Problems*, Springer-Verlag, New York, 1984.
192. Motzkin, T.S. and Schoenberg, I.J., The relaxation method for linear inequalities, *Canadian Journal of Mathematics*, 6(1954)393-404.
193. Nagao, M. and Matsuyama, T., Edge preserving smoothing, *Computer Graphics and Image Processing*, 9(1979)394-407.
194. Nagaraja, G. and Krishna, G., An algorithm for the solution of linear inequalities, *IEEE Transactions on Computers*, 23(1974)421-427.
195. Narula, S.C. and Wellington, J.F., An efficient algorithm for the MSAE and the MMAE regression problems, *SIAM Journal on Scientific and Statistical Computing*, 9(1988)717-727.
196. Nie, Y.Y. and Xu, S.R., Determination and correction of an inconsistent system of linear inequalities, *Journal of Computational Mathematics*, 13(1995)211-217.
197. Noble, B., A method for computing the generalized inverse of a matrix, *SIAM Journal on Numerical Analysis*, 3(1966)582-584.
198. Noble, B., *Applied Linear Algebra*, Prentice-Hall, Englewood Cliffs, NJ, 1969.
199. Osborne, M.R. and Watson, G.A., On the best linear Chebyshev approximation, *Computer Journal*, 10(1967)172-177.
200. Outrata, J.V., On the minimum time problem in linear discrete systems with the discrete set of admissible controls, *Kybernetika*, 11(1975)368-374.
201. Pavlidis, T., Waveform segmentation through functional approximation, *IEEE Transactions on Computers*, 22(1973)689-697.

202. Pavlidis, T., Optimal piecewise polygonal L_2 approximation of functions of one and two variables, *IEEE Transactions on Computers*, 24(1975)98-102.
203. Pavlidis, T., The use of algorithms of piecewise approximations for picture processing applications, *ACM Transactions on Mathematical Software*, 2(1976)305-321.
204. Pavlidis, T., Polygonal approximation by Newton's method, *IEEE Transactions on Computers*, 26(1977)800-807.
205. Pavlidis, T. and Horowitz, S.L., Segmentation of plane curves, *IEEE Transactions on Computers*, 23(1974)860-870.
206. Pavlidis, T. and Maika, A.P., Uniform piecewise polynomial approximation with variable joints, *Journal of Approximation Theory*, 12(1974)61-69.
207. Pei, S-C. and Lin, C-N., The detection of dominant points on digital curves by scale-space filtering, *Pattern Recognition*, 25(1992)1307-1314.
208. Perez, J-C. and Vidal, E., Optimum polygonal approximation of digitized curves, *Pattern Recognition Letters*, 15(1994)743-750.
209. Peters, G. and Wilkinson, J.H., The least squares problem and pseudo-inverses, *Computer Journal*, 13(1970)309-316.
210. Phillips, D.L., A technique for the numerical solution of certain integral equations of the first kind, *Journal of ACM*, 9(1962)84-97.
211. Phillips, D.L., A note on best one-sided approximations, *Communications of ACM*, 14(1971)598-600.
212. Phillips, G.M., Algorithms for piecewise straight line approximation, *Computer Journal*, 11(1968)211-212.
213. Pierre, D.A., *Optimization Theory with Applications*, John Wiley & Sons, New York, 1969.
214. Pikaz, A. and Dinstein, I.H., Optimal polygonal approximation of digital curves, *Pattern Recognition*, 28(1995)373-379.
215. Pinar, M.C. and Chen, B., l_1 solution of linear inequalities, *IMA Journal of Numerical Analysis*, 19(1999)19-37.
216. Pinar, M.C., and Elhedhli, S., A penalty continuation method for the l_∞ solution of overdetermined linear systems, *BIT – Numerical Mathematics*, 38(1998)127-150.

217. Pinkus, A.M., *On L^1-Approximation*, Cambridge University Press, London, 1989.
218. Polak, E. and Deparis, M., An algorithm for minimum energy control, *IEEE Transactions on Automatic Control*, 14(1969)367-377.
219. Porter, W.A., *Modern Foundations of System Engineering*, Macmillan, New York, 1966.
220. Powell, M.J.D., The minimax solution of linear equations subject to bounds on the variables, *Proceedings of the Fourth Manitoba Conference on Numerical Mathematics*, Hartnell, B.L. and Williams, H.C. (eds.), Winnipeg, Manitoba, Canada, pp. 53-107, 1975.
221. Powell, M.J.D., *Approximation Theory and Methods*, Cambridge University Press, London, 1981.
222. Press, W.H., Flannery, B.P., Teukolsky, S.A. and Vetterling, W.T., *Numerical Recipes in C, The Art of Scientific Computing*, Second Edition, Cambridge University Press, Cambridge, 1992.
223. Ralston, A., *A First Course in Numerical Analysis*, McGraw-Hill, New York, 1965.
224. Ramer, U., An iterative procedure for the polygonal approximation of plane curves, *Computer Graphics and Image Processing*, 1(1972)244-256.
225. Rannou, F. and Gregor, J., Equilateral polygon approximation of closed contours, *Pattern Recognition*, 29(1996)1105-1115.
226. Ray, B.K. and Ray, K.S., An algorithm for detection of dominant points and polygonal approximation of digitized curves, *Pattern Recognition Letters*, 13(1992)849-856.
227. Ray, B.K. and Ray, K.S., Determination of optimal polygon from digital curve using L_1 norm, *Pattern Recognition*, 26(1993)505-509.
228. Rice, J.R., Tchebycheff approximation in a compact metric space, *Bulletin of American Mathematical Society*, 68(1962)405-410.
229. Rice, J.R., *The Approximation of Functions, Vol. 2*, Addison-Wesley, Reading, MA, 1969.
230. Rice, J.R. and White, J.S., Norms for smoothing and estimation, *SIAM Review*, 6(1964)243-256.

231. Robers, P.D. and Ben-Israel, A., An interval programming algorithm for discrete linear L_1 approximation problems, *Journal of Approximation Theory*, 2(1969)323-336.
232. Robers, P.D. and Robers, S.S., Algorithm 458: Discrete linear L_1 approximation by interval linear programming [E2], *Communications of ACM*, 16(1973)629-631.
233. Roberts, F.D.K. and Barrodale, I., An algorithm for discrete Chebyshev linear approximation with linear constraints, *International Journal for Numerical Methods in Engineering*, 15(1980)797-807.
234. Rosen, J.B., Park, H. and Glick, J., Signal identification using a least L_1 norm algorithm, *Optimization and Engineering*, 1(2000)51-65.
235. Rosen, J.B., Park, H., Glick, J. and Zhang, L., Accurate solution to overdetermined linear equations with errors using L_1 norm minimization, *Computational Optimization and Applications*, 17(2000)329-341.
236. Rosin, P.L., Assessing the behavior of polygonal approximation algorithms, *Pattern Recognition*, 36(2003)505-518.
237. Russell, E., Chiang, L.H. and Braatz, R.D., *Data-Driven Methods for Fault Detection and Diagnosis in Chemical Processes*, Springer-Verlag, London, 2000.
238. Ryan, T.P., *Modern Regression Methods*, John Wiley & Sons, New York, 1997.
239. Salotti, M., Optimal polygonal approximation of digitized curves using the sum of square deviations criterion, *Pattern Recognition*, 35(2002)435-443.
240. Sarkar, D., A simple algorithm for detection of significant vertices for polygonal approximation of chain-coded curves, *Pattern Recognition Letters*, 14(1993)959-964.
241. Sarkar, B., Singh, L.K. and Sarkar, D., A genetic algorithm-based approach for detection of significant vertices for polygonal approximation of digital curves, *International Journal of Image and Graphics*, 4(2004)223-239.
242. Sato, Y., Piecewise linear approximation of plane curves by perimeter optimization, *Pattern Recognition*, 25(1992)1535-1543.

243. Sierksma, G., *Linear and Integer Programming, Theory and Practice,* Second Edition, Marcel Dekker Inc., New York, 2002.
244. Sklansky, J. and Gonzalez, V., Fast polygonal approximation of digitized curves, *Pattern Recognition*, 12(1980)327-331.
245. Sklansky, J. and Michelotti, L., Locally trained piecewise linear classifier, *IEEE Transactions on Pattern Analysis and Machine Intelligence*, 2(1980)101-110.
246. Sklar, M.G., L_∞ norm estimation with linear restrictions on the parameters, *Numerical Functional Analysis and Optimization*, 3(1981)53-68.
247. Sklar, M.G. and Armstrong, R.D., Least absolute value and Chebyshev estimation utilizing least squares results, *Mathematical Programming*, 24(1982)346-352.
248. Sklar, M.G. and Armstrong, R.D., A piecewise linear approximation procedure for Lp norm curve fitting, *Journal of Statistical Computation and Simulation*, 52(1995)323-335.
249. Smith, F.W., Pattern classifier design by linear programming, *IEEE Transactions on Computers*, 17(1968)367-372.
250. Solodov, A.V., *Linear Automatic Control Systems with Varying Parameters*, Blackie, London, 1966.
251. Spath, H., *Mathematical Algorithms for Linear Regression*, Academic Press, English Edition, London, 1991.
252. Sposito, V.A., Kennedy, W.J. and Gentle, J.E., Useful generalized properties of L_1 estimators, *Communications in Statistics-Theory and Methods*, A9(1980)1309-1315.
253. Squire, W., The solution of ill-conditioned linear systems arising from Fredholm equations of the first kind by steepest descents and conjugate gradients, *International Journal for Numerical Methods in Engineering*, 10(1976)607-617.
254. Stark, P.B. and Parker, R.L., Bounded-variable least-squares: An algorithm and applications, *Computational Statistics*, 10(1995)129-141.
255. Stewart, G.W., *Introduction to Matrix Computations*, Academic Press, New York, 1973.
256. Stewart, G.W., The effects of rounding residual on an algorithm for downdating a Cholesky factorization, *Journal of Institute of Mathematics and Applications*, 23(1979)203-213.

257. Stiefel, E., Uber diskrete und lineare Tschebyscheff-approximation, *Numerische Mathematik*, 1(1959)1-28.
258. Stiefel, E., Note on Jordan elimination, linear programming and Tchebycheff approximation. *Numerische Mathematik*, 2(1960)1-17.
259. Stone, H., Approximation of curves by line segments, *Mathematics of Computation*, 15(1961)40-47.
260. Storoy, S., Error control in the simplex technique, *BIT*, 7(1967)216-225.
261. Sun, Y-N. and Huang, S-C., Genetic algorithms for error-bounded polygonal approximation, *International Journal of Pattern Recognition and Artificial Intelligence*, 14(2000)297-314.
262. Taylor, G.D., Approximation by functions having restricted ranges III, *Journal of Mathematical Analysis and Applications*, 27(1969)241-248.
263. Taylor, G.D. and Winter, M.J., Calculation of best restricted approximations, *SIAM Journal on Numerical Analysis*, 7(1970)248-255.
264. Teh, C-H. and Chin, R.T., On the detection of dominant points on digital curves, *IEEE Transactions on Pattern Analysis and Machine Intelligence*, 11(1989)859-872.
265. te Riele, H.J.J., A program for solving first kind Fredholm integral equations by means of regularization, *Computer Physics Communications*, 36(1985)423-432.
266. Thiran, J.P. and Thiry, S., Strict Chebyshev approximation for general systems of linear equations, *Numerische Mathematik*, 51(1987)701-725.
267. Tikhonov, A.N., Solution of incorrectly formulated problems and method of regularization, *Soviet Mathematics*, 4(1963)1035-1038.
268. Tomek, I., Two algorithms for piecewise-linear continuous approximation of functions of one variable, *IEEE Transactions on Computers*, 23(1974)445-448.
269. Torng, H.C., Optimization of discrete control systems through linear programming, *Journal of Franklin Institute*, 278(1964)28-44.

270. Tou, J.T. and Gonzalez, R.C., *Pattern Recognition Principles*, Addison-Wesley, Reading, MA, 1974.
271. Usow, K.H., On L_1 approximation. II. Computation for discrete functions and discretization effects, *SIAM Journal on Numerical Analysis*, 4(1967)233-244.
272. van de Panne, C. and Whinston, A., Simplicial methods for quadratic programming, *Naval Research Logistic Quarterly*, 11(1964)273-302.
273. van de Panne, C. and Whinston, A., Simplex and the dual method for quadratic programming, *Operations Research Quarterly*, 15(1964)355-388.
274. Varah, J.M., On the numerical solution of ill-conditioned linear systems with applications to ill-posed problems, *SIAM Journal on Numerical Analysis*, 10(1973)257-267.
275. Varah, J.M., A practical examination of some numerical methods for linear discrete ill-posed problems, *SIAM Review*, 21(1979)100-111.
276. Varah, J.M., Pitfalls in the numerical solution of linear ill-posed problems, *SIAM Journal on Scientific Statistical Computation*, 4(1983)164-176.
277. Vogel, C.R., Optimal choice of a truncation level for the truncated SVD solution of linear first kind integral equations when data are noisy, *SIAM Journal on Numerical Analysis*, 23(1986)109-134.
278. von Neumann, J. and Goldstine, H.H., Numerical inverting of matrices of high order, *Bulletine of American Mathematical Society*, 53(1947)1021-1099.
279. Wagner, H.M., Linear programming techniques for regression analysis, *Journal of American Statistical Association*, 54(1959)206-212.
280. Wahba, G., Practical approximate solutions to linear operator equations when the data are noisy, *SIAM Journal on Numerical Analysis*, 14(1977)651-667.
281. Wall, K. and Danielsson, P-E., A fast sequential method for polygonal approximation of digitized curves, *Computer Vision, Graphics, and Image Processing*, 28(1984)220-227.

282. Watson, G.A., One-sided approximation and operator equations, *Journal of Institute of Mathematics and Applications*, 12(1973)197-208.

283. Watson, G.A., On the best linear one-sided Chebyshev approximation, *Journal of Approximation Theory*, 7(1973)48-58.

284. Watson, G.A., The calculation of best linear one-sided Lp approximations, *Mathematics of Computation*, 27(1973)607-620.

285. Watson, G.A., The calculation of best restricted approximations, *SIAM Journal on Numerical Analysis*, 11(1974)693-699.

286. Watson, G.A., A multiple exchange algorithm for multivariate Chebyshev approximation, *SIAM Journal on Numerical Analysis*, 12(1975)46-52.

287. Watson, G.A., *Approximation Theory and Numerical Methods*, John Wiley & Sons, New York, 1980.

288. Weisberg, S., *Applied Linear Regression*, Second Edition, John Wiley & Sons, New York, 1985.

289. Weischedel, H.R., A solution of the discrete minimum-time control problem, *Journal of Optimization Theory and Applications*, 5(1970)81-96.

290. Wesolowsky, G.O., A new descent algorithm for the least absolute value regression problem, *Communications in Statistics – Simulation and Computation*, B10(1981)479-491.

291. Westlake, J.R., *A Handbook of Numerical Matrix Inversion and Solution of Linear Equations*, John Wiley & Sons, New York, 1968.

292. Whinston, A., The bounded variable problem – An application of the dual method for quadratic programming, *Naval Research Logistics Quarterly*, 12(1965)173-179.

293. Wilkinson, J.H., Error analysis of floating-point computation, *Numerische Mathematik*, 2(1960)319-340.

294. Wilkinson, J.H., *Rounding Errors in Algebraic Processes*, Prentice-Hall, Englewood Cliffs, NJ, 1963.

295. Wilkinson, J.H., *The Algebraic Eigenvalue Problem*, Clarendon Press, Oxford 1965.

296. Williams, C.M., An efficient algorithm for the piecewise linear approximation of planar curves, *Computer Graphics and Image Processing*, 8(1978)286-293.
297. Wold, S., Ruhe, A., Wold, H. and Dunn III, W.J., The collinearity problem in linear regression. The partial least squares (PLS) approach to generalized inverses, *SIAM Journal on Scientific and Statistical Computing*, 5(1984)735-743.
298. Wolfe, P., Error in the solution of linear programming problems, *Proceedings of a symposium conducted by the MRC and the University of Wisconsin*, Vol. 2, Rall, L.B., (ed.), John Wiley, New York, pp. 271-284, 1966.
299. Yin, P-Y., Algorithms for straight line fitting using k-means, *Pattern Recognition Letters*, 19(1998)31-41.
300. Yin, P-Y., Genetic algorithms for polygonal approximation of digital curves, *International Journal of Pattern Recognition and Artificial Intelligence*, 13(1999)1061-1082.
301. Zhu, Y., and Seneviratne, L.D., Optimal polygonal approximation of digitised curves, *IEEE Proceedings on Vision, Image and Signal Processing*, 144(1997)8-14.

Appendix B

Main Program

```
#include <Windows.h>
#include <stdio.h>
#include <string.h>

#include "LA_Defs.h"

/*-----------------------------------------------------------------
Driver prototypes
-----------------------------------------------------------------*/
void DR_Lone     (void);
void DR_L1       (void);
void DR_Loneside(void);
void DR_Lonebv   (void);
void DR_L1ineq   (void);
void DR_L1pol    (void);
void DR_L1pw1    (void);
void DR_L1pw2    (void);
void DR_Linf     (void);
void DR_Linfside(void);
void DR_Linfbv   (void);
void DR_Chineq   (void);
void DR_Restch   (void);
void DR_Strict   (void);
void DR_Linfpw1  (void);
void DR_Linfpw2  (void);
void DR_Eluls    (void);
void DR_Hhls     (void);
void DR_Mls      (void);
void DR_L2pw1    (void);
void DR_L2pw2    (void);
void DR_Fuel     (void);
void DR_Tmfuel   (void);
void DR_Effort   (void);
void DR_Energy   (void);

/*-----------------------------------------------------------------
Local function prototypes
-----------------------------------------------------------------*/
```

```
void Run_all (void);
void prn_unrecognized (char *pzsInput);
void prn_options (void);

/*-------------------------------------------------------------------
Main program
-------------------------------------------------------------------*/
int main (int argc, char* argv[])
{
    DWORD deltaTick;
    DWORD startTick = GetTickCount ();

    if (argc < 2)
    {
        PRN ("Missing option\n");
        prn_options ();
    }
    else
    {
      strlwr  (argv[1]);
      if      (strncmp (argv[1], "all"     , 3) == 0) Run_all     ();
      else if (strncmp (argv[1], "lone"    , 5) == 0) DR_Lone     ();
      else if (strncmp (argv[1], "l1"      , 5) == 0) DR_L1       ();
      else if (strncmp (argv[1], "l1ineq"  , 5) == 0) DR_L1ineq   ();
      else if (strncmp (argv[1], "lonebv"  , 5) == 0) DR_Lonebv   ();
      else if (strncmp (argv[1], "l1pol"   , 4) == 0) DR_L1pol    ();
      else if (strncmp (argv[1], "l1pw1"   , 5) == 0) DR_L1pw1    ();
      else if (strncmp (argv[1], "l1pw2"   , 5) == 0) DR_L1pw2    ();
      else if (strncmp (argv[1], "linf"    , 5) == 0) DR_Linf     ();
      else if (strncmp (argv[1], "loneside", 5) == 0) DR_Loneside();
      else if (strncmp (argv[1], "linfside", 5) == 0) DR_Linfside();
      else if (strncmp (argv[1], "linfbv"  , 5) == 0) DR_Linfbv   ();
      else if (strncmp (argv[1], "chineq"  , 1) == 0) DR_Chineq   ();
      else if (strncmp (argv[1], "restch"  , 1) == 0) DR_Restch   ();
      else if (strncmp (argv[1], "strict"  , 1) == 0) DR_Strict   ();
      else if (strncmp (argv[1], "linfpw1" , 7) == 0) DR_Linfpw1  ();
      else if (strncmp (argv[1], "linfpw2" , 7) == 0) DR_Linfpw2  ();
      else if (strncmp (argv[1], "eluls"   , 2) == 0) DR_Eluls    ();
      else if (strncmp (argv[1], "hhls"    , 1) == 0) DR_Hhls     ();
      else if (strncmp (argv[1], "l2pw1"   , 5) == 0) DR_L2pw1    ();
      else if (strncmp (argv[1], "l2pw2"   , 5) == 0) DR_L2pw2    ();
      else if (strncmp (argv[1], "mls"     , 1) == 0) DR_Mls      ();
      else if (strncmp (argv[1], "fuel"    , 1) == 0) DR_Fuel     ();
      else if (strncmp (argv[1], "tmfuel"  , 1) == 0) DR_Tmfuel   ();
      else if (strncmp (argv[1], "effort"  , 2) == 0) DR_Effort   ();
```

```
      else if (strncmp (argv[1], "energy"  , 2) == 0) DR_Energy  ();
      else prn_unrecognized (argv[1]);
    }

    deltaTick = GetTickCount () - startTick;

    PRN ("Time to execute: %d msec", deltaTick);
}

/*------------------------------------------------------------------
Local functions
------------------------------------------------------------------*/
void Run_all (void)
{
    DR_Lone     ();
    DR_L1       ();
    DR_L1ineq   ();
    DR_Lonebv   ();
    DR_L1pol    ();
    DR_L1pw1    ();
    DR_L1pw2    ();
    DR_Linf     ();
    DR_Loneside();
    DR_Linfside();
    DR_Linfbv   ();
    DR_Chineq   ();
    DR_Restch   ();
    DR_Strict   ();
    DR_Linfpw1 ();
    DR_Linfpw2 ();
    DR_Eluls    ();
    DR_Hhls     ();
    DR_L2pw1    ();
    DR_L2pw2    ();
    DR_Mls      ();
    DR_Fuel     ();
    DR_Tmfuel   ();
    DR_Effort   ();
    DR_Energy   ();
}

void prn_unrecognized (char *pszInput)
{
    PRN ("Unrecognized option \"%s\"\n", pszInput);
    prn_options ();
```

```
.

}

void prn_options (void)
{
    PRN ("Type \"LA option\":  options are (case-insensitive)\n");
    PRN ("  All      : Run all algorithms\n");
    PRN ("  Lone     : L-ONE Approximation\n");
    PRN ("  L1       : L1 Approximation\n");
    PRN ("  L1lineq  : One-Sided L-One Approximation\n");
    PRN ("  Lonebv   : Bounded Variables L-One Approximation\n");
    PRN ("  L1pol    : L1 Polygonal Approximation\n");
    PRN ("  L1pw1    : Linear L-One Piecewise Approximation (1)\n");
    PRN ("  L1pw2    : Linear L-One Piecewise Approximation (2)\n");
    PRN ("  Linf     : Chebyshev Approximation\n");
    PRN ("  Loneside: L-One Solution of Linear Inequalities\n");
    PRN ("  Linfside: One-Sided Chebyshev Approximation\n");
    PRN ("  Linfbv   : Bounded Variables Chebyshev Approximation\n");
    PRN ("  Chineq   : Chebyshev Solution of Linear Inequalities\n");
    PRN ("  Restch   : Restricted Chebyshev Approximation\n");
    PRN ("  Strict   : Strict Chebyshev Approximation\n");
    PRN ("  Linfpw1 : Linear Chebyshev Piecewise Approximation(1)\n");
    PRN ("  Linfpw2 : Linear Chebyshev Piecewise Approximation(2)\n");
    PRN ("  Eluls    : Least Squares Approximation by "
                       "Gauss LU Decomposition\n");
    PRN ("  Hhls     : Least Squares Approximation by "
                       "Hoseholder's Transformation\n");
    PRN ("  Mls      : Solution of Fredholm Integral Equation of the "
                       "First Kind\n");
    PRN ("  L2pw1    : Least Squares Piecewise Approximation (1)\n");
    PRN ("  L2pw2    : Least Squares Piecewise Approximation (2)\n");
    PRN ("  Fuel     : L-ONE Solution of Underdetermined "
                       "Linear Equations\n");
    PRN ("  Tmfuel   : Bounded/L-One-Bounded Solution of "
                       "Underdetermined Linear Equations\n");
    PRN ("  Effort   : Chebyshev Solution of Underdetermined "
                       "Linear Equations\n");
    PRN ("  Energy   : Bounded Least Squares Solution of"
                       "Underdetermined Linear Equations\n");
    PRN ("NOTE:  You can type the first N unique chararcters in the "
                       "option\n");
    PRN ("        - e.g. \"LA t\"     instead of \"LA Tmfuel\"\n");
    PRN ("               \"LA Loneb\" instead of \"LA LA_Lonebv\"\n");
}
```

Appendix C

Constants, Types and Function Prototypes

```
/*---------------------------------------------------------------
DR_Defs.h

Constants and Type Definitions used by Drivers
---------------------------------------------------------------*/

#ifndef _DR_DEFS_H_
#define _DR_DEFS_H_

/* Constants used only by Driver programs.
   Not visible to LA_ algorithm programs */

/* General Constants */
#define N_ROWS              25
#define M_COLS              250
#define KK_PIECES           50
#define MM_COLS             25
#define MMc_COLS            26
#define NN_ROWS             250

/* Energy + Fuel */
#define Ne_ROWS             25
#define Me_COLS             150

/* Eluls + Hhls */
#define NN2_ROWS            50
#define MM2_COLS            30
#define KK2_COLS            12

#endif
```

```
/*------------------------------------------------------------------
LA_Defs.h

Constants and Type Definitions used by Linear Approximation Library
Functions
------------------------------------------------------------------*/

#ifndef _LA_DEFS_H_
#define _LA_DEFS_H_

#define PRN                           printf

/* Used to identify unused (index 0) vector/matrix elements */
#define NIL                   0.0

typedef int                  *tVector_I;
typedef int                 **tMatrix_I;

typedef double                tNumber_R;
typedef double               *tVector_R;
typedef double              **tMatrix_R;

/* Comment-out the following line for single precision */
#define DOUBLE_PRECISION

#ifdef DOUBLE_PRECISION
#define PREC                  1.0E-16
#define EPS                   1.0E-11
#else /* SINGLE PRECISION */
#define PREC                  1.0E-06
#define EPS                   1.0E-04
#endif

/*------------------------------------------------------------
Linear Approximation Return Code
Provides the result of execution of any algorithm
------------------------------------------------------------*/
typedef enum
{
    /*--------------------------------------
    Intermediate calculation did not fail - never returned externally
    --------------------------------------*/
    LaRcOk                          = 0,
```

```
    /*----------------------------------------
    Algorithm has calculated results
    ----------------------------------------*/
    /* Solution vector was found */
    LaRcSolutionFound                   = 1,

    /* Solution vector is unique */
    LaRcSolutionUnique                  = 2,

    /* Solution vector is most-probably not unique */
    LaRcSolutionProbNotUnique           = 3,

    /* Solution vector is definitely not unique due to rank
       deficiency */
    LaRcSolutionDefNotUniqueRD          = 4,

    /*----------------------------------------
    Algorithm could not calculate results
    ----------------------------------------*/
    /* No feasible solution found, due to one of:
       (1) The problem itself has no solution, or
       (2) Matrix "c" of the system c*a=f is ill-conditioned */
    LaRcNoFeasibleSolution              = -1,

    /* There is no solution because c*a=f is not consistent */
    LaRcInconsistentSystem              = -2,

    /* The calculation terminated because the conditions
         el(i) <= f(i) <= ue(i),   i=1,...,n
       were not satisfied */
    LaRcInconsistentConstraints         = -3,

    /*----------------------------------------
    Errors
    ----------------------------------------*/
    /* Argument(s) out of legal bounds */
    LaRcErrBounds                       = -4,

    /* Null pointer argument */
    LaRcErrNullPtr                      = -5,

    /* Memory allocation error */
    LaRcErrAlloc                        = -6,
} eLaRc;
```

```
/* Macros to simplify/abstract run-time error/return-code handling */
#define GOTO_CLEANUP_RC(r)     { rc = r; goto CLEANUP; }
#define VALIDATE(cond,r)       if (!(cond)) {GOTO_CLEANUP_RC (r);}
#define VALIDATE_BOUNDS(cond) VALIDATE(cond,LaRcErrBounds)
#define VALIDATE_PTRS(cond)    VALIDATE(cond,LaRcErrNullPtr)
#define VALIDATE_ALLOC(cond)   VALIDATE(cond,LaRcErrAlloc)
#define IN_BOUNDS(v,lo,hi)     (((v) >= (lo)) && ((v) <= (hi)))

#endif
```

```
/*-------------------------------------------------------------------
LA_Prototypes.h

Linear Approximation Library Function Prototypes
-------------------------------------------------------------------*/

#ifndef _LA_PROTOTYPES_H
#define _LA_PROTOTYPES_H

#include "LA_Utils.h"

eLaRc LA_L1 (int m, int n, tMatrix_R ct, tVector_R f, int *pIrank,
    int *pIter, tVector_R r, tVector_R a, tNumber_R *pZ);

void LA_l1_part_1 (int m, int n, tMatrix_R ct, tVector_I ic,
    tVector_I ir, tMatrix_R ginv, int *pIrank, int *pIter);

void LA_l1_part_2 (int n, tMatrix_R ct, tVector_R f, tVector_I ic,
    tVector_I ib, tVector_R uf, tVector_R vb, int *pIrank,
    tVector_R r);

void LA_l1_triang_matrix (int m, tVector_R p, int *pIrank);

void LA_l1_vleav (int *pIvo, int *pIout, tNumber_R *pXb,
    tVector_R xp, int *pIrank);

void LA_l1_alfa (int iout, int n, tMatrix_R ct, tVector_I ic,
    tVector_I ib, tVector_R xp, tVector_R t, tVector_R alfa,
    int *pIrank, tVector_R r);

void LA_l1_vent (int ivo, int *pJin, int *pItest, int n,
    tVector_I ic, tVector_R alfa, int *pIrank);

void LA_l1_skip_iters (int iciout, int icjin, tNumber_R xb,
    tNumber_R *pBxb, tNumber_R pivotn, tMatrix_R ct, tVector_I ib,
    tVector_R t, tVector_R vb, int *pIrank);

void LA_l1_update_p (int iout, int jin, tMatrix_R ct, tVector_I ic,
    tVector_R p, int *pIrank);

void LA_l1_update_ginv (int i, int n, tMatrix_R ct, tVector_I ir,
    tVector_R p, tMatrix_R ginv, tVector_R vb, int *pIrank);

void LA_l1_calcul_r (tNumber_R alpha, int n, tMatrix_R ct,
    tVector_R f, tVector_I ic, tVector_I ib, tVector_R uf,
```

```
    tVector_R xp, tVector_R t, tVector_R p, int *pIrank,
    tVector_R r);

void LA_l1_pslv (int id, int *pIrank, tVector_R p, tVector_R vb,
    tVector_R xp);

void LA_l1_gauss_jordn (int iout, int jin, int lj, int *pIrank,
    int n, tMatrix_R ct, tMatrix_R ginv, tVector_I ic);

eLaRc LA_l1_res (int *pIrank, int m, int n, tVector_R r,
    tMatrix_R ginv, tVector_R vb, tVector_I ir, tVector_R xp,
    tVector_R a, tNumber_R *pZ);



eLaRc LA_Lone (int mCols, int nRows, tMatrix_R ct, tVector_R f,
    int *pIrank, int *pIter, tVector_R r, tVector_R a,
    tNumber_R *pZ);

void LA_lone_gauss_jordn (int ipart, int iout, int jin, int nRows,
    tMatrix_R ct, tVector_I icbas, tMatrix_R binv, tVector_R bv,
    tVector_I ibound, int *pIrank, tVector_R r);

void LA_lone_part_1 (int ipart, int nRows, tMatrix_R ct,
    tVector_I icbas, tVector_I irbas, tMatrix_R binv, tVector_R bv,
    tVector_I ibound, int *pIrank, int *pIter, tVector_R r);

void LA_lone_part_2 (int nRows, tMatrix_R ct, tVector_R f,
    tVector_I icbas, tVector_R bv, tVector_I ibound, int *pIrank,
    tVector_R r);

void LA_lone_th (int iout, int nRows, tMatrix_R ct, tVector_I icbas,
    tVector_I ibound, tVector_R th, int *pIrank, tVector_R r);

void LA_lone_vent (int ivo, int *pItest, int *pJin, int nRows,
    tVector_R th);

eLaRc LA_lone_res (int mCols, int nRows, tVector_R f,
    tVector_I icbas, tVector_I irbas, tMatrix_R binv, tVector_R bv,
    int *pIrank, tVector_R r, tVector_R a, tNumber_R *pZ);

void LA_lone_vleav (int *pIvo, int *pIout, int *pIrank,
    tNumber_R *pXb, tVector_R bv);
```

```
eLaRc LA_Loneside (int iside, int m, int n, tMatrix_R ct,
    tVector_R f, int *pIrank, int *pIter, tVector_R r, tVector_R a,
    tNumber_R *pZ);

void LA_loneside_basic_sol (int m, int n, tMatrix_R ct,
    tVector_I irbas, tVector_R bv);

void LA_loneside_part_1 (int m, int n, tMatrix_R ct, tVector_I icbas,
    tVector_I irbas, tMatrix_R binv, tVector_R  bv, int *pIrank,
    int *pIter, tVector_R r);

void LA_loneside_gauss_jordn (int iout, int jin, int m, int n,
    tMatrix_R ct, tVector_I icbas, tMatrix_R binv, tVector_R bv,
    tVector_R r);

eLaRc LA_loneside_res (int iside, int m, int n, tVector_R f,
    tVector_I icbas, tVector_I irbas, tMatrix_R binv, tVector_R bv,
    int *pIrank, tVector_R r, tVector_R a, tNumber_R *pZ);

void LA_loneside_marg_costs (int m, int n, tMatrix_R ct, tVector_R f,
    tVector_I icbas, tVector_R r);

void LA_loneside_vent (int *pIvo, int *pJin, int m, int n,
    tVector_I icbas, tVector_R r);

void LA_loneside_vleav (int jin, int *pIout, int *pItest, int m,
    tMatrix_R ct, tVector_R bv);



eLaRc LA_Lonebv (int m, int n, tMatrix_R ct, tVector_R f,
    tVector_I icbas, tMatrix_R binv, tVector_R bv, tVector_I ibbv,
    tVector_R thbv, int *pIter, tVector_R rbv, tVector_R a,
    tNumber_R *pZ);

void LA_lonebv_part_1 (int m, int n, tMatrix_R ct, tVector_R f,
    tVector_I icbas, tMatrix_R binv, tVector_R bv, tVector_I ibbv,
    tVector_R rbv);

void LA_lonebv_vleav (int *pIvo, int *pIout, tNumber_R *pXb, int m,
    int n, tVector_I icbas, tVector_R bv);

void LA_lonebv_thbv (int ivo, int iout, int m, int n, tMatrix_R ct,
    tVector_I icbas, tMatrix_R binv, tVector_I ibbv, tVector_R thbv,
```

```
    tVector_R rbv);

void LA_lonebv_vent (int ivo, int *pJin, int *pItest, int m, int n,
    tVector_R thbv);

void LA_lonebv_update_bv (int iout, int jout, int jin,
    tNumber_R *pPivot, tNumber_R pivoto, tNumber_R *pXb, int m,
    int n, tMatrix_R ct, tMatrix_R binv, tVector_R bv,
    tVector_I ibbv);

void LA_lonebv_gauss_jordn (int jin, int iout, int m, int n,
    tMatrix_R ct, tVector_I icbas, tMatrix_R binv, tVector_R bv,
    tVector_I ibbv, tVector_R rbv);

void LA_lonebv_res (int m, int n, tVector_R f, tVector_I icbas,
    tMatrix_R binv, tVector_R rbv, tVector_R a, tNumber_R *pZ);



eLaRc LA_L1pol (int m, int n, tNumber_R enorm, tMatrix_R ct,
    tVector_R f, tMatrix_R ctn, tMatrix_R binv, tVector_I icbas,
    tVector_R r, tVector_R a, tVector_I ixl, tVector_R v,
    tVector_R rp, tMatrix_R ap, tVector_R zp, int *pNpiece);

void LA_l1pol_vertic_line (int je, int m, tMatrix_R ct,
    tMatrix_R ctn, tVector_R f, tVector_R r, tMatrix_R binv,
    tVector_R a, tVector_R v, tNumber_R *pPiv);

void LA_l1pol_residuals (int m, int n1, int n2, tMatrix_R ct,
    tVector_R f, tVector_I icbas, tMatrix_R binv, tVector_R r,
    tVector_R a, tNumber_R *pZ, int kase);

void LA_l1pol_gauss_jordn (int n1, int n2, int m, int iout, int jin,
    tMatrix_R ct, tMatrix_R binv, tVector_I icbas);

void LA_l1pol_res (int n1, int n2, int m, tVector_R f, tVector_R r,
    tVector_R a, tMatrix_R binv, tVector_I icbas, tNumber_R *pZ);



eLaRc LA_L1pwl (int m, int n, tNumber_R enorm, int konect,
    tMatrix_R ct, tVector_R f, tVector_I ixl, tVector_I irankp,
    tMatrix_R rp1, tMatrix_R ap, tVector_R zp, int *pNpiece);
```

```
eLaRc LA_L1pw2 (int m, int n, int npiece, tMatrix_R ct, tVector_R f,
    tMatrix_R ap, tVector_R rp2, tVector_R zp, tVector_I ixl);



eLaRc LA_Linf (int m, int n, tMatrix_R ct, tVector_R f, int *pIrank,
    int *pIter, tVector_R r, tVector_R a, tNumber_R *pZ);

void LA_linf_init (int m, int n, tMatrix_R ct, tVector_I icbas,
    tVector_I irbas, tMatrix_R binv, tVector_I ibound, tVector_R r,
    tVector_R a);

void LA_linf_gauss_jordn (int iout, int jin, int kl, int m, int n,
    tMatrix_R ct, tVector_I icbas, tMatrix_R binv, tVector_R bv,
    int *pIter);

void LA_linf_detect_rank (int *pKl, int iout, int jin, int m, int n,
    tNumber_R piv, tMatrix_R ct, tVector_R f, tVector_I icbas,
    tVector_I irbas, tMatrix_R binv, tVector_R bv, tVector_I ibound,
    int *pIrank, int *pIter);

void LA_linf_part_1 (int *pKl, int m, int n, tMatrix_R ct,
    tVector_R f, tVector_I icbas, tVector_I irbas, tMatrix_R binv,
    tVector_R bv, tVector_I ibound, int *pIrank, int *pIter);

void LA_linf_part_2 (int kl, int m, int n, tMatrix_R ct, tVector_R f,
    tVector_I icbas, tMatrix_R binv, tVector_R bv, tVector_I ibound,
    int *pIter);

eLaRc LA_linf_res (int kl, int m, int n, tVector_R f,
    tVector_I icbas, tVector_I irbas, tMatrix_R binv, tVector_R bv,
    tVector_I ibound, tVector_R r, tVector_R a, int irank,
    tNumber_R *pZ);

void LA_linf_part_3 (int kl, int m, int n, tMatrix_R ct, tVector_R f,
    tVector_I icbas, tMatrix_R binv, tVector_R bv, tVector_I ibound,
    tVector_R r, tNumber_R *pZ);

void LA_linf_vent (int *pIvo, int *pJin, int kl, int m, int n,
    tVector_I icbas, tVector_R r, tNumber_R *pZ);

void LA_linf_vleav (int *pItest, int *pIout, int jin, int kl, int m,
    tMatrix_R ct, tVector_R bv);
```

```
eLaRc LA_Linfside (int iside, int m, int n, tMatrix_R ct,
    tVector_R f, int *pIrank, int *pIter, tVector_R r, tVector_R a,
    tNumber_R *pZ);

void LA_linfside_vleav (int *pItest, int *pIout, int jin, int kl,
    int m, tMatrix_R ct, tVector_R bv);

void LA_linfside_init (int iside, int m, int n, tMatrix_R ct,
    tVector_I irbas, tMatrix_R binv, tVector_I ibound, tVector_R a);

void LA_linfside_gauss_jordn (int iout, int jin, int kl, int m,
    int n, tMatrix_R ct, tVector_I icbas, tMatrix_R binv,
    tVector_R bv);

void LA_linfside_detect_rank (tNumber_R piv, int iout, int jin,
    int *pKl, int m, int n, tMatrix_R ct, tVector_R f,
    tVector_I icbas, tVector_I irbas, tMatrix_R binv, tVector_R bv,
    tVector_I ibound, int *pIrank, int *pIter);

void LA_linfside_part_2 (int kl, int m, int n, tMatrix_R ct,
    tVector_R f, tVector_I icbas, tMatrix_R binv, tVector_R bv,
    tVector_I ibound, int *pIter);

void LA_linfside_resid_norm (int kl, int m, int n, tMatrix_R ct,
    tVector_R f, tVector_I icbas, tMatrix_R binv, tVector_R bv,
    tVector_I ibound, tVector_R r, tNumber_R *pZ);

eLaRc LA_linfside_res (int iside, int m, int n, int kl, int iout,
    tVector_R f, tVector_I icbas, tVector_I irbas, tMatrix_R binv,
    tVector_R bv, tVector_I ibound, tVector_R r, tVector_R a,
    int irank, tNumber_R *pZ);

void LA_linfside_vent (int *pIvo, int *pJin, int kl, int m, int n,
    tVector_I icbas, tVector_R r, tNumber_R *pZ);



eLaRc LA_Linfbv (int m, int n, tMatrix_R ct, tVector_R f,
      tVector_I icbas, tVector_I irbas, tMatrix_R binv, tVector_R bv,
    tVector_I ibound, int *pIter, tVector_R r, tVector_R a,
    tNumber_R *pZ);

void LA_linfbv_init (int m, int n, tMatrix_R ct, tVector_I icbas,
```

```
    tVector_I irbas, tMatrix_R binv, tVector_I ibound, tVector_R r,
    tVector_R a);

void LA_linfbv_part_2 (int m, int n, tMatrix_R ct, tMatrix_R binv,
    tVector_R bv, tVector_I ibound);

void LA_linfbv_resid_norm (int m, int n, tMatrix_R ct, tVector_R f,
    tVector_I icbas, tVector_R bv, tVector_R r, tNumber_R *pZ);

void LA_linfbv_vent (int *pIvo, int *pJin, int m, int n,
    tVector_I icbas, tVector_R r, tNumber_R *pZ);

void LA_linfbv_vleav (int jin, int *pIout, int *pItest, int m, int n,
    tMatrix_R ct, tMatrix_R binv, tVector_R bv, tVector_I ibound);

void LA_linfbv_gauss_jordn (int m, int n, int iout, int jin,
    tMatrix_R ct, tVector_I icbas, tMatrix_R binv, tVector_R bv,
    tVector_I ibound);

void LA_linfbv_res (int m, int n, tVector_R f, tVector_I icbas,
    tVector_I irbas, tMatrix_R binv, tVector_I ibound, tVector_R r,
    tVector_R a, tNumber_R *pZ);


eLaRc LA_Mls (int m, int n, tMatrix_R a, tVector_R b,
    tNumber_R toler, tVector_R x, int *pIrank);


eLaRc LA_Restch (int m, int n, tMatrix_R ct, tVector_R f,
    tVector_R el, tVector_R ue, int *pIrank, int *pIter, tVector_R r,
    tVector_R a, tNumber_R *pZ);

eLaRc LA_restch_init (int m, int n, tMatrix_R ct, tVector_R f,
    tVector_R el, tVector_R ue, tVector_I ir, tVector_I ib,
    tVector_I ic, tVector_R g, tMatrix_R ginv, tVector_R a);

void LA_restch_part_1 (tNumber_R *pPiv, int iout, int *pJin, int n,
    tMatrix_R ct, tVector_I ic, tMatrix_R ginv);

void LA_restch_detect_rank (tNumber_R *Ppiv, int *pIout, int *pJin,
    int n, tMatrix_R ct, tVector_I ic, tVector_I ir, tVector_I ib,
    tVector_R g, tVector_R v, tMatrix_R ginv, int *pIrank,
    int *pIter);
```

```
void LA_restch_part_2 (int n, tMatrix_R ct, tVector_I ic,
    tVector_I ib, tVector_R g, tVector_R v, tMatrix_R ginv,
    tVector_R vb, int *pIrank);

void LA_restch_init_p (int m, tVector_R p, int *pIrank);

void LA_restch_update_ginv (int iout, tMatrix_R ct, tVector_I ir,
    tVector_R p, tMatrix_R ginv, tVector_R vb, int *pIrank);

void LA_restch_vent (int *pIvo, int *pJin, tNumber_R *pGgg, int n,
    tVector_R f, tVector_R el, tVector_R ue, tVector_I ic,
    tVector_I ib, tVector_R zc, int *pIrank, tNumber_R *pZ);

void LA_restch_vleav (int jin, int *pIout, int *pItest, tMatrix_R ct,
    tVector_R u, tVector_R v, tVector_R w, tVector_R p,
    tMatrix_R ginv, int *pIrank);

void LA_restch_update_p (int iout, tVector_I ic, tVector_R p,
    int *pIrank);

eLaRc LA_restch_res (int m, tVector_R f, tVector_R el, tVector_R ue,
    tVector_I ic, tVector_I ib, tVector_R v, tVector_R p,
    tVector_R vb, int *pIrank, tVector_R r, tNumber_R *pZ);

void LA_pslvc (int id, int k, tVector_R p, tVector_R b, tVector_R x);

void LA_restch_marg_costs (int m, int n, tMatrix_R ct, tVector_R f,
    tVector_R el, tVector_R ue, tVector_I ic, tVector_I ir,
    tVector_I ib, tVector_R g, tVector_R u, tVector_R v, tVector_R w,
    tVector_R zc, tVector_R p, tMatrix_R ginv, int *pIrank,
    tVector_R r, tVector_R a, tNumber_R *pZ);



eLaRc LA_Strict (int m, int n, tMatrix_R c, tVector_R f, int *pKsys,
    int *pIrank, int *pIter, tVector_R r, tVector_R a, tVector_R z);

void LA_strict_init_1 (int m, int n, tMatrix_R binv, tVector_I ib,
    tVector_I iv, tVector_I irbas, tVector_R v, tVector_R fs,
    tVector_R a, tVector_R r, tVector_R z);

void LA_strict_init_2 (int kl, int m, int n, tMatrix_R c,
    tMatrix_R ct, tVector_I ib, tVector_I ibound,
    tVector_I irbas);
```

```
void LA_strict_part_1 (int *pKl, int kn, int *pMa, int m, int n,
    tMatrix_R c, tMatrix_R ct, tVector_R f, tMatrix_R binv,
    tVector_R bv, tVector_R v, tVector_R fs, tVector_I ib,
    tVector_I ih, tVector_I iv, tVector_I ibound, tVector_I icbas,
    tVector_I irbas, int *pIrank, int *pKsys, int *pIter,
    tVector_R r, tVector_R a, tVector_R z, eLaRc rc);

void LA_strict_map (int kl, int m, tVector_R f, tVector_R fs,
    tVector_I ih, tVector_I icbas, tVector_R r, tNumber_R zz);

void LA_strict_piv (int iout, int *pJin, int n, tNumber_R *pPiv,
    tMatrix_R ct, tVector_I ib);

void LA_strict_swapping (int kl, int iout, int m, int n,
    tMatrix_R ct, tMatrix_R binv, tVector_I ib, tVector_I icbas,
    tVector_I irbas);

void LA_strict_detect_rank (int kl, int *pJin, int m, int n,
    tMatrix_R ct, tVector_R f, tVector_I ib, tVector_I ibound,
    tVector_I icbas);

void LA_strict_part_2 (int kl, int m, int n, tMatrix_R ct,
    tVector_R f, tMatrix_R binv, tVector_R bv, tVector_I ib,
    tVector_I ibound, tVector_I icbas, int *pIter);

void LA_strict_part_3 (int kl, int m, int n, tMatrix_R c,
    tMatrix_R ct, tVector_R f, tMatrix_R binv, tVector_R bv,
    tVector_I ib, tVector_I ibound, tVector_I icbas, tNumber_R *pZz);

void LA_strict_vent (int *pIvo, int *pJin, int kl, int m, int n,
    tMatrix_R c, tVector_I ib, tVector_I icbas, tNumber_R zz);

void LA_strict_vleav (int kl, int *pIout, int jin, int *pItest,
    int m, tMatrix_R ct, tVector_R bv);

void LA_strict_gauss_jordn (int iout, int jin, int kl, int m, int n,
    tMatrix_R ct, tMatrix_R binv, tVector_R bv, tVector_I ib,
    tVector_I icbas);

void LA_strict_uniquens (int *pIvo, int *pIout, int kl, int *pKn,
    int *pLd, int m, int n, tMatrix_R c, tMatrix_R ct, tVector_R f,
    tVector_R bv, tVector_R v, tVector_I ib, tVector_I ih,
    tVector_I ibound, tVector_I icbas, tVector_R r,  tNumber_R zz);
```

```
void LA_strict_eliminat_zz (int *pIout, int kl, int m, tMatrix_R c,
    tMatrix_R binv, tVector_R bv, tVector_I ib, tVector_I ih,
    tVector_I ibound, tVector_I icbas, int *pKsys, tVector_R r,
    tVector_R z, tNumber_R zz);

void LA_strict_gauss_jordn_binv (int iout, int kl, int m,
    tMatrix_R binv, tVector_I iv);

void LA_strict_permute_binv (int kj, int kl, int *pKm, int m,
    tMatrix_R binv, tVector_R bv, tVector_I ih, tVector_I icbas);

void LA_strict_calcul_a (int m, tVector_R fs, tMatrix_R binv,
    tVector_I iv, tVector_I icbas, tVector_I irbas, tVector_R a);

void LA_strict_reduce_sys (int kl, int kln, int *pMa, int m, int n,
    tMatrix_R c, tVector_R f, tMatrix_R binv, tVector_I ib,
    tVector_I iv, tVector_I irbas, tVector_R a);

void LA_strict_modify_binv (int kl, int kj, int m, tMatrix_R binv,
    tVector_R v, tVector_I iv);

void LA_strict_eliminate_ll (int kl, int kln, int *pMa, int ld,
    int m, int n, tMatrix_R c, tVector_R f, tVector_I ib,
    tVector_I iv, tVector_I ibound, tVector_I icbas, tVector_I irbas,
    tVector_R r, tNumber_R zz);

void LA_strict_calcul_r_3 (int kl, int kln, int m, tMatrix_R c,
    tVector_I ib, tVector_I ibound, tVector_I icbas, tVector_R r,
    tNumber_R zz);

void LA_strict_calcul_r_2 (int *pKn, int kln, int m, int n,
    tMatrix_R c, tVector_I ib, tVector_I ibound, tVector_I irbas,
    tVector_R r, tNumber_R zz);

void LA_strict_calcul_r_1 (int m, int n, tMatrix_R c, tVector_I ib,
    tVector_I ibound, tVector_R r, tNumber_R zz);


eLaRc LA_Linfpw1 (int m, int n, tNumber_R enorm, int konect,
    tMatrix_R ct, tVector_R f, tVector_I ixl, tVector_I irankp,
    tMatrix_R rp1, tMatrix_R ap, tVector_R zp, int *pNpiece);
```

```
eLaRc LA_Linfpw2 (int m, int n, int npiece, tMatrix_R ct, tVector_R f,
    tMatrix_R ap, tVector_R rp2, tVector_R zp, tVector_I ixl);



eLaRc LA_Eluls (int ientry, int n, int m, int irhs, tMatrix_R aa,
    tMatrix_R bs, tMatrix_R l, tMatrix_R u, tVector_R diag,
    tVector_R b, int *pIrank, tMatrix_R apsudo, tMatrix_R xs,
    tMatrix_R rres);

void LA_eluls_lu_decomp (int *pIout, int n, int m, tMatrix_R aa,
    tMatrix_R l, tMatrix_R u, tVector_R diag, tVector_I ir,
    tVector_I ic);

void LA_eluls_l_decomp (int n, int m, tMatrix_R tempp, tMatrix_R l,
    tMatrix_R u, int *pIrank);

void LA_eluls_u_decomp (int m, tMatrix_R tempp, tMatrix_R u,
    tVector_R diag, int *pIrank);

void LA_calcul_y (int *pIrank, int n, tMatrix_R t, tVector_R y,
    tVector_R w, tVector_R temp, tMatrix_R l, tVector_R b);

void LA_calcul_x (int *pIrank, int m, tMatrix_R v, tVector_R y,
    tVector_R w, tVector_R temp, tMatrix_R u, tVector_R diag,
    tVector_R x);

void LA_permute_x (int iwish, int ifirst, int isecnd, int n, int m,
    tMatrix_R aa, tMatrix_R bs, tVector_R w, tVector_R x,
    tVector_I ir, tVector_I ic, tMatrix_R apsudo, tMatrix_R xs,
    tMatrix_R rres);

void LA_chols (int irank, tMatrix_R tempp, tMatrix_R el);



eLaRc LA_Hhls (int ientry, int n, int m, int irhs, tMatrix_R aa,
    tMatrix_R bs, tMatrix_R q, tMatrix_R r, tMatrix_R p, int *pIrank,
    tMatrix_R apsudo, tMatrix_R xs, tMatrix_R res);

void LA_hhls_init (int n, int m, tMatrix_R aa, tMatrix_R q,
    tMatrix_R t, tVector_I ic);

void LA_hhls_calcul_q (int *pIout, int n, int m, tMatrix_R q,
    tMatrix_R t, tVector_I ic, tVector_R w, tVector_R dm);
```

```
void LA_hhls_init_p (int n, int m, int *pIrank, tMatrix_R r,
    tMatrix_R t, tMatrix_R p);

void LA_hhls_calcul_r_p (int *pIrank, int m, tMatrix_R r,
    tMatrix_R p, tVector_R w, tVector_R dm);

void LA_hhls_calcul_res (int iwish, int ifirst, int isecnd,
    int *pIrank, int n, int m, tMatrix_R aa, tMatrix_R bs,
    tMatrix_R xs, tMatrix_R apsudo, tMatrix_R q, tMatrix_R r,
    tMatrix_R p, tVector_I ic, tVector_R b, tVector_R x, tVector_R w,
    tVector_R dm, tMatrix_R res);

eLaRc LA_Hhlsro (int n, int m, tMatrix_R c, tVector_R f, tVector_R a,
    tMatrix_R r, tVector_R res, tMatrix_R t, tVector_R dm,
    tVector_R w, tNumber_R *pZ);

eLaRc LA_hhlsro_hh_t (int iend, int n, int m, tMatrix_R t,
    tVector_R dm, tVector_R w);

void LA_hhlsro_x_res (int n, int m, tMatrix_R c, tVector_R f,
    tVector_R a, tMatrix_R r, tVector_R res, tNumber_R *pRho);



eLaRc LA_L2pw1 (int n, int m, tNumber_R enorm, int konect,
    tMatrix_R c, tVector_R f, tVector_I ixl, tMatrix_R rp1,
    tMatrix_R ap, tVector_R zp, int *pNpiece);



eLaRc LA_L2pw2 (int m, int n, int npiece, tMatrix_R c, tVector_R f,
    tMatrix_R ap, tVector_R rp2, tVector_R zp, tVector_I ixl);

void LA_l2pw2_init (int k, int npiece, int n, int m, int *pIs,
    int *pIe, tMatrix_R c, tVector_R f, tMatrix_R cp, tVector_R fp,
    tVector_I ixl);

void LA_pw1_init (int *pNpiece, int is, int ie, int m, tMatrix_R rp1,
    tMatrix_R ap, tVector_R zp);

void LA_pw1_map (int m, int nu, tVector_R r, tVector_R a,
    tNumber_R z, tMatrix_R rp1, tMatrix_R ap, tVector_R zp,
    int *pNpiece);
```

```
eLaRc LA_pw1_prn_rp1 (int konect, int npiece, int n, tVector_I ixl,
    tMatrix_R rp1);

void LA_pw2_init (int k, int npiece, int m, int n, int *pIs,
    int *pIe, tMatrix_R ct, tVector_R f, tMatrix_R ctp, tVector_R fp,
    tVector_I ixl);

eLaRc LA_pw2_prn_rp2 (int npiece, int n, tVector_I ixl,
    tVector_R rp2);



eLaRc LA_Fuel (int m, int n, tMatrix_R c, tVector_R f, int *pIrank,
    int *pIter, tVector_R a, tNumber_R *pZ);

void LA_fuel_init (int m, int n, tVector_I icbas, tVector_I irbas,
    tVector_I ibound, tVector_R a);

eLaRc LA_fuel_part_1 (int iout, int *pJin, int *pKl, int m, int n,
    tMatrix_R c, tVector_R f, tVector_I icbas, tVector_I irbas,
    int *pIrank, int *pIter);

void LA_fuel_part_2 (int kl, int m, int n, tMatrix_R c,
    tVector_I icbas, tVector_R zc);

void LA_fuel_vent (int *pIvo, int *pJin, int kl, int m, int n,
    tVector_I icbas, tVector_R zc);

void LA_fuel_leav (int *pItest, int jin, int *pIout, int kl, int n,
    tMatrix_R c, tVector_R f);

void LA_fuel_gauss_jordn (int iout, int jin, int kl, int m, int n,
    tMatrix_R c, tVector_R f, tVector_I icbas);

eLaRc LA_fuel_res (int kl, int n, tVector_R f, tVector_I icbas,
    tVector_I ibound, tVector_R a, tNumber_R *pZ);



eLaRc LA_Tmfuel (int itf, int m, int n, tMatrix_R c, tVector_R f,
    int *pIrank, int *pIter, tVector_R a, tNumber_R *pZnorm);

void LA_tmfuel_init (int m, int n, tVector_I icbs,  tVector_I irbs,
    tVector_I ibnd, tVector_I kbnd, tVector_I ib,  tVector_R zc,
    tVector_R a);
```

```
eLaRc LA_tmfuel_part_1 (int iout, int *pJin, int m, tMatrix_R c,
    tVector_R f, tVector_I icbs, tVector_I irbs, int *pIrank,
    int *pIter);

void LA_tmfuel_marg_costs (int m, tMatrix_R c, tVector_R f,
    tVector_I icbs, tVector_I ibnd, tVector_I kbnd, tVector_R zc,
    int *pIrank);

void LA_tmfuel_vleav (int *pIrank, int *pIout, int *pIvo,
    tVector_R f, tNumber_R *pXb);

void LA_tmfuel_th_tu (int itf, int m, int ivo, int iout, int *pJout,
    tMatrix_R c, tVector_I icbs, tVector_I ibnd, tVector_I kbnd,
    tVector_R th, tVector_R tu, tVector_R zc);

void LA_tmfuel_vent (int m, int *pJin, int ivo, int *pItest,
    tVector_R th, tVector_R tu);

void LA_tmfuel_swap (int itf, int *pIrank, int *pJin, tMatrix_R c,
    tVector_I ibnd, tVector_I kbnd, tVector_I ib, tVector_R th,
    tVector_R tu, tVector_R zc);

void LA_tmfuel_cascade (int iout, int jin, int *pJout, tNumber_R xb,
    tNumber_R *pPivot, tNumber_R pivoto, tMatrix_R c, tVector_R f,
    tVector_I ibnd, int *pIrank);

void LA_tmfuel_test (int iout, int jin, int *pItest, tNumber_R *pXb,
    tNumber_R pivot, tVector_R f, tVector_I icbs, tVector_R th);

void LA_tmfuel_gauss_jordn (int iout, int jin, int *pIrank, int m,
    tMatrix_R c, tVector_R f);

eLaRc LA_tmfuel_res (int itf, int m, tVector_R f, tVector_I icbs,
    tVector_I ibnd, tVector_I kbnd, tVector_I ib, tVector_R zc,
    int *pIrank, tVector_R a, tNumber_R *pZnorm);


eLaRc LA_Effort (int m, int n, tMatrix_R ct, tVector_R f,
    int *pIrank, int *pIter, tVector_R a, tNumber_R *pZ);

void LA_effort_gauss_jordn (int iout, int jin, int m, int n,
    tNumber_R pivot, tMatrix_R ct, tVector_I ic, tVector_I nb);
```

```
eLaRc LA_effort_res (int m, int n, tMatrix_R ct, tVector_I ib,
    tVector_I ic, tVector_I ip, tVector_I nb, tVector_R zc,
    int *pIrank, tVector_R a, tNumber_R *pZ);

void LA_effort_init (int m, int n, tMatrix_R ct, tVector_I ib,
    tVector_I ic, tVector_I ip, tVector_I nb);

void LA_effort_marg_costs (int m, int n, tMatrix_R ct, tVector_R f,
    tVector_I ib, tVector_R zc);

void LA_effort_vent (int *pIvo, int *pJin, int n, tVector_I nb,
    tVector_R zc, tNumber_R *pZ);

void LA_effort_vleav (int *pItest, int jin, int *pIout, int m,
    tMatrix_R ct);


eLaRc LA_Energy (int m, int n, tMatrix_R ct, tVector_R f,
    int *pIrank, int *pIter, tVector_R a, tNumber_R *pZ);

void LA_energy_init (int m, int n, tMatrix_R ct, tVector_I ic,
    tVector_I ir, tVector_I ik, tVector_R a);

eLaRc LA_energy_phase_1 (int m, int n, tMatrix_R ct, tVector_R f,
    tVector_I ic, tVector_R a, int *pIrank, int *pIter);

void LA_energy_phase_2 (int m, int n, tMatrix_R ct, tVector_I ir,
    tVector_I ik, int *pIrank, tVector_R a);

void LA_energy_norm (int m, tVector_R a, tNumber_R *pZ);

void LA_energy_gauss_jordn_e0 (int iout, int m, int n, tMatrix_R ct,
    tVector_R f, tVector_R a);

void LA_energy_gauss_jordn_e (int m, int n, int iout, int nj0,
    tMatrix_R ct, tVector_I ir, tVector_I ik, tVector_R a);

void LA_energy_vent (int *pIvo, int *pJin, int m, int n,
    tVector_I ir, tVector_R a);

void LA_energy_vleav (int *pItest, int jin, int m, int n, int *pIout,
    int j0, tMatrix_R ct, tVector_I ir, tVector_R a);
```

```
void LA_energy_res (int m, int n, tMatrix_R ct, tVector_I ir,
    tVector_R a);

#endif
```

Appendix D

Utilities and Common Functions

```
/*-------------------------------------------------------------------
LA_Utils.h

Linear Approximation Utility Prototypes
-------------------------------------------------------------------*/

#ifndef _LA_UTILS_H_
#define _LA_UTILS_H_

#include <stdlib.h>
#include <stdio.h>
#include <math.h>

#include "LA_Defs.h"

/* Function prototypes */
void      LA_check_rank_def   (int n, int irank);

void      prn_la_rc           (eLaRc rc);
void      prn_dr_bnr          (char *pszMsg);
void      prn_algo_bnr        (char *pszMsg);
void      prn_example_delim   (void);

tVector_I alloc_Vector_I      (int nElems);
void      swap_elems_Vector_I (tVector_I v, int elemA, int elemB);
void      prn_Vector_I        (int *v, int nElems);
void      free_Vector_I       (int *v);

tVector_R alloc_Vector_R      (int nElems);
void      swap_elems_Vector_R (tVector_R v, int elemA, int elemB);
void      prn_Vector_R        (tNumber_R *v, int nElems);
void      prn_Vector_R_nDec   (tNumber_R *v, int nElems, int nDec);
void      prn_Vector_R_exp    (tNumber_R *v, int nElems);
void      free_Vector_R       (tNumber_R *v);

tMatrix_R alloc_Matrix_R      (int nRows, int mCols);
tMatrix_R init_Matrix_R       (tNumber_R *c, int nRows, int mCols);
void      swap_rows_Matrix_R  (tMatrix_R m, int elemA, int elemB);
```

```
void      prn_Matrix_R       (tMatrix_R, int, int);
void      free_Matrix_R      (tMatrix_R, int nRows);
void      uninit_Matrix_R    (tNumber_R **m);

#endif

/*-------------------------------------------------------------------
LA_Utils.c

Linear Approximation Utility Functions
-------------------------------------------------------------------*/

#include <malloc.h> /* malloc () */
#include <memory.h> /* memcpy () */
#include <string.h>

#include "LA_Utils.h"

/*-------------------------------------------------------------------
Interpretation of results functions
-------------------------------------------------------------------*/

void LA_check_rank_def(int n, int irank)
{
    if (irank < n)
    {
        PRN ("System is rank deficient\n");
    }
}

/*-------------------------------------------------------------------
Print functions
-------------------------------------------------------------------*/

void prn_la_rc (eLaRc rc)
{
    PRN("---------- RC = ");

    switch (rc)
    {
        case LaRcSolutionFound:
            PRN ("Solution Found");                     break;
        case LaRcSolutionUnique:
            PRN ("Solution is Unique");                 break;
        case LaRcSolutionProbNotUnique:
```

```
            PRN ("Solution is probably not Unique");  break;
        case LaRcSolutionDefNotUniqueRD:
            PRN ("Solution is definitely not Unique"
                 " due to Rank deficiency");          break;
        case LaRcNoFeasibleSolution:
            PRN ("No Feasible Solution");             break;
        case LaRcInconsistentSystem:
            PRN ("Inconsistent System");              break;
        case LaRcInconsistentConstraints:
            PRN ("Inconsistent Constraints");         break;
        case LaRcErrBounds:
            PRN ("ERROR: Argument out of bounds");    break;
        case LaRcErrNullPtr:
            PRN ("ERROR: NULL pointer argument");     break;
        case LaRcErrAlloc:
            PRN ("ERROR: Memory allocation failure"); break;
        case LaRcOk: /* Should never return this */
        default:
            PRN ("ERROR: Unknown");                   break;
    }

    PRN(" ----------\n");
}

void prn_dr_bnr (char *pszMsg)
{
PRN("\n==========================================================\n");
PRN("===== Executing Driver:  %s", pszMsg);
PRN("\n==========================================================\n");
}

void prn_algo_bnr (char *pszMsg)
{
PRN("\n----------------------------------------------------------\n");
PRN("----- Executing Algorithm:  %s", pszMsg);
PRN("\n----------------------------------------------------------\n");
}

void prn_example_delim (void)
{
    PRN("----------\n");
}
```

```
/*------------------------------------------------------------------
Vector Functions
------------------------------------------------------------------*/

/* Allocate an integer vector */
tVector_I alloc_Vector_I (int nElem)
{
    int *v;
    if (nElem <= 0) return NULL;
    v = (int *)malloc (nElem * sizeof (int));
    if (!v) return NULL;
    memset (v, nElem * sizeof (int), 0);
    return v - 1;
}

void swap_elems_Vector_I (tVector_I v, int idxA, int idxB)
{
    int temp = v[idxA];
    v[idxA] = v[idxB];
    v[idxB] = temp;
}

void prn_Vector_I (int *v, int nElems)
{
    int i;
    for (i = 1; i <= (nElems - 1); i++)
        PRN ("%d\t", v[i]);
    PRN ("%d\n", v[nElems]);
}

void free_Vector_I (int *v)
{
    if (v == NULL) return;
    free ((char *) (v+1));
}

/* Allocate a real vector */
tNumber_R *alloc_Vector_R (int nElem)
{
    tNumber_R *v;
    if (nElem <= 0) return NULL;
    v = (tNumber_R *)malloc (nElem * sizeof (tNumber_R));
    if (!v) return NULL;
    memset (v, nElem * sizeof (tNumber_R), 0);
    return v - 1;
```

```
}

void swap_elems_Vector_R (tVector_R v, int idxA, int idxB)
{
    tNumber_R temp = v[idxA];
    v[idxA] = v[idxB];
    v[idxB] = temp;
}

void prn_Vector_R (tNumber_R *v, int nElems)
{
    int i;
    for (i = 1; i <= nElems; i++)
        PRN ("% 7.3f ", v[i]);
    PRN ("\n");
}

/* Print to nDec decimal places */
void prn_Vector_R_nDec (tNumber_R *v, int nElems, int nDec)
{
    char szFormat[] = "% 5.pf ";
    int i;
    szFormat[4] = (char)((nDec % 10) + '0');
    for (i = 1; i <= (nElems - 1); i++)
        PRN (szFormat, v[i]);
    PRN (szFormat, v[nElems]);
    PRN ("\n");
}

/* Print vector in exponent format */
void prn_Vector_R_exp (tNumber_R *v, int nElems)
{
    int i;
    for (i = 1; i <= (nElems - 1); i++)
        PRN ("% 1.3E\t", v[i]);
    PRN ("% 1.3E\n", v[nElems]);
}

void free_Vector_R (tNumber_R *v)
{
    if (v == NULL) return;
    free ((char *) (v+1));
}
```

```
/*-----------------------------------------------------------------
Matrix Functions
-----------------------------------------------------------------*/

/* Allocate a real matrix */
tNumber_R **alloc_Matrix_R (int nRows, int mCols)
{
    int i;
    tNumber_R **m;
    if (nRows <= 0 || mCols <= 0) return NULL;
    m = (tNumber_R **)malloc (nRows * sizeof (tNumber_R *));
    if (!m) return NULL;

    m -= 1;

    for (i = 1; i <= nRows; i++)
    {
        m[i] = (tNumber_R *)
               malloc (mCols * sizeof (tNumber_R));
        if (!m[i]) return NULL;
        memset (m, mCols * sizeof (tNumber_R), 0 );
        m[i] -= 1;
    }

    return m;
}

/* Allocate matrix and initialize it with contents of matrix "c" */
tNumber_R **init_Matrix_R (tNumber_R *c, int nRows, int mCols)
{
    int i, j;
    tNumber_R **m = (tNumber_R **)
                    malloc (nRows * sizeof (tNumber_R *));
    if (!m) return NULL;

    m -= 1;

    for (i = 0, j = 1; i <= nRows - 1; i++, j++)
        m[j] = c + mCols * i - 1;
    return m;
}

void swap_rows_Matrix_R (tMatrix_R m, int idxA, int idxB)
{
    tVector_R temp = m[idxA];
```

```
    m[idxA] = m[idxB];
    m[idxB] = temp;
}

void prn_Matrix_R (tNumber_R **m, int nRows, int mCols)
{
    int i;
    for (i = 1; i <= nRows; i++) prn_Vector_R (m[i], mCols);
}

void free_Matrix_R (tNumber_R **m, int nRows)
{
    int i;
    if (m == NULL) return;
    for (i = nRows; i > 0; i--)
        if (m[i] != NULL) free ((char *)(m[i] + 1));
    free ((char *)(m + 1));
}

void uninit_Matrix_R (tNumber_R **m)
{
    free ((char *)(m + 1));
}
```

```
/*-------------------------------------------------------------------
LA_PwCmn.h

Commonly-Used Piecewise Linear Approximation Functions
-------------------------------------------------------------------*/

#include "LA_Prototypes.h"

/*-------------------------------------------------------------------
Initializing the data for "npiece" for piecewise approximation
programs
-------------------------------------------------------------------*/
void LA_pwl_init (int *pNpiece, int is, int ie, int m, tMatrix_R rpl,
    tMatrix_R ap, tVector_R zp)
{
    int i, j, ji;

    /* Initializing the data for "npiece" */
    zp[*pNpiece] = 0.0;

    for (j = is; j <= ie; j++)
    {
        ji = j - is + 1;
        rpl[*pNpiece][ji] = 0.0;
    }

    for (i = 1; i <= m; i++)
    {
        ap[*pNpiece][i] = 0.0;
    }
}

/*-------------------------------------------------------------------
Mapping the data for piecewise approximation programs LA_L1pwl() and
LA_Linfpwl()
-------------------------------------------------------------------*/
void LA_pwl_map (int m, int nu, tVector_R r, tVector_R a,
    tNumber_R z, tMatrix_R rpl, tMatrix_R ap, tVector_R zp,
    int *pNpiece)
{
    int j;

    zp[*pNpiece] = z;

    for (j = 1; j <= m; j++)
```

```
    {
        ap[*pNpiece][j] = a[j];
    }

    for (j = 1; j <= nu; j++)
    {
        rp1[*pNpiece][j] = r[j];
    }
}

/*-------------------------------------------------------------------
Initializing L1pw2 or LA_Linfpw2
-------------------------------------------------------------------*/
void LA_pw2_init (int k, int npiece, int m, int n, int *pIs,
    int *pIe, tMatrix_R ct, tVector_R f, tMatrix_R ctp,
    tVector_R fp, tVector_I ixl)
{
    int i, j, ji, jj, kp1;

    jj = n/npiece;
    ixl[k] = (k-1) * jj + 1;
    *pIs = ixl[k];
    if (k == npiece) *pIe = n;
    if (k != npiece)
    {
        kp1 = k + 1;
        ixl[kp1] = k * jj + 1;
        *pIe = ixl[kp1] - 1;
    }
    for (j = *pIs; j <= *pIe; j++)
    {
        ji = j - *pIs + 1;
        fp[ji] = f[j];
        for (i = 1; i <= m; i++)
        {
            ctp[i][ji] = ct[i][j];
        }
    }
}

/*-------------------------------------------------------------------
Printing the residual vectors for the "npiece" segments for
LA_L1pw1(), LA_Linfpw1() and LA_L2pw1()
-------------------------------------------------------------------*/
eLaRc LA_pw1_prn_rp1 (int konect, int npiece, int n, tVector_I ixl,
```

```
    tMatrix_R rp1)
{
    tVector_R   w = alloc_Vector_R (n);
    int         i, j, j1, j2, j3, jj = 0;

    eLaRc       rc = LaRcOk;
    VALIDATE_ALLOC (w);

    /* For connected piecewise approximation */
    if (konect == 1)
    {
        for (i = 1; i <= npiece-1; i++)
        {
            j1 = ixl[i];
            j2 = ixl[i + 1];
            for (j = j1; j <= j2; j++)
            {
                jj = j2 - j1 + 1;
                j3 = j - j1 + 1;
                w[j3] = rp1[i][j3];
            }
            prn_Vector_R (w, jj);
        }
        j1 = ixl[npiece];
        j2 = n;
        for (j = j1; j <= j2; j++)
        {
            jj = j2 - j1 + 1;
            j3 = j - j1 + 1;
            w[j3] = rp1[npiece][j3];
        }
        prn_Vector_R (w, jj);
    }

    /* For disconnected piecewise approximation */
    if (konect != 1)
    {
        for (i = 1; i <= npiece-1; i++)
        {
            j1 = ixl[i];
            j2 = ixl[i + 1] - 1;
            for (j = j1; j <= j2; j++)
            {
                jj = j2 - j1 + 1;
                j3 = j - j1 + 1;
```

```
                w[j3] = rp1[i][j3];
            }
            prn_Vector_R (w, jj);
        }
        j1 = ixl[npiece];
        j2 = n;
        for (j = j1; j <= j2; j++)
        {
            jj = j2 - j1 + 1;
            j3 = j - j1 + 1;
            w[j3] = rp1[npiece][j3];
        }
        prn_Vector_R (w, jj);
    }

CLEANUP:

    free_Vector_R (w);

    return rc;
}

/*-------------------------------------------------------------------
Printing the residual vectors for the "npiece" segments
for LA_L1pw2(), LA_Linfpw2() and LA_L2pw2()
-------------------------------------------------------------------*/
eLaRc LA_pw2_prn_rp2 (int npiece, int n, tVector_I ixl,
    tVector_R rp2)
{
    tVector_R   w = alloc_Vector_R (n);
    int         i, j, j1, j2, j3, jj = 0;

    eLaRc       rc = LaRcOk;
    VALIDATE_ALLOC (w);

    for (i = 1; i <= npiece-1; i++)
    {
        j1 = ixl[i];
        j2 = ixl[i + 1] - 1;
        for (j = j1; j <= j2; j++)
        {
            jj = j2 - j1 + 1;
            j3 = j - j1 + 1;
            w[j3] = rp2[j];
        }
```

```
        prn_Vector_R (w, jj);
    }
    j1 = ixl[npiece];
    j2 = n;
    for (j = j1; j <= j2; j++)
    {
        jj = j2 - j1 + 1;
        j3 = j - j1 + 1;
        w[j3] = rp2[j];
    }
    prn_Vector_R (w, jj);

CLEANUP:

    free_Vector_R (w);

    return rc;
}
```

Index

D

E

F

G

H

I

K

L

M

S

T

U

V

W